南京大学大理科丛书

NJU Crosd-Disciplinary Science Series

Foundations of Modern Applied Mathematics

近代应用数学基础

苏维宜 著

Su Weiyi

清華大學出版社

Tsinghua University Press

北 京

Bei Jing

内 容 简 介

本书结合例题，系统地介绍集合论、近世代数、点集拓扑、泛函分析、分布理论、微分几何等近代应用数学的基本内容及其在自然科学研究中的应用。书中强调对近代数学概念的理解和对重要论证方法的思路分析，以帮助读者掌握数学推理的基本思维方法，学会把近代应用数学中的重要定理和方法应用到本专业的具体问题中去。

本书可作为物理、天文、化学、地学、生物、计算机等专业学习相关课程的教材或参考书，也可供相关领域科研人员阅读参考。

图书在版编目(CIP)数据

近代应用数学基础＝Foundations of Modern Applied Mathematics：英文/苏维宜著. —北京：清华大学出版社，2024.7

(南京大学大理科丛书)

ISBN 978-7-302-62082-2

Ⅰ. ①近…　Ⅱ. ①苏…　Ⅲ. ①应用数学－英文　Ⅳ. ①O29

中国版本图书馆 CIP 数据核字(2022)第 195092 号

责任编辑：朱红莲
封面设计：常雪影
责任校对：欧　洋
责任印制：刘　菲

出版发行：清华大学出版社
网　　址：https://www.tup.com.cn，https://www.wqxuetang.com
地　　址：北京清华大学学研大厦 A 座　　**邮　　编**：100084
社 总 机：010-83470000　　**邮　　购**：010-62786544
投稿与读者服务：010-62776969，c-service@tup.tsinghua.edu.cn
质量反馈：010-62772015，zhiliang@tup.tsinghua.edu.cn
印 装 者：三河市龙大印装有限公司
经　　销：全国新华书店
开　　本：185mm×230mm　　**印　　张**：28　　**字　　数**：562 千字
版　　次：2024 年 7 月第 1 版　　**印　　次**：2024 年 7 月第1 次印刷
定　　价：99.00 元

产品编号：092096-01

Preface

The new century, the 21st century, has come. It indicates that the rapid development of science and technology, as well as productive activities require a new level of ability of scientists and technicians. It not only asks for profound knowledge and practical ability, but also for lofty ideals, excellent morality, and exquisite thinking. Especially, the great new era asks for extensive modern knowledge, smart and creative thoughts, sensitive and flexible using ability of mathematics from those members who are working in numerous areas of natural science. Clearly, the traditional course advanced calculus could not meet these new requirements.

On the other hand, various excellent concepts, valuable theories, powerful methods in modern applied mathematics are permeating into lots of scientific fields deep. From abstract theory to reality objects, from top to base, modern mathematics and natural sciences are united closely, indeed. Thus, a course of modern applied mathematics must be presented in universities after that of advanced calculus. Hence, it is extremely urgent nowadays to write a textbook of *foundations of modern applied mathematics*.

The contents of modern applied mathematics are very wide, the theory of it is very deep, and the knowledge included is very powerful. We mainly aim to prepare the basic knowledge and technical abilities of modern applied mathematics for undergraduate students who major in natural sciences, such as physics, astronomy, computer science, chemistry, geology, geography, biology, as well as life science. Thus, at the beginning of the 1990s, a course of *modern applied mathematics* was set after that of the advanced mathematics for Kuang Yaming Honors School of Nanjing University. It took one semester, five class hours per week. Its contents were mainly Lebesgue integral and differential geometry. Its results were beneficial, indeed. Then, we arranged and moved the content of Lebesgue integral into the textbook *Advanced Mathematics* published in 2003, and began to write *Foundations of Modern Applied Mathematics*. Before accomplishing the manuscript, we have printed lecture notes twice in 2007 and 2009, and used them as teaching material for 10 years in Kuang Yaming Honors School of Nanjing University and modified them continuously.

The contents of this book are arranged as follows:

Basic knowledge of set theory and modern algebra, introduced in Chapter 1. The first part of this chapter is set theory, including concepts, operations and important properties of sets and mappings between sets. The second part is modern algebra, a main branch of modern mathematics, in which the operation structure of sets is described and studied. It contains several main structures of sets, such as groups, fields, and linear spaces, and particularly how to generate certain new groups from a given group, such as subgroup, product group, and quotient group. Then, several useful groups, such as transformation groups, permutation groups, circulate groups; and their properties are concerned.

Linear spaces and linear transformations, set out in Chapter 2. These are the continuing contents of linear algebra in advanced mathematics, and also are the parts of modern linear algebra involving the structures of orthogonal geometry and skew geometry which play important roles in modern mathematics and modern physics. In the point of view of operation structures in sets, we guide our students to recognize the significance and importance of linear spaces and linear transformations based on a higher level of spaces and transforms. Finally, the theory of multilinear algebra is presented as the essential knowledge of tensor theory and differential geometry.

Point set topology, a corner stone of modern mathematics, is presented in Chapter 3. Both topological and operation structures of sets are essential and intrinsic. A variety of deeper properties of sets can be described by topological structures. A set endowed topological structure is called topological space, this concept comes from reality and Euclidean space. It is highly abstract, and implies both abundant ideas and dedicated methods of modern mathematics. We suggest our students to understand abstract definitions with visual(形象化的、栩栩如生的) examples. Specifically, we emphasize how to generate some new topological spaces from given spaces, such as sub-topological space, product topological space, and quotient topological space; what are the topological structures of these new spaces? The other two concepts are very significant and very useful: a continuous mapping between spaces and the compactness of topological spaces. The classification, separability, as well as connectivity of topological spaces are presented at the end of this chapter.

Functional analysis, an indispensable part of modern applied mathematics, presented in Chapter 4. Its contents are metric space theory, linear operator theory, and linear functional theory. Firstly, about metric space theory, we prove the completion theorem and show certain useful properties of complete metric spaces. Furthermore, various kinds of compactness in metric spaces, such as the compactness, countable compactness, sequential compactness, accumulative compactness and local compactness are defined exactly. Then, two criteria of sequential compactness in normed linear spaces are listed. The Schauder base of Banach spaces, as a generalized concept of finite base of the linear space in Chapter 2, and the orthogonal expansion in Hilbert spaces, as a generalized method of Fourier series in advanced mathematics, both are presented clearly. Secondly, for linear operator theory, we prove the three famous theorems of bounded linear operators on Banach spaces: open mapping theorem, inverse operator theorem and closed graph theorems, moreover, we prove the uniform boundness principle (resonance theorem), and analyze those excellent ideas, methods and proofs. Thirdly, spectrum theory of bounded linear operators plays a role in many scientific areas. It is an important content of the functional analysis, we list carefully basic concepts and useful properties, with enlightening examples. At the end of this chapter, on linear functional theory, we discuss mainly about the conjugate spaces of Banach and Hilbert spaces, as well as the conjugate operators of bounded linear operators in both spaces, including the famous Hermitian operator.

Distribution theory, a quite new direction in the cross-discipline of scientific areas, appeared and was completed at the 1950s, spread out in Chapter 5. From Fourier transform of $L^p(\mathbb{R})$, $1\leqslant p\leqslant 2$, Fourier transform of Schwartz function class, up to Fourier transform of Schwartz distributions, we arrange these contents in details. At the end of this chapter, the newest development of harmonic analysis — wavelet transform, is displayed with multi-signal analysis and applied algorithms. The significance of this chapter is to recognize Fourier transform from the point of view of "distribution theory" in height. We emphasize that the famous Dirac δ function in physics turns out to be a distribution with compact support in seminormed distribution space $E^*(\mathbb{R})$, and this δ is the unit element of normed operator algebra $(L^p(\mathbb{R}), +, \alpha\cdot, \|\cdot\|_{L^p(\mathbb{R})}, *)$, but $\delta\notin L^p(\mathbb{R})$, $1\leqslant p\leqslant\infty$. It is certain that the new idea and new results in distribution theory could bring a new sense and effect to the "δ-function", elegant and mystical, and has puzzled people's mind for a long time.

Calculus on manifolds, not only the theme of our book, but also the essential base of differential geometry and Riemann geometry, organized in Chapter 6. Taking materials from [3] by the great master of mathematics, S. S. Chern, we start from basic concepts of smooth manifold, cotangent space, tangent space, vector field, tensor algebra, to **exterior differential form** on **an exterior differential form space**, in detail. Then, we show the definitions of **exterior differentiation** of an exterior differential form, and **integration** of an exterior differential form on a directed smooth manifold. Applied examples in Euclidean space as patterns are pointed out. The contents and concepts in this chapter are highly abstract and quite difficult to understand. We give models by three-dimensional Euclidean space to help readers for establishing deep-going and essential mathematics thought: from special cases to general ones, from concrete cases to abstract ones, from finite cases to infinite ones, and from theoretical cases to applied ones. These are the essence of modern applied mathematics.

Complimentary knowledge, disposed in the last chapter, including the useful variational calculus and some important theorems, such as Stone-Weierstrass theorem, implicit mapping theorem, inverse mapping theorem, as well as the fixed point theorem on Banach spaces. Moreover, the Haar integral is introduced since it is needed in many natural science areas.

This is a self-contained textbook with a wide span of knowledge, and its contents cover almost all of modern applied mathematics needed by research works in various natural sciences. We have committed for several years to modify and replenish our teaching materials by practice, and to inspire abstract ideas by thinking in terms of images, to analyze difficult concepts by geometric ocular demonstration, to arrange certain questions and exercises for deepening derstanding. All these efforts are fruitful, thus, forming a complete textbook used independently for undergraduates, or as reference materials for other readers concerned.

Before publishing, this textbook was used by Professors Y. Z. Sun and H. Qiu in Kuang Yaming Honors School of Nanjing University. Professors D. X. Lu and W. Xu, as well as M. W. Xiao gave many very valuable suggestions; particularly, D. X. Lu pays his great attention to this Preface. Deputy Editor-in-Chief L. Shi of Tsinghua University Press suggested lots of accurate editorial ideas; Dr. M. Chen has spent energy to check the manuscript, and given pertinent opinions. Heart-felt gratitude is given to all of them. Many thanks to Dr. J. Nuzum and Mrs. C. Nuzum, my closed America friends, for helping the English of this Preface.

Su Weiyi

Feb. 2024 in Nanjing University

CONTENTS

Chapter 1

Set, Structure of Operation on Set

Set theory is the foundation of modern mathematics. We start from its basic knowledge showing the following frame, firstly.

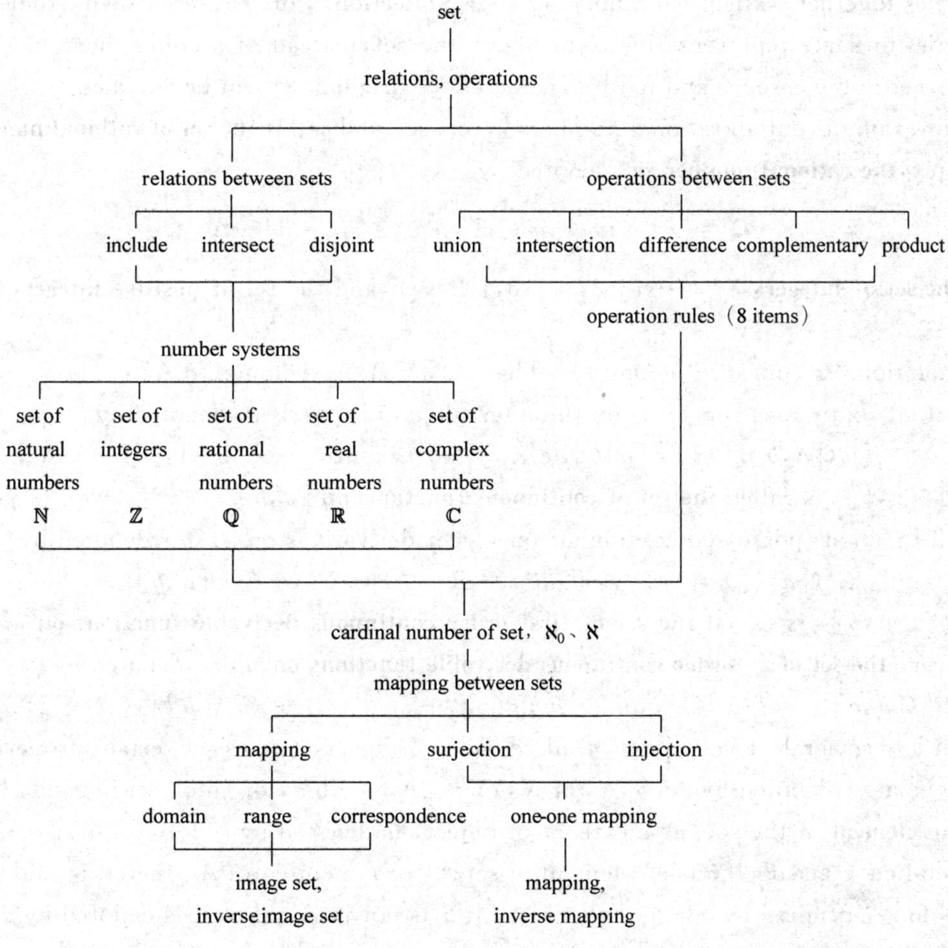

1.1 Sets, the Relations and Operations between Sets

Set theory was constructed at the later period of 19 century by German mathematician G. Cantor. It has developed as an important and fundamental branch of mathematics, rapidly, and permeated through lots of scientific fields as a mathematical tool. There are abundant bibliography and references. We introduce some basic knowledge of set theory in our course, and refer to [1], [2].

1.1.1 Relations between sets

An important thought of mathematics is: put observed objects those having certain properties together, called "a family", or "a collection", or "a set", then studies the properties of some representative elements in the set, instead of studing those individual objects, hence the essence and nature characters of original set can be revealed.

For example, put all rational numbers in one set, and call it **the set of rational numbers**, or simply, **the rational number set**, denoted by

$$\mathbb{Q}=\left\{x: x=\frac{q}{p}, q \in \mathbb{Z}, p \in \mathbb{Z}^{+}\right\}, \quad a \in A$$

with **the set of integers** $\mathbb{Z}=\{\cdots,-2,-1,0,1,2,\cdots\}$, and **the set of positive integers** $\mathbb{Z}^{+}=\{1,2,\cdots\}$.

Here notation "$\in$" means "belong to". That is "$a \in A \Leftrightarrow a$ belongs to A".

Put all continuous functions on the interval $[a,b]$ together, denoted by

$$C([a,b])=\{f: [a,b] \rightarrow \mathbb{R}, f(x) \text{ is continuous on } [a,b]\};$$

Then $C([a,b])$ is called **the set of continuous functions on** $[a,b]$.

All functions possessed continuous one order derivatives on $[a,b]$, denoted by

$$C^{1}([a,b])=\{f(x): [a,b] \rightarrow \mathbb{R}, f(x), f'(x) \in C[a,b]\};$$

Then $C^{1}([a,b])$ is called **the set of first order continuous derivable functions on** $[a,b]$. Moreover, **the set of k-order continuous derivable functions on** $[a,b]$, denoted by

$$C^{k}([a,b])=\{f(x): [a,b] \rightarrow \mathbb{R}, f(x), \cdots, f^{(k)}(x) \in C[a,b]\}, \quad k \in \mathbb{N}.$$

Set In general, a collection of all considered objects possessed certain properties is said to be **a set**, denoted by $A, B, \cdots$, or $X, Y, \cdots$. A member, or an object in a set, is said to be an **element** of the set; A member, or object, is denoted by a, $b\cdots$, or x, y, $\cdots$.

Notation $\in$ and $\notin$ If a is an element of set A, or a member of A, then it is said that a **belongs to** A, denoted by $a \in A$; otherwise, if a is not in A, then it is denoted by $a \notin A$.

"Set" is not only an abstract and profound mathematical idea, but also a very useful and exquisite mathematical tool; It plays role not only in the natural science and social science field, but also in the medical science and humanities science field. It is a main concept、important thought idea, as well as main tool in our book too.

1. Number Systems

positive integer number set $\mathbb{Z}^+=\{x: x=1,2,3,\cdots\}$;

natural number set $\mathbb{N}=\{x: x=0,1,2,3,\cdots\}$;

integer number set $\mathbb{Z}=\{x: x=0,\pm 1,\pm 2,\cdots\}$;

rational number set $\mathbb{Q}=\left\{x: x=\frac{q}{p}, q\in\mathbb{Z}, p\in\mathbb{Z}^+\right\}$;

real number set $\mathbb{R}=\{x: x \text{ is real}\}$;

complex number set $\mathbb{C}=\{z: z=a+\mathrm{i}b,\ a,b\in\mathbb{R}\}$, $\mathrm{i}=\sqrt{-1}$.

2. Relations between sets

We consider the position relations between sets.

Include Let A and B be two sets. If each element a in A is also in B, i. e., "$a\in A$ implies $a\in B$", then A is said to be **included in** B, denoted by $A\subseteq B$, or B **includes** A, denoted by $B\supseteq A$; and we say that A **is a subset of** B.

If A is included in B, but $A\neq B$, then A is said to be a **proper subset of** B, denoted by $A\subset B$.

For example, $\mathbb{Z}^+\subset\mathbb{N}\subset\mathbb{Z}\subset\mathbb{Q}\subset\mathbb{R}\subset\mathbb{C}$; $\{x\in\mathbb{N}: x^2=1\}=\{1\}\subset\{x\in\mathbb{Z}: x^2=1\}=\{1,-1\}$.

Equal If $A\subseteq B$, and $B\subseteq A$, then we say that A **equals** B, denoted by $A=B$.

Intersect Let A and B be two sets. If there are common elements of A and B, then we say that A **intersects** B (in weak sense). It is clear that "include" and "equal" are special cases of "intersect (in weak sense)".

In general, A **intersects** B (in strong sense) means: A and B have common elements, and there is at least one element $a\in A$, but $a\notin B$; also there is at least one element $b\in B$, but $b\notin A$. For example, $A=\{1,2,3,4\}$ intersects $B=\{1,3,5,7\}$ (in strong sense).

Empty set If a set A does not contain any element, then it is called **empty set**, or, A is **empty**, denoted by $A=\varnothing$.

For example, $\{x\in\mathbb{N}: x^2=-1\}=\varnothing$, and $\{z\in\mathbb{C}: z^2=-1\}\neq\varnothing$.

Disjoint Let A and B be two sets. If A and B do not have common element, i. e., their common part is an empty set $\varnothing$, then we say that A **disjoints** B, or A **and** B **disjoint each other**.

For example, the odd number set $A = \{1,3,5,\cdots\}$ and the even number set $B = \{2,4,6,\cdots\}$ disjoint each other.

1.1.2 Operations between sets

1. Union, intersection, difference, complimentary and product of sets

Union of sets Let A and B be two sets. The union set $A \cup B$ of A and B, or union of A and B is defined by(Fig. 1.1.1(a).)

$$A \cup B = \{x : x \in A \text{ or } x \in B\},$$

Intersection of sets Let A and B be two sets. The intersection $A \cap B$ of A and B, or simply intersection of A and B, is defined by(Fig. 1.1.1(b).)

$$A \cap B = \{x : x \in A \text{ and } x \in B\},$$

Usually, it is more general and useful to consider the union $\bigcup\limits_{\lambda \in \Lambda} A_\lambda$ and intersection $\bigcap\limits_{\lambda \in \Lambda} A_\lambda$ of set family $\{A_\lambda : \lambda \in \Lambda\}$ with index set Λ. For example, if $\Lambda = \{1,2,\cdots,n\}$, $n \in \mathbf{N}$, then they are $\bigcup\limits_{j=1}^{n} A_j$ and $\bigcap\limits_{j=1}^{n} A_j$; and if $\Lambda = \mathbf{N}$, they are $\bigcup\limits_{j \in \mathbf{N}} A_j$ and $\bigcap\limits_{j \in \mathbf{N}} A_j$ $\left(\text{or } \bigcup\limits_{j=1}^{\infty} A_j, \bigcap\limits_{j=1}^{\infty} A_j\right)$ respectively.

Difference of sets Let A and B be two sets. The difference $A \setminus B$ of A and B, or for simply, difference of A and B is defined by(Fig. 1.1.1(c).)

$$A \setminus B = \{x : x \in A \text{ and } x \notin B\},$$

Symmetric difference For two sets A and B, the symmetric difference $A \triangle B$ is defined by

$$A \triangle B = (A \setminus B) \cup (B \setminus A).$$

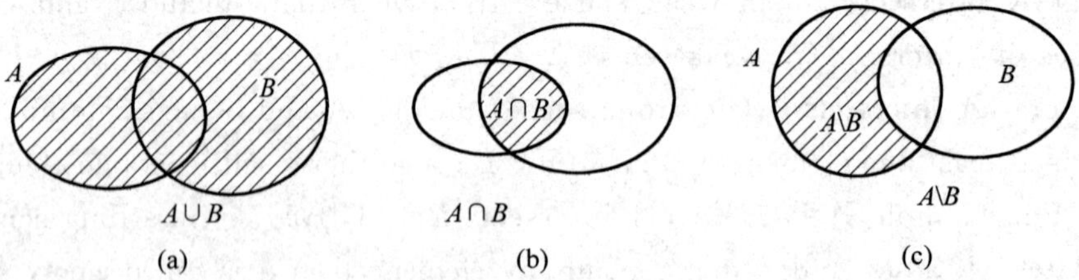

Fig. 1.1.1 Union, intersection and difference

Complementary of sets Let X be a basic set, $A \subset X$ a subset of X. The difference set $X \setminus A$ is said to be **the complementary** of A about X(Fig. 1.1.2(a)), denoted by $A^C = X \setminus A$, or by $\complement A$.

Product of sets Let A and B be two sets. The product set $A \times B$ of A and B is defined by(Fig. 1. 1. 2(b).)

$$A \times B = \{(x, y): x \in A, y \in B\}.$$

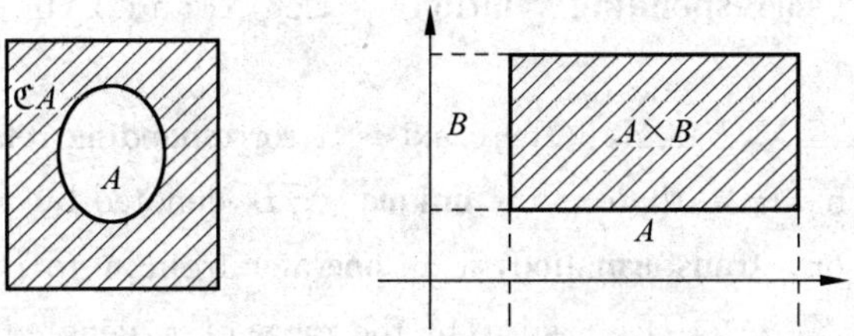

(a) ∁A (b) A×B

Fig. 1. 1. 2 Complementary and product

The product of n sets is denoted by

$$\prod_{j=1}^{n} A_j \equiv A_1 \times A_2 \times \cdots \times A_n = \{(x_1, x_2, \cdots, x_n): x_j \in A_j, j = 1, 2, \cdots, n\}.$$

2. Laws of operations

(1) $A \cap B \subset A \subset A \cup B, A \cap B \subset B \subset A \cup B$;

(2) $A \cap B = B \cap A, A \cup B = B \cup A$;

(3) $A \cap (B \cap C) = A \cap (B \cap C), A \cup (B \cup C) = A \cup (B \cup C)$;

(4) $A \cap (A \cup B) = A \cup (A \cap B) = A$;

(5) $A \cap (B \cup C) = (A \cap B) \cup (A \cap C)$, $A \cup (B \cap C) = (A \cup B) \cap (A \cup C)$;

(6) $\complement(\complement A) = A, \complement(A \cap B) = \complement A \cup \complement B, \complement(A \cup B) = \complement A \cap \complement B$;

(7) $A \cap \left(\bigcap_{j \in I} A_j\right) = \bigcup_{j \in I} (A \cap A_j)$, $A \cup \left(\bigcap_{j \in I} A_j\right) = \bigcap_{j \in I} (A \cup A_j)$;

(8) $\complement\left(\bigcup_{j \in I} A_j\right) = \bigcap_{j \in I} \complement A_j$, $\complement\left(\bigcap_{j \in I} A_j\right) = \bigcup_{j \in I} \complement A_j$. (de Morgan formulas)

1.1.3 Mappings between sets

"Function" is the most important concept and plays crucial role in advanced mathematics, recall that: the domain D_f of a function f is contained in the real number set $\mathbb{R}$ for one variable, or contained in $\mathbb{R}^n$ for several variables; The range R_f is always contained in $\mathbb{R}$. However, variables appearing in modern science and technology are differing, the corresponding relations between two variables, as well as their domains and ranges, all overstep far from Euclidean spaces $\mathbb{R}$ and $\mathbb{R}^n$. Thus, we will consider more general sets instead to be domains and ranges, and give new ideas and methods to describe the concepts of corresponding relations in the place of classical functions.

1. Mappings

Definition 1.1.1 (corresponding relation) Let X and Y be two sets (may be the same, or different).

For **given subsets** $A \subseteq X$, $B \subseteq Y$, if there exists a **corresponding relationship**, denoted by $f: A \to B$, such that for each $x \in A$, there is the **unique** $y \in B$, denoted by $y = f(x)$, then $f: A \to B$ is said to be a **mapping**, or a **transformation**, or an **operator** from A to B. Set A is said to be the **domain** of f, denoted by $\mathfrak{D}_f$; set B is said to be the **range** of f, denoted by $\mathfrak{R}_f$; the set $f(A) = \{f(x): x \in A\} \subseteq B$ is said to be **the image set** of f, denoted by $\text{im}(f) = f(A)$.

For a subset $C \subseteq B$, the set $f^{-1}(C) = \{x \in A: f(x) \in C\} \subseteq A$ is said to be the **inverse image set** of f.

Surjection If $f(A) = B$, then f is said to be a surjection from A onto B, i.e., for each $y \in B$, there is at least one $x \in A$ satisfying $f(x) = y \in Y$. Then, the image set $f(A)$ of f is B, that is, $\text{im}(f) = B$.

Injection If a mapping $f: A \to B$ is one-one corresponding from A into B, i.e.

$$f(x_1) = f(x_2) \quad \text{implies} \quad x_1 = x_2,$$

then f is said to be an infection, or an **one-one mapping** (from A to B).

Note 1 a one-one mapping $f: A \to B$ may not be necessarily a surjection.

Inverse mapping For an injective mapping $f: A \to B$, its inverse mapping $f^{-1}: B \to A$ defined by "$f(x) = y$ implies $f^{-1}(y) = x$"

Note 2 Distinguishing an inverse mapping $f^{-1}: B \to A$ of f with an inverse image set $f^{-1}(B) \subset X$ of f.

Example 1.1.1 Let $A = \{a, b, c\}$, $B = \{1, 2, 3, 4\}$. Then the mapping $f: A \to B$ with

$$f(a) = 1, \quad f(b) = 2, \quad f(c) = 3$$

is one-one from A into B, but it is not surjective.

Example 1.1.2 Take $A = B = \mathbb{Z} = \{0, \pm 1, \pm 2, \pm 3, \cdots\}$, then the mapping $f: A \to B$ with $f(n) = n + 1$ is one-one from $\mathbb{Z}$ onto $\mathbb{Z}$, and it is a surjective too.

Definition 1.1.2 (compound mapping) Let A, B, C be given sets. For given mappings $f: A \to B$ and $g: B \to C$, if the mapping $h: A \to C$ satisfies

$$z = h(x) = g(f(x)), \quad x \in A,$$

then h is said to be **a compound mapping of f and g**, or **a composition mapping of f and g**, denoted by $h = g \circ f$; and $h = g \circ f: A \to C$.

Note 3 in this definition the image set $f(A)$ must be contained in B which is the domain of mapping $g: B \to C$, but is not necessarily equal to B.

Example 1.1.3 Let $A = C[a,b]$, and mapping $J: C[a,b] \to \mathbb{R}$ be determined by Riemann integral: $J(f) = \int_a^b f(x)\mathrm{d}x$. If $f \in C[a,b]$, then $J(f)$ can be regarded as a function of f, called a **functional on** $C[a,b]$.

We will come back to this concept (i. e. functional) in the Chapter 4.

2. Cardinal numbers of sets

"Element number of a set" shows "how many elements in the set". However, it is just suitable for those sets in which there are finite elements. Still, how to measure "element number" is an important problem in many natural science field.

Finite set If a set contains finite elements

$$A = \{a_1, a_2, \cdots, a_n\}, \quad n \in \mathbb{Z}^+,$$

where $n \in \mathbb{Z}^+$ is a determined, finite positive integer, then n is said to be the element **number of** A, and A is said to be a **finite set.**

Infinite set If element number of A is not finite, i. e., A is not a finite set, then A is said to be an **infinite set.**

For example, $\{1,2,3,4\}$ and $\mathbb{Z}_2 = \{0,1\}$ are finite sets; $\mathbb{Z}^+$, $\mathbb{N}$, $\mathbb{Z}$, $\mathbb{Q}$, $\mathbb{R}$ and $\mathbb{C}$ are infinite sets.

It is clear that "element number" is no meaning for infinite sets. The positive integer set $\mathbb{Z}^+ = \{1,2,\cdots\}$ and even number set $\mathbb{O} = \{2,4,6,\cdots\}$ contain infinite elements, both are infinite set. So, the word "element number" does not make sense for them.

To show "how many (or how much) elements" for the infinite set, we introduce an important concept by motivated by the essential property of the finite set: if the numbers of two finite sets A and B equal each other, then the elements of A and B can be established a one-one relationship.

Definition 1.1.3 (equivalence) If there exists a corresponding relationship $\varphi: A \to B$ between sets A and B, such that φ is an injection from A onto B, and also a surjection, i. e., φ is one-one mapping from A onto B with $B = \varphi(A)$, then A and B are said to be **equivalent**, or, A is **equivalent** to B, denoted by $A \sim B$; and φ is said to be **equivalent relation** between A onto B.

By this definition, **we have: two finite sets are equivalent, if and only if they have same numbers of elements.**

For infinite sets, we have different phenomenon: $\mathbb{Z}^+ = \{1,2,\cdots\}$ and $\mathbb{O} = \{2,4,6,\cdots\}$ do not have same "element number", the "element number" of $\mathbb{O}$ is just a half of that of $\mathbb{Z}^+$, but they are equivalent (give the equivalence relationship $\varphi: \mathbb{Z}^+ \to \mathbb{O}$ by readers, please). So does the rational number set $\mathbb{Q}$ with positive integer set $\mathbb{Z}^+$.

Definition 1.1.4 (cardinality) If two infinite sets A and B are equivalent, then they are said to have same **cardinality.** The cardinality ("card", for short) of A is denoted by $\overline{\overline{A}}$, or card $A=\overline{\overline{A}}$.

The number of elements of a finite set is also said to be its **cardinality.**

The cardinality of a set can be compared with that of its subsets.

Comparison of cardinalities For a proper subset $A\subset B$ of infinite set B, with $A\neq B$.

(1) If A is a finite set, then $\overline{\overline{A}}<\overline{\overline{B}}$;

(2) If A is an infinite set, and $A\sim B$, then $\overline{\overline{A}}=\overline{\overline{B}}$;

(3) If A is an infinite set, and A is not equivalent to B, then $\overline{\overline{A}}<\overline{\overline{B}}$, or say $\overline{\overline{B}}>\overline{\overline{A}}$.

For number sets, **we agree on**: the positive integer set $\mathbb{Z}^+$ has a minimum cardinality, denoted by $\aleph_0$. A set is said to be **countable** if its cardinality is $\aleph_0$. All sets which are equivalent to $\mathbb{Z}^+$ have the cardinality $\aleph_0$. Clearly, $\mathbb{Z}^+\sim\mathbb{N}\sim\mathbb{Z}\sim\mathbb{Q}$, thus, we have

$$\text{card } \mathbb{Z}^+=\text{card } \mathbb{N}=\text{card } \mathbb{Z}=\text{card } \mathbb{Q}=\aleph_0.$$

For interval $[0,1]$, it contains all rational and irrational numbers in $[0,1]$, in fact, there is no corresponding relation between $[0,1]$ and $\mathbb{Z}^+$ (see [6], [14]), **we agree on**: the cardinality of $[0,1]$ is $\aleph$. It holds $\aleph_0<\aleph$. We have

$$\text{card}[a,b]=\text{card}\,[a,b)=\text{card}(a,b]=\text{card}(a,b)=\text{card } \mathbb{R}=\aleph.$$

1.2 Structures of Operations on Sets

The position relations and the operation relations between two sets are discussed in the above section, whereas operations between elements in a set A are considered now, motivated by the familiar example $\mathbb{R}$, real number set, in which there are addition, subtraction, multiplication, division, ... What means "structure of operations of sets"? How to endow operations to a set? We discuss in this section, and introduce some quite new concepts compared with in Advanced Calculus, such as groups, rings, fields, ideals, modulo and linear spaces.

A set A endowed operations between its elements is said **to possess a structure of operations**, or **to have a structure of algebra.**

1.2.1 Groups, rings, fields, and linear spaces

1. Groups

We begin with the familiar real number set $\mathbb{R}$. For any $x, y\in\mathbb{R}$, sum $x+y$ can be

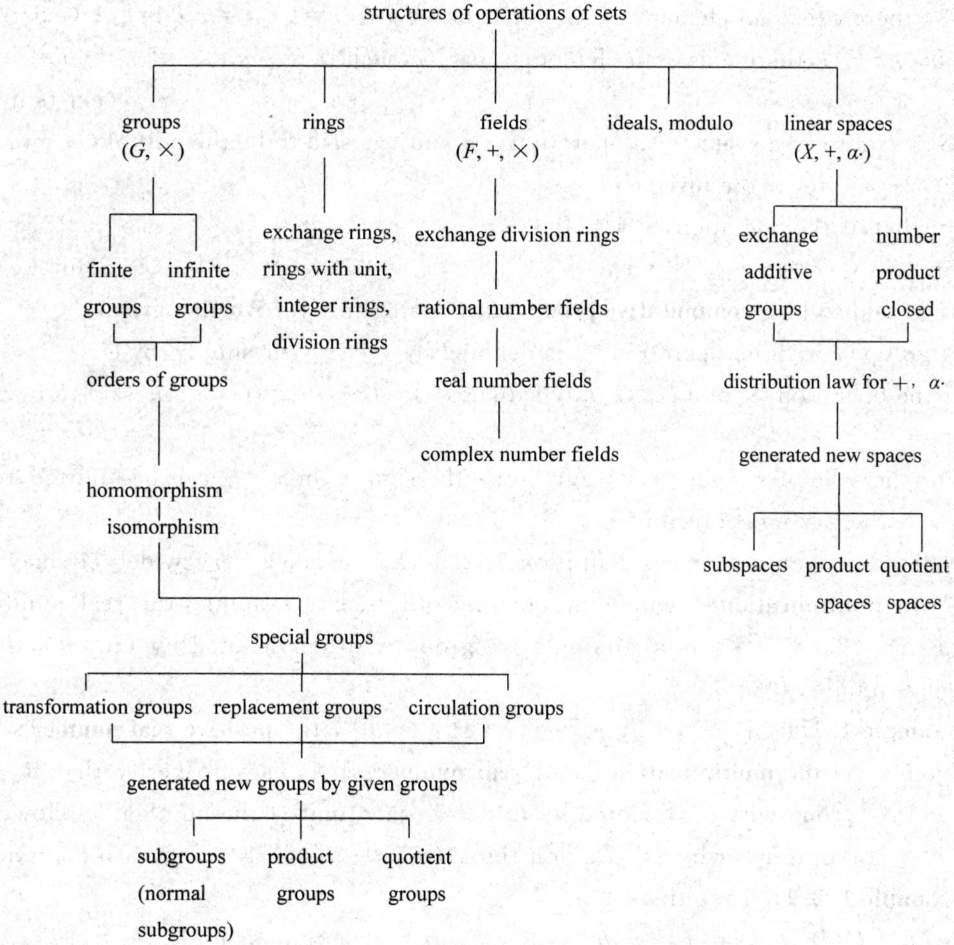

regarded as a result of "operation + of x and y". This operation + has the following familiar properties:

Closed property $x, y \in \mathbb{R} \Rightarrow x + y \in \mathbb{R}$;

Combination law $x, y, z \in \mathbb{R} \Rightarrow (x+y)+z = x+(y+z)$;

unit element there exists "unit" element $0 \in \mathbb{R}$, such that $\forall x \in \mathbb{R} \Rightarrow x + 0 = 0 + x = x$;

inverse element $\forall x \in \mathbb{R}$, there exists an inverse element $(-x) \in \mathbb{R}$, such that $x + (-x) = 0$.

Then $\mathbb{R}$ is called **a group** under operation +.

Abstractly, we introduce the concept of the group.

Definition 1.2.1 (group) If there is an **operation** on a given set G, denoted by $\cdot$, satisfying

(1) $x, y \in G \Rightarrow x \cdot y \in G$; (closed property)

(2) $x, y, z \in G \Rightarrow (x \cdot y) \cdot z = x \cdot (y \cdot z)$; (combination law)

(3) there exists an element $1 \in G$, such that $\forall x \in G \Rightarrow 1 \cdot x = x \cdot 1 = x$; (exists unit)

(4) $\forall x \in G$, there exists an element $x^{-1} \in G$, such that $x \cdot x^{-1} = x^{-1} \cdot x = 1$, (exists inverse)

then G is said to be a **group** with operation $\cdot$; and 1 is said to be the **unit** of G; moreover, $x^{-1} \in G$ is said to be the **inverse of** x.

Further, if the operation $\cdot$ satisfies

(5) $x, y \in G \Rightarrow x \cdot y = y \cdot x$, (commutation law)

then G is said to be **a commutative group**, with operation $\cdot$, or **Abelian group**;

A group G with its operation $\cdot$ is denoted by $(G, \cdot)$, or simply, by G.

If the operation $\cdot$ of a set G only satisfies (1), (2), then $(G, \cdot)$ is said to be a **semi-group**.

For the sake of simplicity, we agree on: the sign $\cdot$ in $x \cdot y$ can be omitted, turning into xy, if there is no confusion.

Note An operation $\cdot$ in definition 1.2.1 can be taken very widely, it may be an addition, a multiplication, a compound, or any others. For example, the real number set $\mathbb{R} = \{x: -\infty < x < +\infty\}$ is a commutative group with $+$, denoted by $(\mathbb{R}, +)$, the unit element is number 0.

Example 1.2.1 Denoted by $\mathbb{R}^+ = \{x \in \mathbb{R}: x > 0\}$, the **positive real number set**, take operation $\cdot$ as **the multiplication** $\times$ **of real numbers.** It is easy to verify that $\mathbb{R}^+$ is an commutative group with $\times$, denoted by $(\mathbb{R}^+, \times)$, and unit is the number 1. However, $\mathbb{R}$ with $\times$ is not a group (why ?). Do you think $(\mathbb{R}\setminus\{0\}, \times)$ is a group with operation $\times$?

Example 1.2.2 Take the set

$$C([a,b]) = \{f: [a,b] \to \mathbb{R}, \quad f(x) \text{ is continuous on } [a,b]\},$$

define operation $\cdot$ as **the addition** $+$ **of functions**, i. e., $(f+g)(x) = f(x) + g(x)$, then the set $(C([a,b]), +)$ is called **the continuous function set**, it is an Abelian group.

Example 1.2.3 The set

$$L(\mathbb{R}, \mathbb{R}) = \{T: T(x) = ax + b, a, b \in \mathbb{R}, a \neq 0\}$$

is that of all linear mappings from $\mathbb{R}$ onto $\mathbb{R}$, where $T(x)$ is called **affine transformation.** Define operation $\cdot$ as **the compound** $\circ$ **of functions**, then the set $(L(\mathbb{R}, \mathbb{R}), \circ)$ is a group.

In fact, for $S, T \in L(\mathbb{R}, \mathbb{R})$, let $S(x) = ax + b$, $T(x) = cx + d$, $a \neq 0$, $c \neq 0$. Then

$$(T \circ S)(x) = T(S(x)) = c(S(x)) + d = c(ax + b) + d = (ca)x + (cb + d),$$

thus $T \circ S \in L(\mathbb{R}, \mathbb{R})$ by $a, c \neq 0$. This shows that the closed property (1) in Definition 1.2.1 holds.

It is clear that the combination law (2) holds.

For (3), the identity mapping $I: x \to x$ with $I(x) = x$, is the unit element with $a = 1$

and $b=0$. Thus (3) holds.

Then for (4), $\forall S(x)=ax+b\in L(\mathbb{R},\mathbb{R}), a\neq 0$, the inverse S^{-1} of S is

$$S^{-1}(x)=\frac{x}{a}-\frac{b}{a}\in L(\mathbb{R},\mathbb{R}).$$

Hence, $L(\mathbb{R},\mathbb{R})$ is a group; however, it is not commutative group, since the (5) fails.

Example 1.2.4 $(\mathbb{R},+),(\mathbb{Q},+),(\mathbb{Z},+)$ all are groups with $+$ as the addition of real numbers, and all are Abelian groups.

Example 1.2.5 (1) $(\mathbb{R}\setminus\{0\},\times)$, $(\mathbb{R}^+,\times)$ are Abelian groups with $\times$ as the multiplication of real numbers.

(2) The **semi-group of positive integers** with multiplication, $(\mathbb{Z}^+,\times)$.

(3) The p-**semi-group of positive integers** with multiplication, $(\mathbb{N}_p,\times)=(\{np: n\in\mathbb{Z}^+\},\times)$.

(4) The p-**adic finite group** $(\{0,1,\cdots,p\},\oplus)$, where $p\geqslant 2$ is an integer, $\oplus$ is the operation mod p: $x\oplus y=x+y\pmod p$.

(5) The **Cayley finite group** $G=(\{1,\mathrm{i},-1,-\mathrm{i}\},\otimes)$, with operation $\otimes$ given by **the Cayley table**

$\otimes$	1	i	−1	−i
1	1	i	−1	−i
i	i	−1	−i	1
−1	−1	−i	1	i
−i	−i	1	i	−1

(6) The **nonsingular n-complex matrix group**

$$(GL(n,\mathbb{C}),\times)=(\{A=[\alpha_{jk}]_{n\times n}: \alpha_{jk}\in\mathbb{C}, \det A\neq 0\},\times),$$

where $\times$ is **the multiplication of matrix**; det A is the determinant of matrix A;

The **nonsingular** n-real matrix group

$$(GL(n,\mathbb{R}),\times)=(\{A=[\alpha_{jk}]_{n\times n}: \alpha_{jk}\in\mathbb{R}, \det A\neq 0\},\times).$$

When $n\geqslant 2$, they are non-Abelian groups, and $(GL(n,\mathbb{R}),\times)\subset(GL(n,\mathbb{C}),\times)$.

(7) The **group of nonsingular n-complex matrix with determinant 1**

$$(SL(n,\mathbb{C}),\times)=(\{A\in GL(n,\mathbb{C}): \det A=1\},\times);$$

The **group of nonsingular n-real matrix with determinant 1**

$$(SL(n,\mathbb{R}),\times)=(\{A\in GL(n,\mathbb{R}): \det A=1\},\times).$$

When $n\geqslant 2$, they are non-Abelian groups, and $(SL(n,\mathbb{R}),\times)\subset(SL(n,\mathbb{C}),\times)$.

(8) The **rotation group by a square** $ABCD$ **revolving around its center**

$$\mathfrak{R}=\left\{\alpha: \alpha=0,\frac{1}{2}\pi,\pi,\frac{3}{2}\pi\right\},$$

where α is the revolving angle, and **the operation** $\circ$ is defined by: $\alpha_1 \circ \alpha_2 \Rightarrow$ rotate α_2 firstly, then rotate α_1; Thus $(\mathfrak{R}, \circ)$ is the rotation group.

(9) The **group of roots of unit circle**

$$\Omega_p = \left\{\exp\left(\frac{\mathrm{i}2k\pi}{p}\right): k = 0, 1, 2, \cdots, p-1\right\},$$

with the operation of **multiplication of complexes** $\times$, and $p \geqslant 2$ is an integer, then $(\Omega_p, \times)$ is a group.

Example 1.2.6 (subgroup) For a subset $H \subset G$ of given group $(G, \cdot)$, if H is also a group with the operation $\cdot$ of G, then H is called a **subgroup of** G.

For example, $GL(n, \mathbb{R})$ is a subgroup of $GL(n, \mathbb{C})$; and $SL(n, \mathbb{R})$ is a subgroup of $SL(n, \mathbb{C})$, i,e, $SL(n, \mathbb{R}) \subset SL(n, \mathbb{C})$.

2. Rings, fields

Definition 1.2.2 (ring, field) If we define two operations on a given set R, denoted by **addition** $+$ **and multiplication** $\times$, respectively, then

1) **Ring**—If the operations $+$ and $\times$ on R satisfy

(1) R is an Abelian group with $+$; the unit of $+$ is 0, called **zero element** of R;

(2) R is a semi-group with $\times$; i. e., holding the following

$$x, y \in R \Rightarrow x \times y \in R; \quad x, y, z \in R \Rightarrow (x \times y) \times z = x \times (y \times z);$$

(3) $+$ and $\times$ on R hold the following the distributive laws

$$x, y, z \in R \Rightarrow x \times (y + z) = x \times y + x \times z, \quad (y + z) \times x = y \times x + z \times x,$$

then R is said to be a **ring** with $+$ and $\times$, denoted by $(R, +, \times)$.

2) **commutative ring**—For a given ring $(R, +, \times)$, if

(4) R satisfies the commutative law with $\times$, i. e., $x, y \in R \Rightarrow x \times y = y \times x$, then $(R, +, \times)$ is said to be an **commutative ring**; thus, an exchange ring satisfies (1)-(4).

3) **Ring with unit element** for $\times$—For a given ring $(R, +, \times)$, if

(5) R has a unit element of $\times$, denoted by 1, i. e., $\forall x \in R \Rightarrow 1 \times x = x \times 1 = x$, then $(R, +, \times)$ is said to be a **ring with unit** 1; thus, a ring with unit 1 satisfies (1)-(3), (5); a commutative ring with unit 1 satisfies (1)-(5).

Note In a ring $(R, +, \times)$ with unit 1, each element is not necessary to have an inverse about the operator $\times$, for example, in the integer ring $(\mathbb{Z}, +, \times)$, only two elements $+1$ and -1 have inverses about $\times$, but other integers do not have inverse.

4) **Integer ring**—If $(R, +, \times)$ is a commutative ring with unit satisfying

(6) there is no "zero factor" in R, then $(R, +, \times)$ is said to be an **integer ring**; Thus, an integer ring satisfies (1)-(6).

zero factor of a ring is defined as: For a given ring $(R, +, \times)$, if $a \neq 0, b \neq 0, a, b \in R$, imply that $a \times b = 0$, then a is called **a left zero factor about** b; and b is called **a right zero factor about** a;

If R is a commutative ring, then left and right zero factors both are called **zero factors.**

5) **Division ring** (or **body**)—If a given ring $(R, +, \times)$ with unit satisfies

(7) R contains at least one nonzero element, i. e., $\exists a \in R, a \neq 0$;

(8) each nonzero element $x \in R$ $(x \neq 0)$ has an inverse element $x^{-1} \in R$ with $x^{-1}x = 1$, then $(R, +, \times)$ is said to be **a division ring**(body). Thus, a division ring satisfies, (1)-(3), (5), (7) and (8).

6) **Field**—A commutative division ring is said to be a **field**, denoted by $(F, +, \times)$.

We have the following propositions.

Proposition 1.2.1 (1) A division ring R does not have zero factor;

(2) If R is a division ring, then $\{R \backslash \{0\}, \times\}$ is a group with $\times$.

Proposition 1.2.2 (1) A field F does not have zero factor;

(2) A field F makes $F \backslash \{0\}$ is a commutative group with multiplication.

By the above Propositions, a field satisfies (1)-(8).

An equivalent definition of the field If a given set F has two operations $+$ and $\times$, such that both $(F, +)$ and $(F \backslash \{0\}, \times)$ are commutative groups; moreover, $+$ and $\times$ satisfy the distributive laws; then $(F, +, \times)$ is said to be a field.

Example 1.2.7 $(\mathbb{R}, +, \times)$ is a familiar ring, it is a field also, where $+$ and $\times$ are **the addition and multiplication of real numbers** in $\mathbb{R}$.

In fact, the set $\mathbb{Z}$, $\mathbb{Q}$, $\mathbb{R}$ and $\mathbb{C}$ are rings with usual $+$ and $\times$; all of them are commutative rings.

Example 1.2.8 **Dyadic Galois field** $\mathbb{Z}_2 = \{0, 1\}$, its operations $\oplus$ and $\otimes$ are defined by

$\oplus$	0	1
0	0	1
1	1	0

$\otimes$	0	1
0	0	0
1	0	1

As we know, this field is very useful in the computer science and physics.

3. Ideals, and principal ideals

Definition 1.2.3 (ideal) If $(R, +, \times)$ is a ring, and its subset J of ring R satisfies

(1) $(J, +)$ is a commutative group with the addition $+$, the zero element is $0 \in J$;

(2) If $a\in J, b\in R$, then $a\times b\in J, b\times a\in J$;

Then J is said to be an **ideal** of R.

The condition (1) is equivalent to that $a, b\in J\Rightarrow a-b\in J$.

Definition 1.2.3′ If one element $a\in R$ in a given ring $(R,+,\times)$ generates an ideal, denoted by $\langle a\rangle=\{r\,a: r\in R\}$, then the $\langle a\rangle$ is said to be a **principal ideal** generated by $a\in R$.

If $(R,+,\times)$ is a ring, and all ideals of it are principal ideals, then $(R,+,\times)$ is said to be **a principal ideal integer ring.**

Example 1.2.9 The integer number ring $(\mathbb{Z},+,\times)$ is a principal ideal integer ring, since any ideal $J\subset\mathbb{Z}$ in $\mathbb{Z}$ is generated by the minimal positive integer in J.

Example 1.2.10 A set consisted of one-variable monic polynomials (their coefficients of first terms are 1) with coefficients in given field F, denoted by $F[x]$, is a commutative ring with unit, and is an integer ring, is a principal ideal integer ring too.

4. Modules over a ring

Definition 1.2.4 (module over a ring) Let $(R,+,\times)$ be a commutative ring with unit 1. A set M is said to be **a module over ring** R, if there are two operations $(+)_M$ and $(\alpha\cdot)_M$ on M, such that.

"addition" $(+)_M$: $(u,v)\in M\times M\Rightarrow u(+)_M v\in M$;

"module product" $(\alpha\cdot)_M$: $(r,u)\in R\times M\Rightarrow ru\equiv r(\alpha\cdot)_M u\in M$,

satisfying

(1) $(M,(+)_M)$ is a commutative group with $(+)_M$, the zero element is $0_M\in M$;

(2) for $u,v\in M$ and $r,s\in R$, the distribution law of $(+)_M$ and $(\alpha\cdot)_M$ holds,

$$r(\alpha\cdot)_M(u(+)_M v)=\{r(\alpha\cdot)_M u\}(+)_M\{r(\alpha\cdot)_M v\};$$

$$(r(+)_M s)(\alpha\cdot)_M u=\{r(\alpha\cdot)_M u\}(+)_M\{s(\alpha\cdot)_M u\}.$$

Moreover, the combination law of $\times$ and $(\alpha\cdot)_M$ holds,

$$(r\times s)(\alpha\cdot)_M u=r(\alpha\cdot)_M(s(\alpha\cdot)_M u);\quad u\in M, 1\in R\Rightarrow 1(\alpha\cdot)_M u=u,$$

Then, **a module over ring** $(R,+,\times)$ is denoted by $(M,+_M,(\alpha\cdot)_M)$.

Example 1.2.11 The module M on number field $\mathbb{F}$ ($\mathbb{R}$ or $\mathbb{C}$) is the familiar linear space; i.e., **a module** M **over number fields** $\mathbb{F}$ is a (real or complex) linear space.

Example 1.2.12 The module M over the integer ring $\mathbb{Z}$ is an Abelian group.

Example 1.2.13 The module M over a principal ideal integer ring $F[x]$ on a field F, is called an $F[x]$-module, where $F[x]$ is the set of all monic polynomials $m(x)$, (see Example 1.2.10).

We have more examples:

If $(R,+,\times)$ is a ring, then the set $\mathfrak{M}_{m\times n}(R)$ of all $m\times n$ matrices is an R-modulus; Its operations $(+)_M$ and $(\alpha\cdot)_M$ are the addition and number product of matrices.

If R is $F[x]$ in Example 1. 2. 13, then the set $\mathfrak{M}_{m\times n}(F[x])$ is the set of all $m\times n$ matrices whose elements are polynomials, called a $F[x]$-module.

If $(R,+,\times)$ is a ring, then R itself is **an R-module**, i. e., the set $R(R)$ is a special R-module.

5. Linear spaces

"Linear space" is a very important operation structure. In a linear space, there are two operations, the addition and the number multiplication. Let us recall the familiar 3-dimensional vector space (linear space) $\mathbb{R}^3$, firstly.

$$\mathbb{R}^3=\{x=(x_1,x_2,x_3):\ x_1,x_2,x_3\in\mathbb{R}\}.$$

The "**addition** $+$" and "**number product** $\alpha\cdot$" (or "scalar product") are defined as

$$x+y=(x_1,x_2,x_3)+(y_1,y_2,y_3)=(x_1+y_1,x_2+y_2,x_3+y_3),x,y\in\mathbb{R}^3;$$

$$\alpha\cdot x=\alpha(x_1,x_2,x_3)=(\alpha x_1,\alpha x_2,\alpha x_3),\quad \alpha\in\mathbb{R},\quad x\in\mathbb{R}^3.$$

It is clear that the "$+$" satisfies the Definition 1. 2. 1, thus $(\mathbb{R}^3,+)$ is an Abelian group; For the number product "$\alpha\cdot$" satisfies the distribution law and combination law. Consider $(\mathbb{R}^3,+)$ as a motivation, we turn to define "linear space" (or "vector space").

Definition 1. 2. 5 (linear space) Let X be a given set, its elements be $x,y,\cdots\in X$, and $\mathbb{F}$ be a number field, either real number field $\mathbb{F}=\mathbb{R}$, or complex number field $\mathbb{F}=\mathbb{C}$. If there are two operations, the addition $+$ and the number multiplication $\alpha\cdot$, $\alpha\in\mathbb{F}$, satisfying the following properties:

(1) X is an Abelian group with operation $+$, i. e.,

① $x,y\in X\Rightarrow x+y$; (closed property)

② $x,y,z\in X\Rightarrow (x+y)+z=x+(y+z)$; (combination law)

③ $x,y\in X\Rightarrow x+y=y+x$; (commutative law)

④ It exists unique zero element $0\in X$, s. t. $\forall x\in X\Rightarrow x+0=x$; (zero element)

⑤ $\forall x\in X$, there exists unique inverse $(-x)\in X$, s. t. $x+(-x)=0$. (inverse element)

(2) "Uumber product $\alpha\cdot$" for $\alpha\in F$ on X satisfies

① $x\in X,\alpha\in\mathbb{F}\Rightarrow\alpha\cdot x\in X$; $x\in X,1\in\mathbb{F}\Rightarrow 1\cdot x=x$; (closed property)

② $x,y\in X,\alpha,\beta\in\mathbb{F}\Rightarrow\alpha\cdot(x+y)=\alpha\cdot x+\alpha\cdot y$,

$(\alpha+\beta)\cdot x=\alpha\cdot x+\beta\cdot x$; (distribution law)

③ $x\in X,\alpha,\beta\in\mathbb{F}\Rightarrow(\alpha\beta)\cdot x=\alpha\cdot(\beta\cdot x)=\beta\cdot(\alpha\cdot x)$. (combination law)

Then X is said to be **a linear space on number field F**, simply, **linear space** (or **vector space**); its element is said to be **a vector.**

We agree on: omitting the sign $\cdot$ in $\alpha\cdot$, i. e. simplifying $\alpha\cdot x$ as $\alpha\cdot x$, if there is no confusion.

$\mathbb{R}^3$ is a linear space on $\mathbb{F}=\mathbb{R}$, called **real linear space**, or called **3-dimensional Euclidean space.** In general, $\mathbb{R}^n=\{x=(x_1,x_2,\cdots,x_n): x_1,x_2,\cdots,x_n\in\mathbb{R}\}$ is an n-**dimensional linear space**, or n-dimensional Euclidean space.

There are more examples of linear spaces.

Example 1. 2. 14 $C([a,b])=\{f: [a,b]\to\mathbb{R}, f(x)$ is continuous on $[a,b]\}$.

Define the addition and the number product of $f,g\in C[a,b]$ on $[a,b]$:

$$(f+g)(x)=f(x)+g(x),\quad x\in[a,b];$$

$$(\alpha f)(x)=\alpha f(x),\quad x\in[a,b],\alpha\in\mathbb{R},$$

then $C[a,b]$ is a linear space on $\mathbb{R}$, called **the continuous function space on** $[a,b]$.

We will meet lots of linear spaces here affer, they play very important roles in modern mathematics and many other scientific areas.

1. 2. 2 Group theory, some important groups

A group $G\equiv(G,\cdot)$, defined in Definition 1. 2. 1, is regarded as an algebraic system. Its operation $\cdot$, so-called "multiplication", can be endowed with very general meanings satisfying the properties listed in Definition 1. 2. 1. Then G becomes a group, and $(G,\cdot)$ has the structure of groups. The structure of groups is one kind of operation structures; and the structures of rings, fields, linear spaces are all operation structures. We concentrate our mind on learning susb structure of groups in this subsection.

1. Finite groups and infinite group

The numbers of elements of a group $(G,\cdot)$ may be finite, or infinite.

Definition 1. 2. 6 (finite group, infinite group) Let $(G,\cdot)$ be a group. If the element number of G is a positive integer, then G is said to **a finite group**; otherwise, it is said to be **an infinite group.**

The examples 1. 2. 6, (4) and (5) are finite groups. The numbers of them is p and 4, respectively.

Both finite and infinite groups have the following important properties.

Theorem 1. 2. 1 Let $(G,\cdot)$ be a group, then the operation $\cdot$ satisfies the eliminated law:

(1) for $a,x,x'\in G$, if $ax=ax'$, then $x=x'$;

(2) for $a, y, y' \in G$, if $ya = y'a$, then $y = y'$.

Proof Only prove (1), by

$$ax = ax' \Rightarrow a^{-1}(ax) = a^{-1}(ax') \Rightarrow (a^{-1}a)x = (a^{-1}a)x' \Rightarrow x = x'.$$

Definition 1.2.7 (degree of finite group G, order of element $a \in G$) Let $(G, \cdot)$ be a group with unit element $e \in G$. If G is a finite group, then the number of elements of G is said to be **the degree of group G**.

For an element $a \in G$, the minimal positive integer $m \in \mathbf{Z}^+$ satisfying $a^m = e$, is said to be **the order of element a in G**, and a is said to be m-**order**. If there is no such m, then a is said to be **infinite order**.

If operation of group is addition $+$, then $a^m = \underbrace{a + a + \cdots + a}_{m}$.

Example 1.2.15 Let G be the set consisted by the 3 roots of $x^3 = 1$. Take operation $\cdot$ as the multiplication of complex numbers, then $(G, \cdot)$ is a group.

Proof It is easy to check that $G = \left\{ e = 1, \varepsilon_1 = \dfrac{-1+\sqrt{-3}}{2}, \varepsilon_2 = \dfrac{-1-\sqrt{-3}}{2} \right\}$ is a group with $\times$. And $e^{-1} = e, \varepsilon_1^{-1} = \varepsilon_2, \varepsilon_2^{-1} = \varepsilon_1$.

Moreover, e is $m = 1$-order, ε_1 is $m = 3$-order, ε_2 is $m = 3$-order.

2. Homomorphisms and isomorphisms

Definition 1.2.8 (homomorphism, isomorphism) Let $(G, (\cdot)_G)$ and $(H, (\cdot)_H)$ be groups, $f: G \to H$ be a mapping from G to H. For $\alpha, \beta \in G$, If the images $f(\alpha), f(\beta) \in H$ satisfy

$$f(\alpha (\cdot)_G \beta) = f(\alpha)(\cdot)_H f(\beta),$$

then f is said to be **a homomorphism mapping** from G to H, or simply, f is a homomorphism. If f is one-one, onto homomorphism, i. e., injection and surjection homomorphism, then f is said to be **an isomorphism mapping** from G onto H, or simply, f is an isomorphism.

In other words, a homomorphism $f: G \to H$ is a mapping that keeps the operations of the groups.

Example 1.2.16 The familiar example of isomorphism is log: $\mathbf{R}^+ \to \mathbf{R}$, it is an isomorphism from group $(\mathbf{R}^+, \times)$ to $(\mathbf{R}, +)$, keeping operations: $\log(x \times y) = \log x + \log y$.

Example 1.2.17 Let $G, \tilde{G}$ be two groups, $\cdot$, $\tilde{\cdot}$ be the operations of them, $e, \tilde{e}$ be unit elements of them, respectively. Let $f: G \to \tilde{G}$, $f(x) = \tilde{e}$, $\forall x \in G$. Then f is a homomorphism from G to $\widetilde{G}$, since for $z=x\cdot y\in G$, we have $f(x\cdot y)=f(z)=\tilde{e}=\tilde{e}\ \tilde{\cdot}\ \tilde{e}=f(x)\ \tilde{\cdot}\ f(y)$.

Example 1.2.18 Let $(\mathbb{R}, +)$ and $(G, \times)$ be two groups, where $G=\{z\in\mathbb{C}: |z|=1\}$ be the unit circle in $\mathbb{C}$ with $\times$. Suppose that mapping $\varphi: \mathbb{R}\to\mathbb{C}$ is $\varphi(x)=e^{ix}$, then φ is a homomorphism from $\mathbb{R}$ to $\mathbb{C}$.

Theorem 1.2.2 Let G and $\widetilde{G}$ be two groups, a mapping $f: G\to\widetilde{G}$ be a surjection and homomorphism, then the image of unit e of G, i.e., $f(e)$, is the unit $\tilde{e}=f(e)$ of $\widetilde{G}$. Moreover, the image $f(a^{-1})$ of the inverse a^{-1} of any $a\in G$, is the inverse $(f(a))^{-1}$ of image $f(a)$ of a, i.e., $f(a^{-1})=(f(a))^{-1}$.

The proof of Theorem is left as the exercise.

We introduce some kinds of important groups, such as the transformation groups (non-commutative groups), the replacement groups (commutative groups), and the circulate groups, they are generally useful widely.

3. Transformation groups

Definition 1.2.9 (transformation of set) Let A be a given set (finite, or infinite). A mapping from A to itself $\tau: A\to A$ with $a\in A\to\tau(a)\in A$ is said to be **a transformation on** A.

Let $\mathbb{S}=\{\tau: A\to A: \tau$ is one-one, onto$\}$ be the set of all injection and surjection mapping from A onto A. Endow the operation $\circ$ on $\mathbb{S}$: $\forall \tau_1, \tau_2\in\mathbb{S}$, operation $\circ$ is the compound of τ_1 and τ_2,

$$\tau_1\circ\tau_2(a)=\tau_1(\tau_2(a)),\quad \forall a\in G.$$

Thus, $(\mathbb{S}, \circ)$ is a group, and is called **a transformation group on set** A.

The transformation group $(\mathbb{S}, \circ)$ on A exists, nature. The compound operation $\circ$ satisfies the closed property, and combination law, exists unit (the identity mapping), as well as exists the inverse (transformation); However, $(\mathbb{S}, \circ)$ is not commutative group.

Example 1.2.19 Let $A=\mathbb{R}^2$. The transformation τ_θ is defined by that τ_θ rotates the Cartesian coordinate system in $\mathbb{R}^2$ around origin $O(0,0)$ with angle θ, then the set $\mathbb{S}=\{\tau_\theta: \mathbb{R}^2\to\mathbb{R}^2\}$ is a transformation group. τ_θ is called **a rotation by angle** θ.

Proof For $\tau_\varphi, \tau_\psi, \tau_\chi\in\mathbb{S}$, then

(1) $\tau_\varphi\circ\tau_\psi=\tau_{\varphi+\psi}$;

(2) $(\tau_\varphi\circ\tau_\psi)\circ\tau_\chi=\tau_\varphi\circ(\tau_\psi\circ\tau_\chi)$;

(3) $\tau_0\in\mathbb{S}$;

(4) $(\tau_\theta)^{-1}=\tau_{-\theta}$.

Example 1.2.20 Let $A=\mathbb{R}^2$; τ denote a rotation on $\mathbb{R}^2$, t **a translation** on $\mathbb{R}^2$. The set of all rotations and translations on $\mathbb{R}^2$, denoted by $\mathbb{T}$, endowed operation "compound" $\circ$, then $\mathbb{T}$ constrcts a transformation group on $A=\mathbb{R}^2$. It is a non-commutative group.

In fact, take translation t on $A=\mathbb{R}^2$: t translates $(0,0)$ to $(1,0)$; then take rotation τ, it rotates around $(0,0)$ with $\theta=\frac{\pi}{2}$. We have

$$t\circ\tau:(0,0)\to(1,0),\quad \tau\circ t:(0,0)\to(0,1),$$

clearly, $t\circ\tau\neq\tau\circ t$.

The transformation group has the following important property.

Theorem 1.2.3 Any group $(G,\cdot)$ is isomorphic with a transformation group.

Proof Let $(G,\cdot)$ be the group, $G=\{a,b,c,\cdots\}$. Take any $x\in G$, define a mapping depending on x,

$$\tau_x: g\to x\cdot g,\quad \forall g\in G$$

and denote $\tau_x(g)=x\cdot g$: then τ_x is a mapping on G. Hence, $\forall x\in G$ determines a transformation on G. Denote $\widetilde{G}=\{\tau_x: x\in G\}$. Thus,

$$\varphi: x\to\tau_x\in\widetilde{G},\quad \forall x\in G$$

is a surjection from G onto $\widetilde{G}$. Moreover, if $x,y\in G, x\neq y$, then $xg\neq yg$, $\forall g\in G$, by the elimination law. We conclude that $\varphi: x\to\tau_x$ is an injection, so it is a one-one mapping.

Then, we check that $\varphi: x\to\tau_x$ is isomorphism, $\tau_{x\cdot y}=\tau_x\circ\tau_y$. Take any $g\in G$, we deduce:

$$\tau_{x\cdot y}(g)=(x\cdot y)g=x(yg)=x(\tau_y(g))=\tau_x(\tau_y(g))=(\tau_x\circ\tau_y)(g).$$

The unit is the mapping $\varphi: e\to\tau_e$. So, $G=\{a,b,c,\cdots\}\overset{\text{iso.}}{\longleftrightarrow}\widetilde{G}=\{\tau_x: x\in G\}$.

4. Replacement groups

The replacement group is one of important groups with lots of applications in theoretical science areas and applied science areas.

Definition 1.2.10 (replacement, replacement group) Let $\alpha_1,\alpha_2,\cdots,\alpha_n$ be n given elements, $n\in\mathbb{Z}^+$; denoted by $\Omega=\{\alpha_1,\alpha_2,\cdots,\alpha_n\}$. Consider the set

$$A=\{(\alpha_{k_1}\ \alpha_{k_2}\ \cdots\ \alpha_{k_n}): \alpha_{k_j}\in\Omega,\quad j=1,2,\cdots,n\},$$

i.e., an element in A is as $(\alpha_{k_1}\ \alpha_{k_2}\cdots\alpha_{k_n})$, where $k_1,k_2,\cdots,k_n$ is **a permutation** of $1,2,\cdots,n$.

(1) **Replacement** — An "one-one" mapping from A to itself, $\tau: A\to A$, is said to be **a replacement on** A; or simply, **a replacement.**

A replacement $\tau: A\to A$ on A is denoted by

$$\tau=\begin{pmatrix}\alpha_1 & \alpha_2 & \cdots & \alpha_n\\ \alpha_{k_1} & \alpha_{k_2} & \cdots & \alpha_{k_n}\end{pmatrix},$$

where $k_1, k_2, \cdots, k_n$ is **a permutation of** $1, 2, \cdots, n$; it means that: a replacement τ maps element $(\alpha_1\ \alpha_2 \cdots \alpha_n) \in A$ to element $(\alpha_{k_1}\ \alpha_{k_2} \cdots \alpha_{k_n}) \in A$; or precisely,

$$\tau: \alpha_1 \xrightarrow{\text{maps to}} \alpha_{k_1}, \alpha_2 \xrightarrow{\text{maps to}} \alpha_{k_2}, \cdots, \alpha_n \xrightarrow{\text{maps to}} \alpha_{k_n}.$$

(2) n**-replacement group** — We define the set S_n (all replacements on A):

$$S_n=\left\{\tau=\begin{pmatrix}1 & 2 & \cdots & n\\ k_1 & k_2 & \cdots & k_n\end{pmatrix}: 1 \leqslant k_j \leqslant n, k_i \neq k_j, \text{if } i \neq j\right\};$$

Endow with operation "composition": $\tau, \nu \in S_n \Rightarrow \tau \circ \nu \in S_n$; then $(S_n, \circ)$ constructs a group, it is called **the n-replacement group**, or called **the symmetric group of n-degree.**

(3) **Operation on** $(S_n, \circ)$ —

① **Representations of a replacement**: $\forall \tau \in S_n$ has $n!$-representations

$$\tau=\begin{pmatrix}1 & 2 & 3 & \cdots & n\\ k_1 & k_2 & k_3 & \cdots & k_n\end{pmatrix}=\begin{pmatrix}2 & 1 & 3 & \cdots & n\\ k_2 & k_1 & k_3 & \cdots & k_n\end{pmatrix}=\begin{pmatrix}3 & 2 & 1 & \cdots & n\\ k_1 & k_2 & k_1 & \cdots & k_n\end{pmatrix}$$

$$=\begin{pmatrix}4 & 2 & 3 & 1 & \cdots & n\\ k_4 & k_2 & k_3 & k_1 & \cdots & k_n\end{pmatrix}=\begin{pmatrix}5 & 2 & 3 & 4 & 1 & \cdots & n\\ k_5 & k_2 & k_3 & k_4 & k_1 & \cdots & k_n\end{pmatrix}=\cdots$$

$$=\begin{pmatrix}n & 2 & 3 & 4 & \cdots & 1\\ k_n & k_2 & k_3 & k_4 & \cdots & k_1\end{pmatrix}.$$

They represent the same element $\tau=\begin{pmatrix}1 & 2 & 3 & \cdots & n\\ k_1 & k_2 & k_3 & \cdots & k_n\end{pmatrix}$, all of them means that:

$$\tau: 1 \xrightarrow{\text{replace by}} k_1, 2 \xrightarrow{\text{replace by}} k_2, \cdots, n \xrightarrow{\text{replace by}} k_n.$$

For example, for $n=3$, there are 6 elements in S_3

$$S_3=\left\{\begin{pmatrix}1 & 2 & 3\\ 1 & 2 & 3\end{pmatrix}, \begin{pmatrix}1 & 2 & 3\\ 1 & 3 & 2\end{pmatrix}, \begin{pmatrix}1 & 2 & 3\\ 2 & 1 & 3\end{pmatrix}, \begin{pmatrix}1 & 2 & 3\\ 2 & 3 & 1\end{pmatrix}, \begin{pmatrix}1 & 2 & 3\\ 3 & 1 & 2\end{pmatrix}, \begin{pmatrix}1 & 2 & 3\\ 3 & 2 & 1\end{pmatrix}\right\},$$

each element has 6 forms, for example, $\tau=\begin{pmatrix}1 & 2 & 3\\ 2 & 1 & 3\end{pmatrix}$ has:

$$\tau=\begin{pmatrix}1 & 2 & 3\\ 2 & 3 & 1\end{pmatrix}=\begin{pmatrix}1 & 3 & 2\\ 2 & 1 & 3\end{pmatrix}=\begin{pmatrix}2 & 1 & 3\\ 3 & 2 & 1\end{pmatrix}=\begin{pmatrix}2 & 3 & 1\\ 3 & 1 & 2\end{pmatrix}=\begin{pmatrix}3 & 1 & 2\\ 1 & 2 & 3\end{pmatrix}=\begin{pmatrix}3 & 2 & 1\\ 1 & 3 & 2\end{pmatrix}.$$

We use $\tau=\begin{pmatrix}1 & 2 & \cdots & n\\ k_1 & k_2 & \cdots & k_n\end{pmatrix}$, $(1 \leqslant k_j \leqslant n, k_i \neq k_j, \text{if } i \neq j)$ to represent an element, i. e. , a replacement in **the replacement group** S_n.

② **Operation on replacement group** S_n: $\circ$ is compound of replacements τ_1 and τ_2

$\tau_2 \circ \tau_1$: perform τ_1 first, then τ_2:

$$\tau_1=\begin{pmatrix}1 & 2 & \cdots & n\\ j_1 & j_2 & \cdots & j_n\end{pmatrix},\quad \tau_2=\begin{pmatrix}1 & 2 & \cdots & n\\ k_1 & k_2 & \cdots & k_n\end{pmatrix}$$

$$\Rightarrow\quad \tau_2\circ\tau_1=\begin{pmatrix}1 & 2 & \cdots & n\\ k_1 & k_2 & \cdots & k_n\end{pmatrix}\circ\begin{pmatrix}1 & 2 & \cdots & n\\ j_1 & j_2 & \cdots & j_n\end{pmatrix}$$

$$\equiv\begin{pmatrix}1 & 2 & \cdots & n\\ k_1 & k_2 & \cdots & k_n\end{pmatrix}\begin{pmatrix}1 & 2 & \cdots & n\\ j_1 & j_2 & \cdots & j_n\end{pmatrix}\equiv\tau_2\tau_1$$

It is easy to check that S_n is a group with operation $\circ$.

③ **The rules of operation on replacement group**:

(i) write down $\tau_2\circ\tau_1$

$$\tau_2\circ\tau_1=\begin{pmatrix}1 & 2 & \cdots & n\\ k_1 & k_2 & \cdots & k_n\end{pmatrix}\begin{pmatrix}1 & 2 & \cdots & n\\ j_1 & j_2 & \cdots & j_n\end{pmatrix};$$

(ii) replace the first row $(1\quad 2\quad \cdots\quad n)$ in τ_2 by the second row $(j_1\quad j_2\quad \cdots\quad j_n)$ in τ_1. Then the second row in τ_2 is as $(k_1\quad k_2\quad \cdots\quad k_n)\to(k'_1\quad k'_2\quad \cdots\quad k'_n)$; i. e.,

$$\tau_2\to\tau'_2=\begin{pmatrix}j_1 & j_2 & \cdots & j_n\\ k'_1 & k'_2 & \cdots & k'_n\end{pmatrix}.$$

Thus we have

$$\tau_2\circ\tau_1\to\tau'_2\circ\tau_1=\begin{pmatrix}j_1 & j_2 & \cdots & j_n\\ k'_1 & k'_2 & \cdots & k'_n\end{pmatrix}\begin{pmatrix}1 & 2 & \cdots & n\\ j_1 & j_2 & \cdots & j_n\end{pmatrix}$$

$$\equiv\begin{pmatrix}1 & 2 & \cdots & n\\ k'_1 & k'_2 & \cdots & k'_n\end{pmatrix}.$$

For example, when $n=3$, take $\tau_2\circ\tau_1=\begin{pmatrix}1 & 2 & 3\\ 1 & 3 & 2\end{pmatrix}\begin{pmatrix}1 & 2 & 3\\ 2 & 1 & 3\end{pmatrix}$; then

$$\tau_2=\begin{pmatrix}1 & 2 & 3\\ 1 & 3 & 2\end{pmatrix}\xrightarrow{\text{replace by}}\tau'_2=\begin{pmatrix}2 & 1 & 3\\ 3 & 1 & 2\end{pmatrix};$$

Thus,

$$\tau_2\circ\tau_1=\begin{pmatrix}1 & 2 & 3\\ 1 & 3 & 2\end{pmatrix}\begin{pmatrix}1 & 2 & 3\\ 2 & 1 & 3\end{pmatrix}\to\tau'_2\circ\tau_1=\begin{pmatrix}2 & 1 & 3\\ 3 & 1 & 2\end{pmatrix}\begin{pmatrix}1 & 2 & 3\\ 2 & 1 & 3\end{pmatrix}=\begin{pmatrix}1 & 2 & 3\\ 3 & 1 & 2\end{pmatrix};$$

We have $\quad\tau_2\circ\tau_1=\begin{pmatrix}1 & 2 & 3\\ 3 & 1 & 2\end{pmatrix}$.

Note The replacement group is a non-commutative group. For example,

$$\begin{pmatrix}1 & 2 & 3\\ 1 & 3 & 2\end{pmatrix}\begin{pmatrix}1 & 2 & 3\\ 2 & 1 & 3\end{pmatrix}=\begin{pmatrix}2 & 1 & 3\\ 3 & 1 & 2\end{pmatrix}\begin{pmatrix}1 & 2 & 3\\ 2 & 1 & 3\end{pmatrix}=\begin{pmatrix}1 & 2 & 3\\ 3 & 1 & 2\end{pmatrix};$$

However, we have

$$\begin{pmatrix}1&2&3\\2&1&3\end{pmatrix}\begin{pmatrix}1&2&3\\1&3&2\end{pmatrix}=\begin{pmatrix}1&3&2\\2&3&1\end{pmatrix}\begin{pmatrix}1&2&3\\1&3&2\end{pmatrix}=\begin{pmatrix}1&2&3\\2&3&1\end{pmatrix}.$$

Theorem 1.2.4 *The n-replacement group S_n has $n!$ elements.*

Proof From the course in elementary mathematics, we use the concept of arrangement.

Proposition 1.2.3 *Let S_n be n-replacement group. For two special replacements*

$$\tau_1=\begin{pmatrix}j_1&\cdots&j_k&j_{k+1}&\cdots&j_n\\j_1^{(1)}&\cdots&j_k^{(1)}&j_{k+1}&\cdots&j_n\end{pmatrix},\quad \tau_2=\begin{pmatrix}j_1&\cdots&j_k&j_{k+1}&\cdots&j_n\\j_1&\cdots&j_k&j_{k+1}^{(2)}&\cdots&j_n^{(2)}\end{pmatrix},$$

we have

$$\tau_2\circ\tau_1=\begin{pmatrix}j_1&\cdots&j_k&j_{k+1}&\cdots&j_n\\j_1^{(1)}&\cdots&j_k^{(1)}&j_{k+1}^{(2)}&\cdots&j_n^{(2)}\end{pmatrix}.$$

Proof We rew rite τ_1 by

$$\tau_1:(j_1\ \cdots\ j_k\ j_{k+1}\ \cdots\ j_n)\to(j_1^{(1)}\ \cdots\ j_k^{(1)}\ j_{k+1}\ \cdots\ j_n),$$

keep $j_{k+1},\cdots,j_n$, but for $l\leqslant k$, it holds

$$j_l^{(1)}=j_l,\quad l\leqslant k.$$

Thus, if $l\leqslant k$, then

$$\tau_2(\tau_1(j_l))=\tau_2(j_l^{(1)})=\tau_2(j_l)=j_l^{(1)};$$

if $l>k$, then

$$\tau_2(\tau_1(j_l))=\tau_2(j_l)=j_l^{(2)}.$$

So that $\tau_2\circ\tau_1=\begin{pmatrix}j_1&\cdots&j_k&j_{k+1}&\cdots&j_n\\j_1^{(1)}&\cdots&j_k^{(1)}&j_{k+1}^{(2)}&\cdots&j_n^{(2)}\end{pmatrix}$.

We have another representation of a replacement group defined as follows, called k-recurring replacement.

Definition 1.2.11 (k-**recurring replacement**) A replacement $\tau\in S_n$ in the replacement group S_n is said to be a k-**recurring replacement**, if τ transforms a_{i_1} to a_{i_2}, a_{i_2} to a_{i_3}, $\cdots$, $a_{i_{k-1}}$ to a_{i_k}, a_{i_k} to a_{i_1}, and keeps others.

In k-recurring replacement group S_n, an element $(a_{i_1}\ \cdots\ a_{i_k}\ a_{i_{k+1}}\ \cdots\ a_{i_n})$ is denoted by $(i_1\ i_2\ \cdots\ i_k)$, or by $(i_1\ i_2\ \cdots\ i_k)=(i_2\ i_3\ \cdots\ i_k\ i_1)=\cdots=(i_k\ i_1\ \cdots\ i_{k-1})$.

A replacement is not necessary a recurring one, for example, $\tau=\begin{pmatrix}1&2&3&4\\2&1&4&3\end{pmatrix}$ is not a recurring replacement in S_4, since τ is: $1\to2, 2\to1, 3\to4, 4\to3$; if it would be a recurring replacement, it must be 4-recurring one. But "$2\to1, 4\to3$" is not ordered.

However, $\tau=\begin{pmatrix}1&2&3&4\\2&1&4&3\end{pmatrix}$ is a compound of two 4-recurring replacements; i.e.,

$$\tau=\begin{pmatrix}1&2&3&4\\2&1&4&3\end{pmatrix}=\begin{pmatrix}1&2&3&4\\1&2&4&3\end{pmatrix}\begin{pmatrix}1&2&3&4\\2&1&3&4\end{pmatrix}=(3\quad 4)(1\quad 2).$$

Definition 1.2.12 (non-conjoint recurring, exchange) If two recurring replacements do not have common number, then they are said to be **non-conjoint recurring.** A 2-recurring replacement is said to be **an exchange.**

Theorem 1.2.5 In the n-replacement group S_n, any replacement $\tau\in S_n$ can be expressed as a compound of several non-conjoint recurring replacements; Moreover, each recurring replacement can be expressed as a compound of several exchanges.

As a conclusion, each replacement can be expressed as a compound of several exchanges.

Proof By induction.

In S_4, there are $4!=24$ elements, by recurring replacements:

(1);

(1 2), (3 4), (1 3), (2 4), (1 4), (2 3);

(1 2 3), (1 3 2), (1 3 4), (1 4 3), (1 2 4), (1 4 2), (2 3 4), (2 4 3);

(1 2 3 4), (1 2 4 3), (1 3 2 4), (1 3 4 2), (1 4 2 3), (1 4 3 2);

(1 2)(3 4), (1 3)(2 4), (1 4)(2 3).

In S_5, $\begin{pmatrix}1&2&3&4&5\\1&2&3&4&5\end{pmatrix}=(1)=(2)=(3)=(4)=(5)$;

$$\begin{pmatrix}1&2&3&4&5\\5&1&2&3&4\end{pmatrix}=(1\ 2\ 3\ 4\ 5)=(2\ 3\ 4\ 5\ 1)=(3\ 4\ 5\ 1\ 2)=(4\ 5\ 1\ 2\ 3)=(5\ 1\ 2\ 3\ 4);$$

$$\begin{pmatrix}1&2&3&4&5\\2&3&1&4&5\end{pmatrix}=(1\ 2\ 3)=(2\ 3\ 1);\ \cdots.$$

Theorem 1.2.6 Let S_n be an n-replacement group. Since each replacement can be expressed as a compound of several exchanges, thus the numbers of those exchanges do not change, and the odd or even of the numbers of those exchanges do not change also. Thus we conclude that: each replacement can be decomposed as a compound of odd interchanges, or even interchanges; one and only one of these two cases must be appeared.

Definition 1.2.13 (odd and even replacement) Let S_n be an n-replacement group. If $\tau\in S_n$ can be represented as a compound by odd numbers of replacements, then τ is said to be **an odd replacement**; if a replacement can be represented as a compound by even numbers of replacement, then it is said to be **an even replacement.**

We sometimes represent an odd or even replacement by the following forms.

Let $S=\{1,2,\cdots,n\}$ be a set having n elements, the n-permutation group on S denoted by

$$\mathfrak{T}(n) \equiv S_n = \{\sigma : \sigma\{1,2,\cdots,n\} = \{\sigma(1),\cdots,\sigma(n)\}\}.$$

Then $\mathfrak{T}(n)$ has $n!$ elements.

Define an operation $\circ$ on $\mathfrak{T}(n)$: for $\sigma,\tau \in \mathfrak{T}(n)$, $\tau\circ\sigma$ is the compound of σ and τ.

Thus, $(\mathfrak{T}(n),\circ)$ with the operation $\circ$ forms a **permutation group**, the unit of $\mathfrak{T}(n)$ is the identity mapping $I(s)=s, s\in\mathfrak{T}(n)$; and $\sigma\in\mathfrak{T}(n)\Rightarrow\tau^{-1}\in\mathfrak{T}(n)$.

Moreover, $\forall\sigma\in\mathfrak{T}(n)$, if $\sigma=\begin{pmatrix}1 & 2 & \cdots & n\\ \sigma(1) & \sigma(2) & \cdots & \sigma(n)\end{pmatrix}$, we take $\sigma(j)-\sigma(k)$, $k<j$, and $k, j\in\mathfrak{T}(n)$,

then the product $\prod\limits_{1\leqslant k<j\leqslant r}(\sigma(j)-\sigma(k))$ has a formula

$$J(\sigma)=\prod_{1\leqslant k<j\leqslant r}(\sigma(j)-\sigma(k))=\pm(2!)(3!)\cdots((n-1)!).$$

Definition 1.2.13′ (odd and even permutations) If the product $J(\sigma)$ is a positive number, then σ is said to be an **even permutation**; And if this product $J(\sigma)$ is a negative number, then σ is said to be an **odd permutation.**

$$J(\sigma)=\begin{cases}>0, & \text{then } \sigma \text{ is even permutation}\\ <0, & \text{then } \sigma \text{ is odd permutation}\end{cases}$$

5. Circulating groups

Definition 1.2.14 (circulating group) Let G be a group. If there exists an element $a\in G$, such that each element $x\in G$ can be expressed as a power of $a\in G$; i.e., $\exists a\in G$, such that for each element $x\in G$, $\exists m\equiv m(x)\in\mathbf{Z}^+$ satisfying $x=a^m$; then G is said to be a **circulating group**; this a is said to be the **generator of** G, and the circulating group is denoted by $G=(a)$.

For example, the p-adic group $G=\{0,1,\cdots,p-1\}$ with addition mod p, is a circulating group.

Theorem 1.2.7 The construction of a circulating group $G=(a)$ is determined fully by the order of its generator $a\in G$:

(1) if the order of a is infinite, then G is isomorphic with the integer number group $(\mathbf{Z},+)$;

(2) if the order of a is a finite positive integer $p\in\mathbf{Z}^+$, then G is isomorphic with the group $(\mathbf{Z}_p,\oplus)$.

Proof (1) If the order of a is infinite: We prove that $G\leftrightarrow(\mathbf{Z},+)$.

① We claim that $a^k=a^h\Leftrightarrow h=k$, firstly

"$\Rightarrow$" It is clear that: $h=k$ implies $a^k=a^h$.

"$\Leftarrow$" We prove that: $a^k = a^h$ implies $h = k$.

In fact, if $a^k = a^h$ would imply, say, $h > k$. Then we have

$$a^k = a^h \quad \Rightarrow \quad e = a^k a^{-k} = a^h a^{-k} = a^{h-k}.$$

This contradicts with that the infinite order of a, thus $h = k$.

Then, we have proved that $a^k = a^h \Leftrightarrow h = k$.

② We claim that $G \leftrightarrow (\mathbf{Z}, +)$, secondly

Since the transformation $\varphi: a^k \to k$ is an one-one mapping from G onto $\mathbf{Z}$; and $a^h a^k = a^{h+k} \leftrightarrow h + k$, so we have $G \leftrightarrow \mathbf{Z}$.

Then, (1) is proved.

(2) If the order of a is finite, $m \in \mathbf{Z}^+$, i. e., $a^m = e$: We prove that $G \leftrightarrow (\mathbf{Z}_m, \oplus)$

① We prove $a^k = a^h \Leftrightarrow m \mid h - k$, firstly

"$\Rightarrow$" It is clear that $a^h = a^k \Rightarrow h - k = mq + r$, where $0 \leqslant r \leqslant m - 1$, so that

$$e = a^{h-k} = a^{mq+r} = a^{mq} a^r = e a^r = a^r,$$

by the definition of order, we get $r = 0$. This implies $m \mid h - k$.

"$\Leftarrow$" If $m \mid h - k$, then $\exists\, q \in \mathbf{Z}^+$, such that

$$h - k = mq \Rightarrow h = k + mq,$$

thus

$$a^h = a^{k+nq} = a^k a^{nq} = a^k (a^n)^q = a^k e^q = a^k;$$

so we have $a^h = a^k$.

Combining the above two steps, we have got $a^k = a^h \Leftrightarrow m \mid h - k$.

② We claim that $G \leftrightarrow (\mathbf{Z}_m, \oplus)$, secondly

By $a^k \to [k]$ with $[k] \in \mathbf{Z}_m$, where $\mathbf{Z}_m = \{0, 1, \cdots, m-1\}$ is the mod m addition group, and the addition is $\oplus: x \oplus y = x + y \pmod m$. Thus

$$a^h a^k = a^{h+k} \to [h+k] = [h] + [k],$$

so that

$$G \leftrightarrow \mathbf{Z}_m = \{0, 1, \cdots, m-1\}.$$

And we conclude that $G \leftrightarrow (\mathbf{Z}_m, \oplus)$.

Then, (2) is proved.

Theorem 1.2.8 The circulating group $G = (a)$ is a commutative group.

The proof is left to readers.

1.2.3 Subgroups, product groups, quotient groups

We have shown some important concrete groups in 1.2.2, whereas certain new important and significant groups will be constructed in this section, such as, subgroup, product groups, quotient groups, they will play important roles in modern mathematics, modern physics, and all modern scientific areas.

1. Subgroups, normal subgroups

Definition 1.2.15 (subgroup, embedding mapping) Let $(G, \cdot)$ be a group, $S \subset G$ a subset of G. If S is a group with operation $\cdot$, denoted by $(S, \cdot)$, then it is said to be **a subgroup** of $(G, \cdot)$.

It is an important mapping between subgroup $S \subset G$ and G, called **embedding mapping**, $I: S \to G$, with $I(x) = x$, $\forall x \in S$, i.e., **the identity mapping** from S into G.

Theorem 1.2.9 The imbedding mapping $I: S \to G$ is a homomorphism from subgroup $(S, \cdot)$ into group $(G, \cdot)$.

The proof is left as an exercise.

Example 1.2.21 $(\mathbb{R}, +)$ is a group, and $(\mathbb{Q}, +)$, $(\mathbb{Z}, +)$ both are subgroups of $(\mathbb{R}, +)$.

Example 1.2.22 Let group $G = S_3$. Then $H = \left\{\begin{pmatrix}1 & 2 & 3\\1 & 2 & 3\end{pmatrix}, \begin{pmatrix}1 & 2 & 3\\2 & 1 & 3\end{pmatrix}\right\}$ is a subgroup of S_3.

Proof H has the operation "compound" as that in G, and this compound is closed in H. By

$$S_3 = \left\{\begin{pmatrix}1 & 2 & 3\\1 & 2 & 3\end{pmatrix}, \begin{pmatrix}1 & 2 & 3\\1 & 3 & 2\end{pmatrix}, \begin{pmatrix}1 & 2 & 3\\2 & 1 & 3\end{pmatrix}, \begin{pmatrix}1 & 2 & 3\\3 & 2 & 1\end{pmatrix}, \begin{pmatrix}1 & 2 & 3\\3 & 1 & 2\end{pmatrix}, \begin{pmatrix}1 & 2 & 3\\2 & 3 & 1\end{pmatrix}\right\}, \text{ and}$$

$\begin{pmatrix}1 & 2 & 3\\1 & 2 & 3\end{pmatrix}$ is the unit.

The subset $H = \left\{\begin{pmatrix}1 & 2 & 3\\1 & 2 & 3\end{pmatrix}, \begin{pmatrix}1 & 2 & 3\\2 & 1 & 3\end{pmatrix}\right\}$ contains two elements: one is the unit $\begin{pmatrix}1 & 2 & 3\\1 & 2 & 3\end{pmatrix}$ of H; the other one is $\begin{pmatrix}1 & 2 & 3\\2 & 1 & 3\end{pmatrix}$, its inverse is itself by $\begin{pmatrix}1 & 2 & 3\\2 & 1 & 3\end{pmatrix}\begin{pmatrix}1 & 2 & 3\\2 & 1 & 3\end{pmatrix} = \begin{pmatrix}1 & 2 & 3\\1 & 2 & 3\end{pmatrix}$; hence H is a subgroup of S_3.

Theorem 1.2.10 A subset $H \subset G$ is a subgroup of group G, if and only if

(1) $a, b \in H \Rightarrow ab \in H$;

(2) $a \in H \Rightarrow a^{-1} \in H$.

These statements are equivalent to:

A subset $H \subset G$ of group G is a subgroup, iff $a, b \in H \Rightarrow ab^{-1} \in H$.

Definition 1.2.16 (normal subgroup, coset and center) Let $(G, \cdot)$ be a group. If a subgroup $N \subset G$ of G satisfies that $\forall g \in G$ implies $gN = Ng$, then N is said to be **a normal subgroup**; or **an invariant subgroup of** G.

The sets $gN = \{gn : n \in N\}$ and $Ng = \{ng : n \in N\}$ are said to be **a left** and **a right coset of** N **about** G, respectively. If $gN = Ng$, then it is said to be **a coset** of N **about** G, or simply, **a coset of** N.

A normal subgroup $C \subset G$ of group G is said to be **a center**, if it satisfies

$$\forall c \in C, \quad \forall g \in G \Rightarrow cg = gc.$$

Theorem 1.2.11 If N is a subgroup of group G, then N is a normal subgroup, iff

$$aNa^{-1} = N, \quad \forall a \in G.$$

Proof **Necessity** If H is a normal subgroup, then $\forall a \in G$ implies $aN = Na$, $\forall a \in G$; thus indicate $aNa^{-1} = (aN)a^{-1} = (Na)a^{-1} = N(aa^{-1}) = N$. The necessity is proved.

Sufficiency If $aNa^{-1} = N$, $\forall a \in G$, then $Na = (aNa^{-1})a = (aN)(a^{-1}a) = (aN)e = aN$, thus N is a normal subgroup.

The numbers of cosets of normal subgroup N about G can be finite or infinite.

Definition 1.2.17 (index number of a normal subgroup) If the number of cosets of a normal subgroup N about group G is a finite integer j, then this j is said to be **the index number of** N **in** G.

Theorem 1.2.12 If G is a finite group, $H \subset G$ is a normal subgroup of G, then the order of normal subgroup H, say n, and its index number in G, say j, satisfy $N = nj$. Moreover, for any $a \in G$, the order of a is a factor of the order of G.

Proof Let the order of G be N, the order of H be n, and the index number of H in G be j, all the three be finite integers. Then G is divided by H into j cosets, and N must satisfy $N = nj$.

Example 1.2.23 The group S_3 has a subgroup $H = \{(1), (1\ 2)\}$, but it is not a normal subgroup of S_3, since

$$(1\ 3)\ H = \{(1\ 3), (1\ 2\ 3)\}, \quad H\ (1\ 3) = \{(1\ 3), (1\ 3\ 2)\}.$$

2. Product groups

Definition 1.2.18 (product group, project mapping) Let $(G_1, (\cdot)_1)$ and $(G_2, (\cdot)_2)$ be two groups. Define **the product set** of G_1 and G_2 as

$$G_1 \times G_2 = \{(g_1, g_2) : g_1 \in G_1, g_2 \in G_2\}.$$

For $(a_1, a_2) \in G_1 \times G_2$ and $(b_1, b_2) \in G_1 \times G_2$, the operation $\cdot$ in $G_1 \times G_2$ is defined by

$$(a_1, a_2)(b_1, b_2) \equiv (a_1, a_2) \cdot (b_1, b_2) = (a_1 (\cdot)_1 b_1, a_2 (\cdot)_2 b_2) \equiv (a_1 b_1, a_2 b_2).$$

It is easy to check that $(G_1 \times G_2, \cdot)$ is a group, called **the product group of** G_1 **and** G_2.

There is an important mapping $\mathrm{Pr}_1 : G \to G_1$ between the product group $G = G_1 \times G_2$

and group G_1, called **the projective mapping**; There is **such a mapping** $Pr_2: G \to G_2$ for product group $G = G_1 \times G_2$ and group G_2, they satisfy

$$Pr_1(g_1, g_2) = g_1, \quad \text{for } (g_1, g_2) \in G;$$

$$Pr_2(g_1, g_2) = g_2, \quad \text{for } (g_1, g_2) \in G,$$

and are homomorphism mappings from G to G_1 and G_2, respectively (see exercise).

3. Quotient groups

Definition 1.2.19 (equivalent class, quotient set) Let $(G, (\cdot)_G)$ be a group, $N \subset G$ be a normal subgroup of G. For $a, b \in G$, if $a^{-1}b \in N$, then b is said to be **equivalent to** a **about normal subgroup** N, or simply, b **is equivalent to** a, denoted by $b \sim a$.

It is clear that, $a^{-1}b \in N$ iff $b \in aN$.

The set of all elements in G which are equivalent to $a \in G$, denoted by $[a]$, i. e.,

$$[a] = \{b \in G: b \sim a\} \equiv aN.$$

Then $[a]$ is said to be **an equivalent class of** a; also $[a]$ is said to be **a coset of** a.

Hence, we have a new "combination" of elements in G, i. e., regard an equivalent class $[a]$ of a by equivalent relation "$\sim$" as "a new element". The set of all equivalent classes, denoted by

$$G/\sim \equiv G/N = \{[a]: a \in G\},$$

called **the quotient set by equivalent relation**$\sim$, or simply, **quotient set.**

Theorem 1.2.13 Let $N \subset G$ be a normal subgroup of group G, and $G/\sim \equiv \{[a]: a \in G\}$ be the quotient set with equivalent relation$\sim$. Then

(1) $a \sim a$, i. e., $a \in [a]$; (self-reciprocity)

(2) $a \sim b \Leftrightarrow b \sim a$, i. e., $b \in [a] \Leftrightarrow a \in [b]$; (symmetry)

(3) $a \sim b, b \sim c \Rightarrow a \sim c$, i. e., $b \in [a], c \in [b] \Rightarrow c \in [a]$; (transitivity)

Moreover,

(4) $a \in [b] \Rightarrow [a] = [b]$, i. e., $a \sim b \Rightarrow [a] = [b]$;

(5) $[a] \neq [b] \Rightarrow [a] \cap [b] = \varnothing$, i. e., $[a] \cap [b] \neq \varnothing \Rightarrow [a] = [b]$.

The proof is left as an exercise.

Definition 1.2.20 (quotient group, quotient mapping) In the quotient set $G/\sim \equiv \{[a]: a \in G\}$, an operation $\cdot$ is defined by

$$[a][b] \equiv [a] \cdot [b] = [a(\cdot)_G b] \equiv [ab],$$

then the quotient set $G/\sim$ consisted of quotient classes with the operation $\cdot$ forms a group, called **a quotient group of group** G **about normal subgroup** N, or simply, **a quotient group**, denoted by

$$(G/\sim, \cdot) \equiv (G/\sim \equiv \{[a]: a \in G\}, [a] \cdot [b] = [ab]).$$

A canonical mapping $\pi: a \to \pi(a)=[a]$ is said to be **a quotient mapping** from group G onto quotient group $G/\sim$.

We invite readers to verify that: the quotient mapping $\pi: a \to \pi(a)=[a]$ is a homomorphism mapping form group G onto quotient group $G/\sim$, and it is a surjection.

Example 1.2.24 (residue class group) Let $(\mathbb{Z},+)$ be the integer group. Take positive integer $m \geqslant 2$, denote the set $N=\{mk: k \in \mathbb{Z}\}$, then $(N,+)$ is a normal subgroup of $(\mathbb{Z},+)$. (Why?)

By a normal subgroup $(N,+)$ of $(\mathbb{Z},+)$, we make equivalent sets, i.e., cosets of $(\mathbb{Z},+)$,

$$[k]=\{mk: k \in \mathbb{Z}\}=kN,$$

then $\mathbb{Z}_m \equiv \mathbb{Z}/\sim=\{[k]: k \in \mathbb{Z}\}$ is the set of all equivalent classes of group $(\mathbb{Z},+)$ about normal subgroup $(N,+)$, i.e., the quotient set

$$\mathbb{Z}_m=\mathbb{Z}/N=\{[0],[1],\cdots,[m-1]\} \equiv \{0,1,\cdots,m-1\},$$

it is a quotient group endowed with the addition mod m, $x \oplus y = x+y \pmod{m}$, and contained m elements, called **the residue class group.**

Example 1.2.25 Let S_3 be the 3-replacement group, i.e.,

$$S_3=\{(1),(1\ 2),(1\ 3),(2\ 3),(1\ 2\ 3),(1\ 3\ 2)\},$$

take the subgroup $H=\{(1),(1\ 2)\}$, it is not normal, so that there are no coset, but it has right cosets and left cosets.

All right cosets—

$H(1)=\{(1),(1\ 2)\}$, $H(1\ 3)=\{(1\ 3),(1\ 3\ 2)\}$, $H(2\ 3)=\{(2\ 3),(1\ 2\ 3)\}$;

that is $H(1\ 2)=\{(1\ 2),(1)\}$, $H(1\ 3\ 2)=\{(1\ 3\ 2),(1\ 3)\}$, $H(1\ 2\ 3)=\{(1\ 2\ 3),(2\ 3)\}$;

All left cosets—

$(1)H=\{(1),(1\ 2)\}$, $(1\ 3)H=\{(1\ 3),(1\ 2\ 3)\}$, $(2\ 3)H=\{(2\ 3),(1\ 3\ 2)\}$;

that is $(1\ 2)H=\{(1\ 2),(1)\}$, $(1\ 2\ 3)H=\{(1\ 2\ 3),(1\ 3)\}$, $(1\ 3\ 2)H=\{(1\ 3\ 2),(2\ 3)\}$.

However, $(1\ 3)H \neq H(1\ 3)$, $(2\ 3)H \neq H(2\ 3)$; $(1\ 3\ 2)H \neq H(1\ 3\ 2)$; $(1\ 2\ 3)H \neq H(1\ 2\ 3)$; and only holding.

$(1)H=H(1)$, $(1\ 2)H=H(2\ 1)$.

Theorem 1.2.14 A group G is homomorphic with the quotient group G/N about a normal subgroup $N \subset G$.

Proof Let G be a group, N be a normal subgroup of G. Then the quotient mapping $\pi: a \to [a]$, $a \in G$, between group G and quotient group G/N, i.e., $a \to aN$, $a \in G$, is a surjection from G onto $G/N \equiv \{[a]: a \in G\}$.

Moreover, $\forall\, a, b \in G$ implies $ab \to abN=(aN)(bN)$, thus, $a \to aN$, $a \in G$, is a surjection homomorphism, this tells that the group G and quotient group G/N are homomorphic to each other.

Exercise 1

1. Prove the operation laws of sets (1)-(6).

2. Prove: the cardinality of rational number set $\mathbb{Q}$ is $\aleph_0$.

3. Let X, Y be two basic sets, and $A, B \subset X$ and $C, D \subset Y$ be their subsets, respectively. If $f: X \to Y$ is a mapping from X to Y, the sets $f(A) = \{f(x), x \in A\}$、$f^{-1}(C) = \{x \in A: f(x) \in C\}$ are the image set of A and the inverse image set of B. Prove the following:

(1) if $A \subset B \subset X$, then $f(A) \subset f(B)$; if $C \subset D \subset Y$, then $f^{-1}(C) \subset f^{-1}(D)$;

(2) $f(A \cup B) = f(A) \cup f(B)$; generally, $f(\bigcup_{\lambda \in L} A_\lambda) = \bigcup_{\lambda \in L} f(A_\lambda)$;

$f(A \cap B) \subset f(A) \cap f(B)$, if f is an injection, then $f(A \cap B) = f(A) \cap f(B)$;

$f(A - B) \supset f(A) - f(B)$, if f is an injection, then $f(A - B) = f(A) - f(B)$;

(3) $f^{-1}(C \cup D) = f^{-1}(C) \cup f^{-1}(D)$; generally, $f^{-1}(\bigcup_{\lambda \in L} C_\lambda) = \bigcup_{\lambda \in L} f^{-1}(C_\lambda)$;

$f^{-1}(C \cap D) = f^{-1}(C) \cap f^{-1}(D)$; generally, $f^{-1}(\bigcap_{\lambda \in L} C_\lambda) = \bigcap_{\lambda \in L} f^{-1}(C_\lambda)$;

$f^{-1}(C \setminus D) = f^{-1}(C) \setminus f^{-1}(D)$;

(4) $X \setminus (\bigcup_{\lambda \in L} A_\lambda) = \bigcap_{\lambda \in L} (X \setminus A_\lambda)$, $X \setminus (\bigcap_{\lambda \in L} A_\lambda) = \bigcup_{\lambda \in L} (X \setminus A_\lambda)$;

(5) $(\bigcup_{\lambda \in L} A_\lambda) \cap (\bigcup_{\mu \in M} B_\mu) = \bigcup_{(\lambda, \mu) \in L \times M} (A_\lambda \cap B_\mu)$,

$(\bigcap_{\lambda \in L} A_\lambda) \cup (\bigcap_{\mu \in M} B_\mu) = \bigcap_{(\lambda, \mu) \in L \times M} (A_\lambda \cap B_\mu)$;

(6) $f^{-1}(f(A)) \supset A$; $f(f^{-1}(C)) \subset C$;

(7) if $f: A \to C$ is an injective, then $f^{-1}(f(A)) = A$;

(8) if $f: A \to B$ is a surjective, then $f(f^{-1}(B)) = B$.

4. Prove: if each element $x \in G$ of a group G satisfies $x^2 = e$, then G is a commutative group.

5. The order of each element in a finite group is a finite number.

6. Let $G = \mathbb{R}$, and the set of all one-one transformations from G onto G denote by S, as well as by $\tau: a \to a' = \tau(a), a \in G$.

Define the operation $\circ$ as a compound: $\tau_1, \tau_2 \in S \Rightarrow \tau_1[\tau_2(a)] = (\tau_1 \circ \tau_2)(a), a \in G$.

Prove that $(S, \circ)$ is a group.

7. Find those elements in S_3 which can't exchange with $\begin{pmatrix} 1 & 2 & 3 \\ 2 & 3 & 1 \end{pmatrix}$.

8. Prove Theorems 1.2.8 and 1.2.10; Prove that: in Theorem 1.2.9, the embedding mapping I is an isomorphism from subgroup S onto the image set $I(S)$.

9. If G is a circulate group, and $\widetilde{G}$ is homomorphic with G, prove that: $\widetilde{G}$ is a circulate group.

10. Let G be a circulate group with infinite order, and $\widetilde{G}$ be any circulate group, prove that: $\widetilde{G}$ and G are homomorphic.

11. Prove that: for positive integer $m \geqslant 2$, the set $N = \{mk: k \in \mathbb{Z}\}$ is a normal subgroup with the usual addition of real numbers.

12. A Group $(G, \cdot)$ has two subgroups G_1, G_2, define so-called the direct product $G_1 \times G_2$ of $(G, \cdot)$:

(1) Ask for that G_1, G_2 are normal subgroups;

(2) $G = G_1 \times G_2 = \{g_1 \cdot g_2: g_1 \in G_1, g_2 \in G_2\}$, $G_1 \cap G_2 = \{e\}$; where $\{e\}$ is the unit of G, it is a special subgroup with only one element. Prove that: $(G, \cdot)$ is the direct product of subgroups G_1 and G_2, if and only if

(a) $\forall g \in G$ can be expressed uniquely as $g = g_1 g_2$, where $g_1 \in G_1, g_2 \in G_2$;

(b) $g_1 g_2 = g_2 g_1$, for any $g_1 \in G_1$ and $g_2 \in G_2$.

Chapter 2 Linear Spaces and Linear Transformations

Algebra is the earliest one of numerous mathematical branches which were known by human, from arithmetic in primary school, to elementary algebra in middle school, including "figure symbolization","solving algebraic equation (system)". Until the 1950's, "algebraic operations" was studied, and combined with "equation theory", formed new viewpoint, such that the general theory of classical algebra and algebraic operations is coincident with modern algebra. Nowadays, **modern algebra** and **linear algebra** both are basic knowledge in foundation of modern mathematics and applications.

Main objects studied **in the modern algebra** are various sets endowed with some "operation structure" or "algebraic structure", such as groups, rings, bodies, fields, ideals, moduli…. "Algebraic structure" represents the inherent formation and essential properties of sets; An important method to study algebraic structure is to show "its representation"; Moreover, "action of one algebraic structure on the other" is an important aim in studying such structure. Indeed, certain properties and expressions of operation structures, as well as the properties of mappings between two sets with operation structures, are studied in the modern algebra. We have presented parts of main contents of modern algebra in the Chapter 1.

We now turn to introduce **the linear algebra**, one of many branches of mathematical science; its main objects are **linear spaces** (or vector spaces) and moduli (generalized concept of linear space), as well as **linear transformations on them**; further, some related problems, such as bilinear mappings, quadratic functions… are studied.

The elementary knowledge of linear algebra is important and indispensible in modern applied mathematics and other scientific areas. It is sure that our readers are familiar with the concepts and rules of operations for determinants and matrices and so on. We refer to [11], [13] for this Chapter.

We show a following frame:

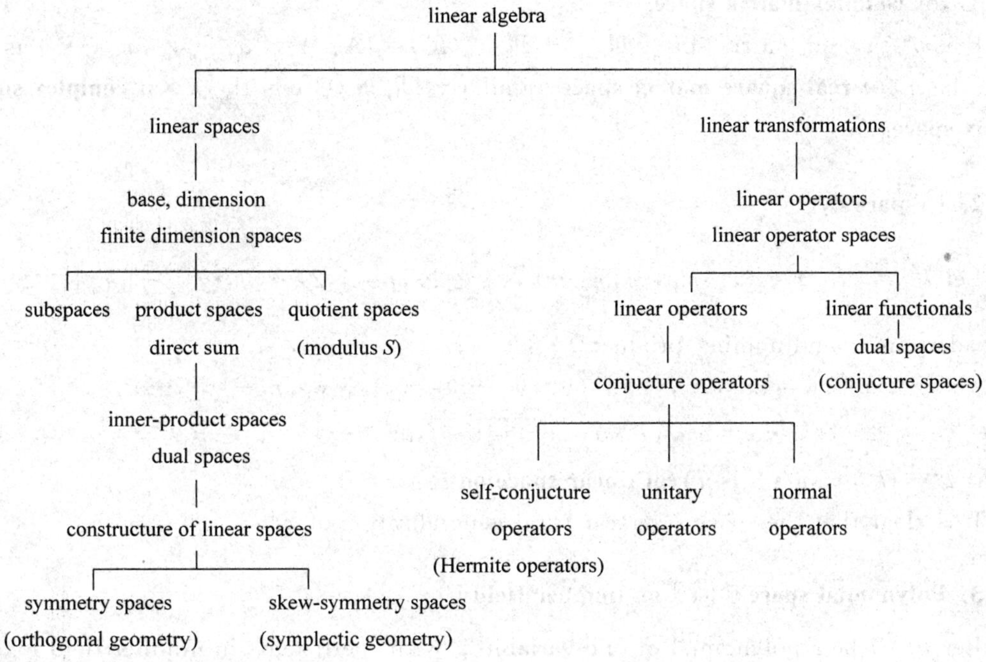

2.1 Linear Spaces

2.1.1 Examples

The definition of linear space has been given in Definition 1.2.5, and two examples were shown there: n-dimensional Euclidean space $\mathbb{R}^n$ and the continuous function space $C([a,b])$ on interval $[a,b]$. We now present more important examples.

1. Matrix spaces

The set of all $m\times n$-real matrices

$$\mathfrak{M}_{m\times n}(\mathbb{R})=\{[a_{ij}]: 1\leqslant i\leqslant m, 1\leqslant j\leqslant n, a_{ij}\in\mathbb{R}\}$$

with the addition $A+B$ of matrices for A, B, and the number product $\alpha\cdot A\equiv\alpha A$ for matrix A with $\alpha\in\mathbb{R}$:

$$A=[a_{ij}], B=[b_{ij}]\in\mathfrak{M}_{m\times n}(\mathbb{R})\Rightarrow A+B=[a_{ij}+b_{ij}]\in\mathfrak{M}_{m\times n}(\mathbb{R});$$

$$A=[a_{ij}]\in\mathfrak{M}_{m\times n}(\mathbb{R}), \alpha\in\mathbb{R}\Rightarrow\alpha A=[\alpha\, a_{ij}]\in\mathfrak{M}_{m\times n}(\mathbb{R}).$$

The $(\mathfrak{M}_{m\times n}(\mathbb{R}),+,\alpha\cdot)$, or simply by $\mathfrak{M}_{m\times n}(\mathbb{R})$, is a linear space on $\mathbb{R}$, and is said to be **the matrix space on real number field** $\mathbb{R}$, or **real matrix space.** If $a_{ij}\in\mathbb{C}$, $\alpha\in\mathbb{C}$, then $\mathfrak{M}_{m\times n}(\mathbb{C})\equiv(\mathfrak{M}_{m\times n}(\mathbb{C}),+,\alpha\cdot)$ is said to be **the matrix space on complex number**

field $\mathbb{C}$,or **complex matrix space.**

For $m=n$, real matrix space $\mathfrak{M}_{n\times n}\equiv\mathfrak{M}_{n\times n}(\mathbb{R})=\{A: A=[a_{ij}]_{n\times n}, a_{ij}\in\mathbb{R}\}$ is said to be the $n\times n$-**real square matrix space.** Similarly, $\mathfrak{M}_{n\times n}(\mathbb{C})$ is the $n\times n$-**complex square matrix space.**

2. l^2 space

Let $l^2=\left\{x: x=(x_1, x_2, \cdots, x_n \cdots), x_j\in\mathbb{R}, j=1,2,\cdots,n,\cdots, \sum_{j=1}^{+\infty}|x_j|^2<+\infty\right\}$ with addition $+$ and number product $\alpha\cdot$:

$$x+y=(x_1, x_2, \cdots)+(y_1, y_2, \cdots)=(x_1+y_1, x_2+y_2, \cdots),$$

$$\alpha x\equiv\alpha\cdot x=\alpha\cdot(x_1, x_2, \cdots)=(\alpha x_1, \alpha x_2, \cdots), \quad \alpha\in\mathbb{R}.$$

Then, $l^2\equiv(l^2, +, \alpha\cdot)$ **is a real linear space on** $\mathbb{R}$.

It is clear that the space l^2 is a natural generalization of $\mathbb{R}^2$.

3. Polynomial space $\mathbb{F}[x]$ on number field $\mathbb{F}$

Let $p(x)$ be a polynomial of one-variable x with coefficients in number field $\mathbb{F}$,i. e.

$$p(x)=a_nx^n+a_{n-1}x^{n-1}+\cdots+a_1x+a_0, \quad a_j\in\mathbb{F}.$$

Denote the set of all polynomials of one variable on field $\mathbb{F}$ by

$$\mathbb{F}[x]=\{p: p(x)=a_nx^n+a_{n-1}x^{n-1}+\cdots+a_1x+a_0, \quad a_j\in\mathbb{F}, n\in\mathbb{N}, j=1,2,\cdots,n\}.$$

Define the operations of addition $+$ and number product $\alpha\cdot$ on $\mathbb{F}[x]$:

For $p, q\in\mathbb{F}[x], \alpha\in\mathbb{F}$,

$$(p+q)(x)=p(x)+q(x)$$
$$=(a_n+b_n)x^n+(a_{n-1}+b_{n-1})x^{n-1}+\cdots+(a_1+b_1)x+(a_0+b_0),$$
$$(\alpha\cdot p)(x)\equiv\alpha p(x)=\alpha a_nx^n+\alpha a_{n-1}x^{n-1}+\cdots+\alpha a_1x+\alpha a_0.$$

We can verify that $\mathbb{F}[x]\equiv(\mathbb{F}[x], +, \alpha\cdot)$ forms a linear space on $\mathbb{F}$, and $\mathbb{F}[x]$ is said to be a **polynomial space on number field** $\mathbb{F}$.

4. *L*-integrable space $L^1(E)$, and *L*-square integrable space $L^2(E)$

The set of all Lebesgue integrable functions on a measurable set $E\subset\mathbb{R}$ is denoted by

$$L^1(E)=\left\{f: \int_E|f(x)|\,\mathrm{d}m<+\infty\right\};$$

and the set of all Lebesgue square integrable functions on a measurable set $E\subset\mathbb{R}$ is denoted by

$$L^2(E)=\left\{f: \int_E|f(x)|^2\,\mathrm{d}m<+\infty\right\}.$$

Define the operations addition $+$ and number product $\alpha \cdot$ on above two sets by

$$(f+g)(x)=f(x)+g(x),\quad x\in E;\quad (\alpha\cdot f)(x)=\alpha\cdot f(x),\quad x\in E,\quad \alpha\in\mathbb{F}.$$

Under the operations $+$, $\alpha\cdot$, $L^1(E)$ is said to be **the Lebesgue integrable function space**; $L^2(E)$ is said to be **the Lebesgue square integrable function space.** Both are linear spaces.

2.1.2 Bases of linear spaces

1. Some elementary concepts

1) **Linearly independent subsets**

Let $(X,+,\alpha\cdot)$ be a linear space on number field $\mathbb{F}$, and $S\subset X$ be a subset of X.

Linear independence of n elements $a_1,a_2,\cdots,a_n\in S$: If n numbers $r_1,r_2,\cdots,r_n\in\mathbb{F}$ satisfy that "$r_1a_1+r_2a_2+\cdots+r_na_n=0$ implies $r_1=\cdots=r_n=0$", then these n elements $a_1,a_2,\cdots,a_n\in S$ are said to be **linearly independent**; Otherwise, to be **linear dependence.**

Linearly independent subset: if n elements $a_1,a_2,\cdots,a_n\in S$ for $\forall n\in\mathbb{Z}^+$ are linearly independent, then S is said to be **a linearly independent subset in linear space X**; otherwise, to be **a linearly dependent subset in X**.

2) **X generated by subset $V\subset X$**

If X is a linear space on number field $\mathbb{F}$, and $V\subset X$ is a nonempty subset; moreover, if element $x\in X$ can be expanded as a linear combination of certain finite elements of V, i.e., $\forall x\in X$, there exist $a_j\in V$ and $r_j\in\mathbb{F}$, $j=1,2,\cdots,m$, $m\equiv m(x)\in\mathbb{Z}^+$, such that $x=r_1a_1+r_2a_2+\cdots+r_ma_m$, then the linear space X is said to be **generated by subset V**; and V is said to be **a generating set of X**.

Without loss of generality, we suppose that **a generating set V of X is a nonempty linearly independent set.**

A nonempty linearly independent set with finite linear combinations of $S\subset X$ forms a linear subspace of X, denoted by

$$\langle S\rangle\equiv\operatorname{span}(S)=\left\{a=\sum_{j=1}^{m}r_ja_j:\ a_j\in S,r_j\in\mathbb{F},j=1,2,\cdots,m,m\in\mathbb{Z}^+\right\}.$$

Thus, "$\varnothing\neq S\subset X\Rightarrow\langle S\rangle\subset X$", and $\langle S\rangle$ is **a linear space on $\mathbb{F}$**.

3) **Isomorphism between two linear spaces**

Let X_1 and X_2 be linear spaces on number set $\mathbb{F}$. If there exists an one-one and surjective mapping $T: X_1\to X_2$, such that

$$T(x_1+x_2)=T(x_1)+T(x_2),\quad x_1,x_2\in X_1,$$

$$T(\alpha x_1)=\alpha T(x_1),\quad x_1\in X_1,\quad \alpha\in\mathbb{F},$$

then T is said to be an isomorphism from X_1 onto X_2, or of X_1 and X_2; moreover, X_1

and X_2 are said to be isomorphic spaces, or simply, X_1 and X_2 are isomorphic.

Note An isomorphic mapping is **an one-one, surjective mapping, and keeps the operations of the linear space.** Recall the definition of isomorphism between two groups in section 1.2, and give a comparison of these two definitions of isomorphisms.

2. Bases of linear spaces

Definition 2.1.1 (base of a linear space) Let $(X, +, \alpha \cdot)$ be a linear space on number field $\mathbb{F}$. If a linearly independent subset $S \subset X$ generates X, i.e., $X = \text{span}(S)$, then S is said to be **a base of** X.

Theorem 2.1.1 Any nonempty linear space X on $\mathbb{F}$ has a base.

Proof Let $X \neq \{0\}$ be a linear space. Denote the set of all linearly independent subsets of X by

$$\mathfrak{B} = \{S \subset X : S \neq \varnothing \text{ is linearly independent subset of } X\}.$$

It is clear that $\mathfrak{B} \neq \varnothing$, since $S = \{a \in X : a \neq 0\} \in \mathfrak{B}$ is the element in $\mathfrak{B}$, which contains only one element a, and it is linearly independent subset of X.

Now for set $\mathfrak{B}$, endowing "order" by "$S_1 \subset S_2 \Leftrightarrow S_1$ is contained in S_2", $\mathfrak{B}$ becomes **a partial order set** (its definition see [6]).

Take a "chain", a full order set of linearly independent subsets in $\mathfrak{B}$ as

$$S_1 \subset S_2 \subset \cdots \subset S_k \subset \cdots, \ S_k \in \mathfrak{B},$$

then the union $U = \bigcup\limits_{k \geqslant 1} S_k$ is in $\mathfrak{B}$, and is an upper bound of chain $S_1 \subset S_2 \subset \cdots \subset S_k \subset \cdots$. So that, by the Zorn Lemma ([6]), $\mathfrak{B}$ has the maximal element $\widetilde{S} \in \mathfrak{B}$, i.e., $\widetilde{S}$ is a linearly independent subset, and any subset $V \subset X$ that contains $\widetilde{S}$ is linearly dependent of X. This implies that $\widetilde{S}$ generates X, since if it would be not, then there exists $x \in X \backslash \widetilde{S}$, such that x is not a linear combination of elements of $\widetilde{S}$, and $\widetilde{S} \cup \{x\}$ is a linearly independent subset of $\mathfrak{B}$, also $\widetilde{S}$ is a proper subset of $\widetilde{S} \cup \{x\}$, i.e., $\widetilde{S} \underset{\neq}{\subset} \widetilde{S} \cup \{x\}$. This is in a contradiction to the maximal element $\widetilde{S} \in \mathfrak{B}$. The proof is complete.

Definition 2.1.2 (dimension of a linear space) Let $(X, +, \alpha \cdot)$ be a linear space on $\mathbb{F}$, and S be a base of X. The cardinality $\overline{\overline{S}} = \text{card}(S)$ of S is said to be **the dimension of** X, denoted by $\dim X = \text{card}(S)$.

If $\dim X = \text{card}(S)$ is an integer $n = \text{card}(S)$, $n \in \mathbf{Z}^+$, then X is said to be an **n-dimensional linear space**. For example, $(\mathbb{R}^n, +, \alpha \cdot)$ has an orthogonal base

$$\mathfrak{B} \equiv \{\underbrace{(1,0,0,\cdots,0)}_{n}, \underbrace{(0,1,0,\cdots,0)}_{n}, \cdots, \underbrace{(0,0,0,\cdots,0,1)}_{n}\},$$

which contains n linearly independent vectors, so that $\mathbb{R}^n$ is a linear space with finite dimension n.

In this book we agree on: A finite non-negative integer n-dimensional linear space is said to be n-**dimensional linear space.**

Proposition 2.1.1 The definition 2.1.2 of the dimension of a linear space is reasonable and exact.

Proof By three steps:

(1) We prove that: If a linear space X on number field $\mathbb{F}$ has two linearly independent subsets $V=\{v_1, v_2, \cdots, v_m\}$ and $S=\{s_1, s_2, \cdots, s_n\}$, as well as $\langle S\rangle = X$ and $n=\text{card}(S)$. Then it holds $m \leqslant n$.

In fact, $S=\{s_1, s_2, \cdots, s_n\}$ generates X, thus, $\forall x \in X \Rightarrow x = \sum_{j=1}^{n} \alpha_j s_j, \alpha_j \in \mathbb{F}, j=1, 2, \cdots, n$.

List a system "$s_1, s_2, \cdots, s_n;\ v_1, v_2, \cdots, v_m$". Rewrite it as "$v_m, s_1, s_2, \cdots, s_n;\ v_1, v_2, \cdots, v_{m-1}$".

Since $S=\{s_1, s_2, \cdots, s_n\}$ generates X, then v_m can be expressed as a linear combination of $s_1, s_2, \cdots, s_n$, and thus we can move out one element, say s_j, from $s_1, s_2, \cdots, s_n$, such that the set $\{v_m, s_1, \cdots, s_{j-1}, s_{j+1}, \cdots, s_n\}$ still generates X, and denoted it by $\{s_1, \cdots, \hat{s}_j, \cdots, s_n\}$, where $\hat{s}_j$ represents that s_j has been moved out. Hence we have a new set

$$\{v_m, s_1, \cdots, \hat{s}_j, \cdots, s_n;\ v_1, \cdots, v_{m-1}\}.$$

Repeating the above process, we get a new set $\{v_{m-1}, v_m, s_1, \cdots, \hat{s}_j, \cdots, s_n;\ v_1, \cdots, v_{m-2}\}$ again. Similarly, we move out s_k from the set, and get new set $\{v_{m-1}, v_m, s_1, \cdots, \hat{s}_j, \cdots, \hat{s}_k, \cdots, s_n;\ v_1, \cdots, v_{m-2}\}$, such that it still generates X.

Continue this process, above "the moving out method" is said to be **Steinitz replacement.** The result of Steinitz replacement is: After each replacement, the new set still generates X.

If all $s_1, s_2, \cdots, s_n$ in S are moved out firstly, i.e., $n < m$, then elements $v_{m-n}, v_{m-n+1}, \cdots, v_{m-1}, v_m, \hat{s}_1, \cdots, \hat{s}_n$ forms a proper subset $\widetilde{V} \equiv \{v_{m-n}, v_{m-n+1}, \cdots, v_{m-1}, v_m, \hat{s}_1, \cdots, \hat{s}_n\}$ of $V=\{v_1, v_2, \cdots, v_m\}$, and $\widetilde{V}$ generates the linearly space X, this contradicts the linearly independent property of V, because $v_{m-n-1}, v_{m-n-2}, \cdots, v_1 \notin \widetilde{V}$, but they can be expressed linearly by elements in $\widetilde{V}(\subset V)$, this implies that V is not a linearly independent set, and so that $m \leqslant n$.

(2) Change the assumptions for S and V, i.e., suppose that $S=\{s_1, s_2, \cdots, s_n\}$ is linearly independent set, and $V=\{v_1, v_2, \cdots, v_m\}$ with $m=\text{card}(V)$ generates X, such

that $X = \mathrm{span}(V)$, then it holds $n \leqslant m$, similarly by (1).

(3) If $S = \{s_1, s_2, \cdots, s_n\}$ and $V = \{v_1, v_2, \cdots, v_m\}$ all are the generating sets of X, thus they are linearly independent sets, so that $m = n$.

By (1)-(3) we conclude that the definition of base of a linear space X is reasonable and exactl.

Theorem 2.1.2 Let$(X, +, \alpha \cdot)$ be an n-dimensional linear space on number field $\mathbb{F}$, $S \subset X$ be a subset, then the following statements are equivalent:

(1) S is a base of X;

(2) each element $x \in X$ can be represented uniquely by

$$x = r_1 a_1 + \cdots + r_n a_n, \quad r_j \in \mathbb{F}, \quad a_j \in S, \quad j = 1, 2, \cdots, n;$$

(3) S is the minimal generating set of X, i. e. , S is contained in all generating sets of X;

(4) S is the maximal linear independent set of X, i. e. , S contains all linearly independent subsets of X.

Important note The structure of an n-dimensional linear space is connected with its number field.

Theorem 2.1.3 Let$(X, +, \alpha \cdot)$ be an n-dimensional linear space on number field $\mathbb{F}$, then

(1) X is isomorphic with $\mathbb{F}^n$; i. e. , the n-dimensional linear space X on real number set $\mathbb{R}$ is isomorphic with $\mathbb{R}^n$; the n-dimensional linear space X on complex number set $\mathbb{C}$ is isomorphic with $\mathbb{C}^n$.

(2) Two linear spaces X and Y on same number field $\mathbb{F}$ are isomorphic, if and only if their dimensions are equal each other.

Proof Only need to prove (1).

To prove that $(X, +, \alpha \cdot)$ is isomorphic with $\mathbb{F}^n$, take a base $\mathfrak{B} = \{b_1, \cdots, b_n\}$ of X, then $\forall x \in X$, $\exists (\alpha_1, \cdots, \alpha_n) \in \mathbb{F}^n$, such that $x = \alpha_1 b_1 + \cdots + \alpha_n b_n$, and by the column vector expression, $[x]_{\mathfrak{B}} \equiv \begin{bmatrix} \alpha_1 \\ \vdots \\ \alpha_n \end{bmatrix}$, this $[x]_{\mathfrak{B}}$ is said to be **the coordinate** of $x \in X$ in $\mathfrak{B}$. Define a mapping $\varphi_{\mathfrak{B}}: X \to \mathbb{F}^n$, such that

$$\varphi_{\mathfrak{B}}(x) = [x]_{\mathfrak{B}} \leftrightarrow (\alpha_1, \cdots, \alpha_n) \in \mathbb{F}^n, \quad x \in X.$$

It is clear that $\varphi_{\mathfrak{B}}$ is a one-one, linear surjective from X onto $\mathbb{F}^n$. Furthermore, $\varphi_{\mathfrak{B}}$ is an isomorphic mapping from X onto $\mathbb{F}^n$, since it keeps operations:

$$\varphi_{\mathfrak{B}}(x + y) = [x + y]_{\mathfrak{B}} = \begin{bmatrix} \alpha_1 + \beta_1 \\ \vdots \\ \alpha_n + \beta_n \end{bmatrix}_{\mathfrak{B}} = \begin{bmatrix} \alpha_1 \\ \vdots \\ \alpha_n \end{bmatrix}_{\mathfrak{B}} + \begin{bmatrix} \beta_1 \\ \vdots \\ \beta_n \end{bmatrix}_{\mathfrak{B}} = \varphi_{\mathfrak{B}}(x) + \varphi_{\mathfrak{B}}(y),$$

$$\varphi_{\mathfrak{B}}(\alpha x)=[\alpha x]_{\mathfrak{B}}=\begin{bmatrix}\alpha\alpha_1\\ \vdots\\ \alpha\alpha_n\end{bmatrix}_{\mathfrak{B}}=\alpha\begin{bmatrix}\alpha_1\\ \vdots\\ \alpha_n\end{bmatrix}_{\mathfrak{B}}=\alpha\varphi_{\mathfrak{B}}(x).$$

Thus, from the isomorphism point of view, n-dimensional linear space on $\mathbb{F}$ is unique, i. e., $\mathbb{F}^n$.

Example 2.1.1 The Euclidean space $(\mathbb{R}^n,+,\alpha\cdot)$ is an n-dimensional linear space, with the base

$$\{e_1=\underbrace{(1,0,0,\cdots,0,0)}_{n},e_2=\underbrace{(0,1,0,\cdots,0,0)}_{n},\cdots,e_n=\underbrace{(0,0,0,\cdots,0,1)}_{n}\}.$$

Example 2.1.2 The set of all $m\times n$-real matrices on $\mathbb{R}$ forms a linear space with $+$ and $\alpha\cdot$ of matrices

$$\mathfrak{M}_{m\times n}=\{A=[a_{ij}]:1\leqslant i\leqslant m,1\leqslant j\leqslant n,a_{ij}\in\mathbb{R}\},$$

with the base

$$\mathfrak{B}_{m\times n}=\{[e_{ij}]:1\leqslant i\leqslant m,1\leqslant j\leqslant n\},$$

where $e_{ij}=[a_{ks}]_{m\times n},a_{ks}=\begin{cases}1, & k=i,s=j\\ 0, & \text{others}\end{cases}$. We have $\dim\mathfrak{M}_{m\times n}=mn$.

Example 2.1.3 The set of all polynormals with order $\leqslant n$ (including zero polynormal) on $\mathbb{R}$, denoted by

$$\mathfrak{P}_n=\{p_n(x):p_n(x)=a_nx^n+\cdots+a_1x+a_0,a_j\in\mathbb{R},\ j=0,1,2,\cdots,n\},$$

has a base $\mathfrak{B}=\{1,x,x^2,\cdots,x^n\}$. We have $\dim\mathfrak{P}_n=n+1$.

2.1.3 Subspaces and product/direct-sum/quitient spaces

1. Subspaces

Let $(X,+,\alpha\cdot)$ be a linear space on number field $\mathbb{F}$, and $S\subset X$ be a nonempty subset of X. How to construct a linear subspace on S, denoted by $(S,+,\alpha\cdot)\equiv(S,(+)_S,(\alpha\cdot)_S)$, from the linear space $(X,+,\alpha\cdot)\equiv(X,(+)_X,(\alpha\cdot)_X)$?

Subset $S\subset X$ is a subset of X, and $S\neq\varnothing$,

addition $(+)_S$— $x,y\in S\Rightarrow x(+)_Sy=x(+)_Xy$,

number product $(\alpha\cdot)_S$— $x\in S$, $\alpha\in\mathbb{F}\Rightarrow(\alpha\cdot)_Sx=(\alpha\cdot)_Xx$.

If set S is a linear space on $\mathbb{F}$ with $(+)_S,(\alpha\cdot)_S$, then $(S,(+)_S,(\alpha\cdot)_S)$ is said to be **a linear subspace of** $(X,(+)_X,(\alpha\cdot)_X)$, or simply, S is **a subspace of** X.

To verify that whether a subset $S\subset X$ forms a linear subspace, it only needs to check that two operations $(+)_X$ and $(\alpha\cdot)_X$ are closed in subset S.

Embedding mapping The mapping $I: S\to X$ from subspace S onto space X with

$I(x)=x, x\in S$, is said to be **an embedding mapping.**

The embedding mapping from S onto X is an isomorphism one, it is an important bridge connected S with X.

Theorem 2.1.4 Let $(X, +, \alpha\cdot)$ be an n-dimensional linear space on number field $\mathbb{F}$, and $S\subset X, T\subset X$ be two subspaces. Then

$$\dim(S)+\dim(T)=\dim(S+T)+\dim(S\cap T).$$

Specially, if $S\cap T=\{0\}$, then

$$\dim(S)+\dim(T)=\dim(S+T)=\dim X=n.$$

2. Product spaces

How to construct a linear space, denoted by $(X\times Y, +, \alpha\cdot)\equiv(X\times Y, (+)_{X\times Y}, (\alpha\cdot)_{X\times Y})$, called a product space, from two linear spaces $(X, (+)_X, (\alpha\cdot)_X)$, $(Y, +_Y, (\alpha\cdot)_Y)$ on $\mathbb{F}$?

Product set $X\times Y=\{(x,y): x\in X, y\in Y\}$ is said to be a product set of X and Y.

addition $(+)_{X\times Y}$

$$(x_1,y_1), (x_2,y_2)\in X\times Y\Rightarrow(x_1,y_1)(+)_{X\times Y}(x_2,y_2)=(x_1(+)_X x_2, y_1(+)_Y y_2),$$

number product $(\alpha\cdot)_{X\times Y}$

$$(x,y)\in X\times Y,\quad \alpha\in\mathbb{F}\Rightarrow(\alpha\cdot)_{X\times Y}(x,y)=((\alpha\cdot)_X x, (\alpha\cdot)_Y y).$$

It is easy to verify that $(X\times Y, +, \alpha\cdot)\equiv(X\times Y, (+)_{X\times Y}, (\alpha\cdot)_{X\times Y})$ is a linear space on $\mathbb{F}$ with operations $(+)_{X\times Y}$ and $(\alpha\cdot)_{X\times Y}$, called **the product space of** $(X, (+)_X, (\alpha\cdot)_X)$ **and** $(Y, (+)_Y, (\alpha\cdot)_Y)$, or simply, **a product space of** X and Y.

There are important bridges connected the product space $(X\times Y, (+)_{X\times Y}, (\alpha\cdot)_{X\times Y})$ with $(X, (+)_X, (\alpha\cdot)_X)$, $(Y, (+)_Y, (\alpha\cdot)_Y)$: two project mappings.

Project mappings the mapping $\mathrm{Pr}_1: X\times Y\to X$ with $\mathrm{Pr}_1(x,y)=x$, and mapping $\mathrm{Pr}_2: X\times Y\to Y$ with $\mathrm{Pr}_2(x,y)=y$, are said to be **the project mappings** from $X\times Y$ into X and Y, respectively.

It is easy to verify that Pr_1 and Pr_2 are homomorphic mappings form $X\times Y$ into X and Y, respectively.

The concept of product space of two linear spaces can be generalized to that of m linear spaces

$$(X_1\times\cdots\times X_m, +, \alpha\cdot)\equiv(X_1\times\cdots\times X_m, (+)_{X_1\times\cdots\times X_m}, (\alpha\cdot)_{X_1\times\cdots\times X_m}).$$

And there are m product mappings:

$$\mathrm{Pr}_j: X_1\times\cdots\times X_m\to X_j, \mathrm{Pr}_j(x_1,\cdots,x_m)=x_j,\quad j=1,2,\cdots,m.$$

3. Direct-sum spaces

The other linear space—direct-sum linear space can be generated by two linear spaces.

Let $(X,+,\alpha\cdot)\equiv(X,(+)_X,(\alpha\cdot)_X)$ and $(Y,+,\alpha\cdot)\equiv(Y,(+)_Y,(\alpha\cdot)_Y)$ be two linear spaces on number field $\mathbb{F}$ with $X\cap Y=\{0\}$.

direct sum For two linear spaces X and Y with $X\cap Y=\{0\}$, the set $X\oplus Y=\{(x,y): x\in X, y\in Y\}$ is called a direct sum of X and Y.

operations For $(+)_{X\oplus Y}$ and $(\alpha\cdot)_{X\oplus Y}$ on $X\oplus Y$:

$(x_1,y_1),(x_2,y_2)\in X\oplus Y\Rightarrow(x_1,y_1)(+)_{X\oplus Y}(x_2,y_2)=(x_1(+)_X x_2, y_1(+)_Y y_2)$,

$(x,y)\in X\oplus Y, \alpha\in\mathbb{F}\Rightarrow(\alpha\cdot)_{X\oplus Y}(x,y)=((\alpha\cdot)_X x,(\alpha\cdot)_Y y)$.

Denote $(X\oplus Y,+,\alpha\cdot)\equiv(X\oplus Y,(+)_{X\oplus Y},(\alpha\cdot)_{X\oplus Y})$, then $(X\oplus Y,(+)_{X\oplus Y},(\alpha\cdot)_{X\oplus Y})$ forms a linear space, called **a direct-sum space on** $\mathbb{F}$ of $(X,+,\alpha\cdot)$ and $(Y,+,\alpha\cdot)$ with $X\cap Y=\{0\}$, simply, **a direct-sum space.**

The direct-sum space can be generizaled to m linear spaces

$$(X_1\oplus\cdots\oplus X_m,+,\alpha\cdot)\equiv(X_1\oplus\cdots\oplus X_m,(+)_{X_1\oplus\cdots\oplus X_m},(\alpha\cdot)_{X_1\oplus\cdots\oplus X_m})$$

for $X_i\cap X_j=\{0\}, i\neq j$.

The direct sum leads to an important concept "complementary set".

Definition 2.1.3 (complementary set) If $S\subset X$ and $T\subset X$ are linear spaces on number field $\mathbb{F}$, and $X=S\oplus T$ with $S\cap T=\{0\}$. Then, sets S and T are said to be **complementary**, denoted by $S=\mathfrak{C}T$ and $T=\mathfrak{C}S$.

Theorem 2.1.5 Let $(X,+,\alpha\cdot)$ be n-dimensional linear space on number field $\mathbb{F}$, and $S\subset X$ be a linear subspace, then $\dim X=\dim(S)+\dim(\mathfrak{C}S)$.

Example 2.1.4 Let $(\mathbb{R}^n,+,\alpha\cdot)$ be n-dimensional linear space, and $(\mathbb{R}^m,+,\alpha\cdot)$ be m-dimensional linear subspace of $\mathbb{R}^n$ with $m<n$. Then $(\mathbb{R}^{n-m},+,\alpha\cdot)$ is a linear subspace of $\mathbb{R}^n$, and $\mathbb{R}^n$ is the direct-sum $\mathbb{R}^n=\mathbb{R}^m\oplus\mathbb{R}^{n-m}$.

Example 2.1.5 In the square matrix space $\mathfrak{M}_{n\times n}$, the subset

$$W=\left\{[a_{ij}]_{n\times n}\in\mathfrak{M}_{n\times n}: \sum_{j=1}^{n}a_{1j}=0\right\}$$

forms a linear subspace of $\mathfrak{M}_{n\times n}$.

It is clear that: "addition of matrices" and "number product of matrices" are closed in W; the zero matrix $O=[0]_{n\times n}\in W$, thus W is a linear subspace of $\mathfrak{M}_{m\times n}$. Since the $n-1$ elements in the first row of W are linearly independent, so that $\dim W=n^2-1$.

4. Quotient spaces

From a linear space $(X,+,\alpha\cdot)$ on number field $\mathbb{F}$ and an equivalent relation $\sim$ on the space X, we can generate a so-called quotient space.

Equivalent relation $\sim$ For $x, y\in X$, elements x and y are said to be equivalent, denoted by $x\sim y$, if

(1) $x \sim x$, (self-reciprocity)

(2) $x \sim y \Rightarrow y \sim x$, (symmetry)

(3) $x \sim y, y \sim z \Rightarrow x \sim z$. (transitivity)

Equivalent class $[x]$ The set $[x]=\{y \in X: y \sim x\}$ is said to be an equivalent class of $x \in X$, also, $[x]$ is called **a coset of** x.

Quotient set $X/\sim$ The set $X/\sim \equiv \{[x]: x \in X\}$ is said to be a quotient set about linear space X with equivalent relation $\sim$.

Operations For $(+)_{X/\sim}$ and $(\alpha \cdot)_{X/\sim}$ on $X/\sim$:

$$[x],[y] \in X/\sim \Rightarrow [x](+)_{X/\sim}[y]=[x(+)_X y],$$

$$[x] \in X/\sim, \quad \alpha \in \mathbb{F} \Rightarrow (\alpha \cdot)_{X/\sim}([x])=[(\alpha \cdot)_X x].$$

It is easy to verify that $X/\sim$ is a linear space on $\mathbb{F}$ with $(+)_{X/\sim}$ and $(\alpha \cdot)_{X/\sim}$, denoted by

$$(X/\sim, +, \alpha \cdot) \equiv (X/\sim, (+)_{X/\sim}, (\alpha \cdot)_{X/\sim}),$$

and called **the quotient space** of linear space $(X, +, \alpha \cdot)$ with respect to equivalent relation $\sim$, or simply, **a quotient space.**

Quotient mapping the mapping $\pi: x \to \pi(x)=[x]$ is said to be **quotient mapping from** X **onto quotient space** $X/\sim$, or simply, **quotient mapping.** The quotient mapping π is an onto homomorphic one.

Example 2.1.6 In $\mathbb{R}^3$ and its subspace $\mathbb{R}^2 \subset \mathbb{R}^3$, define the mapping $T: \mathbb{R}^3 \to \mathbb{R}^2$, such that $\forall x=(x_1, x_2, x_3) \in \mathbb{R}^3 \to T(x)=(x_1, x_2, 0) \in \mathbb{R}^2$. Then, the set

$$\operatorname{im}(T)=T(\mathbb{R}^3)=\mathbb{R}^2$$

is said to be **the image of mapping** T; and the set

$$\ker(T)=T^{-1}(\{0\})$$

is said to be **the kernel of mapping** T.

We have $\ker(T)=\{x=(x_1, x_2, x_3) \in \mathbb{R}^3: T(x)=0\}=\{(0,0,x_3) \in \mathbb{R}^3: x_3 \in \mathbb{R}\}$.

Consider the equivalent relation $\sim$: $x, y \in \mathbb{R}^3 \Rightarrow x-y \in \ker(T)$. Then, we have

$$\mathbb{R}^3/\sim=\operatorname{im}(T), \quad \text{and} \quad \mathbb{R}^2=\mathbb{R}^3/\sim.$$

5. Congruence on modulus S

The concept of congruence on modulus S is a generalization of quotient spaces.

Definition 2.1.4 (congruence on modulus S**)** Let $(X, +, \alpha \cdot)$ be a linear space on field F, and $S \subset X$ be a subspace. If $u, v \in X \Rightarrow u-v \in S$, then u and v are said to be **a congruence on modulus** S, denoted by

$$u=v \pmod{S}.$$

The set of all congruence on modulus S, denoted by

$$[v]=\{u \in X: u=v \ (\mathrm{mod} S)\},$$

is called **a congruence class of v on modulus S in linear space X**, simply, **a congruence class on modulus S**; or $[v]$ is said to be a coset of v on modulus S in linear space X, or simply, a coset of v on modulus S.

Congruence on modulus S is an equivalent relation $\sim$, called the equivalent relation on modulus S, then linear space X can be divided into certain "block" $[v]$.

6. Quotient spaces on modulus S of X

Definition 2.1.5 (quotient space on modulus S) Let $(X,+,\alpha\cdot)$ be a linear space on field F, and $S\subset X$ be a linear subspace, $\sim$ be the equivalent relation on modulus S of X. The set of all congruence classes on modulus S, denotes by

$$X/\sim \ \equiv X/S=\{[v]: v \in X\}=\{[v]=v+S: v \in X\}.$$

Endow the following operations as addition and number product with F on $X/\sim$:

$$[u]+[v]=(u+S)+(v+S)=(u+v)+S, \quad u,v \in X,$$

$$\alpha[u]=\alpha(u+S)=\alpha u+S, \quad u \in X, \quad \alpha \in F,$$

thus, $X/\sim=X/S$ forms a linear space on number field F, called **a quotient space on modulus S of X**.

2.1.4 Inner product spaces

The most important and useful linear space is so-called inner product space.

1. Definition of inner product spaces

Definition 2.1.6 (real inner product, real inner product space) Let $(X,+,\alpha\cdot)$ be a linear space on real number set $\mathbb{R}$. If there exists a mapping $(x, y): X\times X\to\mathbb{R}$, satisfying

(1) $\forall x\in X\Rightarrow(x,x)\geqslant 0$; $(x,x)=0\Leftrightarrow x=0$, (positive definiteness)

(2) $(x,y)=(y,x)$, (symmetry)

(3) $(\alpha x+\beta y,z)=\alpha(x,z)+\beta(y,z)$, (linear)

$(x,\alpha y+\beta z)=\alpha(x,y)+\beta(x,z)$. (bilinear)

Then (x,y) is said to be an **inner product** of x and y on X; the space $(X,+,\alpha\cdot)$ is said to be a **real inner product space**, denoted by $(X,+,\alpha\cdot,(x,y))$; and the mapping $(x,y): X\times X\to\mathbb{R}$ is said to be **a bilinear form.**

Definition 2.1.7 (complex inner product, complex inner product space) Let $(X,+,\alpha\cdot)$ be a linear space on complex number set $\mathbb{C}$. If there exists a mapping (x,y): $X\times X\to\mathbb{C}$, satisfying

(1) $\forall x \in X \Rightarrow (x,x) \geqslant 0$; $(x,x)=0 \Leftrightarrow x=0$, (positive definiteness)

(2) $(x,y)=\overline{(y,x)}$, (conjugate symmetry)

(3) $(\alpha x+\beta y,z)=\alpha(x,z)+\beta(y,z)$, (linearity)

$(x,\alpha y+\beta z)=\bar{\alpha}(x,y)+\bar{\beta}(x,z)$. (conjugate linearity)

Then (x, y) is said to be **a complex inner product of** x, y **on** X, and $(X,+,\alpha \cdot,(x,y))$ is said to be **a complex inner product space.**

Note The mapping (x,y): $X \times X \to \mathbb{C}$ is not a bilinear form in the Definition 2.1.7.

Example 2.1.7 The n-dimensional Euclidean space $(\mathbb{R}^n,+,\alpha \cdot)$, with

$$\mathbb{R}^n=\{x=(x_1,\cdots,x_n): x_j \in \mathbb{R}, 1 \leqslant j \leqslant n\},$$

is the most familiar inner product space, and the inner product of $\mathbb{R}^n$ is defined by

$$(x,y)=\sum_{j=1}^{n} x_j y_j, \quad x,y \in \mathbb{R}^n.$$

Example 2.1.8 $l^2=\left\{x=(x_1,x_2,\cdots,x_n,\cdots): \sum_{j=1}^{+\infty} x_j^2<+\infty\right\}$ is a real inner product space, and the inner product is defined by

$$(x,y)=\sum_{j=1}^{+\infty} x_j y_j, \quad x,y \in l^2.$$

Example 2.1.9 $L^2([a,b])=\left\{f: \left(\int_{[a,b]} |f(x)|^2 \mathrm{d}x\right)^{\frac{1}{2}}<+\infty\right\}$ is a real inner product space, and the inner product is defined by

$$(f,g)=\int_{[a,b]} f(x) g(x) \mathrm{d}x, \quad f,g \in L^2([a,b]).$$

2. Lengths and distances in an inner product space

There are operations in a linear space $(X,+,\alpha \cdot)$, but no concepts of "length" for elements in the linear space. However, in an inner product space, the inner product (x,y): $X \times X \to \mathbb{R}$ (or $\mathbb{C}$) can help us to introduce **length** of an element, such that linear spaces have more deep senses, more important structures and more wide applications.

Definition 2.1.8 (length, distance) Let $(X,+,\alpha \cdot,(x,y))$ be an inner product space. Then, $\|x\|=\sqrt{(x,x)}$ is said to be **the length of** $x \in X$; and $d(x,y)=\sqrt{(x-y,x-y)}$ is said to be **the distance between** x **and** y **of elements** $x,y \in X$.

Recall that we have learned some concepts in the course of *Advance Mathematics*, such as "open set", "closed set", "neighborhood", "limit", "continuity", etc. Now for an inner product space, we can consider to define some concepts such that an inner product space can be endowed so-called "topological structure", a more essential structure of sets.

In the next chapter, the details will be shown about "topological spaces", and here, some useful properties of inner product space are listed.

Theorem 2.1.6 An inner product (x, y) satisfies

(1) Cauchy inequality $|(x,y)| \leqslant \|x\| \cdot \|y\|, x, y \in X$;

(2) Triangle inequality $\|x+y\| \leqslant \|x\| + \|y\|, x, y \in X$;

(3) Three-point distance inequality $d(x,y) \leqslant d(x,z) + d(z,y), x, y, z \in X$;

(4) Parallelogram law $\|x+y\|^2 + \|x-y\| = 2\|x\|^2 + 2\|y\|^2, x, y \in X$.

The proofs are left to exercises.

2.1.5 Dual spaces

1. Mappings on linear spaces

The "linear spaces and moduli", "linear transformations between linear spaces" are main subjects in linear algebra. In fact, to study structures of linear spaces, "linear transformations between linear spaces" are absolutely necessary. Again, homomorphisms and isomorphisms between linear spaces play key roles.

Recall that homomorphism and isomorphism between **two groups (or two fields)** defined in section 1.2.2, these two mappings keep operations of two groups. This motivates that a homomorphism (or isomorphism) can bring structures of one set to those of the other one. Similarly, we will see that homomorphisms and isomorphisms between two linear spaces have similar properties.

Homomorphism between groups— $(G_1, (\cdot)_{G_1}), (G_2, (\cdot)_{G_2})$ are groups, a mapping $f: (G_1, (\cdot)_{G_1}) \to (G_2, (\cdot)_{G_2})$ satisfies

$$f(x(\cdot)_{G_1} y) = f(x)(\cdot)_{G_2} f(y), \quad x, y \in G_1.$$

Homomorphism between fields— $(F_1, (+)_{F_1}, (\times)_{F_1}), (F_2, (+)_{F_2}, (\times)_{F_2})$ are fields, a mapping $f: (F_1, (+)_{F_1}, (\times)_{F_1}) \to (F_2, (+)_{F_2}, (\times)_{F_2})$ satisfies

$$f(x(+)_{F_1} y) = f(x)(+)_{F_2} f(y), \quad x, y \in F_1, \quad f(x), f(y) \in F_2;$$

$$f(x(\times)_{F_1} y) = f(x)(\times)_{F_2} f(y), \quad x, y \in F_1, \quad f(x), f(y) \in F_2.$$

We consider "homomorphism between linear spaces" to generalize the above definitions.

Definition 2.1.9 (linear mapping) Let $(X, (+)_X, (\alpha \cdot)_X)$ and $(Y, (+)_Y, (\alpha \cdot)_Y)$ be linear spaces on number field $\mathbb{F}$. A transformation $T: X \to Y$ is called **a linear mapping** from X to Y, if it satisfies

$$T(\alpha x_1(+)_X\beta x_2)=\alpha T(x_1)(+)_Y\beta T(x_2),\quad x_1,x_2\in X,\quad \alpha,\beta\in\mathbb{F}. \qquad (*)$$

A linear mapping from $(X,(+)_X,(\alpha\cdot)_X)$ to $(Y,(+)_Y,(\alpha\cdot)_Y)$, as a matter of fact, is a homomorphism mapping, since we rewrite $T:\ (X,(+)_X,(\alpha\cdot)_X)\to(Y,(+)_Y,(\alpha\cdot{}_Y))$, and $(*)$ can be split as

$$T(x_1(+)_Xx_2)=T(x_1)(+)_YT(x_2),\quad x_1,x_2\in X,\quad y_1=T(x_1),\quad y_2=T(x_2)\in Y;$$

$$T((\alpha\cdot)_Xx)=(\alpha\cdot)_YT(x),\quad \alpha\in\mathbb{F},\quad x\in X,\quad y=T(x)\in Y.$$

So, a linear mapping between two linear spaces is a **homomorphism** between the two spaces of $(X,(+)_X,(\alpha\cdot)_X)$ and $(Y,(+)_Y,(\alpha\cdot)_Y)$.

It is clear that the concept of "linear mapping" is a special case of "mapping"; and the concept of "function" in the course of advanced mathematics is the prototype of "mapping". In general, it is suitable to understanding "mapping" as "a corresponding relationship between two sets". Thus, "function", "injection", "surfection", "linear mapping", "homomorphism", and "isomorphism", all are "mappings". More mappings between two sets, or two groups, or two linear spaces, ..., will appear in the study of many natural sciences. Therefore, we have:

For notation "$f:\mathfrak{D}_f\to\mathfrak{R}_f$", f is **a corresponding relationship**; $\mathfrak{D}_f(\subset X)$ is said to be **domain of** f; $\mathfrak{R}_f(\subset Y)$ is said to be **range of** f; they satisfy "for each $x\in\mathfrak{D}_f$, there exists one and only one $y\in\mathfrak{R}_f$ corresponding to x, denoted by $y=f(x)$.

If $\mathfrak{D}_f=[a,b]\subset\mathbb{R}$, $\mathfrak{R}_f\subset\mathbb{R}$, then f is said to be a **function**;

If $\mathfrak{D}_f\subset\mathbb{R}^n$, $\mathfrak{R}_f\subset\mathbb{R}$, then f is said to be a **function of n-variables**;

If $\mathfrak{D}_f\subset\mathbb{R}^n$, $\mathfrak{R}_f\subset\mathbb{R}^n$, then f is said to be a **vector-valued function of n-variables**;

If $\mathfrak{D}_f\subset X$, $\mathfrak{R}_f\subset\mathbb{R}$, then f is said to be a **functional**;

If $\mathfrak{D}_f\subset X$, $\mathfrak{R}_f\subset Y$, then f is said to be a **mapping, transformation, or operator.** Where X, Y are groups, or fields, or linear spaces, or more general sets.

2. Dual spaces

We introduce a quite-new-type linear space, its elements are linear functionals on a linear space X, called **dual space of** X, or **conjugate space of** X. We agree on that the number filed $\mathbb{F}$ is real number field $\mathbb{R}$, or complex number field $\mathbb{C}$.

1) **Linear functionals on linear spaces, dual spaces**

Definition 2.1.10 (linear functional) let $(X,+,\alpha\cdot)$ be a linear space on $\mathbb{F}$. If $f: X\to\mathbb{F}$ satisfies

$$f(\alpha x+\beta y)=\alpha f(x)+\beta f(y),\quad x,y\in X,\quad \alpha,\beta\in\mathbb{F}$$

then f is said to be **a linear fractional on** X.

Denote the set of all linear functionals on X by

$$X^* = \{f : f \text{ is a functional on } X\}.$$

The operations addition $(+)_{X^*}$ and number product $(\alpha\cdot)_{X^*}$ on X^* are defined by

$$f, g \in X^* \Rightarrow (f(+)_{X^*} g)(x) = f(x)(+)_{\mathbb{F}} g(x), \quad x \in X;$$

$$f \in X^*, \quad \alpha \in \mathbb{F} \Rightarrow ((\alpha\cdot)_{X^*} f)(x) = (\alpha\cdot)_{\mathbb{F}} f(x) = \alpha f(x), \quad x \in X.$$

Then $(X^*, (+)_{X^*}, (\alpha\cdot)_{X^*})$ forms a linear space on $\mathbb{F}$; X^* is said to be the **dual space of** X, or is called the **conjugate space of** X (we use both "dual space" and "conjugate space" in this book).

Example 2.1.10 Let $(X, +, \alpha\cdot)$ be the 3-dimensional real Euclidean space $\mathbb{R}^3$. Then $(\mathbb{R}^3)^* = \mathbb{R}^3$, where sign "=" means that two linear spaces are isomorphic, i.e., there exists one-one isomorphic mapping which is a surjection and keeps operations of spaces.

In fact, take any $y \in \mathbb{R}^3$, define a functional $T_y: \mathbb{R}^3 \to \mathbb{R}$, satisfying

$$T_y(x) = (x, y) = \sum_{j=1}^{3} x_j y_j, \quad x \in \mathbb{R}^3, y \in \mathbb{R}^3,$$

where $x(x_1, x_2, x_3) \in \mathbb{R}^3$, $y(y_1, y_2, y_3) \in \mathbb{R}^3$, and (x, y) is the inner product of $\mathbb{R}^3$.

It is easy to verify: $\forall y \in \mathbb{R}^3$, it holds

$$T_y(\alpha x + \beta x') = \alpha T_y(x) + \beta T_y(x'), \quad \alpha, \beta \in \mathbb{R}, x, x' \in \mathbb{R}^3.$$

Thus, $\mathbb{R}^3 \subset (\mathbb{R}^3)^*$.

Conversely, we prove that $(\mathbb{R}^3)^* \subset \mathbb{R}^3$. If $f: \mathbb{R}^3 \to \mathbb{R}$ is any functional on $\mathbb{R}^3$, i.e., $f \in (\mathbb{R}^3)^*$, then $f(x + x') = f(x) + f(x')$ and $f(\alpha x) = \alpha f(x)$ determine f as a linear function of x_1, x_2, x_3, thus it must be a form $f(x) = f(x_1, x_2, x_3) = \sum_{k=1}^{3} c_k x_k$ with $c_k \in \mathbb{R}$, $k = 1, 2, 3$.

So, any linear functional $f \in (\mathbb{R}^3)^*$ is corresponding to a triplet $c_k \in \mathbb{R}$, $k = 1, 2, 3$, i.e., an element in $\mathbb{R}^3$, or, $f \in (\mathbb{R}^3)^* \longleftrightarrow (c_1, c_2, c_3) \in \mathbb{R}^3$. This implies that $(\mathbb{R}^3)^* \subset \mathbb{R}^3$. Hence, $(\mathbb{R}^3)^*$ is isomorphic to $\mathbb{R}^3$.

Definition 2.1.11 (**annihilator of a linear functional**) Let $(X, +, \alpha\cdot)$ be a linear space on number field $\mathbb{F}$, and $A \subset X$ be a subspace of X. The subset $A^0 = \{f \in X^* : f(A) = 0\}$ in dual space X^* is said to be an **annihilator of** A, where $f(A)$ is the image of A.

Annihilator $A \subset X$ has the following properties:

(1) $A^0 \subset X^*$ is a linear subspace of dual space X^*, whether set $A \subset X$ is a linear subspace of X, or not;

when $A \subset X$ is a linear subspace of X and $\dim X = n$, then $\dim A + \dim A^0 = n$;

(2) If $A \subset B \subset X$, then $B^0 \subset A^0$;

(3) If $\dim X = n$, and X^{**} is regarded as identifying with X, then for any subset $A \subset X$, it holds $(A^0)^0 = \text{span}(A)$; if A is a linear subspace of X, then $(A^0)^0 = A$;

(4) If $\dim X = n$, and A, B are linear subspaces, then

$$(A \cap B)^0 = A^0 \cup B^0, \quad (A + B)^0 = A^0 \cap B^0;$$

(5) If $X = A \oplus B$ is the direct sum of linear subspaces A and B, then it holds $A^* \approx B^0$, that is, A^* is identified with B^0; and holds $X^* = (A \oplus B)^* = A^0 \oplus B^0$.

2) **Second dual of linear spaces**

Definition 2.1.12 (second dual of linear space X) Let $(X, +, \alpha\cdot)$ be a linear space on $\mathbb{F}$, and the dual space of X be $(X^*, (+)_{X^*}, (\alpha\cdot)_{X^*})$. Then X^* is a linear space on $\mathbb{F}$.

May define the dual space of X^* as $(X^*)^* \equiv X^{**}$, called **the second dual of X**. The third dual, fourth dual, ..., can be defined, similarly.

About the dual space of a finite-dimensional linear space, we have the following theorem.

Theorem 2.1.7 Let $(X, +, \alpha\cdot)$ ba n-dimensional linear space on $\mathbb{F}$, i. e. $\dim X = n$, then

(1) $\dim X = \dim X^* = \dim X^{**} = n$;

(2) X, X^*, X^{**} are isomorphic; and there is a corresponding relationship between $x \in X$ and $x^{**} \in X^{**}$, which satisfies

$$x^{**}(w) = w(x), \quad \forall\, w \in X^*. \tag{2.1.1}$$

Proof By assumption, $\dim X = n$. Let $\mathfrak{B} = \{e_1, \cdots, e_n\} \subset X$ be a base of X. Define n linear functionals on X, $e_k^* \in X^*$, $k = 1, 2, \cdots, n$, satisfying

$$e_k^*(e_j) = \delta_{jk} = \begin{cases} 1, & k = j \\ 0, & k \neq j \end{cases}, \quad k, j = 1, 2, \cdots, n, \tag{2.1.2}$$

where δ_{jk} is said to be **Kronecker symbol**, or simply, **Kronecker.**

It is easy to verify that $\mathfrak{B}^* = \{e_1^*, \cdots, e_n^*\} \subset X^*$ is a base of X^*, and $\mathfrak{B}^*$ is said to be **the dual base of $\mathfrak{B}$.**

Similarly, define the set of linear functionals on X^*, denoted by $\mathfrak{B}^{**} = \{e_1^{**}, \cdots, e_n^{**}\} \subset$

X^{**}, satisfying

$$e_i^{**}(e_k^*)=\delta_{ki}=\begin{cases}1, & i=k\\ 0, & i\neq k\end{cases}, \quad i,k=1,2,\cdots,n, \tag{2.1.3}$$

and prove that $\mathfrak{B}^{**}=\{e_1^{**},\cdots,e_n^{**}\}\subset X^{**}$ is a base of X^{**}, then $\mathfrak{B}^{**}$ is said to be **the second dual of $\mathfrak{B}$.**

Thus we conclude that $\dim X=\dim X^*=\dim X^{**}=n.$

To prove(2), we define the mapping from X onto X^*,

$$\tau_1: x=\sum_{j=1}^{n}\alpha_j e_j\in X\to x^*=\sum_{j=1}^{n}\alpha_j e_j^*\in X^*, \quad \alpha_j\in\mathbb{F}, \quad j=1,2,\cdots,n.$$

It is clear that $\tau_1: X\to X^*$ is an isomorphism, called **a natural isomorphism from X onto X^*.**

Denote $x=\sum_{j=1}^{n}\alpha_j e_j\in X$ by $[x]_{\mathfrak{B}}=\begin{bmatrix}\alpha_1\\ \vdots\\ \alpha_n\end{bmatrix}$, and $x^*=\sum_{j=1}^{n}\alpha_j e_j^*\in X^*$ by $[x^*]_{\mathfrak{B}^*}=\begin{bmatrix}\alpha_1\\ \vdots\\ \alpha_n\end{bmatrix}$, then we have that X is isomorphic with X^*; and

$$[x]_{\mathfrak{B}}=\begin{bmatrix}\alpha_1\\ \vdots\\ \alpha_n\end{bmatrix}\overset{\tau_1}{\longleftrightarrow}[x^*]_{\mathfrak{B}^*}=\begin{bmatrix}\alpha_1\\ \vdots\\ \alpha_n\end{bmatrix}.$$

Similarly, define the mapping from X^* onto X^{**},

$$\tau_2: x^*=\sum_{j=1}^{n}\alpha_j e_j^*\in X^*\to x^{**}=\sum_{j=1}^{n}\alpha_j e_j^{**}\in X^{**}, \quad \alpha_j\in\mathbb{F}, \quad j=1,2,\cdots,n,$$

then τ_2 is **a natural isomorphism from X^* onto X^{**}**; and

$$[x^*]_{\mathfrak{B}^*}=\begin{bmatrix}\alpha_1\\ \vdots\\ \alpha_n\end{bmatrix}\overset{\tau_2}{\longleftrightarrow}[x^{**}]_{\mathfrak{B}^{**}}=\begin{bmatrix}\alpha_1\\ \vdots\\ \alpha_n\end{bmatrix}.$$

Let us consider the relationship between X and X^{**}. Let $\tau=\tau_2\circ\tau_1: X\to X^{**}$, and $x\in X$ have an expression $x=\sum_{j=1}^{n}\alpha_j e_j\in X$ in $\mathfrak{B}$. What is the expression of $\tau_2(\tau_1(x))$ in $\mathfrak{B}^{**}$? By

$$\tau(x)=\tau_2(\tau_1(x))=\tau_2(x^*)=x^{**}, \tag{2.1.4}$$

take any $w\in X^*$, and suppose that w has an expression in $\mathfrak{B}^*$, i. e.,

$$w=\sum_{i=1}^{n}\beta_i e_i^* \in X^*. \tag{2.1.5}$$

Since (2.1.2), $w(e_j)=\sum_{i=1}^{n}\beta_i e_i^*(e_j)=\beta_j, j=1,2,\cdots,n$, substitute it into (2.1.5), then $w=\sum_{i=1}^{n}w(e_i)e_i^*$ holds. Thus, for any $x^{**}=\sum_{j=1}^{n}\alpha_j e_j^{**}\in X^{**}$ and any $w=\sum_{i=1}^{n}\beta_i e_i^*=\sum_{i=1}^{n}w(e_i)e_i^*\in X^*$, it holds

$$\begin{aligned}x^{**}(w)&=\Big(\sum_{j=1}^{n}\alpha_j e_j^{**}\Big)(w)=\sum_{j=1}^{n}\alpha_j e_j^{**}(w)=\sum_{j=1}^{n}\alpha_j e_j^{**}\Big(\sum_{i=1}^{n}w(e_i)e_i^*\Big)\\&=\sum_{j=1}^{n}\sum_{i=1}^{n}\alpha_j w(e_i)e_j^{**}(e_i^*)=\sum_{j=1}^{n}\alpha_j w(e_j)=w\Big(\sum_{j=1}^{n}\alpha_j e_j\Big)=w(x).\end{aligned}$$

This is $[x^{**}]_{\mathfrak{B}^{**}}=\begin{bmatrix}\alpha_1\\ \vdots\\ \alpha_n\end{bmatrix}\overset{\tau_2\circ\tau_1}{\longleftrightarrow}[x]_{\mathfrak{B}}=\begin{bmatrix}\alpha_1\\ \vdots\\ \alpha_n\end{bmatrix}$. Hence, we get the following important corresponding relation

$$x^{**}=\sum_{j=1}^{n}\alpha_j e_j^{**}\in X^{**}\overset{\tau_2\circ\tau_1}{\longleftrightarrow}x=\sum_{j=1}^{n}\alpha_j e_j\in X,$$

and (2.1.1) holds. We induce the following table.

linear space X base $\mathfrak{B}=\{e_1,\cdots,e_n\}$	dual space X^* dual base $\mathfrak{B}^*=\{e_1^*,\cdots,e_n^*\}$	second dual space X^{**} second dual base $\mathfrak{B}^{**}=\{e_1^{**},\cdots,e_n^{**}\}$
$e_k^*(e_j)=\delta_{jk}$		$e_i^{**}(e_k^*)=\delta_{ki}$
natural isomorphism $\tau_1: x=\sum_{j=1}^{n}\alpha_j e_j\in X$ $\to x^*=\sum_{j=1}^{n}\alpha_j e_j^*\in X^*$	natural isomorphism $\tau_2: x^*=\sum_{j=1}^{n}\alpha_j e_j^*\in X^*$ $\to x^{**}=\sum_{j=1}^{n}\alpha_j e_j^{**}\in X^{**}$	natural isomorphism $\tau=\tau_2\circ\tau_1: x=\sum_{j=1}^{n}\alpha_j e_j\in X$ $\to x^{**}=\sum_{j=1}^{n}\alpha_j e_j^{**}\in X^{**}$

2.1.6 Structures of linear spaces

The structures of linear space $(X,+,\alpha\cdot)$ have very closed connection with its dual space, a very important "bridge" is so-called "bilinear form", which is bilinear functional defined on $X\times X$, and can determine the structures of $(X,+,\alpha\cdot)$.

1. Bilinear forms, quadratic forms

Definition 2.1.13 (bilinear form on linear space, quadratic form) Let $(X, +, \alpha \cdot)$ be a linear space on $\mathbb{F}$. A mapping $f: X\times X \to \mathbb{F}$ is called **a bilinear form on** X, if f is linear about two variables, i.e.

$$f(\alpha x+\beta y, z)=\alpha f(x, z)+\beta f(y, z), \quad x, y, z \in X, \quad \alpha, \beta \in \mathbb{F};$$

and

$$f(x, \alpha y+\beta z)=\alpha f(x, y)+\beta f(x, z), \quad x, y, z \in X, \quad \alpha, \beta \in \mathbb{F}.$$

A bilinear form $(x, y): X\times X \to \mathbb{F}$ is also called **a quadratic form on** $X\times X$.

Definition 2.1.14 (symmetric and skew-symmetric bilinear forms) Let $(X, +, \alpha \cdot)$ be a linear space on $\mathbb{F}$. A bilinear form $(x, y): X\times X \to \mathbb{F}$ is said to be **symmetric**, if

$$\forall x, y \in X \Rightarrow (x, y)=(y, x);$$

it is said to be **skew-symmetric**, if

$$\forall x, y \in X \Rightarrow (x, y)=-(y, x).$$

Definition 2.1.15 (metric linear space, nonsingular metric linear space) If there exists a bilinear form $(x, y): X\times X \to \mathbb{F}$ on linear space $(X, +, \alpha \cdot)$, then space $(X, (x, y)) \equiv ((X, +, \alpha \cdot), (x, y))$ is said to be **a metric linear space**; and the bilinear form (x, y) is said to be **a metric on the metric linear space** $(X, (x, y))$.

For metric linear space $(X, (x, y))$, if "$\forall y \in X, (x, y)=0 \Rightarrow x=0$", then $(X, (x, y))$ is said to be **a nonsingular metric linear space.**

2. Orthogonal geometry and symplectic geometry on linear spaces

Definition 2.1.16 (orthogonal geometry and symplectic geometry) For a nonsingular metric linear space $(X, (x, y))$, we have the following concepts:

(1) If $(x, y)=0$, then two elements $x \in X$ and $y \in X$ are said to be **orthogonal**, denoted by $x \perp y$. Two linear subspaces $V \subset X$ and $W \subset X$ are said to be orthogonal, if for any $x \in V$ and any $y \in W$, it holds $x \perp y$, denoted by $V \perp W$.

(2) If $(x, y)=(y, x)$, then $(X, (x, y))$ is said to **be a symmetric metric linear space**, or X is said to be **an orthogonal geometry**, or a **symmetric geometry on** $\mathbb{F}$.

(3) If $(x, y)=-(y, x)$, then $(X, (x, y))$ is said to be **a skew-symmetric metric linear space**, or X is said to be **a skew-symmetric geometry**, or **a symplectic geometry on** $\mathbb{F}$.

3. Riesz representative theorem

A representative theorem of linear functionals on a finite-dimensional nonsingular metric linear space is called Riesz representative theorem.

Theorem 2.1.8 Let $(X,(x,y))$ be a finite-dimensional, nonsingular metric linear space on $\mathbb{F}$. Then for each $f\in X^*$, there exists unique $x\in X$, satisfying "$\forall v\in X$, the action of functional f on $v\in X$ can be expressed as $f(v)\equiv(v,x)$, where (v,x) is a metric on X (a bilinear form on X)".

Proof Take $x\in X$, define a mapping $\varphi_x: X\to\mathbb{F}$ by $\varphi_x(v)=(v,x)$. It is clear that, φ_x is a functional on X, i.e., $\varphi_x\in X^*$ satisfies

$(\varphi_x,\alpha v_1+\beta v_2)=(\alpha v_1+\beta v_2,x)=\alpha(v_1,x)+\beta(v_2,x)$ for $x\in X,\alpha,\beta\in\mathbb{F}$, $v_1,v_2\in X^*$.

Moreover, define a mapping $\tau: X\to X^*$ satisfying $\tau(x)=\varphi_x$.

Then we turn to prove that $\tau: X\to X^*$ is linear. For $x,y\in X$, $\alpha,\beta\in\mathbb{F}$, we prove

$$\tau(\alpha x+\beta y)=\alpha\tau(x)+\beta\tau(y). \tag{2.1.6}$$

In fact, for any $v\in X$, $x,y\in X$, $\alpha,\beta\in\mathbb{F}$, by

$$\begin{aligned}(\tau(\alpha x+\beta y),v)&=\varphi_{\alpha x+\beta y}(v)=(v,\alpha x+\beta y)=\alpha(v,x)+\beta(v,y)\\&=\alpha\varphi_x(v)+\beta\varphi_y(v)=(\alpha\varphi_x+\beta\varphi_y)(v)=(\alpha\tau(x)+\beta\tau(y))(v),\end{aligned}$$

this implies the linearity of mapping τ, so that (2.1.6) holds.

Finally, by nonsingularity of space $(X,(x,y))$, then set $\{x\in X: \varphi_x=0\}=\{x\in X: (v,x)=0, \forall v\in X\}$ contains only one element $\{x\in X: \varphi_x=0\}=\{0\}$, this implies that $\tau: X\to X^*$ is a one-one and surjective mapping, so it is isomorphic from X onto X^*. Thus, if X is a finite-dimension linear space, then the linear functional $\varphi_x\in X^*$ on X has a form $\varphi_x(v)=(v,x)$. The proof is complete.

Riesz representative theorem tells us: The linear functional on a finite-dimensional nonsingular metric linear space $(X,+,\alpha\cdot,(x,y))$, has and only has one form, i.e., bilinear form on X as (v,x).

4. Structures of orthogonal geometry and symplectic geometry on linear spaces

To study structures of linear spaces, we start from the transformation matrix between linear spaces.

1) **Transformation matrix between linear spaces**

(1) **Transformation formula for two coordinate systems of $\mathbb{R}^3$**

Recalling in *advance mathematics*, the orthogonal coordinate system with origin zero in $\mathbb{R}^3$, is denoted by $Oxyz$, and a new coordinate system with origin zero in $\mathbb{R}^3$, is denoted by $Ox_1y_1z_1$.

Let $\mathfrak{B}=\{Oxyz: i,j,k\}$ and $\mathfrak{C}=\{Ox_1y_1z_1: i_1,j_1,k_1\}$, with $\{i,j,k\}$, $\{i_1,j_1,k_1\}$ as orthogonal unit vectors on two systems $\mathfrak{B}$ and $\mathfrak{C}$, respectively. They are two bases of $\mathbb{R}^3$, satisfying the following table.

	i_1	j_1	k_1
i	α_1	β_1	γ_1
j	α_2	β_2	γ_2
k	α_3	β_3	γ_3

The transformation formula of two bases is

$$\begin{cases} i_1 = i\cos\alpha_1 + j\cos\alpha_2 + k\cos\alpha_3 \\ j_1 = i\cos\beta_1 + j\cos\beta_2 + k\cos\beta_3 \\ k_1 = i\cos\gamma_1 + j\cos\gamma_2 + k\cos\gamma_3 \end{cases}, \tag{2.1.7}$$

where the **transformation matrix** $C = \begin{bmatrix} \cos\alpha_1 & \cos\alpha_2 & \cos\alpha_3 \\ \cos\beta_1 & \cos\beta_2 & \cos\beta_3 \\ \cos\gamma_1 & \cos\gamma_2 & \cos\gamma_3 \end{bmatrix}$ is orthogonal, $CC^{\mathrm{T}} = I$, with C^{T} **the transposition** of C. Let $\delta = \begin{bmatrix} i \\ j \\ k \end{bmatrix}$, $\delta_1 = \begin{bmatrix} i_1 \\ j_1 \\ k_1 \end{bmatrix}$, then the formula (2.1.7) is as $\delta_1 = C\delta$.

Using notations

$$\mathfrak{B} = \{i, j, k\} \leftrightarrow M_{\mathfrak{B}} \equiv \begin{bmatrix} b_1 \\ b_2 \\ b_3 \end{bmatrix} \in \mathfrak{M}_{3\times1}, \quad \mathfrak{C} = \{i_1, j_1, k_1\} \leftrightarrow M_{\mathfrak{C}} \equiv \begin{bmatrix} c_1 \\ c_2 \\ c_3 \end{bmatrix} \in \mathfrak{M}_{3\times1},$$

then, $\mathfrak{B} = \{i, j, k\} \xrightarrow{C} \mathfrak{C} = \{i_1, j_1, k_1\}$. (Where we regard the 3×3 matrix $\mathfrak{B}$ as a vector $\mathfrak{B} \leftrightarrow M_{\mathfrak{B}} \in \mathfrak{M}_{3\times1}$, this brings convenience in high-dimensional cases.) Denote $C \equiv M_{\mathfrak{B}\mathfrak{C}}$, then we call $M_{\mathfrak{B}\mathfrak{C}}: M_{\mathfrak{B}} \to M_{\mathfrak{C}}$ **a transformation matrix.** Thus, the formula $\delta_1 = C\delta$ in (2.1.7), i. e. , **the transformation formula of two bases of** $\mathbb{R}^3$, becomes

$$M_{\mathfrak{C}} = M_{\mathfrak{B}\mathfrak{C}} M_{\mathfrak{B}}. \tag{2.1.8}$$

To determine the transformation matrix $C \equiv M_{\mathfrak{B}\mathfrak{C}} \in \mathfrak{M}_{3\times3}$, denote $C \equiv [C_1 \quad C_2 \quad C_3]$, and

$$[C(M_{\mathfrak{B}})]_{\mathfrak{C}} = CM_{\mathfrak{B}}.$$

Take $e_1 = \begin{bmatrix} 1 \\ 0 \\ 0 \end{bmatrix}$, $e_2 = \begin{bmatrix} 0 \\ 1 \\ 0 \end{bmatrix}$, $e_3 = \begin{bmatrix} 0 \\ 0 \\ 1 \end{bmatrix}$ in $\mathfrak{B} = \mathfrak{N} = \{e_1, e_2, e_3\}$, then

$$[C(e_j)]_{\mathfrak{C}} = Ce_j = C_j, \quad j = 1, 2, 3$$

Hence, **the transformation matrix of two bases** $\mathfrak{B}$ and $\mathfrak{C}$ of $\mathbb{R}^3$ is

$$M_{\mathfrak{B}\mathfrak{C}} \equiv [C(e_j)]_{\mathfrak{C}}. \tag{$*$}$$

(2) **Transformation formula for coordinate systems of** n**-dimensional linear space** X to m**-dimensional linear space** Y

The result in (1) can be generalized to that of in n-dimensional linear X to m-dimensional linear space Y.

Let dimensions of X, Y be $\dim X = n$, $\dim Y = m$, respectively; the base $\mathfrak{B}$ of X be $\begin{bmatrix} b_1 \\ b_2 \\ \vdots \\ b_n \end{bmatrix} \equiv M_B$, and the base $\mathfrak{C}$ of Y be $\begin{bmatrix} c_1 \\ c_2 \\ \vdots \\ c_m \end{bmatrix} \equiv M_{\mathfrak{C}}$. Thus, we have $\mathfrak{B} \leftrightarrow M_B \in \mathfrak{M}_{n\times 1}$, and $\mathfrak{C} \leftrightarrow M_{\mathfrak{C}} \in \mathfrak{M}_{m\times 1}$. If linear transformation $\tau_{\mathfrak{BC}}: \mathfrak{B} \to \mathfrak{C}$ transforms the base $\mathfrak{B}$ of X to the base $\mathfrak{C}$ of Y, then we have

$$\tau_{\mathfrak{BC}} \leftrightarrow M_{\mathfrak{BC}} \in \mathfrak{M}_{m\times n}; \tag{2.1.9}$$

thus, **the transformation formula of bases of X and Y** is

$$M_{\mathfrak{C}} = M_{\mathfrak{BC}} M_{\mathfrak{B}}. \tag{2.1.10}$$

Similarly, as (*) in (1), **the transformation matrix of** $\mathfrak{B}$ and $\mathfrak{C}$ becomes

$$M_{\mathfrak{BC}} = [[\tau_{\mathfrak{BC}}(e_1)]_{\mathfrak{C}} \cdots [\tau_{\mathfrak{BC}}(e_n)]_{\mathfrak{C}}]. \tag{* *}$$

(3) **Transformation formula of coordinates for points in two coordinate systems of $\mathbb{R}^3$**

In $\mathbb{R}^3$, a transformation of two coordinate systems $\mathfrak{B}$ and $\mathfrak{C}$ of $\mathbb{R}^3$, denotes by $\tau_{\mathfrak{BC}}: \mathfrak{B} \to \mathfrak{C}$. Now, a transformation formula of coordinates for a point $P \in \mathbb{R}^3$ with respect to $\mathfrak{B}$ and $\mathfrak{C}$ is deduced as follows.

By $\mathfrak{B} = \{i, j, k\} \leftrightarrow \delta \equiv \begin{bmatrix} b_1 \\ b_2 \\ b_3 \end{bmatrix} \in \mathfrak{M}_{3\times 1}$, the coordinate of point P in $\mathfrak{B}$ is denoted by $(x_1, x_2, x_3) \equiv [x]_{\mathfrak{B}}$; and by $\mathfrak{C} = \{i_1, j_1, k_1\} \leftrightarrow \delta_1 \equiv \begin{bmatrix} c_1 \\ c_2 \\ c_3 \end{bmatrix} \in \mathfrak{M}_{3\times 1}$, the coordinate of P in $\mathfrak{C}$ is denoted by $(x_1', x_2', x_3') \equiv [x]_{\mathfrak{C}}$. Since $\tau_{\mathfrak{BC}}: \mathfrak{B} \to \mathfrak{C}$ is an orthogonal transformation, thus the transformation matrix $M_{\mathfrak{BC}}: \delta \to \delta_1$ ($M_{\mathfrak{C}} = M_{\mathfrak{BC}} M_B$) satisfies $M_{\mathfrak{BC}} (M_{\mathfrak{BC}})^{\mathrm{T}} = I$ with the transposed matrix $(M_{\mathfrak{BC}})^{\mathrm{T}} = (M_{\mathfrak{BC}})^{-1}$. Note, that the length of $\overrightarrow{OP}$ is invariant under the orthogonal transformation $\tau_{\mathfrak{BC}}: \mathfrak{B} \to \mathfrak{C}$, and so does the inner product of two vectors, thus take $[x]_{\mathfrak{B}} \equiv \begin{bmatrix} x_1 \\ x_2 \\ x_3 \end{bmatrix} \in \mathfrak{M}_{3\times 1}$ and $\delta \equiv \begin{bmatrix} b_1 \\ b_2 \\ b_3 \end{bmatrix} \in \mathfrak{M}_{3\times 1}$ in $\mathfrak{B}$; as well as $[x]_{\mathfrak{C}} \equiv \begin{bmatrix} x_1' \\ x_2' \\ x_3' \end{bmatrix} \in \mathfrak{M}_{3\times 1}$ and $\delta_1 \equiv \begin{bmatrix} c_1 \\ c_2 \\ c_3 \end{bmatrix} \in \mathfrak{M}_{3\times 1}$ in $\mathfrak{C}$. By $[x]_{\mathfrak{C}} \cdot \delta_1 = [x]_{\mathfrak{B}} \cdot \delta$, we deduce that:

$$[x]_{\mathfrak{C}} \cdot \delta_1 = [x]_{\mathfrak{B}} \cdot \delta \quad (\text{since } \delta = (M_{\mathfrak{BC}})^{-1}\delta_1 = (M_{\mathfrak{BC}})^{\mathrm{T}}\delta_1)$$
$$\Rightarrow [x]_{\mathfrak{C}} \cdot \delta_1 = [x]_{\mathfrak{B}} \cdot \{(M_{\mathfrak{BC}})^{\mathrm{T}}\delta_1\} \quad (\text{multiply } (\delta_1)^{\mathrm{T}} \text{ on the right})$$
$$\Rightarrow \{[x]_{\mathfrak{C}} \cdot \delta_1\}(\delta_1)^{\mathrm{T}} = \{[x]_{\mathfrak{B}} \cdot (M_{\mathfrak{BC}})^{\mathrm{T}}\delta_1\} \cdot (\delta_1)^{\mathrm{T}} \Rightarrow [x]_{\mathfrak{C}} = M_{\mathfrak{BC}}[x]_{\mathfrak{B}}.$$

Then, we have **the Transformation formula of coordinates for a point in two coordinate systems of $\mathbb{R}^3$**:

$$[x]_{\mathfrak{C}} = M_{\mathfrak{BC}}[x]_{\mathfrak{B}}. \tag{2.1.11}$$

We conclude that the transformation matrix $M_{\mathfrak{BC}}$ is not only that of $M_{\mathfrak{B}}(\leftrightarrow\mathfrak{B})$ to $M_{\mathfrak{C}}(\leftrightarrow\mathfrak{C})$, i.e., $M_{\mathfrak{C}} = M_{\mathfrak{BC}}M_{\mathfrak{B}}$ (see (2.1.10)), but also is the transformation matrix of $[x]_{\mathfrak{B}}$ to $[x]_{\mathfrak{C}}$, i.e., $[x]_{\mathfrak{C}} = M_{\mathfrak{BC}}[x]_{\mathfrak{B}}$ (see (2.1.11)).

(4) **Transformation formula of coordinates for points in coordinate systems of n-dimensional linear space X to m-dimensional linear space Y**

The result in (3) can be generalized to that of n-dimensional linear space X to m-dimensional linear space Y.

Under the assumption of (2), we suppose that the coordinate of $x \in X$ in the base $\mathfrak{B} \subset X$ is $[x]_{\mathfrak{B}} \equiv \begin{bmatrix} x_1 \\ \vdots \\ x_n \end{bmatrix} \in \mathfrak{M}_{n\times 1}$, and the coordinate of $y \in Y$ in the base $\mathfrak{C} \subset Y$ is $[y]_{\mathfrak{C}} \equiv \begin{bmatrix} y_1 \\ \vdots \\ y_m \end{bmatrix} \in \mathfrak{M}_{m\times 1}$. Thus, $\tau_{\mathfrak{BC}}: \mathfrak{B} \to \mathfrak{C}$ implies $\tau_{\mathfrak{BC}}: X \to Y$ with $\tau_{\mathfrak{BC}}(x) = y$, and

$$\tau_{\mathfrak{BC}}([x]_{\mathfrak{B}}) = [y]_{\mathfrak{C}}, \tag{2.1.12}$$

or, by the transformation matrix $M_{\mathfrak{BC}}$ of $\tau_{\mathfrak{BC}}$, it is

$$[y]_{\mathfrak{C}} = M_{\mathfrak{BC}}[x]_{\mathfrak{B}}. \tag{2.1.13}$$

2) **Finite-dimensional nonsingular metric linear spaces**

We have the following theorem about an n-dimensional nonsingular metric linear space.

Theorem 2.1.9 Let $(X, +, \alpha\cdot, (x, y))$ be a finite-dimensional ($\dim X = n$), nonsingular metric linear space on $\mathbb{F}$, and $\mathfrak{B} = \{b_1, \cdots, b_n\}$ be a base of X as

$$\mathfrak{B} = \{[b_{11} \ \cdots \ b_{1n}]^{\mathrm{T}}, \cdots, [b_{n1} \ \cdots \ b_{nn}]^{\mathrm{T}}\} = \left\{ \begin{bmatrix} b_{11} \\ b_{12} \\ \vdots \\ b_{1n} \end{bmatrix}, \cdots, \begin{bmatrix} b_{n1} \\ b_{n2} \\ \vdots \\ b_{nn} \end{bmatrix} \right\},$$

with the transposed matrix $[\cdot]^{\mathrm{T}}$ of $[\cdot]$. Suppose that $[\mathfrak{B}] \equiv [b_{jk}]_{n\times n}$ is represented as $[\mathfrak{B}] = [(b_j, b_k)]_{n\times n}$ with inner product (b_j, b_k) of b_j and b_k, called the corresponding matrix determined by bilinear form (x, y), or simply, the corresponding matrix. Then,

(1) Bilinear form (x, y) is symmetric, if and only if $[\mathfrak{B}] = [(b_j, b_k)]_{n\times n}$ is

symmetric; i. e. , $b_{jk}=b_{kj}$, $1\leqslant j,k\leqslant n$.

(2) Bilinear form (x,y) is skew-symmetric, if and only if the corresponding matrix $[\mathfrak{B}]=[(b_j,b_k)]_{n\times n}$ is skew-symmetric; i. e. , $(b_j,b_j)=0$, $(b_j,b_k)=-(b_k,b_j)$, $1\leqslant j\neq k\leqslant n$.

Proof We express $x\in X$ and $y\in X$ in the base $\mathfrak{B}=\{b_1,\cdots,b_n\}$ as $x=\sum_{j=1}^{n}x_jb_j$ and $y=\sum_{j=1}^{n}y_jb_j$, respectively; and denote $[x]_{\mathfrak{B}}=\begin{bmatrix}x_1\\ \vdots\\ x_n\end{bmatrix}$, $[y]_{\mathfrak{B}}=\begin{bmatrix}y_1\\ \vdots\\ y_n\end{bmatrix}$. Then, in the base $\mathfrak{B}$, we have

$$(x,y)=\Big(\sum_{j=1}^{n}x_jb_j,\sum_{k=1}^{n}y_kb_k\Big)=\sum_{j=1}^{n}x_j\Big(b_j,\sum_{k=1}^{n}y_kb_k\Big)=\sum_{j=1}^{n}\sum_{k=1}^{n}x_jy_k(b_j,b_k).$$

Thus, for any $x,y\in X$, if $(x,y)=(y,x)$, then $(b_j,b_k)=(b_k,b_j)$, $1\leqslant j,k\leqslant n$, so that $[\mathfrak{B}]=[(b_j,b_k)]_{n\times n}$ is symmetric matrix; and vice versa.

Conversely, if $(x,y)=-(y,x)$, then $(b_j,b_k)=-(b_k,b_j)$, $1\leqslant j,k\leqslant n$, this implies that $(b_j,b_j)=0$, and $(b_j,b_k)=-(b_k,b_j)$, $1\leqslant j\neq k\leqslant n$, so that $[\mathfrak{B}]$ is skew-symmetric matrix; and vice versa.

3) **Decomposition theorem of orthogonal geometry**

Theorem 2. 1. 10 Let $(X,+,\alpha\cdot,(x,y))$ be a finite n-dimensional, nonsingular, symmetric, metric linear space on number field $\mathbb{F}$. Then there exists an orthogonal base $\mathfrak{B}=\{b_1,\cdots,b_n\}$ with $(b_j,b_k)=\begin{cases}(b_j,b_j)\neq 0, & k=j\\ 0, & k\neq j\end{cases}$, $j,k=1,2,\cdots,n$, such that X can be decomposed as the orthogonal direct sum

$$X=S_1\underline{\oplus}\cdots\underline{\oplus}S_n.$$

Where: ① Sign $\underline{\oplus}$ means the orthogonal direct sum, i. e. , $S_j\perp S_k$, $S_j\oplus S_k$, $j\neq k$, and $\forall x\in X$ can be expressed as $x=\sum_{j=1}^{n}s_j$, $s_j\in S_j$, $j=1,2,\cdots,n$.

② By b_j, $(b_j,b_j)=a_j$, $j=1,2,\cdots,n$, it generates an one-dimensional linear subspace S_j, $j=1,2,\cdots,n$, that is $S_j=\{x\in X: x=\alpha\, b_j, b_j\in\mathfrak{B}\}$.

③ In the orthogonal base $\mathfrak{B}=\{b_1,\cdots,b_n\}$ determined by the bilinear form (x,y) of X, the corresponding matrix $[\mathfrak{B}]=\begin{bmatrix}(b_1,b_1) & \cdots & (b_1,b_n)\\ \vdots & \ddots & \vdots\\ (b_n,b_1) & \cdots & (b_n,b_n)\end{bmatrix}$ is nonsingular orthogonal one: $|[\mathfrak{B}]|\neq 0$, and $[\mathfrak{B}]^{\mathrm{T}}[\mathfrak{B}]=[(b_j,b_j)]_{n\times n}$.

When $\mathbb{F}=\mathbb{C}$ (it is an algebraic closed field, i. e., all n-order polynomials with coefficients in $\mathbb{F}$ can be decomposed into a product of one-order linear factors), there exists a normal orthogonal base $\mathfrak{N}$, such that the corresponding matrix is

$$[\mathfrak{B}]=\begin{bmatrix}1 & & \\ & \ddots & \\ & & 1\end{bmatrix}. \tag{2.1.14}$$

When $\mathbb{F}=\mathbb{R}$ (it is not algebraic closed field), there exists an orthogonal base

$$\mathfrak{B}=\{b_1=(b_{11},\cdots,b_{1n}),\cdots,b_n=(b_{n1},\cdots,b_{nn})\}$$

with $(b_j,b_k)=\begin{cases}(b_j,b_j)\neq 0, & k=j\\ 0, & k\neq j\end{cases}$, and $(b_j,b_j)=\begin{cases}1, & 1\leqslant j\leqslant j_0\\ -1, & j_0<j\leqslant n\end{cases}$, such that the corresponding matrix is

$$[\mathfrak{B}]=\begin{bmatrix}1 & & & & & \\ & \ddots & & & & \\ & & 1 & & & \\ & & & -1 & & \\ & & & & \ddots & \\ & & & & & -1\end{bmatrix}. \tag{2.1.15}$$

④ If P is an n-order nonsingular symmetric matrix, then there exists n-order nonsingular matrix Q, such that

$$P=Q^{\mathrm{T}}[\mathfrak{B}]Q=Q^{\mathrm{T}}\begin{bmatrix}a_1 & & & \\ & a_2 & & \\ & & \ddots & \\ & & & a_n\end{bmatrix}Q, \tag{2.1.16}$$

where $(b_j,b_j)=a_j$ is as in ②: if $\mathbb{F}=\mathbb{C}$, then $[\mathfrak{B}]$ is given by (2.1.14); if $\mathbb{F}=\mathbb{R}$, then $[\mathfrak{B}]$ is given by (2.1.15).

4) **Decomposition theorem of symplectic geometry**

Theorem 2.1.11 Let $(X,+,\alpha\cdot,(x,y))$ be a finite n-dimensional, nonsingular, skew-symmetric metric linear space on number field $\mathbb{F}$. Then there exists an orthogonal base $\mathfrak{B}=\{b_1,\cdots,b_n\}$, such that X can be decomposed as the orthogonal direct sum

$$X=H_1\underline{\oplus}\cdots\underline{\oplus}H_k.$$

Where: ① Sign $\underline{\oplus}$ means the orthogonal direct sum, i. e., $H_j\perp H_l$, $H_j\oplus H_l$, $j\neq l$, and $\forall x\in X$ can be expressed by $x=\sum_{j=1}^{k}h_j$, $h_j\in H_j$, $j=1,2,\cdots,k$.

② H_j is 2-dimensional skew-symmetric metric linear subspace, the corresponding matrix is $\begin{bmatrix} 0 & 1 \\ -1 & 0 \end{bmatrix}$; thus, the dimension of a nonsingular, skew-symmetric, metric linear space is certainly a even number.

③ There exists an orthogonal base $\mathfrak{B}=\{b_1,\cdots,b_n\}$, such that the corresponding matrix of (x,y) is

$$[\mathfrak{B}]=\begin{bmatrix} 0 & 1 & 0 & 0 & \cdots & 0 & 0 \\ -1 & 0 & 0 & 0 & \cdots & 0 & 0 \\ 0 & 0 & 0 & 1 & \cdots & 0 & 0 \\ 0 & 0 & -1 & 0 & \cdots & 0 & 0 \\ \vdots & \vdots & \vdots & \vdots & \ddots & \vdots & \vdots \\ 0 & 0 & 0 & 0 & \cdots & 0 & 1 \\ 0 & 0 & 0 & 0 & \ddots & -1 & 0 \end{bmatrix}.$$

④ If P is an n-order nonsingular skew-symmetric matrix, then there exists an n-order nonsingular matrix Q, such that

$$P=Q^{\mathrm{T}}[\mathfrak{B}]Q=Q^{\mathrm{T}}\begin{bmatrix} 0 & 1 & 0 & 0 & \cdots & 0 & 0 \\ -1 & 0 & 0 & 0 & \cdots & 0 & 0 \\ 0 & 0 & 0 & 1 & \cdots & 0 & 0 \\ 0 & 0 & -1 & 0 & \cdots & 0 & 0 \\ \vdots & \vdots & \vdots & \vdots & \ddots & \vdots & \vdots \\ 0 & 0 & 0 & 0 & \cdots & 0 & 1 \\ 0 & 0 & 0 & 0 & \ddots & -1 & 0 \end{bmatrix}Q,$$

thus, a nonsingular, skew-symmetric matrix certainly is even order.

5) **Sylvester's law of inertia**

Theorem 2.1.12 Let $(X,+,\alpha\cdot,(x,y))$ be a finite n-dimensional nonsingular metric linear space on number field $\mathbf{F}$.

(1) If (x,y) is symmetric, then there exists an orthogonal base of X, denoted by $\mathfrak{B}=\{b_1,\cdots,b_n\}$, such that for any $x,y\in X$, it holds

$$(x,y)=a_1\xi_1\eta_1+a_2\xi_2\eta_2+\cdots+a_n\xi_n\eta_n. \tag{2.1.17}$$

If (x,y) is skew-symmetric, then there exists an orthogonal base of X, denoted by $\mathfrak{B}=\{b_1,\cdots,b_n\}$, such that for any $x,y\in X$, it holds

$$(x,y)=\xi_1\eta_2-\xi_2\eta_1+\cdots+\xi_{n-1}\eta_n-\xi_n\eta_{n-1}. \tag{2.1.18}$$

In (2.1.17), (2.1.18), we have $[x]_{\mathfrak{B}}=\begin{bmatrix}\xi_1\\ \xi_2\\ \vdots\\ \xi_n\end{bmatrix}$, $[y]_{\mathfrak{B}}=\begin{bmatrix}\eta_1\\ \eta_2\\ \vdots\\ \eta_n\end{bmatrix}$, $(b_j,b_j)=a_j, j=1,2,\cdots,n$.

(2) If $\mathbb{F}=\mathbb{C}$, then there exists an orthonormal basis of X, denoted by

$$\mathfrak{N}=\{b_1=(b_{11},\cdots,b_{1n}),\cdots,b_n=(b_{n1},\cdots,b_{nn})\},$$

with $(b_j,b_k)=\begin{cases}1, & k=j\\ 0, & k\neq j\end{cases}$, such that for any $x,y\in X$, it holds

$$(x,y)=\xi_1\eta_1+\xi_2\eta_2+\cdots+\xi_n\eta_n. \tag{2.1.19}$$

If $\mathbb{F}=\mathbb{R}$, then there exists an orthogonal basis of X, denoted by

$$\mathfrak{N}=\{b_1=(b_{11},\cdots,b_{1n}),\cdots,b_n=(b_{n1},\cdots,b_{nn})\},$$

with $(b_j,b_k)=\begin{cases}(b_j,b_j)\neq 0, & k=j\\ 0, & k\neq j\end{cases}$, and $(b_j,b_j)=\begin{cases}1, & 1\leqslant j\leqslant j_0\\ -1, & j_0<j\leqslant n\end{cases}$, such that for any $x,y\in X$,

$$(x,y)=\xi_1\eta_1+\cdots+\xi_k\eta_k-\xi_{k+1}\eta_{k+1}-\cdots-\xi_n\eta_n. \tag{2.1.20}$$

(3) Specially, for a quadratic form (x,x), if $\mathbb{F}=\mathbb{C}$, then $(x,x)=\xi_1^2+\xi_2^2+\cdots+\xi_n^2$; if $\mathbb{F}=\mathbb{R}$, then $(x,x)=\xi_1^2+\cdots+\xi_k^2-\xi_{k+1}^2-\cdots-\xi_n^2$.

2.2 Linear Transformations

2.2.1 Linear operator spaces

1. Linear operator on linear spaces

Let $X\equiv(X,(+)_X,(\alpha\cdot)_X)$, $Y\equiv(Y,(+)_Y,(\alpha\cdot)_Y)$ be two linear spaces on number field $\mathbb{F}$.

Definition 2.2.1 (linear operator, linear operator space) Let $T: X\to Y$ be a mapping of $(X,(+)_X,(\alpha\cdot)_X)$ to $(Y,(+)_Y,(\alpha\cdot)_Y)$. If T satisfies

$$T(\alpha x_1+\beta x_2)=\alpha T(x_1)+\beta T(x_2)=\alpha y_1+\beta y_2,\quad x_1,x_2\in X, y_1,y_2\in Y, \alpha,\beta\in\mathbb{F},$$

then T is said to be **a linear operator from X to Y**, or simply, **a linear transformation**, or **a linear mapping.**

Space of linear operators denote all linear operators from X to Y by

$$L(X,Y)=\{T: X\to Y \text{ is the linear operator from } X \text{ to } Y\},$$

it is the set of linear operators from X to Y. The elements in $L(X,Y)$ are denoted by T, S, R, $\cdots$. We define operations on $L(X,Y)$:

addition $+\equiv(+)_{L(X,Y)}$:

$$T,S\in L(X,Y)\Rightarrow (T(+)_{L(X,Y)}S)(x)=T(x)(+)_Y S(x),\quad \forall x\in X;$$

number product $\alpha\cdot\equiv(\alpha\cdot)_{L(X,Y)}$:

$$T\in L(X,Y),\quad \alpha\in\mathbb{F}\Rightarrow((\alpha\cdot)_{L(X,Y)}T)(x)=(\alpha\cdot)_Y T(x),\quad \forall x\in X,\quad \alpha\in\mathbb{F}.$$

Then $L(X,Y)$ is a linear space on $\mathbb{F}$, called **the linear operator space from X to Y**, or simply, **the linear operator space.**

Particularly, if Y is $\mathbb{F}$ ($\mathbb{R}$ or $\mathbb{C}$), then $L(X,\mathbb{F})$ is called **a linear functional space**, an element $T \in L(X,\mathbb{F})$ turns out a linear functional. **The dual space** $X^* = L(X,\mathbb{F})$ of X (see Definition 2.1.10), is also called **the conjugate space of** X.

There is an isomorphism theorem for linear operator space $L(X,Y)$.

Theorem 2.2.1 Let X, Y **be linear spaces on** $\mathbb{F}$ with $\dim X = n$, $\dim Y = m$, respectively. Then

$$L(X,Y) \overset{\text{iso.}}{\longleftrightarrow} \mathfrak{M}_{m\times n} \equiv \mathfrak{M}_{m\times n}(\mathbb{F}),$$

where $\mathfrak{M}_{m\times n}(\mathbb{F})$ is the $m\times n$-matrix space on $\mathbb{F}$. Moreover, the isomorphic mapping is

$$T \in L(X,Y) \leftrightarrow A = [T]_{\mathfrak{B}\mathfrak{C}} = [a_{ij}]_{1\leqslant i\leqslant m, 1\leqslant j\leqslant n}.$$

Proof Let $\mathfrak{B} = \{b_1, \cdots, b_n\}$ be a base of X, each $x \in X$ has the coordinate in $\mathfrak{B}$ as $[x]_{\mathfrak{B}} = \begin{bmatrix}\alpha_1\\ \vdots \\ \alpha_n\end{bmatrix}$; and $\mathfrak{C} = \{c_1, c_2, \cdots, c_m\}$ be a base of Y, each $y \in Y$ has the coordinate in $\mathfrak{C}$ as $[y]_{\mathfrak{C}} = \begin{bmatrix}\beta_1\\ \vdots \\ \beta_m\end{bmatrix}$. Suppose that linear operator $T \in L(X,Y)$ transforms $\mathfrak{B}$ to $\mathfrak{C}$, i. e., $T: \mathfrak{B} \to \mathfrak{C}$, and $T(x) = y \in Y$.

To determine T, let $[T(x)]_{\mathfrak{C}} = A[x]_{\mathfrak{B}} = [y]_{\mathfrak{C}}$ by $[T(x)]_{\mathfrak{C}} = [y]_{\mathfrak{C}}$. Next, we determine the matrix $A = [A_1 \ \cdots \ A_n]$ with $A_j = \begin{bmatrix}a_{1j}\\ \vdots \\ a_{mj}\end{bmatrix}$, $j = 1, 2, \cdots, n$. Take $\mathfrak{B}$ as the orthonormal base

$$\mathfrak{N} = \{b_1 = \underbrace{(1,0,\cdots,0)}_{n}, \cdots, b_n = \underbrace{(0,\cdots,0,1)}_{n}\},$$

i. e., $[b_1]_{\mathfrak{B}} = [e_1]_{\mathfrak{B}}, \cdots, [b_n]_{\mathfrak{B}} = [e_n]_{\mathfrak{B}}$, substitute into $[T(x)]_{\mathfrak{C}} = A[x]_{\mathfrak{B}}$, we get

$$[T(b_1)]_{\mathfrak{C}} = \begin{bmatrix}(T(b_1))_1\\ \vdots \\ (T(b_1))_m\end{bmatrix} = A[b_1]_{\mathfrak{B}} = [A_1 \ \cdots \ A_n][b_1]_{\mathfrak{B}} = [A_1 \ \cdots \ A_n]\begin{bmatrix}1\\0\\ \vdots \\ 0\end{bmatrix} = A_1.$$

Similarly, we have $[T(b_j)]_{\mathfrak{C}}$, $j = 2, \cdots, n$. Then $[T(b_j)]_{\mathfrak{C}} = \begin{bmatrix}(T(b_j))_1\\ \vdots \\ (T(b_j))_m\end{bmatrix}_{\mathfrak{B}} = A[b_j]_{\mathfrak{B}} = A_j$.

Thus, $A_j = [T(b_j)]_{\mathfrak{C}}, j=1,2,\cdots,n$, it follows

$$A \equiv [T]_{\mathfrak{BC}} = [A_1 \ \cdots \ A_n] = [[T(b_1)]_{\mathfrak{C}} \cdots [T(b_n)]_{\mathfrak{C}}]_{m\times n}.$$

Hence, $T \in L(X,Y) \leftrightarrow A \equiv [T]_{\mathfrak{BC}} \in \mathfrak{M}_{m\times n}$. We conclude that $T \in L(X,Y)$ is corresponding to a matrix $[T]_{\mathfrak{BC}} \in \mathfrak{M}_{m\times n}$. Moreover, for any $x \in X$ and its image $T(x) \in Y$, the coordinates $[x]_{\mathfrak{B}} = [\alpha_1 \ \cdots \ \alpha_n]^{\mathrm{T}}$ and $[T(x)]_{\mathfrak{C}} = [\beta_1 \ \cdots \ \beta_m]^{\mathrm{T}}$ hold the relationship

$$[T(x)]_{\mathfrak{C}} = [T]_{\mathfrak{BC}}[x]_{\mathfrak{B}},$$

where $A = [T]_{\mathfrak{BC}}: \mathfrak{B} \to \mathfrak{C}$. So $T \in L(X,Y)$ determines a linear transformation from $\mathbb{F}^n$ to $\mathbb{F}^m$, i. e., $A = [T]_{\mathfrak{BC}} \in \mathfrak{M}_{m\times n}$, with $T \in L(X,Y) \to A = [T]_{\mathfrak{BC}} \in \mathfrak{M}_{m\times n} \leftrightarrow L(\mathbb{F}^n, \mathbb{F}^m)$.

To prove $L(X,Y)$ and $\mathfrak{M}_{m\times n}$ are isomorphic, we define a mapping $\varphi: L(X,Y) \to \mathfrak{M}_{m\times n}$ by

$$\varphi(T) = A = [T(b_j)]_{\mathfrak{C}}, \quad \forall\, T \in L(X,Y).$$

Linearity— In fact, $\forall\, \alpha, \beta \in \mathbb{F}$, $\forall\, T, S \in L(X,Y)$, by the definition of φ, we have

$$\varphi(\alpha T + \beta S) = [\alpha T(b_j) + \beta S(b_j)]_{\mathfrak{C}} = \alpha [T(b_j)]_{\mathfrak{C}} + \beta [S(b_j)]_{\mathfrak{C}} = \alpha\varphi(T) + \beta\varphi(S).$$

Surjection—For any $B \in \mathfrak{M}_{m\times n}$, we have

$$B = [B_1 \ \cdots \ B_n], \quad B_j = \begin{bmatrix} \beta_{1j} \\ \vdots \\ \beta_{mj} \end{bmatrix}, j = 1,2,\cdots,n.$$

In the base $\mathfrak{B} = \{b_1, \cdots, b_n\}$, we select one $T \in L(X,Y)$, with $[T(b_j)]_{\mathfrak{C}} = B_j$, $j=1,2,\cdots,n$, so that for any element $B \in \mathfrak{M}_{m\times n}$, there is one $T \in L(X,Y)$ corresponding to it. Thus φ is a surjection.

Injection— If $A = [T]_{\mathfrak{BC}} = 0$, then $[T(b_j)]_{\mathfrak{C}} = 0, j=1,2,\cdots,n$; and $T(b_j) = 0$ for all $j = 1,2,\cdots,n$. It implies that $T = 0$, and φ is injective. Then,

$$L(X,Y) \xleftrightarrow{\varphi} \mathfrak{M}_{m\times n} \equiv \mathfrak{M}_{m\times n}(\mathbb{F}).$$

The proof is complete.

Note By $T \in L(X,Y) \leftrightarrow [T]_{\mathfrak{BC}} \in \mathfrak{M}_{m\times n}$, **the transformation formula between two bases** is

$$M_{\mathfrak{C}} = [T]_{\mathfrak{BC}} M_{\mathfrak{B}} \quad (\text{see}(2.1.10)).$$

The transformation formula between two points is

$$[T(x)]_{\mathfrak{C}} = [T]_{\mathfrak{BC}}[x]_{\mathfrak{B}}, \quad \forall\, x \in X \quad (\text{see}(2.1.12)).$$

$[\mathrm{T}]_{\mathfrak{BC}} \in \mathfrak{M}_{m\times n}$ is said to be **the representation matrix of operator** $T \in L(X,Y)$.

A special circumstance of the above theorem is as follows.

Theorem 2.2.2 Let X be a linear space on $\mathbb{F}$ with $\dim X = n$. Then for two bases $\mathfrak{B}$ and $\mathfrak{C}$ of X, the representation matrix of operator $T \in L(X,X)$ is

$$[T]_{\mathfrak{CC}}=M_{\mathfrak{BC}}[T]_{\mathfrak{BB}}M_{\mathfrak{BC}}^{-1},$$

where $M_{\mathfrak{BC}}\equiv[[b_1]_{\mathfrak{C}}\cdots[b_n]_{\mathfrak{C}}]_{n\times n}$ is determined by the orthonormal base of X, denoted by

$$\mathfrak{N}=\left\{e_1=\begin{bmatrix}1\\0\\\vdots\\0\end{bmatrix},\cdots,e_n=\begin{bmatrix}0\\\vdots\\0\\1\end{bmatrix}\right\}.$$

We have the generalized case of Theorem 2.2.1 in three linear spaces.

Theorem 2.2.3 Let X, Y, Z be linear spaces on $\mathbb{F}$, with $\dim X=n$, base $\mathfrak{B}=\mathfrak{B}_X$; $\dim Y=m$, base $\mathfrak{C}=\mathfrak{C}_Y$; and $\dim Z=l$, base $\mathfrak{D}=\mathfrak{D}_Z$. Then for $T\in L(X,Y)$, $S\in L(Y,Z)$, it holds

$$[S\circ T]_{\mathfrak{BD}}=[S]_{\mathfrak{CD}}[T]_{\mathfrak{BC}},$$

where, $[T]_{\mathfrak{BC}}\in\mathfrak{M}_{m\times n}$, $[S]_{\mathfrak{CD}}\in\mathfrak{M}_{l\times m}$, $[S\circ T]_{\mathfrak{BD}}\in\mathfrak{M}_{l\times n}$.

Proof By $x\in X\Rightarrow T(x)\in Y\Rightarrow(S\circ T)(x)=S(T(x))\in Z$, we have

$$[T(x)]_{\mathfrak{C}}=[T]_{\mathfrak{BC}}[x]_{\mathfrak{B}},\quad [S(T(x))]_{\mathfrak{D}}=[S]_{\mathfrak{CD}}[T(x)]_{\mathfrak{C}}.$$

Thus, substitute $[T(x)]_{\mathfrak{C}}=[T]_{\mathfrak{BC}}[x]_{\mathfrak{B}}$ into the latter formula, it follows

$$[S(T(x))]_{\mathfrak{D}}=[S]_{\mathfrak{CD}}[T(x)]_{\mathfrak{C}}=[S]_{\mathfrak{CD}}[T]_{\mathfrak{BC}}[x]_{\mathfrak{B}},$$

this is $[(S\circ T)(x)]_{\mathfrak{D}}=[S]_{\mathfrak{CD}}[T]_{\mathfrak{BC}}[x]_{\mathfrak{B}}$, and thus, $[S\circ T]_{\mathfrak{BD}}=[S]_{\mathfrak{CD}}[T]_{\mathfrak{BC}}$.

2. Images, kernels, and nullities of linear operators

Definition 2.2.2 (image, kernel, and nullity) Let X, Y be linear spaces on $\mathbb{F}$. For any operator $T\in L(X,Y)$, the set $\operatorname{im}(T)=\{T(x): x\in X\}$ is called the **image** of T; the set $\ker(T)=\{x\in X: T(x)=0\}$ is called **the kernel** of T; and $\dim(\ker(T))$ is called **the nullity** of T, denoted by $\operatorname{null}(T)$. Moreover, the dimension of image $\operatorname{im}(T)$ is called **the rank** of T, i. e., $\operatorname{rank}(T)=\dim(\operatorname{im}(T))$.

We list the following four theorems, and refer to [13].

Theorem 2.2.4 (rank and nullity) If $T\in L(X,Y)$, then $\dim(\operatorname{im}(T))+\dim(\ker(T))=\dim X$.

Theorem 2.2.5 (first isomorphic theorem) (1) If $T\in L(X,Y)$, then the quotient $X/\ker(T)$ is isomorphic with $\operatorname{im}(T)$, i. e., $X/\ker(T)\overset{\text{iso.}}{\longleftrightarrow}\operatorname{im}(T)$; (2) If X_1 is a linear subspace of X, then the quotient space X/X_1 is isomorphic with the complimentary linear space $\complement X_1$, i. e., $X/X_1\overset{\text{iso.}}{\longleftrightarrow}\complement X_1$.

Theorem 2.2.6 (Second isomorphic theorem) If X_1, X_2 are two linear subspaces of X, then the quotient space $(X_1+X_2)/X_1$ is isomorphic with quotient space $X_2/$

$(X_1 \cap X_2)$, i. e.

$$(X_1 + X_2)/X_1 \overset{\text{iso.}}{\longleftrightarrow} X_2/(X_1 \cap X_2).$$

Theorem 2.2.7(Third isomorphic theorem) If X_1, X_2 are two linear subspaces of X, and $X_1 \subset X_2$, then the quotient space $(X/X_1)/(X_2/X_1)$ of X/X_1 and X_1/X_2 is isomorphic with X/X_2, i. e.

$$(X/X_1)/((X_2/X_1)) \overset{\text{iso.}}{\longleftrightarrow} X/X_2.$$

2.2.2 Conjugate operators of linear operators

1. Definition of conjugate operator

The conjugate operator is connected closely with the linear operator, it plays an important role in modern mathematics.

Definition 2.2.3 (conjugate operator) Let X, Y be two linear spaces on number field $\mathbb{F}$; $T: X \to Y$ be a linear mapping from X to Y. The mapping $T^* \in L(Y^*, X^*)$ is said to be a **conjugate operator of** $T \in L(X, Y)$, if for any $x \in X$ and $f \in Y^*$, it satisfies

$$\langle T^*(f), x\rangle = \langle f, T(x)\rangle, \quad \forall f \in Y^*, \quad \forall x \in X, \tag{2.2.1}$$

where $T^*(f) \equiv \langle T^*, f\rangle$ represents the action of $T^* \in L(Y^*, X^*)$ on $f \in Y^*$; the domain of T^* is Y^*, the range of T^* is X^*, thus, $T^*: f \in Y^* \xrightarrow{T^* \in L(Y^*, X^*)} T^*(f) \in X^* = L(X, \mathbb{F})$ (Fig. 2.2.1).

Note The left-hand side of equality (2.2.1) represents that $T^*(f): X \to \mathbb{F}$ acts on $x \in X$, so that it is in $\mathbb{F}$, and $\langle T^*(f), x\rangle \in \mathbb{F}$, it makes sense. The right side of (2.2.1) is $\langle f, T(x)\rangle$, $\forall f \in Y^*$, $\forall x \in X$, it represents that each $f \in Y^* = L(Y, \mathbb{F})$ acts on $y = T(x) \in Y$, denoted by

$$\langle f, y\rangle = \langle f, T(x)\rangle, \quad \forall f \in Y^*, \quad y = T(x) \in Y, \quad \forall x \in X.$$

This implies that $\forall f \in Y^*$, $\forall x \in X$, it holds $\langle f, T(x)\rangle \in \mathbb{F}$. So that $\langle f, T(x)\rangle \in \mathbb{F}$ is well-defined, and fits the left-hand side $\langle T^*(f), x\rangle \in \mathbb{F}$.

If there is another operator $S \in L(Y^*, X^*)$ with "$\langle S(f), x\rangle = \langle f, T(x)\rangle$ for $\forall f \in Y^*$, $\forall x \in X$", then S is the conjugate operator of $T \in L(X, Y)$, i. e., $S = T^*$. This means the conjugate operator T^* of T is well-defined and unique. (2.2.1) often rewrites as

$$T^*(f)(x) = f(T(x)), \quad \forall f \in Y^*, \forall x \in X.$$

Theorem 2.2.8 Let X, Y be linear spaces on number field $\mathbb{F}$. Then the conjugate operators T^* and S^* of linear operators $T \in L(X, Y)$ and $S \in L(X, Y)$ are in the linear space $L(Y^*, X^*)$, i. e., $T, S \in L(X, Y) \Rightarrow T^*, S^* \in L(Y^*, X^*)$, and

(1) $T, S \in L(X, Y) \Rightarrow (T+S)^* = T^* + S^*$;

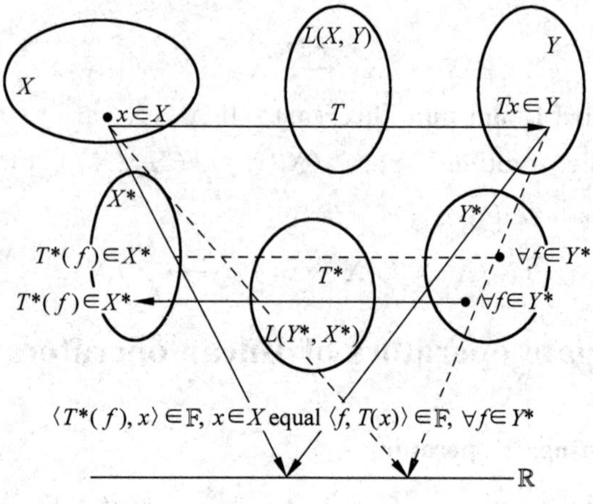

Fig. 2. 2. 1 Conjugate operator

(2) $T\in L(X,Y),\alpha\in\mathbb{F}\Rightarrow(\alpha T)^*=\alpha T^*$;

(3) $T\in L(X,Y),U\in L(Y,Z)\Rightarrow(U\circ T)^*=T^*\circ U^*\in L(Z^*,X^*)$;

(4) $T\in L(X,Y)$, and T^{-1} exists with $T^{-1}\in L(Y,X)$, then $(T^{-1})^*=(T^*)^{-1}$;

(5) If X,Y are finite-dimensional linear spaces, then the conjugate operator $T^*\in L(Y^*,X^*)$ of $T\in L(X,Y)$ satisfies $(T^*)^*=T$;

(6) If X, Y are finite-dimensional linear spaces with $\dim X=n$, $\dim Y=m$, respectively, then "the representation matrix of $T\in L(X,Y)$ is $A\in\mathfrak{M}_{m\times n}$" implies that "the conjugate operator $T^*\in L(Y^*,X^*)$ of T has the representation matrix $A^{\mathrm{T}}\in\mathfrak{M}_{n\times m}$, it is the transposed matrix of A"; and holds $\operatorname{rank}(T)=\operatorname{rank}(T^*)=\operatorname{rank}(A)=\operatorname{rank}(A^{\mathrm{T}})$.

Proof (1) and (2) hold clearly by definitions.

To prove (3), take any $f\in Z^*$, we prove

$$\langle(U\circ T)^*(f),x\rangle=\langle f,U(T(x))\rangle,\quad x\in X$$

The left- hand side is $\langle(U\circ T)^*(f),x\rangle=\langle f,(U\circ T)(x)\rangle$, $\forall f\in Z^*$, $\forall x\in X$, by definition. Also by definition, we continue the above equality that

$$\langle f,(U\circ T)(x)\rangle=\langle f,U(T(x))\rangle=\langle U^*(f),T(x)\rangle=\langle T^*(U^*(f)),x\rangle;$$

then, the right-hand side of the above last equality is $\langle(T^*\circ U^*)(f),x\rangle$, and this implies (3).

For (4), take $U=T^{-1}$ in (3), then, since $T^{-1}\in L(Y,X)$ exists by assumption, so $I=(I)^*=(T^{-1}\circ T)^*=T^*\circ(T^{-1})^*$, this implies (4).

For (5), since X,Y are finite dimension, thus $X\cong X^{**}$, $Y\cong Y^{**}$ (by Theorem 2. 1. 7), so that

$$T\in L(X,Y)\Rightarrow T^*\in L(Y^*,X^*)\Rightarrow(T^*)^*\in L((X^*)^*,(Y^*)^*)=L(X,Y).$$

We only need to prove that $\forall x \in X$ implies $\langle T, x\rangle = \langle T^{**}, x\rangle$.

In fact, for any $x^{**} \in X^{**}$, we have $x^{**} \in X^{**} \leftrightarrow x \in X$, and this x is determined by

$$\langle f, x\rangle = \langle x^{**}, f\rangle, \quad \forall f \in Y^*, \tag{2.2.2}$$

thus, for $T^{**} \in L(X^{**}, Y^{**})$ and for any $f \in Y^*$, it follows that

$$\langle T^{**}(x^{**}), f\rangle \overset{\text{def}}{=} \langle x^{**}, T^*(f)\rangle \overset{\text{def}}{=} \langle x^{**}, f(T)\rangle \xlongequal{(2.2.2)} \langle f(T), x\rangle \overset{\text{def}}{=} \langle T(x), f\rangle,$$

so that $T^{**}(x^{**}) = T(x)$. Moreover, since $X \cong X^{**}$, the equality $T^{**}(x^{**}) = T(x)$ implies that "T^{**} maps $x^{**} \in X^{**}$" and "T maps $x \in X$" are the same value, so $x^{**} \leftrightarrow x$ implies $T^{**} = T$.

Finally, for (6), let $\mathfrak{B}_X = \{b_1, \cdots, b_n\}$, $\mathfrak{C}_Y = \{c_1, \cdots, c_m\}$ be the bases of X, Y, respectively; the dual bases be $\mathfrak{B}^*_{X^*} = \{b_1^*, \cdots, b_n^*\}$, $\mathfrak{C}^*_{Y^*} = \{c_1^*, \cdots, c_m^*\}$. Thus, a linear operator $T \in L(X, Y) \leftrightarrow [T]_{\mathfrak{BC}} \in \mathfrak{M}_{m\times n}$ and conjugate operator $T^* \in L(Y^*, X^*) \leftrightarrow [T^*]_{\mathfrak{C}^*\mathfrak{B}^*} \in \mathfrak{M}_{n\times m}$.

By section 2.2.1, we have

$$[T]_{\mathfrak{BC}} = ([T(b_1)]_{\mathfrak{C}} \quad \cdots \quad [T(b_n)]_{\mathfrak{C}})$$

and

$$[T^*]_{\mathfrak{C}^*\mathfrak{B}^*} = ([T^*(c_1^*)]_{\mathfrak{B}^*} \quad \cdots \quad [T^*(c_m^*)]_{\mathfrak{B}^*}),$$

compare the elements of above two matrices, we have $[T^*]_{\mathfrak{C}^*\mathfrak{B}^*} = ([T]_{\mathfrak{BC}})^{\mathrm{T}}$. The proof of Theorem 2.2.8 is complete.

Example 2.2.1 Let X, Y be linear spaces on $\mathbb{R}$, with $\dim X = n$, $\dim Y = m$. Then $\dim X^* = n$, $\dim Y^* = m$, and $L(X, Y) \leftrightarrow \mathfrak{M}_{m\times n}$, $L(Y^*, X^*) \leftrightarrow \mathfrak{M}_{n\times m}$.

By Theorem 2.2.8, if the representation matrices of $T \in L(X, Y)$ and $T^* \in L(Y^*, X^*)$ are $A_{m\times n}$ and $\mathfrak{B}_{m\times n}$, respectively, i.e.

$$T \in L(X, Y) \leftrightarrow A_{m\times n} \in \mathfrak{M}_{m\times n}, \quad T^* \in L(Y^*, X^*) \leftrightarrow B_{n\times m} \in \mathfrak{M}_{n\times m},$$

then, $B_{n\times m} = (A_{m\times n})^{\mathrm{T}}$ is the **transposed matrix** of $A_{m\times n}$.

2. Equivalent definitions of conjugate operators on inner product spaces

We suppose that X, Y, Z are inner product spaces on number field $\mathbb{F}$ in this subsection.

1) **Riesz representative theorem**

Riesz representative theorem on inner product space X with dual space X^* has the following form (compare with **Theorem** 2.1.8).

Theorem 2.2.9 **Let** $(X, (x, x'))$ be an inner product space, (x, x') be the inner product on X. Let X^* be the dual space of X. Then for any $f \in X^*$, there exists unique

$x \in X$ corresponding to f, such that for any $v \in X$, the "action" of linear functional f on $v \in X$ can be represented by $f(v) \equiv (v, x)$, where (v, x) is the inner product of X.

2) **Two definitions of the conjugate operator**

We consider the other definition of a conjugate operator on an inner product space, it is equivalent to that one defined in Definition 2.2.3.

Let X be an inner product space, and $f \in X^* = L(X, \mathbb{F})$ be a linear functional. By Theorem 2.2.9, it follows that

$$\forall f \in X^* \Rightarrow \exists ! x \in X, \quad \text{s.t.} \quad f(v) = (v, x)_X, \quad \forall v \in X,$$

thus, $f \in X^* \leftrightarrow x \in X$ is a one-one mapping, denoted by $J: f \in X^* \to x \in X$ with $J(f) = x$. Then, substitute into $f(v)$, we have

$$\langle f, v \rangle = (v, x) \Rightarrow \langle f, v \rangle = f(v) = (v, J(f)).$$

$J(f)$ has the following properties: for $\forall v \in X, f, g \in X^*, \alpha, \beta \in \mathbb{F}$, we have

$$\begin{aligned}(v, J(\alpha f + \beta g))_X &= \langle \alpha f + \beta g, v \rangle = \alpha \langle f, v \rangle + \beta \langle g, v \rangle = \langle \alpha f, v \rangle + \langle \beta g, v \rangle \\ &= (\alpha f, v)_X + (\beta g, v)_X = (v, \bar{\alpha} J(f))_X + (v, \bar{\beta} J(g))_X \\ &= (v, \bar{\alpha} J(f) + \bar{\beta} J(g))_X,\end{aligned}$$

this implies $J(\alpha f + \beta g) = \bar{\alpha} J(f) + \bar{\beta} J(g)$, so that $J: X^* \to X$ is a linear mapping when X is a real inner product space, or a conjugate linear one when X is a complex inner product space. Moreover, it is a surjection and injection mapping. Thus, the mapping $J: X^* \to X$ is "a conjugate isomorphism mapping" from X^* onto X, where we agree that "conjugate isomorphism" means "isomorphism" when X is a real inner product space; and it means "conjugate isomorphism" when X is a complex one.

By using the inner product (x, x') of inner product space $(X, (x, x'))$, the inner product on X^* can be defined as

$$(f, g)_{X^*} = (J^{-1} f, J^{-1} g)_X, \quad f, g \in X^*. \tag{2.2.3}$$

It is easy to verify that $(X^*, (f, g)_{X^*})$ is an inner product space.

By $X^* \overset{J_X}{\longleftrightarrow} X, Y^* \overset{J_Y}{\longleftrightarrow} Y$, we have $L(Y^*, X^*) \leftrightarrow L(Y, X)$. Thus, in (2.2.1), we may change $T^* \in L(Y^*, X^*)$ to $T' \in L(Y, X)$, and change $\forall f \in Y^*$ to $\forall y \in Y$ in the sense of isomorphism, it follows

$$(T'(y), x)_X = (y, T(x))_Y, \quad \forall y \in Y, \quad \forall x \in X.$$

Definition 2.2.4 (conjugate operator) Let X, Y be inner product spaces on number field $\mathbb{F}$, and $T: X \to Y$ be a linear mapping from X to Y. For $T \in L(X, Y)$, an operator $T' \in L(Y, X)$ is said to be **a conjugate operator of** T, if it satisfies

$$(T'(y), x) = (y, T(x)), \quad \forall y \in Y, \quad \forall x \in X. \tag{2.2.4}$$

Where $(T'(y),x)$ is the inner product of X, and $(y,T(x))$ is the inner product of Y.

3) **Relationship between two definitions of the conjugate operator**

Let $T^*: Y^* \to X^*$ and $T': Y \to X$ be the conjugate operators on inner product space $(X,(x,x'))$ defined by (2.2.1) and (2.2.4), respectively. If $\varphi_1: X^* \leftrightarrow X$, $\varphi_2: Y^* \leftrightarrow Y$ are isomorphism mappings, we define a mapping $\sigma: Y^* \leftrightarrow X^*$ satisfying $\sigma=(\varphi_1)^{-1}T'\varphi_2: Y^* \to X^*$; it is a linear mapping. If $\forall x \in X$ with $(\varphi_1)^{-1}(x)=x^* \in X^*$, then $\varphi_1(x^*)=x$, so that we have a diagram in Fig. 2.2.2, it holds

$$((\varphi_1)^{-1}(x),v)=(x^*,v)=(v,\varphi_1(x^*)), \quad \forall v \in X, \forall x^* \in X^*.$$

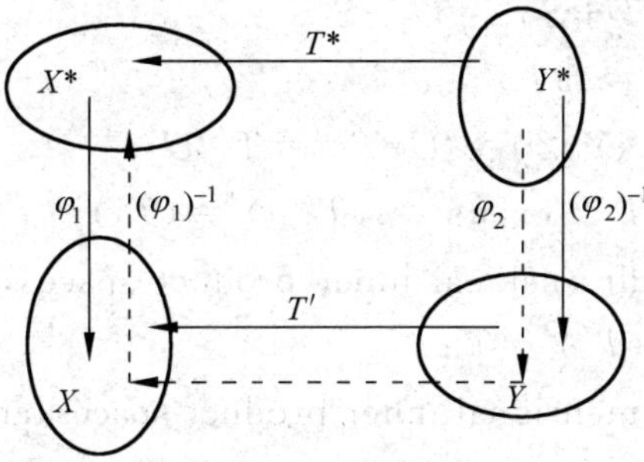

Fig. 2.2.2 T^* and T'

Hence, $\forall f \in Y^*$, $\forall v \in X$, it follows that

$$(\sigma(f),v)=[(\varphi_1)^{-1}T'\varphi_2(f)](v)=(\varphi_1)^{-1}[T'\varphi_2(f)](v)=(v,T'\varphi_2(f))$$
$$=(T(v),\varphi_2(f))=f(T(v))=(T^*(f),v).$$

This implies $\sigma=T^*$, i. e., $T^*=(\varphi_1)^{-1}T'\varphi_2$. Then T^* and T' are isomorphic, the Fig. 2.2.2 is an exchange graph.

Usually, in a finite dimensional inner product space, people prefer to use T' as a conjugate operator of T for convenience.

Next, we consider the relationship between representation matrices of T and T'.

If $\mathfrak{B}=\{b_1,\cdots,b_n\}$ and $\mathfrak{C}=\{c_1,\cdots,c_m\}$ are orthonormal bases of inner product spaces X and Y, respectively, and if for each $T \in L(X,Y)$, it holds

$$[T]_{\mathfrak{B}\mathfrak{C}}=[[T(b_1)]_{\mathfrak{C}} \cdots [T(b_n)]_{\mathfrak{C}}] \equiv [\beta_{ij}]_{m\times n}$$

with $\beta_{ij}=(T(b_i),c_j)$. Moreover, for $T' \in L(Y,X)$, it holds

$$[T']_{\mathfrak{C}\mathfrak{B}}=[[T'(c_1)]_{\mathfrak{B}} \cdots [T'(c_m)]_{\mathfrak{B}}] \equiv [\gamma_{ij}]_{n\times m}$$

with $\gamma_{ij}=(T'(c_i),b_j)$. Then,

$$\gamma_{ij}=(T'(c_i),b_j)=\overline{(b_j,T'(c_i))}=\overline{(T(b_j),c_i)}=\overline{\beta_{ji}}, \quad i=1,2,\cdots,m, \quad j=1,2,\cdots,n,$$

so that $[T']_{\mathfrak{C}\mathfrak{B}}=(\overline{[T]_{\mathfrak{B}\mathfrak{C}}})^{\mathrm{T}}$. Then we conclude that:

In finite dimensional inner product spaces, if the representation matrix of operator T:

$X \to Y$ is $[T]_{\mathfrak{BC}} = [a_{jk}]_{m\times n}$, then its conjugate operator $T^* = T' : Y \to X$ has the representation matrix $[T']_{\mathfrak{CB}} = \overline{([T]_{\mathfrak{BC}})^{\mathrm{T}}} = \overline{[a_{kj}]}_{n\times m}$, **it is the conjugate transposed matrix of** $[T]_{\mathfrak{BC}}$.

We summarize the properties of conjugate operators on inner product spaces as follows.

Theorem 2.2.10 Let X, Y be two inner product spaces on number field $\mathbb{F}$. Then the conjugate operator $T' \in L(Y, X)$ defined by Definition 2.2.4, of a linear operator $T \in L(X, Y)$, has the following properties:

(1) $T, S \in L(X, Y) \Rightarrow (T+S)' = T' + S'$;

(2) $T \in L(X, Y), \alpha \in \mathbb{F} \Rightarrow (\alpha T)' = \bar{\alpha}\, T'$;

(3) $T \in L(X, Y), U \in L(Y, Z) \Rightarrow (U \circ T)' = T' \circ U'$;

(4) $T \in L(X, X)$, and T^{-1} exists $\Rightarrow (T^{-1})' = (T')^{-1}$;

(5) If X, Y are finite dimensional inner product spaces, then $T \in L(X, Y)$ has the conjugate operator T' with $(T')' = T$;

(6) If X, Y are finite dimensional inner product spaces, and $\dim X = n$, $\dim H = m$; if the representation matrix of $T \in L(X, Y)$ is $A \in \mathfrak{M}_{m\times n}$, then the conjugate operator $T' \in L(Y, X)$ has the representation matrix B being the conjugate transposed matrix of $A \in \mathfrak{M}_{m\times n}$, i.e., $B = (\bar{A})^{\mathrm{T}} \in \mathfrak{M}_{n\times m}$, and

$$\operatorname{rank}(T) = \operatorname{rank}(T') = \operatorname{rank}(A) = \operatorname{rank}(B).$$

3. Special conjugate operators on inner product spaces

Definition 2.2.5 (three kinds of special conjugate operators) Let $(X, +, \alpha\cdot, (x, y))$ be an inner product space on number field $\mathbb{F}$, and $T \in L(X, X)$ be a linear operator from X to X with the conjugate operator $T' \in L(X, X)$, satisfying (see (2.2.4)):

$$(T'(y), x) = (y, T(x)), \quad \forall x, y \in X.$$

(1) If $T' = T$, then T is said to be **a self-conjugate operator**, or **Hermitian operator**;

(2) If T is an one-one mapping, and $T' = T^{-1}$, then T is said to be **a unitary operator**;

(3) If $TT' = T'T$, then T is said to be **a normal operator.**

Correspondingly, for $T \in L(X, X) \leftrightarrow A \in \mathfrak{M}_{n\times n}$, and $T' \in L(X, X) \leftrightarrow B \in \mathfrak{M}_{n\times n}$, we have:

When $A \in \mathfrak{M}_{n\times n}(\mathbb{C})$ **is a complex matrix**, and the conjugate transposed matrix of A is $\bar{A}^{\mathrm{T}}$,

(1)′ If $\bar{A}^{\mathrm{T}} = A$, then A is said to be **a self-conjugate matrix**, or **Hermitian matrix**;

(2)′ If $\bar{A}^{\mathrm{T}} = -A$, then A is said to be **a skew-Hermitian matrix**;

(3)′ If A is an invertible matrix, and $\overline{A}^{\mathrm{T}}=A^{-1}$, then A is said to be **a unitary matrix**;

(4)′ If $A\overline{A}^{\mathrm{T}}=\overline{A}^{\mathrm{T}}A$, then A is said to be **a normal matrix.**

When $A\in\mathfrak{M}_{n\times n}(\mathbb{R})$ is **a real matrix**, and the transposed matrix of A is A^{T},

(1)″ If $A^{\mathrm{T}}=A$, then A is said to be **a symmetric matrix**;

(2)″ If $A^{\mathrm{T}}=-A$, then A is said to be **a skew-symmetric matrix**;

(3)″ If A is an invertible matrix, and $A^{\mathrm{T}}=A^{-1}$, then A is said to be **an orthogonal matrix.**

It is easy to see that if $\mathbb{F}=\mathbb{R}$, then a self-conjugate matrix turns out a symmetric matrix $A=A'$; a unitary matrix turns out an orthogonal matrix $A^{-1}=A'$.

Example 2. 2. 2 Hermitian operators and Hermitian matrices play very important roles in modern physics. The following examples come from physical study and experiments.

$I=\begin{bmatrix}1 & 0\\ 0 & 1\end{bmatrix}$, by $I^{\mathrm{T}}=\begin{bmatrix}1 & 0\\ 0 & 1\end{bmatrix}^{\mathrm{T}}=\begin{bmatrix}1 & 0\\ 0 & 1\end{bmatrix}=I$, it is a Hermitian matrix (symmetric matrix);

$A=\begin{bmatrix}0 & 1\\ 1 & 0\end{bmatrix}$, by $A^{\mathrm{T}}=\begin{bmatrix}0 & 1\\ 1 & 0\end{bmatrix}^{\mathrm{T}}=\begin{bmatrix}0 & 1\\ 1 & 0\end{bmatrix}=A$, it is a Hermitian matrix (symmetric matrix);

$B=\begin{bmatrix}1 & 0\\ 0 & -1\end{bmatrix}$, by $B^{\mathrm{T}}=\begin{bmatrix}1 & 0\\ 0 & -1\end{bmatrix}^{\mathrm{T}}=\begin{bmatrix}1 & 0\\ 0 & -1\end{bmatrix}=B$, it is a Hermitian matrix (symmetric matrix), called Pauli matrix;

$C=\begin{bmatrix}0 & 1\\ -1 & 0\end{bmatrix}$, by $C^{\mathrm{T}}=\begin{bmatrix}0 & 1\\ -1 & 0\end{bmatrix}^{\mathrm{T}}=\begin{bmatrix}0 & -1\\ 1 & 0\end{bmatrix}=-C$, it is a skew-symmetric matrix;

$D=\begin{bmatrix}0 & \mathrm{i}\\ \mathrm{i} & 0\end{bmatrix}$, by $D=\overline{\begin{bmatrix}0 & \mathrm{i}\\ \mathrm{i} & 0\end{bmatrix}}^{\mathrm{T}}=\begin{bmatrix}0 & -\mathrm{i}\\ -\mathrm{i} & 0\end{bmatrix}=-\begin{bmatrix}0 & \mathrm{i}\\ \mathrm{i} & 0\end{bmatrix}=-D$, it is a skew-Hermitian matrix;

$E=\begin{bmatrix}0 & -\mathrm{i}\\ \mathrm{i} & 0\end{bmatrix}$, by $E=\overline{\begin{bmatrix}0 & -\mathrm{i}\\ \mathrm{i} & 0\end{bmatrix}}^{\mathrm{T}}=\begin{bmatrix}0 & -\mathrm{i}\\ \mathrm{i} & 0\end{bmatrix}=E$, it is a Hermitian matrix.

In the modern physics, Schrodinger equation

$$\mathrm{i}h\frac{\mathrm{d}}{\mathrm{d}t}\begin{bmatrix}\psi_1\\ \psi_2\end{bmatrix}=\begin{bmatrix}H_{11} & H_{12}\\ H_{21} & H_{22}\end{bmatrix}\begin{bmatrix}\psi_1\\ \psi_2\end{bmatrix}$$

contains Hermitian matrix $H_{nm}=H_{mn}^{*}$, $m,n=1,2$.

The three-order matrices

$$\begin{bmatrix}0&1&0\\1&0&1\\0&1&0\end{bmatrix},\quad \begin{bmatrix}0&-\mathrm{i}&0\\\mathrm{i}&0&-\mathrm{i}\\0&\mathrm{i}&0\end{bmatrix},\quad \begin{bmatrix}1&0&0\\0&0&0\\0&0&-1\end{bmatrix}$$

appear in the modern physics.

The following three theorems are useful and important.

Theorem 2.2.11 Let $(X,+,\alpha\cdot,(x,y))$ be an inner product space on $\mathbb{F}$, and T,S be in $L(X,X)$ with $T',S'\in L(X,X)$. We have

(1) If T is a self-conjugate operator, then $((T,y),x)=(y,(T,x))$, $\forall x,y\in X$;

(2) If T,S are self-conjugate operators, then, so is $T+S$;

(3) If T is a self-conjugate operator, then, so is αT for $\alpha\in\mathbb{R}$;

(4) If T is a self-conjugate operator, and is invertible, then, so is T^{-1};

(5) If T is a self-conjugate operator, then $\forall x\in X$ implies $((T,x),x)\in\mathbb{R}$ (is real);

(6) If T is a self-conjugate operator, then $\forall x\in X$, $((T,x),x)=0$ implies $T=0$.

Theorem 2.2.12 Let $(X,+,\alpha\cdot,(x,y))$ be an inner product space on number field $\mathbb{F}$, and T,S be in $L(X,X)$. The following statements hold:

(1) If T is a unitary operator, then $\forall x,y\in X$ imply $((T,x),y)=(x,(T^{-1},y))$;

(2) If T is a unitary operator, then, so is T^{-1};

(3) If T,S are unitary operators, then, so is $T\circ S$;

(4) T is a unitary operator, if and only if T is a preserving inner product and surjective mapping, i. e., $\forall x,y\in X\Rightarrow((T,x),(T,y))=(x,y)$;

(5) If $\dim X=n$, then T is a unitary operator, if and only if T maps an orthonormal base of X to the other orthonormal base of X.

Correspondingly, for matrices, let $T\in L(X,X)\leftrightarrow A\in\mathfrak{M}_{n\times n}$. The following hold:

(1)′ A is a unitary matrix, if and only if all its column vectors consist of an orthonormal base of $\mathbb{C}^n$; also, if and only if all its line vectors consist of an orthonormal base of $\mathbb{C}^n$;

(2)′ If A is a unitary matrix, then $|\det(A)|=1$;

(3)′ If A is an orthogonal matrix, then $\det(A)=\pm 1$.

Theorem 2.2.13 Let $(X,+,\alpha\cdot,(x,y))$ be an inner product space on number field $\mathbb{F}$, and $T\in L(X,X)$ be a normal operator. We have

(1) For $x\in X$, $T(x)=0$ implies $T'(x)=0$;

(2) $\forall k\in\mathbb{N}$, $k>0$, then $T^k(x)=0$ implies $T(x)=0$;

(3) If $\forall x\in X$, $T(x)=\lambda x$, then $T'(x)=\bar{\lambda}x$.

4. Algebras, operator algebras

In linear space $(X,+,\alpha\cdot)$, we define a new operation, multiplication $\times$ between its elements, such that it has new structure, called algebra.

Definition 2.2.6 (algebra) Let $X \equiv (X, +, \alpha \cdot)$ be a linear space on number field $\mathbb{F}$. We define an operation on X between $x \in X$ and $y \in X$, called "multiplication", denoted by $x \times y$, or simply, by xy, satisfying

(1) $x, y \in X \Rightarrow x \times y \in X$; (closed property)

(2) $x, y, z \in X \Rightarrow (x \times y) \times z = x \times (y \times z)$; (combination law)

(3) $x, y, z \in X \Rightarrow x \times (y + z) = x \times y + x \times z, (y + z) \times x = y \times x + z \times x$; (distribution law)

(4) $x, y \in X,\ \alpha \in \mathbb{F} \Rightarrow \alpha \cdot (x \times y) = (\alpha \cdot x) \times y = x \times (\alpha \cdot y)$. (combination law)

Then $(X, +, \alpha \cdot, \times)$ is said to be **an algebra** on **number field** $\mathbb{F}$, or simply, **algebra.**

Note (1) and (2) imply that X is a semi-group with the multiplication $\times$.

Moreover, if (1)-(4) and the following (5) hold

(5) there exists the unit element $I \in X$ of operation $\times$, such that $\forall x \in X$ implies

$$I \times x = x \times I;$$

Then $(X, +, \alpha \cdot, \times)$ is said to be **an algebra with the unit on number field** $\mathbb{F}$.

Furthermore, if (1)-(5) and the following (6) hold,

(6) $x, y \in X$ imply $x \times y = y \times x$,

then $(X, +, \alpha \cdot, \times)$ is said to be **a commutative algebra with unit on number field** $\mathbb{F}$.

Example 2.2.3 $C([a, b])$: the set of all continuous functions on interval $[a, b]$

The addition, number product, multiplication on $C([a, b])$ are as follows:

$$f, g \in C([a, b]) \Rightarrow (f + g)(x) = f(x) + g(x), \quad x \in [a, b];$$
$$f, g \in C([a, b]) \Rightarrow (\alpha f)(x) = \alpha f(x), \quad \alpha \in \mathbb{R}, x \in [a, b];$$
$$f, g \in C([a, b]) \Rightarrow (f \times g)(x) = f(x) g(x), \quad x \in [a, b].$$

And $(C([a, b]), +, \alpha \cdot, \times)$ is a commutative algebra with unit $I(x) \equiv 1 \in C([a, b])$ on number field $\mathbb{F}$.

Example 2.2.4 $L^1([a, b])$: the set of all Lebesgue integrable functions on interval $[a, b]$

The addition, number product on $L^1([a, b])$ are the same as those on $C([a, b])$. The multiplication is defined by "convolution", i.e.

$$f, g \in L^1([a, b]) \Rightarrow (f * g)(x) = \int_{[a, b]} f(x - t) g(t) \mathrm{d}t,$$

then $(L^1([a, b]), +, \alpha \cdot, *)$ is **a commutative algebra without the unit element** on number field $\mathbb{F}$. The fact that this algebra $(L^1([a, b]), +, \alpha \cdot, *)$ does not have unit element, will be proved in Chapter 5.

Example 2.2.5 $(L(X, X), +, \alpha \cdot)$: the set of all linear operators from X to X

The addition, number product are defined as follows:

$$T,S \in L(X,X) \Rightarrow (T+S)(x)=T(x)+S(x), \quad x \in X;$$

$$T \in L(X,X), \alpha \in \mathbb{F} \Rightarrow (\alpha T)(x)=\alpha T(x), \quad x \in X.$$

The multiplication is defined by "composition", i. e.

$$T,S \in L(X,X) \Rightarrow (S \circ T)(x)=S(T(x)), \quad x \in X.$$

Then $(L(X,X),+,\alpha\cdot,\circ)$ is a non-commutative algebra with unit element $I(x)=x$, $\forall x \in X$, called **the operator algebra on X.**

2.2.3 Multilinear algebra

The multilinear algebra is an important part of linear algebra, it has very wide applications. We introduce the multilinear mappings, the multilinear mapping space $L(V_1 \times V_2 \times \cdots \times V_r, Z)$ and the multilinear functional space $L(V_1 \times V_2 \times \cdots \times V_r, \mathbb{F})$ with r linear spaces $V_1, V_2, \cdots, V_r$ on the same number field $\mathbb{F}$, as well as the dual spaces (conjugate spaces). These are basic knowledge of modern mathematics, specially, used in the differential geometry.

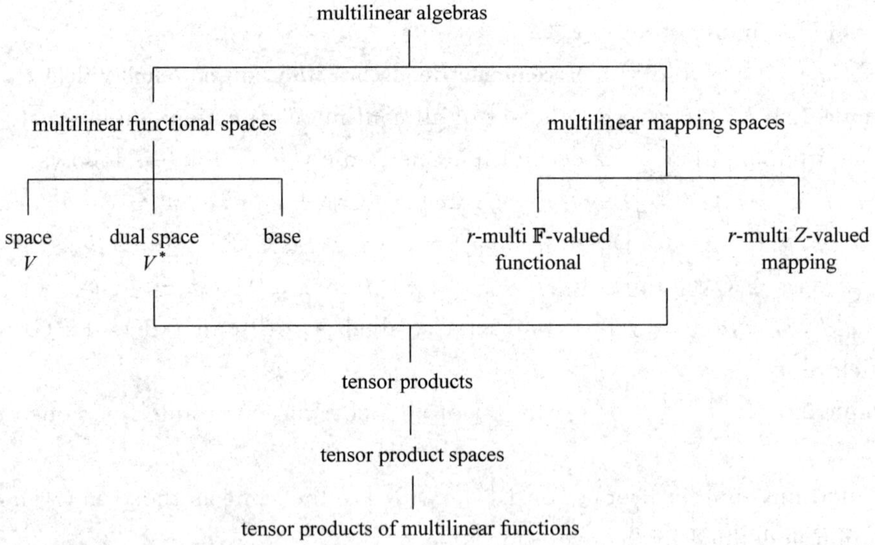

1. Spaces V, V^*; bases and dual bases

Let $V, W, \cdots$ be linear spaces on number field $\mathbb{F}$; and $V^*, W^*, \cdots$ be the dual spaces (i. e., the linear functional spaces of $V, W, \cdots$), respectively.

If $\mathfrak{A}=\{a_1, \cdots, a_n\}$ is a base of n-dimensional linear space V; and each element $v \in V$

is expressed as $v=\sum_{j=1}^{n} v_j a_j$, $v_j \in \mathbb{F}$, $1 \leqslant j \leqslant n$, denoted by $[v]_{\mathfrak{A}} \equiv \begin{bmatrix} v_1 \\ \vdots \\ v_n \end{bmatrix}$, called **coordinate** of $v \in V$ in the base $\mathfrak{A}=\{a_1,\cdots,a_n\}$. Moreover, for $f \in V^*$, we have

$$f(v)=\sum_{j=1}^{n} v_j f(a_j) \equiv \sum_{j=1}^{n} v_j f_j. \tag{2.2.5}$$

Thus, a linear functional $f \in V^*$ is determined by $\begin{bmatrix} f(a_1) \\ \vdots \\ f(a_n) \end{bmatrix} \equiv \begin{bmatrix} f_1 \\ \vdots \\ f_n \end{bmatrix}$ which are the values of f on $\mathfrak{A}=\{a_1,\cdots,a_n\}$. Take a system $\{a_1^*,\cdots,a_n^*\}$ in dual space V^* of V satisfying

$$a_j^*(a_k)=\delta_{kj}, \quad 1 \leqslant j,k \leqslant n, \tag{2.2.6}$$

where $\delta_{kj}=\begin{cases} 1, & j=k \\ 0, & j \neq k \end{cases}$. Then

$$a_j^*(v)=a_j^*\left(\sum_{k=1}^{n} v_k a_k\right)=v_j, \quad 1 \leqslant j \leqslant n.$$

Substitute $v_j=a_j^*(v)$ in (2.2.5), it follows

$$\forall v \in V, v=\sum_{j=1}^{n} v_j a_j \Rightarrow f(v)=\sum_{j=1}^{n} v_j f(a_j)=\sum_{j=1}^{n} a_j^*(v) f(a_j)=\sum_{j=1}^{n} f_j a_j^*(v).$$

This means that a linear functional $f \in V^*$ has an expression $f=\sum_{j=1}^{n} f_j a_j^*$, with $f_j=f(a_j)$, $1 \leqslant j \leqslant n$, and this expression is unique, thus $\{a_j^*, 1 \leqslant j \leqslant n\}$ becomes a base of V^*, called a dual base of base $\mathfrak{A}=\{a_j, 1 \leqslant j \leqslant n\}$, denoted by $\mathfrak{A}^*=\{a_j^*, 1 \leqslant j \leqslant n\}$.

Theorem 2.2.14 For a finite dimensional linear space V with $n=\dim V$ on number field $\mathbb{F}$, its dual space V^* is a linear space on $\mathbb{F}$, and holds

(1) $\dim V=\dim V^*=n$;

(2) If $\mathfrak{A}=\{a_j, 1 \leqslant j \leqslant n\}$ is a base of V, then the dual base $\mathfrak{A}^*=\{a_j^*, 1 \leqslant j \leqslant n\}$ of $\mathfrak{A}=\{a_j, 1 \leqslant j \leqslant n\}$ is a base of dual space V^* with $a_j^*(a_k)=\delta_{kj}$, $1 \leqslant j,k \leqslant n$;

(3) $(V^*)^*=V$.

2. Multilinear mapping space

Definition 2.2.7 (**$\mathbb{F}$-valued functional on $V \times V^*$**) Let V be a finite dimensional linear space on number field $\mathbb{F}$, and V^* be the dual space of V with $\dim V=\dim V^*=n$. For $v^* \in V^*$, we define **the action of** v^* **on** v as

$$v^*(v)=\langle v^*,v\rangle,\quad \forall v\in V,$$

where $\langle v^*,v\rangle\in\mathbb{F}$ is a function defined on $V\times V^*$, valued in $\mathbb{F}$, and is linear about both v and v^*, that is, bilinear form

$$\begin{cases}\langle v^*,\alpha_1v_1+\alpha_2v_2\rangle=\alpha_1\langle v^*,v_1\rangle+\alpha_2\langle v^*,v_2\rangle\\ \langle\alpha_1v_1^*+\alpha_2v_2^*,v\rangle=\alpha_1\langle v_1^*,v\rangle+\alpha_2\langle v_2^*,v\rangle\end{cases},\quad \alpha_1,\alpha_2\in\mathbb{F}$$

holds, with $v,v_1,v_2\in V, v^*,v_1^*,v_2^*\in V^*$. The mapping $\langle v^*,v\rangle$: $V^*\times V\to\mathbb{F}$ is said to be **an $\mathbb{F}$-valued linear functional on $V^*\times V$.**

Remark 1 As usual $\langle v^*,v\rangle$ and $\langle v,v^*\rangle$ are regarded as the same, $\langle v,v^*\rangle\equiv\langle v^*,v\rangle$, in finite dimensional linear spaces, reasonably. Since $V\leftrightarrow V^*\leftrightarrow V^{**}$ hold for finite dimensional spaces, and $\dim V=\dim V^*=\dim V^{**}$. Moreover, $V^*\leftrightarrow V^{**}\leftrightarrow V(\cong(V^{**})^*)$; the corresponding $v\leftrightarrow v^{**}$ is given by

$$\langle v^*,v\rangle=\langle v^{**},v^*\rangle,\quad \forall v^*\in V^*.$$

Thus, substitute $v\leftrightarrow v^{**}$ in the later, it follows $\langle v^*,v\rangle=\langle v,v^*\rangle\Leftrightarrow\langle v,v^*\rangle=\langle v^*,v\rangle$. Hence $v\in V$ can be regarded as a linear functional on V^*, and $\langle v,v^*\rangle$ is the action of v on $v^*\in V^*$ with the same value $\langle v^*,v\rangle$.

Remark 2 The action $\langle v,\cdot\rangle$ is an $\mathbb{F}$-valued linear functional on V^*; conversely, any $\mathbb{F}$-valued functional φ: $V^*\to\mathbb{F}$ on V^* can be represented by $\langle v,\cdot\rangle$. And for a given φ: $V^*\to\mathbb{F}$, let $v=\sum_{j=1}^{n}\varphi(a_j^*)a_j$, then $\forall v^*\in V^*$ implies $\langle v,v^*\rangle=\varphi(v^*)$. We see the duality of V and V^* again.

Remark 3 In Definition 2.2.7, $\mathbb{F}$-valued linear functional can be generalized to "$\mathbb{F}$-valued multilinear functional".

Definition 2.2.8 ($\mathbb{F}$-valued r-multilinear functional, Z-valued r-multilinear mapping)

(1) **$\mathbb{F}$-valued bilinear functional** Let V,W be linear spaces on number field $\mathbb{F}$. If a functional f: $V\times W\to\mathbb{F}$ is linear about both variables, i. e.

$$\begin{cases}f(\alpha_1v_1+\alpha_2v_2,w)=\alpha_1f(v_1,w)+\alpha_2f(v_2,w),\\ f(v,\alpha_1w_1+\alpha_2w_2)=\alpha_1f(v,w_1)+\alpha_2f(v,w_2),\end{cases}\quad \alpha_1,\alpha_2\in\mathbb{F},$$

with $v,v_1,v_2\in V, w,w_1,w_2\in W$, then f is said to be **an $\mathbb{F}$-valued bilinear functional on $V\times W$.** The set of all $\mathbb{F}$-valued bilinear functional on $V\times W$ is denoted by

$$L(V,W;\mathbb{F})\equiv L(V\times W,\mathbb{F}).$$

Endow operations on $L(V,W;\mathbb{F})$:

addition: $f,g\in L(V,W;\mathbb{F})\Rightarrow(f+g)(v,w)=f(v,w)+g(v,w)$;

number product: $f\in L(V,W;\mathbb{F}),\ \alpha\in\mathbb{F}\Rightarrow(\alpha f)(v,w)=\alpha f(v,w)$;

with $v\in V, w\in W$, then $(L(V,W;\mathbb{F}),+,\alpha\cdot)$ is said to be **a bilinear functional space.**

(2) Z-**valued bilinear mapping** Let V, W, Z be linear spaces on number field $\mathbb{F}$. If a mapping $f: V\times W \to Z$ is linear about both variables, i. e.

$$\begin{cases} f(\alpha_1 v_1+\alpha_2 v_2, w)=\alpha_1 f(v_1, w)+\alpha_2 f(v_2, w), \\ f(v, \alpha_1 w_1+\alpha_2 w_2)=\alpha_1 f(v, w_1)+\alpha_2 f(v, w_2), \end{cases} \quad \alpha_1, \alpha_2 \in \mathbb{F},$$

with $v, v_1, v_2 \in V, w, w_1, w_2 \in W$, then f is called a Z-**valued bilinear mapping on** $V\times W$. The set of all Z-valued bilinear mappings on $V\times W$ is denoted by

$$L(V, W; Z) \equiv L(V\times W, Z).$$

Endow operations on $L(V, W; Z)$:

addition: $f, g \in L(V, W; Z) \Rightarrow (f+g)(v, w)=f(v, w)+g(v, w)$;

number product: $f\in L(V, W; Z), \alpha \in \mathbb{F} \Rightarrow (\alpha f)(v, w)=\alpha f(v, w)$;

with $v\in V, w\in W$, then $(L(V, W; Z), +, \alpha\cdot)$ is said to be **a bilinear mapping space.**

(3) $\mathbb{F}$-**valued** r-**multilinear functional** Let $V_1, \cdots, V_r$ be linear spaces on $\mathbb{F}$. If a functional $f: V_1\times\cdots\times V_r \to \mathbb{F}$ satisfies: $f(v_1, \cdots, v_r), v_j \in V_j, j=1,2,\cdots,r$, is linear for every variable, then f is called $\mathbb{F}$-**valued** r-**multilinear functional on** $V=V_1\times\cdots\times V_r$. The set of all $\mathbb{F}$-valued r-multilinear functional on $V=V_1\times\cdots\times V_r$ is denoted by $L(V_1, \cdots, V_r; \mathbb{F}) \equiv L(V_1\times\cdots\times V_r, \mathbb{F})$. With the following operations

$$f, g \in L(V_1, \cdots, V_r; \mathbb{F}) \Rightarrow (f+g)(v_1, \cdots, v_r)=f(v_1, \cdots, v_r)+g(v_1, \cdots, v_r);$$

$$f\in L(V_1, \cdots, V_r; \mathbb{F}), \alpha \in \mathbb{F} \Rightarrow (\alpha f)(v_1, \cdots, v_r)=\alpha f(v_1, \cdots, v_r);$$

this $(L(V_1, \cdots, V_r; \mathbb{F}), +, \alpha\cdot)$ becomes an $\mathbb{F}$-**valued** r-**multilinear functional space.**

(4) Z-**valued** r-**multilinear mapping** Let $V_1, \cdots, V_r, Z$ be linear spaces on $\mathbb{F}$. If a mapping $f: V_1\times\cdots\times V_r \to Z$ satisfies: $f(v_1, \cdots, v_r)$ is linear for every variable $v_j \in V_j$, $j=1,2,\cdots,r$, then f is called a Z-**valued** r-**multilinear mapping on** $V_1\times\cdots\times V_r$. The set of all Z-valued r-multilinear mappings on $V=V_1\times\cdots\times V_r$ is denoted by $L(V_1, \cdots, V_r; Z) \equiv L(V_1\times\cdots\times V_r, Z)$. With the following operations

$$f, g \in L(V_1, \cdots, V_r; Z) \Rightarrow (f+g)(v_1, \cdots, v_r)=f(v_1, \cdots, v_r)+g(v_1, \cdots, v_r);$$

$$f\in L(V_1, \cdots, V_r; Z), \alpha \in \mathbb{F} \Rightarrow (\alpha f)(v_1, \cdots, v_r)=\alpha f(v_1, \cdots, v_r).$$

This $(L(V_1, \cdots V_r; Z), +, \alpha\cdot)$ becomes a Z-**valued** r-**multilinear mapping space.**

3. Isomorphism between multilinear mapping spaces

For a multilinear mapping space $L(V_1, \cdots, V_r; Z)$, the isomorphism theorem is very useful. We only list the theorem of linear mapping space $L(V; Z)$.

1) **Isomorphism theorem of linear mapping space** $L(V, Z)$

Let V, Z be linear spaces on number field $\mathbb{F}$ with dimensions $n=\dim V, m=\dim Z$, and with bases $\mathfrak{A}=\{a_1, \cdots, a_n\}$ of V, $\mathfrak{B}=\{b_1, \cdots, b_m\}$ of Z, respectively.

Consider a mapping $f \in L(V,Z)$: $x \in V \to y = f(x) \in Z$. Suppose that $f(x)=y$ maps $\mathfrak{A}=\{a_1,\cdots,a_n\}$ to $\mathfrak{B}=\{b_1,\cdots,b_m\}$ by

$$b_j = \sum_{k=1}^{n} f_{jk}\, a_k, \quad j=1,2,\cdots,m, \tag{2.2.7}$$

denoted by $A=A_{m\times n}=[f_{jk}]_{1\leqslant j\leqslant m,1\leqslant k\leqslant n}\equiv M_{\mathfrak{A}\mathfrak{B}}$, and $\mathfrak{A}=\begin{bmatrix} a_1 \\ a_2 \\ \vdots \\ a_n \end{bmatrix}\equiv M_{\mathfrak{A}}$, $\mathfrak{B}=\begin{bmatrix} b_1 \\ b_2 \\ \vdots \\ b_m \end{bmatrix}\equiv M_{\mathfrak{B}}$,

then (2.2.7) can be rewritten as $M_{\mathfrak{B}}\equiv M_{\mathfrak{A}\mathfrak{B}}M_{\mathfrak{A}}$. By Theorem 2.2.1, we have

Linear space $L(V,Z)$ **is isomorphic with** $m\times n$**-matrix space** $\mathfrak{M}_{m\times n}$ **on F**; i.e.,

$$L(V,Z) \overset{\text{iso.}}{\longleftrightarrow} \mathfrak{M}_{m\times n}.$$

2) **Decomposition of bilinear mapping space** $L(V,W;Z)$

We show that the bilinear mapping space $L(V,W;Z)$ is corresponding to linear space $L(Y,Z)$, that is, we will construct a linear space Y, such that $f=g\circ h$: $V\times W\to Z$.

Let V,W be linear spaces on $\mathbb{F}$ with $n=\dim V$, $m=\dim W$, and bases $\mathfrak{A}=\{a_1,\cdots,a_n\}$, $\mathfrak{B}=\{b_1,\cdots,b_m\}$, respectively; Z be a linear space on $\mathbb{F}$. We will decompose a bilinear mapping $f\in L(V,W;Z)$ into a transposed mapping of bilinear mapping $h\in L(V\times W,Y)$ and linear mapping $g\in L(Y,Z)$, where Y is depending on V and W, and satisfying the exchange graph Fig. 2.2.3.

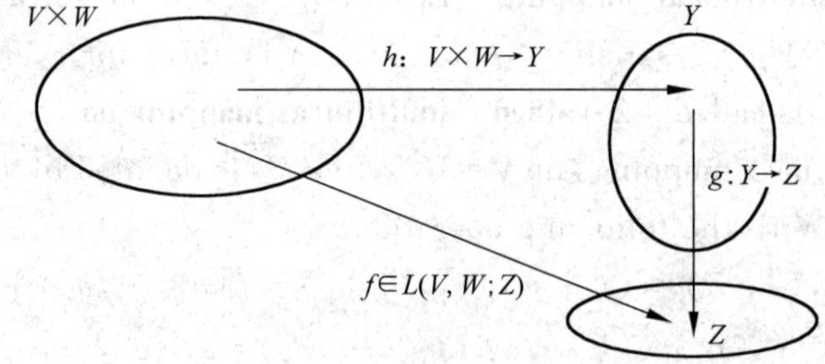

Fig. 2.2.3 Exchange graph

4. Tensor products and tensor product spaces

To construct the space Y, we use the method of tensor product and tensor product space.

1) **Tensor products in** V^* **and** W^*

Definition 2.2.9 (**tensor product of** $v^*\in V^*$ **and** $w^*\in W^*$, **tensor product space of** V^* **and** W^*) Let V,W be linear spaces, the dual spaces be V^*, W^*, respectively; and $v^*\in V^*=L(V,\mathbb{F})$, $w^*\in W^*=L(W,\mathbb{F})$.

(1) Tensor product of elements $v^* \in V^*$ and $w^* \in W^*$

Let $v^* \otimes w^*: V \times W \to \mathbb{F}$ be a mapping satisfying

$$v^* \otimes w^*(v,w) = v^*(v)w^*(w) = \langle v,v^* \rangle \langle w,w^* \rangle, \quad v \in V, w \in W, \tag{2.2.8}$$

with $\langle v,v^* \rangle$ defined in Definition 2.2.7. Then $v^* \otimes w^*$ is said to be **a tensor product of $v^* \in V^*$ and $w^* \in W^*$**. It is clear that $v^* \otimes w^*$ is a bilinear functional on $V \times W$, i. e., $v^* \otimes w^* \in L(V,W; \mathbb{F})$. Hence, the operation "$\otimes$" can be regarded as a bilinear mapping from linear space $V^* \times W^*$ to linear space $L(V,W; \mathbb{F})$, i. e.

$$\otimes: (v^*, w^*) \in (V^*, W^*) \equiv V^* \times W^* \to v^* \otimes w^* \in L(V,W; \mathbb{F}).$$

(2) Tensor product of spaces V^* and W^*

For the linear space spanning by all tensor products $v^* \otimes w^*$ with $v^* \in V^*$ and $w^* \in W^*$, denoted by

$$V^* \otimes W^* \equiv \operatorname{span}\{v^* \otimes w^*: v \in V^*, w^* \in W^*\}, \tag{2.2.9}$$

we define addition and number product operation as follows: $\forall f, g \in V^* \otimes W^*, \alpha \in \mathbb{F}$,

$$(f+g)((v,w)) = f((v,w)) + g((v,w)), \quad (v,w) \in V \times W;$$

$$(\alpha \cdot f)((v,w)) = \alpha f((v,w)), \quad (v,w) \in V \times W.$$

Then, $V^* \otimes W^*$ becomes a linear space on number field $\mathbb{F}$, called **a tensor product space of V^* and W^***.

Note In the tensor product space $V^* \otimes W^*$, each element is a linear combination of tensor product $v^* \otimes w^*$ with $v^* \in V^*$ and $w^* \in W^*$; the element $v^* \otimes w^*$ is a "single form".

2) **The expression of tensor product $v^* \otimes w^*$**

Let $\mathfrak{A} = \{a_k, 1 \leqslant k \leqslant n\}$ be a base of V with $n = \dim V$, and $\mathfrak{A}^* = \{a_j^*, 1 \leqslant j \leqslant n\}$ be the dual base of $\mathfrak{A}$ in V^*, satisfying $a_j^*(a_k) = \delta_{kj}$, $1 \leqslant j, k \leqslant n$. Moreover, let $\mathfrak{B} = \{b_k, 1 \leqslant k \leqslant m\}$ be a base of W with $m = \dim W$, and $\mathfrak{B}^* = \{b_j^*, 1 \leqslant j \leqslant m\}$ be the dual base of $\mathfrak{B}$ in W^*, satisfying $b_j^*(b_k) = \delta_{kj}$, $1 \leqslant j, k \leqslant m$.

Since $\otimes: (v^*, w^*) \in V^* \times W^* \to v^* \otimes w^* \in L(V,W; \mathbb{F})$, we have **the expression**

$$v^* \otimes w^* = \sum_{j,k} (v^*(a_j) w^*(b_k))(a_j^* \otimes b_k^*) \tag{2.2.10}$$

of $v^* \otimes w^*$; and it is easy to prove that $\{a_j^* \otimes b_k^*: 1 \leqslant j \leqslant n, 1 \leqslant k \leqslant m\}$ is a base of the tensor product space $V^* \otimes W^*$ on $\mathbb{F}$ with $n \times m = \dim(V^* \times W^*)$; and by the mapping

$$(v^*, w^*) \in V^* \times W^* \to v^* \otimes w^* \in L(V,W; \mathbb{F})$$

such that $V^* \otimes W^* \overset{\text{iso.}}{\longleftrightarrow} L(V,W; \mathbb{F})$.

3) **Tensor products in V and W**

To define the tensor product in linear spaces V and W, we consider $V=(V^*)^*$ and $W=(W^*)^*$, by similar methods in 1) and 2), we define the tensor product of $v\in V$ and $w\in W$, denoted by $v\otimes w$. Moreover, for the tensor product space $V\otimes W$, we have

$$V\otimes W \overset{\text{iso.}}{\longleftrightarrow} L(V^*, W^*; \mathbb{F}).$$

Also, $\{a_j\otimes b_k: 1\leqslant j\leqslant n, 1\leqslant k\leqslant m\}$ is a base of $V\otimes W$, and $V\otimes W$ is a linear space on $\mathbb{F}$ with $n\times m=\dim(V\times W)$.

4) **Duality of tensor product spaces**

Let $\langle v\otimes w, v^*\otimes w^*\rangle=\langle v, v^*\rangle\langle w, w^*\rangle$, and

$$\langle a_j\otimes b_k, a_l^*\otimes b_s^*\rangle=\delta_{jl}\delta_{ks}=\begin{cases}1, & (j,k)=(l,s)\\ 0, & (j,k)\neq(l,s)\end{cases}, \tag{2.2.11}$$

thus $\{a_j\otimes b_k: 1\leqslant j\leqslant n, 1\leqslant k\leqslant m\}$ and $\{a_j^*\otimes b_k^*: 1\leqslant j\leqslant n, 1\leqslant k\leqslant m\}$ are bases of $V\otimes W$ and $V^*\otimes W^*$, respectively; moreover, they are dual to each other. This implies that

$$V^*\otimes W^*=(V\otimes W)^*.$$

Theorem 2.2.15 Let $h: V\times W\to V\otimes W$ be bilinear mapping

$$h(v,w)=v\otimes w: V\times W\to V\otimes W,$$

then for any bilinear mapping $f: V\times W\to Z$, there exists an unique linear mapping $g: V\otimes W\to Z$, such that $f=g\circ h: V\times W\to Z$.

Proof For any bilinear mapping $f: V\times W\to Z$, we construct a linear mapping $g: V\otimes W\to Z$, such that the base $\{a_j\otimes b_k: 1\leqslant j\leqslant n, 1\leqslant k\leqslant m\}$ of $V\otimes W$ is expressed as

$$g(a_j\otimes b_k)=f(a_j, b_k),\quad 1\leqslant j\leqslant n, 1\leqslant k\leqslant m.$$

Thus, for any $v=\sum_{j=1}^{n} v_j a_j\in V$ and $w=\sum_{k=1}^{m} w_k b_k\in W$, it follows that

$$g(v\otimes w)=\sum_{j,k} v_j w_k\, g(a_j\otimes b_k)=\sum_{j,k} v_j w_k\, f(a_j, b_k)=f(v,w),$$

hence $g: V\otimes W\to Z$ is determined uniquely. The linearity of g is clear. The proof is complete.

We have the following table 2.2.1.

Table 2.2.1

space	base	dual space	dual base
V	$\mathfrak{A}=\{a_1,\cdots,a_n\}$, $\dim V=n$	V^*	$\mathfrak{A}^*=\{a_1^*,\cdots,a_n^*\}$, $\dim V^*=n$
W	$\mathfrak{B}=\{b_1,\cdots,b_m\}$, $\dim V=m$	W^*	$\mathfrak{B}^*=\{b_1^*,\cdots,b_m^*\}$, $\dim W^*=m$
$V\otimes W$	$\{a_j\otimes b_k: 1\leqslant j\leqslant n, 1\leqslant k\leqslant m\}$ $\dim(V\otimes W)=n\times m$	$V^*\otimes W^*$	$\{a_j^*\otimes b_k^*: 1\leqslant j\leqslant n, 1\leqslant k\leqslant m\}$ $\dim(V^*\otimes W^*)=n\times m$

Theorem 2.2.16 Linear space $L(V,W;Z)$ and $L(V\otimes W,Z)$ are isomorphic:

$$L(V,W;Z)\xleftrightarrow{\text{iso.}}L(V\otimes W,Z).$$

Proof Define a mapping $\varphi: L(V\otimes W,Z)\to L(V,W;Z)$ satisfying

$$\varphi(g)=g\circ h,\quad g\in L(V\otimes W,Z),$$

with $h(v,w)=v\otimes w: V\times W\to V\otimes W$, and $\varphi(g)=g\circ h\in L(V,W;Z)$, since

$$\varphi(g)=g\circ h:(v,w)\in V\times W\xrightarrow{h}h(v,w)\in V\otimes W\xrightarrow{g}Z.$$

We prove that $\varphi: L(V\otimes W,Z)\to L(V,W;Z)$ is an isomorphic mapping.

For $(v,w)\in V\times W$, take $h:(v,w)\to v\otimes w$, then:

(1) $\varphi: g\in L(V\otimes W,Z)\to g\circ h\in L(V,W;Z)$; Fig. 2.2.4 shows that $\varphi: g\to g\circ h$

$$\varphi: g\in L(V\otimes W,Z)\to g\circ h\in L(V,W;Z);$$

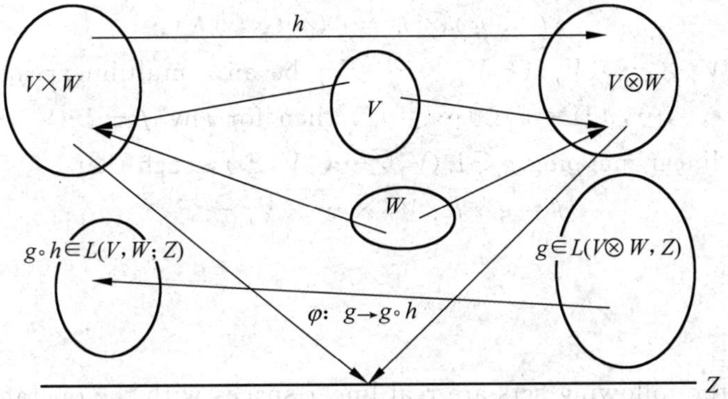

Fig. 2.2.4 Isomorphism

(2) $\varphi: g\to g\circ h$ is a one-one mapping

$$g_1\neq g_2\in L(V\otimes W,Z)\Rightarrow g_1\circ h\neq g_2\circ h\in L(V,W;Z);$$

(3) $\varphi: g\to g\circ h$ keeps the operations:

① $g_1,g_2\in L(V\otimes W,Z)\Rightarrow\varphi(g_1+g_2)=(g_1+g_2)\circ h=g_1\circ h+g_2\circ h=\varphi(g_1)+\varphi(g_2)$, since $\forall(v,w)\in V\times W$ implies $\varphi(g_1+g_2)(v,w)=((g_1+g_2)\circ h)(v,w)=(g_1+g_2)(h(v,w))=g_1(h(v,w))+g_2(h(v,w))=(g_1\circ h)(v,w)+(g_2\circ h)(v,w)=(\varphi(g_1)+\varphi(g_2))(v,w)$;

② $g\in L(V\otimes W,Z),\alpha\in\mathbb{F}\Rightarrow\varphi(\alpha g)=(\alpha g)\circ h=\alpha(g\circ h)=\alpha\varphi(g)$, since $\forall(v,w)\in V\times W$ implies $\varphi(\alpha g)(v,w)=((\alpha g)\circ h)(v,w)=(\alpha g)(h(v,w))=\alpha g(h(v,w))=\alpha(g\circ h)\cdot(v,w)=\alpha\varphi(g)(v,w)$. The proof is complete.

5) **Properties of tensor products**

Definition 2.2.10 (**tensor product of multilinear functionals**) For multilinear functionals $f\in L(V_1,\cdots,V_r;\mathbb{F})$ and $g\in L(W_1,\cdots,W_s;\mathbb{F})$, **the tensor product of** f **and**

g is defined by

$$(f\otimes g)(v_1,\cdots,v_r,w_1,\cdots,w_s)=f(v_1,\cdots,v_r)g(w_1,\cdots,w_s),$$

with $(v_1,\cdots,v_r)\in V_1\times\cdots\times V_r$, $(w_1,\cdots,w_s)\in W_1\times\cdots\times W_s$. Thus, the tensor product $f\otimes g$ is an $\mathbf{F}$-valued $(r+s)$-multilinear functional on the space $V_1\times\cdots\times V_r\times W_1\times\cdots\times W_s$, and the tensor product operator $\otimes$ is a multilinear mapping form $L(V_1,\cdots,V_r;\mathbf{F})\times L(W_1,\cdots,W_s;\mathbf{F})$ to $L(V_1,\cdots,V_r,W_1,\cdots,W_s;\mathbf{F})$.

Furthermore, for linear spaces $V_1,\cdots,V_r$ on number field $\mathbf{F}$, the tensor product space $V_1\otimes\cdots\otimes V_r$ can be defined and hold $V_1\otimes\cdots\otimes V_r\overset{\text{iso.}}{\longleftrightarrow}L(V_1^*,\cdots,V_r^*;\mathbf{F})$, $V_1^*\otimes\cdots\otimes V_r^*\overset{\text{iso.}}{\longleftrightarrow}L(V_1,\cdots,V_r;\mathbf{F})$.

Theorem 2.2.17 Tensor product operation $\otimes$ satisfies the combination law: for $f\in L(V_1,\cdots,V_r;\mathbf{F})$, $g\in L(W_1,\cdots,W_s;\mathbf{F})$, $h\in L(U_1,\cdots,U_t;\mathbf{F})$, it holds

$$(f\otimes g)\otimes h=f\otimes(g\otimes h).$$

Moreover, let $h: V_1\times\cdots\times V_r\to V_1\otimes\cdots\otimes V_r$ be an r-multilinear mapping defined by tensor product $h(v_1,\cdots,v_r)=v_1\otimes\cdots\otimes v_r$, then for any $f\in L(V_1,\cdots,V_r;\mathbf{F})$, there exists the unique linear mapping $g\in L(V_1\otimes\cdots\otimes V,\mathbf{F})$, such that

$$f=g\circ h: V_1\times\cdots\times V_r\to\mathbf{F}.$$

Exercise 2

1. Whether the following sets are real linear spaces with the operations:

(1) The set of all polynomials with order n $(n\geqslant 1)$ and real coefficients, with the addition and real number product of polynomials;

(2) The set of all n-order real symmetry matrices (skew-symmetry matrices, upper-triangular matrices), with the addition and real number product of matrices;

(3) The set of all vectors on plane $\mathbf{R}^2$, with the usual addition of vectors, and real number product defined by $k\circ\alpha=0$, $k\in\mathbf{R}$, $\alpha\in\mathbf{R}^2$;

(4) The set of all positive real numbers $\mathbf{R}^+$, with the addition: $a\oplus b=ab$, and the number product: $k\circ a=a^k$, $a\in\mathbf{R}^+$, $k\in\mathbf{R}$.

2. Prove that: The dimension of a linear space in definition 2.1.2 is reasonable. Find the dimension of the linear space given in 1(4), and seek a base of this linear space.

3. Let $A\in\mathfrak{M}_{n\times n}$, prove that the set of all matrices in $\mathfrak{M}_{n\times n}$ which can exchange with A is a linear subspace of $\mathfrak{M}_{n\times n}$, denoted by $C(A)$. Moreover, when $A=I$ (the unit of $\mathfrak{M}_{n\times n}$), find $C(A)$ and $\dim(C(A))$.

4. Prove: the sum V_1+V_2 is a direct sum, if and only if $V_1\cap V_2=\{0\}$.

5. Prove the Theorem 2.1.2. for $\mathbb{F}=\mathbb{R}$.

6. Let σ be a linear mapping from n-dimensional linear space $(X,+,\alpha\cdot)$ on $\mathbb{F}$ to itself. Prove that: (1) If $f\in X^*$ is a linear functional on X, then, so is $f\circ\sigma$. (2) Define a mapping $\sigma^*: V^*\to V^*$ from X^* to X^* with $\sigma^*: f\to f\circ\sigma$. Then σ^* is a linear mapping from V^* to itself.

7. For $\mathfrak{B}=\{e_1,\cdots,e_n\}$, $\mathfrak{B}^*=\{e_1^*,\cdots,e_n^*\}$, $\mathfrak{B}^{**}=\{e_1^{**},\cdots,e_n^{**}\}$ defined in the Theorem 2.2.1, prove that: $\mathfrak{B}^*$, $\mathfrak{B}^{**}$ are the bases of X^*, X^{**}, respectively. Moreover, if $x\to x^{**}$ is the natural mapping from X onto X^{**}, prove that: $x^{**}(w)=w(x)$, $\forall w\in X^*$.

8. Prove that: a bilinear functional $f(x,y)$ on a linear space $(X,+,\alpha\cdot)$ is skew-symmetric, if and only if $f(x,x)=0$, $\forall x\in X$.

9. Let V be the n-dimensional Euclidean space $\mathbb{R}^n$ with inner product (x,y). For a fix vector a in V, we define a functional a^*: $\forall b\in V\Rightarrow a^*(b)=(a,b)$ on V. Prove that: the mapping $a\to a^*$ from V to V^* is an isomorphism. Moreover, $\mathbb{R}^n \xleftrightarrow{a\to a^*} (\mathbb{R}^n)^*$.

10. Let σ be a linear mapping from V to itself, and $\sigma^{k-1}(\xi)\neq 0$, but $\sigma^k(\xi)=0$. Prove that: for $k>1$, $\xi,\sigma(\xi),\cdots,\sigma^{k-1}(\xi)$ are linearly independent.

11. Let V be a linear space on complex field $\mathbb{C}$ with dimension $n\geqslant 2$, and $f(x,y)$ be a bilinear symmetric functional on V. Prove that: (1) There exists non-zero vector $\xi\in V$, such that $f(\xi,\xi)=0$; (2) If $f(x,y)$ is non-degenerate, then there exists ξ, η, independently, satisfying $f(\xi,\eta)=1$, $f(\xi,\xi)=f(\eta,\eta)=0$.

12. Show the relationship between two definitions of conjugate operators for a finite dimensional inner space.

13. Prove the bilinear property of a tensor product "$\otimes$" for the space $V^*\times W^*$ to $L(V,W;F)$.

14. Let V be a linear space on number field $\mathbb{F}$, and $f_1,f_2,\cdots,f_k$ be linear functionals on V. Prove that: (1) The set $W=\{v\in V: f_j(v)=0, 1\leqslant j\leqslant k\}$ is a linear subspace of V, called the annihilator space for $f_1,f_2,\cdots,f_k$ in V. (2) Any linear subspace of V is an annihilator space for some linear functionals.

Chapter 3 Basic Knowledge of Point Set Topology

A frame about basic knowledge of point set topology is shown below:

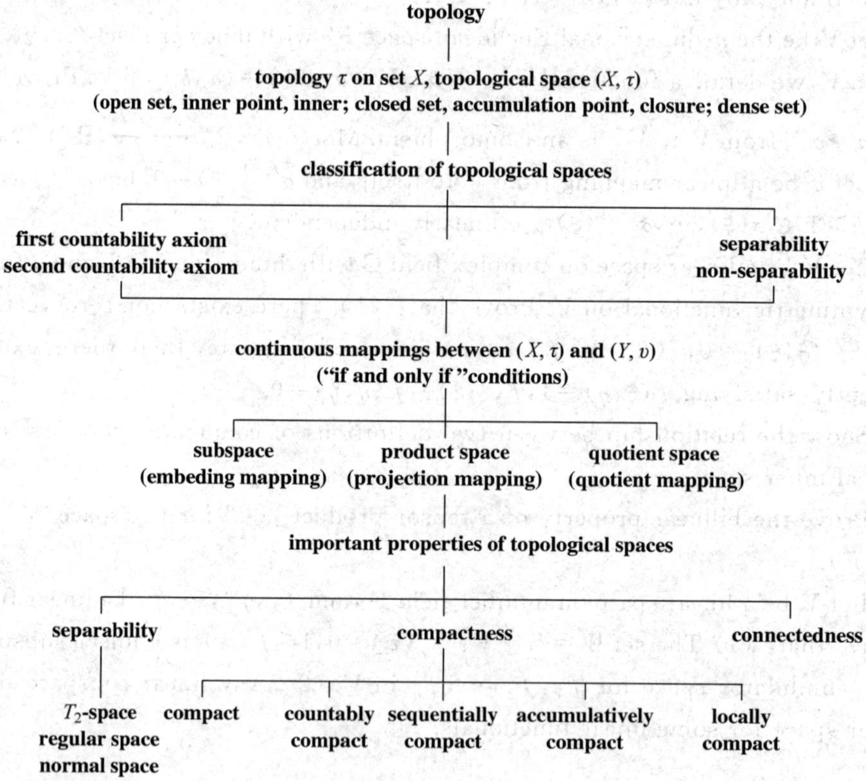

To describe the relationship between positions (far or near) for elements in a set, we need to endow so-called topological structure to the set, such that it turns to **a topological space.** Furthermore, we need the knowledge of topological structures which play essential role in modern mathematics and other sciences. The main references to this chapter are [2], [10], [12], [14], and [17].

Three-dimensional Euclidean space $\mathbb{R}^3$ is the most familiar for humility. In fact, one-dimensional space $\mathbb{R}$ is a straight line, two-dimensional space $\mathbb{R}^2$ is a plane, and three-dimensional space $\mathbb{R}^3$ models the physical world, or to a greater extent, the cosmos.

It is well known that $\mathbb{R}$, $\mathbb{R}^2$, $\mathbb{R}^3$ are linear spaces on number field $\mathbb{R}$ with addition "$+$" and number product "$\alpha\cdot$", $\alpha\in\mathbb{R}$, these are the **operation structure** of them. However, they have another important structure, that is **topological structure**.

Firstly, for any $x, y\in\mathbb{R}$, **a distance** of x and y is defined by $\rho(x,y)=|x-y|$, it means the relationship of far or near between x and y. Similarly, in $\mathbb{R}^2$ and $\mathbb{R}^3$, we have the distances

$$\rho(x,y)=\sqrt{(x_1-y_1)^2+(x_2-y_2)^2},\quad x=(x_1,x_2),y=(y_1,y_2)\in\mathbb{R}^2$$

and

$$\rho(x,y)=\sqrt{(x_1-y_1)^2+(x_2-y_2)^2+(x_3-y_3)^2},$$
$$x=(x_1,x_2,x_3),y=(y_1,y_2,y_3)\in\mathbb{R}^3,$$

respectively.

More examples can be given for non-Euclidean spaces, such as for the set

$$X=\{x: x=(x_1,\cdots,x_n,\cdots),x_j\in\mathbb{R},j\in\mathbb{Z}^+\},$$

it has a subset $l^2=\left\{x\in X: \sum_{j=1}^{+\infty}|x_j|^2<+\infty\right\}\subset X$. We define a "distance"

$$\rho(x,y)=\sqrt{\sum_{j=1}^{+\infty}(x_j-y_j)^2},\quad x,y\in l^2.$$

There are the realistic senses of $\rho(x,y)$ in $\mathbb{R}$, $\mathbb{R}^2$, $\mathbb{R}^3$, since $\rho(x,y)$ is a **distance** between x and y. However, there is no realistic meaning for so-called "distance" in abstract sets. What are the essential characteristic properties of distances of $\rho(x,y)=|x-y|$, $\rho(x,y)=\sqrt{(x_1-y_1)^2+(x_2-y_2)^2}$ and $\rho(x,y)=\sqrt{(x_1-y_1)^2+(x_2-y_2)^2+(x_3-y_3)^2}$? So that we may generalize the concept of distances to that of abstract sets? First of all, we need catch the essential properties of "distance", then define a kind of sets, which is so-called "metric space", and study structure properties of these sets, as well as some special cases of the metric spaces, such as normed spaces, inner product spaces. These are main aims in the point-set topology.

3.1 Metric Spaces, Normed Linear Spaces

3.1.1 Metric spaces

What are the essential properties of distances $\rho(x,y)$ of Euclidean spaces $\mathbb{R}$, $\mathbb{R}^2$, $\mathbb{R}^3$?

They are following three properties:

(1) Non-negativity: $\rho(x,y)\geqslant 0$, moreover, $x=y\Leftrightarrow\rho(x,y)=0$;

(2) Symmetry: $\rho(x,y)=\rho(y,x)$, the distance from A to B equals that from B to A;

(3) Trigonometric inequality: $\rho(x,y)\leqslant\rho(x,z)+\rho(z,y)$, a famous theorem in the elementary geometry.

Thus, we define "distance" for abstract sets by generalizing above three essential properties.

Definition 3.1.1 (distance, metric space) Let X be a set, denote elements of X by x, y, $\cdots\in X$. If for any two elements x and y in X, there exists a real number $\rho(x,y)$ satisfying the following properties:

(1) $\rho(x,y)\geqslant 0$; $\rho(x,y)=0\Leftrightarrow x=y$; (non-negativity)

(2) $\rho(x,y)=\rho(y,x)$; (symmetry)

(3) $\rho(x,y)\leqslant\rho(x,z)+\rho(z,y)$. (triangle inequality)

Then $\rho(x,y)$ is said to be **a distance of** x **and** y. The set X is said to be **a metric space endowed with the distance** $\rho(x,y)$ **as its topological structure**, or simply, a metric space.

We show more useful metric spaces except the metric spaces $\mathbb{R}$, $\mathbb{R}^2$, $\cdots$, $\mathbb{R}^n$.

1) l^2 **space**

For the given set

$$l^2=\left\{x=(x_1,\cdots,x_n,\cdots):x_j\in\mathbb{R},j=1,2,\cdots,n,\cdots,\sum_{j=1}^{+\infty}|x_j|^2<+\infty\right\},\tag{3.1.1}$$

define the distance in l^2: for $x\in l^2$ and $y\in l^2$, let

$$\rho(x,y)=\sqrt{\sum_{j=1}^{+\infty}(x_j-y_j)^2}.\tag{3.1.2}$$

It is easy to verify that $\rho(x,y)=\sqrt{\sum_{j=1}^{+\infty}(x_j-y_j)^2}$ satisfies (1)-(3) in the Definition 3.1.1, thus, it is a distance on l^2, such that l^2 becomes a metric space, denoted by $(l^2,\rho(x,y))$, or simply, l^2.

2) **Continuous function space** $C([a,b])$ **on** $[a,b]$

The set of all real-valued continuous functions in interval $[a,b]$ (see 1.1.1), denote by

$$C([a,b])=\{f: f(x) \text{ is continuous in } [a,b]\}.\tag{3.1.3}$$

For $f,g,h,\cdots\in C([a,b])$, we define

$$\rho(f,g)=\max_{x\in[a,b]}|f(x)-g(x)|,\tag{3.1.4}$$

it is easy to verify that $\rho(f,g)$ is a distance on $C([a,b])$, such that $C([a,b])$ becomes a metric space, called the **continuous function space on** $[a,b]$.

3) ***L*-integrable function space** $L^1(E)$, ***L*-square integrable function space** $L^2(E)$

The sets

$$L^1(E)=\left\{f: \int_E |f(x)|\,\mathrm{d}m<+\infty\right\}, \quad L^2(E)=\left\{f: \int_E |f(x)|^2\,\mathrm{d}m<+\infty\right\} \tag{3.1.5}$$

are all L-integrable functions and all L-square integrable functions on an L-measurable set E, respectively. Define

$$\rho(f,g)=\int_E |f(x)-g(x)|\,\mathrm{d}m, \quad \rho(f,g)=\left\{\int_E |f(x)-g(x)|^2\,\mathrm{d}m\right\}^{\frac{1}{2}} \tag{3.1.6}$$

on $L^1(E)$ and $L^2(E)$, respectively. We may check that they are distances satisfying (1)-(3) in Definition 3.1.1.

We check that $\rho(f,g)=\left\{\int_E |f(x)-g(x)|^2\,\mathrm{d}m\right\}^{\frac{1}{2}}$ satisfies (1)-(3), so that $L^2(E)$ is a metric space.

For (1), take any $f,g\in L^2(E)$, then by the definition of Lebesgue integral, it holds

$$\rho(f,g)=\left\{\int_E |f(x)-g(x)|^2\,\mathrm{d}m\right\}^{\frac{1}{2}}\geqslant 0.$$

Thus, "$f=g$ implies $\rho(f,g)=0$" is evident. Conversely, $\rho(f,g)=\left\{\int_E |f(x)-g(x)|^2\,\mathrm{d}m\right\}^{\frac{1}{2}}$ $=0$ implies "$f=g$, a. e." (this means "f equals g at almost all $x\in E$", i. e., $f=g$ holds for $x\in E\setminus E_0$ with $mE_0=0$, where mE_0 is the Lebesgue measure of E_0), by the famous theorem in "real-variable functions". We agree on: "$f=g$ in $L^2(E)$ iff $f=g$, a. e. $x\in E$". Thus (1) holds.

For (2), the symmetry property is clear.

For (3), firstly, by the Hölder inequality: " if $f,g\in L^2(E)$, then

$$\int_E |f(x)g(x)|\,\mathrm{d}m\leqslant\left\{\int_E |f(x)|^2\,\mathrm{d}m\right\}^{\frac{1}{2}}\left\{\int_E |g(x)|^2\,\mathrm{d}m\right\}^{\frac{1}{2}}\text{"}; \tag{3.1.7}$$

we get the Minkovski inequality: " if $f,g\in L^2(E)$, then

$$\left\{\int_E |f(x)+g(x)|^2\,\mathrm{d}m\right\}^{\frac{1}{2}}\leqslant\left\{\int_E |f(x)|^2\,\mathrm{d}m\right\}^{\frac{1}{2}}+\left\{\int_E |g(x)|^2\,\mathrm{d}m\right\}^{\frac{1}{2}}\text{"}. \tag{3.1.8}$$

So that (3) holds.

Recall that these examples have been endowed certain operation structures in Chapter 1, so they all are linear spaces; and now they have been endowed distances as topological structures.

So, they become metric spaces, denoted by $(l^2, +, \alpha\cdot, \rho_{l^2})$, $(C([a,b]), +, \alpha\cdot, \rho_{C([a,b])})$, and $(L^1(E), +, \alpha\cdot, \rho_{L^1(E)})$, $(L^2(E), +, \alpha\cdot, \rho_{L^2(E)})$, or simply, $l^2, C([a,b]), L^1(E), L^2(E)$.

However, a metric space does not need to have operation structure, for example, as follows.

4) On set $A=\{a,b,c,d\}$, we define $\rho(x,y)=\begin{cases}1, & x\neq y\\ 0, & x=y\end{cases}$, then A becomes a metric space (A,ρ_A), but there is no operation between its elements.

3.1.2 Normed linear spaces

Euclidean spaces $\mathbb{R}, \mathbb{R}^2, \cdots, \mathbb{R}^n$ not only have operation structures—linear space, but also have topological structures—metric space; so there are $l^2, C([a,b]), L^1(E), L^2(E)$. However, as we will see, topological structures are not only endowed by distances, but also by other schemes, for example, endowed so called "norm" that is a generalized "length" in Euclidean spaces.

Definition 3.1.2 (normed linear space) Let $(X, +, \alpha\cdot)$ be a linear space on number field $\mathbb{F}$. If for any $x\in X$, there exists an non-negative real number corresponding to x, denoted by $\|x\|_X$, satisfying

(1) $\|x\|_X\geqslant 0$, $\|x\|_X=0\Leftrightarrow x=0$ ($x=0$ is zero of X); (non-negativity)

(2) $\|\alpha x\|_X=|\alpha|\,\|x\|_X$, $\alpha\in\mathbb{F}$; (absolutely homogeneity)

(3) $\|x+y\|_X\leqslant\|x\|_X+\|y\|_X$, $x, y\in X$. (triangle inequality)

Then $\|x\|_X$ is said to be **a norm** of $x\in X$; and X is said to be **a normed linear space on $\mathbb{F}$**, or simply, **a normed space.** A linear space X is endowed with norm $\|x\|_X$ such that it has a topological structure, denoted by $(X, +, \alpha\cdot, \|x\|_X)$; or simply, $(X, \|x\|_X)$.

$\mathbb{R}^n, l^2, C([a,b]), L^1(E), L^2(E)$ are normed linear spaces, the norms of them are:

$x\in\mathbb{R}^n$ $\quad \|x\|_{\mathbb{R}^n}=\sqrt{x_1^2+x_2^2+\cdots x_n^2}$, $n\in\mathbb{N}$;

$x\in l^2$ $\quad \|x\|_{l^2}=\sqrt{\sum_{j=1}^{+\infty}x_j^2}$;

$f\in C([a,b])$ $\quad \|f\|_{C([a,b])}=\max_{x\in[a,b]}|f(x)|$;

$f\in L^1(E)$ $\quad \|f\|_{L^1(E)}=\int_E|f(x)|\,\mathrm{d}m$;

$f\in L^2(E)$ $\quad \|f\|_{L^2(E)}=\left\{\int_E|f(x)|^2\,\mathrm{d}m\right\}^{\frac{1}{2}}$.

Every normed linear space $(X, \|\cdot\|_X)$ is a metric space, since for any elements x, $y\in X$, define $\rho(x,y)\,\|x-y\|_X$, then (X,ρ) is a metric space. However, a metric space is not necessarily a normed space, since it is not necessarily having operation structure, so that it is not necessarily a linear space.

For an inner product space $(X, +, \alpha \cdot, (x, y))$ on $\mathbb{F}$ (see section 2.1.4), we may define a norm such that it becomes normed linear space. In fact, $\| x \|_X = \{(x, x)\}^{\frac{1}{2}}, x \in X$, is a norm and $(X, +, \alpha \cdot, (x, y))$ becomes normed linear space $(X, +, \alpha \cdot, \| x \|_X)$. However, conversely, many normed linear spaces could not become inner product linear spaces, for example, $C([a, b])$ and $L^1(E)$ do not have inner product, they are normed linear spaces, but not inner product spaces.

3.2 Topological Spaces

3.2.1 Some definitions in topological spaces

As is well known, metric space, normed linear space, inner-product space and Euclidean space have topological structures, so that we can determine the positions of elements in them, and the far or near relationships between two elements in these spaces. There is an interesting and important problem: Do we have some more general topological structures on sets than that of the above listed? Yes, there are many new topological structures appeared in modern science, on one hand; and on the other hand, some quite new, deep and wide research topics are connected closely with more general topological structures of sets in the point of view of modern mathematics. Thus, a very elegant and delicate branch of mathematics, called "general topology" or "point set topology", undertakes the task to study and devise topological structures of more and more sets. Nowadays, the point set topology plays a quite important role in modern science and technology.

1. Topological structures

Definition 3.2.1 (topology, topological space) Let X be a non-empty set, $X \neq \varnothing$, τ be a family of some subsets of X. If τ satisfies:

(1) $X, \varnothing \in \tau$;

(2) $G_1, G_2 \in \tau \Rightarrow G_1 \cap G_2 \in \tau$; (finite intersection property)

(3) $G_\lambda \in \tau, \lambda \in \Lambda \Rightarrow \bigcup_{\lambda \in \Lambda} G_\lambda \in \tau$, Λ is an index set. (any union property)

Then τ is said to be **a topology on** X, or, τ determines **a topological structure on** X. Any $A \in \tau$ is said to be **an open set** of X, so that τ is said to be **an open set family of** X, and (X, τ)

is said to be **a topological space.**

The meaning of Definition 3.2.1: (1) Indicating that the set X and empty set $\varnothing$ both are open sets; (2) Indicating that the intersection of finite open sets is an open set; (3) Indicating that the union of any open sets (finite, countable infinite, uncountable infinite) is an open set.

We agree on that **empty set $\varnothing$ is contained in any set**, i.e., $\forall A \subset X \Rightarrow \varnothing \subset A$.

As examples, metric space, normed linear space, inner product space, and Euclidean space, all are topological spaces. Here we **emphasize** that a topological space does not need endow any operation structure, for instance, a metric space does not need to have an operation structure, only need has a topological structure.

We take an example to determine the topology τ on a metric space.

Let (X, ρ) be a metric space. Take a point $x \in X$ and a real number $r > 0$, then, the set

$$B(x, r) = \{y \in X: \rho(x, y) < r\} \tag{3.2.1}$$

is said to be **an open ball with center x, radius r**; or simply, **an open ball.**

Open set: A set $G \subset X$ in metric space (X, ρ) is said to be **an open set**, if for any point $x \in G$, there exists a ball $B(x, r) = \{y \in X: \rho(x, y) < r\}$ such that $B(x, r) \subset G$ (Fig. 3.2.1).

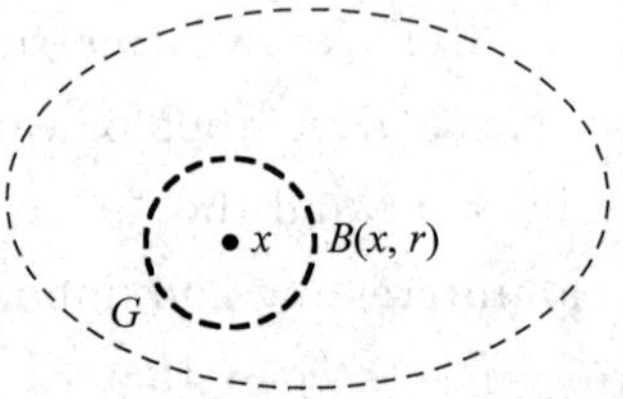

Fig. 3.2.1 An open ball in a metric space

In metric space (X, ρ), the set of all open sets, called **an open set family**, denoted by

$$\tau = \{G \subset X: G \text{ is open set in } (X, \rho)\},$$

satisfies (1)-(3) in Definition 3.2.1. Since

(1) X in metric space (X, ρ) is an open set: $\forall x \in X$, $\exists r > 0$ (not need unique), such that $B(x, r) = \{y \in X: \rho(x, y) < r\} \subset X$, then $X \in \tau$. Moreover, by appointment, $\varnothing \in \tau$.

(2) Let G_1, G_2 be open sets in (X, ρ), then by definition, $\exists r_1$, s.t. $B(x, r_1) \subset G_1$; $\exists r_2$, s.t. $B(x, r_2) \subset G_2$. Thus, there are two cases: ① $G_1 \cap G_2 = \varnothing$, then $G_1 \cap G_2 = \varnothing \in \tau$; ② $G_1 \cap G_2 \neq \varnothing$, take $r = \min\{r_1, r_2\}$, such that $\forall x \in G_1 \cap G_2 \Rightarrow B(x, r) \subset G_1 \cap G_2$; this implies $G_1 \cap G_2 \in \tau$. Thus, the property (2) holds for finite open sets:

$$G_1, \cdots, G_n \in \tau \Rightarrow \bigcap_{j=1}^{n} G_j \in \tau, \quad n \in \mathbf{N}.$$

(3) Let $G_\lambda, \lambda \in \Lambda$, be open sets in metric space (X, ρ), Λ be the index set, finite, countably infinite, or uncountably infinite. Let $G = \bigcup_{\lambda \in \Lambda} G_\lambda$. We prove $G = \bigcup_{\lambda \in \Lambda} G_\lambda \in \tau$:

$$\forall x \in G \Rightarrow \exists \lambda_0 \in \Lambda, \text{s. t. } x \in G_{\lambda_0} \subset \bigcup_{\lambda \in \Lambda} G_\lambda$$

$$\Rightarrow \exists r_0, \quad \text{s. t. } B(x, r_0) \subset G_{\lambda_0} \text{ by } G_{\lambda_0} \text{ open}$$

$$\Rightarrow \exists r_0, \quad \text{s. t. } B(x, r_0) \subset G_{\lambda_0} \subset G = \bigcup_{\iota \in \Lambda} G_\iota \quad \Rightarrow \quad G \text{ is open.}$$

Hence, τ is a topology on metric space (X, ρ), so that, by definition, a metric space (X, ρ) is a topological space.

Left as exercise, please show the open set family τ in a normed linear space $(X, \|x\|_X)$, inner product space $(X, (x, y))$, Euclidean space $\mathbb{R}^n$, respectively. Moreover, prove that the open set family τ which you give is a topology such that (X, τ) is a topological space.

The other set class in a topological space (X, τ), called **a closed set family**, is important as the same as an open set family.

Definition 3.2.2 (closed set) Let (X, τ) be a topological space, $A \subset X$ be a subset of X. If the complementary set $A^c \equiv \complement A \equiv X \setminus A$ of A is an open set, i. e., $A^c \in \tau$, then A is said to be **a closed set.**

For example, in $\mathbb{R}^1 \equiv \mathbb{R}$, ① an interval $[a, b]$ is a closed set; ② the union of finite intervals $\bigcap_{j=1}^{k} [a_j, b_j]$, $k \in \mathbb{Z}^+$, is a closed set.

The properties of closed sets:

Theorem 3.2.1 Let (X, τ) be a topological space. We have

(1) $X, \varnothing$ are closed sets;

(2) F_1, F_2 are closed sets $\Rightarrow F_1 \cup F_2$ is a closed set; (finite union property)

(3) $F_\lambda, \lambda \in \Lambda$, are closed sets $\Rightarrow \bigcap_{\lambda \in \Lambda} F_\lambda$ is a closed set. (any intersection property)

Proof For (1), since $X^c = X \setminus X = \varnothing$, by appointment that $\varnothing \in \tau$ is an open set, so that X is a closed set; the same reason implies that $\varnothing$ is a closed set.

For (2), by de Morgan formula $X \setminus \left\{\bigcup_{\lambda \in \Lambda} A_\lambda\right\} = \bigcap_{\lambda \in \Lambda} (X \setminus A_\lambda)$, it follows that

F_1, F_2 are closed sets $\Rightarrow X \setminus (F_1 \cup F_2) = (X \setminus F_1) \cap (X \setminus F_2)$ with open sets $X \setminus F_1$ and $X \setminus F_2 \Rightarrow X \setminus (F_1 \cup F_2)$ is the intersection of $X \setminus F_1$ and $X \setminus F_2$, thus $X \setminus (F_1 \cup F_2)$ is an open set $\Rightarrow F_1 \cup F_2$ is a closed set.

For (3), by de Morgan formula $X \setminus \left\{\bigcap_{\lambda \in \Lambda} A_\lambda\right\} = \bigcup_{\lambda \in \Lambda} (X \setminus A_\lambda)$, it follows that

$F_\lambda, \lambda \in \Lambda$, are closed sets $\Rightarrow X \setminus \left\{\bigcap_{\lambda \in \Lambda} F_\lambda\right\} = \bigcup_{\lambda \in \Lambda} (X \setminus F_\lambda)$ with $X \setminus F_\lambda$ is an open set $\Rightarrow$ $X \setminus \left\{\bigcap_{\lambda \in \Lambda} F_\lambda\right\}$ is the union of open set family $X \setminus F_\lambda, \lambda \in \Lambda$, so $X \setminus \left\{\bigcap_{\lambda \in \Lambda} F_\lambda\right\}$ is an open set $\Rightarrow$

the set $\bigcap_{\lambda \in \Lambda} F_\lambda$ is a closed set.

2. Neighborhoods of points

Definition 3.2.3 (neighborhood) Let (X, τ) be a topological space. For a point $x \in X$, its subset $A \subset X$ is said to be **a neighborhood of** x, if there exists an open set $G \in \tau$, such that $x \in G \subset A$ (Fig. 3.2.2).

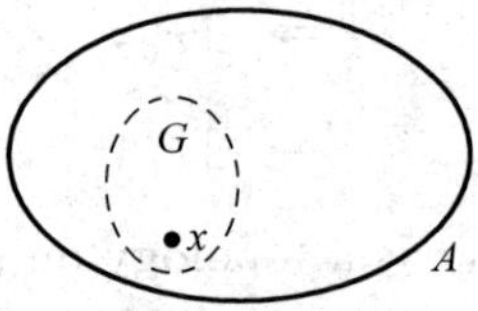

Fig. 3.2.2 A neighborhood of $x \in X$

For a topological space (X, τ), the set of all neighborhoods of $x \in X$ is denoted by

$$\mathfrak{A}_x = \{A \subset X: A \text{ is a neighborhood of } x \in A\},$$

and $\mathfrak{A}_x$ is said to be **a neighborhood system of** $x \in X$.

It is easy to see that any open set $O \in \tau$ that contains $x \in X$ is a neighborhood of x, called an **open neighborhood of** x.

Theorem 3.2.2 Let (X, τ) be a topological space, $\mathfrak{A}_x$ be a neighborhood system of $x \in X$, we have

(1) $A_1, A_2 \in \mathfrak{A}_x \Rightarrow A_1 \cap A_2 \in \mathfrak{A}_x$; (finite-intersection property)

(2) $A_\iota \in \mathfrak{A}_x, \iota \in \Lambda \Rightarrow \bigcup_{\iota \in \Lambda} A_\iota \in \mathfrak{A}_x$; (any union property)

(3) $\forall A \in \mathfrak{A}_x \Rightarrow \exists B \in \mathfrak{A}_x$, s.t. ① $B \subset A$; ② $\forall y \in B \Rightarrow B \in \mathfrak{A}_y$;

(4) $G \in \tau$ is open set $\Leftrightarrow \forall x \in G$ implies $G \in \mathfrak{A}_x$.

Note If a family $\mathfrak{A} \equiv \{\mathfrak{A}_x : x \in X\}$ on a set X is given as a neighborhood system of each $x \in X$, then it will determine a topology on X, such that $(X, \mathfrak{A})$ becomes a topological space.

3. Inner point and the inner; external point and the external; boundary point and the boundary

Let (X, τ) be a topological space (Fig. 3.2.3), we define:

1) **Inner point of set** $A \subset X$: for a point $x \in A$, if there exists open set $G \in \tau$, such that $x \in G \subset A$, then x is said to be **an inner point of** A.

Inner of set $A \subset X$: the set of all inner points of A is said to be **the inner of** A, denoted by A° (or A°).

2) **External point of set** $A \subset X$: for a point $x \in A$, if $x \in (A^c)^\circ$, i.e., x is in the inner part of complementary of A, then x is said to be **an external point of** A.

External part of set $A \subset X$: the set of all external points of A is said to be **the external part of** A, or simply, the external.

3) **Boundary point of set** $A \subset X$: if in any open set $G \in \tau$ that contains $a \in X$, there exist both inner and external points of A, then a is said to be **a boundary point of** A.

Boundary of set $A \subset X$: the set of all boundary points of A is said to be **the boundary of** A, denoted by ∂A.

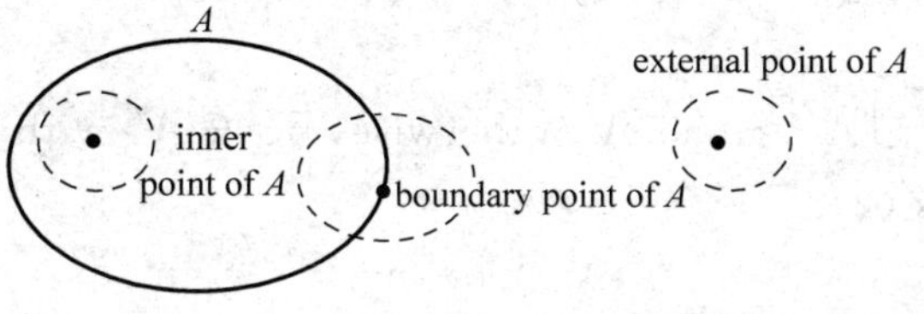

Fig. 3. 2. 3 Inner point, external point, boundary point of A

4. Accumulation point and isolated point of a set, divided set and closure of a set

1) **Accumulation point of a set** $A \subset X$: if for any open set $G \in \tau$ that contains $a \in G$ with $a \in X$, implies $(G \setminus \{a\}) \cap A \neq \varnothing$, then a is called **an accumulation point of** A, or **a limit point of** A;

Isolated point of a set $A \subset X$: if $a \in A$, but a is not accumulation point of A, then a is called **an isolated point of** A.

2) **Divided set of set** $A \subset X$: the set of all accumulation points of A is called **a divided set of** A, denoted by A'.

3) **Closure of set** $A \subset X$: the union of A and A' is called **a closure** of A, denoted by $\bar{A}$, i. e., $\bar{A} = A \cup A'$.

If $A \subset X$ is a closed set and has no isolated point, then A is called **a complete set.**

Note Both boundary points and accumulation points of a set A can belong to A, or do not belong to A; However, an isolated point certainly belongs to A.

We have the following properties for closed sets and closures:

Theorem 3. 2. 3 Let (X, τ) be a topological space. the following equivalences hold:

(1) A is a closed set $\Leftrightarrow$ (2) $A' \subset A$ $\Leftrightarrow$ (3) $A = A \cup A'$ $\Leftrightarrow$ (4) $\bar{A} = A$.

Proof For (1)$\Leftrightarrow$(2):

A is closed $\Rightarrow A^c$ is open $\Rightarrow \forall x \in A^c, \exists G \in \tau$, and $x \in G \subset A^c$, s. t. $G \cap A = \varnothing \Rightarrow (G \setminus \{x\}) \cap A = \varnothing$ (by $\forall x \notin A$) $\Rightarrow \forall x \in A^c$ is not accumulation point of A; i. e., $x \notin A'$ implies $A' \subset A$ (since "$\forall y \in A^c \Rightarrow y \notin A'$" implies "$\forall y \in A' \Rightarrow y \notin A^c \Rightarrow y \in A$") $\Rightarrow$ (1) implies (2);

Conversely, $A' \subset A \Rightarrow (A')^c \supset A^c \Rightarrow \forall x \in A^c$ implies $x \in (A')^c$, thus $x \notin A', x \notin A \Rightarrow \exists G_x \in \tau$, s. t. $(G_x \setminus \{x\}) \cap A = \varnothing \Rightarrow (G_x \setminus \{x\}) \cap A = G_x \cap A = \varnothing$ (by $x \notin A$) $\Rightarrow A^c =$

$\bigcup_{x\in A^c}\{x\}\subset \bigcup_{x\in A^c} G_x \subset A^c \Rightarrow A^c = \bigcup_{x\in A^c} G_x$ is a union of open sets $\Rightarrow A$ is closed $\Rightarrow$ (2) **implies** (1).

For (2)$\Leftrightarrow$(3).

$A'\subset A \Rightarrow A\cup A'\subset A\cup A = A \Rightarrow$ by $A\subset A\cup A'$, it follows $A=A\cup A' \Rightarrow$ (2) **implies** (3);

Conversely, $A=A\cup A' \Rightarrow A'\subset A$ (otherwise, $\exists x\in A'$ with $x\notin A$, this contradicts $A'\subset A$) $\Rightarrow$ (3) **implies** (2).

For (3)$\Leftrightarrow$(4):

$\overline{A}=A \Leftrightarrow A\cup A'=A$ (by definition).

Example 3.2.1 (1) In one-dimensional Euclidean space $\mathbb{R}$, for set $A=(-a,a]$, the closure of A is $\overline{A}=[-a,a]$; the inner of A is $A^\circ=(-a,a)$; the boundary of A is $\partial A=\{-a,a\}$; the external of A is $(-\infty,a]\cup(a,+\infty)$; the divided set of A is $A'=[-a,a]$.

(2) If $A_n=\left(\frac{1}{n},2\right)$, then $\overline{A}_n=\left[\frac{1}{n},2\right]$, $\bigcup_{n=1}^{+\infty}\overline{A_n}=\bigcup_{n=1}^{+\infty}\left[\frac{1}{n},2\right]=(0,2]$, and

$$\overline{\bigcup_{n=1}^{+\infty} A_n}=\overline{\bigcup_{n=1}^{+\infty}\left(\frac{1}{n},2\right)}=\overline{(0,2)}=[0,2].$$

5. Dense sets

Definition 3.2.4 (dense set) Let (X,τ) be a topological space, $A\subset X$ be a subset of X.

(1) If the closure of A is X, i.e., $X=\overline{A}$, then A is said to be **a dense subset of** X; or say that subset A is dense in X.

For a subset $B\subset X$ of X, if $B\subseteq\overline{A}$, then A is said to be dense in B (or $A\subset B$, or not);

(2) If the closure of A has empty inner part, i.e., $(\overset{\circ}{\overline{A}})=\varnothing$, then A is said to be **a nowhere dense subset of** X.

Example 3.2.2 The rational number set $\mathbb{Q}$ is dense in real number set $\mathbb{R}$, $\mathbb{R}=\overline{\mathbb{Q}}$; The integer number set $\mathbb{Z}$ is nowhere dense in $\mathbb{R}$ (because the closure of $\mathbb{Z}$ does not contain any finite real number, so that $\mathbb{Z}'=\varnothing$, and then $\overline{\mathbb{Z}}$ has empty inner part).

Theorem 3.2.4 Let (X,τ) be a topological space. If a subset $A\subset X$ is dense in X, i.e., $X=\overline{A}$, then, $\forall x\in X=\overline{A}$ implies that every open set $U\in\tau$ with $x\in U$ satisfies $U\cap A\neq\varnothing$.

The interpretation of Theorem 3.2.4 is: if $X=\overline{A}$, then any open neighborhood $U\in\tau$

of each $x \in X$ has non-empty intersection with A.

Take $A=\mathbb{Q}$, $X=\mathbb{R}$, then for each $x \in \mathbb{R}$, any open neighborhood of x, for instance, for $U=\left(x-\frac{1}{n}, x+\frac{1}{n}\right)$, $n=1,2,\cdots$, it holds

$$\left(x-\frac{1}{n}, x+\frac{1}{n}\right) \cap \mathbb{Q} \neq \varnothing, \quad n=1,2,\cdots,$$

this means that there exists a sequence of rational numbers

$$q_n \in \left(x-\frac{1}{n}, x+\frac{1}{n}\right) \cap \mathbb{Q} \neq \varnothing, \quad n=1,2,\cdots,$$

such that $\lim\limits_{n \to +\infty} q_n = x$.

In fact, for a topological space (X, τ), many properties can be transferred to its dense subsets.

3.2.2 Classification of topological spaces

In this section, we give an elementary classification of topological spaces. Later in the section 3.4, the classification from the point of view of intrinsic properties about topological spaces will be shown.

1. Topological bases

1) **Topological base of a topological space** (X, τ)

Definition 3.2.5 (**base of topological space** (X, τ)) Let (X, τ) be a topological space. If $\mathfrak{B}$ is an open set family of τ, and for each open set $G \in \tau$, and each point $x \in G$, there exists a set $B \in \mathfrak{B}$ satisfying $x \in B \subset G$, then $\mathfrak{B}$ is said to be **a topological base of** τ, or **a base of topology** τ (Fig. 3.2.4).

The following are a diagram and a characteristic property of a topological base.

$$\forall G \in \tau, \forall x \in G,$$
$$\exists B \in \mathfrak{B}, \text{ s. t. } x \in B \subset G$$

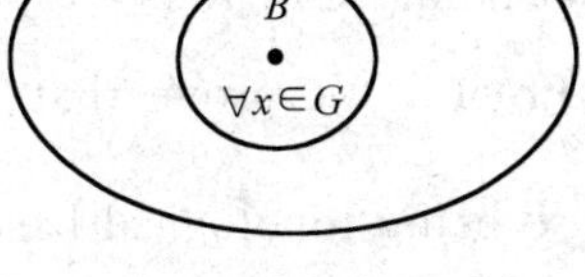

Fig. 3.2.4 Topological base

Theorem 3.2.5 Let (X, τ) be a topological space. Then, $\mathfrak{B} \subset \tau$ is a topological base of (X, τ), if and only if, each open set $G \in \tau$ is a union of some open sets in $\mathfrak{B}$, i. e., $\forall G \in \tau$, there exists open sub set family $\mathfrak{B}_0 \subset \mathfrak{B}$, satisfying

$$G = \bigcup_{B \in \mathfrak{B}_0} B \quad (\text{or } G = \bigcup \{B : B \in \mathfrak{B}_0\}).$$

Proof $\Rightarrow$ Let $\mathfrak{B} \subset \tau$ be a topological base of (X, τ), thus $\forall G \in \tau$, $\forall x \in G$, $\exists B \in \mathfrak{B}$, such that $x \in B \subset G$. Thus, $G = \bigcup_{x \in B} \{x\} \subset \bigcup_{x \in B} B$. However, since each B is contained in G, thus $G = \bigcup_{x \in B} \{x\} \subset \bigcup_{x \in B} B \subset G$, hence $G = \bigcup_{x \in B} B$. This implies that $\forall G \in \tau$ can be expressed as a union of some open sets of $\mathfrak{B}$.

$\Leftarrow$ For a topological space (X, τ), if an open set subfamily $\mathfrak{B} \subset \tau$ satisfies: $\forall O \in \tau$, $\exists \mathfrak{B}_0 \subset \mathfrak{B}$, such that $G = \bigcup_{B \in \mathfrak{B}_0} B$, we prove that this open set subfamily $\mathfrak{B} \subset \tau$ is a topological base of (X, τ).

In fact, $\forall G \in \tau$, $\forall x \in G$, we have $G = \bigcup_{B \in \mathfrak{B}_0} B$ by assumption. Thus, $B \in \mathfrak{B}_0 \subset \mathfrak{B}$ such that $x \in B \subset G$, this shows that the open set subfamily $\mathfrak{B} \subset \tau$ is a topological base.

The proof is finished.

Note By Definition 3.2.5, a topological base $\mathfrak{B}$ is a part of the open set family τ of the space (X, τ), so that $\mathfrak{B}$ can be regarded as "basic open sets" in (X, τ), thus, the study for a topological space (X, τ) can be deduced to that for $(X, \mathfrak{B})$.

For example, for $\mathbb{R} = (-\infty, +\infty)$, if $(\mathbb{R}, \tau)$ is a topology of $\mathbb{R}$, and

$$\tau = \{G \subset \mathbb{R} : G \text{ is open set of } \mathbb{R}\},$$

then, it is easy to verify that $\mathfrak{B} = \{(a, b) : -\infty < a < b < +\infty\}$ is a topological base of $\mathbb{R}$.

To construct a topological base, we have the following theorem.

Theorem 3.2.6 Let X be a set, $\mathfrak{B}$ be a subset family in X. If $\mathfrak{B}$ satisfies

(1) $\bigcup_{B \in \mathfrak{B}} B = X$;

(2) $B_1, B_2 \in \mathfrak{B} \Rightarrow \forall x \in B_1 \cap B_2$, $\exists B \in \mathfrak{B}$, s. t. $x \in B \subset B_1 \cap B_2$;

then there exists unique family $\tau = \left\{O \subset X : \exists \mathfrak{B}_0 \subset \mathfrak{B}, \text{ s. t. } O = \bigcup_{B \in \mathfrak{B}_0} B\right\}$ as a topology of X with topological base $\mathfrak{B}$, such that (X, τ) becomes a topological space. Conversely, if a subset family $\mathfrak{C}$ of X is a topological base, then $\mathfrak{C}$ must satisfy the conditions (1) and (2).

Proof We prove that $\tau = \left\{O \subset X : \exists \mathfrak{B}_0 \subset \mathfrak{B}, \text{ s. t. } O = \bigcup_{B \in \mathfrak{B}_0} B\right\}$ is a topology on X with $\mathfrak{B}$ being topological base.

Firstly, we prove that τ satisfies (1), (2), (3) in the Definition 3.2.1:

(1) $X, \varnothing \in \tau$: by condition (1), we have $X = \bigcup_{B \in \mathfrak{B}} B \in \tau$; and by our appointment, $\varnothing \in \tau$.

(2) $A_1, A_2 \in \tau \Rightarrow A_1 \cap A_2 \in \tau$: by two steps.

① If $A_1 = B_1 \in \mathfrak{B}$, $A_2 = B_2 \in \mathfrak{B}$, then $B_1 \cap B_2 \in \tau$.

In fact, the condition (2) implies that "$\forall x \in B_1 \cap B_2$, $\exists B_x \in \mathfrak{B}$, s. t. $x \in B_x \subset B_1 \cap$

B_2". On the other hand, $B_1 \cap B_2 = \bigcup_{x \in B_1 \cap B_2} \{x\} \subset \bigcup_{x \in B_1 \cap B_2} B_x \subset B_1 \cap B_2$, thus $B_1 \cap B_2 = \bigcup_{x \in B_1 \cap B_2} B_x$, and $B_1 \cap B_2 \in \tau$ by the construction of τ.

② If $A_1, A_2 \in \tau$, then $A_1 \cap A_2 \in \tau$.

Since $A_1 \in \tau \Rightarrow A_1 = \bigcup_{B_1 \in \mathfrak{B}_1} B_1$ and $A_2 \in \tau \Rightarrow A_2 = \bigcup_{B_2 \in \mathfrak{B}_2} B_2$. Then

$$A_1 \cap A_2 = \Big(\bigcup_{B_1 \in \mathfrak{B}_1} B_1\Big) \cap \Big(\bigcup_{B_2 \in \mathfrak{B}_2} B_2\Big) = \bigcup_{\substack{B_1 \in \mathfrak{B}_1, \\ B_2 \in \mathfrak{B}_2}} B_1 \cap B_2.$$

By ①, if $B_1, B_2 \in \mathfrak{B}$, then $B_1 \cap B_2 = \bigcup_{x \in B_1 \cap B_2} B_x \in \tau$, thus $A_1 \cap A_2 = \bigcup_{\substack{B_1 \in \mathfrak{B}_1, \\ B_2 \in \mathfrak{B}_2}} \bigcup_{x \in B_1 \cap B_2} B_x \in \tau$.

(3) $A_\alpha \in \tau, \alpha \in \Lambda \Rightarrow \bigcup_{\alpha \in \Lambda} A_\alpha \in \tau$.

By $A_\alpha \in \tau \Rightarrow \exists B_\beta^\alpha \in \mathfrak{B}_\beta \subset \mathfrak{B}$, s. t. $B_\alpha = \bigcup_{B_\beta^\alpha \in \mathfrak{B}_\beta} B_\beta^\alpha$. Thus, $\bigcup_{\alpha \in \Lambda} A_\alpha = \bigcup_{\alpha \in \Lambda} \bigcup_{B_\beta^\alpha \in \mathfrak{B}_\beta} B_\alpha^\beta \equiv \bigcup_{B_\gamma \in \mathfrak{B}_\gamma} B_\gamma \in \tau$. Then, $(X, \tau) \equiv \big(X, \{O \subset X: \exists \mathfrak{B}_0 \subset \mathfrak{B}, \text{s. t. } O = \bigcup_{B \in \mathfrak{B}_0} B\}\big)$ becomes a topological space; and by definition, $\mathfrak{B}$ is a topological base of $\tau = \{O \subset X: \exists \mathfrak{B}_0 \subset \mathfrak{B}, \text{s. t. } O = \bigcup_{B \in \mathfrak{B}_0} B\}$.

Next, we prove that the set family $\mathfrak{B}$ satisfying (1), (2) of Theorem 3. 2. 6 determines a unique topology τ described in the above (3).

In fact, suppose that there is another topology $\tilde{\tau}$ with base $\mathfrak{B}$, then, by definition of topological base, we have

$$\forall A \in \tilde{\tau} \Rightarrow A = \bigcup_{B \in \mathfrak{B}_A} B \Rightarrow A \in \tau.$$

So that $\tilde{\tau} \subset \tau$. Conversely, since $\mathfrak{B}$ is a base of $\tilde{\tau}$, thus $\mathfrak{B} \subset \tilde{\tau}$. Take an open set $A \in \tau$, then A is a union of some sets in $\mathfrak{B}$. But since $\mathfrak{B}$ is a topological base of $\tilde{\tau}$, this implies that A is a union of some sets in $\tilde{\tau}$. Hence $A \in \tilde{\tau}$, and so that $\tau \subset \tilde{\tau}$. Combine both the above, it follows that $\tilde{\tau} = \tau$.

Finally, we prove that if a subset family $\mathfrak{C}$ of X is a topological base of topology of X, then $\mathfrak{C}$ satisfies (1) and (2).

In fact, let a subset family $\mathfrak{C}$ of X is the topological base of some topology τ^* of X. Then, by $X \in \tau^*$, we have $X = \bigcup_{C \in \mathfrak{C}_\delta \subset \mathfrak{C}} C = \bigcup_{C \in \mathfrak{C}} C$, where $\mathfrak{C}_\delta$ is a subset family of $\mathfrak{C}$, thus $\mathfrak{C}$ satisfies (1). Moreover, for $C_1, C_2 \in \mathfrak{C} \subset \tau^*$, we know C_1, C_2 are open sets in X; Then for $x \in C_1 \cap C_2$, an open set that contains x, so it is an open neighborhood of x. This implies by the definition of neighborhood that there exists $W_x \in \mathfrak{C}$, satisfying $x \in W_x \subset C_1 \cap C_2$, this implies that $\mathfrak{C}$ satisfies (2). The proof is finished.

Remark 1 Theorem 3.2.6 tells that: to construct a topology on a set, we only need to give a set family $\mathfrak{B}$ satisfying conditions (1), (2) in the theorem.

Remark 2 For a topological space (X, τ), suppose that $\mathbb{S} \subset \tau$ is a subset family of τ, and $\mathfrak{B}_0$ is the subset family generated by $\mathbb{S}$ as

$$\mathfrak{B}_0 = \left\{B = \bigcap_{j=1}^{n} S_j : \varnothing \neq S_j \in \mathbb{S} \subset \tau, 1 \leqslant j \leqslant n;\ n \in \mathbb{N}\right\}, \tag{3.2.2}$$

i.e., the elements in $\mathfrak{B}_0$ are those finite intersections of non-empty subsets of $\mathbb{S}$. If $\mathfrak{B}_0$ is a topological base of (X, τ), then $\mathbb{S}$ is said to be **"a subbase" contained in topological base** $\mathfrak{B}_0$, or simply, $\mathbb{S}$ is called **a subbase of topology** τ, or **a subbase of topological space** (X, τ).

Remark 3 A given set can be endowed different topologies, and they may have quite different properties. See the following example:

Example 3.2.3 We have three different topologies on $\mathbb{R}$: the usual topology (or standard topology), the lower topology, the K-topology.

The usual topology $(\mathbb{R}, \tau)$: $\mathfrak{B} = \{(a, b): -\infty < a < b < +\infty\}$ is the topological base of τ, denoted by $(\mathbb{R}, \mathfrak{B})$; Clearly, τ has a subbase $\mathbb{S} = \{\{(-\infty, a): a \in \mathbb{R}\} \cup \{(b, +\infty): b \in \mathbb{R}\}\}$.

The lower topology $(\mathbb{R}, \tau')$: $\mathfrak{B}' = \{[a, b): -\infty < a < b < +\infty\}$ is the topological base of τ', denoted by $(\mathbb{R}, \mathfrak{B}')$.

The K-topology $(\mathbb{R}, \tau'')$: $\mathfrak{B}'' = \left\{(a, b) - \left\{1, \frac{1}{2}, \frac{1}{3}, \cdots\right\} : -\infty < a < b < +\infty\right\}$ is the topological base of τ'', denoted by $(\mathbb{R}, \mathfrak{B}'')$.

$\mathfrak{B}, \mathfrak{B}', \mathfrak{B}''$ satisfy the conditions in Theorem 3.2.6, thus they are topological bases of $\mathbb{R}$, such that $\mathbb{R}$ has three different topological structures.

2) **First countability axiom, second countability axiom**

Definition 3.2.6 (countability axiom) Let (X, τ) be a topological space. If X has a countably topological base $\mathfrak{B}$, then (X, τ) is said to satisfy **the second countability axiom.** If for every point $x \in X$, there exists a countably topological base $\mathfrak{B}_x$ of x in X, then (X, τ) is said to satisfy **the first countability axiom.**

Example 3.2.4 One-dimensional real Euclidean space $(\mathbb{R}, \tau)$ satisfies both the first and the second countability axiom.

In fact, for an open set $O \subset \mathbb{R}$ in $(\mathbb{R}, \tau)$, the n $\forall x \in O$, there exists $\eta > 0$, such that $(x - \eta, x + \eta) \subset O$. Thus the open subset family of τ can be taken as

$$\mathfrak{B}_x = \{B = (x - a_x, x + b_x) : a_x, b_x \in \mathbb{Q}\},$$

with $B=(x-a_x, x+b_x)\subset(x-\eta, x+\eta)\subset O$, here $\mathbb{Q}$ is the rational number set. Hence, $x\in B=(x-a_x, x+b_x)\subset O$, so that $(\mathbb{R}, \tau)$ satisfies the first countability axiom.

Moreover, $\mathfrak{B}=\{\mathfrak{B}_x : x\in\mathbb{Q}\}$ is a countably topological base of $(\mathbb{R}, \tau)$, so that $(\mathbb{R}, \tau)$ satisfies the second countability axiom.

It is not difficult to prove that **any metric space (X, ρ) satisfies the first countability axiom.** The proof is left to exercise.

A topological space (X, τ) satisfying the second countability axiom, then it satisfies the first countability axiom; but the inverse assertion is not necessarily true. Moreover, there are topological spaces which do not satisfy both the first and the second countability axioms.

2. Separability of topological spaces

1) **Definitions of separability of topological spaces**

Definition 3.2.7 (separability) Let (X, τ) be a topological space. If there exists at most countable set $A=\{a_1, a_2, \cdots, a_n, \cdots\}\subset X$, such that A is dense in X, i.e., $X=\overline{A}$, then X is said to be **separable**; otherwise, **unseparable.**

The Euclidean space $\mathbb{R}^n$ is separable, since $\mathbb{R}^n=\overline{\mathbb{Q}^n}$.

Example 3.2.5 The continuous function space $C[a, b]$ is a topological space, and it is a metric space with distance $\rho(f, g)=\max\limits_{x\in[a,b]}|f(x)-g(x)|$. We prove that it is separable.

In fact, the set of polynormials with rational number coefficients

$$P_0=\{p_n : p_n(x)=r_nx^n+\cdots+r_1x+r_0, x\in[a, b], r_j\in\mathbb{Q}\}$$

is a countable dense subset in $C([a, b])$. Since by **Bernstein (Бернштейн) Theorem**: "for any $f\in C([a, b])$, there exists polynormial sequence $\{p_n(x)\}_{n\geqslant 1}\subset P_0$ with rational coefficients, such that $p_n(x)$ is convergent uniformly to $f(x)$ on $[a, b]$". Then $C([a, b])=\overline{P_0}$. So that the metric space $C([a, b])$ is separable.

Similarly, by Weierstrass Theorem (see section 7.2), we can prove that for $1\leqslant p<+\infty$, it holds $L^p([a, b])=\overline{P_0}$, thus $L^p([a, b])$ is separable.

The following example is an unseparable topological space.

Example 3.2.6 The bounded function space on $[0, 1]$, denoted by

$$\mathbb{B}([0,1])=\{f : [0,1]\to\mathbb{R}, \exists M>0, \text{s. t. } |f(x)|\leqslant M\},$$

with distance $\rho(f, g)=\sup\limits_{x\in[0,1]}|f(x)-g(x)|$, is a topological space (metric space), it is unseparable.

In fact, if $A=\{f_n\in\mathbb{B}([0,1]) : n\in\mathbb{Z}^+\}$ is any countable subset of $\mathbb{B}([0,1])$, we prove that A is not dense in $\mathbb{B}([0,1])$. Only need to prove that: there exists $g\in\mathbb{B}([0,1])$, such that $g\notin\bar{A}$.

Take $[0,1]=\left[0,\frac{1}{2}\right)\cup\left[\frac{1}{2},\frac{2}{3}\right)\cup\cdots\cup\left[1-\frac{1}{n},1-\frac{1}{n+1}\right)\cup\cdots$, and a decomposition of it could be

$$[0,1]=\{1\}\cup\left(\bigcup_{n=1}^{+\infty}\left[1-\frac{1}{n},1-\frac{1}{n+1}\right)\right).$$

Thus, for $f_n\in A, n\in\mathbb{Z}^+$, we construct a function $g:[0,1]\to\mathbb{R}$, satisfying

$$g(t)=\begin{cases}2, & t\in\left[1-\frac{1}{n},1-\frac{1}{n+1}\right),\ \forall t'\in\left[1-\frac{1}{n},1-\frac{1}{n+1}\right), |f_n(t')|\leqslant 1\\ 0, & t\in\left[1-\frac{1}{n},1-\frac{1}{n+1}\right), \exists t''\in\left[1-\frac{1}{n},1-\frac{1}{n+1}\right), |f_n(t'')|>1\\ 0, & t=1\end{cases},\tag{3.2.3}$$

then $g\in\mathbb{B}([0,1])$, and $\forall n\in\mathbb{Z}^+$, it holds $\rho(f_n,g)=\sup\limits_{x\in[0,1]}|f_n(x)-g(x)|\geqslant 1$. The ball

$$B(g,1)=\{f\in\mathbb{B}([0,1]):\rho(f,g)<1\}$$

and $A=\{f_n\in\mathbb{B}([0,1])\}$ have empty intersection: $A\cap B(g,1)=\varnothing$, so that $g\notin\bar{A}$, this implies that $\bar{A}\underset{\neq}{\subset}\mathbb{B}([0,1])$, and also that $\mathbb{B}([0,1])$ is an unseparable topological space.

2) **Separable and unseparable topological spaces**

By definition in 1), we can classify topological spaces into separable and unseparable ones.

3.3 Continuous Mappings on Topological Spaces

3.3.1 Mappings between topological spaces, continuity of mappings

1. Continuous mappings

The study of continuous mapping between topological spaces is one of main goals in point set topology, since a continuous mapping keeps lots of properties of the topological space. To define the continuity of a mapping from one topological space (X,τ) to the other one (Y,υ), we are motivated by a necessary and sufficient condition in the courses of *advance calculus* and *theory of real variable functions*: "a function $f:\mathbb{R}\to\mathbb{R}$ is continuous, if and only if the inverse image set $f^{-1}(G)\subset\mathbb{R}$ of open set $G\subset\mathbb{R}$ is an open set."

Definition 3.3.1 (**continuity of a mapping**) Let (X,τ), (Y,υ) be topological spaces, $f: X \to Y$ be a mapping from X to Y. Then, f is said to be **a continuous mapping** from X to Y, if any open set $O \in \upsilon$ in Y implies that the image set $G = f^{-1}(O) \in \tau$ is open in X; i. e., the inverse image set $G = f^{-1}(O)$ of the open set $O \in \upsilon$ in Y is open in X.

The following theorem contains equivalent properties of a continuous mapping.

Theorem 3.3.1 Let (X,τ), (Y,υ) be topological spaces, $f: X \to Y$ be a mapping from X to Y. Then the following statements are equivalent:

(1) f is a continuous mapping from X to Y;

(2) An inverse image $f^{-1}(O)$ of an open set $O \in \upsilon$ in Y is an open set in X;

(3) An inverse image $f^{-1}(F)$ of a closed set $F \in \upsilon$ in Y is a closed set in X;

(4) For any point $x \in X$ and the corresponding $y = f(x) \in Y$, it holds

$$\forall V_y \in \upsilon \text{ with } y \in V_y,\ \exists U_x \in \tau,\ \text{s. t. } f(U_x) \subset V_y;$$

(5) An inverse image $f^{-1}(V)$ of an open neighborhood V of $y \in Y$ in Y is an open neighborhood of $x \in X$ in X with $f(x) = y$;

(6) $B \subset Y$ implies $\overline{f^{-1}(B)} \subset f^{-1}(\bar{B})$;

(7) $A \subset X$ implies $f(\bar{A}) \subset \overline{f(A)}$.

Proof (1)⇔(2) By definition.

(2)⇔(3) Take any closed set $F \subset Y$ in Y, by definition, $Y \backslash F \subset Y$ is an open set, i. e., $F^c = Y \backslash F \in \upsilon$. Thus, by (2), $f^{-1}(Y \backslash F) \in \tau$. Since $f^{-1}(Y \backslash F) = f^{-1}(Y) \backslash f^{-1}(F) = X \backslash f^{-1}(F)$, and $f^{-1}(Y \backslash F)$ is an open set in X, thus $f^{-1}(F)$ is a closed set in X, this is (3); And vise versa. Then, (2) ⇔ (3) is proved.

To prove (2)⇔(4), we deduce as follows.

(2) ⇒ (4) Let $x \in X$, $y = f(x) \in Y$. Since $\forall V_y \in \upsilon$, $y = f(x) \in V_y$, by (2), $f^{-1}(V_y)$ is an open set in X, i. e., $f^{-1}(V_y) \in \tau$, and $x \in f^{-1}(V_y)$; so that $f^{-1}(V_y)$ is an open set containing x, that is, $\exists U_x \in \tau$, s. t. $U_x \subset f^{-1}(V_y)$; hence, $f(U_x) \subset f(f^{-1}(V_y)) \subset V_y$, this is (4).

(4) ⇒ (2) For $O \in \upsilon$, we prove that $f^{-1}(O) \in \tau$ is an open set. In fact, $\forall x \in f^{-1}(O)$, it holds the corresponding $y = f(x) \in f(f^{-1}(O)) \subset O$. For this $y \in O \in \tau$, by (4), $\exists U_x \in \tau$, $x \in U_x$, s. t. $f(U_x) \subset O$, this implies that $x \in U_x \subset f^{-1}(f(U_x)) \subset f^{-1}(O)$; hence it implies that $f^{-1}(O)$ is an open set in X, i. e., $f^{-1}(O) \in \tau$, this is (2). Then, (2) ⇔ (4) is proved.

To prove (2)⇔(5), we deduce as follows.

(2) ⇒ (5) For $y \in Y$, there is $x \in X$, such that $y = f(x)$. Then, $\forall U_y \in \upsilon$, $y = f(x) \in$

U_y, by (2); thus $f^{-1}(U_y)$ is a neighborhood of $x\in X$; i. e., (5) holds.

(5) $\Rightarrow$(2) $\forall O\in \upsilon$, for any $y\in O$, then O is an open neighborhood of y, by (5), $f^{-1}(O)\subset X$ is an open neighborhood of $x\in f^{-1}(O)\subset X$ with $x\in X$ satisfying $y=f(x)$. Thus, for this $f^{-1}(O)$, we have $\exists G\subset X, G\in\tau$, s. t. $x\in G\subset f^{-1}(O)$; hence $f^{-1}(O)$ is an open set in X, then $f^{-1}(O)\in\tau$, this is (2). Then, (2) $\Leftrightarrow$(5) is proved.

Next, we prove (3) $\Leftrightarrow$ (6).

(3) $\Rightarrow$(6) $\forall B\subset Y\Rightarrow f^{-1}(\bar{B})\subset X$ is a closed set (by(3)) $\Rightarrow$ $\overline{f^{-1}(B)}\subset f^{-1}(\bar{B})$ (since $B\subset\bar{B}$ implies $f^{-1}(B)\subset f^{-1}(\bar{B})$, again implies $\overline{f^{-1}(B)}\subset\overline{f^{-1}(\bar{B})}$) $\Rightarrow$ (6) holds;

(6) $\Rightarrow$(3) Each closed subset $F\subset Y\Rightarrow \overline{f^{-1}(F)}\subset f^{-1}(\bar{F})=f^{-1}(F)$ (by(6)) $\Rightarrow$ $\overline{f^{-1}(F)}=f^{-1}(F)$ $\Rightarrow$ $f^{-1}(F)$ is a closed set $\Rightarrow$ (3) holds. Then, (3) $\Leftrightarrow$ (6) is proved.

Finally, we prove (6) $\Leftrightarrow$ (7).

(6) $\Rightarrow$(7) $\forall A\subset X$ implies $f(A)\subset Y\Rightarrow f(A)\subset\overline{f(A)}\Rightarrow$ $\overline{f^{-1}(f(A))}\subset f^{-1}(\overline{f(A)})$ (by(6)) $\Rightarrow f(\bar{A})\subset\overline{f(A)}$ (since $A\subset f^{-1}(f(A))$ implies $\bar{A}\subset\overline{f^{-1}(f(A))}$, again implies $f(\bar{A})\subset f(\overline{f^{-1}(f(A))})\subset f(f^{-1}(\overline{f(A)}))\subset\overline{f(A)}$) $\Rightarrow$ (7) holds;

(7) $\Rightarrow$(6) $\forall B\subset Y$ implies $f^{-1}(B)\subset X$ $\Rightarrow$ take $A\equiv f^{-1}(B)\subset X$ in (7), then $f(\overline{f^{-1}(B)})\subset\overline{f(f^{-1}(B))}\subset\bar{B}$ (by $f(f^{-1}(B))\subset B\subset\bar{B}$, thus $\overline{f(f^{-1}(B))}\subset\bar{B}$) $\Rightarrow$ $f^{-1}(f(\overline{f^{-1}(B)}))\subset f^{-1}(\bar{B})$ (by $f^{-1}(f(\overline{f^{-1}(B)}))=\overline{f^{-1}(B)}$) $\Rightarrow$ $\overline{f^{-1}(B)}\subset f^{-1}(\bar{B})\Rightarrow$ (6) holds. Thus, (6)$\Leftrightarrow$(7) is proved.

The proof of the theorem is complete.

In Example 3. 2. 3, three topologies on Euclidean space $\mathbb{R}$ are given. We show that, for the first topology $\mathfrak{B}$ and the second one $\mathfrak{B}'$, the identity mapping I: $(\mathbb{R}, \mathfrak{B})\rightarrow(\mathbb{R}, \mathfrak{B}')$ with $I(x)=x$ is not continuous. Because for any open set $O=[a, b)\in\mathfrak{B}'$, its inverse image $I^{-1}([a, b))=[a, b)$ is not an open set in topology $\mathfrak{B}$, so that by Theorem 3. 3. 1, the mapping I: $(\mathbb{R}, \mathfrak{B})\rightarrow(\mathbb{R}, \mathfrak{B}')$ is discontinuous.

Definition 3. 3. 2 (**homeomorphism, open** (**closed**) **mapping**) Let (X, τ), (Y, υ) be topological spaces. If there exists a mapping f: $X\rightarrow Y$ from X to Y, such that f is one-one, f and f^{-1} are continuous, i. e., f and f^{-1} both are **single-valued** and **continuous**, then f is said to be **a homeomorphism** from X to Y; also say that: X and Y are **topological homeomorphic**.

A property is said to be **a topological invariant property** if it is invariant under a topological homeomorphism; or, a quantity which is invariant under topological homeomorphism is said to be **a topological invariant quantity**.

If a mapping $f: X \to Y$ maps open (closed) sets in X to open (closed) sets in Y, then f is said to be **an open (closed) mapping from X to Y**.

We have the following equivalent theorem for homeomorphic, open (closed) mappings.

Theorem 3.3.2 Let (X,τ), (Y,υ) be topological spaces. If $f: X \to Y$ is a one-one and surjective mapping from X onto Y (injective and surjective), then the following two statements are equivalent:

(1) f is a homeomorphism from X onto Y;

(2) f is a continuous, open (closed) mapping from X onto Y.

Proof In fact, by continuity of f^{-1}, we have that

$$\forall U_x \in \tau \text{ implies } [(f^{-1})^{-1}](U_x) = f(U_x) \in \upsilon;$$

Hence, f maps an open set to an open set, i. e., f is an open mapping; And vise versa. Similarly, for a closed set.

Theorem 3.3.3 Let (X,τ), (Y,υ), (Z,ω) be three topological spaces. If $f: X \to Y$ is a continuous mapping from X to Y and $g: Y \to Z$ is a continuous mapping from Y to Z, then the compound mapping $g \circ f: X \to Z$ is continuous from X to Z.

Proof Let any $x_0 \in X$ with $y_0 = f(x_0) \in Y$, and $z_0 = g(y_0) = g(f(x_0)) \in Z$. We prove that compound mapping $g \circ f: X \to Z$ is continuous.

$\forall W_{z_0} \in \omega$, by continuity of $g: Y \to Z$, then $\exists V_{y_0} \in \upsilon$, s. t. $g(V_{y_0}) \subset W_{z_0}$. Moreover, by continuity of $f: X \to Y$, for the above $V_{y_0} \in \upsilon$, hence $\exists U_{x_0} \in \tau$, s. t. $f(U_{x_0}) \subset V_{y_0}$. Thus, $\forall W_{z_0} \in \omega$, then $\exists U_{x_0} \in \tau$, s. t. $g(f(U_{x_0})) \subset g(V_{y_0}) \subset W_{z_0}$, this implies that $g \circ f: X \to Z$ is continuous at x_0. Moreover, since x_0 is any point of X, so the continuity of compound mapping $g \circ f: X \to Z$ on X is proved.

Example 3.3.1 Let τ_1, τ_2 be two different topological structures on X, the mapping $I: (X,\tau_1) \to (X,\tau_2)$ be identical with $I(x) = x, x \in X$. Then $I: X \to X$ is a continuous mapping, if and only if $\tau_2 \subset \tau_1$.

Proof By the necessary and sufficient condition in Theorem 3.3.1(2), $\forall O \in \tau_2$, $\exists G \in \tau_1$, s. t. $I(G) \subset O$; but $I(G) = G$, thus $G \subset O$ implies $O \in \tau_1$, this is $\tau_2 \subset \tau_1$. And vise versa.

2. Limits

Recall that in *advanced calculus*, the concept of "limit" is introduced firstly, and then that of "continuity". However, we introduce the concept of "continuity" firstly, and then that of "limit" in our course *foundation of modern applied mathematics*.

There are several ways to define "limit", such as by "net", or "filter". We start from so-called "sequence", and then "limit".

Definition 3.3.3 (**sequence, limit**) Let (X,τ) be a topological space. A mapping S: $j\in\mathbf{Z}^+\to x_j\in X, j\in\mathbf{Z}^+$, is said to be **a sequence** in X, denoted by $\{x_j\}_{j\in\mathbf{Z}^+}$, or for simply, by $\{x_j\}$. If there exists $x\in X$, such that for every open set $G\in\tau$ containing x with $x\in G$, satisfies: "$\exists N\in\mathbf{Z}^+$, s. t. $j>N$ implies $x_j\in G$", then x is said to be **a limit of sequence** $\{x_j\}_{j\in\mathbf{Z}^+}$, also say that **the sequence** $\{x_j\}_{j\in\mathbf{Z}^+}$ **converges to** x, denoted by $\lim\limits_{j\to+\infty} x_j=x$.

Definition 3.3.4 (**subsequence**) Let (X,τ) be a topological space. Let S: $\mathbf{Z}^+\to X$ and S_1: $\mathbf{Z}^+\to X$ be two sequences in (X,τ), denoted by $\{x_j\}_{j\in\mathbf{Z}^+}$ and $\{x_k\}_{k\in\mathbf{Z}^+}$, respectively. If there exists a strictly increasing sequence W: $j\in\mathbf{Z}^+\to k=W(j)\in\mathbf{Z}^+$ (i. e., for $\forall j_1, j_2\in\mathbf{Z}^+$ with $j_1<j_2$ implies $k_1=W(j_1)<k_2=W(j_2)$), such that $x_k=x_{W(j)}$, then $\{x_k\}_{k\in\mathbf{Z}^+}$ is said to be **a subsequence of** $\{x_j\}_{j\in\mathbf{Z}^+}$. Or, "$\forall j\in\mathbf{Z}^+\Rightarrow W(j)=W_j\in\mathbf{Z}^+$, $x_{W(j)}\in\{x_j\}_{j\in\mathbf{Z}^+}$". On the other words, the j-point in subsequence $\{x_{W(j)}\}_{j\in\mathbf{Z}^+}$ is just the $W(j)$-point in sequence $\{x_j\}_{j\in\mathbf{Z}^+}$. A subsequence is always denoted by $\{x_{j_k}\}_{k\in\mathbf{Z}^+}$ or $\{x_{j_k}\}$.

Theorem 3.3.4 Let (X,τ) be a topological space, $\{x_j\}$ be a sequence in X.

(1) If $\{x_j\}$ is a constant sequence, i. e., $\forall x_j=x\in X$, then $\lim\limits_{j\to+\infty} x_j=\lim\limits_{j\to+\infty} x=x$;

(2) If $\{x_j\}$ converges to $x\in X$, then any subsequence $\{x_{j_k}\}$ of $\{x_j\}$ converges to x;

(3) If (X,τ), (Y,υ) are two topological spaces, f: $X\to Y$ is a continuous mapping from X to Y, and a sequence $\{x_j\}$ in X converges to $x\in X$, then the sequence $\{f(x_j)\}$ in Y converges to $f(x)\in Y$.

Proof (1), (2) are clear.

We prove(3), that is to prove "$\lim\limits_{j\to+\infty} x_j=x$ implies $\lim\limits_{j\to+\infty} f(x_j)=f(x)$".

For $y=f(x)$, $x\in X$, take $V_y\in\upsilon$ as an open neighborhood of $y=f(x)$, by the continuity of mapping f: $X\to Y$, thus $\exists U_x\in\tau$, s. t. $f(U_x)\subset V_y$. On the other hand, by $\lim\limits_{j\to+\infty} x_j=x\in U_x$, we have: $\exists N\in\mathbf{Z}^+$, s. t. $j>N$ implies $x_n\in U_x$. Then, $j>N$ implies $f(x_j)\in f(U_x)\subset V_y$. This is $\lim\limits_{j\to+\infty} f(x_j)=f(x)$.

Remark 1 In the *advance calculus*, the inverses of (2), (3) in above theorem are true, i. e., they are necessary and sufficient. For example, the (3) is:

"Let (X,ρ) be a metric space, (Y,υ) be a topological space, f: $X\to Y$ be a mapping from X to Y. Then f is continuous, **iff** the sequence $\{x_j\}$ in X converges to $x\in X$ implies the sequence $\{f(x_j)\}$ in Y converges to $f(x)\in Y$".

However, in a general topological space, the inverse of (3) is not necessarily true.

Remark 2 For a topological space (X,τ), if $\{x_j\}_{j\in\mathbf{Z}^+}$ is a sequence in X, then it may have limit, or may not have. Moreover, limit, if exists, is not necessaryly unique. For example, if (X,τ) is a topological space containing at least two points. Take $\tau=\{X,\varnothing\}$, then in (X,τ), any given sequence $\{x_n\}\subset X$, it holds $\lim\limits_{n\to+\infty} x_n=x$ for every $x\in X$. This means that in $\tau=\{X,\varnothing\}$, any sequence may have more than two different limits.

3.3.2 Subspaces, product spaces, quotient spaces

1. Comparison of topologies, coarser and finer topologies

Definition 3.3.5 (coarser topology, finer topology) Let τ_1,τ_2 be topologies on a given set X, such that (X,τ_1), (X,τ_2) to be "two" topological spaces on the same set X. If $\tau_1\subset\tau_2$, i.e., regarded as sets, τ_1 is contained in τ_2, then τ_1 is said to be **coarser** (or **weaker**) **than** τ_2; or τ_2 is **finer** (or **stronger**) **than** τ_1.

We have **two trivial topologies** on a set X: ① (X,τ_1) with $\tau_1=\{X,\varnothing\}$; ② (X,τ_2) with $\tau_2=\mathfrak{F}(X)=\{A: A\subset X\}$. This τ_1 is **the coarsest topology** on X (or the weakest); And τ_2 is **the finest topology** on X (or the strongest).

For the usual topology $(\mathbb{R},\tau)$ with base $\mathfrak{B}=\{(a,b): -\infty<a<b<+\infty\}$, it is clear that τ is a **non-trivial topology on** $\mathbb{R}$, and $\tau_1=\{\mathbb{R},\varnothing\}$ is the coarsest topology on $\mathbb{R}$, as well as $\tau_2=\mathfrak{F}(\mathbb{R})=\{A: A\subset\mathbb{R}\}$ is the finest one on $\mathbb{R}$, respectively.

2. Subspaces

Definition 3.3.6 (subspace of a topological space) Let (X,τ) be a topological space, X_0 be a subset of X. Denote

$$\tau_0=\{G\subset X_0: G=O\cap X_0, \forall O\in\tau\}, \tag{3.3.1}$$

it is easy to verify that τ_0 satisfies (1)-(3) in Definition 3.2.1, thus τ_0 is an open set family on X_0 such that (X_0,τ_0) becomes a topological space (Fig. 3.3.1). Then, (X_0,τ_0) is said to be **a subspace** of (X,τ); τ_0 is called **the relative topology** of τ, or **the subtopology of** τ.

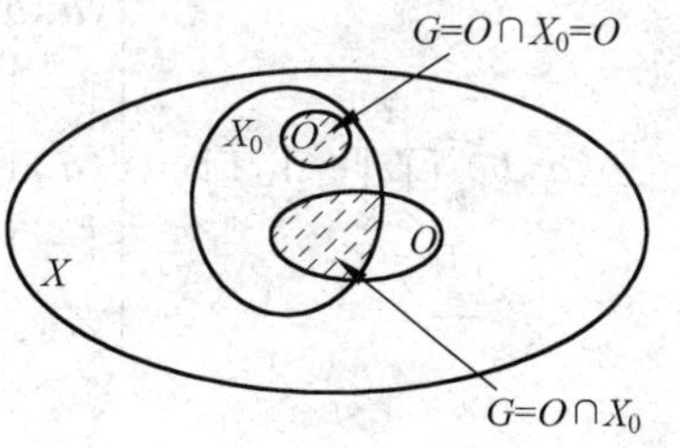

Fig. 3.3.1 Subspace

If $\mathfrak{B}$ is a topological base of topological space (X,τ), then $\mathfrak{B}_0=\{B\cap X_0: B\in\mathfrak{B}\}$ is a topological base of the subspace (X_0,τ_0).

Example 3.3.2 $\mathbb{R}$, $\mathbb{R}^2$ are subspaces of $\mathbb{R}^3$. For $(\mathbb{R}^3,\tau_3)$, we take the base

$$\mathfrak{B}_3=\left\{G: G=\left\{(x_1,x_2,x_3)\in\mathbb{R}^3: \sqrt{x_1^2+x_2^2+x_3^2}<q\right\}, q\in\mathbb{Q}\right\};$$

for $(\mathbb{R}^2, \tau_2)$, there is a subbase in $\mathbb{R}^2$:

$$\mathfrak{B}_2 = \{G: G = \{(x_1, x_2, 0) \in \mathbb{R}^2: \sqrt{x_1^2 + x_2^2} < q\}, q \in \mathbb{Q}\};$$

for $(\mathbb{R}^1, \tau_1)$, there is a subbase in $\mathbb{R}$:

$$\mathfrak{B}_1 = \{G: G = \{(x_1, 0, 0) \in \mathbb{R}^1: |x_1| < q\}, q \in \mathbb{Q}\}.$$

The following is a structure theorem of subtopology.

Theorem 3.3.5 Let (X, τ) be a topological space, $X_0 \subset X$ be a subset of X. Then, for the subspace (X_0, τ_0),

(1) $G \subset X_0$ is an open set in $X_0 \Leftrightarrow$ there exists an open set $O \in \tau$ in X, such that $G = O \cap X_0$;

(2) $C \subset X_0$ is a closed set in $X_0 \Leftrightarrow$ there exists a closed set $F \subset X$ in X, such that $C = F \cap X_0$;

(3) If $\bar{C}_{X_0}$ is the closure of subset $C \subset X_0$ with respect to X_0, then $\bar{C}_{X_0} = \bar{C}_X \cap X_0$, where $\bar{C}_X$ is the closure of $C \subset X$ with respect to X;

(4) Subset $A \subset X_0$ satisfies $\mathfrak{C}_{X_0} A = (\mathfrak{C}_X A) \cap X_0$, i. e., the complementary $\mathfrak{C}_{X_0} A = X_0 \setminus A$ of subset $A \subset X_0$ with respect to subspace X_0 is equal to the intersection of complementary $\mathfrak{C}_X A = X \setminus A$ about X with subset X_0 (Fig. 3.3.2).

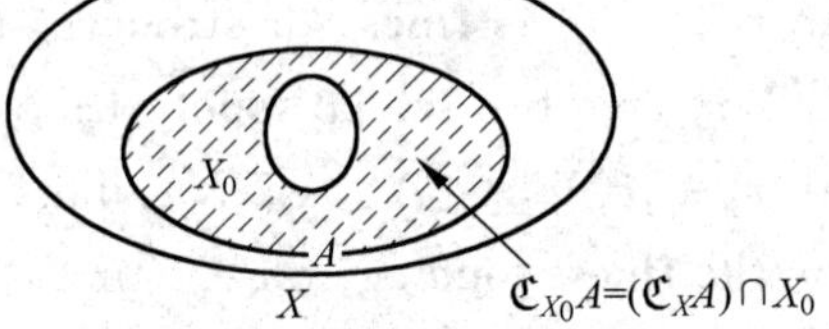

Fig. 3.3.2 Complementary set

Example 3.3.3 Let $(\mathbb{R}, \tau)$ be the one-dimensional Euclidean space. Determine the subtopology of subset $X_0 = [-1, 1]$ of $\mathbb{R}$.

Take $\mathfrak{B} = \{(a, b): -\infty < a < b < +\infty\}$ as a topological base of $(\mathbb{R}, \tau)$. Then, take

$$\mathfrak{B}_{[-1,1]} = \{G: G = (a, b) \cap [-1, 1], -\infty < a < b < +\infty\}$$

as the topological base of $X_0 = [-1, 1]$. We have $(a, b) \cap [-1, 1]$ in $\mathfrak{B}_{[-1,1]}$:

$$(a, b) \cap [-1, 1] = \begin{cases} (a, b), & a \in (-1, 1),\ b \in (-1, 1), \\ [-1, b), & a \notin [-1, 1],\ b \in (-1, 1), \\ (a, 1], & a \in (-1, 1),\ b \notin [-1, 1], \\ [-1, 1], & a, b \notin [-1, 1],\ (a, b) \cap [-1, 1] \neq \varnothing, \\ \varnothing, & a, b \notin [-1, 1],\ (a, b) \cap [-1, 1] = \varnothing. \end{cases} \tag{3.3.2}$$

Thus, in the subtopology $(X_0, \tau_0) = ([-1, 1], \tau_0)$, the open sets are quite different from those of in $\mathbb{R}$. However, if $G \subset X_0$, then $G \in \tau$ implies $G \in \tau_0$.

The **identity mapping** $I: X_0 \to X$ with $x = I(x), x \in X_0 \subset X$, plays an important role in the *point set topology*. The following theorem of continuity of $I: X_0 \to X$ holds.

Theorem 3.3.6 Let (X, τ) be a topological space, (X_0, τ_0) be the subspace of

(X,τ). Then the identity mapping

$$I: X_0 \to X,\ I(x)=x,\quad x \in X_0 \subset X \tag{3.3.3}$$

is one-one, continuous mapping from X_0 into X; Moreover, the subtopology (X_0,τ_0) is the coarsest topology on $X_0 \subset X$ such that the identity mapping $I: X_0 \to X$ is continuous.

Proof $\tau_0=\{G\subset X_0: G=O\cap X_0,\quad \forall O\in\tau\}$ is the subtopology on (X_0,τ_0), and $I: X_0\to X$ satisfies: $x\in X_0 \Rightarrow I(x)\in X$. Thus, $\forall O\in\tau$ implies $I^{-1}(O)=O\cap X_0$ by the property of I. So, $I^{-1}(O)$ is an open set in X_0, i. e., $I^{-1}(O)\in\tau_0$. This implies that $I: X_0\to X$ is a continuous mapping from X_0 into X.

Moreover, if there is an other topology (X_0,υ) on subset X_0, such that $I: (X_0,\upsilon)\to(X,\tau)$ is continuous, then, υ must contain all open sets in $\tau_0=\{G\subset X_0: G=O\cap X_0, O\in\tau\}$, that is $\tau_0\subset\upsilon$, and this means that τ_0 is the coarsest topology for keeping the continuity of $I: X_0\to X$.

3. Product spaces

Definition 3.3.7 (product space of topological spaces) Let (X,τ), (Y,υ) be two topological spaces. The set $X\times Y=\{(x,y): x\in X, y\in Y\}$ is said to be **a product set of X and Y**. Define

$$\mathfrak{B}_{\tau\times\upsilon}=\{G\times O: G\in\tau, O\in\upsilon\} \tag{3.3.4}$$

as a topological base of $X\times Y$, that is, open sets in $\mathfrak{B}_{\tau\times\upsilon}$ are defined as product sets which are as $G\times O$ with open sets $G\in\tau$ and $O\in\upsilon$.

We verify: $\mathfrak{B}_{\tau\times\upsilon}$ satisfies (1), (2) in Theorem 3.2.6. In fact,

(1) Take $G=X\in\tau$, $O=Y\in\upsilon$, thus $X\times Y=G\times O\in\mathfrak{B}_{\tau\times\upsilon}$;

(2) If $A=G_1\times O_1\in\mathfrak{B}_{\tau\times\upsilon}$, $B=G_2\times O_2\in\mathfrak{B}_{\tau\times\upsilon}$, then,

$$A=G_1\times O_1,\ G_1\in\tau, O_1\in\upsilon;\quad B=G_2\times O_2,\ G_2\in\tau, O_2\in\upsilon.$$

So that

$$A\cap B=(G_1\times O_1)\cap(G_2\times O_2)=(G_1\cap G_2)\times(O_1\cap O_2)\in\mathfrak{B}_{\tau\times\upsilon}.$$

Endow the topology $\tau\times\upsilon$ with base $\mathfrak{B}_{\tau\times\upsilon}$ in (3.3.4) to product set $X\times Y$, then $X\times Y$ becomes a topological space, and is said to be **the product space of X and Y with product topology** $\tau\times\upsilon$ of τ and υ, denoted by $(X\times Y,\tau\times\upsilon)$.

Example 3.3.4 $\mathbb{R}^3$ is the product set of $\mathbb{R}^2$ and $\mathbb{R}$, i. e., $\mathbb{R}^3=\mathbb{R}^2\times\mathbb{R}$. The topology of $\mathbb{R}^3$ can be regarded as a product topology $\mathbb{R}^2\times\mathbb{R}$. The concept of product spaces can be generalized to finite spaces:

$$(X_1,\tau_1),\cdots,(X_k,\tau_k)\Rightarrow(X,\tau)=(X_1\times\cdots\times X_k,\tau_1\times\cdots\times\tau_k).$$

Definition 3.3.8 (projective mapping) Let $(X_1,\tau_1),\cdots,(X_k,\tau_k)$ be k topological spaces, and $(X, \tau) = (X_1 \times \cdots \times X_k, \tau_1 \times \cdots \times \tau_k)$ be the product space. The mapping

$$\mathrm{Pr}_j : X \to X_j, \quad \mathrm{Pr}_j(x_1, \cdots, x_k) = x_j, \quad j = 1, 2, \cdots, k \tag{3.3.5}$$

with $(x_1, \cdots, x_k) \in X_1 \times \cdots \times X_k$ is said to be **a projective mapping from X to X_j**, $j = 1, 2, \cdots, k$; And X_j $(j = 1, 2, \cdots, k)$ is said to be **the projective space** of X.

The following theorem describes the continuity of projective mappings.

Theorem 3.3.7 Let (X, τ) be a product space $X = X_1 \times X_2 \times \cdots \times X_k$ with a base in (3.3.4). Then projective mappings $\mathrm{Pr}_j : X \to X_j$, $j = 1, 2, \cdots, k$ are continuous, surjective, and open mappings from X onto X_j, $j = 1, 2, \cdots, k$; Moreover, this topology (X, τ) is the coarsest topology such that every projective mapping Pr_j, $j = 1, 2, \cdots, k$, is continuous.

Proof We only prove for $k = 2$. Let $(X, \tau) = (X_1 \times X_2, \tau_1 \times \tau_2)$, $\mathrm{Pr}_1 : (x_1, x_2) \to x_1$, $\mathrm{Pr}_2 : (x_1, x_2) \to x_2$. And only need to prove for $\mathrm{Pr}_1 : (x_1, x_2) \to x_1$.

Continuity of $\mathrm{Pr}_1 : (x_1, x_2) \to x_1$: take any open set $G_1 \in \tau_1$, then $(\mathrm{Pr}_1)^{-1}(G_1) = G_1 \times X_2$ is an open set in (X, τ), so that for Pr_1, the inverse image $(\mathrm{Pr}_1)^{-1}(G_1) \in \tau$ is an open set $G_1 \times X_2$ in space $X_1 \times X_2$, this implies the continuity of Pr_1.

Surjective for $\mathrm{Pr}_1 : (x_1, x_2) \to x_1$: for any $x_1 \in X_1$, there exists $(x_1, x_2) \in X = X_1 \times X_2$, such that $\mathrm{Pr}_1(x_1, x_2) = x_1$, this means Pr_1 is surjective.

Open mapping for $\mathrm{Pr}_1 : (x_1, x_2) \to x_1$: take any $G = G_1 \times G_2 \in \tau = \tau_1 \times \tau_2$, then $\mathrm{Pr}_1(G) = G_1$, so it is an open mapping.

The coarsest topology for continuity of Pr_1: in fact, Pr_1 is continuous in the topology of $(X_1 \times X_2, \tau_1 \times \tau_2)$, this implies "$\forall G_1 \in \tau_1 \Rightarrow (\mathrm{Pr}_1)^{-1}(G_1) \in \tau$". If there exists another topology $(X, \tilde{\tau})$, such that Pr_1 is continuous, then for the above $G_1 \in \tau_1$, it must have $(\mathrm{Pr}_1)^{-1}(G_1) \in \tilde{\tau}$. So that $\tau \subset \tilde{\tau}$. Thus $(X, \tau) = (X_1 \times X_2, \tau_1 \times \tau_2)$ is the coarsest topology such that projective mapping Pr_1 is continuous. Similarly, for that of Pr_2. The proof is complete.

4. Quotient spaces

Definition 3.3.9 (equivalence) Let (X, τ) be a topological space. Two elements $x \in X$ and $y \in X$ are said to be **equivalent**, denoted by $x \sim y$, if they satisfy an **equivalence relationship** $\sim$ with the following rules:

(1) Reflexive: $x \sim x$;

(2) Symmetric: $x \sim y \Leftrightarrow y \sim x$;

(3) Transitive: $x \sim y$, $y \sim z \Rightarrow x \sim z$.

Denote by $X_a = \{x \in X : x \sim a\}$, the set of all elements in X which are equivalent to $a \in X$, and X_a is said to be **an equivalence class of** $a \in X$, denoted by $X_a \equiv [a]$.

Definition 3.3.10 (quotient space) The set of all equivalence classes is denoted by

$$\widetilde{X}=\{X_a: a\in X\}=\{[a]: a\in X\}, \tag{3.3.6}$$

and $\widetilde{X}$ is said to be **the quotient set of X with respect to equivalence relation** $\sim$, or simply, **the quotient set of X**, denoted by $\widetilde{X}=X/\sim$. Let

$$\pi: a\to[a] \tag{3.3.7}$$

be the mapping from X to $\widetilde{X}$ with $\pi(a)=[a]$. This mapping $\pi: a\to[a]$ is said to be **the quotient mapping** (from X to $\widetilde{X}$). Endow the topology

$$\widetilde{\tau}=\{\widetilde{O}\subset\widetilde{X}: \pi^{-1}(\widetilde{O})\in\tau\} \tag{3.3.8}$$

to quotient set $\widetilde{X}$, i. e., "open set $\widetilde{O}\subset\widetilde{X}$ in $\widetilde{\tau}$ satisfies that the inverse image $\pi^{-1}(\widetilde{O})$ of $\widetilde{O}$ is open set in τ". Thus, $\widetilde{X}$ becomes a topological space, called **the quotient space of topological space X with respect to equivalence relation** $\sim$ or simply, **a quotient space**, denoted by $\widetilde{X}=X/\sim$.

We verify that $\widetilde{\tau}$ in (3.3.8) satisfies three conditions in Definition 3.2.1.

In fact, by definition of quotient mapping $\pi: a\to[a]$, we have

(1) $\pi^{-1}(\widetilde{X})=X\in\tau$, $\pi^{-1}(\varnothing)=\varnothing\in\tau$, thus $\widetilde{X},\varnothing\in\widetilde{\tau}$;

(2) $\forall\widetilde{O}_1,\widetilde{O}_2\in\widetilde{\tau}\Rightarrow\pi^{-1}(\widetilde{O}_1),\pi^{-1}(\widetilde{O}_2)\in\tau\Rightarrow\pi^{-1}(\widetilde{O}_1)\cap\pi^{-1}(\widetilde{O}_2)\in\tau\Rightarrow\pi^{-1}(\widetilde{O}_1\cap\widetilde{O}_2)=\pi^{-1}(\widetilde{O}_1)\cap\pi^{-1}(\widetilde{O}_2)\in\tau\Rightarrow\widetilde{O}_1\cap\widetilde{O}_2\in\widetilde{\tau}$;

(3) $\forall\widetilde{O}_\alpha\in\widetilde{\tau},\alpha\in\Lambda\Rightarrow\pi^{-1}(\widetilde{O}_\alpha)\in\tau,\alpha\in\Lambda\Rightarrow\bigcup_{\alpha\in\Lambda}\pi^{-1}(\widetilde{O}_\alpha)\in\tau\Rightarrow\pi^{-1}\left(\bigcup_{\alpha\in\Lambda}\widetilde{O}_\alpha\right)=\bigcup_{\alpha\in\Lambda}\pi^{-1}(\widetilde{O}_\alpha)\in\tau\Rightarrow\bigcup_{\alpha\in\Lambda}\widetilde{O}_\alpha\in\widetilde{\tau}$.

The verification is complete.

In the usual topological space $(\mathbb{R},\tau)$, we define "equivalence relation" by

$$\text{“}\sim\text{”}=\{x,y\in\mathbb{R}: x,y\in\mathbb{Q}, \text{or } x,y\notin\mathbb{Q}\},$$

i. e., $x,y\in\mathbb{R}$, $x\sim y\Leftrightarrow x,y\in\mathbb{Q}$, or $x,y\notin\mathbb{Q}$. Then, all rational numbers are equivalent, and all irrational numbers are equivalent. Hence, all real numbers in $\mathbb{R}$ are classified into two classes, the rational number class, denoted by $[r]$, and the irrational number class, denoted by $[s]$. In the quotient space $\mathbb{R}/\sim$, there are just 4 elements, $\tau_{\mathbb{R}/\sim}=\{\mathbb{R}/\sim,\varnothing,[r],[s]\}$, and $(\mathbb{R}/\sim,\tau_{\mathbb{R}/\sim})$ is the quotient space.

The following theorem is quite important that describes the continuity of quotient mappings.

Theorem 3.3.8 Let $(\widetilde{X},\widetilde{\tau})$ be the quotient space of topological space (X,τ) with respect to equivalence relation $\sim$. Then quotient mapping $\pi: X \to \widetilde{X}$ is surjective, continuous from X onto $\widetilde{X}$; moreover, the quotient topology is the finest topology such that the quotient mapping $\pi: X \to \widetilde{X}$ is continuous.

Proof (1) The quotient mapping $\pi: X \to \widetilde{X}$ is surjective and continuous:

$\forall [x] \in \widetilde{X} = X/\sim$, there exists at least one $x \in X$, such that $\pi(x) = [x]$, thus $\pi: X \to \widetilde{X}$ is surjective; moreover, by the definition of quotient topology, $\forall U \in \widetilde{\tau}$ implies $\pi^{-1}(U) \in \tau$, this is "the inverse image of open sets in $\widetilde{\tau}$ is open sets in τ", so that $\pi: X \to \widetilde{X}$ is continuous.

(2) The quotient topology $\widetilde{\tau}$ is the finest topology such that $\pi: X \to \widetilde{X}$ is continuous:

Let $(\widetilde{X}, \upsilon)$ be the other topology on $\widetilde{X}$, such that quotient mapping $\pi: X \to \widetilde{X}$ is continuous. Then, $\forall A \in \upsilon \Rightarrow \pi^{-1}(A) \in \tau$. However, the quotient topology $\widetilde{\tau}$ contains all open sets $O \subset \widetilde{X}$ satisfying $\pi^{-1}(O) \in \tau$, thus $A \subset \widetilde{X}$ with "$A \in \upsilon \Rightarrow \pi^{-1}(A) \in \tau$" implies $A \in \widetilde{\tau}$, so that implies $\upsilon \subset \widetilde{\tau}$. This is: any other topology υ on $\widetilde{X}$ which makes quotient mapping π continuous, must be contained in $\widetilde{\tau}$, i. e., $\upsilon \subset \widetilde{\tau}$. This means the quotient topology $\widetilde{\tau}$ of $\widetilde{X}$ is the finest topology such that π is continuous. The proof is complete.

Example 3.3.5 Let $p \in \mathbf{Z}^+$ be a prime, and let $p\mathbf{Z} = \{\cdots, -2p, -p, 0, p, 2p, \cdots\}$. We define the equivalence relation $\sim$ in $\mathbf{Z}$: $a, b \in \mathbf{Z}, a \sim b \Leftrightarrow a - b \in p\mathbf{Z}$. Thus,

$$\mathbf{Z}/\sim = \{0, 1, \cdots, p-1\}. \tag{3.3.9}$$

Example 3.3.6 Let $I^2 = [0,1] \times [0,1]$. The equivalence relation is defined as

$$\sim : \{(x, y) \in I^2 \times I^2 : x = y, \text{or } x_1, y_1 \in \{0,1\}, x_2 = y_2\}, \tag{3.3.10}$$

then the quotient space $I^2/\sim$ is as follow.

The equivalence relation in $I^2 = [0,1] \times [0,1]$ is: for $x = (x_1, x_2), y = (y_1, y_2) \in I^2$, we have

$$x = y \text{ or } x_1, y_1 \in \{0,1\}, \quad x_2 = y_2.$$

By equivalence relation, the points α, β on OB $(x_1, y_1 = 0)$, AC $(x_1, y_1 = 1)$ $(x_2 = y_2)$, respectively, are in the same class; and for $x \in I^2$ which is not on OB and AC, its equivalent class is $[x] = \{x\}$. Thus, the elements in quotient set $I^2/\sim$, can be regarded as a circle gluing points of $(0, y)$ and $(1, y)$ which are the elements in I^2 and on the sides OB and AC, respectively. Endow quotient topology, we get the quotient topological space $I^2/\sim$. Moreover, $I^2/\sim$ is isomorphic with the pipe

$$\{(x_1, x_2, x_3) \in \mathbb{R}^3 : x_1^2 + x_2^2 = 1, 0 \leqslant x_3 \leqslant 1\} = S^1 \times I.$$

Finally, the mapping $f: I^2 \to S^1 \times I \subset \mathbb{R}^3$ with $(t, s) \to f(t, s) = ((\cos 2\pi t, \sin 2\pi t), s)$ and

$(t,s)\in I^2$, $f(t,s)=((\cos 2\pi t,\sin 2\pi t),s)\in S^1$. Clearly, $f: I^2 \to S^1\times I\subset\mathbb{R}^3$ is a continuous and surjective mapping. We can prove that there exists an isomorphic mapping $f^*: I^2/\sim \to S^1\times I$ from $I^2/\sim$ onto $S^1\times I$, satisfying $f=f^*\circ\pi: I^2\to S^1\times I$.

If we glue the points $(0,y)$, $(1,1-y)$ on OB, AC in I^2, respectively, then we get a Möbius strip(Fig. 3. 3. 3).

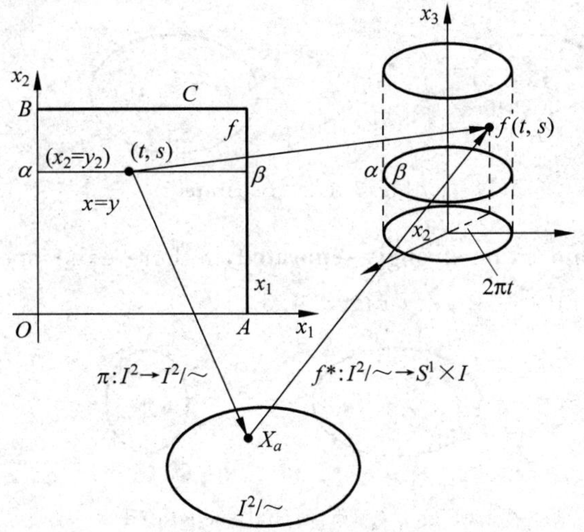

Fig. 3. 3. 3 Equivalence class

3. 4 Important Properties of Topological Spaces

3. 4. 1 Separation axioms of topological spaces

The separability is one of the essential properties of topological spaces, it plays an important role in studies of continuity, limit, compactness, etc. Also it is a principle for the classification of topological spaces. We will show the separation axioms of points and sets, as well as various types of topological spaces with certain separation axioms in this subsection.

Definition 3. 4. 1 (separated for points) Let (X,τ) be a topological space. For any two points $x,y\in X$ with $x\neq y$.

(1) x and y are said to be **weakly separated**, if for at least one point, say x, there exists an open set $G\in\tau$, such that $x\in G$, but $y\notin G$; see Fig. 3. 4. 1.

(2) x and y are said to be **separated**, if there exists an open set $G_1\in\tau$ with $x\in G_1$ such that $y\notin G_1$; and there exists an open set $G_2\in\tau$ with $y\in G_2$, too, such that $x\notin G_2$. see Fig. 3. 4. 2.

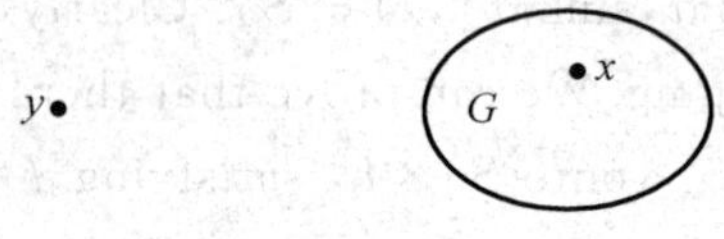

Fig. 3. 4. 1 Weakly separated

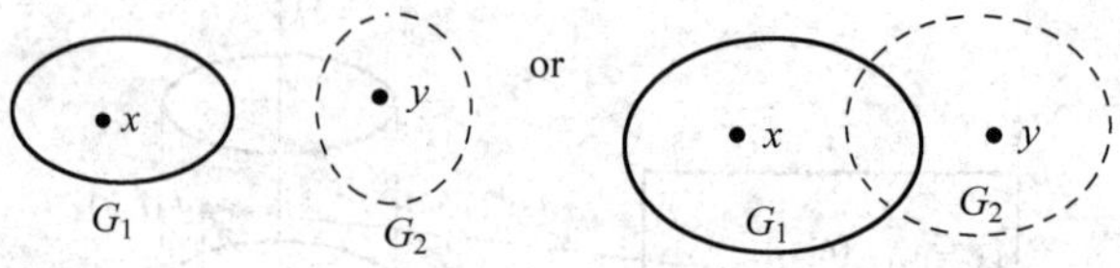

Fig. 3. 4. 2 Separated

(3) x and y are said to be **strongly separated**, if there exist open sets $G_1, G_2 \in \tau$ with $x \in G_1$, $y \in G_2$, and $G_1 \cap G_2 = \varnothing$. see Fig. 3. 4. 3.

Fig. 3. 4. 3 Strongly separated

Similarly, the weakly separated, separated, and strongly separated for two sets $A \subset X$, $B \subset X$ with $A \cap B = \varnothing$ (A, B are said to be disjoint) can be defined. For example, two sets $A \subset X$, $B \subset X$ are said to be **weakly separated**, if at least one of them possesses an open neighborhood not intersected with the other one.

Two sets $A \subset X$, $B \subset X$ are said to be **separated**, if A has an open neighborhood disjoint from B; and B has an open neighborhood disjoint from A. Or, A and B are separated if and only if $A \cap \overline{B} = \overline{A} \cap B = \varnothing$.

Two non-empty sets $A \subset X$, $B \subset X$ are said to be **strongly separated**, if $\overline{A} \cap \overline{B} = \varnothing$.

Definition 3. 4. 2 (separation axiom) Let (X, τ) be a topological space.

(1) T_0 **axiom**: If any two points $x, y \in X$, $x \neq y$, are **weakly separated**, then X is said to **satisfy the T_0 axiom**; or X is called **a T_0-type topological space**, or simply, **T_0-space.**

(2) T_1 **axiom**: If any two points $x, y \in X$, $x \neq y$, are **separated**, then X is said to **satisfy T_1 axiom**; or X is called **a T_1-type topological space**, or simply, **T_1-space.**

(3) T_2 **axiom**: If any two points $x, y \in X$, $x \neq y$, are **strongly separated**, then X is said to **satisfy T_2 axiom**; and X is called **a T_2-type topological space**, or simply, **T_2-space**, usually, it is called **a Hausdorff space.**

(4) **Regular axiom**: If any non-empty closed set $F \subset X$, $F \neq \varnothing$ and any point $x \in X$, $x \notin F$, are strongly separated, then X is said to **satisfy the regular axiom**, and X is called **a**

regular space.

(5) **Normal axiom**: If any non-empty closed subsets $F_1, F_2 \subset X, F_1, F_2 \neq \varnothing$ in X with empty intersection $F_1 \cap F_2 = \varnothing$ are strongly separated, then X is said to **satisfy the normal axiom**, and X is called **a normal space.**

(6) T_3 **axiom**: If X is a T_1-type space satisfying the regular axiom, then X is said to **satisfy** T_3 **axiom**, and X is called **a** T_3**-type topological space**, or simply, T_3**-space.**

(7) T_4 **axiom**: If X is a T_1-type space satisfying the normal axiom, then X is said to **satisfy** T_4 **axiom**, and X is called **a** T_4**-type topological space**, or simply, T_4**-space.**

We have the following table:

$$
\begin{array}{ccccccccc}
T_0 & \longleftarrow & T_1 & \longleftarrow & T_2 & \longleftarrow & T_3 & \longleftarrow & T_4 \\
 & & & & & & \updownarrow & & \updownarrow \\
 & & & & & & \text{regular} + T_1 & & \text{normal} + T_1
\end{array}
$$

Example 3.4.1 Let (X, τ) be a topological space.

Non-T_0-space If X has at least 2 points, then $\tau = \{X, \varnothing\}$ is the coarsest topology of X, and (X, τ) is not a T_0-space.

T_0-space but non-T_1-space If $X = \{a, b: a \neq b\}$, $\tau = \{X, \varnothing, \{a\}\}$, then (X, τ) is a T_0-space, but it is not a T_1-space.

T_1-space but non-T_2-space If $X = \{a_1, a_2, \cdots, a_n, \cdots\}$ is a countable set, **the co-finite topology**

$$\tau = \{X, \varnothing, \{O \subset X: \complement O \text{ is a finite subset}\}\},$$

then (X, τ) is a T_1-space, but is not a T_2-space.

Regular and normal space but non-T_0-space If $X = \{1,2,3\}, \tau = \{X, \varnothing, \{1\}, \{2,3\}\}$, then (X, τ) is a regular and normal space, but it is not a T_0-space, and is not a T_1-space or a T_2-space.

Normal but not regular space If $X = \{1,2,3\}$, $\tau = \{X, \varnothing, \{1\}, \{2\}, \{1,2\}\}$, then (X, τ) is a normal space, but not a regular space.

Regular but not normal space If $X = \{x = (x_1, x_2) \in \mathbb{R}^2: x_2 \geqslant 0\}$,

$$\tau = \{X, \varnothing, \{B(x, \varepsilon): x \in X, \varepsilon \in (0, x_2)\} \cup \{B(x, x_2) \cup \{(x_1, 0)\}: x \in X\}\},$$

then (X, τ) is a regular space, but not a normal space.

T_2-space but non-regular, non-normal space If $X = \mathbb{R}$, $K = \left\{\frac{1}{n}: n \in \mathbb{Z}^+\right\}$, and τ_u is the usual topology of $\mathbb{R}$, denoted by $\tau = \{G \backslash A: A \in \tau_u\}$, then (X, τ) is a non-regular, non-normal T_2-space.

Metric spaces, normal linear spaces, inner product spaces, Euclidean spaces, all are T_4-spaces.

For representations of various topological spaces, we show the following four theorems.

Theorem 3.4.1 (T_0**-space**) (X,τ) is a T_0-type topological space, if and only if for $x,y\in X, x\neq y$, it holds $\overline{\{x\}}\neq\overline{\{y\}}$.

Proof Necessity If X is a T_0-space, then $\forall x,y\in X, x\neq y$, $\exists G\in\tau$, s. t. $y\notin G$; this implies that $G\cap\{y\}=\varnothing$, thus $x\notin\overline{\{y\}}$, and hence $\overline{\{x\}}\neq\overline{\{y\}}$.

Sufficiency If $\forall x,y\in X, x\neq y$, by sufficient condition, $\overline{\{x\}}\neq\overline{\{y\}}$. Thus, either $\overline{\{x\}}\backslash\overline{\{y\}}\neq\varnothing$, or $\overline{\{y\}}\backslash\overline{\{x\}}\neq\varnothing$. So that: ① If $\overline{\{x\}}\backslash\overline{\{y\}}\neq\varnothing$, then $x\notin\overline{\{y\}}$ (otherwise, if $x\in\overline{\{y\}}$, then $\{x\}\subset\overline{\{y\}}$, thus $\overline{\{x\}}\subset\overline{\{y\}}$ and thus $\overline{\{x\}}\backslash\overline{\{y\}}=\varnothing$), this shows that there exists an open set $\complement\overline{\{y\}}$ of x with $x\in\complement\overline{\{y\}}$ which does not contain y; ② If $\overline{\{y\}}\backslash\overline{\{x\}}\neq\varnothing$, by the same reason, there exists an open set of y which does not contain x. Thus, X is a T_0-type space.

Theorem 3.4.2 (T_1**-space**) (X,τ) is a T_1-type topological space, if and only if all single points are closed sets in X.

Proof Necessity If X is a T_1-space, then $\forall y\in X, y\neq x$, $\exists G_y\in\tau$, s. t. $x\notin G_y$, so that we have $G_y\cap\{x\}=\varnothing$, and $y\notin\overline{\{x\}}$, this implies $\overline{\{x\}}=\{x\}$, and it means a single point is a closed set.

Sufficiency If each single point in X is a closed set, then, $\complement\{x\}$, $\complement\{y\}$ are open sets, and $\complement\{x\}$ contains y, $\complement\{y\}$ contains x, respectively; this implies X is a T_1-space.

Theorem 3.4.3 (**uniqueness theorem of the limit in** T_2**-space**) If (X,τ) is a T_2-type topological space, then any convergent sequence $\{x_n\}$ in X has and only has a unique limit.

Proof Let (X,τ) be a T_2-type space. Take a convergent sequence $\{x_n\}\subset X$, and suppose that there are two limits

$$\lim_{n\to+\infty} x_n = x \quad \text{and} \quad \lim_{n\to+\infty} x_n = y,$$

with $x\neq y$. Since X is T_2-type, thus there exist open neighborhoods U of x, V of y, respectively, with $U\cap V=\varnothing$. By definition of the limit, "$\lim_{n\to+\infty} x_n = x$, $\exists N_x>0$, such that $n>N_x$ implies $x_n\in U$" and "$\lim_{n\to+\infty} x_n = y$, $\exists N_y>0$, such that $n>N_y$ implies $x_n\in V$". Take $N=\max\{N_x,N_y\}$, then $n>N$ implies $x_n\in U$ and $x_n\in V$. So that $U\cap V\neq\varnothing$. This contradicts $U\cap V=\varnothing$. We get $x=y$, and for a convergent sequence $\{x_n\}\subset X$, its limit $\lim_{n\to+\infty} x_n$ is unique.

Theorem 3.4.4 (**representation of a normal space**) (X,τ) is a normal space, if and only if for any closed set $F\subset X$ and any open set $G\subset X$ with $F\subset G$, there exists an open set $O\in\tau$, such that $F\subset O\subset\bar{O}\subset G$.

Proof Necessity If X is a normal space, and $F\subset X$ is a closed set, $G\subset X$ is an open set with $F\subset G$. Then, $\mathfrak{C}G$ is a closed set with $\mathfrak{C}G\subset\mathfrak{C}F$, so that $F\cap\mathfrak{C}G=\varnothing$.

Now, for two disjoint closed sets F and $\mathfrak{C}G$, there exist an open set O with $F\subset O$, and an open set W with $\mathfrak{C}G\subset W$, respectively, since X is a normal space. So that $O\cap W=\varnothing$, thus this implies $O\subset\mathfrak{C}W$, and then

$$F\subset O\subset\bar{O}\subset\overline{\mathfrak{C}W}=\mathfrak{C}W\subset\mathfrak{C}(\mathfrak{C}G)=G,$$

we conclude that $F\subset O\subset\bar{O}\subset G$.

Sufficiency For any two closed sets $F_1\subset X$ and $F_2\subset X$ with $F_1\cap F_2=\varnothing$, then $G=\mathfrak{C}F_2$ is an open set which contains F_1, i. e. , $F_1\subset G$. By the assumption of thest Theorem, there exists an open set O, such that

$$F_1\subset O\subset\bar{O}\subset G=\mathfrak{C}F_2.$$

Then, for the two disjoint closed sets F_1, F_2, there exists an open set O containing F_1, $O\supset F_1$; and there exists an open set $\mathfrak{C}\bar{O}$ containing F_2, $\mathfrak{C}\bar{O}\supset F_2$; moreover, $O\cap\mathfrak{C}\bar{O}=\varnothing$. This implies X is a normal space.

For normal space, we have the following two famous theorems.

Theorem 3.4.5 (Urysohn lemma) (X,τ) is a normal space, if and only if for any two disjoint closed sets A, B in X, there exists a real valued continuous function $f: X\to\mathbb{R}$ on X, satisfies

$$f(x)|_{x\in A}=0,\quad f(x)|_{x\in B}=1, \tag{3.4.1}$$

with $0\leqslant f(x)\leqslant 1$, $\forall x\in X$ (Fig. 3.4.4).

Proof Sufficiency We prove (X,τ) is a normal space.

For any two disjoint closed sets A, B, if there exists a real continuous function f: $X\to\mathbb{R}$, with $0\leqslant f(x)\leqslant 1$ and $f(x)|_{x\in A}=0$, $f(x)|_{x\in B}=1$, then we construct two sets

$$U=\left\{x\in X: f(x)<\frac{1}{2}\right\},\quad V=\left\{x\in X: f(x)>\frac{1}{2}\right\}.$$

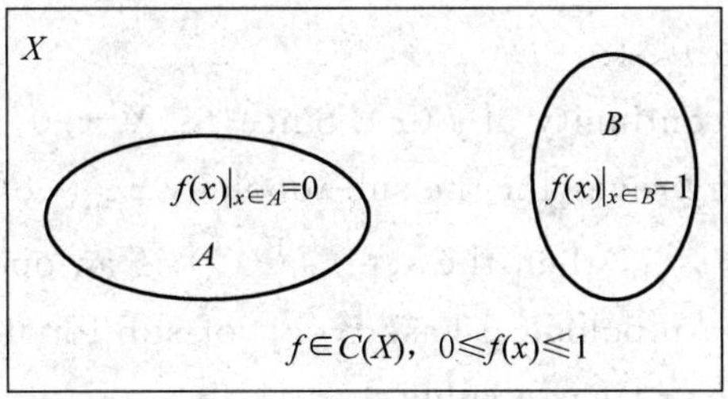

Fig. 3.4.4 Urysohn lemma

They are open sets by the continuity of $f: X\to\mathbb{R}$, and $U\cap V=\varnothing$. Clearly, $A\subset U, B\subset V$, then (X,τ) is a normal space.

Necessity If (X,τ) is a normal space, for two disjoint closed sets A, B, we construct a function $f: X\to\mathbb{R}$, real and continuous, $0\leqslant f(x)\leqslant 1$ and $f(x)|_{x\in A}=0, f(x)|_{x\in B}=1$.

(X,τ) is a normal space $\Rightarrow A\cap B=\varnothing$ implies $A\subset \complement B=X\backslash B \Rightarrow$ for $A\subset\complement B$, $\exists U_{\frac{1}{2}}\subset X$, s. t. $A\subset U_{\frac{1}{2}}\subset\overline{U_{\frac{1}{2}}}\subset\complement B$, by Theorem 3. 4. 4 $\Rightarrow$ for $A\subset U_{\frac{1}{2}}$ and $\overline{U_{\frac{1}{2}}}\subset\complement B$, $\exists U_{\frac{1}{4}}$ and $\exists U_{\frac{3}{4}}$, respectively, s. t.

$$A\subset U_{\frac{1}{4}}\subset\overline{U_{\frac{1}{4}}}\subset U_{\frac{1}{2}}\subset\overline{U_{\frac{1}{2}}}\subset U_{\frac{3}{4}}\subset\overline{U_{\frac{3}{4}}}\subset\complement B,$$

by Theorem 3. 4. 4.

Continue the process, define $U_0=A, U_1=\complement B$, then for any rational number

$$r\in\mathbb{Q}_0=\left\{\frac{m}{2^n}: m=0,1,\cdots,2^n, n\in\mathbb{Z}^+\right\},$$

there exists an open set U_r. We agree on "$\forall r>1, U_r=X$", and then we have

$$U_0\subset U_{\frac{1}{2^n}}\subset\overline{U_{\frac{1}{2^n}}}\subset\cdots\subset U_{\frac{m-1}{2^n}}\subset\overline{U_{\frac{m-1}{2^n}}}\subset\cdots\subset U_{\frac{2^n-1}{2^n}}\subset\overline{U_{\frac{2^n-1}{2^n}}}\subset U_1\subset U_r=X.$$

They satisfy:

(1) $r,s\in\mathbb{Q}_0, r<s\Rightarrow\overline{U_r}\subset U_s$;

(2) $\forall r\in\mathbb{Q}_0\Rightarrow A=U_0\subset U_r\subset\overline{U_r}\subset\complement B\Rightarrow B\subset\complement(\overline{U_r})$;

(3) $\forall x\in X\Rightarrow x\in B$, or $x\in\complement B\Rightarrow x\in B$, or $\exists r_x\in\mathbb{Q}_0$, s. t. $x\in\overline{U}_{r_x}$.

Now we define the real-valued function $f: X\to\mathbb{R}$ satisfying

$$f(x)=\begin{cases}\inf\{r\in\mathbb{Q}_0: x\in U_r\}, & x\in X\backslash B\\ 1, & x\in B\end{cases}. \tag{3.4.2}$$

It has the following properties:

① $\forall x\in A=U_0\subset U_r$, $\inf\{r: x\in U_r, \forall r\in\mathbb{Q}_0\}=0\Rightarrow f(x)=0, \forall x\in A$;

② $\forall x\in B\Rightarrow f(x)=1$;

③ $\forall x\in X\Rightarrow 0\leqslant f(x)\leqslant 1$.

Finally, we prove the continuity of $f(x)$. Since $f: X\to[0,1]$ is a mapping from X to $[0,1]\subset\mathbb{R}$, we only need to prove: for the sub-topology $\tau_{[0,1]}$ of subspace $([0,1],\tau_{[0,1]})$, and for any open set $O\in\tau_{[0,1]}$, then the set $f^{-1}(O)$ is an open set in X. Also, we only need to prove the case for a topological base $\mathbb{B}_{[0,1]}$ of sub-topology $\tau_{[0,1]}$.

By Example 3. 2. 3, $(\mathbb{R},\tau)$ has a subbase

$$\mathbb{S}=\{\{(-\infty,a): a\in\mathbb{R}\}\cup\{(b,+\infty): b\in\mathbb{R}\}\},$$

so that subspace $([0,1],\tau_0)$ has a subbase $\mathbb{S}_{[0,1]}=\{\{(a,1]: a\in[0,1)\}\cup\{[0,b):$

$b\in(0,1]\}\}$. To verify the continuity of $f: X\to[0,1]$, we prove: $\forall O\in\mathbb{S}_{[0,1]}$, the set $f^{-1}(O)$ is an open set in X.

It is enough to consider for $a\in[0,1)$ and $b\in(0,1]$, the inverse image sets $f^{-1}((a,1])$ and $f^{-1}([0,b))$.

If $a\in[0,1)$, $f^{-1}((a,1])\Leftrightarrow f(x)\in(a,1]\Leftrightarrow a<f(x)\leqslant 1\Leftrightarrow a<\inf\{r\in\mathbb{Q}_0: x\in U_r\}$; or $x\in B\Leftrightarrow\exists r\in\mathbb{Q}_0$, s. t., $r>a$, and $x\notin\overline{U_r}$ $(x\in\mathfrak{C}(\overline{U_r}))$; or $x\in B\Rightarrow x\in f^{-1}((a,1])=\{\bigcup_{r>a,r\in\mathbb{Q}_0}\mathfrak{C}(\overline{U_r})\}\cup B=\{\bigcup_{r>a,r\in\mathbb{Q}_0}\mathfrak{C}(\overline{U_r})\}$ (by $\forall r\in\mathbb{Q}_0, B\subset\mathfrak{C}(\overline{U_r})$) $\Rightarrow f^{-1}((a,1])$ is a union of the open set family $\{\mathfrak{C}(\overline{U_r}): r\in\mathbb{Q}_0, r>a\}\Rightarrow f^{-1}((a,1])$ is an open set in (X,τ).

If $b\in(0,1]$, $x\in f^{-1}([0,b))\Leftrightarrow f(x)\in[0,b)\Leftrightarrow 0\leqslant f(x)<b\Leftrightarrow\inf\{r\in\mathbb{Q}_0: x\in U_r\}<b\Leftrightarrow\exists r\in\mathbb{Q}_0$, s. t. $r<b$, $\forall x\in U_r\Rightarrow x\in f^{-1}([0,b))=\bigcup_{r<b,r\in\mathbb{Q}_0}U_r\Rightarrow f^{-1}([0,b))$ is a union of the open set family $\{U_r: r\in\mathbb{Q}_0, r<b\}\Rightarrow$ the set $f^{-1}([0,b))$ is an open set in (X,τ). The proof is complete.

The sense of this theorem: The connection between geometry property (topological type of topological spaces) and analytical property (existence of the continuous function on topological spaces) is revealed in **Urysohn lemma** by the construction method. Also, the combination of geometry and algebra methods is shown in this lemma.

Theorem 3.4.6 (Tietze extension theorem) (X,τ) is a normal space, if and only if for any closed set A in X and any real-valued continuous functions $f: A\to\mathbb{R}$ on A, there exists a continuous function $F: X\to\mathbb{R}$ on X, satisfying $F(x)|_{x\in A}=f(x)$.

For the proof, we refer to [12], [14].

3.4.2 Connectivity of topological spaces

The connectivity has very visual geometry sense.

For example, the set $[1,2)\cup(2,3]$ is divided into two parts, "split" or "break up"; but the set $[1,2)\cup[2,3]=[1,3]$ is as one part, "not split", or "connected together".

Definition 3.4.3 (disconnected) Let (X,τ) be a topological space, $A\subset X$, $B\subset X$ be two non-empty subsets of X. If they have disjoint open neighborhoods, i. e., there exist G_1 and $G_2\in\tau$ with $A\subset G_1$, $B\subset G_2$, such that $G_1\cap G_2=\varnothing$, then A and B are said to be **disconnected.** Or equivalently, **two non-empty sets $A\subset X$, $B\subset X$ are disconnected iff**

$$(A\cap\overline{B})\cup(\overline{A}\cap B)=\varnothing. \tag{3.4.3}$$

Note The condition $(A\cap\overline{B})\cup(\overline{A}\cap B)=\varnothing$ is equivalent to both $A\cap\overline{B}=\varnothing$ and $\overline{A}\cap B=\varnothing$ holding, simultaneously, see Fig. 3.4.5; it means that: sets A, B are disjoint, and A does not contain any accumulative points of B (i. e. , $A\cap\overline{B}=\varnothing$), as well as B does not contain any accumulative points of A (i. e. , $\overline{A}\cap B=\varnothing$), too.

Definition 3. 4. 4 (**connected property of a space**) Let (X,τ) be a topological space. If there exist non-empty sets $A, B\subset X$, such that A and B are disconnected each other, and $A\cup B=X$, then X is said to be **a disconnected topological space**; Otherwise, **a connected topological space.**

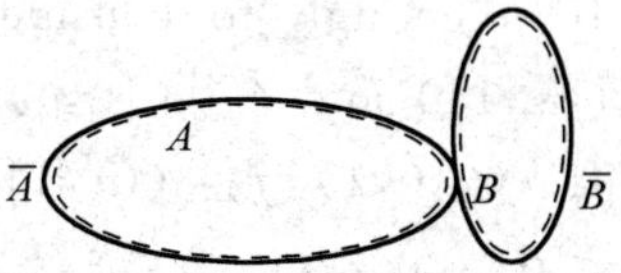

Fig. 3. 4. 5 $A\cap\overline{B}=\varnothing, \overline{A}\cap B=\varnothing$

Thus, $[1,2)$, $(2,3]$ are disconnected; and $[1,2)$, $[2,3]$ are connected.

A non-empty subset $A\subset X$ of a topological space (X,τ) is said to be **a disconnected subset**, if A is a disconnected topological space when it is regarded as a subspace.

We have the equivalent theorem as follows:

Theorem 3. 4. 7 Let (X,τ) be a topological space. Then the following statements are equivalent:

(1) X is a disconnected space;

(2) there exist two nonempty closed sets $A,B\subset X$ in X, such that $A\cap B=\varnothing, A\cup B=X$;

(3) there exist two nonempty open sets $G,O\subset X$ in X, such that $G\cap O=\varnothing, G\cup O=X$;

(4) there exists a nonempty proper subset $A\subset X$, which is both open and closed.

Proof (1)⇒(2) Let X be a connected space. By definition, there exist two non-empty separated subsets A,B in X with $A\cap\overline{B}=\varnothing, \overline{A}\cap B=\varnothing$, s. t. $A\cup B=X$
$\Rightarrow \complement(A\cup B)=\complement X=\varnothing \Rightarrow A\cap B=\varnothing$ by

$$\overline{B}=\overline{B}\cap X=\overline{B}\cap(A\cup B)=(\overline{B}\cap A)\cup(\overline{B}\cap B)=(\overline{B}\cap A)\cup B=\varnothing\cup B=B,$$

$$\overline{A}=\overline{A}\cap X=\overline{A}\cap(A\cup B)=(\overline{A}\cap A)\cup(\overline{A}\cap B)=A\cup(\overline{A}\cap B)=A\cup\varnothing=A.$$

Thus A and B are closed sets with $A\cap B=\varnothing, A\cup B=X \Rightarrow$ (2) holds.

(2) ⇒(3) If there exist two non-empty separated closed sets A,B in X with $A\cap\overline{B}=\varnothing, \overline{A}\cap B=\varnothing$, s. t. $A\cup B=X, A\cap B=\varnothing$
$\Rightarrow \complement A\cap\complement B=\varnothing, \complement A\cup\complement B=X\Rightarrow$ there exist two non-empty open sets $G=\complement A$ and $O=\complement B$ with $G\cap O=\varnothing$ and $G\cup O=X \Rightarrow$ (3) holds.

(3) ⇒(4) If (3) holds, then the open set O satisfying condition (3) is also a closed set by $G\cup O=X$ and $O=\complement G \Rightarrow$ there exists a set which is both open and closed in $X \Rightarrow$ (4) holds.

(4) ⇒(1) If there exists a non-empty proper subset A which is both open and closed in $X \Rightarrow$ Let $B=\complement A$. Then A,B are non-empty proper subsets which are both open and

closed in X with $A\cup B=X, A\cap B=\varnothing$.

$\Rightarrow \bar{A}\cap B=A\cap B=\varnothing, A\cap \bar{B}=A\cap B=\varnothing$, i. e., two disjoint non-empty closed subsets are disconnected $\Rightarrow$ (1) holds.

The proof of this theorem is complete.

A disconnected topological space can be defined by equivalent definitions (by Theorem 3.4.7):

Definition 3.4.4′ Let (X, τ) be a topological space. If there exists non-empty proper subset $A\subset X$, which is both open and closed, then (X, τ) is said to be **a disconnected space.**

Definition 3.4.4″ Let (X, τ) be a topological space. If only X and empty set $\varnothing$ both are open and closed, then (X, τ) is said to be **a connected space.** And if (X, τ) is not connected, then it is said to be a disconnected space.

Example 3.4.2 One-dimensional Euclidean space $(\mathbb{R}, \tau)$ is a connected one; The rational number set $\mathbb{Q}$ is a disconnected subset in $\mathbb{R}$.

We prove the assertions in this example by three steps.

First Prove the connectedness of $(\mathbb{R}, \tau)$. Suppose that $(\mathbb{R}, \tau)$ would be disconnected space, then by Theorem 3.4.7 (2), there exist non-empty closed sets A, B in $\mathbb{R}$ with $A\cap B=\varnothing, A\cup B=\mathbb{R}$. Take any $a\in A$ (then $a\notin B$), $b\in B$ (then $b\notin A$); suppose that $a<b$, without loss of generality.

Let $\tilde{A}=A\cap[a, b], \tilde{B}=B\cap[a, b]$. Then $\tilde{A}, \tilde{B}$ are non-empty closed sets with $a\notin\tilde{B}$, $b\notin\tilde{A}$ and $\tilde{A}\cap\tilde{B}=\varnothing, \tilde{A}\cup\tilde{B}=[a, b]$. This means that $[a, b]$ is a disconnected subset. But this is a contradiction since an interval $[a, b]$ is a connected set (see the second step). Then we have proved that $(\mathbb{R}, \tau)$ is a connected space.

Second Prove the connectivity of $[a, b]$. Since $\tilde{A}$ is a bounded closed set with upper bound b, then $\forall x\in\tilde{A}$ satisfies $x\leqslant b$. For the supremum $\tilde{b}=\sup\tilde{A}$, we have $\tilde{b}\in\tilde{A}$ (since $\tilde{A}$ is closed) and $\tilde{b}<b$ (since $b\notin\tilde{A}$). So that the interval $(\tilde{b}, b]\subset\tilde{B}$ (since on one hand, by $b\in\tilde{B}$; and on the other hand, $\forall x\in(\tilde{b}, b)\Rightarrow x\notin\tilde{A}$, by $\tilde{b}=\sup\tilde{A}$, and $\tilde{A}$ is closed implies $x\in\tilde{B}$). Thus, $\tilde{b}\in\tilde{B}$ (by $\tilde{B}$ is closed), and then $\tilde{b}\in\tilde{A}\cap\tilde{B}$, this contradicts to $\tilde{A}\cap\tilde{B}=\varnothing$. Hence, $[a, b]$ is a connected set.

Third Prove the disconnectivity of $\mathbb{Q}$. Since for any irrational number $s\in\mathbb{R}\setminus\mathbb{Q}$, the set $(-\infty, s)\cap\mathbb{Q}$ is an open and closed set in $\mathbb{Q}$, simultaneously, since $(-\infty, s)\cap\mathbb{Q}=(-\infty, s]\cap\mathbb{Q}$ (note, $(-\infty, s]$ is a closed and, also an open set in $\mathbb{R}$, by $\mathbb{R}\setminus(-\infty, s]=(s, +\infty)$). Hence, $\mathbb{Q}$ is regarded as a subspace of $\mathbb{R}$, and a disconnected topological space.

The proof is complete.

"The connectivity" is an invariant property under continuous mappings.

Theorem 3.4.8 (1) Let (X,τ) be a connected topological space, (Y,υ) be a topological space, and $f: X \to Y$ be a continuous mapping. Then $f(X) \subset Y$ is a connected subset in Y; (2) Let (X_j,τ_j), $j=1,2,\cdots,m$, be connected topological spaces. Then the product space $X = X_1 \times \cdots \times X_m$ is a connected topological space.

Proof (1) If $f(X)$ would be a disconnected subset in Y, then there exist two disjoint non-empty open sets $A, B \subset Y$, such that $f(X) = A \cup B$. Thus, the inverse images $f^{-1}(A), f^{-1}(B)$ are non-empty, and $f^{-1}(A) \in \tau, f^{-1}(B) \in \tau$. Moreover, it holds

$$\begin{aligned}
&[(f^{-1}(A)) \cap \overline{(f^{-1}(B))}] \cup [\overline{(f^{-1}(A))} \cap (f^{-1}(B))] \\
&\subset [(f^{-1}(A)) \cap (f^{-1}(\bar{B}))] \cup [(f^{-1}(\bar{A})) \cap (f^{-1}(B))] \\
&= f^{-1}[(A \cap \bar{B}) \cup (\bar{A} \cap B)] = \varnothing,
\end{aligned}$$

this implies that $f^{-1}(A)$ and $f^{-1}(B)$ are two disjoint non-empty subsets; and it holds

$$f^{-1}(A) \cup f^{-1}(B) = f^{-1}(A \cup B) = f^{-1}(X) = X,$$

so that X is a disconnected space. However, this contradicts the connectivity of (X,τ). Thus, (1) is proved.

(2) It is clear.

Definition 3.4.5 (connectivity of two points, component) Let (X,τ) be a topological space. For two points $x, y \in X$, if there exists a non-empty connected subset $A \subset X$, such that $x, y \in A$, then x and y are said to be **connected**; or x **is connected with** y.

It is clear that the connectivity of two points is an equivalence relation (reflection, symmetry, transition).

In topological space (X,τ), for a point $x \in X$, the union set of all connected subsets which contain x is said to be **the connected component of** x.

For every point $x \in X$, if the connected component of x is a single point set $\{x\} \subset X$, then (X,τ) is said to be **a totally disconnected topological space.**

For example, endow the dyadic topology τ to $G = [0, +\infty)$, then (G,τ) is a totally disconnected topological space.

In fact, let

$$G = \{x = (x_{-s}, \cdots, x_{-1}, x_0, x_1, x_2, \cdots) : x_j \in \{0,1\}, j = -s, -s+1, \cdots, s \in \mathbb{N}\}.$$

Define the addition $\oplus$, mod2, term by term, on G; and define the open base of $0 \in G$ as $\tau = \{G_k : k \in \mathbb{Z}\}$ with

$$G_k = \{y \in G: y = (y_k, y_{k+1}, \cdots), y_k \neq 0, y_j \in \{0,1\}, j \geqslant k+1\},$$

then (G,τ) is a topological space, and it is totally disconnected.

Definition 3.4.6 (path-connectivity) Let (X, τ) be a topological space. If $\forall x, y \in X$, there exists a continuous mapping $\sigma: [0,1] \to X$ from interval $[0,1] \subset \mathbb{R}$ to X, such that $\sigma(0) = x, \sigma(1) = y$, then X is said to be **a path-connected topological space**; The continuous mapping $\sigma: [0,1] \to X$ is said to be **a path in** X; The image $\sigma([0,1]) \subset X$ is said to be **an arc (or curve) in** X; And x, y are said to be **the origin point and final point of** σ, respectively.

We emphasize the following properties for connectivity and path-connectivity:

(1) Path-connected space is connected one, but the inverse is not true;

(2) Path-connectivity is topologically invariant;

(3) In Euclidean spaces, the connectivity of open sets is equivalent to the path-connectivity of open sets.

Definition 3.4.7 (locally connected) Let (X, τ) be a topological space. For $x \in X$, denote the open set family $\tau_x = \{G \in \tau : x \in G\}$. If for any $G \in \tau_x$, there exists a connected subset $V \in \tau_x$ with $V \subset G$, then (X, τ) is said to be **locally connected at** x; If (X, τ) is locally connected at every point $x \in X$, then (X, τ) is said to be **a locally connected space.**

We point the following:

(1) The locally connected component of a point can be defined in a locally connected space;

(2) Connectivity and local connectivity are independent each other;

(3) The locally path-connected space and locally path-connected component can be defined;

(4) The local connectivity is a topological invariant; the local path-connectivity is invariant under continuous mappings.

3.4.3 Compactness of topological spaces

Compactness is the most important concept in topology, as well as in other modern science areas.

compactness

compact | countably compact | sequentially compact | accumulatively compact | locally compact

We concentrate our mind on the concepts of compact and locally compact.

1. Compactness

Borel covering theorem in *advance calculus* state is: "for any closed interval $[\alpha, \beta]$ in $\mathbb{R}$, $-\infty < \alpha < \beta < +\infty$, if $\mathfrak{D} = \{(a_t, b_t)_{t \in \Lambda} : -\infty < a_t < b_t < +\infty\}$ with index set Λ is an

open covering of $[\alpha,\beta]$, i. e., $[\alpha,\beta]\subset\bigcup_{t\in\Lambda}(a_t,b_t)$, then there exist finite $t_1,\cdots,t_n\in\Lambda$, such that $[\alpha,\beta]\subset\bigcup_{j=1}^{n}(a_{t_j},b_{t_j})$ with $t_j\in\Lambda, j=1,\cdots,n$". This Borel theorem is not only a powerful tool that can transfer an infinite process to a finite one, but also an important mathematical thinking and idea, such that a lot of abstract mathematical problems can be solved by this theorem. Moreover, it is also a strong and suitable motivation of the compactness in the topology.

1) **Compact and criterions**

Definition 3.4.8 (compact) Let (X,τ) be a topological space. An open set subfamily $\Lambda\subset\tau$ of τ is said to be **an open covering of** X, if $\bigcup_{A\in\Lambda}A=X$. And topological space (X,τ) is said to be **a compact space**, if for any open covering Λ of X, there exists a finite subcovering $\{O_1,\cdots,O_n\}\subset\Lambda, n\in\mathbb{N}$, of X; i. e.

"any open covering $\bigcup_{A\in\Lambda}A=X\Rightarrow\exists\{O_1,\cdots,O_n\}\subset\Lambda, n\in\mathbb{N}$ with $\bigcup_{j=1}^{n}O_j=X$".

If a subset $A\subset X$ regarded as a subtopological space of X is compact, then it is said to be **a compact set in** X.

For a compact set, we have the following equivalent theorem.

Theorem 3.4.9 Let (X,τ) be a topological space, $A\subset X$ be a compact subset in (X,τ), if and only if for any open covering $\mathfrak{U}=\{U\in\tau\}$ of A in X, there exists a finite subcovering $\widetilde{\mathfrak{U}}\subset\mathfrak{U}$ of A in X.

Proof Necessity Let $\mathfrak{U}=\{U\in\tau\}$ be any open covering of A in $(X,\tau)\Rightarrow$ take a subset family of $\mathfrak{U}$: $\mathfrak{P}_A=\{U\cap A: U\in\mathfrak{U}\}$, it is an open covering of A in subspace $(A,\tau_A)\Rightarrow$ by the assumption of the necessity, A is a compact space in the subtopology, thus there exists finite covering $\{U_j\cap A: U_j\in\mathfrak{U}\}_{j=1}^{n}$, s. t. $\bigcup_{j=1}^{n}(U_j\cap A)$ is a finite covering of A in the subtopological space $(A,\tau_A)\Rightarrow\widetilde{\mathfrak{U}}=\{U_j\}_{j=1}^{n}$ is the finite subfamily of $\mathfrak{U}$ in topological space (X,τ), and $\bigcup_{j=1}^{n}U_j\supset\bigcup_{j=1}^{n}(U_j\cap A)\supset A\Rightarrow$ for any open covering $\mathfrak{U}=\{U\in\tau\}$ of A in (X,τ), there exists a finite subcovering $\widetilde{\mathfrak{U}}=\{U_j\}_{j=1}^{n}$ of A in $X\Rightarrow$ the necessity is proved.

Sufficiency Let $\mathfrak{U}_A=\{V: V\in\tau_A\}$ be any open covering of A in subspace $(A,\tau_A)\Rightarrow$ $\forall V\in\tau_A$, $\exists U\in\tau$, s. t. $V=U\cap A$, then take all $U\in\tau$ with $V=U\cap A\in\mathfrak{U}_A$, and collect these $U\in\tau$, such that $\mathfrak{U}=\{U\in\tau: U\cap A=V\in\mathfrak{U}_A\}\Rightarrow\mathfrak{U}$ is an open covering of A in (X,τ), i. e.,

$A \subset \bigcup_{V \in \mathfrak{U}_A} V = \bigcup_{U \in \mathfrak{U}} (U \cap A) \Rightarrow$ there exists a finite sub-covering of $\mathfrak{U}$, say, $\widetilde{\mathfrak{U}} = \{U_j : U_j \in \mathfrak{U}\}_{j=1}^n$, s. t. $A \subset \bigcup_{j=1}^n U_j$ by assumption of the sufficiency $\Rightarrow A \subset (\bigcup_{j=1}^n U_j) \cap A = \bigcup_{j=1}^n (U_j \cap A)$, denoted by $U_j \cap A = V_j$, $1 \leqslant j \leqslant n \Rightarrow$ for any open covering $\mathfrak{U}_A = \{V : V \in \tau_A\}$ of A in subspace (A, τ_A), there exists a finite sub-covering $\widetilde{\mathfrak{U}}_A = \{V_j\}_{j=1}^n$ of A, i. e., $A \subset \bigcup_{j=1}^n V_j \Rightarrow$ the sufficiency is proved.

Example 3. 4. 3 The interval $[a, b]$ in $\mathbb{R}$ is a compact set, but $\mathbb{R}$ itself is not a compact space. The cuboid $[a_1, b_1; \cdots; a_n, b_n]$, $-\infty < a_i < b_i < +\infty$, in $\mathbb{R}^n$ is a compact set, but $\mathbb{R}^n$ is not a compact space. Moreover, for any non-empty finite set X with a topology τ, space (X, τ) is a compact one.

We only prove: the interval $[0,1]$ in $\mathbb{R}$ is a compact set.

Take an open covering $\mathfrak{U} = \{V = U \cap [0,1] : U \in \tau\}$ of $[0,1]$ in the usual topology $(\mathbb{R}, \tau)$ of $\mathbb{R}$, and $\mathfrak{B} = \{(a,b) : -\infty < a < b < +\infty\}$ is a base of τ. Let $\mathfrak{B}_0 = \{(a,b) \cap [0,1] : (a,b) \in \mathfrak{B}\}$, and let

$$\mathfrak{P} = \{x \in [0,1] : \exists \mathfrak{U}_0 \subset \mathfrak{U} \text{ as finite sub-covering of } [0,x]\}. \tag{3.4.4}$$

We prove: (1) $\mathfrak{P} \neq \varnothing$; (2) $\mathfrak{P}$ is an open set in $[0,1]$; (3) $\mathfrak{P}$ is a closed set in $[0,1]$.

(1) is clear. For (2): by definition, $\forall x \in \mathfrak{P}$, $\exists \mathfrak{U}_0 = \{V_1, \cdots, V_n\} \subset \mathfrak{U}$, s. t. $[0,x] \subset \bigcup_{j=1}^n V_j$. Then there are two cases.

① If $x = 1 \in \mathfrak{P} \Rightarrow [0,x] = [0,1]$, and $[0,1]$ is an open set in the sub-space $([0,1], \tau') \Rightarrow x = 1$ is an inner point of $\mathfrak{P}$. Similarly, if $x = 0 \in \mathfrak{P}$. then 0 is an inner point of $\mathfrak{P}$.

② If $x \in \mathfrak{P}$ and $0 < x < 1 \Rightarrow$ by $0 < x < 1$ and by $\mathfrak{U}$, an open covering of $[0,1]$, thus $\exists V_0$, s. t. $x \in V_0 \in \mathfrak{U} \Rightarrow$ since $V_0 = U_0 \cap [0,1]$ is an open set in $[0,1]$, thus x is an inner point of $V_0 \Rightarrow \exists \varepsilon_1 > 0$, s. t. $[x, x+\varepsilon_1) \subset (x-\varepsilon_1, x+\varepsilon_1) \cap [0,1] \subset V_0$. Then, by $x \in \mathfrak{P}$, $\exists V_1, \cdots, V_n \in \mathfrak{U}$, s. t. $[0,x] \subset \bigcup_{j=1}^n V_j \Rightarrow \forall \varepsilon\ (0 < \varepsilon < \varepsilon_1)$, s. t., $[0, x+\varepsilon] = [0,x] \cup [x, x+\varepsilon_1) \subset (\bigcup_{j=1}^n V_j) \cup V_0 \Rightarrow x + \varepsilon \in \mathfrak{P}$, and $[0, x+\varepsilon) \subset \mathfrak{P}$ is an open set containing x, thus x is an inner point of $\mathfrak{P} \Rightarrow \mathfrak{P}$ is an open set in $[0,1]$.

To prove (3), we prove $\complement_{[0,1]}\mathfrak{P} = [0,1] \setminus \mathfrak{P}$ is an open set, then $\mathfrak{P}$ is a closed set in $[0,1]$.

Take $x \in \complement_{[0,1]}\mathfrak{P} = [0,1] \setminus \mathfrak{P} \Rightarrow [x, 1] \subset \complement_{[0,1]}\mathfrak{P} \Rightarrow x > 0$ (by $0 \in \mathfrak{P}$) $\Rightarrow \exists \widetilde{V} \in \mathfrak{U}$, s. t.

$x\in\widetilde{V}$($\mathfrak{U}$ is open covering of $[0,1]$) $\Rightarrow$ $\widetilde{V}$ is open set in $[0,1]$, $\exists\varepsilon>0$, s. t. $(x-\varepsilon,x]\subset(x-\varepsilon,x+\varepsilon)\cap[0,1]\subset\widetilde{V}\Rightarrow(x-\varepsilon,x]\cap\mathfrak{P}=\varnothing$ (otherwise, if $(x-\varepsilon,x]\cap\mathfrak{P}\neq\varnothing\Rightarrow\exists z\in(x-\varepsilon,x]\cap\mathfrak{P}\Rightarrow z\in\mathfrak{P}\Rightarrow\exists$ finite subset family $\mathfrak{U}'\subset\mathfrak{U}$, s. t. $\mathfrak{U}'$ covers $[0,z]$ $\Rightarrow$ the open covering $\{\widetilde{V}\}\cup\mathfrak{U}'$ covers $[0,z]\cup(x-\varepsilon,x]=[0,x]\Rightarrow x\in\mathfrak{P}$, this contradicts $x\in\complement_{[0,1]}\mathfrak{P}$) $\Rightarrow$ $(x-\varepsilon,x]\subset\complement_{[0,1]}\mathfrak{P}\Rightarrow(x-\varepsilon,1]=(x-\varepsilon,x]\cup[x,1]\subset\complement_{[0,1]}\mathfrak{P}\Rightarrow x$ is an inner point of $\complement_{[0,1]}\mathfrak{P}\Rightarrow\complement_{[0,1]}\mathfrak{P}$ is an open set $\Rightarrow\mathfrak{P}$ is a closed set.

Summary: $\mathfrak{P}$ is an open and a closed subset in $[0,1]$, and $\mathfrak{P}\neq\varnothing$. However, $[0,1]$ is a connected set (see Example 3. 4. 2), it could not contain subset $\mathfrak{P}$, unless $\mathfrak{P}=[0,1]\Rightarrow 1\in\mathfrak{P}\Rightarrow\exists$ finite subset family $\mathfrak{U}'$ of $\mathfrak{U}$, such that $\mathfrak{U}'\supset[0,1]\Rightarrow[0,1]$ is a compact set.

Example 3. 4. 4 Let τ be the topology of (X,τ). If ① τ is the coarsest topology of (X,τ); ② τ is the co-finite topology, when X is an infinite set; then prove that (X,τ) is a compact space.

In fact, ① It is clear that (X,τ) with the coarsest topology $\tau=\{X,\varnothing\}$ is a compact space.

② Let X be an infinite set, $\tau=\{G\subset X: \complement G$ is a finite subset$\}$ be the co-finite topology. For any open covering $\Lambda\subset\tau$ of X, denoted by $X=\bigcup\limits_{O\in\Lambda}O$ with $O\neq\varnothing$, then $\complement O$ is a finite subset in X, i. e., $\complement O=\{a_1,\cdots,a_n\}$. Take $O_j\in\Lambda$ such that $a_j\in O_j$, $j=1,2,\cdots,n$, then set $\{O,O_1,\cdots,O_n\}\subset\Lambda$ is a finite covering of X: $X\subset O\cup O_1\cup\cdots\cup O_n$. This implies that co-finite topology τ makes (X,τ) to be a compact space.

To verify the compactness of a topological space (X,τ), we use the following theorem.

Theorem 3. 4. 10 Let $\mathfrak{B}$ be a topological base of topological space (X,τ). If for any open covering $\mathfrak{U}=\{U\in\mathfrak{B}\}$ of X, there exists a finite subcovering $\widetilde{\mathfrak{U}}\subset\mathfrak{U}$ of X, then (X,τ) is a compact space.

Proof Let $\mathfrak{U}=\{U\subset X: U\in\mathfrak{B}\}$ be an open covering of $X\Rightarrow\forall U\in\mathfrak{U}$, $\exists\mathfrak{B}_U\subset\mathfrak{B}$, s. t. $U=\bigcup\limits_{B\in\mathfrak{B}_U}B\Rightarrow$ let $\widetilde{\mathfrak{U}}=\bigcup\limits_{U\in\mathfrak{U}}\mathfrak{B}_U$, thus $\bigcup\limits_{B\in\widetilde{\mathfrak{U}}}B=\bigcup\limits_{B\in\bigcup\limits_{U\in\mathfrak{U}}\mathfrak{B}_U}B=\bigcup\limits_{U\in\mathfrak{U}}\Big(\bigcup\limits_{B\in\mathfrak{B}_U}B\Big)=\bigcup\limits_{U\in\mathfrak{U}}U=X\Rightarrow\widetilde{\mathfrak{U}}$ is an open covering of X consisting of the elements in $\mathfrak{B}\Rightarrow$ by assumption, there exists a finite covering of X, $\{B_j\in\widetilde{\mathfrak{U}}: j=1,2,\cdots,n\}$, s. t. $\bigcup\limits_{j=1}^{n}B_j=X\Rightarrow$ for $B_j\in\widetilde{\mathfrak{U}}$: $j=1,2,\cdots,n$, $\exists U_j\in\mathfrak{U}$, s. t. $B_j\in\mathfrak{B}_{U_j}$, $j=1,2,\cdots,n\Rightarrow B_j\subset U_j\in\mathfrak{U}\Rightarrow\{U_j: j=1,2,\cdots,n\}\subset\mathfrak{U}$ is a finite subset family and $\bigcup\limits_{j=1}^{n}U_j\supset\bigcup\limits_{j=1}^{n}B_j=X\Rightarrow$ The proof is complete.

2) **Relationships between compactness and continuity**

Theorem 3. 4. 11 (1) Let (X,τ), (Y,υ) be topological spaces, $f: X\to Y$ be a continuous mapping. If $A\subset X$ is a compact set in X, then $f(A)\subset Y$ is a compact set in Y. (2) Let (X_j,τ_j), $j=1,2,\cdots,m$, be compact spaces. Then the product space $X=X_1\times\cdots\times X_m$ is a compact space.

Proof (1) Let $A\subset X$ be a compact set in X. For image set $f(A)\subset Y$, take any open covering of $f(A)$, $\mathfrak{U}=\{U\subset Y: U\in\upsilon\}$, such that $f(A)\subset\bigcup\limits_{U\in\mathfrak{U}}U$. By the continuity of f, we have that $\forall U\in\mathfrak{U}$, $f^{-1}(U)\subset X$ is an open set in X, and

$$A\subset f^{-1}(f(A))\subset f^{-1}\Big(\bigcup_{U\in\mathfrak{U}}U\Big)=\bigcup_{U\in\mathfrak{U}}f^{-1}(U) \tag{3.4.5}$$

Thus, $\bigcup\limits_{U\in\mathfrak{U}}f^{-1}(U)$ is an open covering of A. By compactness of A, there exist finite open sets $U_1, U_2, \cdots, U_n\in\mathfrak{U}$, such that the inverse image set $\{f^{-1}(U_1), f^{-1}(U_2),\cdots, f^{-1}(U_n)\}$ forms an open covering of A, i. e., $A\subset\bigcup\limits_{j=1}^{n}f^{-1}(U_j)=f^{-1}\Big(\bigcup\limits_{j=1}^{n}U_j\Big)$. Thus,

$f(A)\subset f\Big(f^{-1}\Big(\bigcup\limits_{j=1}^{n}U_j\Big)\Big)\subset\bigcup\limits_{j=1}^{n}U_j$, this implies that $f(A)$ is a compact set in Y.

(2) Without loss of generality, we prove the theorem for $j=1,2$. Let (X,τ), (Y,υ) be compact spaces. By the definition of product space, $\mathfrak{B}=\{U\times V: U\in\tau, V\in\upsilon\}$ is a base of $(X\times Y,\tau\times\upsilon)$. Take an open covering $\mathfrak{U}=\{U\times V\subset X\times Y: U\times V\in\mathfrak{B}\}$ of space $X\times Y$. We prove that there exists a finite sub-covering of $\widetilde{\mathfrak{U}}$ of $\mathfrak{U}$, such that $\widetilde{\mathfrak{U}}\supset X\times Y$.

In fact, $\forall x\in X$, the subspace $\{x\}\times Y$ is homeomorphic with $Y\Rightarrow$ the compactness of Y implies the compactness of $\{x\}\times Y\Rightarrow$ since $\mathfrak{U}$ is a covering of $X\times Y$, also a covering of compact set $\{x\}\times Y$, thus there exists a finite subcovering of $\mathfrak{U}$, denoted by $\widetilde{\mathfrak{U}}=\{U_x\times V_{y_1},\cdots,U_x\times V_{y_n}\subset\{x\}\times Y: U_x\times V_{y_j}\in\mathfrak{U}\}$ with $(U_x\times V_{y_j})\cap(\{x\}\times Y)\neq\varnothing$, s. t. $(U_x\times V_{y_1})\cup\cdots\cup(U_x\times V_{y_n})=U_x\times(V_{y_1}\cup\cdots\cup V_{y_n})\supset U_x\times Y\Rightarrow$ since $\bigcup\limits_{x\in X}U_x=X$ is compact, $\exists U_{x_1},\cdots,U_{x_m}$ with $U_{x_j}\times V_{y_k}\in\widetilde{\mathfrak{U}}$, s. t. $U_{x_1}\cup\cdots\cup U_{x_m}=X$, and $(U_{x_1}\cup\cdots\cup U_{x_m})\times Y=X\times Y\Rightarrow\{U_{x_j}\times V_{y_k}: U_{x_j}\times V_{y_k}\in\widetilde{\mathfrak{U}}\}\supset X\times Y$, where $1\leqslant j\leqslant m$, $1\leqslant k\leqslant n\Rightarrow X\times Y$ is a compact space. The proof is complete.

3) **Relationships between compact sets and closed sets**

Theorem 3. 4. 12 Let (X,τ) be a topological space.

(1) If X is a compact space, then any closed set $A\subset X$ is a compact set in X;

(2) If X is a T_2-type space, then any compact set $A\subset X$ is a closed set in X;

(3) If X is a T_2-type compact space, then a set $A\subset X$ is compact, if and only if A is a closed set in X.

Proof For (1) Let X be a compact space, $A\subset X$ be a closed set, and $\mathfrak{U}=\{V\subset X: V\in\tau\}$ be an open covering $A\subset\bigcup_{V\in\mathfrak{U}}V$ of A. Since A is closed, then $\complement A$ is an open set. We have

$$\left(\bigcup_{V\in\mathfrak{U}}V\right)\cup\complement A=\left(\bigcup_{V\in\mathfrak{U}}V\right)\cup(X\backslash A)=X,$$

so, $\mathfrak{U}'\equiv\left\{\left(\bigcup_{V\in\mathfrak{U}}V\right)\cup\complement A\right\}$ is an open covering of X. By the compactness of X, there exist finite open sets $B_j\in\mathfrak{U}'$, $j=1,2,\cdots,n$, such that $X=\bigcup_{j=1}^{n}B_j$. Clearly, one of $B_j\in\mathfrak{U}'$ must be $\complement A$, say, $B_n=\complement A$, s. t. $A\subset\{B_j\}_{j=1}^{n-1}\subset\mathfrak{U}$, this means that for any open covering $\mathfrak{U}$ of A, there exists the finite sub-covering $\{B_j\}_{j=1}^{n-1}$ of A, hence A is a compact set.

To prove (2), suppose X is a T_2-type space, $A\subset X$ is a compact set. We prove that A is a closed set. By two steps.

① Any point $x\notin A$ and compact set A are strongly separable.

In fact, since X is T_2-type space, for a point $x\in X$ and any $y\in A$, there exists an open neighborhood $U_y(x)$ containing x, and there exists an open neighborhood V_y containing y, such that $U_y(x)\cap V_y=\varnothing$. Thus, the open set family $\{V_y: y\in A\}$ is an open covering $A\subset\bigcup_{y\in A}V_y$ of A. Hence, there exist finite open sets $V_{y_1}, V_{y_2}, \cdots, V_{y_n}$, such that $A\subset\bigcup_{j=1}^{n}V_{y_j}\equiv V$, by compactness of A. Correspondingly, we have $U_{y_1}(x), U_{y_2}(x), \cdots, U_{y_n}(x)$. Let $U\equiv U(x)=\bigcap_{j=1}^{n}U_{y_j}(x)$. Then U is an open set containing x; moreover, $V=\bigcup_{k=1}^{n}V_{y_k}$ is an open set containing A. We see that $U\cap V_{y_k}=\left(\bigcap_{j=1}^{n}U_{y_j}\right)\cap V_{y_k}=\bigcap_{j=1}^{n}(U_{y_j}\cap V_{y_k})=\varnothing$ and $U\cap V=U\cap\bigcup_{k=1}^{n}V_{y_k}=\bigcup_{k=1}^{n}(U\cap V_{y_k})=\varnothing$. Thus we have proved that: for any point $x\in X$ and compact set A with $x\notin A$, there exists an open set U with $x\in U$, and there exists an open set V with $A\subset V$, such that $U\cap V=\varnothing$.

② A compact set A is closed.

In fact, by ①, there exists an open neighborhood U of x, such that U does not contain any point of A, thus any point $x\notin A$ could not be accumulative point of A. Thus we conclude that all accumulative points of A are contained in A, i. e., $A'\subset A$, and thus A is a closed set. The proof is complete.

(3) is clear by (1),(2).

Remember: in a T_2-type compact space: closed set $\Leftrightarrow$ compact set

in a compact space: closed set $\Rightarrow$ compact set

in a T_2-type space: closed set $\Leftarrow$ compact set

Example 3.4.5 In topological space (X,τ), generally speaking, a compact set is not a closed set.

The first example: If X is a finite set containing at least two points, take τ as the coarsest topology $\tau=\{X,\varnothing\}$, then for any point x in X, the subset $A=X\setminus\{x\}$ is a compact set in X, since it is a finite set; however, it is not a closed set, since there are only two closed sets in τ: X and $\varnothing$.

The second example: If X is an infinite set, take the co-finite topology $\tau_{\text{co-finite}}$ of X, denoted by $(X,\tau_{\text{co-finite}})$, then for any $x\in X$, the subset $A=X\setminus\{x\}$ is compact (see Example 3.4.4), but not closed, since $(X,\tau_{\text{co-finite}})$ is a T_1-space, a single point set $\{x\}$ is a closed set, thus its complimentary set $A=X\setminus\{x\}$ is an open one.

Theorem 3.4.13 Let (X,τ) be a compact space, (Y,υ) be a T_2-type topological space. Then

(1) Continuous mapping $f: X\to Y$ is a closed mapping;

(2) Continuous, one-one mapping $f: X\to Y$ is a homeomorphism mapping.

Proof (1) To prove a continuous mapping $f: X\to Y$ is a closed mapping, take any closed set $A\subset X\Rightarrow A$ is compact (by the compactness of X and by Theorem 3.4.12(1)) $\Rightarrow$ $f(A)\subset Y$ is a compact set in Y (by the continuity of f and Theorem 3.4.11(1)) $\Rightarrow f(A)$ is a closed set in Y (by T_2-type of Y and Theorem 3.4.12(3)) $\Rightarrow f$ is a closed set.

(2) To prove a continuous, one-one mapping $f: X\to Y$ is a homeomorphism mapping, denoted by f^{-1} the inverse mapping of f, and take any closed set $A\subset X\Rightarrow(f^{-1})^{-1}(A)=f(A)$ is a closed set $\Rightarrow$ the inverse image $(f^{-1})^{-1}(A)$ of f^{-1} is a closed set $\Rightarrow f^{-1}$ is continuous $\Rightarrow f: X\to Y$ is bisingle-valued, bicontinuous, thus a homeomorphism mapping. The proof is complete.

4) **Relationships between compactness and separability**

Theorem 3.4.14 Let (X,τ) be a T_2-type topological space. We have

(1) If $A\subset X$ is a compact set, and a point $x\in X$ is not in A, $x\notin A$, then there exist open neighborhoods $U\in\tau$ of A and $V\in\tau$ of x, respectively, such that $U\cap V=\varnothing$ (the strongly separated for point and compact set).

(2) If $A\subset X$ and $B\subset X$ are compact sets with $A\cap B=\varnothing$, then there exist open neighborhoods $U\in\tau$ of A and $V\in\tau$ of B, respectively, such that $U\cap V=\varnothing$ (the strongly

separated for two compact sets).

Proof (1) It has been proved in the Theorem 3.4.12(2).

(2) For compact sets $A\subset X, B\subset X$ with $A\cap B=\varnothing \Rightarrow \forall x\in A$ and compact set B, by (1), $\exists U_x\in\tau$ of x with $x\in U_x$, and $\exists V_x\equiv V_x\in\tau$ of B with $V_x\supset B$, s. t. $U_x\cap V_x=\varnothing$ $\Rightarrow$ since $\{U_x: x\in A\}$ is an open covering of A, and by compactness of A, $\exists$ finite covering $A\subset\bigcup_{j=1}^{n}U_{x_j}$. Correspondingly, $\exists V_{x_j}\supset B, 1\leqslant j\leqslant n$, s. t. $U_{x_j}\cap V_{x_k}=\varnothing, 1\leqslant k, j\leqslant n \Rightarrow$ let $U\equiv\bigcup_{j=1}^{n}U_{x_j}, V=\bigcap_{k=1}^{n}V_{x_k}\Rightarrow A\subset U, B\subset V$ with $U\cap V=\varnothing\Rightarrow$ (2)holds.

Theorem 3.4.15 Let (X,τ) be a topological space.

(1) If X is a T_2-type compact space, then X is a regular space.

(2) If X is a T_2-type compact space, then X is a normal space.

(3) If X is a compact regular space, then X is a normal space.

Proof (1) To prove X is a regular space, we take any point $x\in X$ and any closed set $A\subset X$ with $x\notin A\Rightarrow$ By the compactness of X, closed set A is compact $\Rightarrow$ By T_2-type and the compactness of X, and by Theorem 3.4.14 (1), for $x\notin A$ and compact set A, there exist open neighborhoods U of x and V of A, respectively, s. t. $U\cap V=\varnothing\Rightarrow(X,\tau)$ is a regular space.

(2) To prove X is a normal space, we take any two disjoint closed sets A, B in $X\Rightarrow$ By T_2-type and the compactness of X, and by Theorem 3.4.14(2), there exist two disjoint open neighborhoods U, V of A, B, respectively, s. t. $U\cap V=\varnothing\Rightarrow X$ is a normal space.

(3) Suppose that X is a compact regular space, we prove that it is a normal space. Take any closed set $F\subset X$ and a neighborhood G with $G\supset F$, we prove that there exists open neighborhood $O\in\tau$, s. t. $F\subset O\subset\bar{O}\subset G$, so that by Theorem 3.4.4, X is a normal space.

In fact, by compactness of X and closed set $F\subset X$, thus F is a compact set. Since X is a regular space, thus $\forall x\in F, \exists V_x\in\tau$ as an open neighborhood of x, s. t. $x\in V_x\subset\bar{V}_x\subset G$. Hence, the open set family $\{V_x: x\in F\}$ is an open covering $F\subset\bigcup_{x\in F}V_x$ of F. By the compactness of F, $\exists\{V_1,\cdots,V_n\}$, s. t. $F\subset\bigcup_{j=1}^{n}V_j$. Let $V=\bigcup_{j=1}^{n}V_j$. Then V is an open neighborhood of F. and $F\subset V\subset\bar{V}=\overline{\bigcup_{j=1}^{n}V_j}=\bigcup_{j=1}^{n}\bar{V}_j\subset G$, then, by Theorem 3.4.4, X is a normal space.

Example 3.4.6 A compact normal space is not necessarily a regular one.

Let $X=\{1,2,3\}$, define $\tau=\{X,\varnothing,\{1\},\{2\},\{1,2\}\}$, then (X,τ) is a normal, compact space, but it is not regular.

If closure $\overline{A}$ of $A\subset X$ is a compact set in X, then A is said to be **a relative compact set.**

5) **Compactness in metric spaces**

In a metric space (X,ρ) and Euclidean space $\mathbb{R}^n$, the compactness has more important properties and applications.

Definition 3.4.9 (bounded set) Let (X,ρ) be a metric space. A subset $A\subset X$ is said to be **a bounded set**, if there exists positive real number $M>0$, such that $\rho(x,y)<M$ holds for all $x,y\in A$. If X is bounded, then X is said to be **a bounded metric space.**

Theorem 3.4.16 Let (X,ρ) be a compact metric space. Then

(1) X is a bounded metric space.

(2) If $X\neq\varnothing$, then a continuous function $f: X\to\mathbb{R}$ can arrive its maximum value and minimum value, i.e., $\exists x_{\max}\in X$ and $\exists x_{\min}\in X$, respectively, s.t.

$$f(x_{\max})=\max_{x\in X}\{f(x)\} \text{ and } f(x_{\min})=\min_{x\in X}\{f(x)\}.$$

(3) If (Y,ρ_1) is a metric space, then a continuous mapping $f: X\to Y$ is uniformly continuous.

Proof (1) (X,ρ) is a compact metric space $\Rightarrow$ for the open ball family $\mathfrak{B}=\{B(x,1): x\in X\}$ of open covering of X, there exists finite covering $\{B(x_j,1)\}_{j=1}^n$ of $X\Rightarrow$ Let $M=\max\limits_{1\leqslant j,k\leqslant n}\{\rho(x_j,x_k)+2\Rightarrow \forall x,y\in X, \exists j,k\ (1\leqslant j,k\leqslant n)$, s.t. $x\in B(x_j,1)$, $y\in B(x_k,1)$, and holds $\rho(x,y)\leqslant\rho(x,x_j)+\rho(x_j,x_k)+\rho(x_k,y)<1+(M-2)+1=M\Rightarrow X$ is a bounded space.

(2) $X\neq\varnothing$ is a compact metric space $\Rightarrow f(X)$ is a compact set in $\mathbb{R}$ by continuity of $f\Rightarrow f(X)$ is a bounded set by (1), and a closed set (by Theorem 3.4.12) in $\mathbb{R}\Rightarrow \inf f(X)\leqslant f(x)\leqslant\sup f(X)\Rightarrow$ since $f(X)$ is a closed set, then $\inf f(X)=\min\limits_{x\in X}\{f(x)\}$, and $\sup f(X)=\max\limits_{x\in X}\{f(x)\}\Rightarrow f(x_{\min})=\min\limits_{x\in X}\{f(x)\}$ and $f(x_{\max})=\max\limits_{x\in X}\{f(x)\}$.

(3) Let $f: X\to Y$ be a continuous mapping from compact metric space (X,ρ) to metric space (Y,ρ_1). We prove the uniformly continuity of f, that is, to prove: $\forall\varepsilon>0$, $\exists\delta=\delta(\varepsilon)>0$, s.t. for any $x', x''\in X$ with $\rho(x',x'')<\delta$, it holds $\rho_1(f(x'),f(x''))<\varepsilon$.

By the continuity of f, $\forall x\in X$, $\forall\varepsilon>0$, $\exists\delta=\delta(\varepsilon,x)>0$, s.t. for $\rho(x',x)<\delta$, it holds

$\rho_1(f(x'),f(x))<\frac{\varepsilon}{2}$ $\Rightarrow$ an open ball family $\left\{B\left(x,\frac{\delta(\varepsilon,x)}{2}\right): x\in X\right\}$ is an open covering of (X,ρ) $\Rightarrow$ there exists a finite covering $\left\{B\left(x_j,\frac{\delta(\varepsilon,x_j)}{2}\right): x_j\in X, j=1,2,\cdots,n\right\}$ of X $\Rightarrow$ Let $\delta(\varepsilon)=\min\limits_{1\leqslant j\leqslant n}\{\delta_j\}$, then $\forall x',x''\in X$ with $\rho(x',x'')<\delta$, for $x'\in B\left(x_{j_0},\frac{\delta(\varepsilon,x_{j_0})}{2}\right)$, and then $x''\in B(x_{j_0},\delta(\varepsilon))$, thus $\rho_1(f(x'),f(x''))\leqslant\rho_1(f(x'),f(x_{j_0}))+\rho_1(f(x_{j_0}),f(x''))<\varepsilon$ $\Rightarrow$ $f(x)$ is uniformly continuous on X. The proof is complete.

Theorem 3.4.17 In $\mathbb{R}^n$, set $A\subset\mathbb{R}^n$ is a compact set, if and only if A is a bounded closed set.

Proof We prove the theorem only for $n=1$.

Necessity Let $A\subset\mathbb{R}$ be a compact set $\Rightarrow$ $A\subset\mathbb{R}$ is a bounded set (Theorem 3.4.16), and closed set ($\mathbb{R}$ is a T_2-space, thus a compact set is a closed set).

Sufficiency Let $A\subset\mathbb{R}$ be a bounded closed set $\Rightarrow$ if $A=\varnothing$, then it is compact. If $A\neq\varnothing$, then $\exists M>0$, s. t. $\forall x,y\in A$, it holds $\rho(x,y)<M$ $\Rightarrow$ take any $x_0\in A$, let $M_0=M+\rho(0,x_0)$ with origin $0\in\mathbb{R}\Rightarrow A\subset[-M_0,M_0]$ (by triangle inequality, $\forall x\in A$, it holds $\rho(x,0)\leqslant\rho(x,x_0)+\rho(x_0,0)<M+\rho(x_0,0)=M_0$) $\Rightarrow$ Since $[-M_0,M_0]\xleftrightarrow{\text{homeomor.}}[0,1]$, thus $[-M_0,M_0]$ is a compact space $\Rightarrow$ the closed subset A in compact space $[-M_0,M_0]$ is a compact set. The proof is complete.

2. Locally compactness

Definition 3.4.10 (locally compact) Let (X,τ) be a topological space. If every point $x\in X$ in X has a **compact neighborhood**, i. e., $\forall x\in X$, $\exists$ compact set $W\subset X$, s. t. $x\in\mathring{W}\subset W$, where $\mathring{W}$ is an open neighborhood of $x\in X$, then (X,τ) is said to be **a locally compact topological space.**

If a set $A\subset X$ regarded as a subspace is a locally compact topological space, then A is said to be a locally compact set in X, or simply, **a locally compact set.**

A compact topological space (X,τ) **is a locally compact one**, since $\forall x\in X$, take $W=X$, and regard W as a compact neighborhood of x. Then, $x\in\mathring{W}=\mathring{X}=X=W$, and $W=X$ is compact; this implies that (X,τ) is a locally compact space. But the inverse is not true.

Example 3.4.7 $(\mathbb{R},\tau)$ is a locally compact topological space, but it is not compact.

In fact, $\forall x \in \mathbb{R}$, take $r > 0$, then $\overline{B(x,r)} = \overline{(x-r,x+r)} = [x-r,x+r]$ is a compact set in $\mathbb{R}$ (bounded, closed), and $x \in B(x,r) = \mathring{B}(x,r) \subset \overline{B(x,r)}$, thus $\overline{B(x,r)}$ is a compact neighborhood of x. But $\mathbb{R}$ is not compact, since the open covering $\mathfrak{U} = \{(-n,n) : n \in \mathbb{N}\}$ of $\mathbb{R}$ does not have finite subcovering of $\mathbb{R}$.

For any real number a, the set $(-\infty, a)$ is a locally compact subset in $\mathbb{R}$; and so is $(a, +\infty)$.

Theorem 3.4.18 Let (X,τ) be a locally compact topological space. Then any closed subset $F \subset X$ is a locally compact set in X.

T_2-type locally compact topological space has the following property.

Theorem 3.4.19 Let (X,τ) be a T_2-type locally compact topological space. If $U \in \tau$ is an open set, then for any $x \in U$, there exists an open neighborhood V of x, such that $x \in V \subset \bar{V} \subset U$. Thus, a T_2-type locally compact topological space is a locally compact regular one (Fig. 3.4.6).

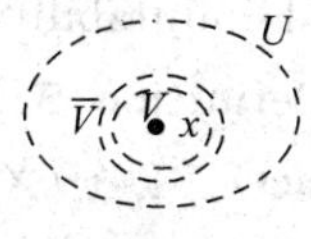

Fig. 3.4.6 Separated

Proof (X,τ) is a T_2-type locally compact space $\Rightarrow \forall x \in U$ with $U \in \tau$, $\exists$ compact set $F \subset X$, s. t. $x \in \mathring{F}$, and F is closed set by T_2-type of $X \Rightarrow (F, \tau_F)$ as a subspace is a T_2-type compact space, so that it is a regular space (by Theorem 3.4.15) $\Rightarrow U \cap \mathring{F}$ is an open neighborhood of x in subspace (F, τ_F), because it is an open neighborhood of x in the space $(X,\tau) \Rightarrow \exists$ an open neighborhood V of x in (F,τ_F) with $x \in V$, s. t. the closure $\bar{V}_F$ of V in space (F,τ_F) satisfies $x \in V \subset \bar{V}_F \subset W = U \cap \mathring{F} \Rightarrow \bar{V}_F = \overline{F \cap V} = \bar{F} \cap \bar{V} = F \cap \bar{V} = \bar{V}$ (by F is closed), this shows that the closure of V in subspace (F,τ_F) equals that of V in the space $(X,\tau) \Rightarrow \bar{V}_F = \bar{V} \subset W = U \cap \mathring{F} \subset F \Rightarrow \bar{V}$ is compact by the compactness of F.

On the other hand, $V \subset W$ implies $V = V \cap W$, thus open set V in (F,τ_F) is also open subset in the open set $W = U \cap \mathring{F}$. However, $W = U \cap \mathring{F}$ is an open set in (X,τ), so that V is an open set in (X,τ). Then we have $x \in V \subset \bar{V}_F = \bar{V} \subset U \cap \mathring{F} \subset U$ for V and $\bar{V}$; so that (X,τ) is a regular topological space.

Urysohn lemma can be generalized to T_2-type locally compact topological spaces.

Theorem 3.4.20 (Urysohn lemma) Let (X,τ) be a T_2-type locally compact topological space, $K \subset X$ be a compact subset, $U \in \tau$ be an open set with $K \subset U$. Then there exists a real-valued continuous function $f: X \to \mathbb{R}$, satisfying $f(x)|_{x \in K} = 1$, $f(x)|_{x \in \complement V} = 0$ with $0 \leqslant f(x) \leqslant 1$, $\forall x \in X$, and V is a compact subset of U, $V \subset U$.

Tietze extension theorem on a T_2-type locally compact topological space is as follows:

Theorem 3.4.21 Let (X, τ) be a T_2-type locally compact topological space, $K \subset X$ be a compact subset. If $f: K \to \mathbb{R}$ is a real-valued continuous function on K, then there exists a continuous function $F: X \to \mathbb{R}$ on X, satisfying $F(x)|_{x \in K} = f(x)$.

3. Other compactness of topological spaces

Definition 3.4.11 (Lindelöff compact) Let (X, τ) be a topological space. If for any open covering of X, there exists a countable subcovering, then X is said to be **Lindelöff compact**; and X is said to be **a Lindelöff space.**

Theorem 3.4.22 If topological space (X, τ) satisfies the second countability axiom, then X is a Lindeloff space; If (X, τ) is a Lindelöff metric space, then X satisfies the second countability axiom.

Definition 3.4.12 (countably compact, sequentially compact and accumulatively compact) Let (X, τ) be a topological space.

(1) If for any countably open covering of X, there exists a finite subcovering of X, then (X, τ) is said to be **a countably compact space.**

(2) If for any sequence in X, there exists a convergent subsequence in X, then (X, τ) is said to be **a sequentially compact space.**

(3) If for any infinite subset in X, there exist the accumulation points of this subset in X, then (X, τ) is said to be **an accumulatively compact space.**

For the above compactness, we have:

Theorem 3.4.23 Let (X, τ) be a topological space.

(1) If (X, τ) is a topological space, then

compact ⇒ countably compact ⇒ accumulatively compact

⇑

sequentially compact

(2) If (X, τ) satisfies the first countability axiom, then

compact ⇒ countably compact ⇒ accumulatively compact

⇕

sequentially compact

(3) If (X, τ) satisfies the first countability axiom and is T_1-type, then

compact ⇒ countably compact ⇔ accumulatively compact

⇕

sequentially compact

(4) If (X,ρ) is a metrlc space, then

compact $\Leftrightarrow$ countably compact $\Leftrightarrow$ accumulatively compact

$\Updownarrow$

sequentially compact

3.4.4 Topological linear spaces

In a topological space (X,τ), it is not necessary to have operation structures. However, we will introduce operation structures to (X,τ) in this subsection, then a new kind of topological spaces will be presented.

Definition 3.4.13 (topological linear space) Let (X,τ) be a topological space. If two operations addition $+$ and number product $\alpha\cdot$, $\alpha\in\mathbb{F}$, are endowed to X, such that $(X,+,\alpha\cdot)$ is a linear space on $\mathbb{F}$, and satisfy

(1) **operation** $+$ is continuous in τ, i.e., mapping $(x,y)\to x+y$ is continuous under topology τ;

(2) **operation** $\alpha\cdot$ is continuous in τ, i.e., mapping $(\alpha,x)\to\alpha x$ is continuous under topology τ; Then X is said to be **a topological linear space**, denoted by $(X,+,\alpha\cdot,\tau)$. Sometimes, **we agree on** adding "Hausdorff space" on a topological linear space.

Normed linear space $(X,\|\cdot\|)$, inner-product space $(X,(\cdot,\cdot))$, Euclidean space $\mathbb{R}^n$ are topological spaces. We have: Euclidean space $\subset$ inner-product space $\subset$ normed linear space $\subset$ topological linear space, metric space $\subset$ topological space

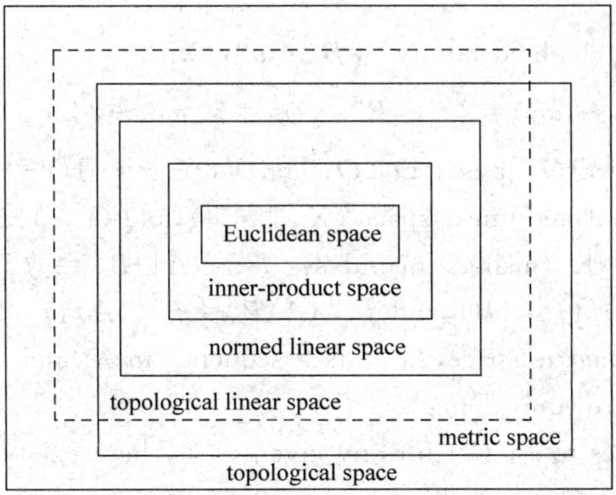

There are various spaces in the scientific areas, we will meet more other kind of spaces in our course.

Exercise 3

1. Let (X,τ) be a topological space, $A\subset X$ be a subset of X. Prove: (1) the closure $\overline{A}=A\cup A'$ has an equivalent definition: "the intersection of all closed subsets containing A, i.e., $\bigcap\limits_{\substack{A\subset F\\ F:\text{ closed}}} F$ with F closed and $A\subset F\subset X$, is called a closure of A, denoted by $\overline{A}=\bigcap\limits_{A\subset F} F$"; (2) the inner $\mathring{A}$ has an equivalent definition: "the inner $\mathring{A}$ of set A is the maximum open set contained in A, i.e., $\mathring{A}=\bigcup\limits_{\substack{G\subset A\\ G:\text{ open}}} G$"; (3) $\overline{A}=\mathring{A}\cup\partial A$.

2. Prove: (1) $A\subset B\Rightarrow A'\subset B'$; (2) $(A\cup B)'=A'\cup B'$; (3) $(A')'\subset A\cup A'$.

3. Prove: (1) $A\subset B\Rightarrow\overline{A}\subset\overline{B}$; (2) $\overline{A\cup B}=\overline{A}\cup\overline{B}$; (3) $\overline{\overline{A}}=\overline{A}$.

4. Prove: any metric space satisfies the first countable axiom.

5. Prove: any topological space satisfying the second countable axiom is separable.

6. Let (X,ρ) be a metric space. Prove: "(X,ρ) has a countable topological base" is equivalent to "(X,ρ) is separable".

7. Prove: the topologies $(\mathbb{R}^n,\tau)$ and $(\mathbb{R}^n,\tilde{\tau})$ of $\mathbb{R}^n$ are equivalent, (1) determined by "ball family"

$$\{B(x,r)\}\text{ with }B(x,r)=\left\{y\in X:\ \sqrt{\sum_{j=1}^{n}(x_j-y_j)^2}<r\right\};$$

(2) determined by "cuboid family" $\{\tilde{B}(x,a)\}$ with

$$\tilde{B}(x,a_1,\cdots,a_n)=\{x\in\mathbb{R}^n:\ |x_1|<a_1,\cdots,\ |x_n|<a_n\}.$$

i.e., prove: $\forall O\in\tau\Rightarrow\exists\tilde{O}\in\tilde{\tau}$, s.t. $\tilde{O}\subset O$; and $\forall\tilde{O}\in\tilde{\tau}\Rightarrow\exists O\in\tau$, s.t. $O\subset\tilde{O}$.

8. If (X,τ) is a normed linear space $(X,\|x\|)$, and (Y,υ) is a metric space (Y,ρ), prove Theorem 3.3.1. (using formulas: $f^{-1}(A\cup B)=f^{-1}(A)\cup f^{-1}(B)$, $f^{-1}(A\cap B)=f^{-1}(A)\cap f^{-1}(B)$, and $f^{-1}(A\backslash B)=f^{-1}(A)\backslash f^{-1}(B)$).

9. If (X,ρ) is a metric space, $\{x_j\}$ is a sequence in X, and $x\in X$ is a point of X. Prove: the following are equivalent:

(1) $\{x_j\}$ converges to x; (2) for any given $\varepsilon>0$, there exists $N\in\mathbb{Z}^+$, such that for $j>N$, it holds $\rho(x_j,x)<\varepsilon$; (3) $\lim\limits_{j\to+\infty}\rho(x_j,x)=0$.

10. Prove: The compound of two continuous functions is continuous.

11. Let $X_0\subset X$ be an open subspace of topological space (X,τ), i.e., (X_0,τ_0) is a subspace of (X,τ), and X_0 is an open subset of X. If $A\subset X_0$ is an open set of X_0, prove:

A is an open set of X.

12. Let (X,τ), (Y,υ) be topological spaces, $X_0\subset X$ be a subset of X, $Y_0\subset Y$ be a subset of Y, respectively. Regard (X_0,τ_0) and (Y_0,υ_0) as two subspaces of (X,τ) and (X,τ), we have $(X_0\times Y_0,\tau_0\times\upsilon_0)$ the product space; regard $X_0\times Y_0$ as a subset of $X\times Y$, we have the subspace $(X_0\times Y_0,(\tau\times\upsilon)_0)$. Prove: the topological spaces $(X_0\times Y_0,\tau_0\times\upsilon_0)$ and $(X_0\times Y_0,(\tau\times\upsilon)_0)$ have the same topological base.

13. Prove: the product topology defined in Definition 3.3.7 is the coarsest topology such that projective mappings are continuous; the quotient topology defined in Definition 3.3.10 is the finest topology such that quotient mapping is continuous.

14. Let $\mathfrak{B}=\{(a,b): -\infty<a<b<+\infty\}$ be a base of usual one-dimensional Euclidean space $(\mathbb{R},\tau)$, and $X=[0,1)\cup\{2\}$. Give a topological base $\mathfrak{B}_0$ of X such that X is a subtopological space of $\mathbb{R}$.

15. Let (X,τ) and (Y,υ) be topological spaces, $A\subset X$ be a closed set of X, $B\subset Y$ be a subset of Y. Prove: $A\times B$ is a closed set in $X\times Y$.

16. Prove: (X,τ) is a regular space, if and only if for any point $x\in X$ in X, and any open set $U\in\tau$ containing x, there exists an open set $O\in\tau$ satisfying $x\in O\subset\bar{O}\subset G$.

17. Let (X,τ) be a topological space. If X is not a connected set, prove: (1) there exist two nonempty closed sets $A,B\subset X$ with $A\cap B=\varnothing$, $A\cup B=X$; (2) there exist two nonempty open sets $G,O\subset X$ with $G\cup O=X$, $G\cap O=\varnothing$.

18. Prove: a metric space is T_2-space (Huasdorff space), also is normal space.

19. Prove: in a T_2-type space (X,τ), if two compact sets A and B are disjoint, then they are separable.

20. Prove: "compact set" has an equivalent definition: in a topological space (X,τ), subset $A\subset X$ is compact, if and only if for any covering of A, there exists a finite sub-covering of A.

21. Prove: in a compact metric space (X,ρ), for any nonempty decreasing closed set sequence $\{F_n\}$: $F_1\supset F_2\supset\cdots\supset F_n\supset\cdots$, if diameter $d(F_n)=\sup\{\rho(x,x'): x,x'\in F_n\}\to 0$, then there exists unique $x\in X$, such that $x\in\bigcap_{n=1}^{+\infty}F_n$.

22. Prove: If a topological space (X,τ) satisfies the second countable axiom, then X is a Lindelöff space; but its inverse is not true. Moreover, for a metric space (X,ρ), X satisfies the second countable axiom, if and only if X is a Lindelöff space.

Chapter 4

Foundation of Functional Analysis

Functional analysis is an important branch of modern mathematics. Its main subjects are **space theory** and **operator theory**(including **linear functional theory**).

The main objects in **space theory** are those spaces with both algebraic and topological structures, such as inner-product space, Hilbert space, normed linear space and Banach space, as well as mappings between these spaces. We give the definitions of these spaces, and their important properties, including bases, dimensions of Banach space, orthogonal expansions in Banach space and in Hilbert space in section 4. 1.

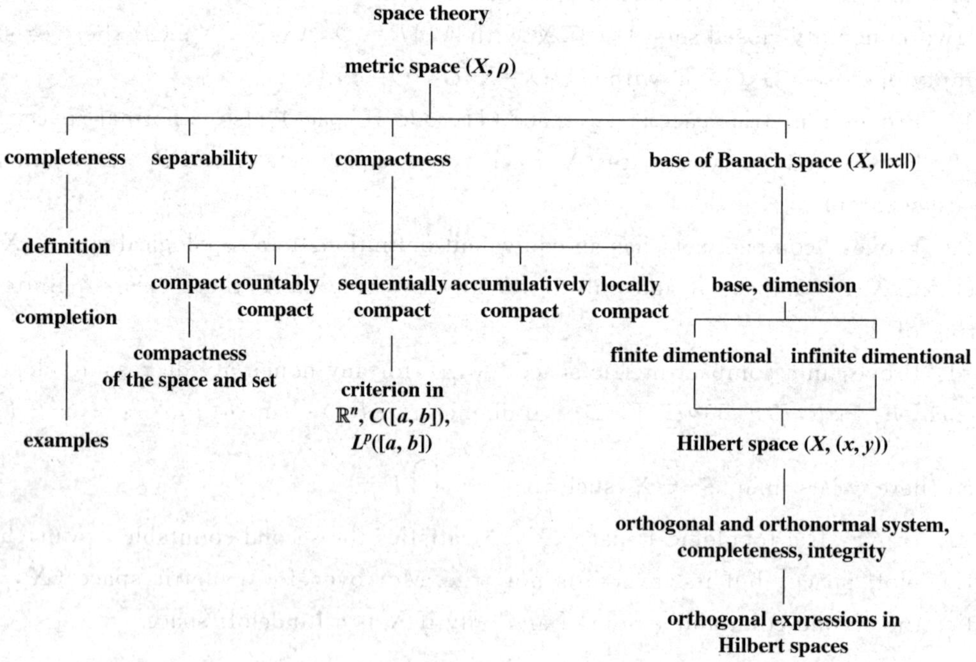

Note that, the metric space does not necessarily possess algebraic structure, however its topological structure is more visual, and easier to understanding, moreover, many sets can be endowed with "distance" possessing reality and visual sense, for instance, Euclidean space, inner-product space, and normed linear space all are metric ones. Thus "metric space" is also one of the main objects in the space theory. We discuss completeness, separability and compactness of such a space.

The **operator theory** and **linear functional theory** will be presented in sections 4.2 and 4.3, respectively. In the operator theory, the linear operators play main roles. Their certain essential properties on normed linear space are shown: open mapping theorem, and inverse-mapping theorem, closed-graph theorem. Then, in bounded linear operator space, various convergences of linear operators sequences and uniform-boundedness theorem are exhibited. Afterwards, the spectrum theory of linear operators is expounded. In the other parts of operator theory — linear functional theory, the conjugate (or dual) space and

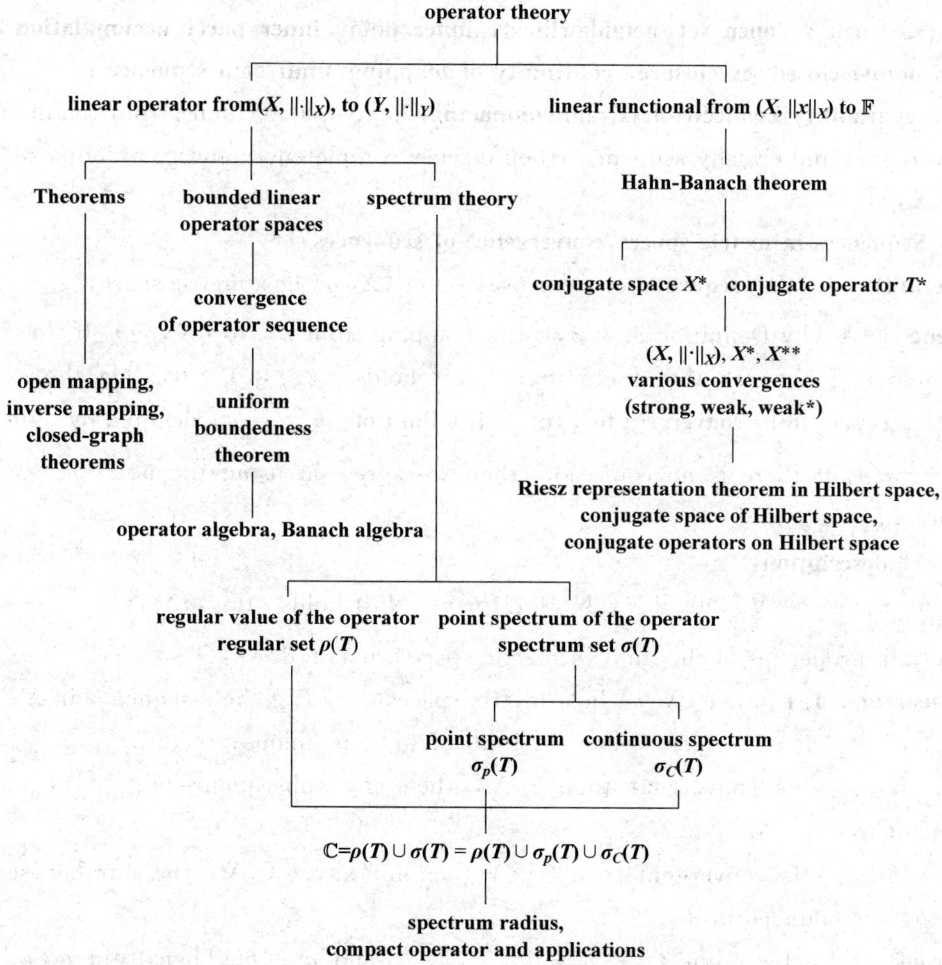

conjugate operator are introduced, and three kinds of convergences, strong, weak and weak* convergences are defined and compared. Special operators appeared in modern physics, such as unitary operators, and Hermitian operators, are presented.

4.1 Metric Spaces

4.1.1 Completion of metric spaces

1. Completeness of metric spaces

A metric space (X, ρ) is a special topological space, by Definition 3.1.1. Examples are as $\mathbb{R}^n$, l^2, $C[a,b]$, $L^1(E)$, $L^2(E)$. Many definitions for such a space can be referred in Chapter 3, such as **open set, neighborhood; inner point, inner part; accumulation point, isolated point; closed set, closure; continuity of mapping, limit of a sequence** in (X, ρ); as well as **separability, connectedness,** and **compactness** of (X, ρ). We start from the definitions of convergence and Cauchy sequence, then discuss completeness and completion of metric space (X, ρ).

1) Sequences in metric spaces, convergence of sequences

Definition 4.1.1 (sequence, convergence) Let (X, ρ) be a metric space, $\{x_n\}_{n \in \mathbf{Z}^+}$ be **a sequence** in X (by Definition 3.3.3, it is a mapping from $\mathbb{Z}^+$ to (X, ρ)). If there exists an element $a \in X$, such that for $n \to +\infty$, it holds $\rho(x_n, a) \to 0$, then the sequence $\{x_n\}_{n \in \mathbf{Z}^+}$ is said to be **convergent to** a, or, a **is a limit of** $\{x_n\}_{n \in \mathbf{Z}^+}$; denoted by $\lim_{n \to +\infty} x_n = a$, or, $x_n \to a$. If there is no confusion, then we agree on using the notation $\{x_n\}$ for sequence $\{x_n\}_{n \in \mathbf{Z}^+}$.

"ε-δ" **description**:

$$\lim_{n \to +\infty} x_n = a \Leftrightarrow \forall \varepsilon > 0, \exists N \in \mathbb{N}, \text{s. t. for } n > N, \text{it holds } \rho(x_n, a) < \varepsilon.$$

Certain properties of the limit in metric space are as follows:

Theorem 4.1.1 Let (X, ρ) be a metric space, $\{x_n\} \subset X$ be a sequence in X.

(1) If $\{x_n\}$ is convergent, then the limit of $\{x_n\}$ is unique;

(2) If $\{x_n\}$ is convergent to $a \in X$, then any subsequence $\{x_{n_k}\}$ of $\{x_n\}$ is convergent to a;

(3) If $\{x_n\}$ is convergent to $a \in X$, then for any $b \in X$, the number sequence $\{\rho(x_n, b)\}$ is bounded in $\mathbb{R}$.

Proof (1) Let $\lim_{n \to +\infty} x_n = a$, $\lim_{n \to +\infty} x_n = b$, and $a \neq b$. Then $\lim_{n \to +\infty} \rho(x_n, a) =$

$\lim\limits_{n\to+\infty}\rho(x_n,b)=0$; thus, $\forall\varepsilon>0$, $\exists N>0$, s. t. for $n>N$, hold $\rho(x_n,a)<\varepsilon$, $\rho(x_n,b)<\varepsilon$. So that for $n>N$, $\rho(a,b)\leqslant\rho(a,x_n)+\rho(x_n,b)<2\varepsilon$. By arbitrary $\varepsilon>0$, we have $\rho(a,b)=0$. This implies $a=b$.

(2) Let $\lim\limits_{n\to+\infty}x_n=a$. Then $\forall\varepsilon>0$, $\exists N>0$, s. t. for $n>N$, it holds $\rho(x_n,a)<\varepsilon$; hence, for any subsequence $\{x_{n_k}\}$, for $n_k>N$, it holds $\rho(x_{n_k},a)<\varepsilon$. This is $\lim\limits_{k\to+\infty}x_{n_k}=a$.

(3) Let $\lim\limits_{n\to+\infty}x_n=a$, then $\forall\varepsilon>0$, $\exists N>0$, s. t. for $n>N$, it holds $\rho(x_n,a)<\varepsilon\Rightarrow$ take $b\in X$ with $\rho(a,b)=\gamma$. Then, for $n>N$, it holds $\rho(x_n,b)\leqslant\rho(x_n,a)+\rho(a,b)<\varepsilon+\gamma\Rightarrow$ Let M be the maximum of $\rho(x_1,b)$, $\cdots$, $\rho(x_n,a)$, $\varepsilon+\gamma$. Then $\rho(x_n,b)<M(n\in\mathbb{Z}^+)$. This implies $\{\rho(x_n,b)\}$ is bounded. The proof is complete.

The continuity of mappings in metric spaces has the following "characteristic" theorem.

Theorem 4.1.2 Let (X,ρ) and (X_1,ρ_1) be two metric spaces, $f: X\to X_1$ be a mapping from X to X_1. Then $f: X\to X_1$ is a continuous mapping, if and only if $\forall\{x_n\}\subset X$ with $\lim\limits_{n\to+\infty}x_n=x$, $x\in X$ implies that sequence $\{f(x_n)\}\subset X_1$ converges to $f(x)$.

Proof Necessity For two metric spaces (X,ρ), (X_1,ρ_1), the continuity of a mapping $f: X\to X_1$ at $x_0\in X$, is equivalent to $\forall\varepsilon>0$, $\exists\delta>0$, s. t. for $\rho(x,x_0)<\delta$, it holds $\rho_1(f(x),f(x_0))<\varepsilon\Rightarrow\forall\{x_n\}\subset X$ with $\lim\limits_{n\to+\infty}x_n=x$, then $\forall\varepsilon>0$, $\exists N>0$, s. t. for $n>N$, it holds $\rho(x_n,x)<\delta\Rightarrow\rho_1(f(x_n),f(x))<\varepsilon$, i. e., $\lim\limits_{n\to+\infty}f(x_n)=f(x)$.

Sufficiency By assumption, $\forall\{x_n\}\subset X$ with $\lim\limits_{n\to+\infty}x_n=x\in X$ implies $\lim\limits_{n\to+\infty}f(x_n)=f(x)$. We use reduction to absurdity. If $\exists x_0\in X$, s. t. $f(x)$ is not continuous at x_0, then $\exists\varepsilon_0>0$, s. t. $\forall\delta=\dfrac{1}{n}$ with $n=1,2,\cdots$, $\exists x_n\in X$, it holds $\rho(x_n,x_0)<\dfrac{1}{n}$. But by $\rho_1(f(x_n),f(x_0))\geqslant\varepsilon_0$, this is a contradiction. So that we have the continuity of $f: X\to X_1$. The proof is complete.

2) Cauchy sequences in metric spaces

Definition 4.1.2 (Cauchy sequence) Let (X,ρ) be a metric space. A sequence $\{x_n\}\subset X$ is said to be **a Cauchy sequence (or a basic sequence) in** X, if for any given $\varepsilon>0$, there exists $N>0$, such that for $m,n>N$, it holds $\rho(x_m,x_n)<\varepsilon$.

3) Completeness in metric spaces and the completion of metric spaces

Definition 4.1.3 (complete metric space) Let (X,ρ) be a metric space. If for any Cauchy sequence $\{x_n\}\subset X$ in (X,ρ), there exists an element $x\in X$, such that it holds $\lim\limits_{n\to\infty}x_n=x$ in (X,ρ), then (X,ρ) is said to be **a complete metric space.**

A subset $A\subset X$ is said to be **a complete set in** (X,ρ), if A regarded as a subspace is a complete metric space.

Example 4.1.1 The rational number set $\mathbb{Q}$ with distance $\rho(x,y)=|x-y|$ is a metric space $(\mathbb{Q},\rho)$, and a subspace of $(\mathbb{R},\rho)$. But $(\mathbb{Q},\rho)$ is not complete.

In fact, for any rational Cauchy sequence $\{r_n\}$, $r_n\in\mathbb{Q}$, if it is convergent under distance $\rho(x,y)=|x-y|$, then the limit may be a rational number $r\in\mathbb{Q}$, or an irrational number $s\notin\mathbb{Q}$. For example, for Cauchy sequence $r_n=1+\frac{1}{n}$, it holds

$$\lim_{n\to+\infty} r_n=\lim_{n\to+\infty}\left(1+\frac{1}{n}\right)=1\in\mathbb{Q};$$

but for Cauchy sequence $(r_n)^n=\left(1+\frac{1}{n}\right)^n$, it holds (see *Advance Mathematics*, [11])

$$\lim_{n\to+\infty}(r_n)^n=\lim_{n\to+\infty}\left(1+\frac{1}{n}\right)^n=e\notin\mathbb{Q}.$$

Example 4.1.2 Let $\mathbb{R}^1\equiv\mathbb{R}$ be a real number set with distance $\rho(x,y)=|x-y|$. Then $(\mathbb{R}^1,\rho)$ is a complete metric space.

In fact, for any Cauchy sequence

$$\{x_n\}\subset\mathbb{R}^1 \tag{4.1.1}$$

in $(\mathbb{R}^1,|x-y|)$, we take $\varepsilon=1$, then $\exists N\in\mathbf{N}$, for $n,m>N$, it holds $|x_n-x_m|<1$. Thus, fix an integer $m_0>N$, then for $n>N$, it holds $|x_n-x_{m_0}|<1$. Thus by $|x_n|<|x_n-x_{m_0}|+|x_{m_0}|<1+|x_{m_0}|$, $\forall n>N$, it follows that $|x_n|\leqslant\max\{|x_1|,\cdots,|x_{m_0}|,|x_{m_0}|+1\}\equiv M$, $\forall n\in N$. This means that the sequence $\{x_n\}$ in (4.1.1) is a bounded sequence in $\mathbb{R}^1$. By Bolzano-Weierstrass Theorem[17], there exists a convergent subsequence $\{x_{n_k}\}_{k\in\mathbf{Z}^+}$ of the bounded sequence $\{x_n\}$ in (4.1.1), and there exists $x\in\mathbb{R}^1$, such that $\lim\limits_{k\to+\infty}x_{n_k}=x\in[-M,M]$.

Next, we prove that the sequence $\{x_n\}$ in (4.1.1) is convergent to this x.

In fact, $\forall\varepsilon>0$, $\exists N_1\in\mathbf{N}$, s.t. for $k>N_1$, it holds $|x_{n_k}-x|<\frac{\varepsilon}{2}$. By definition of the subsequence, we have $n_k\geqslant k$. On the other hand, since $\{x_n\}$ is a Cauchy sequence, $\exists N_2\in\mathbf{N}$, s.t. for $n,m>N_2$, it holds $|x_n-x_m|<\frac{\varepsilon}{2}$. Let $N=\max\{N_1,N_2\}$. Thus $k>N$; and for $n_k>N$, it holds $|x_{n_k}-x|<\frac{\varepsilon}{2}$; so that for $n>N$, it holds $|x_n-x|\leqslant|x_n-x_{n_k}|+|x_{n_k}-x|<\frac{\varepsilon}{2}+\frac{\varepsilon}{2}=\varepsilon$, and then $\lim\limits_{n\to+\infty}x_n=x\in\mathbb{R}^1$. This is the completeness of $\mathbb{R}^1$.

4) Some complete metric spaces

(1) $(\mathbb{R}^n, \rho(x,y))$

$$\mathbb{R}^n = \{x = (x_1, x_2, \cdots, x_n) : x_j \in \mathbb{R},\ j = 1,2,\cdots,n\}, \quad \rho(x,y) = \sqrt{\sum_{j=1}^{n} (x_j - y_j)^2};$$

(2) $(l^p, \rho(x,y)), 1 \leqslant p < +\infty$

$$l^p = \left\{x = (x_1, x_2, \cdots) : \sum_{j=1}^{+\infty} |x_j|^p < +\infty\right\}, \quad \rho(x,y) = \left(\sum_{j=1}^{\infty} |x_j - y_j|^p\right)^{\frac{1}{p}};$$

(3) $(l^\infty, \rho(x,y))$

$$l^\infty = \{x = (x_1, x_2, \cdots) : |x_j| < +\infty, j \in \mathbb{Z}^+\}, \quad \rho(x,y) = \sup_{j \in \mathbb{Z}^+} |x_j - y_j|;$$

(4) $(s, \rho(x,y))$

$$s = \{x = (x_1, x_2, \cdots) : x_j \in \mathbb{R}\}, \quad \rho(x,y) = \sum_{j=1}^{\infty} \frac{1}{2^j} \frac{|x_j - y_j|}{1 + |x_j - y_j|};$$

(5) $(S([a,b]), \rho(x,y))$

$$S([a,b]) = \{f : [a,b] \text{ a.e. finite and measurable}\}, \rho(x,y) = \int_{[a,b]} \frac{|f(t) - g(t)|}{1 + |f(t) - g(t)|} \mathrm{d}t;$$

(6) $(C([a,b]), \rho(x,y))$

$$C([a,b]) = \{x(t) : t \in [a,b], x(t) \text{ is continuous}\}, \quad \rho(x,y) = \max_{t \in [a,b]} |x(t) - y(t)|;$$

(7) $(C^k([a,b]), \rho(x,y)), k \in \mathbb{N}$

$C^k([a,b]) = \{x(t) : x^{(k)}(t) \in C([a,b])\}$, we agree on $C^{(0)}([a,b]) = C([a,b])$,

$$\rho(x,y) = \sum_{j=0}^{k} \max_{t \in [a,b]} |x^{(j)}(t) - y^{(j)}(t)|;$$

(8) $(C^\infty([a,b]), \rho(x,y))$

$$C^\infty([a,b]) = \{x(t) : x^{(k)}(t) \in C([a,b]),\ k \in \mathbb{N}\},$$

$$\rho(x,y) = \sum_{n=0}^{+\infty} \frac{1}{2^n} \frac{\max\limits_{t \in [a,b]} |x^{(n)}(t) - y^{(n)}(t)|}{1 + \max\limits_{t \in [a,b]} |x^{(n)}(t) - y^{(n)}(t)|};$$

(9) $(L^p(E), \rho(x,y)), 1 \leqslant p < +\infty, E \subset \mathbb{R}$ is a Lebesgue measurable set

$$L^p(E) = \left\{x(t) : \int_E |x(t)|^p \mathrm{d}t < +\infty\right\}, \quad \rho(x,y) = \left(\int_E |x(t) - y(t)|^p \mathrm{d}t\right)^{\frac{1}{p}};$$

(10) $(L^\infty(E), \rho(x,y))$, $E \subset \mathbb{R}$ is a Lebesgue measurable set

$$L^\infty(E) = \{x(t) : \operatorname*{esssup}_{t \in E} |x(t)| = \inf_{\substack{mE_0 = 0 \\ E_0 \subset E}} \sup_{t \in E \setminus E_0} |x(t)| < +\infty\},$$

$$\rho(x,y) = \operatorname*{esssup}_{t \in E} |x(t) - y(t)|.$$

5) Completion of metric spaces

To discuss the completion of a metric space, we need to introduce the concept of denseness in a metric space (recall and compare the Definition 3.2.4 in topological space (X,τ)).

Definition 4.1.4 (dense set) Let $A\subset X$ and $B\subset X$ be two subsets in metric space (X,ρ). If $B\subseteq\overline{A}$, then A is said to be **dense** in B. If $B=X$, and holds $X=\overline{A}$, then A is said to be **a dense subset in** X.

Remark **The meaning of the denseness in metric space** (X,ρ): "$B\subseteq\overline{A}$" tells that all points in B are accumulation points of A, i.e., "$\forall x\in B, \forall\varepsilon>0 \Rightarrow \exists y\in A, s.t.\ \rho(x,y)<\varepsilon$"; or equivalently,

$$\forall x \in B \Rightarrow \exists\{y_n\}\subset A,\quad \text{s. t. } \lim_{n\to+\infty}\rho(x,y_n)=0.$$

Definition 4.1.5 (completion of a metric space) For a metric space (X,ρ), the minimum complete metric space (X_0,ρ_0) containing (X,ρ) is said to be **the completion of metric space** (X,ρ), or simply, **the completion of** (X,ρ).

Or, an equivalent definition: For a metric space (X,ρ), if there exists a complete metric space (X_0,ρ_0), such that (X,ρ) is **equidistant isomorphic** with a dense subset $\widetilde{X}_0$ of (X_0,ρ_0), then (X_0,ρ_0) is said to be **the completion of metric space** (X,ρ); i.e., if there exists a complete metric space (X_0,ρ_0) and a dense subset $\widetilde{X}_0$ with $X_0=\overline{\widetilde{X}_0}$, as well as there exists mapping $T: X\to\widetilde{X}_0$, such that $\forall x,y\in X$, holds $\rho_0(Tx,Ty)=\rho(x,y)$. Here, the mapping $T: X\to\widetilde{X}_0$ is called **an equidistant isomorphism**, i.e., it is a one-one, bisingle valued, bicontinuous mapping from X onto $\widetilde{X}_0$, and $\rho_0(Tx,Ty)=\rho(x,y)$.

Theorem 4.1.3 Let (X,ρ) be a metric space. Then there exists the completion (X_0,ρ_0) *of* (X,ρ); Moreover, the completion of (X,ρ) is unique in the equidistant isomorphism sense.

Proof We will construct a complete metric space (X_0,ρ_0) with a dense subset $\widetilde{X}_0\subset X_0$, such that $X=\overline{\widetilde{X}_0}$, and $(\widetilde{X}_0,\rho_0)$ is equidistant isomorphic with (X,ρ). We have six steps.

(1) **Define "equivalent relation"** $\sim$: For two Cauchy sequences $\{x_n\}$ and $\{y_n\}$ in (X,ρ), if

$$\lim_{n\to+\infty}\rho(x_n,y_n)=0,$$

then they are said to be equivalent, denoted by $\{x_n\}\sim\{y_n\}$. It is clear that $\sim$ satisfies self-reciprocity, symmetry property, and transmission property, so it is an equivelant relation. Denote equivalent classes by $\xi=\{x_n\},\eta=\{y_n\},\cdots$, and denote

$$X_0=\{\xi,\eta,\zeta,\cdots\}=\{\{x_n\},\{y_n\},\{z_n\},\cdots\}.$$

(2) **Define "distance"** on $X_0=\{\xi,\eta,\zeta,\cdots\}$: For $\xi=\{x_n\}\in X_0$ and $\eta=\{y_n\}\in X_0$, define

$$\rho_0(\xi,\eta)=\lim_{n\to+\infty}\rho(x_n,y_n). \tag{4.1.1}$$

We prove that ρ_0 is a distance on $X_0=\{\xi,\eta,\zeta,\cdots\}$.

1) The limit in (4.1.1) exists, and is a non-negative real number: We prove that real sequence $\{\alpha_n\}=\{\rho(x_n,y_n)\}$ is a Cauchy sequence, then by the continuity of the real number, there exists a limit $\lim\limits_{n\to+\infty}\alpha_n=\lim\limits_{n\to+\infty}\rho(x_n,y_n)\in\mathbb{R}$; and $\alpha_n\geqslant 0$ implies $\lim\limits_{n\to+\infty}\alpha_n=\alpha\geqslant 0$.

In fact, real number sequence $\{\alpha_n\}=\{\rho(x_n,y_n)\}$ is a Cauchy sequence, since

$$\begin{aligned}|\alpha_n-\alpha_m|&=|\rho(x_n,y_n)-\rho(x_m,y_m)|\\&\leqslant|\rho(x_n,y_n)-\rho(x_m,y_n)|+|\rho(x_m,y_n)-\rho(x_m,y_m)|\\&\leqslant\rho(x_n,x_m)+\rho(y_n,y_m).\end{aligned}$$

Take two Cauchy sequences $\xi=\{x_n\},\eta=\{y_n\}$, i.e., $\forall\varepsilon>0$, $\exists N>0$, s.t. for $n,m>N$, it holds

$$\rho(x_n,x_m)<\frac{\varepsilon}{2},\quad\rho(y_n,y_m)<\frac{\varepsilon}{2}.$$

Thus, $\forall\varepsilon>0$, $\exists N>0$, s.t. for $n,m>N$, it holds $|\alpha_n-\alpha_m|<\varepsilon$. This means $\{\alpha_n\}=\{\rho(x_n,y_n)\}$ is a Cauchy sequence in $\mathbb{R}$.

2) The limit in (4.1.1) is independent of the choice of Cauchy sequence $\{x_n\}$ in ξ: We prove that for two choices $\{x_n\},\{x'_n\}\in\xi,\{y_n\},\{y'_n\}\in\xi$, it holds

$$\rho_0(\xi,\eta)=\lim_{n\to+\infty}\rho(x_n,y_n)=\lim_{n\to+\infty}\rho(x'_n,y'_n).$$

In fact, by

$$\begin{aligned}|\rho(x_n,y_n)-\rho(x'_n,y'_n)|&\leqslant|\rho(x_n,y_n)-\rho(x'_n,y_n)|+|\rho(x'_n,y_n)-\rho(x'_n,y'_n)|\\&\leqslant\rho(x_n,x'_n)+\rho(y_n,y'_n);\end{aligned}$$

and $\{x_n\},\{x'_n\}\in\xi$ imply $\rho(x_n,x'_n)\to 0$; $\{y_n\},\{y'_n\}\in\eta$ imply $\rho(y_n,y'_n)\to 0$, thus the right-hand side of the above tends to zero, thus $\lim\limits_{n\to+\infty}\rho(x_n,y_n)=\lim\limits_{n\to+\infty}\rho(x'_n,y'_n)$.

3) The limit $\rho_0(\xi,\eta)=\lim\limits_{n\to+\infty}\rho(x_n,y_n)$ in (4.1.1) satisfies the three conditions of the distance (left as exercise 7). Hence, (X_0,ρ_0) is a metric space with set $X_0=\{\xi,\eta,\zeta,\cdots\}$, and distance $\rho_0(\xi,\eta)=\lim\limits_{n\to+\infty}\rho(x_n,y_n)$.

(3) **Define a dense subset** $\widetilde{X}_0\subset X_0=\{\xi,\eta,\zeta,\cdots\}$, i.e., $X_0=\overline{\widetilde{X}_0}$: By constructing a constant sequence family $\widetilde{X}_0$.

In fact, for $x \in X$, it is clear that $\{x_n = x\}_{n \geq 1}$ is a Cauchy sequence, called **a constant sequence**, denoted by $\tilde{\xi}_x \equiv \{x, x, \cdots\} \in X_0$. The set of all constant sequences denotes by $\tilde{X}_0 = \{\tilde{\xi}_x : x \in X\} \subset X_0$, we prove $X_0 = \overline{\tilde{X}_0}$.

In fact, $\forall\ \xi = \{x_n\} \in X_0 \Rightarrow \forall \varepsilon > 0, \exists N > 0$, s. t. for $n, m > N$, it holds $\rho(x_n, x_m) < \varepsilon \Rightarrow \forall j > N$, $\exists\ \tilde{\xi}_j \equiv \tilde{\xi}_{x_j} = \{x_j, x_j, \cdots, x_j, \cdots\} \in \tilde{X}_0$, s. t. $\rho_0(\tilde{\xi}_j, \xi) = \lim\limits_{n \to +\infty} \rho(x_j, x_n) = 0 \Rightarrow \lim\limits_{j \to +\infty} \rho_0(\tilde{\xi}_j, \xi) = 0 \Rightarrow \forall \xi = \{x_n\} \in X_0, \exists\ \tilde{\xi}_j \in \tilde{X}_0$, s. t. $\lim\limits_{j \to +\infty} \rho_0(\tilde{\xi}_j, \xi) = 0 \Rightarrow X_0 = \overline{\tilde{X}_0}$.

(4) (X_0, ρ_0) **is a complete metric space**: We prove "for a Cauchy sequence $\{\xi_n\} \subset X_0$, $\exists\ \eta \in X_0$, s. t. $\lim\limits_{n \to +\infty} \rho_0(\xi_n, \eta) = 0$".

In fact, $\forall \varepsilon = \dfrac{1}{n}$, $\forall\ \xi_n \in X_0 = \overline{\tilde{X}_0}$, $\exists\ \tilde{\eta}_n \in \tilde{X}_0$, s. t. $\rho_0(\xi_n, \tilde{\eta}_n) < \dfrac{1}{n}$, for $n = 1, 2, \cdots \Rightarrow$ $\forall n$, a constant sequence $\tilde{\eta}_n = \{y_n, y_n, \cdots, y_n, \cdots\} \in \tilde{X}_0 \subset X_0$ is an equivalent class in $X_0 \Rightarrow \{\tilde{\eta}_n\} \subset X_0$ satisfies $\rho_0(\tilde{\eta}_n, \tilde{\eta}_m) = \rho(y_n, y_m)$, where $y_n \in X$, and $\rho(y_n, y_m) = \rho_0(\tilde{\eta}_n, \tilde{\eta}_m) \leqslant \rho_0(\tilde{\eta}_n, \xi_n) + \rho_0(\xi_n, \xi_m) + \rho_0(\xi_m, \tilde{\eta}_m) < \dfrac{1}{n} + \rho_0(\xi_n, \xi_m) + \dfrac{1}{m} \Rightarrow$ since $\{\xi_n\} \subset X_0$ is a Cauchy sequence, then $\lim\limits_{n, m \to +\infty} \rho_0(\xi_n, \xi_m) = 0 \Rightarrow \lim\limits_{n, m \to +\infty} \rho(y_n, y_m) = 0 \Rightarrow \{y_m\} \subset X$ is a Cauchy sequence $\Rightarrow \{y_m\}$ belongs to some equivalent class $\Rightarrow \exists\ \eta \in X_0$, s. t. $\{y_m\} \in \eta$, and holds $\rho_0(\xi_n, \eta) \leqslant \rho_0(\xi_n, \tilde{\eta}_n) + \rho_0(\tilde{\eta}_n, \eta) < \dfrac{1}{n} + \rho_0(\tilde{\eta}_n, \eta) \Rightarrow \eta \in X_0$ is the equivalent class that contains $\{y_m\} \Rightarrow$ for $\{y_n, y_n, \cdots, y_n, \cdots\} = \tilde{\eta}_n \in \tilde{X}_0$, $\forall n = 1, 2, \cdots$, it holds $\rho_0(\tilde{\eta}_n, \eta) = \lim\limits_{m \to +\infty} \rho(y_n, y_m) = 0 \Rightarrow$ for enough large n, it holds $\rho_0(\tilde{\eta}_n, \eta) < \dfrac{1}{n} \Rightarrow$ there exists $\eta \in X_0$, s. t. $\lim\limits_{n \to +\infty} \rho_0(\xi_n, \eta) = 0 \Rightarrow (X_0, \rho_0)$ is a complete space.

(5) $\tilde{X}_0 \subset X_0$ **and** X **are equidistant**: Define the mapping $T: x \to \xi_x$ with the following properties.

① T is an one-one surjective mapping from X to $\tilde{X}_0$: since $\forall x \in X, \exists !\ \xi_x \in \tilde{X}_0$; conversely, $\forall \xi \in \tilde{X}_0$ is a constant sequence, thus $\xi = \{x_1, x_2, \cdots, x_n, \cdots\}$ is as $x_1 = x_2 = \cdots = x_n = \cdots \equiv x \in X$;

② T is an equidistant mapping from X to $\tilde{X}_0$: since it holds

$$\rho_0(\xi, \eta) = \lim_{n \to +\infty} \rho(x_n, y_n) = \lim_{n \to +\infty} \rho_0(\tilde{\xi}_x, \tilde{\eta}_y) = \rho(x, y);$$

③ T and T^{-1} are continuous: clearly, from ②.

(6) **The completion** (X_0, ρ_0) **of** (X, ρ) **is uniquely determined**: We omit the proof, and

refer to [6].

Example 4.1.3 Let $X=\{x=(\xi_1,\xi_2,\cdots,\xi_k,0,0,\cdots):\xi_j\in\mathbb{R},1\leqslant j\leqslant k,k\in\mathbb{Z}^+\}$. It is a subspace of $l^p=\left\{x=(x_1,x_2,\cdots):\sum_{j=1}^{+\infty}|x_j|^p<+\infty\right\}$, but it is not complete with respect to l^p-distance $\rho(x,y)=\left(\sum_{j=1}^{\infty}|x_j-y_j|^p\right)^{\frac{1}{p}}$.

In fact, if we take a sequence in X

$$x_1=(1,0,0\cdots),x_2=\left(1,\frac{1}{2},0,0\cdots\right),\cdots,x_n=\left(1,\frac{1}{2},\cdots,\frac{1}{2^{n-1}},0,0,\cdots\right),\cdots,$$

it is easy to see that $\{x_n\}\subset X$ is a Cauchy sequence in $\rho(x,y)=\left(\sum_{j=1}^{\infty}|x_j-y_j|^p\right)^{\frac{1}{p}}$, but it does not have limit in subspace X. We see that X is dense in l^p, and l^p is the completion of the space $X=\{x=(\xi_1,\xi_2,\cdots,\xi_k,0,0,\cdots):\xi_j\in\mathbb{R},1\leqslant j\leqslant k,k\in\mathbb{Z}^+\}$.

Example 4.1.4 $C([a,b])$ is a complete metric space with distance $\rho(f,g)=\max_{x\in[a,b]}|f(x)-g(x)|$; And $L^p([a,b])$ is a complete metric space with distance $\rho(x,y)=\left(\int_{[a,b]}|x(t)-y(t)|^p\,\mathrm{d}t\right)^{\frac{1}{p}}$, $1\leqslant p<+\infty$. We prove the completeness of these two important spaces.

Proof (1) For any Cauchy sequence $\{x_n(t):t\in[a,b]\}$ in $C([a,b])$, by definition, $\forall\varepsilon>0,\exists N\in\mathbb{N}$, s.t. for $n,m>N$, it holds $\max_{t\in[a,b]}\{|x_n(t)-x_m(t)|\}<\varepsilon$; i.e., $|x_n(t)-x_m(t)|<\varepsilon$ holds for $t\in[a,b]$, uniform; also, $\lim_{n\to+\infty}x_n(t)$ uniformly converges on $[a,b]$. By a result about "uniformy convergence" in *Advanced Mathematics*, $\exists x(t)\in C([a,b])$, s.t. $\lim_{n\to+\infty}x_n(t)=x(t)$. This implies the completeness of $C([a,b])$.

(2) For any Cauchy sequence $\{x_n(t):t\in E\}$ in $L^p(E)$, and for $k\in\mathbb{Z}^+$, there exists a subsequence $n_k\in\mathbb{Z}^+$, such that $\left\{\int_E|x_{n_k}(t)-x_{n_{k+1}}(t)|^p\,\mathrm{d}t\right\}^{\frac{1}{p}}<\frac{1}{2^k}$, $k=1,2,\cdots$. Then for any measurable subset $e\subset E$ with $me<+\infty$, by Hölder inequality, it holds

$$\int_e|x_{n_k}(t)-x_{n_{k+1}}(t)|\,\mathrm{d}t\leqslant\left\{\int_e|x_{n_k}(t)-x_{n_{k+1}}(t)|^p\,\mathrm{d}t\right\}^{\frac{1}{p}}\left\{\int_e 1^q\,\mathrm{d}t\right\}^{\frac{1}{q}}<\frac{1}{2^k}(me)^{\frac{1}{q}},$$

where $\frac{1}{p}+\frac{1}{q}=1$. Thus series $\sum_{k=1}^{+\infty}\int_e|x_{n_k}(t)-x_{n_{k+1}}(t)|\,\mathrm{d}t$ is convergent. Further, $|x_{n_k}(t)-x_{n_{k+1}}(t)|\geqslant 0$ guarantees (see [6])

$$\sum_{k=1}^{+\infty}\int_e |x_{n_k}(t)-x_{n_{k+1}}(t)|\,\mathrm{d}t=\int_e \sum_{k=1}^{+\infty}|x_{n_k}(t)-x_{n_{k+1}}(t)|\,\mathrm{d}t.$$

Then, series

$$|x_{n_1}(t)|+|x_{n_2}(t)-x_{n_1}(t)|+|x_{n_3}(t)-x_{n_2}(t)|+\cdots$$

converges on the subset e, a. e.; Moreover, since $e\subset E$ has finite measure, thus sum function $x(t)$ is L-measurable; Then the series $x_{n_1}(t)+x_{n_2}(t)-x_{n_1}(t)+x_{n_3}(t)-x_{n_2}(t)+\cdots$ converges to $x(t)$ on E, a. e.; This implies that $\lim\limits_{k\to+\infty}\{x_{n_k}(t)\}=x(t)$ on E, a. e..

Next, we prove: $x(t)\in L^p(E)$, and $\lim\limits_{n\to+\infty}\left\{\int_E |x_n(t)-x(t)|^p\,\mathrm{d}t\right\}^{\frac{1}{p}}=0.$

Since $\{x_n(t):t\in E\}$ is a Cauchy sequence, $\forall\,\varepsilon>0$, $\exists\,N\in\mathbf{N}$, s. t. for $n, n_k>N$, it holds

$$\left\{\int_E |x_{n_k}(t)-x_n(t)|^p\,\mathrm{d}t\right\}^{\frac{1}{p}}<\varepsilon.$$

Fix the n, let $k\to+\infty$, then by a theorem in *The Real Variable Function Course*[6],[17], it follows

$$\begin{aligned}\lim_{k\to+\infty}\left\{\int_E |x_{n_k}(t)-x_n(t)|^p\,\mathrm{d}t\right\}^{\frac{1}{p}}&=\left\{\int_E \lim_{n_k\to+\infty}|x(t)-x_n(t)|^p\,\mathrm{d}t\right\}^{\frac{1}{p}}\\&=\left\{\int_E |x(t)-x_n(t)|^p\,\mathrm{d}t\right\}^{\frac{1}{p}}<\varepsilon.\end{aligned}$$

So that, $\forall\,\varepsilon>0$, $\exists\,N\in\mathbf{Z}^+$, s. t. for $n>N$, it holds $\left\{\int_E |x(t)-x_n(t)|^p\,\mathrm{d}t\right\}^{\frac{1}{p}}<\varepsilon.$

Moreover, by $x(t)-x_n(t)\in L^p(E)$, then $x(t)=\{x(t)-x_n(t)\}+x_n(t)\in L^p(E)$, this implies that $\{x_n(t):t\in E\}\subset L^p(E)$ converges to $x(t)\in L^p(E)$. The completeness of $L^p(E)$ is proved.

We emphasize that $C([a,b])$ is a subspace of $L^2([a,b])$ with distance $\rho(x,y)=\left(\int_{[a,b]}|x(t)-y(t)|^2\,\mathrm{d}t\right)^{\frac{1}{2}}$, but it is not complete under this distance $\rho(x,y)$. The reason is as follows.

Take a point $c\in(a,b)$, and a function $x_n(t)=\arctan n(t-c)$, $t\in[a,b]$, $n\in\mathbf{Z}^+$, we see that this $x_n(t)\in C([a,b])\subset L^2([a,b])$; and the function $x(t)=\begin{cases}\dfrac{\pi}{2}, & c<t\leqslant b\\ 0, & t=c\\ -\dfrac{\pi}{2}, & a\leqslant t<b\end{cases}$

$\in L^2([a,b])$ is the limit of $x_n(t)$ in $L^2([a,b])$ satisfying $\lim\limits_{n\to+\infty}\rho(x_n,x)=\lim\limits_{n\to+\infty}\left(\int_{[a,b]}|x_n(x)-x(x)|^2\mathrm{d}x\right)^{\frac{1}{2}}=0$. Thus $\{x_n(t)\}\subset C([a,b])$ is a Cauchy sequence with the distance $\rho(x,y)=\left(\int_{[a,b]}|x(t)-y(t)|^2\mathrm{d}t\right)^{\frac{1}{2}}$; However, $x(t)$ is not continuous, so that $C([a,b])$ is not complete with distance $\rho(x,y)=\left(\int_{[a,b]}|x(t)-y(t)|^2\mathrm{d}t\right)^{\frac{1}{2}}$.

The above example tells that: the completeness of metric space depends fundamentally on the distance of the space, and has very closed relationship with the distance, for example, there are two distances on X, such that (X,ρ_1) and (X,ρ_2) are two "metric spaces", the first one can be complete, but the second one may not.

Example 4.1.5 Let $P([a,b])$ be the set of all polynomials on $[a,b]$. Consider the completeness of $P([a,b])$ as a subspace with respect to the metric space $(C([a,b]),\rho(x,y))$.

Clearly, with the distance $\rho(x,y)$ of $C([a,b])$, subset $P([a,b])$ is a subspace. However, following sequence

$$p_1(t)=1,\quad p_2(t)=1+\frac{t}{2},\quad \cdots,\quad p_n(t)=1+\frac{t}{2}+\cdots+\frac{t^{n-1}}{2^{n-1}},\quad \cdots$$

in $P([a,b])$ is a Cauchy sequence with the distance of $C([a,b])$, but it does not have limit in $P([a,b])$ (why?). On the other hand, $P([a,b])$ is dense in $C([a,b])$, thus $C([a,b])$ is the completion of $P([a,b])$.

Since normed linear space, inner-product space, and Euclidean space all are metric ones, thus the discussions above are available for them. Then, we can use the completion theory of metric spaces to them.

A complete normed linear space is called **Banach space**, a complete inner-product space is called **Hilbert space**.

2. Important properties of complete metric spaces

Theorem 4.1.4 Let (X,ρ) be a complete metric space. Then any nonempty closed set in X is a complete subspace; Conversely, any complete subspace of X is a closed subspace of X.

Proof **Necessity** Since in nonempty closed subset F of complete metric space (X,ρ), any Cauchy sequence $\{x_n\}\subset F$ converges to $x_0\in X$, and $\lim\limits_{n\to+\infty}x_n=x_0\in\bar{F}=F$, thus $x_0\in F$ implies that (F,ρ) is a complete subspace.

Sufficiency Suppose (E,ρ) is a complete subspace of complete metric space (X,ρ).

To prove E is a closed subset of X, take any $a\in\bar{E}$, then $\exists\{x_n\}\subset E$, s. t. $\lim\limits_{n\to+\infty} x_n=a \Rightarrow \{x_n\}\subset E$ converges by definition $\Rightarrow \{x_n\}$ is a Cauchy sequence $\Rightarrow \exists b\in E$, s. t. $\lim\limits_{n\to+\infty} x_n = b$, by the completeness of $E \Rightarrow$ any two limits $b\in E$ and $a\in\bar{E}$ are equal, since the metric space is T_2-type $\Rightarrow a\in E \Rightarrow \bar{E}=E\Rightarrow (E,\rho)$ is a closed subspace.

We now work out so-called **fixed-point theorem**, it is very important and useful, as well as has extensive and wide applications.

Theorem 4. 1. 5 For a mapping $T: X\to X$ from complete metric space (X,ρ) to itself, if $\forall x,y\in X$, holds inequality

$$\rho(T(x),T(y))\leqslant\theta\,\rho(x,y) \tag{4.1.2}$$

with $0\leqslant\theta<1$, then there exists unique fixed point $x_0\in X$ in X satisfying $T(x_0)=x_0$. A mapping T satisfying (4. 1. 2) is said to be a compression mapping.

Proof Note that firstly, (4. 1. 2) implies the continuity of T:

In fact, $\forall x\in X$, $\forall\varepsilon>0$, $\exists\delta>0$, s. t. for $\rho(x,y)<\delta$, it holds $\rho(T(x),T(y))<\varepsilon$ $\left(\text{if } \theta=0, \text{take } \delta=\varepsilon;\ \text{if } 0<\theta<1, \text{take } \delta=\frac{\varepsilon}{\theta}\right)$; So that $T: X\to X$ is continuous at $x\in X$, and then it is continuos on X.

Then, take any point $x_1\in X$, and let $x_2=T(x_1), x_3=T(x_2),\cdots,x_{n+1}=T(x_n),\cdots$. We prove that this $\{x_n\}\subset X$ is a Cauchy sequence. In fact,

$$\rho(x_1,x_2)=\rho(T(x_0),T(x_1))\leqslant\theta\,\rho(x_0,x_1)=\theta\,\rho(x_0,T(x_0)),$$

$$\rho(x_2,x_3)=\rho(T(x_1),T(x_2))\leqslant\theta\,\rho(x_1,x_2)=\theta^2\,\rho(x_0,T(x_0)),$$

……

$$\begin{aligned}\rho(x_n,x_{n+1})&=\rho(T(x_{n-1}),T(x_n))\leqslant\theta\,\rho(x_{n-1},x_n)=\theta\,\rho(T(x_{n-2}),T(x_{n-1}))\\&\leqslant\theta^2\,\rho(x_{n-2},x_{n-1})\leqslant\cdots\leqslant\theta^n\,\rho(x_0,T(x_0)),\quad n=1,2,\cdots.\end{aligned}$$

Thus, $\forall p\in\mathbb{N}$, if $n\to+\infty$, it holds

$$\begin{aligned}\rho(x_n,x_{n+p})&\leqslant\rho(x_n,x_{n+1})+\rho(x_{n+1},x_{n+2})+\cdots+\rho(x_{n+p-1},x_{n+p})\\&\leqslant\{\theta^n+\theta^{n+1}+\cdots+\theta^{n+p-1}\}\,\rho(x_0,T(x_0))=\frac{\theta^n(1-\theta^p)}{1-\theta}\rho(x_0,T(x_0))\\&\leqslant\frac{\theta^n}{1-\theta}\rho(x_0,T(x_0))\to0,\quad\left(\lim_{n\to+\infty}\theta^n=0 \text{ by } 0\leqslant\theta<1\right).\end{aligned}$$

Hence, $\{x_n\}$ is a Cauchy sequence in X; Then by the completeness of X, there exists $x_0\in X$, such that $\lim\limits_{n\to+\infty} x_n=x_0$.

Take limit at both sides: $\lim\limits_{n\to+\infty} x_{n+1}=\lim\limits_{n\to+\infty} T(x_n)$, then by the continuity of T, we have

$$x_0 = \lim_{n\to+\infty} x_{n+1} = \lim_{n\to+\infty} T(x_n) = T(\lim_{n\to+\infty} x_n) = T(x_0),$$

thus $x_0 = T(x_0)$, so x_0 is a fixed point of T.

Finally, we prove the uniqueness of the fixed point: If there exist two fixed points x_0, y_0 of T, then

$$\rho(x_0, y_0) = \rho(T(x_0), T(y_0)) \leqslant \theta\, \rho(x_0, y_0),$$

because $0 \leqslant \theta < 1$, the above inequality implies $\rho(x_0, y_0) = 0$, and thus $x_0 = y_0$. The uniqueness is proved.

Example 4.1.6 The integral equation

$$x(t) = f(t) + \lambda \int_a^b K(t,s)x(s)\mathrm{d}s \tag{4.1.3}$$

is said to be **a Fredholm integral equation**, where $f \in L^2([a,b])$ is given, λ is a parameter, a given function $K(x,s)$ with $x, s \in [a,b]$ satisfying $\int_a^b \int_a^b |K(t,s)|^2 \mathrm{d}t\, \mathrm{d}s < +\infty$ is said to be **a kernel of Fredholm integral equation**, or simply, **a kernel**; $x(t) \in L^2([a,b])$ is **unknown function** of equation (4.1.3). Then, Fredholm integral equation has unique solution $x = x(t) \in L^2([a,b])$ for enough small $|\lambda|$.

Proof To use fixed-point Theorem 4.1.5, we define a mapping $T: L^2([a,b]) \to L^2([a,b])$ by

$$(T(x))(t) \equiv T(x(t)) = f(t) + \lambda \int_a^b K(t,s)x(s)\mathrm{d}s.$$

Then Minkovski inequality gives the estimation

$$\int_a^b \left| \int_a^b K(t,s)x(s)\mathrm{d}s \right|^2 \mathrm{d}t \leqslant \int_a^b \left[\int_a^b |K(t,s)|^2 \mathrm{d}s \int_a^b |x(s)|^2 \mathrm{d}s \right] \mathrm{d}t$$
$$= \int_a^b \int_a^b |K(t,s)|^2 \mathrm{d}s\, \mathrm{d}t \int_a^b |x(s)|^2 \mathrm{d}s < +\infty,$$

Thus, T maps $x(t)$ from $L^2([a,b])$ to $L^2([a,b])$.

For enough small $|\lambda|$, we take $\theta \equiv |\lambda| \left\{ \int_a^b \int_a^b |K(t,s)|^2 \mathrm{d}s\mathrm{d}t \right\}^{\frac{1}{2}} < 1$. Hence, by Hölder inequality, it holds

$$\rho(T(x), T(y)) = \left\{ \int_a^b \left| \left\{ f(t) + \lambda \int_a^b K(t,s)x(s)\mathrm{d}s \right\} - \left\{ f(t) + \lambda \int_a^b K(t,s)y(s)\mathrm{d}s \right\} \right|^2 \right\}^{\frac{1}{2}}$$
$$= |\lambda| \left\{ \int_a^b \left| \int_a^b K(t,s)(x(s) - y(s))\, \mathrm{d}s \right|^2 \mathrm{d}t \right\}^{\frac{1}{2}}$$
$$\leqslant |\lambda| \left\{ \int_a^b \left(\int_a^b |K(t,s)|^2 \mathrm{d}t \right) \mathrm{d}s \right\}^{\frac{1}{2}} \cdot \left\{ \int_a^b |x(s) - y(s)|^2 \mathrm{d}s \right\}^{\frac{1}{2}}$$
$$= |\lambda| \left\{ \int_a^b \left(\int_a^b |K(t,s)|^2 \mathrm{d}t \right) \mathrm{d}s \right\}^{\frac{1}{2}} \cdot \rho(x,y) = \theta\, \rho(x,y),$$

this shows that T is compression mapping, and by Theorem 4.1.5, there exists unique function $x(t)$ in $L^2([a,b])$ satisfying Fredholm integral equation (4.1.3).

Example 4.1.7 Let (X,ρ) be a metric space, $T: X\to X$ be a mapping from X to X. If there exist a constant $\theta, 0\leqslant\theta<1$, and an integer $n_0\in\mathbb{Z}^+$, such that it holds

$$\rho(T^{n_0}x, T^{n_0}y)\leqslant\theta\,\rho(x,y),\quad \forall(x,y)\in X\times X,$$

where T^n is **the nth-iterated of** T, i.e., $T^n(x)=T(T^{n-1}(x))$. Then there exists unique fixed point of T in X.

Proof The n_0^{th}-iterated T^{n_0} of T satisfies $\rho(T^{n_0}x, T^{n_0}y)\leqslant\theta\,\rho(x,y)$, i.e., satisfies (4.1.2). By the fixed-point Theorem 4.1.5, the mapping T^{n_0} exists unique fixed point $x_0\in X$. **This x_0 is also a fixed point of** T, since $T^{n_0}(x_0)=x_0\Rightarrow T^{n_0}(T(x_0))=T^{n_0+1}(x_0)=T(T^{n_0}(x_0))=T(x_0)\Rightarrow T(x_0)$ is a fixed point of $T^{n_0}\Rightarrow T(x_0)=x_0$, since the uniqueness of fixed point of $T^{n_0}\Rightarrow x_0$ is a fixed point of T.

Moreover, **this x_0 is the unique fixed point of** T, because: if there is the other one x_1 such that $T(x_1)=x_1\Rightarrow T^{n_0}(x_1)=T^{n_0}(T(x_1))=T^{n_0+1}(x_1)=\underbrace{T\cdots T}_{n_0+1}(x_1)=\underbrace{T\cdots T}_{n_0}(T(x_1))=\underbrace{T\cdots T}_{n_0}(x_1)=\underbrace{T\cdots T}_{n_0-1}(T(x_1))=\cdots=T(x_1)=x_1\Rightarrow x_1$ is a fixed point of $T^{n_0}\Rightarrow x_0=x_1$.

Example 4.1.8 The integral equation

$$x(t)=f(t)+\lambda\int_a^t K(t,s)x(s)\mathrm{d}s \tag{4.1.4}$$

is said to be **a Volterra integral equation**, where $f\in C([a,b])$ is given, λ is a parameter; $K(x,s)$ with $a\leqslant x\leqslant b, a\leqslant s\leqslant t$, is a given continuous function called **a kernel**; $x(t)\in C([a,b])$ is unknown function of equation (4.1.4), and the integral $\int_a^t K(t,s)x(s)\mathrm{d}s$ in (4.1.4) is with a varying upper limit t. Then Volterra integral equation (4.1.4) has a unique solution $x=x(t)\in C([a,b])$ for any constant λ and any continuous function $f\in C([a,b])$.

Proof Define a mapping $T: C([a,b])\to C([a,b])$ by

$$(T(x))(t)=f(t)+\lambda\int_a^t K(t,s)x(s)\mathrm{d}s. \tag{4.1.5}$$

Then for any $x_1(t), x_2(t)\in C([a,b])$, it holds

$$\begin{aligned}|(T(x_1))(t)-(T(x_2))(t)|&=|\lambda|\left|\int_a^t K(t,s)[x_1(s)-x_2(s)]\mathrm{d}s\right|\\&\leqslant|\lambda|M(t-a)\max_{t\in[a,b]}|x_1(t)-x_2(t)|\\&=|\lambda|M(t-a)\rho(x_1,x_2),\end{aligned} \tag{4.1.6}$$

where $M=\max\{|K(t,s)|: s\in[a,t], t\in[a,b]\}$, and $\rho(x_1,x_2)=\max\limits_{t\in[a,b]}|x_1(t)-x_2(t)|$ is a distance between x_1 and x_2 in space $C([a,b])$.

We show that the mapping T defined in (4.1.5) satisfies the conditions in Example 4.1.7, and prove

$$\rho(T^{n_0}x, T^{n_0}y)\leqslant\theta\,\rho(x,y),\quad n_0\in\mathbf{Z}^+.$$

In fact, when $n=1$, (4.1.6) implies

$$|(T(x_1))(t)-(T(x_2))(t)|\leqslant|\lambda|M(t-a)\rho(x_1,x_2);$$

Suppose $n=k$, it holds

$$|(T^k(x_1))(t)-(T^k(x_2))(t)|\leqslant|\lambda|^kM^k\frac{(t-a)^k}{k!}\rho(x_1,x_2),$$

then we prove that it holds for $n=k+1$. Since for $n=k+1$,

$$\begin{aligned}|(T^{k+1}(x_1))(t)-(T^{k+1}(x_2))(t)|&=|\lambda|\left|\int_a^t K(t,s)[T^kx_1(s)-T^kx_2(s)]\,\mathrm{ds}\right|\\&\leqslant|\lambda|^{k+1}M^{k+1}\frac{1}{k!}\left\{\int_a^t(s-a)^k\,\mathrm{ds}\right\}\rho(x_1,x_2)\\&=|\lambda|^{k+1}M^{k+1}\frac{(t-a)^{k+1}}{(k+1)!}\rho(x_1,x_2),\end{aligned}$$

thus by induction, the mapping T in (4.1.5) holds for all $n\in\mathbf{Z}^+$

$$|(T^n(x_1))(t)-(T^n(x_2))(t)|\leqslant|\lambda|^nM^n\frac{(t-a)^n}{n!}\rho(x_1,x_2).$$

Then, choose $n\in\mathbf{Z}^+$ large enough, such that $\theta=|\lambda|^nM^n\dfrac{(t-a)^n}{n!}<1$ for any constant λ, so that there exists a unique function $x(t)\in C([a,b])$ satisfying Volterra integral equation (4.1.4), by the result in Example 4.1.7.

Recall that in *advanced mathematics*, an initial-valued problem of differential equation $\begin{cases}\dfrac{\mathrm{d}y}{\mathrm{d}x}=f(x,y)\\ y|_{x=x_0}=y_0\end{cases}$ has a unique integral curve passing through initial point (x_0,y_0), when Lipschitz condition of y is satisfied:

$$|f(x,y)-f(x,y')|\leqslant M|y-y'|$$

with constant M. The method of proof used there: Transform the initial problem to a Volterra integral equation

$$y(x)=y_0+\int_{x_0}^{x}f(t,y(t))\,\mathrm{d}t$$

with a varying upper limit x, then **the successive approximation method** is used. However, in this course, we define the mapping

$$(T(y))(x) \equiv T(y(x)) = y_0 + \int_{x_0}^{x} f(t, y(t))\,dt, \tag{4.1.7}$$

and by

$$\begin{aligned}\rho(T(y_1), T(y_2)) &= \max_{|x-x_0|\leqslant\delta} \left| \int_{x_0}^{x} [f(t, y_1(t)) - f(t, y_2(t))]\,dt \right| \\ &\leqslant \max_{|x-x_0|\leqslant\delta} \left| \int_{x_0}^{x} M|y_1(t) - y_2(t)|\,dt \right| \\ &\leqslant M\delta \max_{|x-x_0|\leqslant\delta} |y_1(t) - y_2(t)| \\ &= M\delta\, \rho(y_1, y_2),\end{aligned}$$

take $\delta < 1$ with $K\delta < 1$, then by $T: C([x_0-\delta, x_0+\delta]) \to C([x_0-\delta, x_0+\delta])$ in (4.1.7), and by the result in Example 4.1.8, there exists a unique integral curve passing though initial point (x_0, y_0).

The other example in the course of *Linear Algebra*: for a matrix $[a_{ij}]_{n\times n} \in \mathfrak{M}_{n\times n}$ with $\sum_{i=1}^{n}\sum_{j=1}^{n} |a_{ij}|^2 < 1$, then the equation system

$$\xi_i - \sum_{j=1}^{n} a_{ij}\xi_j = b_i, \quad i = 1, 2, \cdots, n,$$

has unique solution $\xi^0 = [\xi_i^0]_{n\times 1} \in \mathfrak{M}_{n\times 1}$ for unknown vector $\xi = [\xi_j]_{n\times 1} \in \mathfrak{M}_{n\times 1}$, and a given nonzero $b = [b_i]_{n\times 1} \in \mathfrak{M}_{n\times 1}$.

Proof Let $T: \mathfrak{M}_{n\times 1} \to \mathfrak{M}_{n\times 1}$ be a linear mapping $(T\xi)_i = \sum_{j=1}^{n} a_{ij}\xi_j + b_i, i = 1, 2, \cdots, n$, from the complete metric space $\mathfrak{M}_{n\times 1}$ ($\xleftrightarrow{\text{iso.}} \mathbb{R}^n$) to itself. Then, for $x = [x_j]_{n\times 1} \in \mathfrak{M}_{n\times 1}$, $y = [y_j]_{n\times 1} \in \mathfrak{M}_{n\times 1}$,

$$\begin{aligned}\rho(Tx, Ty) &= \left\{\sum_{i=1}^{n}\left[\sum_{j=1}^{n} |a_{ij}(x_j - y_j)|\right]^2\right\}^{\frac{1}{2}} \leqslant \left\{\sum_{i=1}^{n}\sum_{j=1}^{n} |a_{ij}|^2 \sum_{j=1}^{n} |x_j - y_j|^2\right\}^{\frac{1}{2}} \\ &\leqslant \left\{\sum_{i=1}^{n}\sum_{j=1}^{n} |a_{ij}|^2\right\}^{\frac{1}{2}} \rho(x, y).\end{aligned}$$

Let $\theta = \sum_{i=1}^{n}\sum_{j=1}^{n} |a_{ij}|^2 < 1$, then $T: \mathfrak{M}_{n\times 1} \to \mathfrak{M}_{n\times 1}$ is a compression mapping, thus there exists unique solution $x^0 = [x_i^0]_{n\times 1} \in \mathfrak{M}_{n\times 1}$.

4.1.2 Compactness in metric spaces

We have introduced various compactness in a topological space (X, τ) in Chapter 3: compact, countably compact, sequentially compact, accumulatively compact, and locally compact, these four sorts of compactnesses are closed related with the topological structures. Now, suppose (X,τ) is a metric space (X,ρ), then we have all these compactnesses in (X,ρ) since it is a special topological space; moreover, we have more important, interesting and useful properties for these compactnesses in (X,ρ).

1. Totally bounded sets

We have known the definition of the boundedness of a subset $A\subset X$ in Definition 3.4.9 in a metric space (X,ρ), as well as known the relationships between boundedness and compactness in Theorems 3.4.16 and 3.4.17. Now, we introduce the concept of "totally bounded".

Definition 4.1.6 (totally bounded set) Let (X,ρ) be a metric space. A subset $A\subseteq X$ is said to be **a totally bounded set**, if for any given $\varepsilon>0$, there exists a finite subset $B=\{x_1,\cdots,x_n\}\subset X$, s.t. $\forall x\in A$, $\exists x_j\in B$, it holds $\rho(x,x_j)<\varepsilon$. And B is said to be **a finite ε-net of** A.

It is clear that "totally boundedness" is a generalization of "boundedness" (Figs. 4.1.1 and 4.1.2).

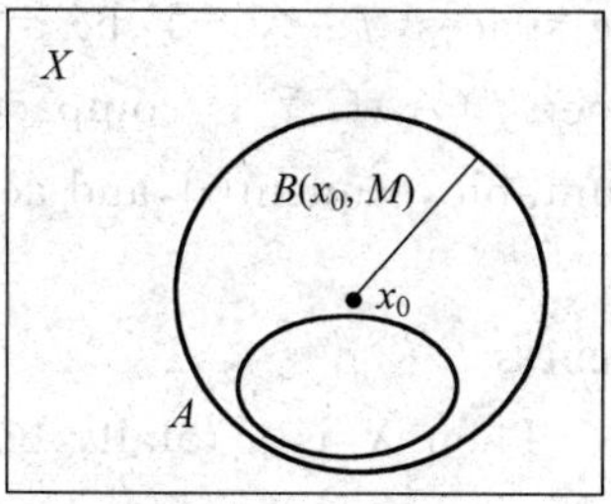

Fig. 4.1.1 Bounded set A

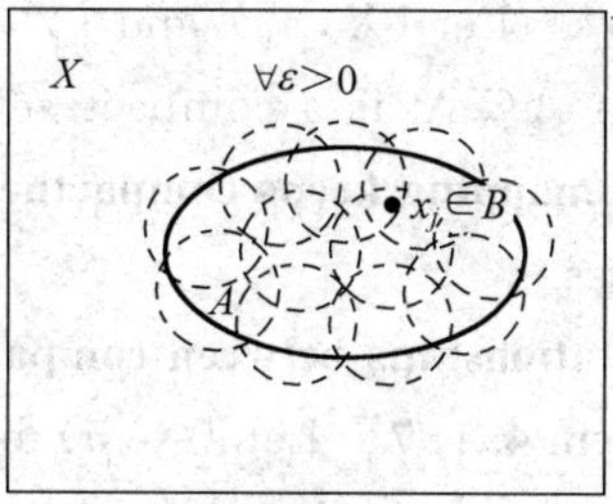

Fig. 4.1.2 Finite ε-net

2. Compactness in metric spaces

1) Definitions of the compactness

Definition 4.1.7 Let (X,ρ) be a metric space.

(1) **Compactness** Any open covering of X exists a finite subcovering (motivated by "Heine-Borel finite covering theorem: Any open covering of closed interval $[a,b]$ exists a finite subcovering.")

(2) **Countable compactness** Any countable covering of X exists a finite subcovering (motivated by "Chinese-boxes theorem: If $[a_1,b_1]\supset\cdots\supset[a_n,b_n]\supset\cdots$, and holds $\lim\limits_{n\to+\infty}(b_n-a_n)=0$, then there exists unique $x_0\in\bigcap\limits_{n=1}^{+\infty}[a_n,b_n]$")

(3) **Sequential compactness** Any infinite sequence in X exists a convergent subsequence with limit $x \in X$ (motivated by "'sequential' Bolzano-Weierstrass theorem: Any infinite sequence in closed interval $[a,b]$ exists a convergent subsequence with limit $x \in [a,b]$").

(4) **Accumulative compactness** Any infinite subset in X exists an accumulation point $x \in X$ (motivated by "'subsetly' Bolzano-Weierstrass theorem: Any infinite subset in closed interval $[a,b]$ exists an accumulation point $x \in [a,b]$")

(5) **Local compactness** Any point $x \in X$ exists a compact closure neighborhood in X (motivated by "Any point $x \in \mathbb{R}^n$ has an open ball neighborhood with compact closure in $\mathbb{R}^n$").

Theorem 4.1.6 If (X,ρ) is a metric space, then:

X compactness $\Leftrightarrow$ X countable compactness $\Leftrightarrow$ X sequential compactness $\Leftrightarrow$ X accumulative compactness $\Rightarrow$ X total boundedness.

If (X,ρ) is a complete metric space, then:

X compactness $\Leftrightarrow$ X countable compactness $\Leftrightarrow$ X sequential compactness $\Leftrightarrow$ X accumulative compactness $\Leftrightarrow$ X total boundedness.

We omit the proof, and refer to [6], [12], [14].

Remark Let (X,ρ_1) and (Y,ρ_2) be two metric spaces, $f: X \to Y$ be a continuous mapping. If $A \subset X$ is a compact set, then the image set $f(A) \subset Y$ is compact. Thus, "**a continuous mapping keeps compactness**", also, keeps countable, sequential, and accumulative compactness.

2) **Relationships between compactness and completeness**

Theorem 4.1.7 Let (X,ρ) be a compact space. Then X is a totally bounded and complete space.

Proof For a metric space (X,ρ), we prove that it is totally bounded space, firstly:

For a given $\varepsilon > 0$, let $B(x,\varepsilon) = \{y \in X: \rho(x,y) < \varepsilon\}$ be an open ball with center $x \in X$ and radius ε. The set of all open balls $\{B(x,\varepsilon): \forall x \in X, \forall \varepsilon > 0\}$ is an open covering of X, $\bigcup_{x \in X} B(x,\varepsilon) = X$. Then, $\exists \{B(x_j,\varepsilon_j)\}_{j=1}^{m}$, such that it is a finite covering of X by compactness; This implies that $\{x_1, x_2, \cdots, x_m\}$ is a finite $\varepsilon = \max_j \{\varepsilon_1, \cdots, \varepsilon_m\}$-net of X, thus X is totally bounded.

Next, prove the completeness of X: ① **the completeness of X implies sequential compactness** — since for any infinite sequence $\{x_n\}$ in X, let $F_n = \overline{\{x_n, x_{n+1}, \cdots, x_{n+p}, \cdots\}}$. Then $\{F_n\}$ is a decreasing closed set sequence in X satisfying "an intersection of any finite closed sets in $\{F_n\}$ is not empty" (this is the condition of "finite-intersection property", see Exercise 16).

For $F_n=\overline{\{x_n,x_{n+1},\cdots,x_{n+p},\cdots\}} \Rightarrow U_n=X\backslash F_n$ is an open set, and $\bigcup_{n=1}^{+\infty} U_n \supset X$ (by $X=\complement(\bigcap_{n=1}^{+\infty} F_n)=\bigcup_{n=1}^{+\infty}\complement F_n=\bigcup_{n=1}^{+\infty} U_n) \Rightarrow \exists \{U_{n_1},\cdots,U_{n_m}\}$ s. t. $\bigcup_{k=1}^{m} U_{n_k}\supset X$ by compactness of $X \Rightarrow$ if $\bigcap_{n=1}^{+\infty} F_n=\varnothing$, then $\bigcap_{k=1}^{m} F_{n_k}=\bigcap_{k=1}^{m}\complement U_{n_k}=\complement(\bigcup_{k=1}^{m} U_{n_k})=\varnothing \Rightarrow$ this contradicts $\bigcap_{k=1}^{m} F_{n_k}\neq\varnothing \Rightarrow$ so that $\bigcap_{n=1}^{+\infty} F_n\neq\varnothing$ implies $\exists x\in\bigcap_{n=1}^{+\infty} F_n\neq\varnothing$, and this x is an accumulation point of $\{x_n\} \Rightarrow \exists \{x_{n_j}\}$ converges to $x\in X \Rightarrow$ any infinite sequence $\{x_n\}$ in X has convergent subsequence $\Rightarrow$ the compactness of X implies the sequential compactness of X.

② **the completeness of** X — take any Cauchy sequence $\{x_n\}\subset X$, then it has accumulation point $x\in F_1=\overline{\{x_n\}}$, by ①. We prove that this accumulation point $x\in X$ is the limit of Cauchy sequence $\{x_n\}$ with $\lim_{n\to+\infty}\rho(x_n,x)=0$: Since $x\in X$ is an accumulation point of Cauchy sequence $\{x_n\}$, then $\forall\varepsilon>0$, $\exists N\in\mathbf{N}$, s. t. for $n, m>N$, it holds $\rho(x_n,x_m)<\varepsilon$; again by definition of the accumulation point, $\forall\varepsilon>0$, $\exists p\geqslant N$, s. t. $\rho(x_p,x)<\varepsilon$; then for $n>N$, it holds $\rho(x_n,x)\leqslant\rho(x_n,x_p)+\rho(x_p,x)<\varepsilon+\varepsilon=2\varepsilon$, so that $\lim_{n\to+\infty}\rho(x_n,x)=0$, and X is a complete space. The Theorem is proved.

3. Compactness of subspaces in metric spaces

Definition 4.1.8 (compactness of set) Let (X,ρ) be a metric space, $A\subset X$ be a subset.

(1) A is said to be **a compact set**, if any open covering of A has a finite subcovering;

(2) A is said to be **a countably compact set**, if any countable open covering of A has a finite subcovering;

(3) A is said to be **a sequentially compact set**, if any sequence in A has the convergent sub-sequence; A is said to be **self-sequentially compact**, if any sequence in A has the convergent sub-sequence with limit in A itself;

(4) A is said to be **an accumulatively compact**, if any infinite subset in A has accumulation points; A is said to be **self-accumulatively compact**, if any infinite subset of A has accumulation point as limit point in A itself.

Note For metric space (X,ρ) itself, the sequential compactness and self-sequential compactness are same, since the limit is in X itself in both cases; and so are accumulative compactness and self-accumulative compactness. However, for subsets of (X,ρ), we have to distinguish these conceptions.

Theorem 4.1.8 (1) Let (X,ρ) be a metric space, $A\subset X$ be a subset of X. Then:

A is compact $\Leftrightarrow$ A is countably compact $\Leftrightarrow$ A is self-sequentially compact $\Leftrightarrow$ A is self-accumulatively compact $\Rightarrow$ A is totally bounded;

(2) Let (X,ρ) be a complete metric space, $A\subset X$ be a subset of X. Then:

A is compact $\Leftrightarrow$ A is countably compact $\Leftrightarrow$ A is self-sequentially compact $\Leftrightarrow$ A is self-accumulatively compact $\Leftrightarrow$ A is totally bounded;

(3) In $\mathbb{R}^n$, for a subset $A\subset\mathbb{R}^n$, then:

A is compact $\Leftrightarrow$ A is countably compact $\Leftrightarrow$ A is self-sequentially compact $\Leftrightarrow$ A is self-accumulatively compact $\Leftrightarrow$ A is bounded closed.

Example 4.1.9 The real-number set $X=\mathbb{R}$ is a non-compact set, since the sequence $\{n\}$ does not contain any convergent subsequence; For $X=\mathbb{R}$, take $F=[a,b]$ with $-\infty<a<b<+\infty$, then F is a compact set, thus it is self-compact; However, $O=(a,b)$ is an accumulatively compact subset, but it is not self-accumulatively compact.

Example 4.1.10 Let $X=L^2[-\pi,\pi]$. The trigonometric system

$$\left\{\frac{1}{\sqrt{2\pi}},\frac{\cos t}{\sqrt{\pi}},\frac{\sin t}{\sqrt{\pi}},\cdots,\frac{\cos nt}{\sqrt{\pi}},\frac{\sin nt}{\sqrt{\pi}},\cdots\right\}$$

is bounded, but it does not have any convergent subsequence since any elements have distance $\sqrt{2}$, so it is not a sequentially compact set.

4. Criterions of the sequential compactness in normed linear spaces

We show the criterions of sequential compactness in $\mathbb{R}^n$, $C([a,b])$ and $L^p([a,b])$, respectively.

1) Criterion of the sequential compactness of sets in $\mathbb{R}^n$

Theorem 4.1.9 In $\mathbb{R}^n$, any subset is sequential compact, if and only if A is a bounded closed set in $\mathbb{R}^n$.

2) Criterion of the sequential compactness of sets in $C([a,b])$

Theorem 4.1.10 (Ascoli -Arzela Theorem) Any subset $A\subset C([a,b])$ is sequential compact, if and only if:

(1) A is "C-uniformly bounded", i. e. $\exists M>0$, s. t.,

$$\forall f\in A\Rightarrow|f(x)|\leqslant M,\quad \forall x\in[a,b];$$

also that is, the set A is uniformly bounded with norm $\|f\|_{C([a,b])}=\max_{x\in[a,b]}|f(x)|$ of $C([a,b])$;

(2) A is "C-equi-continuous", i. e. $\forall\varepsilon>0$, $\exists\delta=\delta(\varepsilon)>0$, s. t.,

$$\forall x_1,x_2\in[a,b],|x_1-x_2|<\delta\Rightarrow|f(x_1)-f(x_2)|<\varepsilon,\quad \forall f\in A.$$

3) Criterion of the sequential compactness of sets in $L^p([a,b])$

Theorem 4.1.11 Any subset $A\subset L^p([a,b])(1<p<+\infty)$ is sequential compact, if and only if:

(1) A is "L^p-uniformly bounded", i. e. $\exists M>0$, s. t.,

$$\forall f\in A\Rightarrow\int_{[a,b]}|f(x)|^p\mathrm{d}x\leqslant M;$$

also that is, the set A is uniformly bounded with norm $\|f\|_{L^p([a,b])}=\left(\int_{[a,b]}|f(x)|^p\mathrm{d}x\right)^{\frac{1}{p}}$ of $L^p([a,b])$;

(2) A is "L^p-equi-continuous", i. e. $\forall\varepsilon>0,\exists\delta=\delta(\varepsilon)>0$, s. t.

$$\forall h,0<h<\delta,\forall f\in A\Rightarrow\left\{\int_{[a,b]}|f_h(x)-f(x)|^p\mathrm{d}x\right\}^{\frac{1}{p}}<\varepsilon,$$

where f_h is defined by $\forall h>0,\forall x\in[a,b]$, $f_h(x)=\frac{1}{2h}\int_{[x-h,x+h]}f(t)\mathrm{d}t$; we agree on that if $x\notin[a,b]$, then $f(x)=0$ ($\delta=\delta(\varepsilon)$ is independent of $h\in(0,\delta)$ and $f\in A$).

For the proofs of Theorems 4.9, 4.1.10 and 4.1.11, we refer to [6].

4) Local compactness of normed linear spaces

Theorem 4.1.12 A normed linear space $(X,\|x\|_X)$ is locally compact, if and only if any bounded closed subset in X is compact.

4.1.3 Bases of Banach spaces

The concept of the base of a linear space $(X,+,\alpha\cdot)$ is introduced in Chapter 2. We consider now a linear space as normed linear space $(X,+,\alpha\cdot,\|x\|_X)$, $\alpha\in\mathbb{F}$, on number field $\mathbb{F}$ ($\mathbb{R}$ or $\mathbb{C}$), also consider Banach space, a complete normed linear space; and construct the base of space $(X,\|x\|_X)\equiv(X,+,\alpha\cdot,\|x\|_X)$, in this subsection.

1. Important Banach spaces

(1) $(\mathbb{R}^n,\|x\|_{\mathbb{R}^n})$:

$$\mathbb{R}^n=\{x=(x_1,x_2,\cdots x_n):x_j\in\mathbb{R},j=1,2,\cdots,n\},\ \|x\|_{\mathbb{R}^n}=\sqrt{\sum_{j=1}^{n}|x_j|^2};$$

(2) $(l^p,\|x\|_{l^p})$, $1\leqslant p<+\infty$:

$$l^p=\left\{x=(x_1,x_2,\cdots):\sum_{j=1}^{+\infty}|x_j|^p<+\infty\right\},\ \|x\|_p\equiv\|x\|_{l^p}=\left(\sum_{j=1}^{\infty}|x_j|^p\right)^{\frac{1}{p}};$$

(3) $(l^{\infty}, \|x\|_{l^{\infty}})$:

$l^{\infty}=\{x=(x_1,x_2,\cdots): |x_j|<+\infty, j\in\mathbb{Z}^+\}$, $\|x\|_{\infty}\equiv\|x\|_{l^{\infty}}=\sup_{j\in\mathbb{Z}^+}|x_j|$;

(4) $(C([a,b]), \|f\|_{C([a,b])})$:

$C([a,b])=\{f$: continuous function on $[a,b]\}$, $\|f\|_{C([a,b])}=\max_{x\in[a,b]}|f(x)|$;

(5) $(C^k([a,b]), \|f\|_{C^k([a,b])}), k\in\mathbb{N}$:

$C^k([a,b])=\{f: f^{(k)}\in C([a,b])\}$, $\|f\|_{C^k([a,b])}=\sum_{j=0}^{k}\max_{x\in[a,b]}|f^{(j)}(x)|$,

we agree on $C^0([a,b])=C([a,b])$;

(6) $(L^p([a,b]), \|f\|_{L^p([a,b])})$: $1\leqslant p<+\infty$

$L^p([a,b])=\left\{f: \int_{[a,b]}|f(x)|^p\,\mathrm{d}x<+\infty\right\}$, $\|f\|_{L^p([a,b])}=\left(\int_{[a,b]}|f(x)|^p\,\mathrm{d}x\right)^{\frac{1}{p}}$;

(7) $(L^{\infty}([a,b]), \|f\|_{L^{\infty}([a,b])})$:

$L^{\infty}([a,b])=\{f: \operatorname{esssup}_{x\in[a,b]}|f(x)|<+\infty\}$, $\|f\|_{L^{\infty}(E)}=\operatorname{esssup}_{x\in[a,b]}|f(x)|$.

2. Finite-dimensional Banach spaces

Definition 4.1.9 (base, dimension of a Banach space) Let $(X, \|x\|_X)$ be a Banach space. If there exists n $(n\in\mathbb{Z}^+)$ elements $e_1,e_2,\cdots,e_n\in X$, independently, such that any $x\in X$ can be expressed uniquely as

$$x=\sum_{j=1}^{n}\xi_j e_j, \xi_j\in\mathbb{F}, \quad j=1,2,\cdots,n,$$

then the set $\{e_1,e_2,\cdots,e_n\}\subset X$ is said to be **a base of** X, and $\xi_1,\xi_2,\cdots,\xi_n\in\mathbb{F}$ is said to be **the coordinates of** $x\in X$ **with respect to base** $\{e_1,e_2,\cdots,e_n\}$; n is said to be **the dimension of** X; and X is said to be **(finite)** n**-dimensional Banach space**, denoted by $\dim X=n$.

If $X=\{0\}$, then **we agree on** that $\dim\{0\}=0$.

If a Banach space $(X, \|x\|_X)$ is not finite dimensional, then it is said to be an **infinite-dimensional Banach space.**

Definition 4.1.10 (topological isomorphism) Let $(X, \|x\|_X)$, and $(Y, \|y\|_Y)$ be two Banach spaces. If they satisfy:

(1) there exists an isomorphism mapping $T: X\to Y$ (one-one, surjection, keeping operations), such that linear spaces X and Y are isomorphic;

(2) $T: X\to Y$ is a bicontinuous mapping, i. e., T, and T^{-1} are continuous;

Then, X **and** Y **are said to be topologically isomorphic**; i. e., they are algebraically isomorphic and topologically homeomorphic.

The following is an isomorphic theorem of n-dimensional Banach spaces.

Theorem 4.1.13 Let n be a positive integer, $n \in Z^+$. Then

(1) any real n-dimensional Banach space $(X, \|x\|_X)$ is topologically isomorphic with $\mathbb{R}^n$; any complex n-dimensional Banach space $(X, \|x\|_X)$ is topologically isomorphic with $\mathbb{C}^n$; So that, in any Banach space, there exists a finite-dimensional Banach subspace; moreover, it is a closed subspace.

(2) a normed linear space $(X, \|x\|_X)$ is a finite-dimensional space, if and only if $(X, \|x\|_X)$ is locally compact.

Proof We only show the idea of the proof, and only prove for the case of the normed space.

For (1), let $(X, \|x\|_X)$ be an n-dimensional linear space. Take base $\{e_1, \cdots, e_n\}$, then $\forall x \in X$, we have $x = \sum_{j=1}^{n} \xi_j e_j$, $\xi_j \in \mathbb{F}$. Let $(\xi_1, \cdots, \xi_n)$ be a point in $\mathbb{F}^n$, and T: $T(x) = (\xi_1, \cdots, \xi_n)$ be a mapping form X to $\mathbb{F}^n$.

We list that T satisfies:

① $T: X \to \mathbb{F}^n$, $T(x) = (\xi_1, \cdots, \xi_n)$ **is an isomorphic linear mapping** form X to $\mathbb{F}^n$;

② With base $\{e_1, \cdots, e_n\}$, $\exists c_1, c_2 > 0$, s.t., for any $x = \sum_{j=1}^{n} \xi_j e_j \in X$, it holds inequalities

$$c_1 \|x\|_X \leqslant \left(\sum_{j=1}^{n} |\xi_j|^2\right)^{\frac{1}{2}} \leqslant c_2 \|x\|_X;$$

and then holds $c_1 \|x - y\|_X \leqslant \|T(x) - T(y)\|_{\mathbb{F}^n} \leqslant c_2 \|x - y\|_X$.

③ By the above inequalities, we get the continuity of T and T^{-1}: since if $\{x_m\} \subset X$ converges to $x_0 \in X$, $\|x_m - x_0\|_X \to 0$, then $\|T(x_m) - T(x_0)\|_{\mathbb{F}^n} \leqslant c_2 \|x_m - x_0\|_X$ implies the continuity of T; Moreover, by $c_1 \|T^{-1}(x) - T^{-1}(y)\|_{\mathbb{F}^n} \leqslant \|x - y\|_X$, so that T^{-1} is continuous.

To prove (2), we need **Riesz lemma**: Let X_0 be a normed linear space, and $X_0 \subsetneq X$ be a proper closed subspace. Then, $\forall \varepsilon > 0$, $\exists x_0 \in X$, $\|x_0\|_X = 1$, s.t. for all $x \in X_0$, it holds $\|x - x_0\|_X \geqslant 1 - \varepsilon$.

The idea of the proof of Riesz lemma: $\exists x_1 \in X - X_0$ by $X_0 \subsetneq X$. Let $d = \inf_{x \in X_0} \|x_1 - x\|_X > 0$ (since $X_0 \subsetneq X$ is a closed subspace). By definition of the "inf": ① $\forall x \in X_0$ implies $\|x_1 - x\|_X \geqslant d$; ② $\forall \eta > 0$, $\exists x_1' \in X_0$, s.t. it holds $\|x_1 - x_1'\|_X < d + \eta$, thus we have: for any given ε with $0 < \varepsilon < 1$, and $\frac{d}{1-\varepsilon} > d$, there exists $x_1' \in X_0$, s.t.

$\| x_1 - x_1' \|_X < d \dfrac{1}{1-\varepsilon}$. Let $x_0 = \dfrac{x_1 - x_1'}{\| x_1 - x_1' \|_X}$ with $\| x_0 \|_X = 1$. Then, $\forall x \in X_0$, it holds

$$\begin{aligned}
\| x - x_0 \|_X &= \left\| x - \frac{x_1 - x'_1}{\| x_1 - x'_1 \|_X} \right\|_X \\
&= \frac{1}{\| x_1 - x'_1 \|_X} \left\| \| x_1 - x'_1 \|_X x - x_1 + x'_1 \right\|_X \\
&= \frac{1}{\| x_1 - x'_1 \|_X} \left\| (\| x_1 - x'_1 \|_X x + x'_1) - x_1 \right\|_X. \qquad (4.1.8)
\end{aligned}$$

Since $x_1' \in X_0$, $x \in X_0$, and X_0 is a linear space, we have $\| x_1 - x_1' \|_X x + x_1' \in X_0$, so that

$$\| (\| x_1 - x'_1 \|_X x + x'_1) - x_1 \|_X \geqslant d.$$

Moreover, by (4.1.8),

$$\| x - x_0 \|_X = \frac{1}{\| x_1 - x'_1 \|_X} \left\| (\| x_1 - x'_1 \|_X x + x'_1) - x_1 \right\|_X \geqslant \frac{d}{\| x_1 - x'_1 \|_X} > 1 - \varepsilon,$$

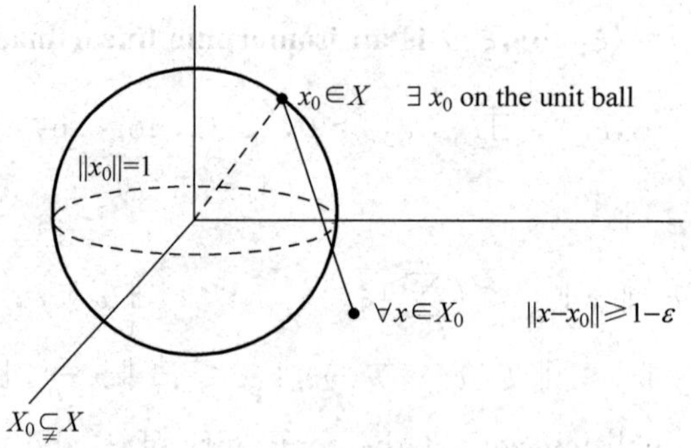

Fig. 4.1.3 Riesz lemma

and then Riesz lemma is proved.

Necessity of (2) By the result in(1), $(X, \| x \|_X)$ is topological isomorphic with $\mathbb{F}^n$, thus the isomorphic mapping maps a bounded closed subset in X to a bounded closed subset in $\mathbb{F}^n$; and vice versa. On the other hand, for a bounded closed set A in $\mathbb{F}^n$, if and only if it is a compact set, so that, the closure of any open ball neighborhood is bounded closed in $(X, \| x \|_X)$, and also compact. This implies $(X, \| x \|_X)$ is a locally compact normed linear space.

Sufficiency of (2) We suppose that $(X, \| x \|_X)$ is not a finite-dimensional locally compact normed linear space. Let $S = \{x \in X: \| x \|_X = 1\}$ be a unit ball shell in $(X, \| x \|_X)$, it is a bounded closed set, and thus is a compact set (Exercise 14).

Take any $x_1 \in S$, denoted by X_1, the linear subspace expanded by x_1, it is a finite-dimensional subspace, and is isomorphic with $\mathbb{R}^1$, so that X_1 is a finite-dimensional proper subspace of X. Then Riesz lemma tells that $\exists x_2 \in S$, such that for all $x \in X_1$, it holds $\| x-x_2 \|_X \geqslant \frac{1}{2}$; also holds $\| x_1-x_2 \|_X \geqslant \frac{1}{2}$ by $x_1 \in X_1$. For x_1 and x_2, denoted by $X_2 = \text{span}\{x_1, x_2\}$, it is a finite-dimensional proper closed subspace. Again by Riesz lemma, $\exists x_3 \in S$, such that for all $x \in X_2$, it holds $\| x-x_3 \|_X \geqslant \frac{1}{2}$. Because $x_1, x_2 \in X_2$, we have $\| x_3-x_1 \|_X \geqslant \frac{1}{2}$, $\| x_3-x_2 \|_X \geqslant \frac{1}{2}$. Continue the process, we get the sequence $x_1, x_2, \cdots, x_k, \cdots \in S$ with $\| x_k-x_l \|_X \geqslant \frac{1}{2}$, $k \neq l$, $k, l \geqslant 1$. But by $\| x_k-x_l \|_X \geqslant \frac{1}{2}$, $k \neq l$, $k, l \geqslant 1$ for $\{x_k\} \subset S$, so that $\{x_k\}$ does not have any subsequence. This contradicts the compactness of S. The proof of the theorem is complete.

3. Infinite-dimensional Banach spaces

Definition 4.1.11 (Schauder base) Let $(X, \| x \|_X)$ be a Banach space. An infinite sequence $\{e_1, \cdots, e_n, \cdots\} = \{e_n\}_{n=1}^{+\infty} \equiv \{e_n\} \subset X$, its any finite elements are linear independent, is said to be **a Schauder base of** X, if $\forall x \in X$ has an expression

$$x = \sum_{n=1}^{+\infty} c_n e_n, \quad c_n \in \mathbb{F}, n \in \mathbb{Z}^+,$$

with series converges in the sense of norm of $(X, \| x \|_X)$, i. e., $\lim\limits_{n \to +\infty} \left\| x - \sum\limits_{k=1}^{n} c_k e_k \right\|_X = 0$; The number sequence $\{c_n\} \subset \mathbb{F}$ is said to be **a coordinate of** $x \in X$ **in base** $\{e_n\}$.

Schauder base is **a countable base**, and a normed linear space (or Banach space) X possessing a Schauder base is said to be **an infinite-dimensional normed linear space** (or **an infinite-dimensional Banach space**).

Example 4.1.11 $(l^p, \| x \|_{l^p})$, $1 \leqslant p < +\infty$.

For $l^p = \left\{ x = (x_1, x_2, \cdots) : \sum\limits_{j=1}^{+\infty} |x_j|^p < +\infty \right\}$, let $\| x \|_{l^p} = \left(\sum\limits_{j=1}^{\infty} |x_j|^p \right)^{\frac{1}{p}}$.

Then $(l^p, \| x \|_{l^p})$ is an infinite-dimensional Banach space.

(1) l^p is a linear space on number field $\mathbb{F} = \mathbb{C}$ or $\mathbb{R}$. The operation "+" closed in l^p can be obtained by Minkowski inequality

$$\left(\sum_{j=1}^{\infty} |x_j + y_j|^p\right)^{\frac{1}{p}} \leqslant \left(\sum_{j=1}^{\infty} |x_j|^p\right)^{\frac{1}{p}} + \left(\sum_{j=1}^{\infty} |y_j|^p\right)^{\frac{1}{p}};$$

and operation "$\alpha\cdot$" is closed, clearly.

(2) $\|x\|_{l^p}=(\sum\limits_{j=1}^{\infty}|x_j|^p)^{\frac{1}{p}}$ is a norm. Since "$\|x\|_{l^p}\geqslant 0, x=0\Leftrightarrow\|x\|_{l^p}=0$" and "$\|\alpha x\|_{l^p}=|\alpha|\ \|x\|_{l^p}$" are clear; the trigonometric inequality $\|x+y\|_{l^p}\leqslant\|x\|_{l^p}+\|y\|_{l^p}$ holds by Minkowski inequality.

(3) l^p is a Banach space. It is complete in norm $\|x\|_{l^p}=(\sum\limits_{j=1}^{\infty}|x_j|^p)^{\frac{1}{p}}$, since about the distance $\rho(x,y)=(\sum\limits_{j=1}^{\infty}|x_j-y_j|^p)^{\frac{1}{p}}$, any Cauchy sequence in l^p converges to "a point" in l^p by the continuity of $\mathbb{R}$.

(4) l^p has Schauder base

$$e_1=(1,0,0,\cdots,0,\cdots), e_2=(0,1,0,\cdots,0,\cdots),\cdots, e_n=(\underbrace{0,\cdots,0,1}_{n},0,\cdots),\cdots.$$

For any $x=(x_1,x_2,\cdots,x_j,\cdots)\in l^p$, it holds $x=\sum\limits_{j=1}^{+\infty}x_je_j$, and $\lim\limits_{k\to+\infty}\|x-\sum\limits_{j=1}^{k}x_je_j\|_{l^p}=0$.

Example 4.1.12 $(L^p([a,b]),\|f\|_p), 1\leqslant p<+\infty$.

For $L^p([a,b])=\left\{f:\int_{[a,b]}|f(x)|^p\mathrm{d}x<+\infty\right\}$, let $\|f\|_p=\left(\int_{[a,b]}|f(x)|^p\mathrm{d}x\right)^{\frac{1}{p}}$. Then $(L^p([a,b]),\|f\|_p)$ is an infinite-dimensional Banach space.

(1) $L^p([a,b])$ is a linear space on number field $\mathbb{F}=\mathbb{C}$ or $\mathbb{R}$: The closed property of linear operations holds by Minkowski inequality $\left(\int_{[a,b]}|f(x)+g(x)|^p\mathrm{d}x\right)^{\frac{1}{p}}\leqslant\left(\int_{[a,b]}|f(x)|^p\mathrm{d}x\right)^{\frac{1}{p}}+\left(\int_{[a,b]}|g(x)|^p\mathrm{d}x\right)^{\frac{1}{p}}$;

(2) $\|f\|_p=\left(\int_{[a,b]}|f(x)|^p\mathrm{d}x\right)^{\frac{1}{p}}$ is a norm:

① For "$\|f\|_p\geqslant 0$, $\|f\|_p=0\Leftrightarrow f\sim 0$" — the nonnegative condition $\|f\|_p\geqslant 0$ is clear. We prove "$\|f\|_p=0\Leftrightarrow f\sim 0$"; i. e., "$\|f\|_p=0\Leftrightarrow f=0$, a. e.", that is, "$\forall x\in[a,b]\backslash E$ implies $f(x)=0$ with $m(E)=0$ on Lebesgue measurable set E.

About "$f\sim 0\Rightarrow\|f\|_p=0$": By $\int_{[a,b]}|f(x)|^p\mathrm{d}x=\int_{[a,b]\backslash E}|f(x)|^p\mathrm{d}x+\int_E|f(x)|^p\mathrm{d}x=0$, the first integral is zero since $f(x)=0$ on $[a,b]\backslash E$; The second one is also zero by $mE=0$.

Conversely, about "$\|f\|_p=0\Rightarrow f\sim 0$": For $n\in\mathbb{Z}^+$, then we have by inequality

$$\int_{[a,b]} |f(x)|^p \mathrm{d}x \geqslant \int_{E(|f|^p \geqslant \frac{1}{n})} |f(x)|^p \mathrm{d}x \geqslant \frac{1}{n} mE\left(|f|^p \geqslant \frac{1}{n}\right),$$

thus $\|f\|_p = 0 \Rightarrow mE\left(|f|_p \geqslant \frac{1}{n}\right) = 0$; And $f \sim 0$ by "$E(f \neq 0) = \bigcup_{n=1}^{+\infty} E\left(|f|^p \geqslant \frac{1}{n}\right)$ implies $mE(f \neq 0) \leqslant \sum_{n=1}^{+\infty} mE\left(|f|^p \geqslant \frac{1}{n}\right) = 0$".

② For "$\|\alpha f\|_p = |\alpha| \|f\|_p$", it is clear;

③ For inequality "$\|f+g\|_p \leqslant \|f\|_p + \|g\|_p$", it is clear by Minkowski inequality.

(3) $L^p([a,b])$ is a Banach space: the norm is $\|f\|_p = \left(\int_{[a,b]} |f(x)|^p \mathrm{d}m\right)^{\frac{1}{p}}$, distance is $\rho(f,g) = \left(\int_{[a,b]} |f(x) - g(x)|^p \mathrm{d}m\right)^{\frac{1}{p}}$. The completeness has been proved in Example 4.1.4.

(4) $L^p([a,b])$, $1 \leqslant p < +\infty$, is an infinite-dimensional Banach space: Since $C([a,b]) \subset L^p([a,b]) \subset L^1([a,b])$, we prove $C([a,b])$ is an infinite-dimensional space. In fact, $\forall n \in \mathbb{Z}^+$, there exists n linearly independent function system, for example, $\{x^k: k=0,1,2,\cdots,n, x \in [a,b]\} \subset C([a,b])$ in $C([a,b])$. If the space $C([a,b])$ is a finite-dimensional Banach space, say, n-dimensional, then any $n+1$ function system in $C([a,b])$ is linearly dependent, however it is impossible. So that $C([a,b])$ is an infinite-dimensional Banach space, and so does $L^p([a,b])$ for $p \geqslant 1$.

4.1.4 Orthogonal systems in Hilbert spaces

The elements of spaces $\mathbb{R}^n$ ($n \in \mathbb{Z}^+$) and l^p ($1 \leqslant p \leqslant +\infty$) are points, whereas elements in spaces $C^k([a,b])$ ($k \in \mathbb{N}$) and $L^p([a,b])$ ($1 \leqslant p \leqslant +\infty$) are functions, so that $C^k([a,b])$, $L^p([a,b])$ are function spaces. Almost all function spaces are infinite dimensional, so that we consider "bases" of function spaces as "function system", and "a function in the space can be expressed by the function system", particularly, in inner-product spaces and Hilbert spaces.

1. Orthogonality in inner-product spaces

We recall some useful properties of an inner-product space $(X, (x,y))$, firstly.

(1) **Linearity and conjugate linearity of inner-product**

If $(X, (x,y))$ is a real inner product space, then inner product (x,y) is linear about x, y, respectively; If it is a complex inner product space, then (x,y) is linear about x, and is conjugate linear about y.

(2) **Inner-product space is normed linear spaces**

If $(X,(x,y))$ is an inner product space, and $\|x\|_X=\sqrt{(x,x)}$, then $(X,\|x\|_X)$ is a normed linear space. And Schwarz inequality $|(x,y)|\leqslant\|x\|_X\|y\|_X$ holds.

(3) **Inner-product is a continuous function of x and y**

If $(X,(x,y))$ is an inner product space, for sequences $\{x_n\}\subset X$, $\{y_n\}\subset X$ with

$$\lim_{n\to+\infty}\|x_n-x\|_X=0,\ \lim_{n\to+\infty}\|y_n-y\|_X=0,\quad x\in X, y\in Y,$$

where $\|z\|_X=\sqrt{(z,z)}$ is the norm of element $z\in X$, then it holds $\lim\limits_{n\to+\infty}|(x_n,y_n)-(x,y)|=0$.

In fact, by the boundedness of a convergent sequence $\{y_n\}$ in X, we have

$$\begin{aligned}|(x_n,y_n)-(x,y)|&\leqslant|(x_n,y_n)-(x,y_n)|+|(x,y_n)-(x,y)|\\&=|(x_n-x,y_n)|+|(x,y_n-y)|\\&\leqslant\|x_n-x\|_X\|y_n\|_X+\|x\|_X\|y_n-y\|_X\\&\leqslant M_1\|x_n-x\|_X+\|x\|_X\|y_n-y\|_X\to0,\end{aligned}$$

this implies the continuity of inner product (x,y).

(4) **Identical equalities in inner product spaces**

If $(X,(x,y))$ is a real inner product space, then the norm $\|x\|_X=\sqrt{(x,x)}$ satisfies the following identical equalities

$$(x,y)=\frac{1}{4}(\|x+y\|_X^2-\|x-y\|_X^2)$$

and

$$\|x+y\|_X^2+\|x-y\|_X^2=2(\|x\|_X^2+\|y\|_X^2);$$

the latter one is called **the parallelogram identity equality in inner product spaces;**

If $(X,(x,y))$ is a complex inner product space, then the identical equality holds

$$(x,y)=\frac{1}{4}(\|x+y\|_X^2-\|x-y\|_X^2+\mathrm{i}\|x+\mathrm{i}y\|_X^2-\mathrm{i}\|x-\mathrm{i}y\|_X^2).$$

(5) **Pythagorean Theorem in inner product spaces**

If in subset $\mathfrak{A}=\{x_1,\cdots,x_k\}\subset X$, any couple elements are orthogonal, then for $x=x_1+\cdots+x_k$, the norm $\|x\|_X=\sqrt{(x,x)}$ satisfies **the generalized Pythagorean Theorem:** $\|x\|_X^2=\|x_1\|_X^2+\cdots+\|x_k\|_X^2$.

The concept of orthogonality in Euclidean spaces can be generalized to Hilbert spaces.

Definition 4.1.12 (orthogonality) Let $(X,(x,y))$ be a Hilbert space. For any couple of non-elements $x,y\in X$, if "$x,y\in X\Rightarrow(x,y)=0$", then x,y are said to be **orthogonal mutually**, denoted by $x\perp y$.

For an element $x \in X$ and a subset $M \subset X$, if x is orthogonal to every element $y \in M$, then x is said to be **orthogonal to** M, denoted by $x \perp M$.

The set of all elements in X which are orthogonal to subset $M \subset X$, is denoted by

$$M^{\perp} = \{x \in X : x \perp y, \forall y \in M\}.$$

Definition 4.1.13 (orthogonal decomposition) Let $(X, (x, y))$ be a Hilbert space, M be a closed subspace of X. If for any $x \in X$, it holds a unique decomposition $x = y + z$ with $y \in M, z \in M^{\perp}$, then $x = y + z$ is said to be **the orthogonal decomposition of** x, this y is said to be **the orthogonal projection of** $x \in X$ **in closed subspace** M.

Theorem 4.1.14 Let $(X, (x, y))$ be a Hilbert space, M be a closed subspace of X. Then for any $x \in X$, there exists a unique orthogonal decomposition: $\forall x \in X, \exists x_0 \in M$, $y \in M^{\perp}$, s. t. $x = x_0 + y$.

Proof Without loss of generality, let $M \underset{\neq}{\subset} X$ be a proper subspace. If $x \notin M$, then $\alpha = \inf\limits_{y \in M} \| x - y \|_X > 0$. Thus, $\exists \{x_n\} \subset M$, s. t. $\alpha = \lim\limits_{n \to +\infty} \| x - x_n \|_X \Rightarrow \dfrac{x_n + x_m}{2} \in M$ implies $\left\| x - \dfrac{x_n + x_m}{2} \right\|_X \geqslant \alpha \Rightarrow \| x_m - x_n \|_X^2 = \| x_m - x - x_n + x \|_X^2 = 2 \| x_m - x \|_X^2 + 2 \| x_n - x \|_X^2 - 4 \left\| x - \dfrac{x_m + x_n}{2} \right\|_X^2 \leqslant 2 \| x_m - x \|_X^2 + 2 \| x_n - x \|_X^2 - 4\alpha^2 \to 2\alpha^2 + 2\alpha^2 - 4\alpha^2 = 0 \Rightarrow \lim\limits_{n, m \to +\infty} \| x_m - x_n \|_X = 0 \Rightarrow M$ is closed, thus complete, so $\exists x_0 \in M$, s. t. $\lim\limits_{n \to +\infty} \| x_n - x_0 \|_X = 0$, and for $x \in X \setminus M$, it holds $\| x - x_0 \|_X = \alpha$.

Next, we prove $(x - x_0) \perp M$. Take any $z \in M, z \neq 0$, for a complex number $\lambda \in \mathbb{C}$, it holds $x_0 + \lambda z \in M$

$\Rightarrow \alpha^2 \leqslant \| x - (x_0 + \lambda z) \|_X^2 = \| x - x_0 \|_X^2 - \bar{\lambda}(x - x_0, z) - \lambda(z, x - x_0) + |\lambda|^2 \| z \|^2$

$\Rightarrow \bar{\lambda}(x - x_0, z) + \lambda(z, x - x_0) - |\lambda|^2 \| z \|^2 \leqslant 0$ (by $\| x - x_0 \|_X = \alpha$)

$\Rightarrow |(x - x_0, z)|^2 + |(z, x - x_0)|^2 - |(x - x_0, z)|^2 \leqslant 0$ (take $\lambda = \dfrac{(x - x_0, z)}{\| z \|^2}$)

$\Rightarrow |(x - x_0, z)|^2 \leqslant 0 \Rightarrow (x - x_0, z) = 0 \Rightarrow (x - x_0) \perp M$ (by $z \in M$ arbitrary)

$\Rightarrow$ let $y = x - x_0$, then $x = x_0 + y$ with $x_0 \in M, y \in M^{\perp} \Rightarrow$ The proof is complete.

Definition 4.1.14 (orthonormal system) Let $(X, (x, y))$ be a Hilbert space. If a subset $\{e_1, e_2, \cdots, e_n, \cdots\} \subset X$ satisfies $(e_i, e_j) = \begin{cases} 1, & i = j \\ 0, & i \neq j \end{cases}$, then $\{e_n\}_{n \geqslant 1}$ is said to be **an orthonormal system.**

Example 4.1.13 In Example 4.1.11, take $p = 2$, we consider

$$l^2=\left\{x=(x_1,x_2,\cdots):\ \sum_{j=1}^{+\infty}|x_j|^2<+\infty\right\}.$$

Define $(x,y)=\sum_{j=1}^{+\infty}x_jy_j$, by Schwarz inequality $(\sum_{j=1}^{\infty}|x_jy_j|)\leqslant(\sum_{j=1}^{\infty}|x_j|^2)^{\frac{1}{2}}\cdot(\sum_{j=1}^{\infty}|y_j|^2)^{\frac{1}{2}}$, then $x,y\in l^2\Rightarrow|(x,y)|<+\infty$; it is easy to verify that (x,y) is an inner product on l^2, a complete inner product space, i. e., a Hilbert space.

Schauder base $\{e_n=(\underbrace{0,\cdots,0,1}_{n},0,\cdots):n\in\mathbf{Z}^+\}$ **of** l^2 is **an orthonormal system in** l^2.

Example 4.1.14 In Example 4.1.12, take $p=2$, we consider $L^2([a,b])=\left\{f:\int_{[a,b]}|f(x)|^2\mathrm{d}x<+\infty\right\}$.

Define $(f,g)=\int_{[a,b]}f(x)\overline{g(x)}\,\mathrm{d}x$. By Schwarz inequality (function type)

$$\int_{[a,b]}|f(x)\overline{g(x)}|\,\mathrm{d}x\leqslant\left(\int_{[a,b]}|f(x)|^2\mathrm{d}x\right)^{\frac{1}{2}}\left(\int_{[a,b]}|g(x)|^2\mathrm{d}x\right)^{\frac{1}{2}},$$

then $f,g\in L^2([a,b])\Rightarrow|(f,g)|<+\infty$, and $f\cdot g\in L^1([a,b])$. Thus, (f,g) is an inner product, and $L^2([a,b])$ is a Hilbert space.

The function family $\left\{\frac{1}{\sqrt{2\pi}}\mathrm{e}^{inx}\right\}_{n\in\mathbf{Z}}$ is **an orthonormal system in** $L^2([-\pi,\pi])$.

Example 4.1.15 The weighted Hilbert space $L^2([a,b],\omega(x))$ with positive measurable weighted function $\omega(x)$. Let

$$L^2([a,b],\omega(x))=\left\{f(x):\int_{[a,b]}\omega(x)|f(x)|^2\,\mathrm{d}x<+\infty\right\},$$

by the generalized Schwarz inequality

$$\int_{[a,b]}\omega(x)|f(x)\overline{g(x)}|\,\mathrm{d}x\leqslant\left(\int_{[a,b]}\omega(x)|f(x)|^2\mathrm{d}x\right)^{\frac{1}{2}}\left(\int_{[a,b]}\omega(x)|g(x)|^2\mathrm{d}x\right)^{\frac{1}{2}},$$

we can prove that $L^2([a,b],\omega(x))$ with **the weighted inner product** $(f,g)=\int_{[a,b]}\omega(x)f(x)\cdot\overline{g(x)}\,\mathrm{d}x<+\infty$ is **a weighted Hilbert space.**

Example 4.1.16 **Chegyshev orthonormal system** in weighted Hilbert space $L^2\left([-1,1],\frac{1}{\sqrt{1-x^2}}\right)$.

Let $L^2\left([-1,1],\frac{1}{\sqrt{1-x^2}}\right)$ be weighted Hilbert space with **weighted function** $\omega(x)=\frac{1}{\sqrt{1-x^2}}$. In this space, there exists an orthonormal system $\{\widetilde{T}_n(x):x\in[-1,1]\}_{n\in\mathbf{N}}$

with $\widetilde{T}_n(x)=\begin{cases}\dfrac{1}{\sqrt{\pi}}T_n(x), & n=0\\ \dfrac{2}{\sqrt{\pi}}T_n(x), & n\neq 0\end{cases}$, generalized by so-called **first-class Chegyshev polynomials** $\{T_n(x)=\cos(n\arccos x): x\in[-1,1]\}_{n\in\mathbf{N}}$.

Example 4.1.17 An orthonormal system in weighted Hilbert space $L^2((-\infty,+\infty),\mathrm{e}^{-x^2})=L^2(\mathbb{R},\mathrm{e}^{-x^2})$.

Clearly, $\{1,x,x^2,\cdots,x^n,\cdots\}\subset L^2(\mathbb{R},\mathrm{e}^{-x^2})$. For weighted function $\omega(x)=\mathrm{e}^{-x^2}$, we have

$$\omega'(x)=-2x\mathrm{e}^{-x^2},\omega''(x)=(4x^2-2)\mathrm{e}^{-x^2},\cdots,\omega^{(n)}(x)=y_n(x)\mathrm{e}^{-x^2},$$

where $y_n(x)$ is n-polynomial of x with the highest coefficient $(-2)^n$. We can prove that $\{y_n(x)\}_{n\in\mathbf{N}}$ **is a polynomial family generalized by the orthogonalization of** $\{1,x,x^2,\cdots,x^n,\cdots\}$. Let $H_n(x)=\dfrac{y_n(x)}{\|y_n\|}$, then $\{H_n(x)\}_{n\in\mathbf{N}}^{+\infty}$ is an orthonormal system in $L^2(\mathbb{R},\mathrm{e}^{-x^2})$.

2. Orthogonal expansions in Hilbert spaces

The concept of the orthogonality in a Hilbert space brings about many important properties and applications, particularly, the orthogonal expansion in a Hilbert space.

1) Fourier expansions in Hilbert spaces

Definition 4.1.15 (Fourier coefficient) Let $(X,(x,y))$ be a Hilbert space, $\{e_n\}\subset X$ be an orthonormal system in X. For any $x\in X$, the number sequence $\{c_n\}$ with

$$c_n=(x,e_n),\quad n=1,2,\cdots \tag{4.1.9}$$

is said to be the **Fourier coefficient** of x with respect to orthonormal system $\{e_n\}$, and number c_n is called the nth-Fourier coefficient of x.

Theorem 4.1.15 Let $(X,(x,y))$ be a Hilbert space, $\{e_n\}\subset X$ be an orthonormal system in X. Then for any $x\in X$, the inequality

$$\sum_{n=1}^{+\infty}|(x,e_n)|^2\leqslant\|x\|_X^2 \tag{4.1.10}$$

holds.

Inequality (4.1.10) is said to be **Bessel inequality**, and $(x,e_n)=c_n$ is the nth-Fourier coefficient of x; $\|x\|_X=\sqrt{(x,x)}$ is the norm of x in Hilbert space $(X,(x,y))$, when it is regarded as a Banach space.

2) Integrity and completeness of orthonormal systems in Hilbert spaces

Definition 4.1.16 (Parseval equality) Let $(X,(x,y))$ be a Hilbert space, $\{e_n\}\subset X$ be an orthonormal system in X. For any $x\in X$, the equality

$$\sum_{n=1}^{+\infty} |(x,e_n)|^2 = \|x\|_X^2 \tag{4.1.11}$$

is said to be **Parseval equality**, where $(x,e_n)=c_n, n=1,2,\cdots$, and $\|x\|_X=\sqrt{(x,x)}$.

Definition 4.1.17 (integrity) Let $(X,(x,y))$ be a Hilbert space, $\{e_n\}\subset X$ be an orthonormal system in X. If for any $x\in X$, Parseval equality holds, then the orthonormal system $\{e_n\}$ is said to possess **the integrity.**

Note **"integrity"** means: a space obeys "the law of conservation of energy "; i. e., **Parseval equality reveals the conservation law of energy.**

Definition 4.1.18 (completeness) Let $(X,(x,y))$ be a Hilbert space, $\{e_n\}\subset X$ be an orthonormal system in X. If for any $x\in X$, it holds that

$$(x,e_n)=0,\quad n=1,2,\cdots, \text{implies } x=0, \tag{4.1.12}$$

then the orthonormal system $\{e_n\}$ is said to possess **the completeness.**

Note **"completeness"** means: for nonzero element $x\in X$, if it is orthogonal to every e_n in $\{e_n\}$, then it must be zero, $x=0$. So that this orthonormal system $\{e_1,e_2,\cdots,e_n,\cdots\}$ is completed, since it contains all nonzero mutually orthogonal elements of X but no loss.

Theorem 4.1.16 Let $(X,(x,y))$ be a Hilbert space, $\{e_n\}\subset X$ be an orthonormal system in X. Then the following four statements are equivalent:

(1) $\{e_n\}$ has integrity;

(2) $\{e_n\}$ has completeness;

(3) The subspace $L\subset X$ expanded by $\{e_n\}$ is dense in X;

(4) $\forall x\in X$, the series $\sum_{n=1}^{+\infty}(x,e_n)e_n$ converges to x in $(X,(x,y))$.

Proof (1) $\Leftrightarrow$(3) —

(1) $\Rightarrow$ (3): Let $\{e_n\}$ have the integrity, i. e., $\forall x\in X$, it holds equality $\sum_{n=1}^{+\infty}|(x,e_n)|^2=\|x\|_X^2$; and this is equivalent to: for $x\in X$, it holds $\left\|\sum_{k=1}^{n}c_ke_k - x\right\|_X \to 0$ with $(x,e_k)=c_k$; Then, this means that the subspace L expanded by $\{e_n\}$ is dense in X.

(3) $\Rightarrow$ (1): Take $x\in X$, and $(x,e_k)=c_k$ $(k=1,2,\cdots)$ in (4.1.9). Since L is dense in X, then for this $x\in X$, and $\forall\varepsilon>0$, $\exists\{\alpha_k\}$, $\exists n>0$, s. t. it holds $\|x-s_n\|_X<\varepsilon$ with $s_n=\sum_{k=1}^{n}\alpha_ke_k$.

We declare that the equality $\| x - s_n \|_X = \left\| x - \sum_{k=1}^{n} \alpha_k e_k \right\|_X$ takes minimum value, if and only if $\alpha_k = c_k$.

In fact, by $(x - \sum_{k=1}^{n} \alpha_k e_k, e_j) = (x, e_j) - (\sum_{k=1}^{n} \alpha_k e_k, e_j) = (x, e_j) - \alpha_j = 0, j = 1, 2, \cdots, n$, then $x - \sum_{k=1}^{n} \alpha_k e_k$ is orthogonal with each e_j, if and only if $\alpha_j = c_j, j = 1, 2, \cdots, n$. Moreover, by the generalized Pythagorean Theorem, it holds

$$\begin{aligned}\left\| x - \sum_{k=1}^{n} \alpha_k e_k \right\|_X^2 &= \left\| x - \sum_{k=1}^{n} c_k e_k + \sum_{k=1}^{n} (c_k - \alpha_k) e_k \right\|_X^2 \\ &= \left\| x - \sum_{k=1}^{n} c_k e_k \right\|_X^2 + \left\| \sum_{k=1}^{n} (c_k - \alpha_k) e_k \right\|_X^2 \\ &= \left\| x - \sum_{k=1}^{n} c_k e_k \right\|_X^2 + \sum_{k=1}^{n} | c_k - \alpha_k |^2 .\end{aligned}$$

Thus, if and only if $\| x - s_n \|_X = \left\| x - \sum_{k=1}^{n} \alpha_k e_k \right\|_X$ arrives the minimum value at $\alpha_k = c_k$; This implies that $\left\| x - \sum_{k=1}^{n} c_k e_k \right\|_X < \varepsilon$. Then, by arbitrary $\varepsilon > 0$ and Parseval equality $\sum_{n=1}^{+\infty} | (x, e_n) |^2 = \| x \|_X^2$ holds.

(1) ⇔ (2)—

(1) ⇒ (2): we prove "$\{e_n\}$ has integrity" implies "$\{e_n\}$ has completeness". Take $x \in X$ with "$(x, e_n) = 0, n = 1, 2, \cdots$". Since $\{e_n\}$ has integrity, then $(x, e_n) = 0$ implies $\| x \|_X^2 = \sum_{n=1}^{+\infty} | (x, e_n) |^2 = \sum_{n=1}^{+\infty} | c_n |^2 = 0$, so that $x = 0$, and the completeness of $\{e_n\}$ holds.

(2) ⇒ (1): If $\{e_n\}$ has completeness, we prove $\{e_n\}$ has integrity. Only need to prove that subspace L spanned by $\{e_n\}$ is dense in X. If not, then $\bar{L} \neq X$, so that $\exists x \in X \setminus \bar{L}$ with $x \neq 0$. Then, $x = x_0 + y$, s. t. $x_0 \in \bar{L}$, $y \perp \bar{L}$ (Theorem 4. 1. 14), and $y \neq 0$. However, $y \perp e_n$ for $n = 1, 2, \cdots$. This contradicts the completeness of $\{e_n\}$, so that $\bar{L} = X$, and then $\{e_n\}$ has integrity by(3)⇔(1).

The equivalences of (4) with (1), (2), (3) are clear.

Theorem 4. 1. 17 Let $(X, (x, y))$ be a Hilbert space with separability. Then there exists an orthonormal system $\{e_n\} \subset X$ in X.

We omit the proof, only show the idea of it: By the separability of space, there exists countable dense set $A=\{x_n\}\subset X$, such that $X=\overline{A}$. Using **Schmidt orthogonalization method** to $A=\{x_n\}$, we construct an orthonormal system $B=\{e_n\}$ such that $(X,(x,y))$ is the space expanded by $B=\{e_n\}$.

Schmidt orthogonalization method: Let $\{x_n\}$ be a countable sequence in inner product space $(X,(x,y))$.

Take $e_1=\dfrac{x_1}{\|x_1\|_X}$, then $\|e_1\|_X=1$—let $y_2=x_2-(x_2,e_1)e_1$, then $y_2\neq 0$, $(y_2,e_1)=0$. Thus, $y_2\perp e_1$;

Take $e_2=\dfrac{y_1}{\|y_1\|_X}$, then $\|e_2\|_X=1, e_2\perp e_1$—let $y_3=x_3-(x_3,e_1)e_1-(x_3,e_2)e_2$, then $y_3\neq 0, (y_3,e_1)=0, (y_3,e_2)=0$. Thus, $y_3\perp e_1, y_3\perp e_2$;

Take $e_3=\dfrac{y_2}{\|y_2\|_X}$, then $\|e_3\|_X=1, e_3\perp e_1, e_3\perp e_2$—let $y_4=x_4-(x_4,e_1)e_1-\cdots$, then $y_4\neq 0,\cdots$; Continue the procedure, we get an orthonormal system $\{e_n\}$.

Theorem 4.1.18 For an infinite-dimensional Hilbert space $(X,(x,y))$, it holds

(1) Any real Hilbert space $(X,(x,y))$ with separability is equidistance isomorphic with the real space l^2; Any complex Hilbert space $(X,(x,y))$ with separability is equidistance isomorphic with the complex space l^2;

(2) All Hilbert spaces with separability are equidistance isomorphic each other.

By using an orthonormal system $\{e_n\}$ of Hilbert space $(X,(x,y))$, for any $x\in X$, it has Fourier expansion with Fourier coefficients $\{c_n\}, c_n=(x,e_n)$, and $\{c_n\}\leftrightarrow l^2$. So that $(X,(x,y))\leftrightarrow l^2$.

Example 4.1.18 Rademarch function system $\{\varphi_n(x)\}_{n\in\mathbf{N}}$ on interval $[0,1]$ with

$$\varphi_0(x)=\begin{cases}1, & x\in[0,2^{-1})\\ -1, & x\in(2^{-1},1)\end{cases},\varphi_n(x)=\varphi_0(2^n x),\quad n=1,2,\cdots$$

it is an orthonormal system on $[0,1]$; but not complete. The completion of $\{\varphi_n(x)\}_{n\in\mathbf{N}}$ is Walsh function system $\{\mathrm{wal}_n(x)\}_{n\in\mathbf{N}}$, with $\mathrm{wal}_0(x)=1, \mathrm{wal}_k(x)=\prod\limits_{(j:\,k'_{-j}=1)}\varphi_j(x), k\in\mathbf{Z}^+$, it is an orthonormal system; here the dyadic expression of $k\in\mathbf{Z}^+$ is $k=(k_{-N+1},k_{-N},\cdots,k_{-1},k_0)$, with $k'_{-j}=k_{-j}\oplus k_{-(j+1)}$.

The orthogonal expansions in Hilbert spaces, including Fourier series, have learned in the course of *Advance Mathematics*. We will learn more contents, such as Fourier Analysis, Laplace transformations, wavelets, in Chapter 5. All these contents are very important and useful in modern applied mathematics and other natural scientific areas.

4.2 Operator Theory

4.2.1 Linear operators on Banach spaces

"space", can be regarded as a "stadium", variables sport on stadiums, and functions sport on the same, or different stadiums. Thus, various kinds of spaces and mappings connected variables with functions are the main objects in mathematics. We study diverse linear operators on Banach spaces and operator spaces in this section.

1. Bounded linear operators on Banach spaces

1) Bounded linear operators

Definition 4.2.1 (linear operator) Let $(X, \| x \|_X)$, $(Y, \| y \|_Y)$ be Banach spaces on number field $\mathbb{F}$, and $D \subseteq X$ be a subspace of X. If a mapping $T: D \to Y$ satisfies:

$$T(x_1 + x_2) = T(x_1) + T(x_2), \quad T(\alpha x) = \alpha T(x),$$

holding for $\forall x_1, x_2, x \in D$, $\forall \alpha \in \mathbb{F}$, then T is said to be **a linear operator from X to Y**, and D is said to be **the domain of T**.

If $D = X$, $Y = \mathbb{R}$ or $\mathbb{C}$, then T is said to be **a linear functional** on X.

Definition 4.2.2 (bounded linear operator) Let $(X, \| x \|_X)$, $(Y, \| y \|_Y)$ be two Banach spaces on number field $\mathbb{F}$, set $D \subseteq X$ be a subspace of X. If mapping $T: D \to Y$ is a linear operator from X to Y, and exists constant $M > 0$, satisfying

$$\| T(x) \|_Y \leqslant M \| x \|_X, \quad \forall x \in D, \tag{4.2.1}$$

then T is said to be **a bounded linear operator** from X to Y; If $Y = \mathbb{F}$, and exists constant $M > 0$, satisfying

$$| T(x) | \leqslant M \| x \|_X, \quad \forall x \in D, \tag{4.2.2}$$

then T is said to be **a bounded linear functional** on X.

Definition 4.2.3 (norm of an operator) Let $(X, \| x \|_X)$, $(Y, \| y \|_Y)$ be Banach spaces on number field $\mathbb{F}$. If mapping $T: D \subseteq X \to Y$ is a bounded linear operator form X to Y, then the infimum of $M > 0$ satisfying the inequality

$$\| T(x) \|_Y \leqslant M \| x \|_X, \quad \forall x \in D,$$

is said to be **the norm of operator T**, denoted by $\| T \|$:

$$\| T \| = \inf\{M: \| T(x) \|_Y \leqslant M \| x \|_X, \forall x \in D\}. \tag{4.2.3}$$

Equivalently, the norm of an operator can be defined by

$$\| T \| = \sup\{ \| T(x) \|_Y: \| x \|_X \leqslant 1, \forall x \in D\}, \tag{4.2.4}$$

or

$$\| T \| = \sup\{ \| T(x) \|_Y : \| x \|_X = 1, \forall x \in D \}. \tag{4.2.5}$$

Example 4.2.1 Determine bounded linear operator $T: \mathbb{R}^n \to \mathbb{R}^n$ from $\mathbb{R}^n$ to $\mathbb{R}^n$. In fact, $T: \mathbb{R}^n \to \mathbb{R}^n$ is given by an $n\times n$-matrix $A = [a_{ij}]_{1\leqslant i,j\leqslant n} \in \mathfrak{M}_{n\times n}$. Let

$$x = (\xi_1, \cdots, \xi_n) \in \mathbb{R}^n \to y = (\eta_1, \cdots, \eta_n) \in \mathbb{R}^n$$

by the transformation $\eta_i = \sum_{j=1}^{n} a_{i,j}\xi_j \ (i=1,\cdots n)$. Where $T(x)=y$ has a form

$$T(x) = [a_{ij}]_{n\times n}[x_i]_{n\times 1} = \begin{bmatrix} a_{11} & \cdots & a_{1n} \\ \vdots & \ddots & \vdots \\ a_{n1} & \cdots & a_{nn} \end{bmatrix} \begin{bmatrix} \xi_1 \\ \vdots \\ \xi_n \end{bmatrix} = \begin{bmatrix} a_{11}\xi_1 + \cdots + a_{1n}\xi_n \\ \vdots \\ a_{n1}\xi_1 + \cdots + a_{nn}\xi_n \end{bmatrix}$$

$$= \Big[\sum_{j=1}^{n} a_{ij}\xi_j\Big]_{n\times 1} = \begin{bmatrix} \eta_1 \\ \vdots \\ \eta_n \end{bmatrix} = y.$$

It is easy to verify that $T(x) = y^{\mathrm{T}} = [\eta_1 \ \cdots \ \eta_n]$ ($y^{\mathrm{T}} = [\eta_1 \ \cdots \ \eta_n]$ is the transposed matrix of $y = \begin{bmatrix} \eta_1 \\ \vdots \\ \eta_n \end{bmatrix}$) is a bounded linear operator from $\mathbb{R}^n$ to $\mathbb{R}^n$, since it is linear, firstly;

Then, it has boundedness by Cauchy inequality

$$\Big(\sum_{i=1}^{n}\eta_i^2\Big)^{\frac{1}{2}} = \Big(\sum_{i=1}^{n}\Big(\sum_{j=1}^{n}a_{i,j}\xi_j\Big)^2\Big)^{\frac{1}{2}} \leqslant \Big(\sum_{i=1}^{n}\sum_{j=1}^{n}a_{i,j}^2\Big)^{\frac{1}{2}}\Big(\sum_{j=1}^{n}\xi_j^2\Big)^{\frac{1}{2}} = \Big(\sum_{i=1}^{n}\sum_{j=1}^{n}a_{i,j}^2\Big)^{\frac{1}{2}} \| x \|_{\mathbb{R}^n},$$

and $\| T \| \leqslant \Big(\sum_{i=1}^{n}\sum_{j=1}^{n}a_{i,j}^2\Big)^{\frac{1}{2}}$.

Example 4.2.2 Let $C([a,b])$ be the space of all continuous functions on $[a,b]$. With norm $\| f \|_{C([a,b])} = \max_{x\in[a,b]} |f(x)|$ of $f \in C([a,b])$, then $(C([a,b]), \| f \|_{C([a,b])})$ is a Banach space. For $f \in C([a,b])$, we define **an integral operator** J

$$J(f)(g) = \int_a^b f(x)g(x)\,\mathrm{d}x, \quad \forall g \in C([a,b]),$$

then $J: C([a,b]) \to \mathbb{R}$ is a **bounded linear functional** on $C([a,b])$ by

$$|J(f)(g)| = \left|\int_a^b f(x)g(x)\,\mathrm{d}x\right| \leqslant (b-a) \| f \|_{C([a,b])} \| g \|_{C([a,b])}.$$

Example 4.2.3 Let $C^k([a,b])$ be the space of all k^{th}-continuous differential functions on $[a,b]$ with norm $\| f \| = \sum_{j=0}^{k} \max_{x\in[a,b]} |f^{(j)}(x)|$ of $f \in C^k([a,b])$, such that $C^k([a,b])$ is a Banach space. Define **the derivation operator** $\frac{\mathrm{d}}{\mathrm{d}x}$

$$\frac{\mathrm{d}}{\mathrm{d}x}: C^k([a,b]) \to C^{k-1}([a,b]), \quad k \geqslant 1,$$

then $\frac{\mathrm{d}}{\mathrm{d}x}$ is a linear operator on $C^k([a,b])$.

However, $\frac{\mathrm{d}}{\mathrm{d}x}$ is not bounded, since there is function sequence $\{f_n(x)\} = \{\sin nx\}_{n=1}^{+\infty} \subset C^1([0,1])$ with $\| f_n(x) \| = 1$, but $\left\| \frac{\mathrm{d}}{\mathrm{d}x} f_n(x) \right\|_{C([0,1])} = \| n\cos nx \|_{C([0,1])} = n \to +\infty$. Thus there is no positive number M satisfying inequality (4.2.2).

So that, a linear operator on a Banach space is not necessary to be bounded.

Theorem 4.2.1 Let $(X, \| x \|_X)$, $(Y, \| y \|_Y)$ be Banach spaces on number field $\mathbb{F}$. Then a mapping from X to Y is a bounded operator, if and only if it is a continuous linear operator from X to Y; also, if and only if it maps a bounded subset in X to a bounded subset in Y.

Proof (1) "a mapping is bounded" $\Leftrightarrow$ "a mapping is continuous"

Necessity Let $T: X \to Y$ be a bounded linear operator, then $\exists M > 0$ satisfying $\| T(x) \|_Y \leqslant M \| x \|_X$ for each $x \in X \Rightarrow \forall x \in X$, take sequence $x_n \in X, n = 1, 2, \cdots$, s.t. $\| x_n - x \|_X \to 0$ implies $\| T(x_n - x) \|_Y \leqslant M \| x_n - x \|_X$, thus $\| T(x_n) - T(x) \|_Y = \| T(x_n - x) \|_Y \to 0 \Rightarrow T: X \to Y$ is continuous.

sufficiency Let $T: X \to Y$ be a linear continuous operator. If T is unbounded, then $\forall n \in \mathbb{N}$, $\exists x_n \in X$ with $x_n \neq 0$, s.t. $\| T(x_n) \|_Y \geqslant n \| x_n \|_X \Rightarrow$ let $y_n = \frac{x_n}{n \| x_n \|_X}$, then $\| y_n \|_X = \left\| \frac{x_n}{n \| x_n \|_X} \right\|_X = \frac{1}{n} \to 0 \Rightarrow T(y_n) \to 0 \ (n \to +\infty)$ for the above $y_n \in Y$, by the continuity of $T \Rightarrow \| T(y_n) \|_Y \to 0 \ (n \to +\infty)$.

On the other hand, by the continuity of T, it holds $T(y_n) = T\left(\frac{x_n}{n \| x_n \|_X}\right) = \frac{1}{n \| x_n \|_X} T(x_n)$, so that

$$\| T(y_n) \|_Y = \left\| T\left(\frac{x_n}{n \| x_n \|_X}\right) \right\|_Y = \left\| \frac{1}{n \| x_n \|_X} T(x_n) \right\|_Y$$
$$= \frac{1}{n \| x_n \|_X} \| T(x_n) \|_Y \geqslant 1,$$

this contradicts $\| T(y_n) \|_Y \to 0 \ (n \to +\infty)$. Hence we conclude that T is bounded.

(2) "a mapping is bounded" $\Leftrightarrow$ "a mapping maps bounded subset to bounded subset"

Necessity Let $T: X \to Y$ be a bounded linear operator, then $\exists M > 0$ satisfying

$\| T(x) \|_Y \leqslant M \| x \|_X$ for each $x \in X \Rightarrow$ take a bounded set $A \subset X$, then $\exists K > 0$, such that $x \in A$ implies $\| x \|_X \leqslant K \Rightarrow$ combine both inequalities, then $\| T(x) \|_Y \leqslant M \| x \|_X \leqslant MK$, $\forall x \in A \Rightarrow$ the image set $T(A)$ is bounded in X.

Sufficiency Let linear operator $T: X \to Y$ maps a bounded subset in X to a bounded subset in $Y \Rightarrow$ the set $S = \{x \in X: \| x \|_X = 1\}$ is a unit sphere in X, and bounded in $X \Rightarrow$ by assumption, $T(S) \subset Y$ is bounded in $Y \Rightarrow \exists M > 0$, s. t. $\forall x \in S = \{x \in X: \| x \|_X = 1\}$ implies $\| T(x) \|_Y \leqslant M \Rightarrow \forall x \in X$ with $x \neq 0$ implies $\frac{1}{\| x \|_X} x \in S \Rightarrow \left\| T\left(\frac{x}{\| x \|_X}\right) \right\|_Y \leqslant M \Rightarrow \forall x \in X$ implies $\| T(x) \|_Y \leqslant M \| x \|_X \Rightarrow T: X \to Y$ is a bounded linear operator.

The proof is complete.

We use both "bounded linear operator" and "continuous linear operator", simultaneously.

2) Bounded linear operator spaces

Definition 4.2.4 (operator space) Let $(X, \| x \|_X)$, $(Y, \| y \|_Y)$ be normed linear spaces on number field $\mathbb{F}$. The set of all bounded linear operators from X to Y is denoted by

$$\mathfrak{B}(X, Y) = \{T: X \to Y, T \text{ is bounded linear operator}\}.$$

The operations $+$ and $\alpha \cdot$ on $\mathfrak{B}(X, Y)$ are endowed by:

addition$+$: $T, S \in \mathfrak{B}(X, Y) \Rightarrow (T + S)(x) = T(x) + S(x)$,

number product $\alpha \cdot$: $T \in \mathfrak{B}(X, Y), \alpha \in \mathbb{F} \Rightarrow (\alpha T)(x) = \alpha T(x)$,

then $\mathfrak{B}(X, Y)$ is a linear space on $\mathbb{F}$; Moreover, any $T \in \mathfrak{B}(X, Y)$ has the norm $\| T \|$, such that $\mathfrak{B}(X, Y)$ is a normed linear space $(\mathfrak{B}(X, Y), \| T \|)$ on $\mathbb{F}$, called **a bounded linear operator space**, or **a continuous linear operator space**.

2. Important properties of bounded linear operators on Banach spaces

There are three important theorems of bounded linear operators on Banach spaces: open-mapping, inverse-operator, and closed-graph theorems. In this subsection, we prove the three for both X and Y being Banach spaces.

1) Open mapping theorem

Theorem 4.2.2 Let $T: X \to Y$ be a bounded linear operator $T \in \mathfrak{B}(X, Y)$ from Banach spaces X onto Y satisfying $T(X) = Y$, i. e., T be surjective. Then T maps an open set in X to an open set in Y; that is,

$$T \in \mathfrak{B}(X, Y) \xrightarrow{T(X) = Y} T \text{ is an open mapping.}$$

Proof By three steps:

(1) Prove the following equivalence

"for open set $W \subset X$, the image $T(W) \subset Y$ is an open set" $\Leftrightarrow$ "$\exists \delta > 0$, s. t. $T(B(0, 1)) \supset$

$U(0,\delta)$" where $B(0,1)$ is open ball with center 0, radius 1 in X, and $U(0,\delta)$ is open ball with center 0, radius δ in Y.

Necessity For open set $W\subset X$, the image set $T(W)\subset Y$ is an open set $\Rightarrow$ take $W=B(0,1)\subset X$, it is an open ball, then the image $T(B(0,1))\subset Y$ is an open set in Y, and $T(0)=0\in T(B(0,1))$ by the linearity of T $\Rightarrow$ there exists $\delta>0$, such that the open ball $U(0,\delta)$ in Y is contained in open set $T(B(0,1))$, i. e., $T(B(0,1))\supset U(0,\delta)$.

Sufficiency Suppose that $\exists\,\delta>0$, s. t. $T(B(0,1))\supset U(0,\delta)$ $\Rightarrow$ for any open set $W\subset X$, since T is surjective, then: $\forall\, y_0\in T(W)$, $\exists\, x_0\in W$, s. t. $y_0=T(x_0)$ $\Rightarrow$ since W is open, $\exists\, r>0$, s. t. $B(x_0,r)\subset W\Rightarrow y_0=T(x_0)\in T(B(x_0,r))\subset T(W)$.

On the other hand, since $T\in\mathfrak{B}(X,Y)$ is continuous, then $\exists\,\varepsilon>0$, s. t. the neighborhood $U(T(x_0),r\varepsilon)$ is contained in $T(B(x_0,r))$ $\Rightarrow$ let $\delta=r\varepsilon>0$, then $\exists\,\delta>0$, s. t. $U(y_0,\delta)\subset T(B(x_0,r))\subset T(W)$ $\Rightarrow$ $\forall\, y_0\in T(W)$, $\exists\,\delta>0$, s. t. $U(y_0,\delta)\subset T(W)\Rightarrow$ $T(W)$ is an open set $\Rightarrow$ (1) is proved.

(2) Prove "$\forall\, B(0,1)\subset X$, $\exists\,\delta>0$, s. t. $T(B(0,1))\supset U(0,\delta)$". By two steps.

First step Prove "$\exists\,\delta>0$, s. t. $\overline{T(B(0,1))}\supset U(0,3\delta)$"

Since T is surjective, thus $Y=T(X)=\bigcup\limits_{n=1}^{+\infty}T(B(0,n))$ $\Rightarrow$ since Y is a Banach space, then there exists at least one $n_0\in\mathbb{Z}^+$, s. t. $\overline{T(B(0,n_0))}$ has non-empty inner part; i. e., $\overline{T(B(0,n_0))}$ has at least one inner point $y_0\Rightarrow$ $\exists\, r>0$, s. t. the neighborhood $U(y_0,r)\subset$ $T(B(0,n_0))\subset\overline{T(B(0,n_0))}$ $\Rightarrow$ $B(0,n_0)\subset X$, then $x\in B(0,n_0)$ implies $(-x)\in B(0,n_0)$, by the symmetry of ball $\Rightarrow$ $T(B(0,n_0))\subset Y$ is symmetry, i. e., $y=T(x)\in T(B(0,n_0))$ implies $(-y)=-T(x)=T(-x)\in T(B(0,n_0))\Rightarrow U(-y_0,r)\subset\overline{T(B(0,n_0))}\Rightarrow U(0,r)\subset\frac{1}{2}U(y_0,r)+\frac{1}{2}U(-y_0,r)\subset\overline{T(B(0,n_0))}$.

by $y\in U(0,r)\Rightarrow y=y_1+y_2$, with $y_1\in\frac{1}{2}U(y_0,r)$, $y_2\in\frac{1}{2}U(-y_0,r)$, then

$$\|y-0\|_Y\leqslant\|y_1-y_0\|_Y+\|y_2-(-y_0)\|_Y\leqslant\frac{r}{2}+\frac{r}{2}=r;$$

by "$U(y_0,r)\subset\overline{T(B(0,n_0))}\Rightarrow\frac{1}{2}U(y_0,r)\subset\frac{1}{2}\overline{T(B(0,n_0))}$";

"$U(-y_0,r)\subset\overline{T(B(0,n_0))}\Rightarrow\frac{1}{2}U(-y_0,r)\subset\frac{1}{2}\overline{T(B(0,n_0))}$"

$$\Rightarrow\frac{1}{2}U(y_0,r)+\frac{1}{2}U(-y_0,r)\subset\overline{T(B(0,n_0))}\Rightarrow\frac{1}{n_0}U(0,r)\subset\overline{T(B(0,1))}$$

by $T(\alpha x)=\alpha T(x)$, and by $U(0,r)\subset\overline{T(B(0,n_0))}=\overline{T(n_0B(0,1))}=n_0\overline{T(B(0,1))}$

$\Rightarrow U\left(0,\frac{r}{n_0}\right)\subset\overline{T(B(0,1))}\Rightarrow$ take $3\delta=\frac{r}{n_0}$, then $\exists\delta>0$, there exists $\delta>0$, such that $U(0,3\delta)\subset\overline{T(B(0,1))}$, i. e., $\overline{T(B(0,1))}\supset U(0,3\delta)$. The first step is proved.

Second step Prove "for the above $\delta>0$, it holds $T(B(0,1))\supset U(0,\delta)$". This is only need to prove: "$\forall y_0\in U(0,\delta)$, $\exists x_0\in B(0,1)$, s. t. $T(x_0)=y_0$". Also, only need to prove: "$\forall y_0\in U(0,\delta)$, solve the equation $T(x)=y_0$, find its solution $x_0\in B(0,1)$. "We prove this step by **the successive approximation method.**

$\forall y_0\in U(0,\delta)$, $\exists x_1\in B\left(0,\frac{1}{3}\right)$, s. t., $\|y_0-T(x_1)\|_Y<\frac{\delta}{3}\Rightarrow$ by the result in first step, it holds $U(0,\delta)\subset\overline{T\left(B\left(0,\frac{1}{3}\right)\right)}$. There are two cases: $U\left(y_0,\frac{\delta}{3}\right)\subset\overline{T\left(B\left(0,\frac{1}{3}\right)\right)}$, or $U\left(y_0,\frac{\delta}{3}\right)\not\subset\overline{T\left(B\left(0,\frac{1}{3}\right)\right)}\Rightarrow$ in first case, $U\left(y_0,\frac{\delta}{3}\right)\subset\overline{T\left(B\left(0,\frac{1}{3}\right)\right)}$, then $\exists x_1\in B\left(0,\frac{1}{3}\right)$, s. t. $\|y_0-T(x_1)\|_Y<\frac{\delta}{3}$; in second case, $U\left(y_0,\frac{\delta}{3}\right)\not\subset\overline{T\left(B\left(0,\frac{1}{3}\right)\right)}$, then $\exists k>3$, s. t. $U\left(y_0,\frac{\delta}{k}\right)\subset\overline{T\left(B\left(0,\frac{1}{3}\right)\right)}$. However, we can prove that in both cases, $\exists x_1\in B\left(0,\frac{1}{3}\right)$, s. t. $\|y_0-T(x_1)\|_Y<\frac{\delta}{k}<\frac{\delta}{3}\Rightarrow$ by the successive approximation method, for $y_0\in U(0,\delta)$, $\exists x_1\in B\left(0,\frac{1}{3}\right)$, s. t., $\|y_0-T(x_1)\|_Y<\frac{\delta}{3}\Rightarrow$ let $y_1=y_0-T(x_1)\in U\left(0,\frac{\delta}{3}\right)$, then $\exists x_2\in B\left(0,\frac{1}{3^2}\right)$, s. t. $\|y_1-T(x_2)\|_Y<\frac{\delta}{3^2}\Rightarrow\cdots\Rightarrow$ let $y_n=y_{n-1}-T(x_n)\in U\left(0,\frac{\delta}{3^n}\right)$, then $\exists x_{n+1}\in B\left(0,\frac{1}{3^{n+1}}\right)$, s. t. $\|y_n-T(x_{n+1})\|_Y<\frac{\delta}{3^{n+1}}\Rightarrow$ continue the steps, it follows $\sum_{n=1}^{+\infty}\|x_n\|_X\leqslant\sum_{n=1}^{+\infty}\frac{1}{3^n}=\frac{1}{2}$. Let $x_0\equiv\sum_{n=1}^{+\infty}x_n$ with $\|x_0\|_X\leqslant\frac{1}{2}$, i. e., $x_0\in B(0,1)\Rightarrow$ Denote by $s_n\equiv\sum_{j=1}^{n}x_j$, then $y_n=y_{n-1}-T(x_n)\in U\left(0,\frac{\delta}{3^n}\right)$ implies $\frac{\delta}{3^n}>$

$$\|y_n\|_Y=\|y_{n-1}-T(x_n)\|_Y=\cdots=\|y_0-T(x_1+\cdots+x_n)\|_Y=\|y_0-T(\sum_{j=1}^{n}x_j)\|_Y\Rightarrow$$

for $n=1,2,3,\cdots$, it holds $\|y_0-T(\sum_{j=1}^{n}x_j)\|_Y<\frac{\delta}{3^n}\Rightarrow\|y_0-T(s_n)\|_Y<\frac{\delta}{3^n}$, $n=1,2,3,\cdots\Rightarrow$ $\lim\limits_{n\to+\infty}T(s_n)=y_0$.

To complete the proof of the second step, we prove the continuity of T, i. e., $y_0=$

$T(x_0)$.

By $\| y_0 - T(x_0) \|_Y = \| y_0 - T(s_n) + T(s_n) - T(x_0) \|_Y \leqslant \| y_0 - T(s_n) \|_Y + \| T(s_n) - T(x_0) \|_Y = \| y_0 - T(s_n) \|_Y + \| T(s_n - x_0) \|_Y \Rightarrow$ for any given $\varepsilon > 0$, by $\| T \| \leqslant M$ and $x_0 = \sum_{n=1}^{+\infty} x_n$, then there exists $N_1 > 0$, s. t. for $n > N_1$, it holds $\| s_n - x_0 \|_X < \frac{\varepsilon}{2M}$; and there exists $N_2 > 0$, s. t. for $n > N_2$, it holds $\| T(s_n) - T(x_0) \|_Y < M \| s_n - x_0 \|_X < \frac{\varepsilon}{2}$; moreover, there exists $N_3 > 0$, s. t. for $n > N_3$ with $\frac{\delta}{3^n} < \frac{\varepsilon}{2} \Rightarrow \forall \varepsilon > 0, \exists N > 0$, s. t. for $n > N$, it holds $\| y_0 - T(s_n) \|_Y < \frac{\delta}{3^n} < \frac{\varepsilon}{2} \Rightarrow$ we have $\| y_0 - T(x_0) \|_Y \leqslant \| y_0 - T(s_n) \|_Y + \| T(s_n - x_0) \|_Y < \varepsilon \Rightarrow y_0 = T(x_0)$. The proof is complete.

2) Inverse-operator theorem

Theorem 4. 2. 3 Let $T: X \to Y$ be a bounded linear operator from X to Y, T be one-one surjective. Then there exists inverse operator T^{-1} of T, and $T^{-1}: Y \to X$ is a bounded linear operator from Y to X; i. e.

$$T \in \mathfrak{B}(X, Y) \xrightarrow{T(X) = Y,\ \text{one-one}} T^{-1} \in \mathfrak{B}(Y, X).$$

Proof By assumption, $T: X \to Y$ is a bounded linear operator from X to Y, and $T \in \mathfrak{B}(X, Y)$ is one-one surjective, thus $T^{-1}: Y \to X$ makes sense, and is one-one surjective. Hence, we only need to prove that T^{-1} is linear, and $T^{-1} \in \mathfrak{B}(Y, X)$.

(1) T^{-1} **is linear**—

$$\begin{aligned} y_1, y_2 \in Y &\Rightarrow \exists x_1, x_2 \in X, \text{s. t. } y_1 = T(x_1), \quad y_2 = T(x_2) \\ &\Rightarrow x_1 = T^{-1}(y_1), \quad x_2 = T^{-1}(y_1) \Rightarrow T^{-1}(y_1) + T^{-1}(y_2) = x_1 + x_2 \\ &\Rightarrow T(T^{-1}(y_1) + T^{-1}(y_2)) = T(x_1 + x_2) = T(x_1) + T(x_2) = y_1 + y_2 \\ &\Rightarrow T^{-1}(y_1) + T^{-1}(y_2) = T^{-1}(y_1 + y_2); \end{aligned}$$

$$\begin{aligned} y \in Y, \alpha \in \mathbb{F} &\Rightarrow \exists x \in X, \text{s. t. } y = T(x) \Rightarrow T^{-1}(y) = x \Rightarrow \alpha T^{-1}(y) = \alpha x \\ &\Rightarrow T(\alpha T^{-1}(y)) = T(\alpha x) = \alpha T(x) = \alpha y \Rightarrow \alpha T^{-1}(y) = T^{-1}(\alpha y). \end{aligned}$$

(2) $T^{-1} \in \mathfrak{B}(Y, X)$—

By $T(B(0,1)) \supset U(0, \delta)$ in the open mapping theorem, we rewrite it as $U(0,1) \subset T\left(B\left(0, \frac{1}{\delta}\right)\right) \Rightarrow T^{-1}(U(0,1)) \subset B\left(0, \frac{1}{\delta}\right) \Rightarrow \forall y \in Y$, $\| y \|_Y < 1$, it holds $\| T^{-1}(y) \|_Y < \frac{1}{\delta} \Rightarrow$ by linearity of the operator, it holds $\forall y \in X$, $\forall \varepsilon > 0$ implies $\| T^{-1}(y) \|_Y <$

$\frac{(1+\varepsilon)}{\delta}\|y\|_Y \Rightarrow$ let $\varepsilon \to 0$, then $\forall y \in Y$, it holds $\|T^{-1}(y)\|_X < \frac{1}{\delta}\|y\|_Y$. $\left(\text{for example}, \delta=\frac{r}{3n_0}\right) \Rightarrow T^{-1} \in \mathfrak{B}(Y, X)$.

3) Closed-graph theorem

Definition 4.2.5 (closed operator) Let $T: \mathfrak{D}_T \to Y$ be a linear operator from $\mathfrak{D}_T \subset X$ to Y. The set $G(T)=\{(x, T(x)) \in X \times Y: x \in \mathfrak{D}_T\}$ is said to be **the graph of operator T in Banach space $X \times Y$**. If $G(T)$ is a closed set in $X \times Y$, then T is said to be **a closed linear operator**, or simply, a **closed operator.**

A closed linear operator has an equivalent definition by the following proposition.

Proposition 4.2.1 Let $T: \mathfrak{D}_T \to Y$ be a linear operator. Then T is a closed operator, if and only if $\forall \{x_n\} \subset \mathfrak{D}_T$ with $x_n \xrightarrow{X} x \in X$ and $T(x_n) \xrightarrow{Y} y \in Y$, then $x \in \mathfrak{D}_T$ and $T(x)=y$; i.e.

$$(x, T(x)) \in G(T)=\mathfrak{D}_T \times Y.$$

Proof Necessity Let $T: \mathfrak{D}_T \to Y$ is a closed linear operator, i.e., $G(T)=\mathfrak{D}_T \times Y$ is a closed set in $X \times Y \Rightarrow$ "$\forall \{x_n\} \subset \mathfrak{D}_T$ with $x_n \xrightarrow{X} x \in X$ and $Tx_n \xrightarrow{X} Tx \in Y$" implies that "with respect to product topology it holds $(x_n, T(x_n)) \in \mathfrak{D}_T \times Y \xrightarrow{X \times Y} (x, T(x)) \in X \times Y$" $\Rightarrow$ by assumption, $G(T)=\mathfrak{D}_T \times Y$ is a closed set in $X \times Y$, then we conclude that $(x, T(x)) \in \mathfrak{D}_T \times Y$.

Sufficiency Let $(x, y) \in \overline{G(T)}=\overline{\mathfrak{D}_T \times Y} \Rightarrow \exists \{x_n\} \subset \mathfrak{D}_T$, such that $(x_n, T(x_n)) \in G(T) \xrightarrow{X \times Y} (x, y) \in X \times Y$, i.e., $\lim\limits_{n \to +\infty}\|(x_n - x, T(x_n)-y)\|_{X \times Y}=0 \Rightarrow$ hold $\|x_n - x\|_X \leqslant \|(x_n - x, T(x_n)-y)\|_{X \times Y} \to 0$ and $\|T(x_n)-y\|_Y \leqslant \|(x_n - x, T(x_n)-y)\|_{X \times Y} \to 0$, i.e., $\{x_n\} \subset \mathfrak{D}_T$ implies $x_n \xrightarrow{X} x \in X$ and $T(x_n) \xrightarrow{Y} y \in Y \Rightarrow$ by assumption of the sufficiency, it holds $(x, T(x)) \in \mathfrak{D}_T \times Y=G(T) \Rightarrow G(T)$ is a closed set in $X \times Y$.

Before proving the closed-graph theorem, we need a concept of "comparision of norms".

Comparision of two norms If there are two norms $\|x\|_1$, $\|x\|_2$ on a linear space X, such that both $(X, \|x\|_1)$, $(X, \|x\|_2)$ are normed linear spaces. **Norm $\|x\|_1$ is said to be stronger than** $\|x\|_2$, if there exists a constant $M>0$, such that $\|x\|_2 \leqslant M\|x\|_1$. Then we define that: **Two norms $\|x\|_1$ and $\|x\|_2$ are said to be equivalent**, if there exist constants $M>0$ and $m>0$, such that $m\|x\|_1 \leqslant \|x\|_2 \leqslant M\|x\|_1$.

Proposition 4.2.2 Let $(X, \|x\|_1)$ and $(X, \|x\|_2)$ be two Banach spaces, and norm $\|x\|_1$ be stronger than $\|x\|_2$. Then $\|x\|_1$ and $\|x\|_2$ are equivalent each other.

Proof Consider the identity mapping $I: X \to X$, it is a linear operator from Banach space $(X, \| x \|_1)$ to Banach space $(X, \| x \|_2)$. Since $\| x \|_1$ is stronger than $\| x \|_2$, then there exists constant $M>0$, such that

$$\| I(x) \|_2 \leqslant M \| x \|_1. \tag{4.2.6}$$

This implies that $I: (X, \| x \|_2) \to (X, \| x \|_1)$ is a continuous mapping. It is clear that I is an injective and surjective mapping, by the inverse-mapping theorem, I^{-1} is injective, surjective, and bounded linear operator, hence, there exists constant $c>0$, such that it holds $\| I^{-1}(x) \|_1 \leqslant c \| x \|_2$. However, $I(x)=x$ and $I^{-1}(x)=x$, so take $m=\frac{1}{c}$, then hold $m \| x \|_1 \leqslant \| x \|_2$. Combine (4.2.6), we get $m \| x \|_1 \leqslant \| x \|_2 \leqslant M \| x \|_1$, then $\| x \|_1$ and $\| x \|_2$ are equivalent each other.

Theorem 4.2.4 (closed graph theorem) Let $T: X \to Y$ be a closed linear operator, and $\mathfrak{D}_T \subseteq X$ be a closed set. Then T is a bounded linear operator; i.e.

$$T \text{ is a closed operator} \xrightarrow{\mathfrak{D}_T \text{ is a closed set}} T \in \mathfrak{B}(X, Y).$$

Proof Since $X, Y, X \times Y$ are Banach spaces, if T is a closed operator, then $G(T)=\mathfrak{D}_T \times Y$ is a closed set in $X \times Y$, it is also a closed linear subspace. This implies that $G(T)=\mathfrak{D}_T \times Y$ is a Banach space; then $(\mathfrak{D}_T, \| x \|_X)$ is a Banach space. We define a new norm $\| x \|_D$ on $\mathfrak{D}_T$ by

$$\| x \|_D = \| x \|_X + \| T(x) \|_Y, \quad \forall x \in \mathfrak{D}_T, \tag{4.2.7}$$

and prove that $(\mathfrak{D}_T, \| x \|_D)$ is a Banach space.

Take a Cauchy sequence $\{x_n\} \subset (\mathfrak{D}_T, \| x \|_D)$, by (4.2.6), it holds

$$\| x_n - x_m \|_D = \| x_n - x_m \|_X + \| T(x_n) - T(x_m) \|_Y \to 0, \quad n, m \to +\infty.$$

Since X and Y are two Banach spaces, $\exists x \in X$, $\exists y \in Y$, s. t. $x_n \xrightarrow{X} x$, $T(x_n) \xrightarrow{Y} y$; moreover, T is a closed operator, then $x \in \mathfrak{D}_T$, $T(x)=y$. Thus, $T(x_n) \xrightarrow{Y} T(x)$. This implies

$$\| x_n - x \|_D = \| x_n - x \|_X + \| T(x_n) - T(x) \|_Y \to 0, \quad n, m \to +\infty.$$

Hence $(\mathfrak{D}_T, \| x \|_D)$ is a Banach space. By (4.2.7), it holds $\| x \|_D = \| x \|_X + \| T(x) \|_Y \geqslant \| T(x) \|_Y$.

On the other hand, it holds $\| x \|_D \geqslant \| x \|_X$. Then by Proposition 4.2.2, $\exists M>0$, such that $\| x \|_D \leqslant M \| x \|_X$. This conclude that for $x \in X_1$, it holds $\| T(x) \|_Y \leqslant \| x \|_D \leqslant M \| x \|_X$, and thus $T: X \to Y$ is a bounded linear operator from X to Y. The closed-graph theorem is proved.

3. Important properties on the operator spaces

In this subsection, we study some important properties of bounded linear operator space $(\mathfrak{B}(X,Y), \|T\|)$, including the uniform boundedness principle, Banach-Steinhaus theorem, and their applications. The spaces in this subsection are supposed as normed linear spaces, or Banach spaces.

1) Convergence in bounded linear operator spaces

We have two kinds of convergence in $(\mathfrak{B}(X,Y), \|T\|)$.

(1) **Convergence in operator norm sense (strong convergence)**

Definition 4.2.6 (convergence of operator sequence in operator norm sense) Let $(X, \|x\|_X)$ and $(Y, \|y\|_Y)$ be two normed linear spaces on number field $\mathbb{F}$, and T, $T_n \in \mathfrak{B}(X,Y)$. If it holds $\lim\limits_{n\to+\infty} \|T_n - T\| = 0$, then the operator sequence $\{T_n\}$ is said to **converge to operator T in operator norm sense**; i. e., the convergence in operator norms $\|T_n - T\|$ of operator sequence $\{T_n - T\}$ in normed linear space $(\mathfrak{B}(X,Y), \|T\|)$; also say that **operator sequence $\{T_n\}$ is strongly convergent to operator T.**

Theorem 4.2.5 A linear operator sequence $\{T_n\}$ converges to linear operator T in operator norm sense, if and only if $\{T_n\}$ converges uniformly on the unit sphere $S = \{x \in X: \|x\|_X = 1\}$ in X.

Proof By equivalence "$\lim\limits_{n\to+\infty} \|T_n - T\| = 0 \Leftrightarrow \lim\limits_{n\to+\infty} \sup\limits_{\|x\|_X=1} \|T_n(x) - T(x)\|_Y = 0$", and equivalence

$$\text{“}\lim_{n\to+\infty} \|T_n(x) - T(x)\|_Y = 0, \quad \forall x \in S \text{ uniformly”}.$$

In fact, this means: "the uniform convergence of $\{T_n\}$ on unit ball S" is equivalent to "$\{T_n\}$ converges in any closed bounded subset in X".

(2) **Convergence in point-wise sense**

Definition 4.2.7 (convergence in point-wise sense of an operator sequence) Let $(X, \|x\|_X)$, and $(Y, \|y\|_Y)$ be two normed linear spaces on number field $\mathbb{F}$, and $T, T_n \in \mathfrak{B}(X,Y)$. If for every $x \in X$, it holds $\lim\limits_{n\to+\infty} T_n(x) = T(x)$, i. e., it holds that

$$\lim_{n\to+\infty} \|T_n(x) - T(x)\|_Y = 0, \quad \forall x \in X,$$

then the operator sequence $\{T_n\}$ is said to **converge to T in point-wise sense**, where the limit is in the norm of Y.

2) Properties of bounded linear operator spaces

Theorem 4.2.6 (completeness of $\mathfrak{B}(X,Y)$) Let $(X, \|x\|_X)$ be a normed linear space, $(Y, \|y\|_Y)$ be a Banach space. Then the bounded linear operator space $(\mathfrak{B}(X,Y), \|T\|)$ is a Banach space with the norm of operator $\|T\|$.

Note If $(X, \|x\|_X)$ and $(Y, \|y\|_Y)$ both are normed linear spaces, $\mathfrak{B}(X,Y)$ is not necessarily a Banach space. However, if $(Y, \|y\|_Y)$ is a Banach space, then $\mathfrak{B}(X,Y)$ must be a Banach space, even $(X, \|x\|_X)$ is or not.

The proof of Theorem 4.2.6 is left as an exercise.

Theorem 4.2.7 (uniform boundedness principle) Let $(X, \|x\|_X)$ be a Banach space and $(Y, \|y\|_Y)$ be a normed linear space, $W=\{T\}$ be a family of bounded linear operators in bounded linear operator space $\mathfrak{B}(X,Y)$. If for every $x\in X$, it holds $\sup\limits_{T\in W}\{\|T(x)\|_Y\}<+\infty$, then $V=\{\|T\|\}$ is a bounded set in $\mathbb{R}$. i.e., $\exists M>0$, such that for all $T\in W$, it holds $\|T\|\leqslant M$.

Analysis—(1) **About the assumption condition** "for bounded linear operator family $W=\{T\}\subset\mathfrak{B}(X,Y)$, if for each $x\in X$, it holds $\sup\limits_{T\in W}\{\|T(x)\|_Y\}<+\infty$" **would be interpreted as** "for bounded linear operator family $W=\{T\}$, for each $x\in X$, there exists $M_x>0$, s.t. for all $T\in W$, it holes $\|T(x)\|_Y\leqslant M_x$ with M_x independent of T"; i.e., "in bounded linear operator family $W=\{T\}$, all $\|T(x)\|_Y$ have the common finite supremum $\sup\limits_{T\in W}\{\|T(x)\|_Y\}$ for every $T\in W$ at each $x\in X$".

About the conclusion of the theorem "$\exists M>0$, s.t. for all $T\in W$, it holds $\|T\|\leqslant M$" **would be interpreted as** "the set $\{\|T\|: T\in W\}$, regarded as a number subset in $\mathbb{R}^+$, is bounded"; i.e., the number set $\{\|T\|: T\in W\}\subset\mathbb{R}^+$ is bounded.

The uniformly bounded principle means: "The norm set $\{\|T(x)\|_Y\}_{x\in X}$ (depending on $x\in X$) of linear operator family $W=\{T\}\subset\mathfrak{B}(X,Y)$ is bounded and point-wise in $x\in X$" implies "the operator family $W=\{T\}$ is uniformly bounded (i.e., the norm set $\{\|T\|: T\in W\}$ is bounded)", so that it is called **the uniform boundedness principle.**

(2) **About the corollary of theorem**: From the uniform boundedness principle, we conclude that: "if linear operator family $W=\{T\}\subset\mathfrak{B}(X,Y)$ satisfies $\sup\limits_{T\in W}\|T\|=+\infty$, then $\exists x_0\in X$, s.t. $\sup\limits_{T\in W}\|T(x_0)\|_Y=+\infty$", so it is also called **the resonance theorem.**

(3) **The idea of proof of theorem**: We prove Theorem 4.2.7: "$\forall x\in X$, $\sup\limits_{T\in W}\{\|T(x)\|_Y\}<+\infty$" implies "$\exists M>0$, s.t. $\forall T\in W$, it holds $\|T(x)\|_Y\leqslant M\|x\|_X$" by the following ideas.

① Since $\forall x\in X$, $\sup\limits_{T\in W}\{\|T(x)\|_Y\}<+\infty \Leftrightarrow \forall x\in X$, $\exists M_x>0$, s.t. $\forall T\in W$, it holds $\|T(x)\|_Y\leqslant M_x\|x\|_X$, we need to look for $\sup\{M_x: x\in X\}$, it is a common number $M>0$ independent of T,

$$\|T(x)\|_Y\leqslant M\|x\|_X,\quad \forall T\in W. \tag{4.2.8}$$

However, take the supremum of $\|T(x)\|_Y\leqslant M_x\|x\|_X$ on both sides, could get nothing. The key idea is to use the assumption of $\sup_{T\in W}\{\|T(x)\|_Y\}<+\infty$: the value $\sup_{T\in W}\{\|T(x)\|_Y\}$ is only dependent on $x\in X$ but independent of $T\in W$; and value $\sup_{T\in W}\{\|T(x)\|_Y\}$ is non-negative for each x, we may use it to characterize each $x\in X$! Let $\|x\|_W=\|x\|_X+\sup_{T\in W}\|T(x)\|_Y$ be a "new norm" of $x\in X$; if we could prove $\|x\|_W$ is an equivalent norm with $\|x\|_X$, then it would hold inequality $c_2\|x\|_X\leqslant\|x\|_W\leqslant c_1\|x\|_X$ with two constants $c_1>0$, $c_2>0$, hence

$$\|T(x)\|_Y\leqslant\sup_{T\in W}\|T(x)\|_Y\leqslant\|x\|_X+\sup_{T\in W}\|T(x)\|_Y=\|x\|_W\leqslant c_1\|x\|_X.$$

② Since the "new norm" satisfies $\|x\|_W=\|x\|_X+\sup_{T\in W}\|T(x)\|_Y\geqslant\|x\|_X$, thus it is stronger than $\|x\|_X$, so that by the Proposition 4.2.2, we only need to prove that $(X,\|x\|_W)$ is a Banach space in the "new norm".

We turn to prove the uniform boundedness principle now.

Proof It has no difficulty to verify that $\|x\|_W$ is a norm on X, and is stronger than $\|x\|_X$ (left to exercise).

Then we prove that $(X,\|x\|_W)$ is a complete space. In fact, take a Cauchy sequence $\{x_n\}\subset X$ in new norm $\|x\|_W$ with $\|x_n-x_m\|_W\to 0$, $n,m\to+\infty$. Then $\|x_n-x_m\|_W=\|x_n-x_m\|_X+\sup_{T\in W}\|T(x_n-x_m)\|_Y\to 0$ implies

$$\lim_{n,m\to\infty}\|x_n-x_m\|_X=0 \text{ and } \lim_{n,m\to\infty}\sup_{T\in W}\|T(x_n-x_m)\|_Y=0.$$

Since $(X,\|x\|_X)$ is a Banach space, then $\exists x\in X$, s. t. $x_n\xrightarrow{X}x$, i. e., $\forall\varepsilon>0$, $\exists N>0$, s. t. for $n>N$, it holds $\|x_n-x\|_X<\varepsilon$ and $\sup_{T\in W}\|T(x_n-x)\|_Y<\varepsilon$ by taking $\lim_{m\to\infty}\sup_{T\in W}\|T(x_n-x_m)\|_Y$.

By virtue of assumption $\sup_{T\in W}\{\|T(x)\|_Y\}<+\infty$ for $x\in X$, we see that, for Cauchy sequence $\{x_n\}\subset X$ and its limit $x\in X$, it holds $\lim_{n\to\infty}\|x_n-x\|_W=\lim_{n\to\infty}\|x_n-x\|_X+\sup_{T\in W}\|T(x_n-x)\|_Y=0$, thus $(X,\|x\|_W)$ is complete.

Finally, Proposition 4.2.1 shows that $\|x\|_W$ and $\|x\|_X$ are equivalent. Then, there exists constant $M>0$, such that

$$\|T(x)\|_Y\leqslant\sup_{T\in W}\|T(x)\|_Y\leqslant\|x\|_W\leqslant M\|x\|_X,\quad \forall x\in X,$$

This implies that $\forall T\in W$, it holds $\|T\|\leqslant M$. The proof is complete.

As an application of the uniform boundedness principle, we show the other useful Banach-Steinhaus Theorem in linear operator space $\mathfrak{B}(X,Y)$.

Theorem 4.2.8 (Banach-Steinhaus theorem) Let $(X,\|x\|_X)$ be a Banach space, $(Y,\|y\|_Y)$ be a normed linear space. If $A\subset X$ is a dense subset in X, an operator sequence $\{T_n\}$ is in $\mathfrak{B}(X,Y)$, and operator T is in $\mathfrak{B}(X,Y)$. Then the limit

$$\lim_{n\to+\infty} T_n(x) = T(x), \quad \forall x \in X \tag{4.2.9}$$

holds, if and only if (1) $\| T_n \|$ is bounded; (2) $\lim\limits_{n\to+\infty} T_n(x)=T(x), \forall x\in A$.

Proof Necessity By (4.2.9), the limit(2) holds.

To prove(1), let $W=\{T_n\}\subset\mathfrak{B}(X,Y)$. Then (4.2.9) implies $\lim\limits_{n\to+\infty} \| T_n(x)-T(x) \|_Y=0, \forall x\in X$. Thus, for any given $\varepsilon>0$, there exists $N>0$, s. t. for $n>N$, it holds

$$\| T_n(x) \|_Y < \| T(x) \|_Y + \varepsilon, \quad \forall x \in X, \quad n=1,2,\cdots,N.$$

This is equivalent to $\| T_n(x) \|_Y < M_x \| x \|_X, \forall x \in X, n=1,2,\cdots,N$; and implies $\sup\limits_{T_n\in W} \| T(x) \|_X<+\infty$ for all $x\in X$. By Theorem 4.2.7, $\forall T_n\in W$, it holds $\| T_n \| \leqslant M$, thus(1) is proved.

Sufficiency Suppose (1) and (2) hold, we prove (4.2.9).

Let $\| T_n \| \leqslant M, \forall n\in \mathbf{N}$, by(1). Then, by $X=\overline{A}$, we have $\forall x\in X, \forall \varepsilon>0, \exists y\in A$, such that $\| x-y \|_X\leqslant\dfrac{1}{4(\| T \| +M)}\varepsilon$; This implies that for all $n>N$, hold

$$\begin{aligned}\| T_n(x)-T(x) \|_Y &\leqslant \| T_n(x)-T_n(y) \|_Y + \| T_n(y)-T(y) \|_Y + \| T(y)-T(x) \|_Y \\ &\leqslant M \| x-y \|_X + \| T_n(y)-T(y) \|_Y + M \| y-x \|_X \\ &< \frac{\varepsilon}{4} + \| T_n(y)-T(y) \|_Y + \frac{\varepsilon}{4} = \frac{\varepsilon}{2} + \| T_n(y)-T(y) \|_Y.\end{aligned}$$

Then, the condition (2) and $y\in A$ imply that it holds $\| T_n(y)-T(y) \|_Y<\dfrac{\varepsilon}{2}$ for n large enough, and then $\forall x\in X, \forall\varepsilon>0, \exists N>0$, s. t. for $n>N$, it holds $\| T_n(x)-T(x) \|_Y \leqslant \varepsilon$. The proof is complete.

The following examples are exquisite applications of the uniform boundness principle.

Example 4.2.4 (convergence of integration formula) A familiar approximation formula of Riemann integral is

$$\int_a^b f(x)\mathrm{d}x \approx \sum_{k=0}^{n} f(x_k)\Delta x_k,$$

with partition $a=x_0<x_1<\cdots<x_{n-1}<x_n=b$. Such a problem has to be considered: What conditions guarantee that the above approximation formula has error tended to zero when $n\to+\infty$?

We rewrite the right-hand side of formula as $\sum\limits_{k=0}^{n} f(x_k)\Delta x_k = \sum\limits_{k=0}^{n} A_k^{(n)} f(x_k^{(n)})$. Then ask: What conditions guarantee $\lim\limits_{n\to+\infty} \sum\limits_{k=0}^{n} A_k^{(n)} f(x_k^{(n)})=\int_a^b f(x)\mathrm{d}x$ holds?

Proposition 4.2.3 (convergence conditions) Let $f(x) \in C([a,b])$ be a continuous function on interval $[a,b]$. Then the formula

$$\lim_{n\to+\infty} \sum_{k=0}^{n} A_k^{(n)} f(x_k^{(n)}) = \int_a^b f(x)\,\mathrm{d}x \tag{4.2.10}$$

holds, if and only if:

(1) there exists a constant $M>0$, such that $\sum_{k=0}^{n} |A_k^{(n)}| \leqslant M, n \in \mathbf{N}$;

(2) the equality $\lim_{n\to+\infty} \sum_{k=0}^{n} A_k^{(n)} f(x_k^{(n)}) = \int_a^b f(x)\,\mathrm{d}x$ holds for each polynormial

$$f(x) = p(x) = a_0 + a_1 x + \cdots + a_m x^m, \quad x \in [a,b],\ m \in \mathbb{Z}^+.$$

Proof ① **Prove** $T_n \in \mathfrak{B}(C([a,b]), \mathbb{R})$— Consider a functional on Banach space $C([a,b])$:

$$T_n(f) = \sum_{k=0}^{n} A_k^{(n)} f(x_k^{(n)}), \quad f \in C([a,b]), x \in [a,b],$$

then, (4.2.10) can be rewritten as

$$\lim_{n\to+\infty} T_n(f) = \int_a^b f(x)\,\mathrm{d}x = T(f), \quad f \in C([a,b]).$$

Hence, $T_n(f): C([a,b]) \to \mathbb{R}$.

It is clear that $T \in \mathfrak{B}(C([a,b]), \mathbb{R})$; and T_n is linear. We prove $T_n \in \mathfrak{B}(C([a,b]), \mathbb{R})$: $\forall f \in C([a,b])$ implies $\| f \| \leqslant M_1$, so that

$$| T_n(f) | = \left| \sum_{k=0}^{n} A_k^{(n)} f(x_k^{(n)}) \right| \leqslant \left(\sum_{k=0}^{n} |A_k^{(n)}|\right) \| f \| \leqslant M_1 \left(\sum_{k=0}^{n} |A_k^{(n)}|\right),$$

we get $\| T_n \| \leqslant M_1 \left(\sum_{k=0}^{n} |A_k^{(n)}|\right) \leqslant M_1 M = M_2$.

On the other hand, $\forall n \in \mathbf{N}$, take $f_n(x) \in C([a,b])$, without loss of generality, we suppose that

$$f_n(x_k^{(n)}) = \mathrm{sgn} A_k^{(n)}, \quad k = 0,1,2,\cdots,n$$

with $\| f_n \| = \sup\{|f_n(x)| : x \in [a,b]\} = 1$. Then, $\| T_n \| \geqslant | T_n(f_n) | = \sum_{k=0}^{n} |A_k^{(n)}|$, and thus

$$\| T_n \| = \sum_{k=0}^{n} |A_k^{(n)}|, \quad n \in \mathbf{N}. \tag{4.2.11}$$

This is $T_n \in \mathfrak{B}(C([a,b]), \mathbb{R})$. Moreover, condition (1) is $\| T_n \| \leqslant M, \forall n \in \mathbf{N}$.

② **Sufficiency** Suppose (1) and (2) hold. Since the set of all polynormials is dense in $C([a,b])$, then

$$\lim_{n\to+\infty}\sum_{k=0}^{n}A_k^{(n)}f(x_k^{(n)})=\int_a^b f(x)\mathrm{d}x$$

holds for every $f\in C([a,b])$.

③ **Necessity** Suppose (4.2.10) holds, i.e.

$$\lim_{n\to+\infty}\sum_{k=0}^{n}A_k^{(n)}f(x_k^{(n)})=\lim_{n\to+\infty}T_n(f)=T(f)=\int_a^b f(x)\mathrm{d}x,\quad \forall f\in C([a,b]),$$

that is

$$\lim_{n\to+\infty}|T_n(f)-T(f)|=0,\quad \forall f\in C([a,b]).$$

Then, $\sup_n |T_n(f)-T(f)|\leqslant M$ (a uniformly convergent series is bounded); By Banach-Steinhaus theorem, it holds $\|T_n\|\leqslant M$. Hence, (1) holds; and with (4.2.11), we get (2).

Example 4.2.5 (convergence of Fourier series) Let $C_{2\pi}$ be the Banach space with norm $\|f\|_{C_{2\pi}}=\max_{x\in[0,2\pi]}|f(x)|$, consisted of all 2π-periodic real continuous functions. For $f\in C_{2\pi}$, Fourier series $f(x)\sim\dfrac{a_0}{2}+\sum_{k=1}^{+\infty}(a_k\cos kx+b_k\sin kx)$ has the $(n+1)$-partial sum

$$S_n(f)(x)=\frac{a_0}{2}+\sum_{k=1}^{n}(a_k\cos nx+b_k\sin nx)$$

$$=\frac{1}{\pi}\int_0^{2\pi}f(x)\left(\frac{1}{2}+\sum_{k=1}^{n}\cos k(t-x)\right)\mathrm{d}t=\int_0^{2\pi}f(x)D_n(t,x)\,\mathrm{d}t,$$

with $D_n(t,x)=\dfrac{\sin\left(n+\frac{1}{2}\right)(t-x)}{\sin\frac{1}{2}(t-x)}$ $(n\in\mathbb{N})$, called **Dirichlet kernel.** Note, $\{S_n(f)\}_{n\in\mathbb{N}}$ with $S_n(f)$: $C_{2\pi}\to C_{2\pi}$, is a linear operator family from $C_{2\pi}$ to $C_{2\pi}$.

We prove that $\forall x_0\in[0,2\pi]$, $\exists f\in C_{2\pi}$, s. t. Fourier series of $f(x)$ is divergent at x_0.

In fact, without loss of generality, regard $[0,2\pi]$ as $[-\pi,\pi]$ with $x_0=0$. Let

$$T_n(f)=\int_{-\pi}^{\pi}f(t)D_n(t,0)\mathrm{d}t\ :\ C_{2\pi}\to\mathbb{R}$$

be a functional sequence on space $C_{2\pi}$ with $D_n(t,0)=\dfrac{1}{2\pi}+\dfrac{1}{\pi}\sum_{k=1}^{n}\cos kt=\dfrac{\sin\left(n+\frac{1}{2}\right)t}{\sin\frac{1}{2}t}$ $(n\in\mathbb{N})$. Clearly, $D_n(t,0)$ is continuous on $[-\pi,\pi]$, thus $\|T_n\|\leqslant M$, and $\|T_n\|=\int_{-\pi}^{\pi}|D_n(t,0)|\,\mathrm{d}t, n\in\mathbb{N}$. Since

$$\int_{-\pi}^{\pi} |D_n(t,0)|\,\mathrm{d}t = \int_0^{2\pi} |D_n(t,0)|\,\mathrm{d}t = \frac{1}{2\pi}\int_0^{2\pi} \frac{\left|\sin\left(n+\frac{1}{2}\right)t\right|}{\left|\sin\frac{1}{2}t\right|}\mathrm{d}t$$

$$= \frac{1}{\pi}\int_0^{\pi} \frac{|\sin(2n+1)t|}{|\sin t|}\mathrm{d}t = \frac{1}{\pi}\int_0^{\pi} \frac{|\sin(2n+1)t|}{\sin t}\mathrm{d}t$$

$$\geqslant \frac{1}{\pi}\int_0^{\pi} \frac{|\sin(2n+1)t|}{t}\mathrm{d}t = \frac{1}{\pi}\int_0^{(2n+1)\pi} \frac{|\sin u|}{u}\mathrm{d}u,$$

the non-absolutely convergence of integral $\int_0^{+\infty} \frac{\sin t}{t}\mathrm{d}t$ implies $\lim_{n\to\infty} \frac{1}{\pi}\int_0^{(2n+1)\pi} \frac{|\sin u|}{u}\mathrm{d}u = +\infty$. Then, $\| T_n \| \to +\infty$. The Banach-Steinhaus theorem shows that, there exists at least one function $f_0 \in C_{2\pi}$, such that the sequence

$$\{T_n(f_0)\} = \left\{T_n(f_0) = \int_{-\pi}^{\pi} f_0(t)D_n(t,0)\,\mathrm{d}t\right\}$$

is divergent, i. e., $\lim_{n\to+\infty} T_n(f_0) = \infty$, since if "$\forall f \in C_{2\pi} \Rightarrow \lim_{n\to+\infty} T_n(f_0) = A$", then "$\| T_n(f) \| \leqslant M$ holds for all $f \in C_{2\pi}$ uniformly", and then $\| T_n \|$ is uniformly bounded, this contradicts $\| T_n \| \to +\infty$.

The result in Example 4.2.5 defecates an misconception: Fourier series of $f \in C_{2\pi}$ is always convergent at $[-\pi,\pi]$.

3) Operator algebra on bounded linear operator space $\mathfrak{B}(X,X) \equiv \mathfrak{B}(X)$

If $(X, \| x \|_X)$ is a normed linear space, $(Y, \| y \|_Y)$ is a Banach space, then the bounded linear operator set $\mathfrak{B}(X,Y)$ with addition $+$, number product $\alpha\cdot$, and operator norm $\| T \|$, becomes a Banach space $(\mathfrak{B}(X,Y), +, \alpha\cdot, \| T \|)$.

If $\mathfrak{B}(X) \equiv \mathfrak{B}(X,X)$, then it has more structure: for two linear operators $T, S \in \mathfrak{B}(X)$, define compound operation $S \circ T$ by $(S\circ T)(x) = S(T(x))$, $\forall x \in X$, then a new structure has defined, called **the operator algebra** $(\mathfrak{B}(X,X), +, \alpha\cdot, \| T \|, S\circ T)$.

We start from a normed linear space $(X, +, \alpha\cdot, \| x \|_X)$ to define operator algebra, then consider $\mathfrak{B}(X)$.

Definition 4.2.8 (**algebra, normed algebra, and Banach algebra**)

(1) **algebra** Let $(X, +, \alpha\cdot)$ be a linear space on number field $\mathbb{F}$. If for any elements $x, y, z, \cdots \in X$, and $\alpha \in \mathbb{F}$, define a "multiplication" $x\circ y$, satisfying

① $x\circ y \in X$, (closed property)

② $(x\circ y)\circ z = x\circ(y\circ z)$, (combineation law)

③ $x\circ(y+z) = x\circ y + x\circ z$, $(y+z)\circ x = y\circ x + z\circ x$, (distribution law)

④ $\alpha(x\circ y) = (\alpha x)\circ y = x\circ(\alpha y)$, (combineation law)

then X is said to be an **algebra**, denoted by $(X, +, \alpha\cdot, x\circ y)$.

Moreover, if an algebra $(X,+,\alpha\cdot,x\circ y)$ satisfies

⑤ $x,y\in X\Rightarrow x\circ y=y\circ x$, (commutative law)

then $(X,+,\alpha\cdot,x\circ y)$ is said to be a **commutative algebra.**

(2) **normed algebra** Let $(X,+,\alpha\cdot,\|x\|_X)$ be a normed linear space on number field $\mathbb{F}$. If for any $x,y,z,\cdots\in X,\alpha\in\mathbb{F}$, define "multiplication" $x\circ y$, satisfies ①-④, and

⑥ $\|x\circ y\|_X\leqslant\|x\|_X\|y\|_X$, (normed inequality)

then it is said to be **a normed algebra**, denoted by $(X,+,\alpha\cdot,\|x\|_X,x\circ y)$; moreover, if $x\circ y$ satisfies commutative law, then $(X,+,\alpha\cdot,\|x\|_X,x\circ y)$ is said to be **a commutative normed algebra**(i. e., a commutative normed algebra satisfies ①-⑥).

Moreover, if a normed algebra $(X,+,\alpha\cdot,\|x\|_X,x\circ y)$ has an element $e\in X$, satisfies

⑦ $e\circ x=x\circ e=x,\forall x\in X$, with $\|e\|_X=1$, (unit element exists)

then $e\in X$ is said to be the **unit element about operator** $\circ$, and $(X,+,\alpha\cdot,\|x\|_X,x\circ y,e)$ is said to be **a normed algebra with unit**; If operation $\circ$ satisfies commutative law, then it is said to be **a commutative normed algebra with unit.**

(3) **Banach algebra** Let $(X,+,\alpha\cdot,\|x\|_X,x\circ y)$ be a normed algebra. If it is a Banach space with norm $\|x\|_X$, then it is said to be **a Banach algebra.**

Correspondingly, commutative Banach algebra, Banach algebra with unit, commutative Banach algebra with unit.

Example 4.2.6 $C([a,b])$ is **a commutative Banach algebra with unit.**

In fact, the operations in $C([a,b])$—

addition $+$: $f,g\in C([a,b])\Rightarrow(f+g)(x)=f(x)+g(x),\forall x\in[a,b]$;

number product $\alpha\cdot$: $f\in C([a,b]),\alpha\in F\Rightarrow(\alpha f)(x)=\alpha f(x),\forall x\in[a,b]$;

multiplication $\cdot$: $f,g\in C([a,b])\Rightarrow(f\cdot g)(x)=f(x)g(x),\forall x\in[a,b]$;

unit element: $f\equiv 1\in C([a,b])\Rightarrow(1\cdot f)(x)=1\cdot f(x)=f(x),\forall x\in[a,b]$;

norm: $f\in C([a,b])\Rightarrow\|f\|_{C[a,b]}=\max\limits_{x\in[a,b]}|f(x)|$;

then, it is easy to see that $(C([a,b]),+,\alpha\cdot,f\cdot g,\|f\|_{C[a,b]},1)$ is a commutative Banach algebra with unit.

Example 4.2.7 $W([0,2\pi])$ is **a commutative Banach algebra.**

In the set $W([0,2\pi])=\left\{f: f(x)=\sum\limits_{k=-\infty}^{+\infty}\xi_k e^{ikx},\sum\limits_{k=-\infty}^{+\infty}|\xi_k|<+\infty,x\in[0,2\pi]\right\}$,

we have

addition$+$: $f,g\in W([0,2\pi])\Rightarrow(f+g)(x)=f(x)+g(x),\forall x\in[0,2\pi]$;

number product $\alpha\cdot$: $f\in W([0,2\pi]),\alpha\in F\Rightarrow(\alpha f)(x)=\alpha f(x),\forall x\in[0,2\pi]$;

multiplication $*$: $f,g\in W([0,2\pi])$, multiplication is defined as "convolution" $*$

$$f(x)=\sum_{k=-\infty}^{+\infty}\xi_k e^{ikx}, g(x)=\sum_{k=-\infty}^{+\infty}\eta_k e^{ikx} \Rightarrow (f*g)(x)=\sum_{n=-\infty}^{+\infty}\sum_{k=-\infty}^{+\infty}\xi_{n-k}\eta_k e^{inx};$$

norm: $f\in W([0,2\pi])\Rightarrow \|f\|_{W([0,2\pi])}=\sum_{k=-\infty}^{+\infty}|\xi_k|$.

Estimate $\|(f*g)(x)\|_{W([0,2\pi])} = \sum_{n=-\infty}^{+\infty}\left|\sum_{k=-\infty}^{+\infty}\xi_{n-k}\eta_k\right| \leqslant \sum_{n=-\infty}^{+\infty}\sum_{k=-\infty}^{+\infty}|\xi_{n-k}||\eta_k| = \sum_{n=-\infty}^{+\infty}\left(\sum_{k=-\infty}^{+\infty}|\xi_{n-k}|\right)|\eta_k| = \|f\|_{W([a,b])}\|g\|_{W([a,b])}$ showing that $(W([0,2\pi]),+,\alpha\cdot, f*g, \|f\|_{W[0,2\pi]})$ is a commutative Banach algebra.

Example 4.2.8 Let $(X,\|x\|_X)$ be a Banach space, and $\mathfrak{B}(X)\equiv\mathfrak{B}(X,X)$ be bounded linear operator space as a Banach space. A multiplication $S\circ T$ of operators $T,S\in\mathfrak{B}(X)$ is defined by the compound of S and T

$$(S\circ T)(x)=S(T(x)),\quad \forall x\in X,$$

with identity operator I as a unit. Then $(\mathfrak{B}(X),+,\alpha\cdot,\|T\|,S\circ T,I)$ is **a non-commutative Banach algebra with unit.**

Example 4.2.9 Let $D=\{z\in\mathbb{C}: |z|\leqslant 1\}$, and $\mathfrak{C}(D)$ be the set of all complex-valued analytic function of complex variables on D. Operations on $\mathfrak{C}(D)$ are similar as in example 4.2.6. Then $\mathfrak{C}(D)$ is **a commutative Banach algebra with unit.**

Example 4.2.10 A Banach algebra $L([a,b])$

Lebesgue integrable function space $L([a,b])$ on interval $[a,b]$, with addition, number product as in Example 4.2.6, norm $\|f\|_{L([a,b])}=\int_{[a,b]}|f(x)|\,dx$, multiplication defined by convolution

$$f,g\in L([a,b])\Rightarrow (f*g)(x)=\int_{[a,b]}f(x-t)g(t)\,dt,$$

such that $(L([a,b]),+,\alpha\cdot,\|f\|,f*g)$ is **a commutative Banach algebra without unit.**

Remark 1 For convolution $(f*g)(x)=\int_{[a,b]}f(x-t)g(t)\,dt$ in $L([a,b])$, **we agree on that:**

"if $f,g\in L([a,b])$, then for $x\notin[a,b]$, we define $f(x)=g(x)=0$".

The closed property of convolution operation can be got by inequality

$$f,g\in L([a,b])\Rightarrow \|f*g\|_{L([a,b])}\leqslant\|f\|_{L([a,b])}\|g\|_{L([a,b])}.$$

Remark 2 To prove that Banach algebra $L([a,b])\equiv(L([a,b]),+,\alpha\cdot,\|f\|_{L([a,b])},f*g)$ is without unit, we need some knowledge of Fourier transformation (see Chapter 5, Exercise 11).

4.2.2 Spectrum theory of bounded linear operators

1. Inverse operator of a linear operator

The concept of inverse operator, or inverse mapping, is familiar. It is one of main objects in the spectrum theory.

Let $(X, \| x \|_X)$ and $(Y, \| y \|_Y)$ be two Banach spaces, $\mathfrak{B}(X, Y)$ be the bounded linear operator space from X to Y.

For an operator $T \in \mathfrak{B}(X, Y)$, if there exists an operator $S: Y \to X$, such that for each $y \in Y$, there exists unique $x \in X$ such that $y = T(x)$, i. e., the compound mapping $S \circ T$ is the unit operator $S \circ T = I: X \to X$, then $S: Y \to X$ is said to be **the inverse operator of** $T: X \to Y$, denoted by $S = T^{-1}$, i. e., the inverse operator $T^{-1}: Y \to X$ is with $T^{-1} \circ T = I: X \to X$. Operator T is said to be **invertible.**

Note 1 It is clear that in the above definition, an invertible operator $T: X \to Y$ is a one-one mapping (bijection). Usually, it is a surjection. And so is its inverse $T^{-1}: Y \to X$.

Note 2 For invertible operators $T_1: X \to Y$ and $T_2: Y \to Z$, the compound operator $T \equiv T_2 \circ T_1: X \to Z$ is invertible, and $(T_2 \circ T_1)^{-1}: Z \to X$ holds $(T_2 \circ T_1)^{-1} = T_1^{-1} \circ T_2^{-1}$.

Note 3 For two invertible operators $T_1: X \to X, T_2: X \to X$, i. e., $T_1, T_2 \in \mathfrak{B}(X) \equiv \mathfrak{B}(X, X)$, then compound operator $T_2 \circ T_1 \in \mathfrak{B}(X)$, and satisfies $(T_2 \circ T_1)^{-1} = T_1^{-1} \circ T_2^{-1}$.

2. Spectrum theorem of linear operators

We start from a finite-dimensional linear space $(X, +, \alpha \cdot)$ on real number fied $\mathbb{R}$ with $\dim X = n$.

Since $\dim X = n$ implies $X \overset{\text{iso}}{\longleftrightarrow} \mathbb{R}^n$ (Theorem 2. 1. 3), so that X is isomorphic with $\mathbb{R}^n$; and $\mathfrak{B}(\mathbb{R}^n) \overset{\text{iso}}{\longleftrightarrow} \mathfrak{M}_{n \times n}$ (Theorem 2. 2. 1), where $\mathfrak{M}_{n \times n}$ is the matrix space, a matrix $A = [a_{ij}]_{n \times n} \in \mathfrak{M}_{n \times n}$ is regarded as an operator $T \leftrightarrow A = [a_{ij}]_{n \times n}$.

The problems of finding eigenvalues and eigenvectors are familiar in *advanced mathematics* course: If an equation $Ax = \lambda x$ (or equivalently, $(A - \lambda I)x = 0$) has nonzero solution, where $x = \begin{bmatrix} x_1 \\ \vdots \\ x_n \end{bmatrix} \in \mathbb{R}^n, \lambda \in \mathbb{R}$, then λ is said to be an **eigenvalue** of matrix $A = [a_{ij}]_{n \times n}$; the set of all eigenvalues is said to be **the spectrum set of** A, or simply, **spectrum set**, or **spectrum.** The eigenvalue is said to be spectrum value, or spectrum point, denoted by $\sigma(A) \equiv \sigma(T) \subset \mathbb{C}$. A complex number $\mu \in \mathbb{C} \backslash \sigma(T)$ is said to be **a regular value of** $A \equiv T$.

Thus, if an operator $T \leftrightarrow A = [a_{ij}]_{n \times n}$ is a matrix, then its spectrum is, as a matter of fact, the set of all eigenvalues of $A = [a_{ij}]_{n \times n}$ with $\sigma(A) = \{\lambda \in \mathbb{C} : (A - \lambda I)x = 0\}$; **The spectrum values** are those nonzero $\lambda \in \sigma(A)$ such that the homogeneous equation $(A - \lambda I)x = 0$ has nonzero solutions; **the regular values** are those $\mu \in \mathbb{C} \backslash \sigma(A)$ such that the non-homogeneous equation $(A - \lambda I)x = b, b \neq 0$, has unique solution.

The operator spectrum theory is very important and useful, particularly, in solving various equations, including algebraic, differential, and functional equations. We show about general operator spectrum theory on Banach spaces in this subsection, and give some examples as applications.

Definition 4.2.9 (regular value, resolvent, spectrum point, and spectrum) Let $X \equiv (X, +, \alpha \cdot, \| x \|_X)$ be a Banach space on number field $\mathbb{F}$. If for any bounded linear operator T on X, i. e., $T \in \mathfrak{B}(X)$, and operator $\lambda I - T \in \mathfrak{B}(X)$ with $\lambda \in \mathbb{C}$, we define:

(1) If there exists inverse operator $(\lambda I - T)^{-1}$ of $\lambda I - T$ with domain X, i. e., the inverse $(\lambda I - T)^{-1} \in \mathfrak{B}(X)$ with $(\lambda I - T)^{-1}: X \to X$, then $\lambda \in \mathbb{C}$ is said to be **a regular value of** T, the operator $R_\lambda = (\lambda I - T)^{-1}$ is said to be **a resolvent of** T; The set of all regular values of T is said to be **the regular set of** T, denoted by $\rho(T)$.

(2) If $\lambda \in \mathbb{C}$ is not a regular value of T, that is, the operator $\lambda I - T$ does not have the inverse, then λ is said to be **a spectrum value** or **spectrum point of** T; The set of all spectrum points of T is said to be **the spectrum set of** T, or **spectrum**, denoted by $\sigma(T)$.

Spectrum points can be recognized in two classes:

① If for $\lambda \in \sigma(T)$, the homogeneous equation $(\lambda I - T)x = 0$ has nonzero solutions, then λ is said to be **an eigenvalue of** T; the corresponding nonzero solutions are said to be **eigenelements** or **eigenvector of** λ; the set of all eigenvalues, denoted by $\sigma_p(T)$, is said to be **the point spectrum of** T; and holds $\sigma_p(T) \subset \sigma(T) \subset \mathbb{C}$;

② If for $\lambda \in \sigma(T)$, the homogeneous equation $(\lambda I - T)x = 0$ has and only has zero solution, that is, only has zero vector corresponding to λ, and the range of operator $\lambda I - T$ is a proper subspace of X, then the set of all such λ is said to be the **continuous spectrum of** T, denoted by $\sigma_c(T)$; every $\lambda \in \sigma_c(T)$ is said to be **a continuous spectrum point**; and holds $\sigma_c(T) \subset \sigma(T) \subset \mathbb{C}$.

Now, we have $\mathbb{C} = \rho(T) \cup \sigma(T) = \rho(T) \cup \{\sigma_p(T) \cup \sigma_c(T)\}$.

Example 4.2.11 In n-dimensional complex space $\mathbb{C}^n$, a trigonometric matrix

$$A = \begin{bmatrix} a_{11} & 0 & \cdots & 0 \\ a_{21} & a_{22} & \cdots & \vdots \\ \vdots & \vdots & \ddots & 0 \\ a_{n1} & \cdots & a_{n,n-1} & a_{nn} \end{bmatrix}, \quad a_{ij} \in \mathbb{C},$$

defines an operator $T: x = (\xi_1, \cdots, \xi_n) \in \mathbb{C}^n \to T(x) = (\eta_1, \cdots, \eta_n) \in \mathbb{C}^n$, such that $\eta_k = \sum_{j=1}^{n} a_{kj}\xi_j\ (k = 1,2,\cdots,n)$, i.e., $T(x) = A[\xi_1 \ \cdots \ \xi_n]^{\mathrm{T}} = \begin{bmatrix} a_{11} & 0 & \cdots & 0 \\ a_{21} & a_{22} & & \vdots \\ \vdots & & \ddots & 0 \\ a_{n1} & \cdots & a_{n\,n-1} & a_{nn} \end{bmatrix} \begin{bmatrix} \xi_1 \\ \xi_2 \\ \vdots \\ \xi_n \end{bmatrix} = \begin{bmatrix} \eta_1 \\ \eta_2 \\ \vdots \\ \eta_n \end{bmatrix}$.

Hence, the operator T is determined by matrix A, and equation

$$0 = (\lambda I - A)x = \begin{bmatrix} \lambda - a_{11} & 0 & \cdots & 0 \\ -a_{21} & \lambda - a_{22} & \cdots & 0 \\ \vdots & \vdots & \ddots & \vdots \\ -a_{n1} & -a_{n2} & \cdots & \lambda - a_{nn} \end{bmatrix} x$$

has nonzero solutions corresponding to $\lambda = a_{11}, \cdots, a_{nn}$, so that these $\lambda = a_{11}, \cdots, a_{nn}$ are eigenvalues of T, and $a_{11}, \cdots, a_{nn} \in \sigma_p(T)$; other $\lambda \in \mathbb{C} \setminus \sigma_p(T) = \rho(T)$ is the regular values of T. We have $\mathbb{C} = \rho(T) \cup \sigma_p(T)$.

Example 4.2.12 Let $C([0,1])$ be the set of all complex-valued continuous functions on $[0,1]$. Consider an operator

$$(Tf)(x) = xf(x), \quad f \in C([0,1]),$$

this $T: f \in C([0,1]) \to T(f)(x) = xf(x) \in C([0,1])$ is a bounded linear operator.

For $\lambda \notin [0,1]$, since

$$((\lambda I - T)(f))(x) = ((\lambda - x)(f))(x), \quad f \in C([0,1]),$$

then $\lambda I - T = \lambda - x$ with resolvent $R_\lambda(T) = (\lambda I - T)^{-1} = \frac{1}{\lambda - x}$, and

$$(R_\lambda f)(x) = ((\lambda I - T)^{-1} f)(x) = \frac{1}{\lambda - x} f(x).$$

This R_λ is defined on $C([0,1])$ with value $\frac{1}{\lambda - x} f(x) \in C([0,1])$, and is a bounded linear operator on $C([0,1])$; moreover, $(R_\lambda f)(x)|_{f(x)=(\lambda I - T)(f)} = ((\lambda I - T)^{-1}(\lambda I - T)(f))(x) = f(x)$, this implies $\lambda \notin [0,1]$ is regular value of T.

For $\lambda \in [0,1]$, since

$$((\lambda I - T)(f))(x) = (\lambda - x)f(x), f \in C([0,1]),$$

if $x = \lambda$, then $\lambda I - T = \lambda - x$ does not have inverse operator. However, $(\lambda I - T)|_{x=\lambda} f(x)$ is zero, then the range of $\lambda I - T$ is a proper subspace of $C([0,1])$ (i.e., $\mathfrak{R}_{\lambda I - T} \subsetneq C([0,1])$), and the homogeneous equation $(\lambda I-T)f(x)=0$ does not have nonzero solution in $C([0,1])$. By definition, $\lambda\in[0,1]$ is the continuous spectrum of T, i.e., $\lambda\in\sigma_c(T)=[0,1]$. Thus, we have $\mathbb{C}=\rho(T)\cup\sigma_c(T)$.

Example 4.2.13 Consider an integral operator on $C([0,1])$

$$(Tf)(x)=\int_0^x f(t)\mathrm{d}t,\quad f\in C([0,1]),$$

here $T: f\in C([0,1])\to T(f)(x)=\int_0^x f(t)\mathrm{d}t\in C([0,1])$ is a bounded linear operator.

If $\lambda\neq 0$, since

$$((\lambda I-T)(f))(x)=\lambda f(x)-\int_0^x f(t)\mathrm{d}t,\quad f\in C([0,1]),$$

then non-homogeneous equation $(\lambda I-T)f(x)=g(x)$ is in the form $\lambda f(x)-\int_0^x f(t)\mathrm{d}t=g(x)$, so that for $\lambda\neq 0$, the above equation is equivalent to $f(x)=\frac{1}{\lambda}g(x)+\frac{1}{\lambda}\int_0^x f(t)\mathrm{d}t$. By knowledge of Volterra integral equation, for any $g(x)\in C([0,1])$, the last equation has unique solution. This shows that for $\lambda\neq 0$, we have

$$(\lambda I-T)f(x)=g(x)\Leftrightarrow f(x)=(\lambda I-T)^{-1}g(x),$$

so the operator $(\lambda I-T)$ is invertible, thus $\forall\lambda\neq 0\Rightarrow\lambda\in\rho(T)$.

If $\lambda=0$, the homogeneous equation $(\lambda I-T)(f(x))|_{\lambda=0}=0$ is equivalent to the integral equation $T(f(x))=-\int_0^x f(t)\mathrm{d}t=0$. By $f(x)\in C([0,1])$, the equation $T(f(x))=-\int_0^x f(t)\mathrm{d}t=0$ implies $f(t)=0$, so that the corresponding homogeneous equation has only zero solution, then $\lambda=0$ is not eigenvalue of T.

On the other hand, any element in the range of operator $(\lambda I-T)|_{\lambda=0}$ has a form as $g(x)=-\int_0^x f(t)dt$ with $g'(x)=-f(x)\in C([0,1])$, and $g(0)=0$. This tells that the range of T is $W=\{g: g\in C^1([0,1]), g(0)=0\}$, and W is a proper subspace of $C([0,1])$, then $\lambda=0\in\sigma_c(T)$; hence $\sigma_c(T)=\{0\}$, $\sigma_p(T)=\varnothing$; We have

$$\sigma(T)=\sigma_c(T),\quad \rho(T)=\mathbb{C}\setminus\{0\},\quad \mathbb{C}=\rho(T)\cup\sigma_c(T).$$

Example 4.2.14 In the complex Lebesgue integrable function space $L^1([0,1])$, we consider an integral operator

$$(Tf)(x)=xf(x)+\int_x^1 f(t)\mathrm{d}t,\quad f\in L^1([0,1]),$$

then operator $T: f\in L^1([0,1])\to T(f)(x)=xf(x)+\int_x^1 f(t)\mathrm{d}t\in L^1([0,1])$ is bounded and linear.

For $\lambda \in (0,1]$, the characteristic function $\chi_{[0,\lambda]}(x)=\begin{cases}1, & x\in[0,\lambda]\\ 0, & x\notin[0,\lambda]\end{cases}$ of interval $[0,\lambda]$ is correspondent to the eigen-vector of λ. In fact, if $0\leqslant x\leqslant\lambda$, it holds

$$(T\chi_{[0,\lambda]})(x)=x\chi_{[0,\lambda]}(x)+\int_x^1\chi_{[0,\lambda]}(t)\mathrm{d}t=x+\int_x^\lambda \mathrm{d}t=\lambda,$$

i. e., $((\lambda I-T)\chi_{[0,\lambda]})(x)=0$. Thus, $(0,1]$ is the point spectrum $\sigma_p(T)=(0,1]$ of T.

For $\lambda\notin[0,1]$, take $g\in L^1([0,1])$, then non-homogeneous equation $((\lambda I-T)f)(x)=g(x)$ deduces to

$$(\lambda-x)f(x)-\int_x^1 f(t)\mathrm{d}t=g(x).$$

Let $h(x)=\int_1^x f(t)\mathrm{d}t$. Then, above integral equation is equivalent to the following Cauchy problem of a differential equation with initial condition $\begin{cases}(\lambda-x)h'(x)+h(x)=g(x)\\ h(1)=0\end{cases}$, and has unique solution. So that if $\lambda\notin[0,1]$, then operator $(\lambda I-T)$ is invertible, $\forall\lambda\notin[0,1]\Rightarrow\lambda\in\rho(T)$.

For $\lambda=0$, by the similar argument in Example 4. 2. 13, we can prove that $\lambda=0$ is not an eigenvalue, thus $\lambda=0\in\sigma_c(T)$.

Finally, it follows $\mathbb{C}=\rho(T)\cup\sigma_p(T)\cup\sigma_c(T)$ with $\sigma_p(T)=(0,1]$, $\sigma_c(T)=\{0\}$.

3. Properties of spectrum sets and regular sets

Theorem 4. 2. 9 Let $X\neq\{0\}$ be a nonempty complex Banach space, $\mathfrak{B}(X)$ be the bounded linear operator space on X. For operator $T\in\mathfrak{B}(X)$, we have

(1) The spectrum set $\sigma(T)\subset\mathbb{C}$ of $T\in\mathfrak{B}(X)$ is a bounded closed set. If λ is an eigenvalue of T, then all corresponding eigenvectors of λ and zero-vector constitute a closed subspace of X, called eigenvector space of λ; If λ_k, $k=1,2,\cdots,n$, are different eigenvalues of T, and eigenvector x_k is correspondent to λ_k, then $x_1,\cdots,x_n$ are linearly independent;

(2) The regular set $\rho(T)\subset\mathbb{C}$ of $T\in\mathfrak{B}(X)$ is an open set; The resolvent operator $R_\lambda=(\lambda I-T)^{-1}$, as a function of λ, is analytic in its domain $\rho(T)$ (i. e., regular set);

(3) For $\lambda\in\mathbb{C}$, if $|\lambda|>\|T\|$, then λ is regular value of T, and holds

$$(\lambda I-T)^{-1}=\sum_{n=0}^{+\infty}\frac{1}{\lambda^{n+1}}T^n,$$

the series converges in operator norm sense, and $\|(\lambda I-T)^{-1}\|\leqslant\dfrac{1}{|\lambda|-\|T\|}$;

(4) For $\lambda\in\mathbb{C}$, if λ is a regular value of T, then for any $\mu\in\mathbb{C}$ with $|\mu-\lambda|<\|(\lambda I-T)^{-1}\|^{-1}$, this μ is a regular value, and inverse operator $(\lambda I-T)^{-1}$ is

$$(\mu I-T)^{-1}=\sum_{n=0}^{+\infty}(-1)^n(\mu-\lambda)^n(\lambda I-T)^{-(n+1)},$$

the series converges in operator norm sense.

4. Properties of invertible operators

Theorem 4.2.10 Let X be a Banach space, $\mathfrak{B}(X)$ be the bounded linear operator space on X. Then

(1) The set of all invertible operators is an open set in $\mathfrak{B}(X)$;

(2) If X is a Banach space containing nonzero elements, then $\forall T\in\mathfrak{B}(X)$ implies $\sigma(T)\neq\varnothing$.

5. Spectral radius

Definition 4.2.10 (spectral radius of a operator) Let $(X,+,\alpha\cdot,\|x\|_X)$ be a Banach space on number field $\mathbb{F}$. For a bounded linear operator $T\in\mathfrak{B}(X)$, non-negative number $r_T=\max\limits_{\lambda\in\sigma(T)}|\lambda|$ is said to be **the spectral radius of** T.

Theorem 4.2.11 Let X be a Banach space, $\mathfrak{B}(X)$ be the bounded linear operator space on X. For $T\in\mathfrak{B}(X)$, its spectral radius is $r_T=\lim\limits_{n\to+\infty}\sqrt[n]{\|T^n\|}$.

6. Compact operator and its properties

Definition 4.2.11 (compact operator) Let $(X,\|x\|_X)$ and $(Y,\|y\|_Y)$ be two Banach spaces on number field $\mathbb{F}$. For a linear operator $T: X\to Y$ with $T\in L(X,Y)$, if T maps any bounded set $E\subset X$ into a sequentially compact set $T(E)\subset Y$, then T is said to be **a compact linear operator**, or simply, **a compact operator**; or said to be **a completely**, or **totally continuous operator.** The set of all compact linear operators from X to Y is denoted by $\mathfrak{C}(X,Y)$.

Note 1 $T\in\mathfrak{C}(X,Y)\Leftrightarrow$ for any bounded set $B\subset X$, the set $\overline{T(B)}$ is compact in Y; i.e., $T(B)$ has compact closure; (A set $A\subset X$ is said to be **relatively compact**, if its closure $\overline{A}\subset X$ is a compact set).

Note 2 $T\in\mathfrak{C}(X,Y)\Leftrightarrow$for any bounded sequence $\{x_n\}\subset X$, the sequence $\{T(x_n)\}$ in Y exists convergent subsequence; i.e., $\{T(x_n)\}$ is sequentially compact.

Theorem 4.2.12 (continuity of compact operator) Let X and Y be two Banach spaces on number field $\mathbb{F}$. If a linear operator $T\in L(X,Y)$ is compact, that is $T\in\mathfrak{C}(X,Y)$, we have

(1) T is a continuous linear operator, i.e., a bounded linear operator, then $\mathfrak{C}(X,Y)\subset\mathfrak{B}(X,Y)$; and $\mathfrak{C}(X,Y)$ is a closed subspace of the topological space $\mathfrak{B}(X,Y)$;

(2) $T\in\mathfrak{B}(X,Y)$ is a compact operator, if and only if T maps the unit closed ball $\{x\in X: \|x\|_X\leqslant 1\}$ in X to a sequentially compact set in Y; i. e., $T(\{x\in X: \|x\|_X\leqslant 1\})\subset Y$ is a sequentially compact subset of Y;

(3) $T\in\mathfrak{B}(X,Y)$ is a compact operator, if and only if T maps any bounded subset E in X to a compact subset $\overline{T(E)}\subset Y$.

Example 4.2.15 Let X and Y be two Banach spaces on number field $\mathbb{F}$. If $T\in\mathfrak{B}(X, Y)$, and the range $T(X)$ is a finite-dimensional subspace of Y, then T is a compact operator, i. e., $T\in\mathfrak{C}(X,Y)$.

Because T is a bounded linear operator, thus T maps bounded set $A\subset X$ to bounded set $T(A)\subset T(X)$; then for $T(X)$ is finite-dimensional, all bounded subset in $T(X)$ is sequentially compact, so that T is a compact operator.

Example 4.2.16 Let $K(t,s)$ be continuous on $\{(t,s)\in\mathbb{R}^2: a\leqslant t\leqslant b, a\leqslant s\leqslant b\}$. Define operator T from $C([a,b])$ to $C([a,b])$ by

$$(Tf)(t)=\int_a^b K(t,s)f(s)\mathrm{d}s,$$

with **kernel** $K(t,s)$ of T. Then, $T: C([a,b])\to C([a,b])$ is a compact operator.

In fact, let $A\subset C([a,b])$ be a bounded linear operator. Then, there exists a constant $M>0$, such that for each $f\in A$, it holds $\|f\|_{C([a,b])}\leqslant M$. Clearly, $(Tf)(t)$ is a bounded linear operator on $C([a,b])$, thus $T(A)$ is an uniformly bounded set in $C([a,b])$. On the other hand, we have

$$\begin{aligned}|(Tf)(t_1)-(Tf)(t_2)| &\leqslant\int_a^b |K(t_1,s)-K(t_2,s)||f(s)|\mathrm{d}s\\ &\leqslant M\int_a^b |K(t_1,s)-K(t_2,s)|\mathrm{d}s.\end{aligned}$$

By the continuity of $K(t,s)$ in the square $\{(t,s)\in\mathbb{R}^2: a\leqslant t\leqslant b, a\leqslant s\leqslant b\}$, it holds that $\forall\varepsilon>0$, $\exists\delta>0$, s. t., for $t_1, t_2\in[a,b]$ with $|t_1-t_2|<\delta$, then $|K(t_1,s)-K(t_2,s)|<\dfrac{\varepsilon}{M(b-a)}$, $\forall s\in[a,b]$. Hence, it implies that $|(Tf)(t_1)-(Tf)(t_2)|<\varepsilon$ for all $f\in A$. This means: the image set $T(A)\subset C([a,b])$ of $A\subset C([a,b])$ is C-equi-continuous, then Theorem 4.1.10 shows that $T(A)$ is sequentially compact, so T is a compact linear operator.

Example 4.2.17 Let $l^2=\left\{x: x=(\xi_1,\xi_2,\cdots), \sum_{j=1}^{+\infty}|\xi_j|^2<+\infty\right\}$; and infinite matrix $T=[a_{i,j}]_{1\leqslant i,j<+\infty}$ with $\sum_{i,j=1}^{+\infty}|a_{i,j}|^2<+\infty$. Define a linear operator $T: x=(\xi_1,\xi_2,\cdots)\to y=(\eta_1,\eta_2,\cdots)$ from l^2 to l^2 by $y=T(x)$ with $\eta_j=\sum_{i=1}^{+\infty}a_{i,j}\xi_i$, $j=1,2,\cdots$. Then $T: l^2\to l^2$ is a compact operator.

In fact, by Cauchy inequality, we have

$$\|y\|_{l^2}^2=\sum_{j=1}^{+\infty}|\eta_j|^2=\sum_{j=1}^{+\infty}\Big|\sum_{i=1}^{+\infty}a_{i,j}\xi_i\Big|^2\leqslant\sum_{j=1}^{+\infty}\Big(\sum_{i=1}^{+\infty}|a_{i,j}|^2\Big)\Big(\sum_{i=1}^{+\infty}|\xi_i|^2\Big)$$
$$=\Big(\sum_{i,j=1}^{+\infty}|a_{i,j}|^2\Big)\|x\|_{l^2}^2,$$

hence, T is a bounded linear operator from l^2 to l^2. Moreover, we prove that T is a compact operator.

Let $A\subset l^2$ be a bounded set, thus $\exists M>0$, s. t., for all $x\in A$, it holds $\|x\|_{l^2}<M$. By $\sum\limits_{i,j=1}^{+\infty}|a_{i,j}|^2<+\infty$, then $\forall\varepsilon>0$, $\exists N>0$, s. t. $\sum\limits_{i=1}^{+\infty}\sum\limits_{j=N+1}^{+\infty}|a_{i,j}|^2<\dfrac{\varepsilon^2}{M^2}$. We deduce

$$\sum_{j=N+1}^{+\infty}|\eta_j|^2=\sum_{j=N+1}^{+\infty}\Big|\sum_{i=1}^{+\infty}a_{i,j}\xi_i\Big|^2\leqslant\sum_{j=N+1}^{+\infty}\Big(\sum_{i=1}^{+\infty}|a_{i,j}|^2\Big)\Big(\sum_{i=1}^{+\infty}|\xi_i|^2\Big)$$
$$\leqslant\frac{\varepsilon^2}{M^2}\|x\|_{l^2}^2\leqslant\varepsilon^2,$$

this means: for the above "$\exists N>0$", if the element $(\eta_1,\cdots,\eta_N,0,0,\cdots)$ is generated by N-coordinates $\eta_1,\cdots,\eta_N$ of operator $T(x)=y$, then the set $B=\{(\eta_1,\cdots,\eta_N,0,0,\cdots)\}$ is a finite ε-net of $T(A)$. However, B is a bounded set in N-dimensional subspace of l^2, so it is sequentially compact. Hence, $T(A)$ has a sequentially compact finite ε-net, this implies that $T(A)$ is a sequentially compact set, and finally, T is a compact operator.

We list some properties of compact operators.

Theorem 4. 2. 13 (compound and conjugate of compact operators) Let X_1, X_2, X_3 be three Banach spaces, $T\in\mathfrak{B}(X_1,X_2)$ and $S\in\mathfrak{B}(X_2,X_3)$ be two operators. If one of T and S is compact operator, say $T\in\mathfrak{C}(X_1,X_2)$, then the compound mapping $S\circ T$ is a compact operator, i. e., $S\circ T\in\mathfrak{C}(X_1,X_3)$; Moreover, $T\in\mathfrak{C}(X_1,X_2)$ implies the conjugate operator $T^*: X_2^*\to X_1^*$ is a compact operator, and holds

$$T\in\mathfrak{C}(X_1,X_2)\Leftrightarrow T^*\in\mathfrak{C}(X_2^*,X_1^*).$$

Theorem 4. 2. 14 (separable range of compact operators) Let X, Y be Banach spaces. If $T\in\mathfrak{C}(X,Y)$ is a compact operator, then the range $\mathfrak{R}(T)$ of T is separable in Y.

Theorem 4. 2. 15 (limit theorem of compact sequence) Let X, Y be Banach spaces, $\{T_n\}\subset\mathfrak{C}(X,Y)$ be compact operator sequence. If $\exists T\in\mathfrak{B}(X,Y)$, such that $\lim\limits_{n\to\infty}\|T_n-T\|=0$, then T is compact operator, $T\in\mathfrak{C}(X,Y)$. i. e., a compact operator sequence $\{T_n\}$ converges to a bounded operator T in operator norm sense, thus implies T is compact.

Theorem 4. 2. 16 (properties of $\lambda I-T$ for a compact operator) Let X be a Banach space, $T\in\mathfrak{C}(X)$ be a compact operator. We have

(1) If $\lambda\neq 0$, then the range $\mathfrak{R}(\lambda I-T)\subset X$ of $\lambda I-T$ is a closed subspace of X;

(2) If $\lambda \neq 0$, then the operator $\lambda I - T$ is a surjective mapping, if and only if $\lambda I - T$ is a injective mapping; Thus, if $\lambda I - T$ is surjective (or injective), then λ is a regular value of T.

Thus, any $\lambda \neq 0$ is a regular value of a compact operator $T \Leftrightarrow \lambda I - T$ is surjective, or injective.

Theorem 4.2.17 (spectrum theorem of compact operators) Let X be a Banach space, $T \in \mathfrak{C}(X)$ be a compact operator. Then the spectrum set of T has the following properties.

(1) $\sigma(T) \subset \mathbb{C}$ is a finite set, or at most a countable set with accumulative point $\lambda = 0$; Thus, any $\lambda \in \sigma(T)$ with $\lambda \neq 0$ is an eigenvalue of T, i.e., $\sigma(T) \backslash \{0\} = \sigma_p(T)$;

(2) If $\lambda \neq 0, \lambda \in \sigma(T), \lambda \in \sigma(T^*)$, then two eigenvector spaces of T and T^* corresponding to the same λ have the same dimensions;

(3) If X is infinite-dimensional, then it holds $0 \in \sigma(T), 0 \in \sigma(T^*)$;

(4) The dimension of eigenvector space $E_\lambda \subset X$ corresponding to eigenvalue $\lambda \in \sigma_p(T)$ is finite;

(5) The eigenvector spaces E_{λ_1} and E_{λ_2} corresponding to different eigenvalues λ_1, $\lambda_1 \in \sigma_p(T), \lambda_1 \neq \lambda_2$ of T and T^* are mutually orthogonal, $E_{\lambda_1} \perp E_{\lambda_2}$;

(6) If a complex number $\lambda \neq 0$ is an eigenvalue of T and T^*, then the equation $(\lambda I - T)x = y$ has solution, if and only if y is orthogonal to the zero space of $\lambda I^* - T^*$; The equation $(\lambda I^* - T^*)f = g$ has nonzero solution, if and only if g is orthogonal to the zero space of $\lambda I - T$.

$\mathfrak{C}(X, Y)$, the class of compact operators, is a very important operator subfamily in the bounded linear operator space $\mathfrak{B}(X, Y)$, it has many applications in such scientific areas as algebraic, operator, differential, and integral equations, and plays an active role.

4.3 Linear Functional Theory

We introduce an other important content: linear functional theory. The main aim of this part is to study the bounded linear functional spaces on normed linear spaces and on inner product spaces, particularly, the bounded linear functional spaces on Banach spaces and Hilbert spaces. Moreover, the bounded linear operators and their conjugate operators play roles in linear functional theory, so that they are main objects here.

The following is a frame of this section.

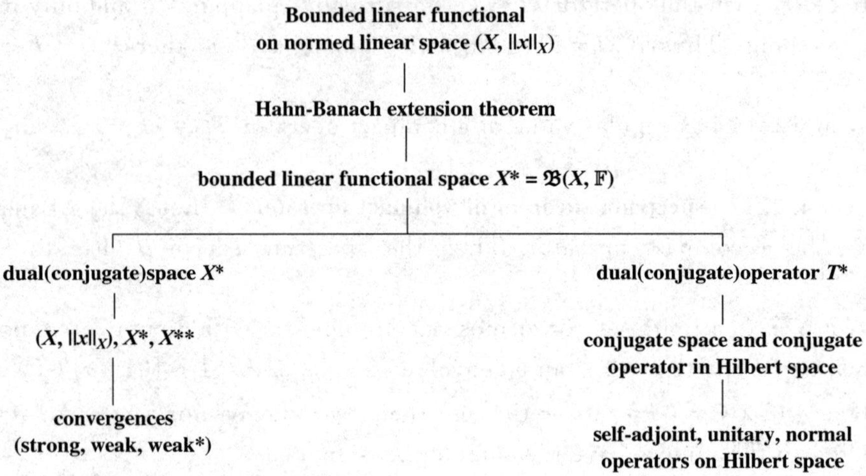

4.3.1 Bounded linear functionals on normed linear spaces

1. Bounded linear functionals

Definition 4.3.1 (linear functional) Let $(X, \| x \|_X)$ be a normed linear space on number field $\mathbb{F}$, simply, a normed space. An operator $T: X \to \mathbb{F}$ is said to be **a linear functional**, if it satisfies

$$T(x+y) = T(x) + T(y), \quad x, y \in X;$$
$$T(\alpha x) = \alpha T(x), \quad x \in X, \alpha \in \mathbb{F}.$$

If a linear functional T is bounded, i. e., $T \in \mathfrak{B}(X, \mathbb{F})$, then it is said to be **a bounded linear functional**, or **a continuous linear functionals** on X.

Theorem 4.3.1 (Hahn-Banach extension theorem) Let $(X, \| x \|_X)$ be a normed space, $G \subset X$ be a subspace of X, and the bounded linear functional space on G be denoted by $\mathfrak{B}(G, \mathbb{F}) \equiv (\mathfrak{B}(G, \mathbb{F}), +, \alpha \cdot, \| T \|_{\mathfrak{B}(G, \mathbb{F})})$. If $T \in \mathfrak{B}(G, \mathbb{F})$ is a bounded linear functional $T: G \to \mathbb{F}$ on G, then T can be extended to whole X as a functional $\widetilde{T}: X \to \mathbb{F}$ with keeping norm and $\widetilde{T} \in \mathfrak{B}(X, \mathbb{F})$; i. e., there exists a bounded linear functional $\widetilde{T}$: $X \to \mathbb{F}$ on X, satisfing

(1) $\widetilde{T}(x)|_{x \in G} = T(x)$; (2) $\| \widetilde{T} \|_{\mathfrak{B}(X, \mathbb{F})} = \| T \|_{\mathfrak{B}(G, \mathbb{F})}$.

Note 1 The proof of this theorem is very delicate, the Zorn lemma and some techniques are used. We refer to [6].

Note 2 By this theorem, we may suppose that the domain of a bounded linear functional $T \in \mathfrak{B}(X, \mathbb{F})$ is space $(X, \| x \|_X)$, without loss of generality.

Note 3 The Fig. 4. 3. 1 is a sketch of this theorem.

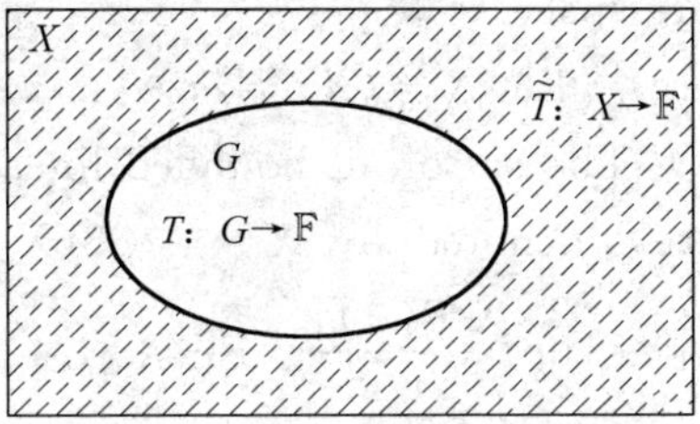

Fig. 4. 3. 1 Hahn-Banach extension theorem

Corollary 1 Let $(X, \|x\|_X)$ be a normed linear space. Then $\forall x_0 \in X, x_0 \neq 0$, there exists a bounded linear functional $f: X \to \mathbb{F}$ on X, such that $\|f\| = 1, f(x_0) = \|x_0\|_X$.

Corollary 2 Let $(X, \|x\|_X)$ be a normed linear space, $G \subset X$ be a subspace, $x_0 \in X$ with

$$d = \rho(x_0, G) = \inf_{x \in G} \|x - x_0\|_X > 0.$$

Then, there exists a bounded linear functional $f: X \to \mathbb{F}$ on X, satisfying $\|f\| = \frac{1}{d}$ with

$$f(x) = \begin{cases} 1, & x = x_0 \\ 0, & x \in G \end{cases}.$$

Example 4. 3. 1 Let $X = \mathbb{R}^2$ with usual norm $\|x\|_{\mathbb{R}^2} = \sqrt{\xi_1^2 + \xi_2^2}$, $x = (\xi_1, \xi_2) \in \mathbb{R}^2$, and $\|x\|_X = |\xi_1| + |\xi_2|$ be an equivalent norm. If for $G = \{x \in \mathbb{R}^2 : x = (\xi_1, 0)\}$, define a function on G: $f(x) = \xi_1$ at $x = (\xi_1, 0) \in G$. Then, $f: G \to \mathbb{F}$ is a bounded linear functional on G with $\|f\| = 1$.

Extend f to whole $\mathbb{R}^2$: take any $\alpha \in [-1, 1]$, define a bounded linear functional on $\mathbb{R}^2$ by $\tilde{f}: \mathbb{R}^2 \to \mathbb{F}$ with $\tilde{f}_\alpha(x) = \xi_1 + \alpha\, \xi_2$, $x = (\xi_1, \xi_2)$. Then $\tilde{f}: \mathbb{R}^2 \to \mathbb{F}$ is an extension of $f: G = \mathbb{R}^1 \to \mathbb{F}$, with $\|\tilde{f}_\alpha\| \geqslant \|f\|$; On the other hand, by

$$|\tilde{f}_\alpha(x)| \leqslant |\xi_1| + |\alpha\, \xi_2| \leqslant |\xi_1| + |\xi_2| = \|x\|_X,$$

this implies $\|\tilde{f}_\alpha\| = \|f\| = 1$.

2. Dual (conjugate) spaces and conjugate operators

1) Dual (conjugate) space of a normed linear space

Recall that we have defined the dual space for a linear space $(X, +, \alpha\cdot)$ which is without any structure of the topology in 2. 1. 5. However, when we study the dual space of a normed linear space, the topological structure of dual space of normed linear space

$(X,+,\alpha\cdot,\|x\|_X)$ must be considered, and the dual space (i. e., the linear functional space) is also called a conjugate space.

Definition 4.3.2 (dual space) Let $X\equiv(X,+,\alpha\cdot,\|x\|_X)$ be a normed linear space on number field $\mathbb{F}$ (or $\mathbb{R}$ or $\mathbb{C}$). The set of all bounded linear functionals on X is called **a dual space** (or **conjugate space**), denoted by $X^*\equiv(\mathfrak{B}(X,\mathbb{F}),+,\alpha\cdot,\|T\|)$, it is a Banach space with norm $\|T\|=\sup\left\{\frac{|T(x)|}{\|x\|_X}: x\in X, x\neq 0\right\}$. We use notation $\langle f,x\rangle\equiv f(x)\in\mathbb{F}$, it means linear functional $f\in X^*$ acts on element $x\in X$. Regard X^* as a Banach space, the second dual space $(X^*)^*\equiv X^{**}$, the third dual space $(X^{**})^*,\cdots$, can be defined, similarly.

Definition 4.3.3 (natural imbedding mapping) Let X be a normed linear space on $\mathbb{F}$, X^* and X^{**} be the dual and second dual space of X, respectively. The mapping $\tau: X\to X^{**}$ satisfying $\langle x^{**},f\rangle=\langle f,x\rangle$, $\forall f\in X^*$, is said to be **a natural imbedding mapping form X to X^{**}**.

If $X^{**}=X$, then X is said to be **a self-reflect space**; If $X^*=X$, then X is said to **be a self-dual (self-conjugate) space, see** Fig. 4.3.2.

For self-reflect space $X=X^{**}$, the natural imbedding mapping $\tau: X\to X^{**}$ is an equimetric isomorphism mapping from X to X^{**}, i. e., $\tau(x)=x^{**}$ with $\langle x^{**},f\rangle=\langle f,x\rangle$, $\forall f\in X^*$, and $\|x\|_X=\|x^{**}\|_{X^{**}}$.

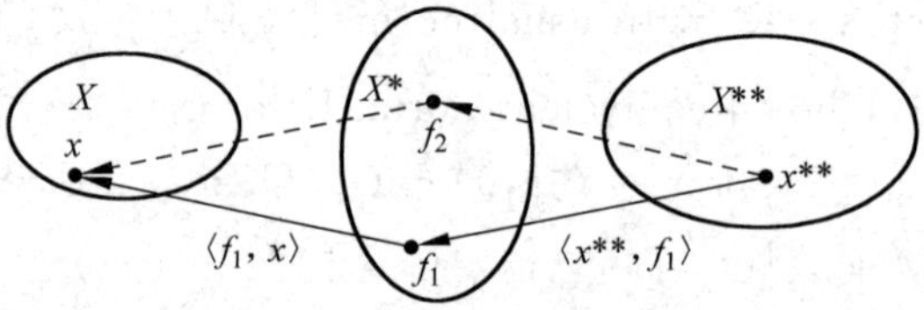

Fig. 4.3.2 Second conjugate, natural imbedding mapping τ form X to X^{**}

2) Some important dual spaces

① The dual space of $\mathbb{R}^n$ is $\mathbb{R}^n$, $n\in\mathbb{Z}^+$, i. e., $(\mathbb{R}^n)^*=\mathbb{R}^n$.

Since $T\in(\mathbb{R}^n)^*\Rightarrow T=\xi=(\xi_1,\cdots,\xi_n)\in\mathbb{R}^n\Rightarrow T(x)=\langle\xi,x\rangle, x\in\mathbb{R}^n$ with $T(x)=\langle\xi,x\rangle=(\xi_1,\cdots,\xi_n)\cdot(x_1,\cdots,x_n)=\sum_{j=1}^{n}\xi_j x_j$, then, $\mathbb{R}^n$ is self-dual, and is self-reflect.

② The dual space of l^p $(1<p<+\infty)$ is the space l^q $(1<q<+\infty)$, with $\frac{1}{p}+\frac{1}{q}=1$, and q is called **the conjugate number of** p, i. e., $(l^p)^*=l^q$. Thus, l^p $(1<p<+\infty)$ is self-reflect, $(l^p)^{**}=l^p$. Moreover, $T\in(l^p)^*=l^q$ is equivalent to

$$T=\xi=(\xi_1,\cdots,\xi_n,\cdots)\Rightarrow T(x)=\langle\xi,x\rangle \text{ with } T(x)=\langle\xi,x\rangle=\sum_{j=1}^{+\infty}\xi_j x_j,\quad \forall x\in l^p.$$

③ The dual space of $L^p(\mathbb{R})$ $(1<p<+\infty)$ is space $L^q(\mathbb{R})$ $(1<q<+\infty)$, with $\frac{1}{p}+\frac{1}{q}=1$, and q is **the conjugate number** of p, i. e., $(L^p(\mathbb{R}))^*=L^q(\mathbb{R})$. And $L^p(\mathbb{R})$ is self-reflect $(L^p(\mathbb{R}))^{**}=L^p(\mathbb{R})$.

Moreover, we have the corresponding relation: $T\in L^q(\mathbb{R})\xleftrightarrow{\text{one-one}} g\in L^q(\mathbb{R})$ by $T\in(L^p(\mathbb{R}))^*=L^q(\mathbb{R}), 1<p<+\infty$ with

$$T_g(f)=\langle g,f\rangle=\int_{\mathbb{R}}g(x)f(x)\mathrm{d}x,\quad \forall f\in L^p(\mathbb{R}).$$

④ If $p=q=2$, we have $(l^2)^*=l^2$, $(L^2(\mathbb{R}))^*=L^2(\mathbb{R})$; i. e. l^2 and $L^2(\mathbb{R})$ both are self-reflect and self-dual Hilbert spaces (complete inner product spaces).

3) Conjugate operators

① Definition of conjugate operators

Definition 4.3.4 (conjugate operator) Let X and Y be two normed linear spaces on number field $\mathbb{F}$, X^*, Y^* be their dual spaces, respectively. For a bounded linear-operator $T\in\mathfrak{B}(X,Y)$ from X to Y, the conjugate operator T^* of T, is defined as the operator $T^*: Y^*\to X^*$ satisfying

$$\langle T^*(f),x\rangle=\langle f,T(x)\rangle,\quad \forall f\in Y^*, \forall x\in X. \tag{4.3.1}$$

The following Fig. 4.3.3 is an exchanged grapy.

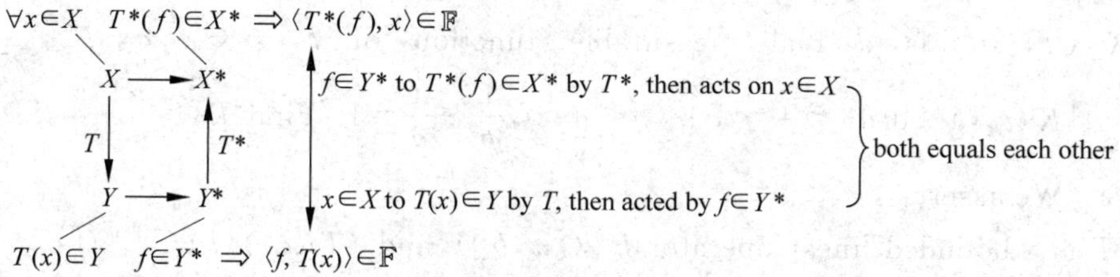

Fig. 4.3.3 Exchanged graph

The right-hand side $\langle T^*(f),x\rangle$ means: operator $T^*\in\mathfrak{B}(Y^*,X^*)$ has domain Y^*, and acts on $f\in Y^*$ denoted by $T^*(f)$; i. e. $T^*: f\in Y^*\xrightarrow{T^*\in\mathfrak{B}(Y^*,X^*)}T^*(f)\in X^*$; so, $T^*(f)\in X^*=\mathfrak{B}(X,\mathbb{F})$, and implies that $T^*(f): X\to\mathbb{F}$ acts on $x\in X$, denoted by $\langle T^*(f),x\rangle$, making sense, and $\langle T^*(f),x\rangle\in\mathbb{F}$. On the other hand, the right-hand side of (4.3.1) makes sense, because $f\in Y^*$, then $f: Y\to\mathbb{F}\to\mathbb{F}$ acts on $y\in Y$, denoted by $\langle f,y\rangle$, for $\forall f\in Y^*$, $\forall y\in Y$; Thus, $\langle f,T(x)\rangle\in\mathbb{F}$, $\forall f\in Y^*$, $\forall x\in X$.

Hence, the definition of conjugate operator $T^*: Y^* \to X^*$ defined in (4.3.1) makes sense. Recall Definition 2.2.3, formula (2.2.1) and Fig. 2.2.1, as well as Definition 4.3.4.

② **Properties of conjugate operators**

Theorem 4.3.2 Let X, Y, Z be normed linear spaces on number field $\mathbb{F}$. Then

(1) $S, T \in \mathfrak{B}(X, Y) \Rightarrow (S+T)^* = S^* + T^*$;

(2) $S \in \mathfrak{B}(X, Y), \alpha \in \mathbb{F} \Rightarrow (\alpha S)^* = \alpha S^*$;

(3) $T \in \mathfrak{B}(X, Y) \Rightarrow T^* \in \mathfrak{B}(Y^*, X^*)$; $\| T^* \| = \| T \|$;

(4) $S \in \mathfrak{B}(X, Y), T \in \mathfrak{B}(Y, Z) \Rightarrow (T \circ S)^* = S^* \circ T^*$;

(5) If $T \in \mathfrak{B}(X, Y)$ has inverse operator $T^{-1} \in \mathfrak{B}(Y, X)$, then so does $T^* \in \mathfrak{B}(Y^*, X^*)$; and hold $(T^*)^{-1} = (T^{-1})^*$, $\| T^{-1} \| = \dfrac{1}{\| T \|}$, $\| (T^*)^{-1} \| = \dfrac{1}{\| T^* \|}$;

(6) If X, Y are finite-dimensional, then $T \in \mathfrak{B}(X, Y) \xleftrightarrow{\text{one-one}} A = [a_{jk}]_{\mathfrak{AC}} \in \mathfrak{M}_{m\times n}(\mathbb{F})$; and $A = [a_{jk}]_{\mathfrak{AC}}$ is called the representation matrix of operator $T \in \mathfrak{B}(X, Y)$. The rank rank(T) of operator T is defined by rank(A) of representation matrix A. Moreover, it holds rank(T)=rank(T^*).

Example 4.3.2 For two Banach spaces $X = \mathbb{R}^n$ and $Y = \mathbb{R}^m$, an operator $T \in L(\mathbb{R}^n, \mathbb{R}^m) \leftrightarrow A \in \mathfrak{M}_{m\times n}$ has conjugate operator $T^* \in L(\mathbb{R}^m, \mathbb{R}^n) \leftrightarrow \overline{A}^{\mathrm{T}} \in \mathfrak{M}_{m\times n}$.

Example 4.3.3 If T is an integral operator with **kernel** $K(t, s)$ on space $X = L^p([a, b]), 1 < p < +\infty$:

$$(T(x))(t) = \int_{[a,b]} K(t,s)x(s)\mathrm{d}s, \quad x \in L^p([a,b]), t \in [a,b],$$

where $K(t, s)$ is a real measurable function of $t, s \in [a, b]$ satisfying $\int_{[a,b]\times[a,b]} |K(t,s)|^q \mathrm{d}t\mathrm{d}s < +\infty$, $1 < q < +\infty$, $\dfrac{1}{p} + \dfrac{1}{q} = 1$. Find T^*.

Solve We assert:

(i) T is a bounded linear operator $L^p([a,b])$, and $(T(x))(t) \in L^q([a,b])$; thus conjugate operator T^* is a bounded linear operator on $L^q([a,b])$, and

$$\langle T^*(f), x \rangle = \langle f, T(x) \rangle, \quad \forall f \in (L^q([a,b]))^*, \quad \forall x \in L^p([a,b]);$$

(ii) $\forall f \in (L^q([a,b]))^*$, $\exists y(t) \in L^q([a,b])$, s. t.

$$\langle f, z \rangle = \int_{[a,b]} y(t)z(t)\mathrm{d}t, \ \forall z \in L^p([a,b]);$$

(iii) T^* is defined by

$$(T^*(f))(s) = \int_{[a,b]} K(t,s)f(t)\mathrm{d}t, \text{for } f \in L^p([a,b]), s \in [a,b]$$

In fact, by (i) and (ii), the conjugate operator T^* of T would satisfy: $\forall f \in$

$(L^q([a,b]))^*$, it holds

$$\langle T^*(f),x\rangle=\langle f,T(x)\rangle=\left\langle f,\int_{[a,b]}K(t,s)x(s)\mathrm{d}s\right\rangle,\quad \forall x\in L^p([a,b]).$$

By Fubini theorem,

$$\begin{aligned}\langle T^*(f),x\rangle&=\left\langle f,\int_{[a,b]}K(t,s)x(s)\mathrm{d}s\right\rangle=\int_{[a,b]}f(t)\left[\int_{[a,b]}K(t,s)x(s)\mathrm{d}s\right]\mathrm{d}t\\&=\int_{[a,b]}x(s)\left[\int_{[a,b]}K(t,s)f(t)\mathrm{d}t\right]\mathrm{d}s\equiv\left\langle\int_{[a,b]}K(t,s)f(t)\mathrm{d}t,x\right\rangle,\end{aligned}$$

it follows

$$T^*(f)(s)=\int_{[a,b]}K(t,s)f(t)\mathrm{d}t,\quad \forall f\in(L^q([a,b]))^*=L^p([a,b]),\quad s\in[a,b].$$

③ **Some special conjugate operators**

Definition 4.3.5 (special conjugate operator) Let $T: X\to X$ be a bounded linear operator from normed linear space X on number field $\mathbb{F}$ to X itself, $T^*\in\mathfrak{B}(X^*)$ be the conjugate operator of $T\in\mathfrak{B}(X)$.

(1) If $T=T^*$, then T is said to be **a self-conjugate operator**, or Hermite operator;

(2) If T is one-one, $T^*=T^{-1}$, then T is said to be **a unitary operator**;

(3) If $T^*T=TT^*$, then T is said to be **a normal operator.**

3. Convergences in Banach spaces

Let $X,Y,\cdots$ be Banach spaces, $X^*,Y^*,\cdots$ be their dual spaces, respectively.

1) Convergence in Banach space X

There are 2 kinds of convergences of sequences in Banach spaces X.

① **Strong convergence**

Definition 4.3.6 (strong convergence) Let $(X,\|x\|_X)$ be a Banach space, $\{x_n\}\subset X$ be a sequence in X. If there exists $x\in X$, such that

$$\lim_{n\to+\infty}\|x_n-x\|_X=0,$$

then $\{x_n\}$ is said to **strongly converge into** x (or $\{x_n\}$ is strongly convergent to x); or $\{x_n\}$ **converges into** x **in the norm sense of** X; and x is said to be **the strong limit of** $\{x_n\}$; denoted by $(s)\lim\limits_{n\to+\infty}x_n=x$.

② **Weak convergence**

Definition 4.3.7 (weak convergence) Let $(X,\|x\|_X)$ be a Banach space on number field $\mathbb{F}$, $\{x_n\}\subset X$ be a sequence in X. If there exist $x\in X$, such that

$$\lim_{n\to+\infty}\langle f,x_n\rangle=\langle f,x\rangle,\quad \forall f\in X^*,$$

where $\langle f, x_n\rangle, \langle f, x\rangle \in \mathbb{F}$, the limit is in norm sense of $\mathbb{R}$ or $\mathbb{C}$; then $\{x_n\}$ is said to **weakly converge into** x; and x is said to be **the weak limit of** $\{x_n\}$; denoted by $(w)\lim\limits_{n\to+\infty} x_n = x$.

③ **Relationships of convergences**

Theorem 4.3.3 Let $(X, \|x\|_X)$ be a Banach space on number field $\mathbb{F}$. We have

(1) If strong limit exists, then it is unique; so is that for weak limit;

(2) $(s)\lim\limits_{n\to+\infty} x_n = x \Rightarrow (w)\lim\limits_{n\to+\infty} x_n = x$;

(3) If $\{x_n\}$ is weakly convergent, then $\{\|x_n\|_X\}$ is a bounded set in $\mathbb{F}$;

(4) If $(X, \|x\|_X)$ is finite-dimensional, then

$$(s)\lim_{n\to+\infty} x_n = x \Leftrightarrow (w)\lim_{n\to+\infty} x_n = x.$$

2) **Convergence in bounded linear operator space $\mathfrak{B}(X, Y)$**

There are 3 kinds of convergences of sequences in bounded linear operator space $\mathfrak{B}(X, Y)$.

① **Strong convergence**

Definition 4.3.8 (strong convergence) Let $(X, \|x\|_X)$, $(Y, \|y\|_Y)$ be Banach spaces, $\{T_n\} \subset \mathfrak{B}(X, Y)$ be a bounded linear operator sequence. If there exists $T \in \mathfrak{B}(X, Y)$, such that

$$\lim_{n\to+\infty} \|T_n - T\| = 0$$

holds, i. e. it is convergent in the operator norm sense of $\mathfrak{B}(X, Y)$, then $\{T_n\}$ is said to **strongly converge into** T (or $\{T_n\}$ is strongly convergent to T); or $\{T_n\}$ **converges into** T **in the norm sense of** $\mathfrak{B}(X, Y)$; and T is said to be **the strong limit of** $\{T_n\}$; denoted by $(s)\lim\limits_{n\to+\infty} T_n = T$. (Compare with Definition 4.2.6.).

② **Pointwise convergence**

Definition 4.3.9 (convergence in pointwise sense) Let $(X, \|x\|_X)$ and $(Y, \|y\|_Y)$ be two Banach spaces, $\{T_n\} \subset \mathfrak{B}(X, Y)$ be a bounded linear operator sequence. If there exists a bounded linear operator $T \in \mathfrak{B}(X, Y)$, such that

$$\lim_{n\to+\infty} T_n(x) = T(x), \quad \forall x \in X,$$

where limit is in norm sense of $(Y, \|y\|_Y)$, then $\{T_n\}$ is said to **pointwise converge into** T or to be pointwise convergent into; and T is said to be **the pointwise limit of** $\{T_n\}$, denoted by $(p)\lim\limits_{n\to+\infty} T_n = T$.

Note that, for space $(Y, \|y\|_Y)$, operator values $T_n(x)$ and $T(x)$ are in space Y, thus pointwise convergence has an equivalent definition:

$$\lim_{n\to+\infty} T_n(x) = T(x), \forall x \in X \Leftrightarrow \lim_{n\to+\infty} \|T_n(x) - T(x)\|_Y = 0, \quad \forall x \in X.$$

(Compare with Definition 4.2.7.)

③ **Weak convergence**

Definition 4.3.10 (weak convergence) Let $(X, \| x \|_X)$ and $(Y, \| y \|_Y)$ be two Banach spaces, $\{T_n\} \subset \mathfrak{B}(X, Y)$ be a bounded linear operator sequence. If there exists $x \in X$, such that

$$\lim_{n \to +\infty} \langle f, T_n(x) \rangle = \langle f, T(x) \rangle, \quad \forall f \in Y^*,$$

where $T_n(x), T(x) \in Y, \forall x \in X$, and the limit is in the norm sense of $\mathbb{R}$ or $\mathbb{C}$, then $\{T_n\}$ is said to **weakly converge into** T; and T is said to be **the weak limit of** $\{T_n\}$, denoted by $(w) \lim_{n \to +\infty} T_n = T$.

④ **Relationships of convergences**

Theorem 4.3.4 Let $(X, \| x \|_X)$ and $(Y, \| y \|_Y)$ be Banach spaces, $\{T_n\} \subset \mathfrak{B}(X, Y)$ be a bounded linear operator. We have

(1) For strong pointwise, and weak limits, if they exist, then they are unique;

(2) $(s) \lim_{n \to +\infty} T_n = T \Rightarrow (p) \lim_{n \to +\infty} T_n = T \Rightarrow (w) \lim_{n \to +\infty} T_n = T$;

(3) If $\{T_n\}$ is weakly convergent, then $\{\| T_n \|\}$ is a bounded set;

(4) if $\mathfrak{B}(X, Y)$ is finite-dimensional, then

$$(s) \lim_{n \to +\infty} T_n = T \Leftrightarrow (p) \lim_{n \to +\infty} T_n = T \Leftrightarrow (w) \lim_{n \to +\infty} T_n = T.$$

The following Fig. 4.3.4 is a draft of the relationship of 3 convergences.

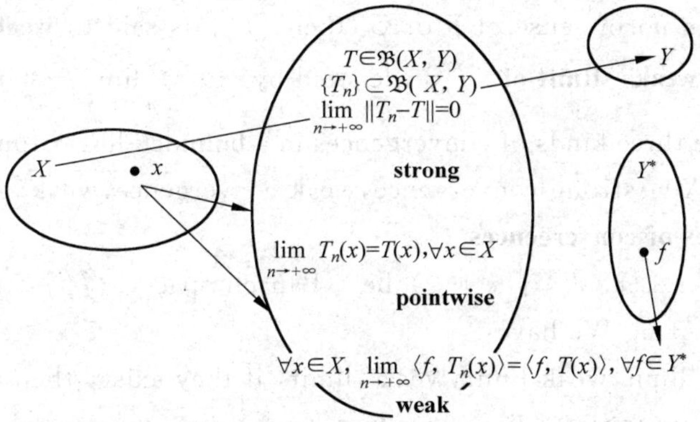

Fig. 4.3.4 Three convergences in $\mathfrak{B}(X, Y)$

3) **Convergence in dual space** X^*

There are 3 kinds of convergences of sequences in dual space $X^* = \mathfrak{B}(X, \mathbb{F})$.

① **Strong convergence**

Definition 4.3.11 (strong convergence) Let $(X, \| x \|_X)$ and $(X^*, \| f \|_{X^*})$ be two Banach spaces, $\{f_n\} \subset X^*$ be a bounded linear functional sequence. If there exists $f \in X^*$, such that

$$\lim_{n\to+\infty} \| f_n - f \|_{X^*} = 0,$$

where the limit is in the norm sense of X^*, then $\{f_n\}$ is said to **strongly converge**, or **be convergent into f in the norm sense $\|f\|_{X^*}$ of X^***; and f is said to be **the strong limit of** $\{f_n\}$; denoted by $(s)\lim\limits_{n\to+\infty} f_n = f$.

② **Weak convergence**

Definition 4.3.12 (weak convergence) Let $(X, \|x\|_X)$, $(X^*, \|x^*\|_{X^*})$, $(X^{**}, \|x^{**}\|_{X^{**}})$ be Banach space, dual space, second dual space, respectively, and $\{f_n\} \subset X^*$ be a functional sequence. If there exists $f \in X^*$, such that

$$\lim_{n\to+\infty} \langle x^{**}, f_n\rangle = \langle x^{**}, f\rangle, \quad \forall x^{**} \in X^{**},$$

where the limit in the norm sense of $\mathbb{R}$ or $\mathbb{C}$, then $\{f_n\}$ is said to **weakly converge into f**; and f is said to be **the weak limit of** $\{f_n\}$, denoted by $(w)\lim\limits_{n\to+\infty} f_n = f$.

③ **Weak* convergence**

Definition 4.3.13 (weak* convergence) Let $(X, \|x\|_X)$ and $(X^*, \|x^*\|_{X^*})$ be a Banach space, its dual space, respectively, and $\{f_n\} \subset X^*$ be a functional sequence. If there exists $f \in X^*$, such that

$$\lim_{n\to+\infty} \langle f_n, x\rangle = \langle f, x\rangle, \quad \forall x \in X,$$

where the limit is in norm sense of $\mathbb{R}$ or $\mathbb{C}$, then $\{f_n\}$ is said to **weak* converge into f**; and f is said to be **weak* limit of** $\{f_n\}$, denoted by $(w^*)\lim\limits_{n\to+\infty} f_n = f$.

Thus, there are three kinds of convergences in a bounded linear functional space $X^* = \mathfrak{B}(X, \mathbb{F})$ (dual of X): strong convergence, weak convergence, weak* convergence.

④ **Relationships of convergences**

Theorem 4.3.5 Let $(X, \|x\|_X)$ be a Banach space, $\{f_n\} \subset X^*$ be a functional sequence, see Fig. 4.3.5. We have:

(1) For strong limit, weak limit, weak* limit, if they exist, then are unique;

(2) $(s)\lim\limits_{n\to+\infty} f_n = f \Rightarrow (w)\lim\limits_{n\to+\infty} f_n = f \Rightarrow (w^*)\lim\limits_{n\to+\infty} f_n = f$;

(3) if $\{f_n\}$ is weak* convergent, then $\{\|f_n\|\} \subset \mathbb{R}$ is a bounded set;

(4) if $(X, \|x\|_X)$ is finite-dimensional, then

$$(s)\lim_{n\to+\infty} f_n = f \Leftrightarrow (w)\lim_{n\to+\infty} f_n = f \Leftrightarrow (w^*)\lim_{n\to+\infty} f_n = f;$$

(5) if $(X, \|x\|_X)$ is self-reflect, then

$$(w)\lim_{n\to+\infty} f_n = f \Leftrightarrow (w^*)\lim_{n\to+\infty} f_n = f.$$

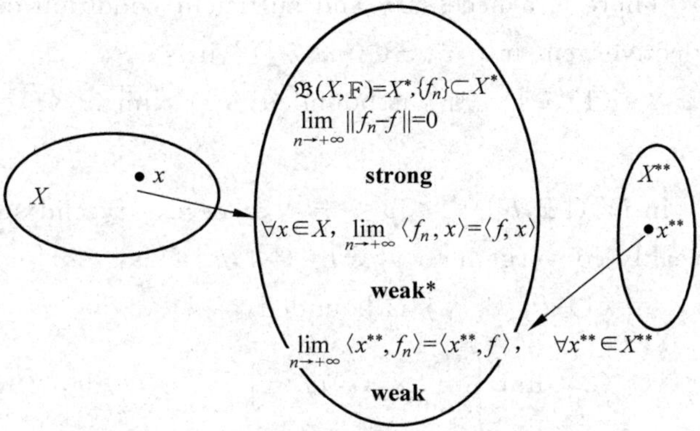

Fig. 4. 3. 5 Three convergences in X^*

4) Convergence in finite-dimension a spaces of X, X^*, X^{**}

If $(X, \|x\|_X)$ is a finite-dimensional Banach space, then $X \leftrightarrow X^* \leftrightarrow X^{**}$, and hold

$$\dim X = \dim X^* = \dim X^{**} = n, \quad n \in \mathbb{N}.$$

The following examples show "meanings of various convergences of a sequence in a same space".

Example 4. 3. 4 For $X = L^p([a,b])$, $1 < p < +\infty$, a sequence $\{x_n\} \subset L^p([a,b])$ converges into $x_0 \in L^p([a,b])$, with the conjugate number $1 < q < +\infty$, and $\frac{1}{p} + \frac{1}{q} = 1$.

(1) Regard as a Banach space $X = L^p([a,b])$

① Strong convergence (in norm sense) $(s)\text{-}\lim\limits_{n \to +\infty} x_n = x_0 \Leftrightarrow \lim\limits_{n \to +\infty} \|x_n - x_0\|_{L^p([a,b])} = 0 \Leftrightarrow \lim\limits_{n \to +\infty} \left\{ \int_{[a,b]} |x_n(t) - x_0(t)|^p \mathrm{d}t \right\}^{\frac{1}{p}} = 0$,

② Weak convergence $(w)\text{-}\lim\limits_{n \to +\infty} x_n = x_0 \Leftrightarrow \lim\limits_{n \to +\infty} \langle f, x_n \rangle = \langle f, x_0 \rangle$, $\forall f \in L^q([a, b]) \Leftrightarrow \lim\limits_{n \to +\infty} \int_{[a,b]} f(t) x_n(t) \mathrm{d}t = \int_{[a,b]} f(t) x_0(t) \mathrm{d}t$, $\forall f \in L^q([a,b])$.

(2) Regard as a dual space $X^* = L^p([a,b])$ of Banach space $X = L^q([a,b])$, $\frac{1}{p} + \frac{1}{q} = 1$

①′ Strong convergence (in norm sense) $(s)\text{-}\lim\limits_{n \to +\infty} x_n = x_0 \Leftrightarrow$ the same as ① in (1)

②′ Weak convergence $(w)\text{-}\lim\limits_{n \to +\infty} x_n = x_0 \Leftrightarrow$ the same as ② in (1), by $X^{**} = L^q([a,b])$

③′ Weak* convergence $(w^*)\text{-}\lim\limits_{n \to +\infty} x_n = x_0 \Leftrightarrow (w)\text{-}\lim\limits_{n \to +\infty} x_n = x_0$

$$\Leftrightarrow \lim_{n \to +\infty} \langle x_n, x^{**} \rangle = \langle x_0, x^{**} \rangle, \ \forall x^{**} \in L^q([a,b])$$

$$\Leftrightarrow \lim_{n \to +\infty} \int_{[a,b]} x_n(t) x^{**}(t) \mathrm{d}t = \int_{[a,b]} x_0(t) x^{**}(t) \mathrm{d}t, \ \forall x^{**} \in L^q([a,b]).$$

Example 4.3.5 There is a necessary and sufficient condition of "sequence $\{x_n\}\subset C([a,b])$ is weakly convergent into $x_0\in C([a,b])$", it is

(w)-$\lim\limits_{n\to+\infty} x_n=x_0 \Leftrightarrow$ (1) $\{\|x_n\|\}$ is bounded; (2) $\lim\limits_{n\to+\infty} x_n(t)=x_0(t)$, for each $t\in[a,b]$.

Example 4.3.6 In $L^p([a,b])$, $1<p<+\infty$, a necessary and sufficient condition of "sequence $\{x_n\}$ is weakly convergent to $x_0\in L^p([a,b])$" is:

(w)-$\lim\limits_{n\to+\infty} x_n=x_0 \Leftrightarrow$ (1) $\{\|x_n\|\}$ is bounded;

$$(2)\ \lim_{n\to+\infty}\int_{[a,c]} x_n(t)\,\mathrm{d}t=\int_{[a,c]} x_0(t)\,\mathrm{d}t,\quad \text{for each } c\in[a,b].$$

4.3.2 Bounded linear functionals on Hilbert spaces

1. Dual spaces, conjugate operators

For an inner product space $(X,(x,y))$ on number field $\mathbb{F}$, it can be regarded as a metric space, or a normed linear space, so that all theories, concepts, theorems in metric and normed spaces can be used for $(X,(x,y))$. We show those properties which are true only for Hilbert spaces in this section.

1) Riesz representation theorem

Riesz representation theorems for finite-dimensional non-singular metric linear spaces and inner product spaces are shown in Theorems 2.1.8 and 2.1.9. However, for a Hilbert space, Riesz representation theorem has the following form.

Theorem 4.3.6 (Riesz representation Theorem) Let $(X,(x,y))$ be a Hilbert space. Then, for any bounded linear functional $f\in X^*$ on X, there exists unique $x\in X$, such that $f(v)=(v,x)$, $\forall v\in X$, with the operator norm $\|f\|=\|x\|_X$ of f.

2) Self-dual and self-reflect properties of Hilbert spaces

Theorem 4.3.7 Let $(X,(x,y))$ be a Hilbert space. Then it is a self-dual and self-reflect space, i. e. $X=X^*=X^{**}$.

Proof By Riesz representation theorem, $f\in X^*$ and $x\in X$ are one-one by a mapping $J: X\to X^*$,

$$J: x\in X\leftrightarrow f\in X^*,\ \text{with } J(x)=f\equiv f_x, \tag{4.3.2}$$

it is an one-one mapping. Thus, for $x,y\in X$ and $\forall v\in X$, we have

$$\begin{aligned}\langle J(x+y),v\rangle &= (v,x+y)=(v,x)+(v,y)\\ &=\langle J(x),v\rangle+\langle J(y),v\rangle=\langle J(x)+J(y),v\rangle,\end{aligned}$$

this implies $J(x+y)=J(x)+J(y)$. On the other hand, $\forall v\in X$, $\forall\alpha\in\mathbb{F}$,

$$\langle J(\alpha x),v\rangle=(v,\alpha x)=\bar{\alpha}(v,x)=\bar{\alpha}\langle J(v),v\rangle=\langle\bar{\alpha}J(x),v\rangle,$$

so that $J(\alpha x)=\bar{\alpha}\,J(x)$.

If $\mathbb{F}=\mathbb{R}$, then $J: X\to X^*$ in (4.3.2) is a linear operator; If $\mathbb{F}=\mathbb{C}$, then $J: X\to X^*$ is a conjugate linear operator. Moreover, $\|J(x)\|_X=\|f\|=\|x\|_X$ implies $J: X\to X^*$ is equidistant. Then we conclude that, X is self-dual, $X=X^*$, in the sense of equidistant isomorphism, so that we may introduce an inner product on X^*. Hence, in the sense of equidistant isomorphism, denote $J^{-1}: X^*\to X$, we have $\forall f,g\in X^*\Rightarrow J^{-1}f$ with $J^{-1}g\in X$. Suppose that $(f,g)_{X^*}=(J^{-1}f,J^{-1}g)_X$, then $(X^*,(f,g)_{X^*})$ is a Hilbert space. This implies that X^* is self-dual in equidistant isomorphism sense, $X^*=(X^*)^*$. Finally, we conclude that $X=X^*=X^{**}$.

3) Conjugate operators on $\mathfrak{B}(X)$

For two real Hilbert spaces $(X,(x_1,x_2))$, $(Y,(y_1,y_2))$, as real inner product spaces, the conjugate operator $T^*\in L(Y^*,X^*)$ of a linear operator $T\in L(X,Y)$ is defined in Definition 2.2.3; and $T'\in L(Y,X)$ is defined in Definition 2.2.4. We discuss the conjugate operators on bounded linear operator space $\mathfrak{B}(X)\equiv\mathfrak{B}(X,X)$ of Hilbert space X in this section.

Definition 4.3.14 (conjugate operator on $\mathfrak{B}(X)$) Let $(X,(x,y))$ be a real Hilbert space, $\mathfrak{B}(X)$ be the bounded linear operator space from X to X, and $T\in\mathfrak{B}(X)$. If an operator $T': X\to X$ satisfies

$$(T'(y),x)=(y,T(x)),\quad \forall y\in X,\forall x\in X,$$

then T' is said to be **a conjugate operator of** T. (compare with Definition 2.2.4).

Note 1 T' is a bounded linear operator, $T'\in\mathfrak{B}(X)$.

Note 2 For Hilbert spaces X and Y, there are two definitions of conjugate operators of a bounded linear operator $T\in\mathfrak{B}(X,Y)$: $T^*\in\mathfrak{B}(Y^*,X^*)$ and $T'\in\mathfrak{B}(Y,X)\leftrightarrow\mathfrak{B}(Y^*,X^*)$. By virtue of the similar method in Chapter 2, we can prove that T^* and T' are topologically isomorphic each other. Particularly, in Hilbert space $(X,(x,y))$, we have that the conjugate operator $T^*\in\mathfrak{B}(X^*)\leftrightarrow\mathfrak{B}(X)$ and conjugate operator $T'\in\mathfrak{B}(X)$ are topologically isomorphic. Then **we agree on** that: in Hilbert space $(X,(x,y))$, operators $T'\in\mathfrak{B}(X)$ and $T^*\in\mathfrak{B}(X^*)$ are regarded as the same (isomorphic), and use the notation T^*, i. e. $T^*\equiv T'\in\mathfrak{B}(X)$ satisfies $(T^*(y),x)=(y,T(x))$, $\forall y\in X,\forall x\in X$.

4) Properties of conjugate operators in Hilbert spaces

Theorem 4.3.8 Let $(X,(x,y))$ be a real Hilbert space. Then the conjugate operator $T^*(\leftrightarrow T')\in\mathfrak{B}(X^*)(\leftrightarrow\mathfrak{B}(X))$ of $T\in\mathfrak{B}(X)$ satisfies

(1) $S,T\in\mathfrak{B}(X)\Rightarrow(S+T)^*=S^*+T^*$;

(2) $S\in\mathfrak{B}(X), \alpha\in\mathbb{R}\Rightarrow(\alpha S)^*=\alpha S^*$;

(3) $S, T\in\mathfrak{B}(X)\Rightarrow(T\circ S)^*=S^*\circ T^*$;

(4) $T\in\mathfrak{B}(X)\Rightarrow\|T^*\|=\|T\|$;

(5) $T\in\mathfrak{B}(X)$, if T^{-1} exists, then $(T^{-1})^*=(T^*)^{-1}$ and $\|(T^{-1})^*\|=\dfrac{1}{\|T^*\|}$;

(6) $T\in\mathfrak{B}(X)\Rightarrow(T^*)^*=T$.

Example 4.3.7 For n-dimensional Euclidean space $X=\mathbb{R}^n$, the following corresponding relation holds for a bounded linear operator $T\in\mathfrak{B}(\mathbb{R}^n)$

$$T\in\mathfrak{B}(\mathbb{R}^n)\leftrightarrow A\in\mathfrak{M}_{n\times n}.$$

If the $n\times m$-matrix A is $A=[\alpha_{ij}]_{n\times n}$, then T is

$$T: x=(\xi_1,\cdots,\xi_n)\in\mathbb{R}^n\to x'=(\xi'_1,\cdots,\xi'_n)\in\mathbb{R}^n$$

with $\xi'_k=\sum_{j=1}^{n}\alpha_{kj}\xi_j$, $k=1,\cdots,n$. Determine the dual operator T^* of T.

Solve $\forall y=(\eta_1,\cdots,\eta_n)\in\mathbb{R}^n$, it holds

$$(x, T^*y)=(Tx, y)=\sum_{k=1}^{n}\left(\sum_{j=1}^{n}\alpha_{kj\xi_j}\right)\bar{\eta}_k=\sum_{j=1}^{n}\xi_j\overline{\sum_{k=1}^{n}\bar{\alpha}_{kj}\eta_k}=\sum_{k=1}^{n}\xi_k\overline{\sum_{j=1}^{n}\bar{\alpha}_{jk}\eta_j},$$

thus, the dual operator T^* of T is determined by **the conjugate transposed matrix** $(\bar{A})^{\mathrm{T}}=([\bar{\alpha}_{ji}]_{n\times n})^{\mathrm{T}}$, i.e.,

$$T^*\in\mathfrak{B}(\mathbb{R}^n)\leftrightarrow(\bar{A})^{\mathrm{T}}\in\mathfrak{M}_{n\times n}.$$

Example 4.3.8 If T is an integral operator on Hilbert space $X=L^2([a,b])$ with kernel $K(t,s)$, i.e.

$$(T(x))(t)=\int_{[a,b]}K(t,s)x(s)\mathrm{d}s,\quad x\in L^2([a,b]), t\in[a,b],$$

where $K(t,s)$ is a real continuous function in $t, s\in[a,b]$ with $\int_{[a,b]\times[a,b]}|K(t,s)|^2\mathrm{d}t\mathrm{d}s<+\infty$. seek for T^*.

Solve We assert:

(1) $T: L^2([a,b])\to L^2([a,b])$ is a bounded linear operator;

(2) By the similar way in Example 4.3.6, we have:

$\Rightarrow\exists y(t)\in L^2([a,b])$, such that $f(z)=\int_{[a,b]}y(t)z(t)\mathrm{d}t$, $\forall z\in L^2([a,b])$

(by definition)

$$\Rightarrow(x, T^*y)=(Tx, y)=\int_{[a,b]}\overline{y(t)}\left[\int_{[a,b]}K(t,s)x(s)\mathrm{d}s\right]\mathrm{d}t\quad\text{(by Fubini theorem)}$$

$$=\int_{[a,b]}x(s)\left[\int_{[a,b]}\overline{K(t,s)}y(t)\mathrm{d}t\right]\mathrm{d}s=\int_{[a,b]}x(t)\left[\int_{[a,b]}\overline{K(s,t)y(s)\mathrm{d}s}\right]\mathrm{d}t$$

$$\Rightarrow (T^* y)(t)=\int_{[a,b]} \overline{K(s,t)}\, y(s)\,ds$$

$\Rightarrow T^*$ is the integral operator with kernel $K^*(t,s)=\overline{K(s,t)}$.

Example 4.3.9 Find conjugate operator T^* of Volterra integral operator T on Hilbert space $X=L^2([0,1])$:

$$(T(x))(t)=\int_{[0,t]} x(s)\,ds,\quad x\in L^2([0,1]),t\in[0,1].$$

Solve The kernel of Volterra integral operator T is

$$K(t,s)=\begin{cases}1, & 0\leqslant t\leqslant 1, 0\leqslant s\leqslant t\\ 0, & 0\leqslant t\leqslant 1, t< s\leqslant 1\end{cases},$$

then by Example 4.3.8, the kernel of conjugate operator T^* is $K^*(t,s)=\overline{K(s,t)}=\begin{cases}1, & 0\leqslant t\leqslant 1, t\leqslant s\leqslant 1\\ 0, & 0\leqslant t\leqslant 1, 0\leqslant s< t\end{cases}$, so that we have

$$(T^*(x))(t)=\int_{[t,1]} x(s)\,ds,\quad x\in L^2([0,1]),t\in[0,1].$$

2. Self-conjugate, unitary, and normal operators in Hilbert spaces

Let $(X,(x,y))$ be a real Hilbert space, $\mathfrak{B}(X)\equiv\mathfrak{B}(X,X)$ be the operator space.

Definition 4.3.15 (self-conjugate, unitary, and normal operators) For $T\in\mathfrak{B}(X)$,

(1) if $T=T^*$, then T is said to be a **self-conjugate operator**, or **Hermite operator** on $(X,(x,y))$;

(2) if T is one-one, and $T^*=T^{-1}$, then T is said to be **a unitary operator** on $(X,(x,y))$;

(3) if $T^*T=TT^*$, then T is said to be a **normal operator** on $(X,(x,y))$.

Example 4.3.10 The operator T is defined on Hilbert space $X=L^2([0,1])$ by

$$(T(x))(t)=tx(t),\quad x\in L^2([0,1]),t\in[0,1].$$

Prove that T is a Hermite operator.

Proof It is clear that the operator T is linear defined on $L^2([0,1])$ with range $L^2([0,1])$; and

$$\int_{[0,1]}|(T(x))(t)|^2\,dt=\int_{[0,1]}|tx(t)|^2\,dt\leqslant\int_{[0,1]}|x(t)|^2\,dt$$
$$=\|x\|^2_{L^2([0,1])}<+\infty$$

implies $T: L^2([0,1])\to L^2([0,1])$ is bounded, so that $T\in\mathfrak{B}(L^2([0,1]))$.

Moreover, (Tx,x) is the inner product of $L^2([0,1])$, thus

$$(Tx,x)=\int_{[0,1]} T(x)(t)\cdot x(t)\,dt=\int_{[0,1]} tx(t)\cdot x(t)\,dt$$

$$=\int_{[0,1]} x(t)\cdot tx(t)\mathrm{d}t=(x,Tx),$$

so that $T: L^2([0,1])\to L^2([0,1])$ is a Hermite operator.

In the matrix space $\mathfrak{M}_{n\times n}(\mathbb{C})$, we have the following theorem.

Theorem 4.3.9 Let $\mathfrak{M}_{n\times n}(\mathbb{C})$ be a matrix space, $T\in\mathfrak{B}(\mathbb{C})$ be a bounded linear operator from $\mathbb{C}^n$ to $\mathbb{C}^n$. Suppose that $T\leftrightarrow A\in\mathfrak{M}_{n\times n}(\mathbb{C})$, $T^*\leftrightarrow B\in\mathfrak{M}_{n\times n}(\mathbb{C})$.

(1) if T is self-conjugate, then $A\in\mathfrak{M}_{n\times n}(\mathbb{C})$ is a self-conjugate matrix, i. e. , $(\bar{A})^{\mathrm{T}}=A$.

(2) if T is unitary, then $A\in\mathfrak{M}_{n\times n}(\mathbb{C})$ is a unitary matrix, i. e. , $(\bar{A})^{\mathrm{T}}=A^{-1}$.

(3) if T is normal, then $A\in\mathfrak{M}_{n\times n}(\mathbb{C})$ is a normal matrix, i. e. , $(\bar{A})^{\mathrm{T}}A=A(\bar{A})^{\mathrm{T}}$.

(4) if T is self-conjugate, then $A\in\mathfrak{M}_{n\times n}(\mathbb{R})$ is a symmetric matrix, i. e. , $A^{\mathrm{T}}=A$.

(5) if T is unitary, then $A\in\mathfrak{M}_{n\times n}(\mathbb{R})$ is an orthogonal matrix, i. e. , $A^{\mathrm{T}}=A^{-1}$.

Theorem 4.3.10 Let $(X,(x,y))$ be a Hilbert space. We have

(1) if $T\in\mathfrak{B}(X)$, then T is a Hermitian operator, if and only if

$$(Tx,y)=(x,Ty),\quad x,y\in X;$$

Also, if and only if for any $x\in X$, the inner product $(Tx,x)\in\mathbb{R}$; then in this case, a Hermitian operator T is called a symmetry operator;

(2) if $T\in\mathfrak{B}(X)$ is a Hermitian operator, and $m=\inf\limits_{\|x\|_X=1}(Tx,x)$, $M=\sup\limits_{\|x\|_X=1}(Tx,x)$, then the norm of T is $\|T\|=\max\{|m|,|M|\}$; and holds

$$-\|T\|(x,x)\leqslant(Tx,x)\leqslant\|T\|(x,x);$$

(3) if $T\in\mathfrak{B}(X)$ is a Hermitian operator, then ①all eigenvalues of T are real; ②two different eigenvectors of T, which are corresponding to two different eigenvalues, are mutually orthogonal;

(4) if $T\in\mathfrak{B}(X)$ is a Hermitian operator, then the spectrum set $\sigma(T)\subset\mathbb{R}$ of T is a closed set.

The above three kinds of operators on Hilbert spaces have many important properties, similar to Theorems 2.2.11—2.2.13; and more other properties; readers can find them in many references.

Exercise 4

1. Let (X,ρ) be a metric space. Prove: $\rho_1(x,y)=\dfrac{\rho(x,y)}{1+\rho(x,y)}$ is a distance on X.

2. The state "ε-δ" language for $\lim\limits_{n\to\infty}x_n=x_0$ is in metric space (X,ρ).

3. Prove: In a metric space (X,ρ), if the limit of a sequence $\{x_n\}$ exists, then it is unique.

4. Let $f: X \to X_1$ be a mapping from metric space (X,ρ) to metric space (X_1,ρ_1). Prove: f is continuous at $x_0 \in X$, if and only if for any sequence $\{x_n\} \subset X$ which is convergent to $x_0 \in X$, it holds that $\{f(x_n)\}$ converges into $f(x_0)$.

5. Prove: any Cauchy sequence in metric space (X,ρ) is convergent. Give anti-example to show the converse proposition is not true.

6. In metric space (X,ρ), two Cauchy sequences $\xi = \{x_n\}$ and $\eta = \{y_n\}$ are called equivalent, if the limit $\lim\limits_{n\to+\infty} \rho(x_n, y_n) = 0$ holds, denoted by "$\xi = \{x_n\} \sim \eta = \{y_n\}$". Prove: "$\xi \sim \eta$" is an equivalence relation.

7. Prove: "$\rho_0(\xi,\eta) = \lim\limits_{n\to+\infty} \rho(x_n, y_n)$" satisfies that (1) if ξ,η are in the same equivalent class, then $\rho_0(\xi,\eta) = 0$; otherwise, $\rho_0(\xi,\eta) \neq 0$; (2) $\rho_0(\xi,\eta)$ satisfies the inequality of distance.

8. Let $X = \{x = (\xi_1, \xi_2, \cdots, \xi_k, 0, 0, \cdots): \xi_j \in \mathbb{R}, 1 \leqslant j \leqslant k, k \in \mathbb{N}\}$. Prove: X is a subspace of l^p, but in the distance $\rho(x,y) = \left(\sum\limits_{j=1}^{\infty} (x_j - y_j)^p\right)^{\frac{1}{p}}$ of l^p, it is not complete; Moreover, prove that l^p is the completion of X.

9. Let $X = \mathbb{Z}$ with distance $d(m,n) = |m-n|$. Prove: $(\mathbb{Z}, d(m,n))$ is a complete metric space.

10. Define a distance $d(x,y) = |\arctan x - \arctan y|$ on $\mathbb{R}$, prove: the space (X,d) is not complete.

11. Prove: $C([a,b])$ with distance $\rho(f,g) = \left(\int_a^b |f(x) - g(x)|^p \mathrm{d}x\right)^{\frac{1}{p}}$ is not complete; Moreover, the space $(L^p([a,b]), \rho(f,g))$ is the completion of $C([a,b])$ in $\rho(f,g)$.

12. Prove: **the closed ball box theorem** in metric space (X,ρ) "If $K_n = \overline{B(x_n, r_n)} \subset X$ is a decreasing closed ball sequence $K_1 \supset K_2 \supset \cdots \supset K_n \supset \cdots$ in X, and $\lim\limits_{n\to+\infty} r_n = 0$ for radius sequence, then there exists unique $x_0 \in X$, such that it holds $x_0 \in \overline{B(x_n, r_n)}$ for any $n \in \mathbb{N}$".

13. Prove: In a metric space (X,ρ), it holds (1) A totally bounded subset A is bounded; (2) A totally bounded subset A has a finite ε-net $B = \{x_1, \cdots, x_n\} \subset A$.

14. Prove: If $(X, \|x\|_X)$ is a locally compact normed linear space, then the unit sphere $S = \{x \in X: \|x\|_X = 1\}$ is a compact set.

15. Prove: The continuous mapping from metric space to metric space keeps compactness.

16. Prove "the finite intersection properties": If $\{F_j\}_{j\in I}$ is a closed set sequence in a

metric space with the property that any finite closed sets have nonempty intersection, then the intersection of the sequence $\bigcap_{j\in I} F_j$ is nonempty, i.e. $\bigcap_{j\in I} F_j \neq \varnothing$".

17. Prove: $\mathbb{N}$ and $\mathbb{Z}$ are not compact subspaces of $\mathbb{R}$; Unit closed interval $[0,1]$ is a compact subspace in $\mathbb{R}$.

18. Prove: For any $x=(\xi_1,\xi_2)\in\mathbb{R}^2$, we have $\|x\|_1=|\xi_1|+|\xi_2|$, $\|x\|_2=(|\xi_1|^2+|\xi_2|^2)^{\frac{1}{2}}$, $\|x\|_3=\max\{|\xi_1|,|\xi_2|\}$ are norms on $\mathbb{R}^2$, and they are equivalent.

19. Let $(X,\|x\|)$ be n-dimensional normed linear space, $\{e_1,\cdots,e_n\}$ be a base of X. For any $x\in X$, let $x=\sum_{j=1}^{n}\xi_j e_j$, $\xi_j\in\mathbb{F}$. Prove: T: $T(x)=(\xi_1,\cdots,\xi_n)$ is an isomorphic mapping from X to $\mathbb{F}^n$.

20. Prove: If $(X,\|x\|_X)$ is a normed linear space, then $(X,\|x\|_X)$ is a locally compact space, if and only if any bounded closed subset in X is compact.

21. Prove: The inner product (x,y) of inner product space $(X,(x,y))$ is a continuous function of x and y.

22. Prove: $x\perp y$, if and only if for all real $a\in\mathbb{R}$, it holds $\|x+ay\|=\|x-ay\|$, in a inner product space $(X,(x,y))$, where $\|x\|$ is determined by inner product (x,y).

23. In $\mathbb{R}^2$, if (a) $M=\{x\in\mathbb{R}^2: x=(\xi_1,\xi_2)\neq 0\}$, (b) $M=\{x_1,x_2\}$ is a linearly independent set in $\mathbb{R}^2$. Look for $M^{\perp}$.

24. Let $(X,(x,y))$ be a Hilbert space, $\{e_n\}\subset X$ be an orthonormal system in X. Prove: for any $x\in X$, it holds Bessel inequality $\sum_{n=1}^{+\infty}|(x,e_n)|^2\leqslant\|x\|^2$.

25. Prove: The operator norm of an operator has the equivalent representations:

$$\|T\|=\sup\{\|T(x)\|_Y:\|x\|_X\leqslant 1,\forall x\in\mathfrak{D}_T\}$$

and

$$\|T\|=\sup\{\|T(x)\|_Y:\|x\|_X=1,\forall x\in\mathfrak{D}_T\}.$$

26. Let T: $X\to Y$ be a linear operator from normed linear space $(X,\|\cdot\|_X)$ to normed linear space $(Y,\|\cdot\|_Y)$. Prove: if T is continuous at $x=0\in X$, then T is continuous on X.

27. Let $C[a,b]$ be the set of all continuous functions on $[a,b]$. With norm $\|f\|=\max_{x\in[a,b]}|f(x)|$, this $C[a,b]$ is a Banach space. Define an operator $J(f)=\int_a^b f(x)\,dx$, $\forall f\in C[a,b]$. Prove: $J: C[a,b]\to\mathbb{R}$ is a continuous linear functional on $C[a,b]$.

28. Let $(X,\|x\|_X)$ be a normed linear space, $(Y,\|y\|_Y)$ be a Banach space. Prove: the bounded linear operator space $(\mathfrak{B}(X,Y),\|T\|)$ is a Banach space with operator norm $\|T\|$.

29. Let $(\mathfrak{B}(X,Y), \|T\|)$ be a bounded linear operator space from Banach space $(X, \|x\|_X)$ to Banach space $(Y, \|y\|_Y)$. If a subset family $W \subset \mathfrak{B}(X,Y)$ satisfies $\sup_{T\in W}\|T(x)\|_Y < +\infty$, prove that: $\|x\|_W = \|x\|_X + \sup_{T\in W}\|T(x)\|_Y$ is a norm on X.

30. Let X, Y be normed linear spaces, $T: X \to Y$ be a closed operator. (1) Prove: the image set $A = T(C)$ of compact subset $C \subset X$ is a closed set in Y; (2) Prove: the inverse-image set $B = T^{-1}(K)$ of compact subset $K \subset Y$ is a closed set in X.

31. Prove: If X, Y are normed linear spaces, a bounded linear operator sequence $\{T_n\} \subset \mathfrak{B}(X,Y)$ uniformly converges into T, then $(s)\lim_{n\to+\infty} T_n = T$.

32. Let $(X, \|\cdot\|)$ be a normed linear space, $Z \subset X$ be a subspace, $T: X \to X$ be a linear operator. If $T(Z) \subset Z$, then Z is called an invariant subspace of T. Prove: the eigenvector space of T is an invariant space of T.

33. Let $\{e_k\}$ be a complete orthonormal sequence in a separable Hilbert space $(H, (x,y))$, an operator $T: \{e_n\} \to \{e_n\}$ is determined by $Te_k = e_{k+1}, k = 1, 2, \cdots$. Look for the extension $\widetilde{T}: H \to H$ of T, and determine an invariant space of T; Moreover, prove that T does not have eigenvalue.

34. Let A be a complex algebra Banach space with unit. If for $x \in A$, there exist y, $z \in A$ with $yx = e, xz = e$. Prove: x is invertible, and holds $y = z = x^{-1}$.

35. Let $(H, (x,y))$ be a Hilbert space, $T: H \to H$ be a bounded linear injective operator with T^{-1} bounded. Prove: $(T^*)^{-1}$ exists, and holds $(T^*)^{-1} = (T^{-1})^*$.

36. Let S, T be linear operators on Hilbert space $(H, (x,y))$. If there exists a unitary operator U on H, such that $S = UTU^{-1} = UTU^*$, then we say "S and T are u-equivalent". Prove: if S and T are u-equivalent, and T is self-conjugate operator, then so is S.

37. Prove: Theorem 4.3.9 holds:

(1) If T is Hermitian operator, then $A \in \mathfrak{M}_{n\times n}(\mathbb{C})$ is a Hermitian matrix, i.e., $(\overline{A})^{\mathrm{T}} = A$, and holds $\|(\overline{A})^{\mathrm{T}}\| = \|A\|$.

(2) If T is an unitary operator, then $A \in \mathfrak{M}_{n\times n}(\mathbb{C})$ is a unitary matrix, i.e., $(\overline{A})^{\mathrm{T}} = A^{-1}$, and holds $\|(\overline{A})^{\mathrm{T}}\| = \|A^{-1}\| = \dfrac{1}{\|A\|}$.

(3) If T is normal operator, then $A \in \mathfrak{M}_{n\times n}(\mathbb{C})$ is a normal matrix, i.e., $(\overline{A})^{\mathrm{T}}A = A(\overline{A})^{\mathrm{T}}$.

(4) If T is a Hermitian operator, then $A \in \mathfrak{M}_{n\times n}(\mathbb{R})$ is a symmetry matrix, i.e., $A^{\mathrm{T}} = A$, and $\|A^{\mathrm{T}}\| = \|A\|$.

(5) If T is a unitary operator, then $A \in \mathfrak{M}_{n\times n}(\mathbb{R})$ is an orthogonal matrix, i.e., $A^{\mathrm{T}} = A^{-1}$, and $\|A^{\mathrm{T}}\| = \|A^{-1}\|$.

Chapter 5

Distribution Theory

Fourier Analysis is a "corner stone" of all scientific areas; as a quantifiable evaluation method, it plays fundamental and important role.

In 1822, French mathematician, physicist J. Fourier published "Heat analysis theory", he decomposed the solutions of heat equations with initial sine waves into "harmonic waves" with different frequencies, and thus established the foundation of Fourier analysis. Then, Fourier analysis was getting developing forward in theory, and had more and more applications, infiltrating into mathematics, physics, astronomy, engineering technology, medicine sciences, etc. Until the distribution theory (or "generalized function") has been appeared in 50s of 20 century, Fourier analysis theory has been into perfection on Euclidean spaces.

In the last decennium of 20 century, "wavelet analysis" has been developing as a new branch of Fourier analysis. Meanwhile, "the non-linear science" with "centers" of chaos, fractal, and soliton has been proposed by many scientifsts. All these show the bright future for the highly developing of Fourier analysis harmonic analysis in 21 century.

For this chapter, we refer to [4], [5], [8], [10], [18].

Its frame is as follows.

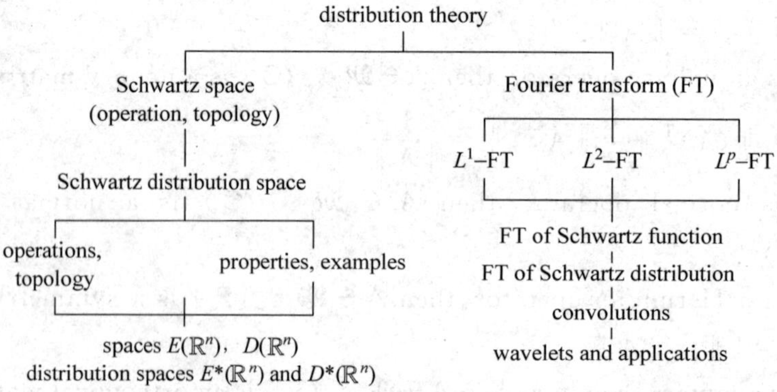

5.1 Schwartz Space, Schwartz Distribution Space

5.1.1 Schwartz space

1. Test function class

Take n-dimensional Euclidean space $\mathbb{R}^n = \{x=(x_1,\cdots,x_n): x_j \in \mathbb{R}, j=1,2,\cdots,n\}$, $n \in \mathbb{Z}^+$, as the basic space; let $\Omega \subseteq \mathbb{R}^n$ be an open set, $f: \Omega \to \mathbb{R}$ be a real-valued function on Ω. And let

$$C^\infty(\Omega) = \{f: \Omega \to \mathbb{R}, \quad f \text{ is any-order continuous derivable on } \Omega\};$$

$$C_C^\infty(\Omega) = \{f \in C^\infty(\Omega): \text{supp } f \text{ is the compact set in } \Omega\}.$$

Definition 5.1.1 (**test function class, Schwartz function class**) The set

$$\mathbb{S}(\mathbb{R}^n) = \{\varphi \in C^\infty(\mathbb{R}^n): |x^\beta \partial^p \varphi(x)| \leqslant M_{\beta,p}, \forall \beta, p \in \mathbb{N}^n\}$$

is said to be **the test function class**, or **Schwartz function class**, where $M_{\beta,p} > 0$ is a constant depending on β, p; and $x^\beta \equiv x_1^{\beta_1} \cdots x_n^{\beta_n}$,

$$x = (x_1,\cdots,x_n) \in \mathbb{R}^n, \quad \beta = (\beta_1,\cdots,\beta_n) \in \mathbb{N}^n;$$

$$\partial^p \varphi(x) \equiv \frac{\partial^p \varphi(x)}{\partial x^p} \equiv \frac{\partial^{p_1+\cdots+p_n} \varphi(x)}{\partial x_1^{p_1} \cdots \partial x_n^{p_n}}, \quad p = (p_1,\cdots,p_n) \in \mathbb{N}^n.$$

2. Schwartz space

We endow operations on Schwartz function class $\mathbb{S}(\mathbb{R}^n)$ such that it is a linear space on number field $\mathbb{F}$

addition $+$: $\varphi, \psi \in \mathbb{S}(\mathbb{R}^n) \Rightarrow (\varphi+\psi)(x) = \varphi(x) + \psi(x), \forall x \in \mathbb{R}^n$;

number product $\alpha\,\cdot$: $\varphi \in \mathbb{S}(\mathbb{R}^n), \alpha \in \mathbb{F} \Rightarrow (\alpha\varphi)(x) = \alpha\varphi(x), \forall x \in \mathbb{R}^n$; thus, $(\mathbb{S}(\mathbb{R}^n), +, \alpha\,\cdot)$ is **a linear space on** $\mathbb{F}$.

Endow $r: X \to [0, +\infty)$, called **a semi-norm**, satisfying

(1) $r(\alpha x) = |\alpha| r(x), \forall x \in X, \alpha \in \mathbb{F}$, (absolutely homogeneity)

(2) $r(x+y) \leqslant r(x) + r(y), \forall x, y \in X$, (sub-additive property)

on $\mathbb{S}(\mathbb{R}^n)$, denoted by $(\mathbb{S}(\mathbb{R}^n), +, \alpha\,\cdot, r)$, then it becomes **a topological linear space** (see Definition 3.4.13). In fact, $\forall \varphi \in \mathbb{S}(\mathbb{R}^n)$, define a mapping $r_{\beta,p}: \mathbb{S}(\mathbb{R}^n) \to [0, +\infty)$ by

$$r_{\beta,p}(\varphi) = \sup_{x \in \mathbb{R}^n} |x^\beta \partial^p \varphi(x)|, \quad \beta, p \in \mathbb{N}^n,$$

then $\{r_{\beta,p}(\varphi): \beta, p \in \mathbb{N}\}$ is a semi-norm family on $\mathbb{S}(\mathbb{R}^n)$, and easy to prove that

$$(\mathbb{S}(\mathbb{R}^n), +, \alpha\,\cdot, \{r_{\beta,p}(\varphi)\}_{\beta,p \in \mathbb{N}^n})$$

is a topological linear space. (compare the definitions of "semi-norm" and "norm", the latter one satisfies "$\| x \| = 0 \Leftrightarrow x = 0$", but the former one does not. However, by virtue of endowing a semi-norm on a set X, such that (X, r) becomes a topological space).

Theorem 5.1.1 (structure of $\mathbb{S}(\mathbb{R}^n)$) The space $(\mathbb{S}(\mathbb{R}^n), +, \alpha \cdot, \{r_{\beta,p}(\varphi)\}_{\beta,p\in\mathbf{N}^n})$ is a T_2-type, complete, self-reflect, metrizable topological space with countable semi-norm family.

Theorem 5.1.2 (equivalent topology of $\mathbb{S}(\mathbb{R}^n)$) For $(\mathbb{S}(\mathbb{R}^n), +, \alpha \cdot, \{r_{\beta,p}(\varphi)\}_{\beta,p\in\mathbf{N}^n})$, its topology structure can be endowed by the following way: for each sequence $\{\varphi_j\} \subset \mathbb{S}(\mathbb{R}^n)$, $\varphi_j \xrightarrow{\mathbb{S}(\mathbb{R}^n)} 0$, if and only if $\forall \beta, p \in \mathrm{N}^n$, $\lim\limits_{j\to+\infty} x^\beta \partial^p \varphi_j(x) = 0, x \in \mathbb{R}^n$, uniformly.

Now, we have more two ways to define topology of a set: ① give "semi-norm family"; ② give "necessary and sufficiency conditions of a zero sequence (sequence with limit zero)".

Example 5.1.1 $C_C^\infty(\mathbb{R}^n) \subset \mathbb{S}(\mathbb{R}^n)$.

Example 5.1.2 $\varphi(x) = \mathrm{e}^{-\alpha|x|^2} \in \mathbb{S}(\mathbb{R}^n), \alpha > 0$.

5.1.2 Schwartz distribution space

1. Schwartz distributions, Schwartz distribution space

The dual (conjugate) space of Schwartz space $\mathbb{S}(\mathbb{R}^n) \equiv (\mathbb{S}(\mathbb{R}^n), +, \alpha \cdot, \{r_{\beta,p}(\varphi)\}_{\beta,p\in\mathbf{P}^n})$, that is, the linear space consisted of all continuous linear functional on $\mathbb{S}(\mathbb{R}^n)$, ("continuity" is in the sense of semi-norm topology of $\mathbb{S}(\mathbb{R}^n)$), denoted by $S^*(\mathbb{R}^n) \equiv (\mathbb{S}(\mathbb{R}^n))^*$, called **Schwartz distribution space**, or **the generalized function space.** Element $T \in \mathbb{S}^*(\mathbb{R}^n)$ is said to be **a Schwartz distribution**, or **generalized function. The action** of a Schwartz distribution $T \in \mathbb{S}^*(\mathbb{R}^n)$ on $\varphi \in \mathbb{S}(\mathbb{R}^n)$ is denoted by $\langle T, \varphi \rangle = T(\varphi)$, $\forall \varphi \in \mathbb{S}(\mathbb{R}^n)$. It is clear that in Schwartz distribution space $(\mathbb{S}^*(\mathbb{R}^n), +, \alpha \cdot)$, the linear operations have been defined: $\forall T_1, T_2 \in \mathbb{S}^*(\mathbb{R}^n)$, the linear operation of $\alpha_1 T_1 + \alpha_2 T_2$ is

$$\langle \alpha_1 T_1 + \alpha_2 T_2, \varphi \rangle = \alpha_1 \langle T_1, \varphi \rangle + \alpha_2 \langle T_2, \varphi \rangle, \quad \forall \varphi \in \mathbb{S}(\mathbb{R}^n), \alpha_1, \alpha_2 \in \mathbf{F}.$$

2. More operations of Schwartz distributions

For Schwartz distributions, the following operations of distributions are always used.

(1) **distribution as a linear functional** $\forall T \in \mathbb{S}^*(\mathbb{R}^n)$,

$$\langle T, \alpha_1\varphi_1 + \alpha_2\varphi_2 \rangle = \alpha_1 \langle T, \varphi_1 \rangle + \alpha_2 \langle T, \varphi_2 \rangle, \quad \forall \varphi_1, \varphi_2 \in \mathbb{S}(\mathbb{R}^n), \alpha_1, \alpha_2 \in \mathbb{R};$$

i. e., each $T \in \mathbb{S}^*(\mathbb{R}^n)$ is a linear functional on Schwartz space $\mathbb{S}(\mathbb{R}^n)$.

(2) **zero distribution** If a distribution $T \in \mathbb{S}^*(\mathbb{R}^n)$ satisfies

$$\langle T,\varphi\rangle=0,\quad \forall\varphi\in \mathbb{S}(\mathbb{R}^n),$$

then $T\in\mathbb{S}^*(\mathbb{R}^n)$ is called a **zero distribution**, denoted by T, $T=0\in\mathbb{S}^*(\mathbb{R}^n)$; In fact, the zero distribution is zero element of linear space $\mathbb{S}^*(\mathbb{R}^n)$.

(3) **equal distributions** If distributions $T_1,T_2\in\mathbb{S}^*(\mathbb{R}^n)$ satisfies

$$\langle T_1-T_2,\varphi\rangle=0,\quad \forall\varphi\in \mathbb{S}(\mathbb{R}^n),$$

i. e., T_1-T_2 is a zero distribution, then distributions T_1 and T_2 are said to **equal each other**; denoted by $T_1=T_2$.

(4) **translation distribution** For distribution $T\in\mathbb{S}^*(\mathbb{R}^n)$ and $a=(a_1,\cdots,a_n)\in\mathbb{R}^n$, if there exists a distribution $U\in\mathbb{S}^*(\mathbb{R}^n)$ satisfying

$$\langle U,\varphi\rangle=\langle T,\varphi_{-a}\rangle,\quad \forall\varphi\in \mathbb{S}(\mathbb{R}^n),$$

then, $U\in\mathbb{S}^*(\mathbb{R}^n)$ is called a **translation distribution of** $T\in\mathbb{S}^*(\mathbb{R}^n)$, with $\varphi_{-a}(x)=\varphi(x+a)$, $\forall\varphi\in\mathbb{S}(\mathbb{R}^n)$, $\forall x\in\mathbb{R}^n$. Denoted $U\in\mathbb{S}^*(\mathbb{R}^n)$ by T_a, or $\tau_a T$, i. e., the translation distribution of $T\in\mathbb{S}^*(\mathbb{R}^n)$ is a distribution $\tau_a T=T_a\in\mathbb{S}^*(\mathbb{R}^n)$ satisfying

$$\langle\tau_a T,\varphi\rangle=\langle T,\tau_{-a}\varphi\rangle,\quad \forall\varphi\in \mathbb{S}(\mathbb{R}^n).$$

(5) **reflecting distribution** For distribution $T\in\mathbb{S}^*(\mathbb{R}^n)$, if there exists a distribution $V\in\mathbb{S}^*(\mathbb{R}^n)$ satisfying

$$\langle V,\varphi\rangle=\langle T,\tilde{\varphi}\rangle,\quad \forall\varphi\in \mathbb{S}(\mathbb{R}^n),$$

then, $V\in\mathbb{S}^*(\mathbb{R}^n)$ is called a **reflection of** $T\in\mathbb{S}^*(\mathbb{R}^n)$, with $\tilde{\varphi}(x)=\varphi(-x)$, and $(-x)=-(x_1,\cdots,x_n)=(-x_1,\cdots,-x_n)$; denoted $V\in\mathbb{S}^*(\mathbb{R}^n)$ by $\tilde{T}$, i. e., the reflecting distribution of $T\in\mathbb{S}^*(\mathbb{R}^n)$ is a distribution $\tilde{T}\in\mathbb{S}^*(\mathbb{R}^n)$ satisfying

$$\langle\tilde{T},\varphi\rangle=\langle T,\tilde{\varphi}\rangle,\quad \forall\varphi\in \mathbb{S}(\mathbb{R}^n).$$

(6) **multiplication of distribution and function** For distribution $T\in\mathbb{S}^*(\mathbb{R}^n)$ and function $f\in\mathbb{S}(\mathbb{R}^n)$, if there exists a distribution $W\in\mathbb{S}^*(\mathbb{R}^n)$ satisfying

$$\langle W,\varphi\rangle=\langle T,f\varphi\rangle,\quad \forall\varphi\in \mathbb{S}(\mathbb{R}^n),$$

then, $W\in\mathbb{S}^*(\mathbb{R}^n)$ is called **a multiplication of** $T\in\mathbb{S}^*(\mathbb{R}^n)$ **and** $f\in\mathbb{S}(\mathbb{R}^n)$, denoted $W\in\mathbb{S}^*(\mathbb{R}^n)$ by fT, i. e., the multiplication of distribution $T\in\mathbb{S}^*(\mathbb{R}^n)$ and function $f\in\mathbb{S}(\mathbb{R}^n)$ is a distribution $fT\in\mathbb{S}^*(\mathbb{R}^n)$ satisfying

$$\langle fT,\varphi\rangle=\langle T,f\varphi\rangle,\quad \forall\varphi\in \mathbb{S}(\mathbb{R}^n).$$

(7) **multiplication of distributions** For distributions $T,S\in\mathbb{S}^*(\mathbb{R}^n)$ satisfying $S\varphi$, $T\varphi\in\mathbb{S}(\mathbb{R}^n)$, if there exists a distribution $W\in\mathbb{S}^*(\mathbb{R}^n)$ satisfying

$$\langle W,\varphi\rangle=\langle T,S\varphi\rangle,\quad \forall\varphi\in \mathbb{S}(\mathbb{R}^n);$$

then, $W\in\mathbb{S}^*(\mathbb{R}^n)$ is called **a multiplication of** $T\in\mathbb{S}^*(\mathbb{R}^n)$ **and** $S\in\mathbb{S}^*(\mathbb{R}^n)$, denoted this $W\in\mathbb{S}^*(\mathbb{R}^n)$ by $S\cdot T\equiv T\cdot S$, i. e., the multiplication of distributions $T,S\in\mathbb{S}^*(\mathbb{R}^n)$ is a distribution $S \cdot T \in \mathbb{S}^*(\mathbb{R}^n)$ satisfying

$$\langle S \cdot T, \varphi \rangle = \langle T, S\varphi \rangle, \quad \forall \varphi \in \mathbb{S}(\mathbb{R}^n).$$

(8) **partials derivatives of distribution** For distribution $T \in \mathbb{S}^*(\mathbb{R}^n)$, if there exists a distribution $Z \in \mathbb{S}^*(\mathbb{R}^n)$ satisfying

$$\langle Z, \varphi \rangle = -\langle T, \partial_{x_1} \varphi \rangle, \quad \forall \varphi \in \mathbb{S}(\mathbb{R}^n),$$

with $\partial_{x_1}\varphi$ the partial derivative of φ with respect to x_1, then $Z \in \mathbb{S}^*(\mathbb{R}^n)$ is called **partial derivative of** $T \in \mathbb{S}^*(\mathbb{R}^n)$ **with respect to** x_1, denoted $Z \in \mathbb{S}^*(\mathbb{R}^n)$ by $\partial_{x_1} T$, i. e., a partial derivative of distribution $T \in \mathbb{S}^*(\mathbb{R}^n)$ with respect to x_1 is a distribution $\partial_{x_1}\varphi \in \mathbb{S}(\mathbb{R}^n)$ satisfying

$$\langle \partial_{x_1} T, \varphi \rangle = -\langle T, \partial_{x_1} \varphi \rangle, \quad \forall \varphi \in \mathbb{S}(\mathbb{R}^n).$$

Since $\varphi \in \mathbb{S}(\mathbb{R}^n)$, thus the right-hand side of above formula makes sense.

Similarly, higher-order partial derivatives can be defined, such as

$$\partial^p T \equiv \frac{\partial^p T}{\partial x^p} \equiv \frac{\partial^{p_1+\cdots+p_n} T}{\partial x_1^{p_1} \cdots \partial x_n^{p_n}}, \quad p = (p_1, \cdots, p_n) \in \mathbb{N}^n.$$

(9) **support of a distribution** Recall the definition of support of a function on $x \in \mathbb{R}^n$: $\operatorname{supp}\varphi = \overline{\{x \in \mathbb{R}^n : \varphi(x) \neq 0\}}$, $\forall \varphi(x) = \varphi(x_1, \cdots, x_n)$, then for a distribution, we have

Definition 5.1.2 (support of a distribution) For a distribution $T \in \mathbb{S}^*(\mathbb{R}^n)$, if $\forall \varphi \in \mathbb{S}(\mathbb{R}^n)$ with $\operatorname{supp}\varphi \subset \Omega$ implies $\langle T, \varphi \rangle = 0$, then $T \in \mathbb{S}^*(\mathbb{R}^n)$ is said to be **equal to zero on open set** $\Omega \subset \mathbb{R}^n$, denoted by $T|_\Omega = 0$.

Let $\{\Omega_\alpha\}_{\alpha \in I}$ be the open set family of all open subsets on which distribution $T \in \mathbb{S}^*(\mathbb{R}^n)$ equals zero, then the union $\bigcup_{\alpha \in I} \Omega_\alpha$ is the largest open set such that $T \in \mathbb{S}^*(\mathbb{R}^n)$ equals zero; and the compliment set $\mathbb{R}^n \setminus \bigcup_{\alpha \in I} \Omega_\alpha$ is said to be **the support of a distribution** T, denoted by $\operatorname{supp}T$, i. e.

$$\operatorname{supp}T = \complement\{x \in \mathbb{R}^n : \exists \text{ open set } U, \quad \text{s. t. } T|_U = 0, x \in U\}.$$

We have: ① $\operatorname{supp}T$ is a closed set in $\mathbb{R}^n$; and equivalently,

$$\operatorname{supp}T = \overline{\{x \in \mathbb{R}^n : T \neq 0 \text{ holds on neighborhood of } x\}};$$

② If $T \in \mathbb{S}^*(\mathbb{R})$, $\varphi \in \mathbb{S}(\mathbb{R})$, and $\operatorname{supp}T \cap \operatorname{supp}\varphi = \varnothing$, then $\langle T, \varphi \rangle = 0$.

3. Topologies of Schwartz distribution space

On linear space $(\mathbb{S}^*(\mathbb{R}^n), +, \alpha \cdot)$, usually, there are three kinds of topologies such that it is a linear space with three topologies.

Definition 5.1.3 (strong topology) Endow the operator norm $\| T \|_{\mathbb{S}^*(\mathbb{R}^n)}$ as a topology on Schwartz distribution space $(\mathbb{S}^*(\mathbb{R}^n), +, \alpha\cdot)$, then $(\mathbb{S}^*(\mathbb{R}^n), +, \alpha\cdot, \| T \|_{\mathbb{S}^*(\mathbb{R}^n)})$ is said to be **a topological linear space with strong topology** (or with norm topology), sometimes, denoted by $(\mathbb{S}^*(\mathbb{R}^n), +, \alpha\cdot, \tau_s)$.

Definition 5.1.4 (weak topology) Endow "zero continuous linear functional sequence" as a topology on Schwartz distribution space $(\mathbb{S}^*(\mathbb{R}^n), +, \alpha\cdot)$; i.e., in $(\mathbb{S}^*(\mathbb{R}^n), +, \alpha\cdot)$, for a continuous linear functional sequence $\{T_k\} \subset \mathbb{S}^*(\mathbb{R}^n)$, define $T_k \to 0 (k \to +\infty)$ by

$$T_k \xrightarrow{\tau_w} 0 \Leftrightarrow \langle \psi, x^\beta \partial^p T_k \rangle \to 0, \quad \forall \psi \in \mathbb{S}^{**}(\mathbb{R}^n), \forall \beta, p \in \mathbb{N}^n,$$

where $x^\beta \partial^p T_k$ is the multiplication of p-order partial derivative $\partial^p T_k$ and function x^β, denoted by $(\mathbb{S}^*(\mathbb{R}^n), +, \alpha\cdot, \tau_w)$, is said to be **a topological linear space with weak topology** on $\mathbb{S}^*(\mathbb{R}^n)$.

Definition 5.1.5 (weak* topology) Endow "zero continuous linear functional sequence" as a topology on Schwartz distribution space $(\mathbb{S}^*(\mathbb{R}^n), +, \alpha\cdot)$, i.e., in $(\mathbb{S}^*(\mathbb{R}^n), +, \alpha\cdot)$, for a continuous linear functional sequence $\{T_k\} \subset \mathbb{S}^*(\mathbb{R}^n)$, define $T_k \to 0 (k \to +\infty)$ by

$$T_k \xrightarrow{\tau_{w^*}} 0 \Leftrightarrow \langle x^\beta \partial^p T_k, \varphi \rangle \to 0, \quad \forall \varphi \in \mathbb{S}(\mathbb{R}^n), \forall \beta, p \in \mathbb{N}^n,$$

denoted by $(\mathbb{S}^*(\mathbb{R}^n), +, \alpha\cdot, \tau_{w^*})$, and is said to be **a topological linear space with weak* topology** on $\mathbb{S}^*(\mathbb{R}^n)$.

Theorem 5.1.3 For the topological structures of $(\mathbb{S}^*(\mathbb{R}^n), +, \alpha\cdot)$, we have

(1) The strong topology is equivalent to: for any bounded set $A \subset \mathbb{S}(\mathbb{R}^n)$, it holds

$$\langle x^\beta \partial^p T_k, \varphi \rangle \to 0, \quad \forall \beta, p \in \mathbb{N}^n, \ \forall \varphi \in A \ \textbf{uniformly};$$

(2) The weak* topology is equivalent to the weak topology;

(3) In the weak* topology, it holds:

$$\forall \varphi_k \xrightarrow{\mathbb{S}(\mathbb{R}^n)} 0 \Leftrightarrow \langle T, x^\beta \partial^p \varphi_k \rangle \to 0, \quad \forall T \in \mathbb{S}^*(\mathbb{R}^n), \quad \forall \beta, p \in \mathbb{N}^n.$$

Since $(\mathbb{S}(\mathbb{R}^n), +, \alpha\cdot, \{r_{\beta,p}(\varphi)\}_{\beta,p \in \mathbf{P}^n})$ is a self-reflect space, thus, the weak topology and weak* topology of space $(\mathbb{S}^*(\mathbb{R}^n), +, \alpha\cdot)$ are equivalent.

4. Examples of Schwartz distributions

Example 5.1.3 $L^1(\mathbb{R}^n) \subset \mathbb{S}^*(\mathbb{R}^n)$.

Since $\forall f \in L^1(\mathbb{R}^n)$, f can be regarded as a distribution, the action is determined by

$$\langle f, \varphi \rangle = \int_{\mathbb{R}^n} f(x)\varphi(x)\mathrm{d}x, \quad \forall \varphi \in \mathbb{S}(\mathbb{R}^n),$$

this $\langle f, \varphi \rangle$ is a continuous linear functional on $\mathbb{S}(\mathbb{R}^n)$, i.e., $\forall f \in L^1(\mathbb{R}^n)$ is a Schwartz distribution on $\mathbb{S}(\mathbb{R}^n)$.

Example 5.1.4 Heaviside function $H(x)=\begin{cases}1, & x>0\\ 0, & x<0\end{cases}$.

It is clear that $\langle H,\varphi\rangle=\int_{[0,+\infty)}\varphi(x)\mathrm{d}x$ is a continuous linear function on $\mathbb{S}(\mathbb{R})$, thus $H(x)\in\mathbb{S}^*(\mathbb{R})$.

Example 5.1.5 Dirac distribution δ.

Dirac distribution δ, a continuous linear functional $\delta\in\mathbb{S}^*(\mathbb{R})$ on $\mathbb{S}(\mathbb{R})$, is defined by

$$\langle\delta,\varphi\rangle=\varphi(0),\quad \forall\varphi\in\mathbb{S}(\mathbb{R}).$$

Linearity of Dirac distribution δ: $\forall\varphi_1,\varphi_2\in\mathbb{S}(\mathbb{R})$

$$\langle\delta,\alpha\varphi_1+\beta\varphi_2\rangle=(\alpha\varphi_1+\beta\varphi_2)(0)=\alpha\varphi_1(0)+\beta\varphi_2(0)=\alpha\langle\delta,\varphi_1\rangle+\beta\langle\delta,\varphi\rangle;$$

Continuity of Dirac distribution δ: $\forall\{\varphi_n\}\subset\mathbb{S}(\mathbb{R})$, $\varphi\in\mathbb{S}(\mathbb{R})$,

$$\varphi_n\xrightarrow{\mathbb{S}(\mathbb{R})}\varphi\Rightarrow|\langle\delta,\varphi_n\rangle-\langle\delta,\varphi\rangle|=|\varphi_n(0)-\varphi(0)|\to 0.$$

We thus obtain a continuous linear functional which could not be expressed as an integral.

Support of Dirac distribution δ: supp $\delta=\{0\}$, by definition, $\langle\delta,\varphi\rangle=\varphi(0)$.

Example 5.1.6 Derivative of Dirac distribution δ.

By the definition of derivatives of distributions, we have

$$\langle\delta',\varphi\rangle=-\langle\delta,\varphi'\rangle=-\varphi'(0),\quad \forall\varphi\in\mathbb{S}(\mathbb{R}).$$

Thus, δ is as a distribution, its derivative δ' is also a distribution, and the action of δ' on $\varphi\in\mathbb{S}(\mathbb{R})$ is $-\varphi'(0)$.

The p-order derivative of Dirac distribution $\delta\in\mathbb{S}^*(\mathbb{R})$ is

$$\langle D^p\delta,\varphi\rangle=(-1)^p\varphi^{(p)}(0),\quad \forall\varphi\in\mathbb{S}(\mathbb{R}), p\in\mathbb{N}.$$

Example 5.1.7 The derivative of Heaviside function $H(x)=\begin{cases}1, & x>0\\ 0, & x<0\end{cases}$.

By definition, it follows

$$\langle H',\varphi\rangle=-\langle H,\varphi'\rangle=-\int_0^{+\infty}\varphi'(x)\mathrm{d}x=\varphi(0)=\langle\delta,\varphi\rangle,\quad \forall\varphi\in\mathbb{S}(\mathbb{R}),$$

thus, $H'=\delta$.

5. Classification of Schwartz distributions

Definition 5.1.6 (regular, singular) For a continuous linear functional $T\in\mathbb{S}^*(\mathbb{R}^n)$ on Schwartz space $\mathbb{S}(\mathbb{R}^n)$, we have:

(1) If the action of $T\in\mathbb{S}^*(\mathbb{R}^n)$ on $\varphi\in\mathbb{S}(\mathbb{R}^n)$ can be expressed as a Lebesgue integral

$$\langle T,\varphi\rangle=\int_{\mathbb{R}^n}f(x)\varphi(x)\mathrm{d}x\ ,\quad \forall\varphi\in\mathbb{S}(\mathbb{R}^n)$$

with $f \in L_{\mathrm{loc}}(\mathbb{R}^n)$, a locally integrable function on $\mathbb{R}^n$ (i. e. f is Lebesgue integrable on any finite cubes in $\mathbb{R}^n$), then T is said to be **a regular distribution**;

(2) If distribution $T \in \mathbb{S}^*(\mathbb{R}^n)$ is not regular, then it is said to be a **singular distribution.**

Since $\mathbb{S}(\mathbb{R}^n) \subset L^p(\mathbb{R}^n) \subset \mathbb{S}^*(\mathbb{R}^n)$, then $f \in \mathbb{S}(\mathbb{R}^n)$ and $f \in L^p(\mathbb{R}^n)$ all determine regular distributions. However, the Dirac δ is a singular distribution.

Remark We explain the rationality for defining derivative of Schwartz distribution in the subsection 2.

In fact, by $f \in \mathbb{S}(\mathbb{R}) \subset \mathbb{S}^*(\mathbb{R})$, take $f \in \mathbb{S}(\mathbb{R})$, then $f' \in \mathbb{S}(\mathbb{R})$, and f, f' both are regular Schwartz distribution. Thus, $\forall \varphi \in \mathbb{S}(\mathbb{R})$, the integration by parts shows:

$$\langle f', \varphi \rangle = \int_{-\infty}^{+\infty} f'(x)\varphi(x)\,\mathrm{d}x = f(x)\varphi(x)\Big|_{-\infty}^{+\infty} - \int_{-\infty}^{+\infty} f(x)\varphi'(x)\,\mathrm{d}x = -\langle f, \varphi' \rangle$$

and implies $\langle f', \varphi \rangle = -\langle f, \varphi' \rangle$. Hence, if a function f would not have derivative, or there is no concept of derivative for distributions, then we may use the formula of usual derivative to define "derivative", such as

$$\langle \partial_{x_1} T, \varphi \rangle = -\langle T, \partial_{x_1} \varphi \rangle, \quad \forall \varphi \in \mathbb{S}(\mathbb{R}^n),$$

compare with $\langle f', \varphi \rangle = -\langle f, \varphi' \rangle$.

5.1.3 Spaces $E(\mathbb{R}^n)$, $D(\mathbb{R}^n)$ and their distribution spaces

We have defined and discussed Schwartz space $\mathbb{S}(\mathbb{R}^n)$, Schwartz distribution space $\mathbb{S}^*(\mathbb{R}^n)$ as well as their properties in 5.1.1 and 5.1.2. However, there are other function spaces and their distribution spaces in mathematics and other scientific fields. We introduce $E(\mathbb{R}^n)$, $D(\mathbb{R}^n)$ and $E^*(\mathbb{R}^n)$, $D^*(\mathbb{R}^n)$ in this section.

1. Space $E(\Omega) \equiv C^\infty(\Omega)$ and its distribution space $E^*(\Omega)$

Let $\Omega \subseteq \mathbb{R}^n$ be an open set in $\mathbb{R}^n$, and $O(0, \cdots, 0) \in \Omega$ be the origin point of $\mathbb{R}^n$; Let

$$E(\Omega) \equiv C^\infty(\Omega) = \{\partial^\beta \varphi \in C(\Omega):\ \forall \beta \in \mathbb{N}^n\}$$

be the set of all any-order continuous differentiable functions on $\Omega \subset \mathbb{R}^n$. Usually, we take $\Omega = \mathbb{R}^n$, and consider $E(\mathbb{R}^n)$, $E^*(\mathbb{R}^n)$.

Endow the following natural topology on $E(\Omega)$: take any compact set sequence $K_j \subset \Omega \subseteq \mathbb{R}^n$, $K_j \subset \mathring{K}_{j+1}$, such that $\bigcup_{j=1}^{+\infty} K_j = \Omega$. For $\varphi \in E(\Omega)$, define semi-norms by

$$r_{m,j}(\varphi) = \sup_{\substack{x \in K_j \\ |\beta| \leqslant m}} |\partial^\beta \varphi(x)|, \quad m \in \mathbb{N}, \quad j \in \mathbb{Z}^+, \quad \beta \in \mathbb{N}^n,$$

then, $\{r_{m,j}(\varphi)\}, m\in\mathbb{N}, j\in\mathbb{Z}^+$ is a **semi-norm family** on $E(\Omega)$, such that $(E(\Omega), +, \alpha\cdot, \{r_{m,j}(\varphi)\})$ is a T_2-type, complete, reflect topological liner space on number field $\mathbb{F}$.

The dual space $E^*(\Omega)$ of $E(\Omega)$ is endowed with w^*-topology: for any functional sequence $\{f_k\}\subset E^*(\Omega)$,

① $f_k \xrightarrow{w^*} 0(k\to+\infty) \Leftrightarrow \langle f_k, \varphi\rangle \to 0(k\to+\infty), \forall \varphi\in E(\Omega)$;

② $f\in E^*(\Omega) \Leftrightarrow \exists c>0, \exists m\geqslant 0$, and exists compact set $K\subset\Omega$, such that

$$|\langle f,\varphi\rangle| \leqslant c \sup_{\substack{x\in K,\\ |\beta|\leqslant m}} |\partial^\beta \varphi(x)|, \quad \forall \varphi\in E(\Omega).$$

Theorem 5.1.4 (structure theorem of $E^*(\mathbb{R}^n)$) All distributions in space $(E^*(\mathbb{R}^n), +, \alpha\cdot, \tau_{w^*})$ have compact supports; and vice versa. i.e., $f\in E^*(\mathbb{R}^n) \Leftrightarrow \operatorname{supp} f = K$ is a compact set.

Example 5.1.8 If $f\in L_{\text{loc}}(\mathbb{R}^n)$, and $\operatorname{supp} f$ is compact, then $f\in E^*(\mathbb{R}^n)$.

For $f\in L_{\text{loc}}(\mathbb{R}^n)$, define a linear functional by:

$$\langle f,\varphi\rangle \equiv \int_{\mathbb{R}^n} f(x)\varphi(x)\,\mathrm{d}x, \quad \forall \varphi\in E(\mathbb{R}^n).$$

It is a linear functional $T\equiv T_f \leftrightarrow f$, and $\operatorname{supp} f$ is compact with $\operatorname{supp} T = \operatorname{supp} f$, thus $f\in E^*(\mathbb{R}^n)$.

Example 5.1.9 Dirac distribution δ: $\delta\in E^*(\mathbb{R})$.

2. Space $D(\Omega)\equiv C_C^\infty(\Omega)$ and its distribution space $D^*(\Omega)$

1) **The inductive limit topology on $D(\Omega)$**

Let $\Omega\subseteq\mathbb{R}^n$ be an open set in $\mathbb{R}^n$, with $O(0,\cdots,0)\in\Omega$ the origin point of $\mathbb{R}^n$; Let

$$D(\Omega) \equiv C_C^\infty(\Omega) = \{\varphi\in C^\infty(\Omega): \operatorname{supp}\varphi = K\subset\Omega \text{ is compact}\}$$

be the set of all any-order continuous differentiable functions with compact supports on $\Omega\subseteq\mathbb{R}^n$.

Endow so-called **the inductive limit topology** τ: For any sequence $\{\varphi_k\}\subset D(\Omega)$, $\varphi_k \xrightarrow{D(\Omega)} 0(k\to+\infty)$, if and only if ① $\exists$ compact set $K\subset\Omega$, such that $\operatorname{supp}\varphi_k\subset K, \forall k\in\mathbb{Z}^+$; ② $\partial^\beta\varphi_k(x)\to 0(k\to\infty)$ on K uniformly for $\forall \beta\in\mathbb{N}^n$.

Then, in the inductive limit topology τ, this $(D(\Omega), +, \alpha\cdot, \tau)$ is a T_2-type, complete, self-reflect topological linear space; However, it is impossible to be metrizable.

2) **The topology on distribution space $D^*(\Omega)$ of $D(\Omega)$**

Endow w^*-topology on $D^*(\Omega)$: For a linear functional sequence $\{f_k\}\subset D^*(\Omega)$, it holds

① $f_k \xrightarrow{w^*} 0(k\to+\infty) \Leftrightarrow \langle f_k,\varphi\rangle \to 0(k\to+\infty), \quad \forall \varphi\in D(\Omega)$;

② $\partial^\beta f_k \xrightarrow{w^*} 0\,(k\to+\infty) \Leftrightarrow \langle \partial^\beta f_k, \varphi\rangle \to 0\,(k\to+\infty),\ \forall \varphi\in D(\Omega)$.

Example 5.1.10 $f\in L_{\text{loc}}(\mathbb{R}^n)$ implies $f\in D^*(\mathbb{R}^n)$.

The linear functional is defined by

$$\langle f,\varphi\rangle = \int_{\mathbb{R}^n} f(x)\varphi(x)\,dx, \quad \forall \varphi \in D(\mathbb{R}^n).$$

Since $\varphi\in D(\mathbb{R}^n)$, then $\operatorname{supp}\varphi$ is compact, this implies the above integral makes sense. And by $f\in L_{\text{loc}}(\mathbb{R}^n)$, the integral determines a bounded linear functional on $D(\mathbb{R}^n)$, this implies that $f\in D^*(\mathbb{R}^n)$.

Example 5.1.11 $\delta\in D^*(\mathbb{R}^n)$.

It is clear that $\langle \delta,\varphi\rangle=\varphi(0)$, $\forall \varphi\in D(\mathbb{R}^n)$ by definition. On the other hand,

$$D(\mathbb{R}^n)\equiv C_C^\infty(\mathbb{R}^n)\subset \mathbb{S}(\mathbb{R}^n)\subset C^\infty(\mathbb{R}^n)\equiv E(\mathbb{R}^n),$$

thus $D^*(\mathbb{R}^n)\supset \mathbb{S}^*(\mathbb{R}^n)\supset E^*(\mathbb{R}^n)$ (left as exercise), and $\operatorname{supp}\delta=\{0\}$ gives $\delta\in E^*(\mathbb{R}^n)\subset D^*(\mathbb{R}^n)$.

5.2 Fourier Transform on $L^p(\mathbb{R}^n)$, $1\leqslant p\leqslant 2$

5.2.1 Fourier transformations on $L^1(\mathbb{R}^n)$

1. Preliminaries

We need the following three inequalities: Hölder, Schwartz, Minkovski inequalities.

Proposition 5.2.1 Let $f\in L^p(\mathbb{R}^n)$, $g\in L^q(\mathbb{R}^n)$, $1\leqslant p,q\leqslant+\infty$, $\frac{1}{p}+\frac{1}{q}=1$. Then Hölder inequality

$$\|fg\|_{L^1(\mathbb{R}^n)}\leqslant \|f\|_{L^p(\mathbb{R}^n)}\|g\|_{L^q(\mathbb{R}^n)}$$

holds. When $p=q=2$, it is Schwartz inequality

$$\|fg\|_{L^1(\mathbb{R}^n)}\leqslant \|f\|_{L^2(\mathbb{R}^n)}\|g\|_{L^2(\mathbb{R}^n)}.$$

Proposition 5.2.2 Let $f,g\in L^p(\mathbb{R}^n)$, $p\geqslant 1$. Then Minkovski inequality

$$\|f+g\|_{L^p(\mathbb{R}^n)}\leqslant \|f\|_{L^p(\mathbb{R}^n)}+\|g\|_{L^p(\mathbb{R}^n)}$$

holds.

For the proofs, we refer to [6], [17].

Proposition 5.2.3 The convolution $f*g(x)=\int_{\mathbb{R}^n} f(t)g(x-t)\,dt$ has the following properties:

(1) if $1\leqslant p,q\leqslant+\infty$, $\frac{1}{p}+\frac{1}{q}=1$, i.e., p and q are said to be conjugate number each

other, then

① $f\in L^p(\mathbb{R}^n), g\in L^q(\mathbb{R}^n), 1\leqslant p,q\leqslant+\infty \Rightarrow f*g\in C(\mathbb{R}^n)$, that is, $f*g(x)$ is a continuous function on $\mathbb{R}^n$, and holds $\|f*g\|_{C(\mathbb{R}^n)}\leqslant\|f\|_{L^p(\mathbb{R}^n)}\|g\|_{L^q(\mathbb{R}^n)}$;

② $f\in L^p(\mathbb{R}^n), g\in L^q(\mathbb{R}^n), 1<p,q<+\infty \Rightarrow f*g\in C_0(\mathbb{R}^n)$, where $C_0(\mathbb{R}^n)=\{f\in C(\mathbb{R}^n): \lim\limits_{|x|\to+\infty} f(x)=0\}$; and $\|f*g\|_{C(\mathbb{R}^n)}\leqslant\|f\|_{L^p(\mathbb{R}^n)}\|g\|_{L^q(\mathbb{R}^n)}$;

(2) $f\in L^1(\mathbb{R}^n)$, $g\in C_0(\mathbb{R}^n)\Rightarrow f*g\in C_0(\mathbb{R}^n)$, and $\|f*g\|_{C(\mathbb{R}^n)}\leqslant\|f\|_{L^1(\mathbb{R}^n)}\|g\|_{C(\mathbb{R}^n)}$;

(3) $f\in L^p(\mathbb{R}^n), 1\leqslant p<+\infty, g\in L^1(\mathbb{R}^n) \Rightarrow f*g\in L^p(\mathbb{R}^n)$, and $\|f*g\|_{L^p(\mathbb{R}^n)}\leqslant\|f\|_{L^p(\mathbb{R}^n)}\|g\|_{L^1(\mathbb{R}^n)}$.

2. Definitions of Fourier transformations

Definition 5.2.1 (Fourier transformation of $f\in L^1(\mathbb{R}^n)$) If $f\in L^1(\mathbb{R}^n)$, then the Fourier transformation of f is defined by

$$f^\wedge(\xi)=\int_{\mathbb{R}^n} f(x)\mathrm{e}^{-\mathrm{i}(x\cdot\xi)}\,\mathrm{d}x,\quad \xi\in\mathbb{R}^n,$$

with $x\cdot\xi=\sum\limits_{j=1}^{n}x_j\xi_j$ is the inner product of $x=(x_1,x_2,\cdots,x_n)\in\mathbb{R}^n$ and $\xi=(\xi_1,\xi_2,\cdots,\xi_n)\in\mathbb{R}^n$.

Note 1 Usually, Fourier transformation is defined as (see [4]) $f^\wedge(\xi)=\dfrac{1}{(\sqrt{2\pi})^n}\displaystyle\int_{\mathbb{R}^n} f(x)\mathrm{e}^{-\mathrm{i}(x\cdot\xi)}\,\mathrm{d}x$, $\xi\in\mathbb{R}^n$; and inverse Fourier transformation (Definition 5.2.2) by $g^\vee(x)=\dfrac{1}{(\sqrt{2\pi})^n}\displaystyle\int_{\mathbb{R}^n} g(\xi)\mathrm{e}^{\mathrm{i}(x\cdot\xi)}\,\mathrm{d}\xi$, $x\in\mathbb{R}^n$, these definitions make some formulas are symmetry. However we define Fourier transformation by $f^\wedge(\xi)=\displaystyle\int_{\mathbb{R}^n} f(x)\mathrm{e}^{-\mathrm{i}(x\cdot\xi)}\,\mathrm{d}x$, $\xi\in\mathbb{R}^n$, and the inverse Fourier transformation by $g^\vee(x)=\dfrac{1}{(2\pi)^n}\displaystyle\int_{\mathbb{R}^n} g(\xi)\mathrm{e}^{\mathrm{i}(x\cdot\xi)}\,\mathrm{d}\xi$, $x\in\mathbb{R}^n$, there is no essential different for two definitions only because the scalar constants are chosen differently in $\mathbb{R}^n$, and most properties of Fourier transformations are in similar forms.

Note 2 Regard $f(t)$ as a signal, t as in "time field" $t\in\mathbb{R}$, then Fourier transformation $f^\wedge(\xi)$ is the continuous frequency spectrum of f, when ξ is in "frequency field" $\mathbb{R}$; whereas, compare with Fourier series, the coefficient sequence $\{c_k\}_{k\in\mathbf{Z}}$ of $f(t)$

is the discrete frequency spectrum of f.

3. Properties of Fourier transformations

We state the theorems in case $n=1$, i. e., in space $L^1(\mathbb{R})\equiv L^1(\mathbb{R}^1)$, and they hold for case $n\,(n\geqslant 1)$ of space $L^1(\mathbb{R}^n)$.

1) Basic operation properties of Fourier transformations

Theorem 5. 2. 1 There are basic operation properties of Fourier transformation of $f\in L^1(\mathbb{R})$:

(1) $[f(\circ-h)]^\wedge(\xi)=\mathrm{e}^{-ih\xi}f^\wedge(\xi),\ \xi\in\mathbb{R},h\in\mathbb{R},$ (FT of translation)

(2) $\tau_h f^\wedge(\xi)=[\mathrm{e}^{i(h\cdot\circ)}f(\circ)]^\wedge(\xi),\ \xi\in\mathbb{R},h\in\mathbb{R},$ (translation of FT)

(3) $[\rho f(\rho\circ)]^\wedge(\xi)=f^\wedge\left(\dfrac{\xi}{\rho}\right),\ \xi\in\mathbb{R},\rho>0,$ (FT of dilation)

(4) $[\overline{f(-\circ)}]^\wedge(\xi)=\overline{f^\wedge(\xi)},\xi\in\mathbb{R}.$ (FT of reflection)

Proof We only deduce formula (4):

$$[\overline{f(-\circ)}]^\wedge(\xi)=(\overline{\tilde{f}(\circ)})^\wedge(\xi)=\int_{\mathbf{R}^n}\overline{f(-x)}\mathrm{e}^{-\mathrm{i}(x\cdot\xi)}\,\mathrm{d}x=\int_{\mathbf{R}^n}\overline{f(t)}\mathrm{e}^{\mathrm{i}(t\cdot\xi)}\,\mathrm{d}t$$

$$=\int_{\mathbf{R}^n}\overline{f(t)\mathrm{e}^{-\mathrm{i}(t\cdot\xi)}}\,\mathrm{d}t=\overline{f^\wedge(\xi)}$$

Others can be proved by the definition of Fourier transformation, left as exercise.

Note that, the formulas (1) and (2) can rewrite as

$$(\tau_h f)^\wedge(\xi)=\mathrm{e}^{-ih\xi}f^\wedge(\xi)\quad\text{and}\quad(\mathrm{e}^{i(h\cdot\circ)}f)^\wedge(\xi)=\tau_h f^\wedge(\xi),$$

respectively.

2) Analytic properties of Fourier transformations

Theorem 5. 2. 2 (1) Fourier transformation $f^\wedge(\xi)$ of $f\in L^1(\mathbb{R})$ is uniformly continuous on $\mathbb{R}$, and $f^\wedge\in C(\mathbb{R})$; (2) Fourier transformation is a continuous linear operator from $L^1(\mathbb{R})$ to $L^\infty(\mathbb{R})$; and it is regarded as an operator $\wedge$: $L^1(\mathbb{R})\to L^\infty(\mathbb{R})$, satisfying norm non-increasing inequality $\|f^\wedge\|_{L^\infty(\mathbf{R})}\leqslant\|f\|_{L^1(\mathbf{R})}$, where $L^\infty(\mathbb{R})=\{f:\mathbb{R}\to\mathbb{R},\ \|f\|_{L^\infty(\mathbf{R})}<+\infty\}$ is the essential bounded function space on $\mathbb{R}$ with norm $\|f\|_{L^\infty(\mathbf{R})}=\operatorname{esssup}\{f(x):x\in\mathbb{R}\}$ called the essential supremum of f.

Proof For $f\in L^1(\mathbb{R})$, its Fourier transformation satisfies $|f^\wedge(\xi+h)-f^\wedge(\xi)|\leqslant\int_{\mathbf{R}}|\mathrm{e}^{-ixh}-1||f(x)|\,\mathrm{d}x$. By $f\in L^1(\mathbb{R})$, it implies $\int_{\mathbf{R}}|\mathrm{e}^{-ixh}-1||f(x)|\,\mathrm{d}x\leqslant M\int_{\mathbf{R}}|f(x)|\,\mathrm{d}x=M\|f\|_{L^1(\mathbf{R})}$. Then, Lebesgue dominated theorem shows that

$$\lim_{|h|\to 0}|f^\wedge(\xi+h)-f^\wedge(\xi)|\leqslant\lim_{|h|\to 0}\int_{\mathbf{R}}|\mathrm{e}^{-ihx}-1||f(x)|\,\mathrm{d}x$$

$$=\int_{\mathbf{R}}\lim_{|h|\to 0}|e^{-ihx}-1||f(x)|\,dx=0$$

holds for all $\xi\in\mathbb{R}$, uniformly. Thus, $f^\wedge(\xi)$ is a uniformly continuous function on $\mathbb{R}$, and $f^\wedge\in C(\mathbb{R})$.

By the definition of Fourier transformation,

$$|f^\wedge(\xi)|\leqslant\int_{\mathbf{R}}|f(x)\,e^{-ix\xi}|\,dx\leqslant\int_{\mathbf{R}}|f(x)|\,dx=\|f\|_{L^1(\mathbf{R})},$$

take the essential supremum, we get $\|f^\wedge\|_{L^\infty(\mathbf{R})}\leqslant\|f\|_{L^1(\mathbf{R})}$, this is the boundedness of operator $\wedge$: $L^1(\mathbb{R})\to L^\infty(\mathbb{R})$.

3) Riemann-Lebesgue lemma

Theorem 5.2.3 (Riemann-Lebesgue lemma) Fourier transformation $f^\wedge(\xi)$ of $f\in L^1(\mathbb{R})$ satisfies $\lim\limits_{|\xi|\to+\infty}f^\wedge(\xi)=0$.

Proof For $f\in L^1(\mathbb{R})$, we write $f^\wedge(\xi)=\frac{1}{2}\int_{\mathbf{R}}\left[f(x)-f\left(x+\frac{\pi}{\xi}\right)\right]e^{-ixh}\,dx$, then by **the continuity in mean** of $f\in L^1(\mathbb{R})$ ($f\in L^1(\mathbb{R})\Rightarrow\lim\limits_{h\to 0}\int_{\mathbf{R}}|f(x+h)-f(x)|\,dx=0$, see [6],[17]), it holds $\lim\limits_{|\xi|\to+\infty}f^\wedge(\xi)=0$.

4) Convolution formulas of Fourier transformations

Theorem 5.2.4 The convolution $f*g(x)=\int_{\mathbf{R}}f(t)g(x-t)\,dt$ of $f,g\in L^1(\mathbb{R})$ is in $L^1(\mathbb{R})$; and satisfies the formula

$$(f*g)^\wedge(\xi)=f^\wedge(\xi)g^\wedge(\xi),\quad \xi\in\mathbb{R}.$$

Proof For $f,g\in L^1(\mathbb{R})$, it holds $\int_{\mathbf{R}}|f(t)g(x-t)|\,dx=|f(t)|\int_{\mathbf{R}}|g(x-t)|\,dt=|f(t)|\,\|g\|_{L^1(\mathbf{R})}$, thus $\int_{\mathbf{R}}|f(t)g(x-t)|\,dx=|f(t)|\,\|g\|_{L^1(\mathbf{R})}\in L^1(\mathbb{R})$. By Fubini theorem, it holds

$$\int_{\mathbf{R}}|(f*g)(x)|\,dx\leqslant\int_{\mathbf{R}}\left|\int_{\mathbf{R}}|f(t)||g(x-t)|\,dt\right|dx$$
$$=\int_{\mathbf{R}}|f(t)|\,dt\int_{\mathbf{R}}|g(x-t)|\,dx=\|f\|_{L^1(\mathbf{R})}\|g\|_{L^1(\mathbf{R})}.$$

This implies $f*g\in L^1(\mathbb{R})$. Moreover, we have

$$(f*g)^\wedge(\xi)=\int_{\mathbf{R}}\left\{\int_{\mathbf{R}}f(t)g(x-t)\,dt\right\}e^{-ix\xi}\,dx$$
$$=\left\{\int_{\mathbf{R}}f(t)e^{-it\xi}\,dt\right\}\left\{\int_{\mathbf{R}}g(x-t)\,e^{-i(x-t)\xi}\,dx\right\}=f^\wedge(\xi)g^\wedge(\xi).$$

The proof is complete.

The sense of Theorem 5. 2. 4: Fourier transformation transfers convolution operation (integral) to pointwise product operation (multiplication).

4. Formulas of Fourier transformations of derivatives

"The function class $AC_{\text{Loc}}(\mathbb{R})$ of all locally absolutely continuous functions" and **"the function class** $AC_{\text{Loc}}^{k-1}(\mathbb{R})$ of all locally absolutely $(k-1)$-order differentiable continuous functions" are defined, respectively

$$AC_{\text{Loc}}(\mathbb{R}) = \left\{f:\ \forall x \in \mathbb{R},\ \exists g \in L_{\text{Loc}}(\mathbb{R}),\ \text{s. t. } f(x) = f(0) + \int_0^x g(u)\,du\right\},$$

$$AC_{\text{Loc}}^{k-1}(\mathbb{R}) = \{f:\ f(x) = a_0 + F(x)\ \forall x \in \mathbb{R}\},$$

with $F(x)$ satisfying: $\exists g \in L_{\text{Loc}}(\mathbb{R}),\ \exists a_0, a_1, \cdots, a_{k-1}$, constants, and

$$F(x) = \int_0^x du_1\left[a_1 + \int_0^{u_1} du_2 + \left[a_2 + \cdots + \int_0^{u_{k-2}} du_{k-2}\left(a_{k-1} + \int_0^{u_{k-1}} g(u_k)\,du_k\right)\cdots\right]\right]du.$$

Theorem 5. 2. 5 For $f \in L^1(\mathbb{R}) \cap AC_{\text{Loc}}(\mathbb{R})$, if $f' \in L^1(\mathbb{R})$, then its Fourier transformation is

$$(f')^\wedge(\xi) = \mathrm{i}\xi f^\wedge(\xi), \quad \xi \neq 0;$$

For $f \in L^1(\mathbb{R}) \cap AC_{\text{Loc}}^{k-1}(\mathbb{R})$, if $f^{(k)} \in L^1(\mathbb{R})$, then its Fourier transformation is

$$(f^{(k)})^\wedge(\xi) = (\mathrm{i}\xi)^k f^\wedge(\xi), \quad \xi \neq 0.$$

We refer to [4] for the proof.

5. Formulas of derivatives of Fourier transformations

Theorem 5. 2. 6 If $f \in L^1(\mathbb{R}), x^k f(x) \in L^1(\mathbb{R})$, then Fourier transformation $f^\wedge$ of f has r-order derivatives $(f^\wedge)^{(r)}$ belonging to $C_0(\mathbb{R})$, and holds formulae

$$(f^\wedge)^{(r)}(\xi) = (-\mathrm{i})^r \int_{\mathbf{R}} x^r f(x)\,\mathrm{e}^{-\mathrm{i}x\xi}\,dx, \quad r = 1, 2, \cdots, k, \xi \in \mathbb{R}.$$

We refer to [4] for the proof.

6. Parseval formula of Fourier transformations

Theorem 5. 2. 7 (1) If $f, g \in L^1(\mathbb{R})$, then Parseval formula holds

$$\int_{\mathbf{R}} f^\wedge(t) g(t)\,\mathrm{d}t = \int_{\mathbf{R}} f(t) g^\wedge(t)\,\mathrm{d}t;$$

(2) If $f, g \in L^1(\mathbb{R}), g^\wedge \in L^1(\mathbb{R})$, then Plancherel-type formula holds

$$\int_{\mathbf{R}} f(t)\overline{g(t)}\,\mathrm{d}t = \frac{1}{2\pi}\int_{\mathbf{R}} f^\wedge(t)\overline{g^\wedge(t)}\,\mathrm{d}t;$$

and for the convolution, it holds

$$f*g(x)=\frac{1}{2\pi}\int_{\mathbf{R}} f^{\wedge}(\xi)g^{\wedge}(\xi)e^{ix\xi}d\xi, \quad \text{a. e. } x\in\mathbb{R}.$$

7. Definitions of inverse Fourier transformations and properties

Definition 5.2.2 (inverse Fourier transformation) The inverse Fourier transformation of a function $g\in L^1(\mathbb{R}^n)$ is defined by

$$g^{\vee}(x)=\frac{1}{(2\pi)^n}\int_{\mathbf{R}^n} g(\xi)e^{i(x\cdot\xi)}d\xi, \quad x\in\mathbb{R}^n,$$

where $x\cdot\xi=\sum_{j=1}^{n}x_j\xi_j$ is the inner product of $x=(x_1,\cdots,x_n)\in\mathbb{R}^n$ and $\xi=(\xi_1,\cdots,\xi_n)\in\mathbb{R}^n$. For $n=1$, the inverse Fourier transformation is $g^{\vee}(x)=\frac{1}{2\pi}\int_{\mathbf{R}} g(\xi)e^{ix\xi}d\xi, x\in\mathbb{R}$.

We state the theorems in case $n=1$, i. e., in space $L^1(\mathbb{R})\equiv L^1(\mathbb{R}^1)$, and they hold for case $n\,(n\geqslant 1)$ of space $L^1(\mathbb{R}^n)$.

Theorem 5.2.8 (inversion formula) (1) If $f, f^{\wedge}\in L^1(\mathbb{R})$, then we have

$$f(x)=\frac{1}{2\pi}\int_{\mathbf{R}} f^{\wedge}(\xi)\, e^{i\xi x}d\xi, \quad \text{a. e. } x\in\mathbb{R};$$

(2) If $f\in L^1(\mathbb{R})\cap C(\mathbb{R})$, $f^{\wedge}\in L^1(\mathbb{R})$, then we have

$$f(x)=\frac{1}{2\pi}\int_{\mathbf{R}} f^{\wedge}(\xi)\, e^{i\xi x}d\xi, \quad \forall x\in\mathbb{R};$$

(3) If $f\in L^1(\mathbb{R})$, and $f^{\wedge}=0$, then we have

$$f(x)=0, \quad \text{a. e. } x\in\mathbb{R};$$

(4) If $f\in L^1(\mathbb{R})$, by introducing "summation factor", then we have

① Cesaro summation factor $\left(1-\frac{|\xi|}{\rho}\right)$ holds

$$\lim_{\rho\to 0}\frac{1}{2\pi}\int_{-\rho}^{\rho}\left(1-\frac{|\xi|}{\rho}\right)f^{\wedge}(\xi)e^{i\xi x}d\xi=f(x), \quad \text{a. e. } x\in\mathbb{R};$$

② Abel summation factor $\exp\left\{-\frac{|\xi|}{\rho}\right\}$ holds

$$\lim_{\rho\to+\infty}\frac{1}{2\pi}\int_{\mathbf{R}}\exp\left\{-\frac{|\xi|}{\rho}\right\}f^{\wedge}(\xi)e^{i\xi x}d\xi=f(x), \quad \text{a. e. } x\in\mathbb{R};$$

③ Gauss summation factor $\exp\left\{-\left(\frac{\xi}{\rho}\right)^2\right\}$ holds

$$\lim_{\rho\to+\infty}\frac{1}{2\pi}\int_{\mathbf{R}}\exp\left\{-\left(\frac{\xi}{\rho}\right)^2\right\}f^{\wedge}(\xi)e^{i\xi x}d\xi=f(x), \quad \text{a. e. } x\in\mathbb{R}.$$

For the proof, we refer to [4].

5.2.2 Fourier transformations on $L^2(\mathbb{R}^n)$

We know that $L^2([-\pi,\pi])\subset L^1([-\pi,\pi])$ holds for compact case $[-\pi,\pi]$, but relationship "$\subset$" is not true for $L^2(\mathbb{R})$ and $L^1(\mathbb{R})$, since $\mathbb{R}$ is locally compact. For example, $f(x)=\dfrac{1}{\sqrt{x}\,(1+|\ln|x||)}\in L^2(\mathbb{R}^n)$, but $f(x)=\dfrac{1}{\sqrt{x}\,(1+|\ln|x||)}\notin L^p(\mathbb{R}^n)$, $p\neq 2$. Thus, in compact case, we only need to define Fourier coefficients (i. e. finite Fourier transformations) $\left\{c_k=\dfrac{1}{\sqrt{2\pi}}\displaystyle\int_{-\pi}^{+\pi}f(x)\mathrm{e}^{-ikx}\,\mathrm{d}x\right\}_{k=-\infty}^{+\infty}$ for $f\in L^1([-\pi,\pi])$, because $\{c_k\}_{k=-\infty}^{+\infty}$ makes sense for $f\in L^2([-\pi,\pi])$; whereas, in locally compact case, we need other idea for $L^2(\mathbb{R})$ since integral $f^\wedge(\xi)=\displaystyle\int_{-\infty}^{+\infty}f(x)\mathrm{e}^{-ix\xi}\,\mathrm{d}x$ may not exist for $f\in L^2(\mathbb{R})$.

1. Two propositions for preparation

Proposition 5.2.4 Let $f\in L^1(\mathbb{R})\cap L^2(\mathbb{R})$. Then $f^\wedge(\xi)\in L^2(\mathbb{R})$, and

$$\|f^\wedge\|_{L^2(\mathbf{R})}=\sqrt{2\pi}\,\|f\|_{L^2(\mathbf{R})}. \tag{5.2.1}$$

Proof Set $g(x)=\overline{f(-x)}$, then $g(x)\in L^1(\mathbb{R})\cap L^2(\mathbb{R})$; Set $h(x)=(f*g)(x)$, by Proposition 5.2.3, (1), ②, (3), we have $h=f*g\in L^1(\mathbb{R})\cap L^2(\mathbb{R})$, thus $h^\wedge$ makes sense, and by Theorem 5.2.8(4), ②, take $x=0$, it holds

$$\lim_{\rho\to+\infty}\frac{1}{2\pi}\int_{\mathbf{R}}\exp\left\{-\frac{|\xi|}{\rho}\right\}h^\wedge(\xi)\,\mathrm{d}\xi=h(0). \tag{5.2.2}$$

Moreover, by convolution formula $(f*g)^\wedge(\xi)=f^\wedge(\xi)g^\wedge(\xi)$ and reflection formula of Fourier transformations, we have

$$g^\wedge(\xi)=[\overline{f(-\circ)}]^\wedge(\xi)=\overline{f^\wedge(\xi)},$$

and hence

$$h^\wedge(\xi)=f^\wedge(\xi)g^\wedge(\xi)=f^\wedge(\xi)\overline{f^\wedge(\xi)}=|f^\wedge(\xi)|^2, \tag{5.2.3}$$

so that the integrand of (5.2.2) is a positive monotonic decreasing function, again, when $\rho\to+\infty$, we have the limit

$$\lim_{\rho\to+\infty}\exp\left\{-\frac{|\xi|}{\rho}\right\}h^\wedge(\xi)=h^\wedge(\xi).$$

Finally, it holds by Lebesgue dominated theorem,

$$h(0)=\frac{1}{2\pi}\int_{\mathbf{R}}\lim_{\rho\to+\infty}\exp\left\{-\frac{|\xi|}{\rho}\right\}h^\wedge(\xi)\,\mathrm{d}\xi=\frac{1}{2\pi}\int_{\mathbf{R}}h^\wedge(\xi)\,\mathrm{d}\xi,$$

that is, $\|h^\wedge\|_{L^1(\mathbf{R})}=\displaystyle\int_{\mathbf{R}}h^\wedge(\xi)\,\mathrm{d}\xi=2\pi h(0)$. This implies $h^\wedge\in L^1(\mathbb{R})$. On the other hand,

$$h(0)=(f*g)(0)=\int_{\mathbf{R}} f(t)g(0-t)\,dt=\int_{\mathbf{R}} f(t)\overline{f(t)}\,dt$$
$$=\int_{\mathbf{R}} |f(t)|^2 dt=\| f\|^2_{L^2(\mathbf{R})};$$

Combine with (5.2.3) $h^\wedge(\xi)=|f^\wedge(\xi)|^2$, it holds

$$\| f^\wedge\|^2_{L^2(\mathbf{R})}=\int_{\mathbf{R}} |f^\wedge(\xi)|^2 d\xi=\int_{\mathbf{R}} h^\wedge(\xi)\, d\xi=2\pi h(0)=2\pi\| f\|^2_{L^1(\mathbf{R})},$$

finally, $\| f^\wedge\|_{L^2(\mathbf{R})}=\sqrt{2\pi}\,\| f\|_{L^2(\mathbf{R})}$.

Note If the definitions of Fourier transformation and inverse Fourier transformation take the symmetric forms, then the norm equality(5.2.1) is $\| f^\wedge\|_{L^2(\mathbf{R})}=\| f\|_{L^2(\mathbf{R})}$.

Proposition 5.2.5 Let $f\in L^2(\mathbb{R})$, and $f_\rho(x)=\begin{cases} f(x), & |x|\leqslant\rho \\ 0, & |x|>\rho\end{cases}$ $(\rho>0)$. Then $f_\rho\in L^1(\mathbb{R})\cap L^2(\mathbb{R})$, and $f_\rho^\wedge\in L^2(\mathbb{R})$, $\forall\rho>0$; Moreover, there exists unique $g\in L^2(\mathbb{R})$, s.t., $\lim\limits_{\rho\to+\infty}\| f_\rho^\wedge-g\|_{L^2(\mathbf{R})}=0$.

Proof By Hölder inequality, $f_\rho\in L^1(\mathbb{R})\cap L^2(\mathbb{R})$, $\forall\rho>0$. Then $f_\rho^\wedge\in L^2(\mathbb{R})$ by Proposition 5.2.4, and $f_{\rho_1}^\wedge-f_{\rho_2}^\wedge$ is Fourier transformation of $f_{\rho_1}-f_{\rho_2}$. Suppose that $\rho_1<\rho_2$, then we have

$$\| f_{\rho_1}^\wedge-f_{\rho_2}^\wedge\|^2_{L^2(\mathbf{R})}=(2\pi)\| f_{\rho_1}-f_{\rho_2}\|^2_{L^2(\mathbf{R})}$$
$$=(2\pi)\int_{(-\rho_2,-\rho_1)} |f(x)|^2 dx+(2\pi)\int_{(\rho_1,\rho_2)} |f(x)|^2 dx.$$

The right-hand side of the above equality tends to zero by $f\in L^2(\mathbb{R})$, this implies $\lim\limits_{\rho_1,\rho_2\to+\infty}\| f_{\rho_1}^\wedge-f_{\rho_2}^\wedge\|_{L^2(\mathbf{R})}=0$.

For any subsequence $\{k_n\}$ $(k_n\in\mathbb{Z}^+)$ of $\rho\in[0,+\infty)$, Fourier transformation sequence $f_{k_n}^\wedge\in L^2(\mathbb{R})$ is a Cauchy sequence in $L^2(\mathbb{R})$; Hence, there exists unique $g\in L^2(\mathbb{R})$, such that $\lim\limits_{\rho\to+\infty}\| f_\rho^\wedge-g\|_{L^2(\mathbf{R})}=0$ by the completeness of $L^2(\mathbb{R})$. The proof of Proposition 5.2.5 is proved.

2. Definition of Fourier transformations of functions in $L^2(\mathbb{R}^n)$

Definition 5.2.3 (**Fourier transformation of** $f\in L^2(\mathbb{R}^n)$) If $f\in L^2(\mathbb{R}^n)$, then **the Fourier transformation of** f is defined by

$$[F^2(f)](\xi)=\overset{(2)}{\lim_{\rho\to+\infty}} f_\rho^\wedge(\xi)=\overset{(2)}{\lim_{\rho\to+\infty}}\int_{(-\rho,\rho)^n} f(x)e^{-i(x\cdot\xi)}dx,\quad \xi\in\mathbb{R}^n,$$

where $x\cdot\xi=\sum\limits_{j=1}^{n}x_j\xi_j$ is the inner product of $x=(x_1,\cdots,x_n)\in\mathbb{R}^n$ and $\xi=(\xi_1,\cdots,\xi_n)\in$

$\mathbb{R}^n$; $\overset{(2)}{\lim\limits_{\rho\to+\infty}}$ is "the limit in mean in L^2-sense".

Usually, denoted by $f^\wedge(\xi) = [F^2(f)](\xi)$, if it is no confusion. Thus, Fourier transformation of $f\in L^2(\mathbb{R}^n)$ is uniquely determined by a function $f^\wedge \equiv [F^2(f)] \in L^2(\mathbb{R}^n)$ satisfying $\lim\limits_{\rho\to+\infty}\left\| F^2(f)(\circ) - \int_{(-\rho,\rho)^n} f(x)e^{-i(\circ,\xi)}\,dx \right\|_{L^2(\mathbb{R}^n)} = 0$, where n-dimensional cube $(-\rho,\rho)^n$ can be replaced by n-dimensional rectangular $(-\rho_1,\rho_1)\times\cdots\times(-\rho_n,\rho_n)$.

3. Properties of $L^2(\mathbb{R}^n)$ -Fourier transformation

The operation properties of $L^2(\mathbb{R})$-Fourier transformation are similar to that in Theorem 5. 2. 1, and will be shown in Theorem 5. 2. 10. We now give the analysis properties.

Theorem 5. 2. 9 (Plancherel theorem) Fourier transformation of $f\in L^2(\mathbb{R})$ is an one-one continuous linear operator from $L^2(\mathbb{R})$ onto $L^2(\mathbb{R})$; that is, for $f\in L^2(\mathbb{R})$, there exists unique function $f^\wedge(\xi) = [F^2(f)](\xi) \in L^2(\mathbb{R})$ satisfying

$$\| f^\wedge \|_{L^2(\mathbf{R})} = \sqrt{2\pi}\, \| f \|_{L^2(\mathbf{R})}.$$

Proof $F^2(f)$ is linear by $F^2(\alpha f + \beta g) = \alpha F^2(f) + \beta F^2(g)$, $f, g\in L^2(\mathbb{R})$, $\alpha,\beta\in\mathbb{R}$. Then, $f^\wedge \equiv F^2(f)$ satisfies $\lim\limits_{\rho\to+\infty} \| f^\wedge - f_\rho^\wedge \|_{L^2(\mathbf{R}^n)} = 0$, and then implies

$$\lim_{\rho\to+\infty} \| f_\rho^\wedge \|_{L^2(\mathbf{R})} = \| f^\wedge \|_{L^2(\mathbf{R})}; \tag{5. 2. 4}$$

On the other hand, by the definition of f_ρ, we have

$$\lim_{\rho\to+\infty} \| f_\rho \|_{L^2(\mathbf{R})} = \| f \|_{L^2(\mathbf{R})}. \tag{5. 2. 5}$$

Moreover, by $f_\rho \in L^1(\mathbb{R}) \cap L^2(\mathbb{R})$, the Proposition 5. 2. 4 gives

$$\| f_\rho^\wedge \|_{L^2(\mathbf{R})} = \sqrt{2\pi}\, \| f_\rho \|_{L^2(\mathbf{R})}, \tag{5. 2. 6}$$

take the limit on both sides of (5. 2. 6), combine with (5. 2. 4), (5. 2. 5), we have $\| f^\wedge \|_{L^2(\mathbf{R})} = \sqrt{2\pi}\, \| f \|_{L^2(\mathbf{R})}$. This is the continuity of operators.

Note If the definitions of Fourier transformations and inverse Fourier transformations take the symmetric forms, then the norm equality is $\| f^\wedge \|_{L^2(\mathbf{R})} = \| f \|_{L^2(\mathbf{R})}$.

5. 2. 3 Fourier transformations on $L^p(\mathbb{R}^n)$, $1<p<2$

For $L^p(\mathbb{R}^n)$ $(1<p<2)$-function, some more preliminaries are needed, such as Riesz-Thorin convex theorem, the types of operators. We omit details, and only state the results for $n=1$.

1. Two propositions for the preparation

Denote $S_{00}(\mathbb{R})$ the set of all **simple functions** on $\mathbb{R}$ (Fig. 5. 2. 1), that is, the functions in $S_{00}(\mathbb{R})$ take **finite values** c_j on **finite measurable sets** $me_j<+\infty$ in $\mathbb{R}$ with **finite sets** e_j, $j=1,2,\cdots,k$, i. e.

$$h\in S_{00}(\mathbb{R})\Rightarrow h(x)=\sum_{j=1}^{k}c_j\chi_{e_j}(x),\quad x\in\mathbb{R},$$

where $-\infty<c_1<c_2<\cdots<c_k<+\infty$, $e_j=\{x\in\mathbb{R}: h(x)=c_j\}$, $0\leqslant me_j<+\infty$, with Lebesgue measure me_j of e_j, the number of e_j, the values of me_j and the values c_j of a simple function all are finite, $j=1,2,\cdots,k\in\mathbb{N}$.

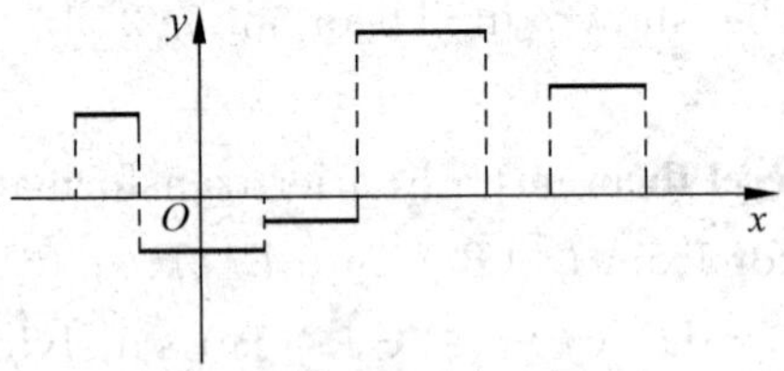

Fig. 5. 2. 1 Simple function

Proposition 5. 2. 6 (properties of $S_{00}(\mathbb{R})$) We have

(1) $S_{00}(\mathbb{R})\subset L^p(\mathbb{R})$, and $S_{00}(\mathbb{R})$ is dense in $L^p(\mathbb{R})$, $1\leqslant p<+\infty$; that is, "$\forall\varepsilon>0$, $\forall f\in L^p(\mathbb{R})\Rightarrow\exists h\in S_{00}(\mathbb{R})$, s. t. $\|f-h\|_{L^p(\mathbf{R})}<\varepsilon$";

(2) If $\operatorname{supp} f$ is a compact set, then there exists sequence $\{h_n\}\subset S_{00}(\mathbb{R})$, s. t. $\lim\limits_{n\to+\infty}\|f-h_n\|_{L^p(\mathbf{R})}=0$, with $\operatorname{supp} g_n\subset\operatorname{supp} f$ for each $n\in\mathbb{Z}^+$;

(3) If $h\in S_{00}(\mathbb{R})$, then $h^\wedge\in L^q(\mathbb{R})$, $\frac{1}{p}+\frac{1}{q}=1$ for $1\leqslant p\leqslant 2$; and

$$\|h^\wedge\|_{L^q(\mathbf{R})}\leqslant\|h\|_{L^p(\mathbf{R})}.$$

Note $h\in S_{00}(\mathbb{R})$ is in $L^p(\mathbb{R})$, $1\leqslant p<+\infty$, so it can be regarded as:

① $h\in S_{00}(\mathbb{R})$, then $h^\wedge\in L^q(\mathbb{R})$, $\frac{1}{p}+\frac{1}{q}=1$, by (3); and $S_{00}(\mathbb{R})\subset L^p(\mathbb{R})$ is dense in $L^p(\mathbb{R})$ by (1); Thus, Fourier transformation of $f\in L^p(\mathbb{R})$, $1<p<2$, belongs to $L^q(\mathbb{R})$, $2<q<+\infty$;

② $h\in L^1(\mathbb{R})$, then $h^\wedge$ is defined by $h^\wedge(\xi)=\int_{\mathbf{R}}h(x)\mathrm{e}^{-\mathrm{i}x\xi}\mathrm{d}x$, and $h^\wedge\in C_0(\mathbb{R})$, satisfying the inequality $\|h^\wedge\|_{L^\infty(\mathbf{R})}\leqslant\|h\|_{L^1(\mathbf{R})}$, we say that "**Fourier transformation** $\wedge$ **is a $(1,\infty)$-type operator**";

③ $h \in L^2(\mathbb{R})$, then $\| h^\wedge \|_{L^2(\mathbf{R})} = \sqrt{2\pi} \| h \|_{L^2(\mathbf{R})}$, we say that **"Fourier transformation $\wedge$ is a (2,2)-type operator"**.

Hence, Fourier transformation regarded as an operator $\wedge$, is "$(1,\infty)$-type" and "$(2,2)$-type" operator on $S_{00}(\mathbb{R})$, by the Riesz-Thorin convex theorem (Proposition 5.2.7), $\wedge$ is a (p, q)-type operator on $S_{00}(\mathbb{R})$, i.e. $\| h^\wedge \|_{L^q(\mathbf{R})} \leqslant \| h \|_{L^p(\mathbf{R})}$, $1 \leqslant p, q < +\infty$, $\frac{1}{p}+\frac{1}{q}=1$.

Proposition 5.2.7 (Riesz-Thorin convex theorem) Let $T: L^p(\mathbb{R}) \to L^r(\mathbb{R})$ $(1 \leqslant p, r \leqslant +\infty)$ be a linear operator, and be (p_0, r_0)-type with norm inequality $\| Tf \|_{L^{r_0}(\mathbf{R})} \leqslant M_0 \| f \|_{L^{p_0}(\mathbf{R})}$, $1 \leqslant p_0, r_0 \leqslant +\infty$; also be (p_1, r_1)-type with norm inequality $\| Tf \|_{L^{r_1}(\mathbf{R})} \leqslant M_1 \| f \|_{L^{p_1}(\mathbf{R})}$, $1 \leqslant p_1, r_1 \leqslant +\infty$. Then T is a (p,r)-type operator with norm inequality $\| Tf \|_{L^r(\mathbf{R})} \leqslant M \| f \|_{L^p(\mathbf{R})}$, $1 \leqslant p, r \leqslant +\infty$, where the parameters satisfying $\frac{1}{p} = \frac{1-t}{p_0} + \frac{1}{p_1}$, $\frac{1}{r} = \frac{1-t}{r_0} + \frac{1}{r_1}$, $\forall t \in [0,1]$, and $M \leqslant M_0^{1-t} M_1^t$.

In Fig. 5.2.2, the coordinate axles are $\frac{1}{p}$ and $\frac{1}{r}$— the points $\left(\frac{1}{p}, \frac{1}{p}\right)$ on the diagonal are corresponding to the (p,p)-type operator; the points $\left(\frac{1}{p}, \frac{1}{q}\right)$, $\left(\frac{1}{p}+\frac{1}{q}=1\right)$ on the anti-diagonal are correspondent to the (p,q)-type operator; and the points $\left(\frac{1}{p}, \frac{1}{r}\right)$ in the shadow part are correspondent to the (p,r)-type; On the other hand, when $p \leqslant r$, points $\left(\frac{1}{p}, \frac{1}{r}\right) \leftrightarrow (p,r)$ appear in the shadow part, thus for $1<p<2$, only can march $2<q<+\infty$; This is the reason why we only can define Fourier transformations for $L^p(\mathbb{R}^n)$-functions with $1 \leqslant p \leqslant 2$, and regarded as operators, Fourier transformations F^p: $L^p(\mathbb{R}^n) \to L^q(\mathbb{R})$ $(1 \leqslant p \leqslant 2)$, only for $2<q<+\infty$.

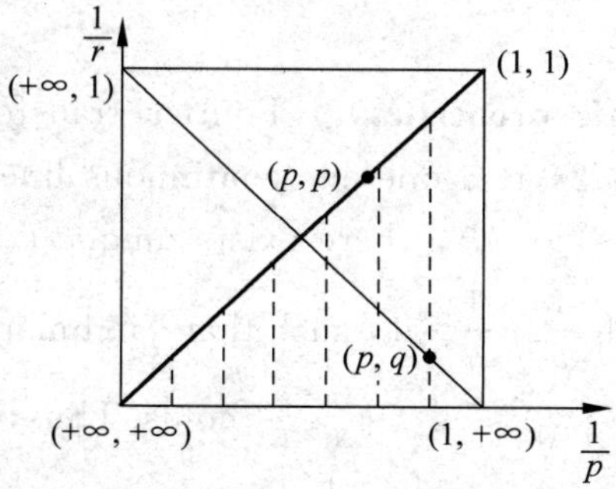

Fig. 5.2.2 Draft of Riesz-Thorin theorem

2. Definition of Fourier transformations of functions in $L^p(\mathbb{R}^n)$ $(1<p<2)$

Definition 5.2.4 Fourier transformation $F^p(f)$ of function $f\in L^p(\mathbb{R}^n)$ with $1<p<2$ is defined as

$$[F^p(f)](\xi)=\underset{\rho\to+\infty}{\overset{(q)}{\lim}} f_\rho^\wedge(\xi)=\underset{\rho\to+\infty}{\overset{(q)}{\lim}}\int_{(-\rho,\rho)^n} f(x)\,\mathrm{e}^{-\mathrm{i}(x\cdot\xi)}\,\mathrm{d}x,\quad \xi\in\mathbb{R}^n,$$

where $x\cdot\xi=\sum_{j=1}^n x_j\xi_j$ is the inner product of $x=(x_1,\cdots,x_n)\in\mathbb{R}^n$ and $\xi=(\xi_1,\cdots,\xi_n)\in\mathbb{R}^n$, $\frac{1}{p}+\frac{1}{q}=1$, "$\underset{\rho\to+\infty}{\overset{(q)}{\lim}}$" means "the Limit in mean in L^q-sense".

So that, Fourier transformation $f^\wedge\equiv F^p(f)$ of $f\in L^p(\mathbb{R}^n)$ with $1<p<2$ is determined by unique function $g\in L^q(\mathbb{R}^n)$ with $g\equiv F^p(f)$ satisfying

$$\lim_{\rho\to+\infty}\left\| F^p(f)(\circ)-\int_{(-\rho,\rho)^n} f(x)\mathrm{e}^{-\mathrm{i}(\circ\cdot\xi)}\,\mathrm{d}x\right\|_{L^q(\mathbb{R}^n)}=0.$$

Where n-dimensional cube $(-\rho,\rho)^n$ can be replaced by any n-dimensional rectangular $(-\rho_1,\rho_1)\times\cdots\times(-\rho_n,\rho_n)$.

3. Properties of Fourier transformations of $L^p(\mathbb{R}^n)$ $(1<p\leqslant 2)$-functions

We denote Fourier transformations of $f\in L^p(\mathbb{R})$ $(1<p\leqslant 2)$-functions by $f^\wedge(\xi)\equiv(F^p(f))(\xi)$, if there is no confusion.

Theorem 5.2.10 (operation properties) Let $f\in L^p(\mathbb{R})$ $(1<p\leqslant 2)$. Then

(1) $(\tau_h f)^\wedge(\xi)=\mathrm{e}^{-\mathrm{i}(h\cdot\xi)}f^\wedge(\xi),\xi\in\mathbb{R},h\in\mathbb{R}$, (FT of translation)

(2) $\tau_h f^\wedge(\xi)=[\mathrm{e}^{\mathrm{i}(h\cdot\circ)}f(\circ)]^\wedge(\xi),\xi\in\mathbb{R},h\in\mathbb{R}$, (translation of FT)

(3) $[\rho f(\rho\circ)]^\wedge(\xi)=f^\wedge\left(\frac{\xi}{\rho}\right),\xi\in\mathbb{R},\rho>0$, (FT of dilation)

(4) $[\overline{f(-\circ)}]^\wedge(\xi)=\overline{f^\wedge(\xi)},\xi\in\mathbb{R}$. (FT of reflection)

When $p=1$, it is Theorem 5.2.1. The proofs can be obtained by using Propositions 5.2.6 and 5.2.7.

Theorem 5.2.11 (analytic properties) Fourier transformation $f^\wedge\equiv[F^p(f)]$ of function $f\in L^p(\mathbb{R})$ $(1<p<2)$ is a one-one continuous linear operator from $L^p(\mathbb{R})$ to $L^q(\mathbb{R})$. For $f\in L^p(\mathbb{R})$ $(1<p<2)$, there exists unique $L^q(\mathbb{R})$ $(2<q<+\infty)$-function $f^\wedge\equiv[F^p(f)]\in L^q(\mathbb{R})$ with $\frac{1}{p}+\frac{1}{q}=1$, such that Titchmarsh inequality (or, norm non-increasing inequality) $\|f^\wedge\|_{L^q(\mathbb{R})}\leqslant\|f\|_{L^p(\mathbb{R})}$ holds. Thus, operator $\wedge\equiv F^p: L^p(\mathbb{R})\to L^q(\mathbb{R})$ is a (p,q)-type $\left(1\leqslant p\leqslant 2,\frac{1}{p}+\frac{1}{q}=1\right)$. Moreover, if $f\in L^1(\mathbb{R})\cap L^p(\mathbb{R})$

$(1<p\leqslant 2)$, then $f^\wedge(\xi)\equiv(F^1 f)(\xi)=(F^p f)(\xi)$, a. e. $\xi\in\mathbb{R}$.

Theorem 5. 2. 12 (convolution formula) For two functions of $f\in L^p(\mathbb{R})$ $(1<p\leqslant 2)$ and $g\in L^1(\mathbb{R})$, the convolution $f*g(x)=\int_{\mathbf{R}} f(t)g(x-t)\,\mathrm{d}t$ satisfies $f*g\in L^p(\mathbb{R})$, and

$$F^p(f*g)(\xi)=(F^p f)(\xi)\cdot(F^1 g)(\xi),\quad \xi\in\mathbb{R},$$

or simply, denoted by

$$(f*g)^\wedge(\xi)=f^\wedge(\xi)g^\wedge(\xi),\quad \xi\in\mathbb{R}.$$

Theorem 5. 2. 13 (FT of derivatives) If $f\in L^p(\mathbb{R})\cap AC_{\mathrm{Loc}}^{r-1}(\mathbb{R})$ $(1<p\leqslant 2)$, and $f^{(r)}\in L^p(\mathbb{R})$, $r\in\mathbb{Z}^+$, then Fourier transformation of $f^{(r)}$ satisfies

$$(f^{(r)})^\wedge(\xi)=(\mathrm{i}\xi)^r f^\wedge(\xi),\quad \text{a. e. } \xi\in\mathbb{R}.$$

Theorem 5. 2. 14 (derivatives of FT) If $f\in L^p(\mathbb{R})$ $(1<p\leqslant 2)$, and $(ix)^r f(x)\in L^1(\mathbb{R})$, $r\in\mathbb{Z}^+$, then the r-order derivative $(f^\wedge)^{(r)}(\xi)$ of L^p-Fourier transformation $f^\wedge$, regarded as $L^q(\mathbb{R})$-function, a. e. exists, and the formula

$$(f^\wedge)^{(r)}(\xi)=(-\mathrm{i})^r(\circ^r f(\circ))^\wedge(\xi),\quad \text{a. e. } \xi\in\mathbb{R}$$

holds.

4. Parseval formula of Fourier transformations

Theorem 5. 2. 15 (1) If $f, g\in L^p(\mathbb{R})$ $(1<p\leqslant 2)$, then Parseval formula $\int_{\mathbf{R}} f^\wedge(t)g(t)\,\mathrm{d}t=\int_{\mathbf{R}} f(t)g^\wedge(t)\,\mathrm{d}t$ holds;

(2) If $f, g\in L^2(\mathbb{R})$, then Plancherel-type formula $\int_{\mathbf{R}} f(t)\overline{g(t)}\,\mathrm{d}t=\frac{1}{2\pi}\int_{\mathbf{R}} f^\wedge(t)\overline{g^\wedge(t)}\,\mathrm{d}t$ holds; Denoted by $(\cdot,\cdot)$ the inner product of $L^2(\mathbb{R})$, then the formula is as $(f,\bar{g})=\frac{1}{2\pi}(f^\wedge,\overline{g^\wedge})$.

Note If the definitions of Fourier transformation and inverse Fourier transformation are in the symmetry forms, Plancherel formula takes form $(f,\bar{g})=(f^\wedge,\overline{g^\wedge})$.

5. Inversion formulas of Fourier transformations

Theorem 5. 2. 16 (1) If $f\in L^p(\mathbb{R})$ $(1<p\leqslant 2)$, then f and its Fourier transformation $f^\wedge$ holds $\lim\limits_{\rho\to+\infty}\left\| f(\circ)-\frac{1}{2\pi}\int_{(-\rho,\rho)} f^\wedge(\xi)\,\mathrm{e}^{\mathrm{i}(\circ\cdot\xi)}\,\mathrm{d}\xi\right\|_{L^p(\mathbf{R})}=0$;

(2) If $f\in L^p(\mathbb{R})$ $(1<p\leqslant 2)$, and $f^\wedge\in L^1(\mathbb{R})$, then the inversion formula of Fourier transformation $f(x)=\frac{1}{2\pi}\int_{\mathbf{R}} f^\wedge(\xi)\mathrm{e}^{\mathrm{i}(x\cdot\xi)}\,\mathrm{d}\xi$, a. e. $x\in\mathbb{R}$, holds; i. e. $f\in L^p(\mathbb{R})$,

$f^\wedge \in L^1(\mathbb{R}) \Rightarrow (f^\wedge)^\vee = f(x)$, a. e. $x \in \mathbb{R}$;

(3) If $f \in L^p(\mathbb{R})$ $(1 < p \leqslant 2)$, and $g, g^\wedge \in L^1(\mathbb{R})$, then the inversion formula of the multiplication of Fourier transformations

$$(f * g)(x) = \frac{1}{2\pi}\int_{\mathbf{R}} f^\wedge(\xi) g^\wedge(\xi) e^{i(x\cdot\xi)} d\xi, \quad x \in \mathbb{R}$$

holds.

The $L^1(\mathbb{R})$-summation factors can be generalized to $f \in L^p(\mathbb{R}), 1 < p \leqslant 2$, cases, then function f itself can be expressed by its Fourier transformation $f^\wedge \equiv F^p f \in L^q(\mathbb{R})$.

Definition 5. 2. 5 (summation factor) An even function $\theta(x) \in L^1(\mathbb{R})$ is said to be a **summation factor**, if it satisfies $\theta^\wedge(\xi) \in L^1(\mathbb{R})$, and $\int_{\mathbf{R}} \theta^\wedge(\xi) d\xi = 2\pi$; Moreover, if a summation factor $\theta(x)$ is a continuous function on $\mathbb{R}$, then $\theta(x)$ is said to be **a continuous summation factor.**

The important summation factors are: (1) **Cesaro** summation factor $\theta(x) = \begin{cases} 1-|x|, & |x| \leqslant 1 \\ 0, & |x| \leqslant 1 \end{cases}$; (2) **Abel** summation factor $\theta(x) = e^{-|x|}$; (3) **Gauss** summation factor $\theta(x) = e^{-x^2}$.

The formula $\frac{1}{2\pi}\int_{\mathbf{R}} f^\wedge(\xi) e^{i(x\cdot\xi)} d\xi$, $x \in \mathbb{R}$, is said to be **a Fourier inversion integral of** $f(x)$, and the formula $U(f, x, \rho) = \frac{1}{2\pi}\int_{\mathbf{R}} \theta\left(\frac{\xi}{\rho}\right) f^\wedge(\xi) e^{i(x\cdot\xi)} d\xi, x \in \mathbb{R}$, is said to be a **$\theta$-sum of Fourier inversion integral.**

Theorem 5. 2. 17 If $f \in L^p(\mathbb{R})(1 \leqslant p \leqslant 2)$, then for $\theta(\xi)$-summation operator, the θ-sum of Fourier inversion integral $U(f, x, \rho) = \frac{1}{2\pi}\int_{\mathbf{R}} \theta\left(\frac{\xi}{\rho}\right) f^\wedge(\xi) e^{i(x\cdot\xi)} d\xi, x \in \mathbb{R}$, satisfies

$$\lim_{\rho \to +\infty} \| U(f, \circ, \rho) - f(\circ) \|_{L^p(\mathbf{R})} = 0;$$

and

$$\| U(f, \circ, \rho) \|_{L^p(\mathbf{R})} \leqslant \| \theta^\wedge \|_{L^1(\mathbf{R})} \| f \|_{L^p(\mathbf{R})}, \quad \rho > 0.$$

On the other hand, if $\theta^\wedge \geqslant 0$, and $\theta^\wedge$ is monotonic decreasing on $[0, +\infty)$, then Fourier inversion integral is θ-summable to f, a. e., that is

$$\lim_{\rho \to +\infty} U(f, x, \rho) = \lim_{\rho \to +\infty} \frac{1}{2\pi}\int_{\mathbf{R}} \theta\left(\frac{\xi}{\rho}\right) f^\wedge(\xi) e^{i(x\cdot\xi)} d\xi = f(x), \quad \text{a. e.}, x \in \mathbb{R}.$$

By virtue of Theorem 5. 2. 17, we may prove Theorem 5. 2. 8 holds for the inversion formulas of Fourier transformation of $L^p(\mathbb{R})(1 \leqslant p \leqslant 2)$-functions.

6. Uniqueness theorem of Fourier transformation

Theorem 5. 2. 18 If $f, g \in L^p(\mathbb{R})(1 \leqslant p \leqslant 2)$, and $f^\wedge(\xi) = g^\wedge(\xi)$, a. e. , then

$$f(x) = g(x), \quad \text{a. e. } x \in \mathbb{R}.$$

The proof is left as exercise.

5. 3 Fourier Transform on Schwartz Distribution Space

5. 3. 1 Fourier transformations of Schwartz functions

We have defined Fourier transformations for $f \in L^p(\mathbb{R}^n)$ $(1 \leqslant p \leqslant 2)$-functions in sections 5. 1 and 5. 2, and shown the properties of them. Now, Fourier transformations of Schwartz functions, Fourier transformations of Schwartz distributions are discussed in this section.

1. Fourier transformations on $\mathbb{S}(\mathbb{R}^n)$

A Schwartz function $\varphi \in \mathbb{S}(\mathbb{R}^n)$ can be regarded as an $L^1(\mathbb{R}^n)$-function by $\mathbb{S}(\mathbb{R}^n) \subset L^1(\mathbb{R}^n)$, thus, we list the definitions of Fourier transformation and inverse Fourier transformation of $\mathbb{S}(\mathbb{R}^n)$-functions, as well as their properties based upon those of $L^1(\mathbb{R}^n)$-functions without proofs, and refer to [4] and [5] to readers who are interested in more knowledge and theory about Fourier analysis.

Definition 5. 3. 1 Fourier transformation of $\varphi \in \mathbb{S}(\mathbb{R}^n)$ is defined by

$$\varphi^\wedge(\xi) = \int_{\mathbb{R}^n} \varphi(x) e^{-i(x \cdot \xi)} dx, \quad \xi \in \mathbb{R}^n,$$

where $x \cdot \xi = \sum_{j=1}^n x_j \xi_j$ is the inner product of $x = (x_1, \cdots, x_n) \in \mathbb{R}^n$ and $\xi = (\xi_1, \cdots, \xi_n) \in \mathbb{R}^n$ (see Definition 5. 2. 1).

The inverse Fourier transformation of $\psi \in \mathbb{S}(\mathbb{R}^n)$ is defined by

$$\psi^\vee(x) = \frac{1}{(2\pi)^n} \int_{\mathbb{R}^n} \psi(\xi) e^{i(x \cdot \xi)} d\xi, \quad x \in \mathbb{R}^n,$$

with $x \cdot \xi = \sum_{j=1}^n x_j \xi_j$ of $x = (x_1, \cdots, x_n) \in \mathbb{R}^n$ and $\xi = (\xi_1, \cdots, \xi_n) \in \mathbb{R}^n$.

2. Properties of Fourier transform on $\mathbb{S}(\mathbb{R}^n)$

We use the following notations in $\mathbb{R}^n$ case.

$|x| = \sqrt{\sum_{j=1}^{n} x_j^2}$, for $x = (x_1, x_2, \cdots, x_n) \equiv (x_j) \in \mathbb{R}^n$;

$x^\beta = x_1^{\beta_1} x_2^{\beta_2} \cdots x_n^{\beta_n}$, for $x = (x_j) \in \mathbb{R}^n, \beta = (\beta_1, \beta_2, \cdots, \beta_n) \equiv (\beta_j) \in \mathbb{N}^n$;

$\alpha! = \alpha_1! \alpha_2! \cdots \alpha_n!$, for $\alpha = (\alpha_1, \alpha_2, \cdots, \alpha_n) \equiv (\alpha_j) \in \mathbb{N}^n$;

$\partial_j \equiv \partial_{x_j} = \frac{\partial}{\partial x_j}, j = 1, 2, \cdots, n$, for $x = (x_j) \in \mathbb{R}^n$;

$D_j = \frac{1}{\mathrm{i}} \partial_j, j = 1, 2, \cdots, n$, with $\mathrm{i} = \sqrt{-1}$;

$D_x^\alpha \equiv D_1^{\alpha_1} \cdots D_n^{\alpha_n} = \left(\frac{1}{\mathrm{i}}\right)^{\alpha_1 + \cdots + \alpha_n} \partial_{x_1}^{\alpha_1} \cdots \partial_{x_n}^{\alpha_n}$, for $x = (x_j) \in \mathbb{R}^n, \alpha = (\alpha_j) \in \mathbb{N}^n$.

1) Basic operation properties of Fourier transformation

Theorem 5.3.1 The operation properties of $\varphi \in \mathbb{S}(\mathbb{R}^n)$-Fourier transformations are:

(1) $(\tau_h \varphi(\circ))^\wedge(\xi) = \mathrm{e}^{-\mathrm{i}(h \cdot \xi)} \varphi^\wedge(\xi), \xi \in \mathbb{R}^n, h \in \mathbb{R}^n$; (FT of translation)

(2) $\tau_h \varphi^\wedge(\xi) = [\mathrm{e}^{\mathrm{i}(h \cdot \circ)} \varphi(\circ)]^\wedge(\xi), \xi \in \mathbb{R}^n, h \in \mathbb{R}^n$; (translation of FT)

(3) $[\varphi(\rho \circ)]^\wedge(\xi) = |\rho|^{-n} \varphi^\wedge(\rho^{-1}\xi), \xi \in \mathbb{R}^n, \rho \neq 0$; (FT of dilation)

(4) $\varphi^\wedge(\rho\xi) = |\rho|^{-n} [\varphi(\rho^{-1} \circ)]^\wedge(\xi), \xi \in \mathbb{R}^n, \rho \neq 0$; (dilation of FT)

(5) $(\widetilde{\varphi}(\circ))^\wedge(\xi) = (2\pi)^n \varphi^\vee(\xi), \xi \in \mathbb{R}^n$; (FT of reflection)

(6) $(\varphi^\wedge(\circ))^\sim(\xi) = (2\pi)^n \varphi^\vee(\xi), \xi \in \mathbb{R}^n$; (reflection of FT)

(7) $(\overline{\varphi}(\circ))^\wedge(\xi) = (2\pi)^n \overline{\varphi^\vee}(\xi), \xi \in \mathbb{R}^n$; (FT of conjugation)

(8) $\overline{\varphi^\wedge}(\xi) = (2\pi)^n (\overline{\varphi})^\vee(\xi), \xi \in \mathbb{R}^n$; (conjugation of FT)

(9) $(D_x^p \varphi(\circ))^\wedge(\xi) = \xi^p \varphi^\wedge(\xi), \xi \in \mathbb{R}^n, p \in \mathbb{P}^n$; (FT of derivative)

(10) $D_\xi^p \varphi^\wedge(\xi) = ((-\circ)^p \varphi(\circ))^\wedge(\xi), \xi \in \mathbb{R}^n, p \in \mathbb{P}^n$; (derivative of FT)

(11) $(\varphi * \psi(\circ))^\wedge(\xi) = \varphi^\wedge(\xi) \psi^\wedge(\xi), \xi \in \mathbb{R}^n$; (FT of convolution)

(12) $(\varphi^\wedge * \psi^\wedge)(\xi) = (2\pi)^n (\varphi(\circ) \cdot \psi(\circ))^\wedge(\xi), \xi \in \mathbb{R}^n$. (convolution of FT)

Proof Properties (1)-(4) can be proved by their definitions (compare with Theorem 5.2.1).

For (5), Fourier transformation $(\widetilde{\varphi})^\wedge$ exists by $\mathbb{S}(\mathbb{R}^n) \subset L^1(\mathbb{R}^n)$, then we deduce for $\varphi \in \mathbb{S}(\mathbb{R}^n)$ that

$$\begin{aligned}(\widetilde{\varphi}(\circ))^\wedge(\xi) &= \int_{\mathbb{R}^n} \varphi(-x) \mathrm{e}^{-\mathrm{i}(x \cdot \xi)} \mathrm{d}x = \int_{\mathbb{R}^n} \varphi(t) \mathrm{e}^{\mathrm{i}(t \cdot \xi)} \mathrm{d}t \\ &= (2\pi)^n \frac{1}{(2\pi)^n} \int_{\mathbb{R}^n} \varphi(t) \mathrm{e}^{\mathrm{i}(t \cdot \xi)} \mathrm{d}t = (2\pi)^n \varphi^\vee(\xi),\end{aligned}$$

since $\varphi \in \mathbb{S}(\mathbb{R}^n) \Rightarrow \widetilde{\varphi} \in \mathbb{S}(\mathbb{R}^n)$.

We prove (6)-(12) only for $\mathbb{R}$ case; and the proofs for those of $\mathbb{R}^n (n > 1)$ are similar.

For (6), we have

$$(\varphi^\wedge(\circ))^\sim(\xi)=\left(\int_{\mathbf{R}}\varphi(x)\mathrm{e}^{-\mathrm{i}x\xi}\,\mathrm{d}x\right)^\sim=\int_{\mathbf{R}}\varphi(x)\mathrm{e}^{\mathrm{i}x\xi}\,\mathrm{d}x$$
$$=(2\pi)^1\frac{1}{(2\pi)^1}\int_{\mathbf{R}}\varphi(x)\mathrm{e}^{\mathrm{i}x\xi}\,\mathrm{d}x=(2\pi)\ \varphi^\vee(\xi).$$

For (7), we have

$$(\overline{\varphi}(\circ))^\wedge(\xi)=\int_{\mathbf{R}}\overline{\varphi}(x)\mathrm{e}^{-\mathrm{i}x\xi}\,\mathrm{d}x=\overline{\int_{\mathbf{R}}\varphi(x)\mathrm{e}^{\mathrm{i}x\xi}\,\mathrm{d}x}$$
$$=(2\pi)^1\frac{1}{(2\pi)^1}\overline{\int_{\mathbf{R}}\varphi(x)\mathrm{e}^{\mathrm{i}x\xi}\,\mathrm{d}x}=(2\pi)\ \overline{\varphi^\vee}(\xi).$$

For (8), we have

$$\overline{\varphi^\wedge}(\xi)=\overline{\int_{\mathbf{R}}\varphi(x)\mathrm{e}^{-\mathrm{i}(x\cdot\xi)}\,\mathrm{d}x}=\int_{\mathbf{R}}\overline{\varphi}(x)\mathrm{e}^{\mathrm{i}x\xi}\,\mathrm{d}x$$
$$=(2\pi)^1\frac{1}{(2\pi)^1}\int_{\mathbf{R}}\overline{\varphi}(x)\mathrm{e}^{\mathrm{i}x\xi}\,\mathrm{d}x=(2\pi)\,(\overline{\varphi})^\vee(\xi).$$

For (9), the definition $\varphi^\wedge(\xi)=\int_{\mathbf{R}}\varphi(x)\mathrm{e}^{-\mathrm{i}x\xi}\,\mathrm{d}x$ gives

$$\xi^p\varphi^\wedge(\xi)=\int_{\mathbf{R}}\xi^p\varphi(x)\mathrm{e}^{-\mathrm{i}x\xi}\,\mathrm{d}x=\int_{\mathbf{R}}\varphi(x)\{(-D_x)^p(\mathrm{e}^{-\mathrm{i}x\xi})\}\,\mathrm{d}x$$
$$=\{\varphi(x)(\mathrm{e}^{-\mathrm{i}x\xi})\}\,|_{-\infty}^{+\infty}-\int_{\mathbf{R}}\{(-D_x)^p\varphi(x)\}\,\mathrm{e}^{-\mathrm{i}x\xi}\,\mathrm{d}x=\int_{\mathbf{R}}\{(D_x^p)\varphi(x)\}\,\mathrm{e}^{-\mathrm{i}x\xi}\,\mathrm{d}x.$$

For (10), we take the derivatives to $\varphi^\wedge(\xi)=\int_{\mathbf{R}}\varphi(x)\mathrm{e}^{-\mathrm{i}x\xi}\,\mathrm{d}x$, $\varphi\in\mathbb{S}(\mathbb{R})$, then the integral by parts, thus

$$D_\xi^p\varphi^\wedge(\xi)=\int_{\mathbf{R}}(-x)^p\varphi(x)\mathrm{e}^{-\mathrm{i}x\xi}\,\mathrm{d}x,$$

so that (10) holds.

Note We may deduce by (9) and (10), and get that $\varphi\in\mathbb{S}(\mathbb{R})$ implies $\varphi^\wedge\in\mathbb{S}(\mathbb{R})$. In fact, $\varphi\in\mathbb{S}(\mathbb{R})$ shows

$$|\xi^\beta\varphi^\wedge(\xi)|=\left|\int_{\mathbf{R}}\{(D_x^\beta)\varphi(x)\}\,\mathrm{e}^{-\mathrm{i}x\xi}\,\mathrm{d}x\right|\leqslant\int_{\mathbf{R}}|(D_x^\beta)\varphi(x)|\,\mathrm{d}x$$

and

$$|\xi^\beta D_\xi^p\varphi^\wedge(\xi)|=\left|\int_{\mathbf{R}}\mathrm{e}^{-\mathrm{i}x\xi}D_x^p\,[(-x)^\beta\varphi(x)]\,\mathrm{d}x\right|$$
$$\leqslant\left|\int_{\mathbf{R}}(1+|x|^2)^{-\frac{n+1}{2}}(1+|x|^2)^{\frac{n+1}{2}}D_x^p\,[(-x)^\beta\varphi(x)]\,\mathrm{d}x\right|$$
$$\leqslant c\sup_{\mathbf{R}}\left|(1+|x|^2)^{\frac{n+1}{2}}D_x^p\,[(-x)^\beta\varphi(x)]\right|,$$

this implies $\varphi^\wedge\in\mathbb{S}(\mathbb{R})$. Hence, we conclude that:

$$\varphi \in \mathbb{S}(\mathbb{R}) \Rightarrow \varphi^\wedge \in \mathbb{S}(\mathbb{R}) \Rightarrow (\varphi^\wedge)^\vee = \varphi = (\varphi^\vee)^\wedge.$$

For (11), Fubini theorem gives

$$(\varphi * \psi(\circ))^\wedge(\xi) = \int_{\mathbb{R}} (\varphi * \psi)(x) e^{-ix\xi} dx = \int_{\mathbb{R}}\int_{\mathbb{R}} e^{-i(x-y)\xi} \varphi(x-y)\psi(y) e^{-iy\xi} dy dx$$

$$= \int_{\mathbb{R}} e^{-ix\xi} \varphi(x) dx \cdot \int_{\mathbb{R}} e^{-iy\xi} \psi(y) dy = \varphi^\wedge(\xi)\psi^\wedge(\xi).$$

By the way, it holds $(\varphi * \psi(\circ))^\vee(\xi) = 2\pi\{\varphi^\vee(\xi)\psi^\vee(\xi)\}$, and holds

$$\varphi, \psi \in \mathbb{S}(\mathbb{R}) \Rightarrow \varphi^\wedge \psi^\wedge \in \mathbb{S}(\mathbb{R}) \Rightarrow (\varphi * \psi)^\wedge \in \mathbb{S}(\mathbb{R}) \Rightarrow \varphi * \psi \in \mathbb{S}(\mathbb{R}).$$

For (12), by the uniqueness of Fourier transformations and (11), we take inverse Fourier transformation on the left-hand side of (12)

$$[(\varphi^\wedge * \psi^\wedge)(\xi)]^\vee(x) = 2\pi(\varphi^\wedge)^\vee(x) \cdot (\psi^\wedge)^\vee(x) = 2\pi\, \varphi(x)\psi(x);$$

then take the inverse Fourier transformation on the right-hand side of (12),

$$((2\pi)(\varphi(\circ) \cdot \psi(\circ))^\wedge(\xi))^\vee(x) = (2\pi)\,[(\varphi(x) \cdot \psi(x))^\wedge]^\vee = 2\pi\, \varphi(x) \cdot \psi(x);$$

thus, (12) holds. The proof is complete.

2) Analytic properties of Fourier transformations

Theorem 5. 3. 2 Fourier transformation $F\varphi \equiv \varphi^\wedge$ of $\varphi \in \mathbb{S}(\mathbb{R}^n)$ has following properties:

(1) Riemann-Lebesgue lemma in $\mathbb{S}(\mathbb{R}^n)$: $\lim\limits_{|\xi| \to \infty} \varphi^\wedge(\xi) = 0$, $\forall \varphi \in \mathbb{S}(\mathbb{R}^n)$;

(2) F: $\varphi \to \varphi^\wedge$ is a continuous linear operator from $\mathbb{S}(\mathbb{R}^n)$ onto $\mathbb{S}(\mathbb{R}^n)$;

(3) F: $\mathbb{S}(\mathbb{R}^n) \to \mathbb{S}(\mathbb{R}^n)$ has continuous linear operator F^{-1}: $\mathbb{S}(\mathbb{R}^n) \to \mathbb{S}(\mathbb{R}^n)$; For Fourier transformation F: $\varphi \to \varphi^\wedge$ and inverse Fourier transformation F^{-1}: $\varphi^\wedge \to \varphi$, it holds inverse formula on $\mathbb{S}(\mathbb{R}^n)$: $\varphi(x) = \dfrac{1}{(2\pi)^n}\displaystyle\int_{\mathbb{R}^n} \varphi^\wedge(\xi) e^{i(\xi \cdot x)} d\xi$; i. e.

$$(\varphi^\wedge(\circ))^\vee(x) = \varphi(x) = (\varphi^\vee(\circ))^\wedge(x), \quad \forall \varphi \in \mathbb{S}(\mathbb{R}^n);$$

thus, F: $\mathbb{S}(\mathbb{R}^n) \to \mathbb{S}(\mathbb{R}^n)$ is isomorphic from $\mathbb{S}(\mathbb{R}^n)$ onto $\mathbb{S}(\mathbb{R}^n)$;

(4) Parseval formula on $\mathbb{S}(\mathbb{R}^n)$:

$$\int_{\mathbb{R}^n} \varphi^\wedge(x)\, \psi(x) dx = \int_{\mathbb{R}^n} \psi^\wedge(x)\varphi(x) dx; \quad (\langle \varphi^\wedge, \psi \rangle = \langle \varphi, \psi^\wedge \rangle)$$

Plancherel-type formula on $\mathbb{S}(\mathbb{R}^n)$:

$$\int_{\mathbb{R}^n} \varphi(x)\, \bar{\psi}(x)\, dx = \frac{1}{(2\pi)^n}\int_{\mathbb{R}^n} \varphi^\wedge(\xi)\, \overline{\psi^\wedge}(\xi)\, d\xi. \quad \left(\langle \varphi, \bar{\psi} \rangle = \frac{1}{(2\pi)^n}\langle \varphi^\wedge, \overline{\psi^\wedge} \rangle\right)$$

Proof For (1), by Riemann-Lebesgue lemma in $L^1(\mathbb{R}^n)$.

For (2), the continuity of F: since $\varphi_j \xrightarrow{\mathbb{S}(\mathbb{R}^n)} \varphi$ implies

$$(\varphi_j)^\wedge(\xi) = \int_{\mathbb{R}^n} \varphi_j(x) e^{-i(x \cdot \xi)} dx \xrightarrow{\mathbb{S}(\mathbb{R}^n)} \varphi^\wedge(\xi) = \int_{\mathbb{R}^n} \varphi(x) e^{-i(x \cdot \xi)} dx.$$

For (3), we prove it only for $\mathbb{R}$. By using Fourier transformation of Gaussian function

$g(x)=e^{-\frac{|x|^2}{2}}$,

$$g^\wedge(\xi)=\int_{\mathbf{R}} e^{-\frac{|x|^2}{2}} e^{-ix\xi}\,dx = e^{-\frac{|\xi|^2}{2}}\int_{\mathbf{R}} e^{-\frac{1}{2}(x+i\xi)^2}\,dx=\sqrt{2\pi}\,e^{-\frac{|\xi|^2}{2}}$$

(Euler integral $\int_{\mathbf{R}} e^{-\frac{t^2}{2}}\,dt=\sqrt{2\pi}$). Then,

$$\begin{aligned}\int_{\mathbf{R}}\varphi^\wedge(\xi)g(\xi)e^{ix\xi}\,d\xi &= \int_{\mathbf{R}} g(\xi)e^{ix\xi}\,d\xi\int_{\mathbf{R}}\varphi(y)e^{-iy\xi}\,dy\\ &=\int_{\mathbf{R}}\int_{\mathbf{R}} e^{i(x-y)\xi}g(\xi)\varphi(y)\,d\xi dy=\int_{\mathbf{R}} g^\wedge(y-x)\varphi(y)\,dy\\ &=\int_{\mathbf{R}} g^\wedge(y)\varphi(x+y)\,dy.\end{aligned}$$

Take $g(\varepsilon\xi)$ $(\varepsilon>0)$ to replace $g(\xi)$, and $\varepsilon^{-1}g^\wedge\left(\frac{y}{\varepsilon}\right)$ to replace $g^\wedge(y)$ in above equality; let $\frac{y}{\varepsilon}=y_1$, then

$$\int_{\mathbf{R}}\varphi^\wedge(\xi)g(\varepsilon\xi)\,e^{ix\xi}\,d\xi=\int_{\mathbf{R}}\varepsilon^{-1}g^\wedge\left(\frac{y}{\varepsilon}\right)\varphi(x+y)\,dy=\int_{\mathbf{R}} g^\wedge(y_1)\varphi(x+\varepsilon y_1)\,dy_1;$$

let $\varepsilon\to 0$, we have

$$g(0)\int_{\mathbf{R}}\varphi^\wedge(\xi)e^{ix\xi}\,d\xi=\varphi(x)\int_{\mathbf{R}} g^\wedge(y_1)\,dy_1.$$

Since $g(0)=1$, and

$$\int_{\mathbf{R}} g^\wedge(y_1)\,dy_1=\sqrt{2\pi}\int_{\mathbf{R}} e^{-\frac{|y_1|^2}{2}}\,dy_1=2\pi,$$

so that the inverse formula $\varphi(x)=\frac{1}{2\pi}\int_{\mathbf{R}}\varphi^\wedge(\xi)e^{ix\xi}\,d\xi$ holds.

The continuity of F^{-1} can be obtained from the above inverse formula.

Then, F: $\mathbb{S}(\mathbb{R}^n)\to\mathbb{S}(\mathbb{R}^n)$ is a bisingle, bicontinuous, keeping operations and a topological isomorphic mapping from $\mathbb{S}(\mathbb{R}^n)$ onto $\mathbb{S}(\mathbb{R}^n)$, and holds $(\varphi^\wedge)^\vee(x)=\varphi(x)=(\varphi^\vee)^\wedge(x)$. (3) is proved.

Finally, Parseval formula is from Fubini Theorem. For the proof of Plancherel-type formula, we refer to [4].

5.3.2 Fourier transformations of Schwartz distributions

1. Fourier transformations on $\mathbb{S}^*(\mathbb{R}^n)$

Definition 5.3.2 Fourier transformation of a distribution $T\in\mathbb{S}^*(\mathbb{R}^n)$ is defined as a distribution, denoted by $T^\wedge$, satisfying

$$\langle T^\wedge, \varphi\rangle = \langle T, \varphi^\wedge\rangle, \quad \forall \varphi \in \mathbb{S}(\mathbb{R}^n);$$

or, equivalently,

$$\langle T^\wedge, \varphi^\vee\rangle = \langle T, \varphi\rangle, \quad \forall \varphi \in \mathbb{S}(\mathbb{R}^n).$$

The inverse Fourier transformation of a distribution $S \in \mathbb{S}^*(\mathbb{R}^n)$ is defined as a distribution, denoted by $S^\vee$, satisfying

$$\langle S^\vee, \varphi\rangle = \langle S, \varphi^\vee\rangle, \quad \forall \varphi \in \mathbb{S}(\mathbb{R}^n);$$

or, equivalently,

$$\langle S^\vee, \varphi^\wedge\rangle = \langle S, \varphi\rangle, \quad \forall \varphi \in \mathbb{S}(\mathbb{R}^n).$$

2. Properties of Fourier transformation on $\mathbb{S}^*(\mathbb{R}^n)$

1) Basic operation properties of FT of Schwartz distributions

Theorem 5.3.3 The basic operation properties of FT of distribution $T \in \mathbb{S}^*(\mathbb{R}^n)$ are:

(1) $(\tau_h T)^\wedge = e^{-i(h\cdot\xi)} T^\wedge, h, \xi \in \mathbb{R}^n$; (FT of translation)

(2) $\tau_h T^\wedge = (e^{i(h\cdot\circ)} T)^\wedge, h \in \mathbb{R}^n$; (translation of FT)

(3) $(\widetilde{T})^\wedge = (2\pi)^n T^\vee$; (FT of reflection)

(4) $(T^\wedge)^\sim = (2\pi)^n T^\vee$; (reflection of FT)

(5) $(D^\alpha T)^\wedge = \xi^\alpha T^\wedge, \xi \in \mathbb{R}^n, \alpha \in \mathbb{N}^n$; (FT of derivative)

(6) $(-D)^\alpha T^\wedge = (\circ^\alpha T)^\wedge, \alpha \in \mathbb{N}^n$. (derivative of FT)

Proof For (1), take $\varphi \in \mathbb{S}(\mathbb{R}^n)$ and $\forall h \in \mathbb{R}^n$, then, $(\tau_h T)^\wedge$ acts on $\varphi \in \mathbb{S}(\mathbb{R}^n)$

$$\langle (\tau_h T)^\wedge, \varphi\rangle = \langle \tau_h T, \varphi^\wedge\rangle = \langle T, \tau_{-h}\varphi^\wedge\rangle;$$

and $e^{-i(h\cdot\xi)} T^\wedge$ acts on $\varphi \in \mathbb{S}(\mathbb{R}^n)$

$$\langle e^{-i(h\cdot\xi)} T^\wedge, \varphi\rangle = \langle T^\wedge, e^{-i(h\cdot\xi)}\varphi\rangle = \langle T, (e^{-i(h\cdot\circ)}\varphi)^\wedge\rangle = \langle T, \tau_{-h}\varphi^\wedge\rangle.$$

The two actions are equal on all $\varphi \in \mathbb{S}(\mathbb{R}^n)$, then we get $(\tau_h T)^\wedge = e^{-i(h\cdot\xi)} T^\wedge$, $\forall h \in \mathbb{R}^n$, $\forall \xi \in \mathbb{R}^n$.

For (2), take $\varphi \in \mathbb{S}(\mathbb{R}^n)$, and $\forall h \in \mathbb{R}^n$, then, $\tau_h T^\wedge$ acts on $\varphi \in \mathbb{S}(\mathbb{R}^n)$

$$\begin{aligned}\langle \tau_h T^\wedge, \varphi\rangle &= \langle T^\wedge, \tau_{-h}\varphi\rangle = \langle T, (\tau_{-h}\varphi)^\wedge\rangle = \langle T, e^{i(h\cdot\xi)}\varphi^\wedge\rangle \\ &= \langle e^{i(h\cdot\xi)} T, \varphi^\wedge\rangle = \langle (e^{i(h\cdot\circ)} T)^\wedge, \varphi\rangle;\end{aligned}$$

This implies $\tau_h T^\wedge = (e^{i(h\cdot\circ)} T)^\wedge, h \in \mathbb{R}^n$, then (2) holds.

For (3), take $\varphi \in \mathbb{S}(\mathbb{R}^n)$, then $(\widetilde{T})^\wedge$ acts on $\varphi \in \mathbb{S}(\mathbb{R}^n)$, and by Theorem 5.3.1(5), we have

$$\langle (\widetilde{T})^\wedge, \varphi\rangle = \langle \widetilde{T}, \varphi^\wedge\rangle = \langle T, \widetilde{\varphi}^\wedge\rangle = \langle T, (2\pi)^n \varphi^\vee\rangle = \langle (2\pi)^n T, \varphi^\vee\rangle;$$

and $(2\pi)^n T^\vee$ acts on $\varphi \in \mathbb{S}(\mathbb{R}^n)$, it implies $\langle (2\pi)^n T^\vee, \varphi\rangle = \langle (2\pi)^n T, \varphi^\vee\rangle$, so this implies that (3) holds, i. e. $(\widetilde{T})^{\wedge}=(2\pi)^n T^{\vee}$.

For (4), the equality $\langle (T^{\wedge})^{\sim},\varphi\rangle=\langle T^{\wedge},\widetilde{\varphi}\rangle=\langle T,(\widetilde{\varphi})^{\wedge}\rangle=\langle T,(2\pi)^n\varphi^{\vee}\rangle=\langle (2\pi)^n T^{\vee},\varphi\rangle$ implies (4).

For (5), take $\varphi\in \mathbb{S}(\mathbb{R}^n)$, and $\alpha=(\alpha_j)\in\mathbb{N}^n$, with $|\alpha|=\sum_{j=1}^{n}\alpha_j$. $(D^{\alpha}T)^{\wedge}$ acts on $\varphi\in\mathbb{S}(\mathbb{R}^n)$,

$$\begin{aligned}\langle (D^{\alpha}T)^{\wedge},\varphi\rangle&=\langle D^{\alpha}T,\varphi^{\wedge}\rangle=(-1)^{|\alpha|}\langle T,D^{\alpha}\varphi^{\wedge}\rangle=(-1)^{|\alpha|}\langle T,(-1)^{|\alpha|}(-D)^{\alpha}\varphi^{\wedge}\rangle\\&=(-1)^{2|\alpha|}\langle T,(-D)^{\alpha}\varphi^{\wedge}\rangle=\langle T,(-D)^{\alpha}\varphi^{\wedge}\rangle=\langle T,(\circ^{\alpha}\varphi)^{\wedge}\rangle;\end{aligned}$$

and $\xi^{\alpha}T^{\wedge}$ acts on $\varphi\in\mathbb{S}(\mathbb{R}^n)$, it implies $\langle \xi^{\alpha}T^{\wedge},\varphi\rangle=\langle T^{\wedge},\xi^{\alpha}\varphi\rangle=\langle T,(\circ^{\alpha}\varphi)^{\wedge}\rangle$, so that it holds $(D^{\alpha}T)^{\wedge}=\xi^{\alpha}T^{\wedge}$, $\xi\in\mathbb{R}^n$, this is (5).

For (6), take $\varphi\in\mathbb{S}(\mathbb{R}^n)$ and $\alpha\in\mathbb{N}^n$, by Theorem 5. 3. 1 (8), we have

$$\langle (\circ^{\alpha}T)^{\wedge},\varphi\rangle=\langle x^{\alpha}T,\varphi^{\wedge}\rangle=\langle T,\xi^{\alpha}\varphi^{\wedge}\rangle=\langle T,(D^{\alpha}\varphi)^{\wedge}\rangle=\langle T^{\wedge},D^{\alpha}\varphi\rangle=\langle (-D)^{\alpha}T^{\wedge},\varphi\rangle,$$

this implies $(-D)^{\alpha}T^{\wedge}=(\circ^{\alpha}T)^{\wedge}$, (6) holds. The proof is complete.

2) Analytic properties of FT of Schwartz distributions

Theorem 5. 3. 4 Fourier transformation $T^{\wedge}$ of distribution $T\in\mathbb{S}^*(\mathbb{R}^n)$ has the following properties:

(1) $F: T\to T^{\wedge}$ is a continuous linear operator from $\mathbb{S}^*(\mathbb{R}^n)$ onto $\mathbb{S}^*(\mathbb{R}^n)$ in the weak* topology of $\mathbb{S}^*(\mathbb{R}^n)$;

(2) $F:\mathbb{S}^*(\mathbb{R}^n)\to\mathbb{S}^*(\mathbb{R}^n)$ has the inverse continuous mapping $F^{-1}:\mathbb{S}^*(\mathbb{R}^n)\to\mathbb{S}^*(\mathbb{R}^n)$; moreover, for $F:\varphi\to\varphi^{\wedge}$ and $F^{-1}:\varphi^{\wedge}\to\varphi$, they hold

$$(T^{\wedge})^{\vee}=T \text{ and } (T^{\vee})^{\wedge}=T$$

on $\mathbb{S}^*(\mathbb{R}^n)$; Thus, Fourier transformation $F:\mathbb{S}^*(\mathbb{R}^n)\to\mathbb{S}^*(\mathbb{R}^n)$ is a topological isomorphism from $\mathbb{S}^*(\mathbb{R}^n)$ onto $\mathbb{S}^*(\mathbb{R}^n)$.

Proof For (1), the linearity of $F:\mathbb{S}^*(\mathbb{R}^n)\to\mathbb{S}^*(\mathbb{R}^n)$ can be obtained by the definition; the continuity of $F:\mathbb{S}^*(\mathbb{R}^n)\to\mathbb{S}^*(\mathbb{R}^n)$ is obtained by the continuity of $F:\mathbb{S}(\mathbb{R}^n)\to\mathbb{S}(\mathbb{R}^n)$ (Theorem 5. 3. 2), thus, the operator $F:\mathbb{S}^*(\mathbb{R}^n)\to\mathbb{S}^*(\mathbb{R}^n)$ is a continuous linear operator from $\mathbb{S}^*(\mathbb{R}^n)$ onto $\mathbb{S}^*(\mathbb{R}^n)$ in the weak*-topology of $\mathbb{S}^*(\mathbb{R}^n)$.

For (2), we prove Fourier transformation $F:\mathbb{S}(\mathbb{R}^n)\to\mathbb{S}(\mathbb{R}^n)$ has inverse mapping $F^{-1}:\mathbb{S}^*(\mathbb{R}^n)\to\mathbb{S}^*(\mathbb{R}^n)$, firstly. In fact, Fourier transformation $F:\mathbb{S}(\mathbb{R}^n)\to\mathbb{S}(\mathbb{R}^n)$ is a topological isomorphism mapping by Theorem 5. 3. 2, thus for each $T\in\mathbb{S}^*(\mathbb{R}^n)$, the action $\langle T,\varphi\rangle$ $(\varphi\in\mathbb{S}(\mathbb{R}^n))$ is also determined by $\varphi^{\wedge}$, and is also a continuous linear functional on space $\{\varphi^{\wedge}:\varphi\in\mathbb{S}(\mathbb{R}^n)\}$, say, g. However, $\{\varphi^{\wedge}:\varphi\in\mathbb{S}(\mathbb{R}^n)\}=\mathbb{S}(\mathbb{R}^n)$, so that g is also a continuous linear functional on $\mathbb{S}(\mathbb{R}^n)$, and holds $\langle T,\varphi\rangle=\langle g,\varphi^{\wedge}\rangle$. This implies $T^{\vee}=g$ by definition. Hence, $F^{-1}T=T^{\vee}$ makes sense, and $F^{-1}T$ can be obtained by $\langle F^{-1}T,\varphi^\wedge\rangle \equiv \langle T^\vee,\varphi^\wedge\rangle = \langle T,\varphi\rangle$. Thus, we obtain that $(T^\wedge)^\vee = T$, $(T^\vee)^\wedge = T$. The continuity and linearity of F^{-1} can be obtained from $\langle F^{-1}T,\varphi^\wedge\rangle = \langle T,\varphi\rangle$, immediately. So that F^{-1}: $\mathbb{S}^*(\mathbb{R}^n) \to \mathbb{S}^*(\mathbb{R}^n)$ is a one-one, onto, continuous linear operator from $\mathbb{S}^*(\mathbb{R}^n)$ onto $\mathbb{S}^*(\mathbb{R}^n)$, and Fourier transformation of Schwartz distribution F: $\mathbb{S}^*(\mathbb{R}^n) \to \mathbb{S}^*(\mathbb{R}^n)$ is a topological isomorphism from $\mathbb{S}^*(\mathbb{R}^n)$ onto $\mathbb{S}^*(\mathbb{R}^n)$. The proof is complete.

Theorem 5.3.5 For distribution $T \in \mathbb{S}^*(\mathbb{R}^n)$ and function $g \in \mathbb{S}(\mathbb{R}^n)$, we have

(1) $\langle T,\bar{g}\rangle = \dfrac{1}{(2\pi)^n}\langle T^\wedge,\overline{g^\wedge}\rangle$;

(2) $\langle T,g\rangle = \langle T^\wedge,g^\vee\rangle$.

Proof For (1), to prove $\langle T,\bar{g}\rangle = \dfrac{1}{(2\pi)^n}\langle T^\wedge,\overline{g^\wedge}\rangle$, take $g \in \mathbb{S}(\mathbb{R}^n)$, and rewrite $\bar{g}$ as

$$\bar{g} = [(\bar{g})^\vee]^\wedge = \left[\frac{1}{(2\pi)^n}\int_{\mathbf{R}^n}\bar{g}(\xi)e^{ix\cdot\xi}d\xi\right]^\wedge \equiv h^\wedge,$$

then,

$$\langle T,\bar{g}\rangle = \langle T,h^\wedge\rangle = \langle T^\wedge,h\rangle = \langle T^\wedge,\frac{1}{(2\pi)^n}\int_{\mathbf{R}^n}\bar{g}(\xi)e^{ix\cdot\xi}d\xi\rangle$$

$$= \left\langle T^\wedge,\frac{1}{(2\pi)^n}\overline{\int_{\mathbf{R}^n}g(\xi)e^{-ix\cdot\xi}d\xi}\right\rangle = \frac{1}{(2\pi)^n}\langle T^\wedge,\overline{g^\wedge}\rangle,$$

This is (1).

For (2), we may get $\langle T,g\rangle = \langle T^\wedge,g^\vee\rangle$ by $\langle T^\wedge,g^\vee\rangle = \langle T,(g^\vee)^\wedge\rangle = \langle T,g\rangle$. Thus, (2) holds.

5.3.3 Schwartz distributions with compact supports

1. Examples of distributions in $E^*(\mathbb{R}^n)$ and $\mathbb{S}^*(\mathbb{R}^n)$

Example 5.3.1 $L^p(\mathbb{R}^n) \subset \mathbb{S}^*(\mathbb{R}^n)$ $(1 \leqslant p < +\infty)$

We have shown in Example 5.1.3 that $L^1(\mathbb{R}^n) \subset \mathbb{S}^*(\mathbb{R}^n)$, now, we show in this example that for each $f \in L^p(\mathbb{R}^n)$ with $1 < p < +\infty$ implies $f \in \mathbb{S}^*(\mathbb{R}^n)$. In fact, the integral

$$\langle f,\varphi\rangle = \int_{\mathbf{R}^n}f(x)\varphi(x)dx, \quad \forall\varphi \in \mathbb{S}(\mathbb{R}^n)$$

defines a linear functional on $\mathbb{S}(\mathbb{R}^n)$. Then by Hölder inequality, the following estimation

$$|\langle f,\varphi\rangle| = \left|\int_{\mathbf{R}^n}f(x)\varphi(x)dx\right| \leqslant \|f\|_{L^p(\mathbf{R}^n)}\|\varphi\|_{L^q(\mathbf{R}^n)}, \quad \forall\varphi \in \mathbb{S}(\mathbb{R}^n) \subset L^q(\mathbb{R}^n)$$

shows the continuity of $\langle f,\varphi\rangle$, $\forall\varphi \in \mathbb{S}(\mathbb{R}^n)$. So that $L^p(\mathbb{R}^n) \subset \mathbb{S}^*(\mathbb{R}^n)$, $1 < p < +\infty$.

Next, we prove $\varphi\in \mathbb{S}(\mathbb{R}^n) \Rightarrow \varphi\in L^p(\mathbb{R}^n), 1<p<+\infty$. It holds

$$\|\varphi\|_{L^p(\mathbf{R}^n)}=\left(\int_{\mathbf{R}^n}|\varphi(x)|^p\mathrm{d}x\right)^{\frac{1}{p}}\leqslant\sup_{\mathbf{R}^n}(1+|x|^2)^{\frac{n+1}{2p}}|\varphi(x)|\left[\int_{\mathbf{R}^n}(1+|x|^2)^{\frac{-n-1}{2}}\mathrm{d}x\right]^{\frac{1}{p}}$$

$$=c\sup_{\mathbf{R}^n}(1+|x|^2)^{\frac{n+1}{2p}}|\varphi(x)|\leqslant c_1<+\infty,$$

this implies $\mathbb{S}(\mathbb{R}^n)\subset L^p(\mathbb{R}^n)\subset \mathbb{S}^*(\mathbb{R}^n)$, and shows $\forall g\in L^p(\mathbb{R}^n)\subset\mathbb{S}^*(\mathbb{R}^n)$, the action $\langle g,\varphi\rangle$ on $\varphi\in\mathbb{S}(\mathbb{R}^n)$ satisfies $|\langle g,\varphi\rangle|\leqslant c_1\sup_{\mathbf{R}^n}(1+|x|^2)^{\frac{n+1}{2p}}|\varphi(x)|$. Thus, $f\in\mathbb{S}(\mathbb{R}^n)$ is a regular Schwartz distribution, however, not necessary having compact support.

Example 5.3.2 The set of all polynomials with constant coefficients is a $\mathbb{S}^*(\mathbb{R}^n)$-distribution.

We only need to consider single item $x^m, m\in\mathbb{N}$. It is clear that $x^m\in L^1_{\text{loc}}(\mathbb{R}^n)$, and

$$\langle x^m,\varphi\rangle=\int_{\mathbf{R}^n}x^m\varphi(x)\mathrm{d}x,\quad \forall\varphi\in\mathbb{S}(\mathbb{R}^n),$$

the integral exists since $\lim\limits_{|x|\to+\infty}x^s\cdot x^m\varphi(x)=0, \forall s\in\mathbb{P}^n$; The linearity and continuity are evident. However, $x^m\in L^1_{\text{loc}}(\mathbb{R}^n)$ does not have compact support.

Example 5.3.3 Dirac distribution $\delta\in E^*(\mathbb{R}^n)\subset\mathbb{S}^*(\mathbb{R}^n)$

The δ-distribution is a Schwartz distribution, defined by $\langle\delta,\varphi\rangle=\varphi(0), \forall\varphi\in\mathbb{S}(\mathbb{R}^n)$, and has compact support with $\operatorname{supp}\delta=\{0\}$.

The translation $\tau_{t_0}\delta$ of δ-distribution

$$\langle\tau_{t_0}\delta,\varphi\rangle=\langle\delta,\tau_{-t_0}\varphi\rangle=\langle\delta,\varphi(t+t_0)\rangle=\varphi(t+t_0)\,|_{t=0}=\varphi(t_0),\ \forall\varphi\in\mathbb{S}(\mathbb{R}^n).$$

Thus, $\tau_{t_0}\delta\in\mathbb{S}^*(\mathbb{R}^n)$. And $\tau_{t_0}\delta\in E^*(\mathbb{R}^n)$, with $\operatorname{supp}\tau_{t_0}\delta=\{t_0\}$, thus, both δ and $\tau_{t_0}\delta$ are singular distribution with compact support.

Example 5.3.4 Fourier transformation $\delta^\wedge$ of Dirac δ-distribution

Evaluate Fourier transformation $\delta^\wedge$ of δ-distribution:

$$\langle\delta^\wedge,\varphi\rangle=\langle\delta,\varphi^\wedge\rangle=\varphi^\wedge(0)=\int_{\mathbf{R}^n}\varphi(x)\mathrm{e}^{-\mathrm{i}x\cdot 0}\mathrm{d}x=\langle 1,\varphi\rangle,\quad\forall\varphi\in\mathbb{S}(\mathbb{R}^n),$$

thus, $\delta^\wedge=1$, and $\delta^\wedge\in L^1_{\text{loc}}(\mathbb{R}^n)$ is a regular distribution, but does not have compact support.

Example 5.3.5 The distribution δ_{t_0} with support $\{t_0\}$ (see Example 5.3.3)

The distribution δ_{t_0} with $\operatorname{supp}\delta_{t_0}=\{t_0\}$ is the translation of δ-distribution, $\delta_{t_0}=\tau_{t_0}\delta$. Its Fourier transformation is

$$(\delta_{t_0})^\wedge=(\tau_{t_0}\delta)^\wedge=\mathrm{e}^{-\mathrm{i}t_0\cdot\xi}\delta^\wedge=\mathrm{e}^{-\mathrm{i}t_0\cdot\xi},\quad\xi\in\mathbb{R}.$$

by Theorem 5.3.3(2). This implies

$$\delta_{t_0} \in E^*(\mathbb{R}^n), \quad (\delta_{t_0})^\wedge \in S^*(\mathbb{R}^n).$$

However, we see that $(\delta_{t_0})^\wedge = e^{-it_0 \cdot \xi}$, $\xi \in \mathbb{R}$, is a regular distribution without compact support.

Example 5.3.6 Fourier transformation of 1

By definition, we have

$$\begin{aligned}\langle 1^\wedge, \varphi \rangle &= \langle 1, \varphi^\wedge \rangle = \int_{\mathbf{R}^n} \varphi^\wedge(\xi) d\xi = \int_{\mathbf{R}^n} \varphi^\wedge(\xi) e^{i(0\cdot\xi)} d\xi \\ &= (2\pi)^n \frac{1}{(2\pi)^n} \int_{\mathbf{R}^n} \varphi^\wedge(\xi) e^{-i(0\cdot\xi)} d\xi = (2\pi)^n (\varphi^\wedge)^\vee(0) \\ &= (2\pi)^n \varphi(0) = (2\pi)^n \langle \delta, \varphi \rangle, \quad \forall \varphi \in S(\mathbb{R}^n),\end{aligned}$$

and have $1^\wedge = (2\pi)^n \delta \in E^*(\mathbb{R}^n)$, a singular distribution with compact support.

In one-dimensional case, it is $1^\wedge = (2\pi)\delta$.

2. Constructions of $E^*(\mathbb{R}^n)$-distributions and $S^*(\mathbb{R}^n)$-distributions

For the function spaces $D(\mathbb{R}^n) \equiv C_C^\infty(\mathbb{R}^n)$, $E(\mathbb{R}^n) \equiv C^\infty(\mathbb{R}^n)$ and Schwartz space $S(\mathbb{R}^n)$, it holds

$$C_C^\infty(\mathbb{R}^n) \subset S(\mathbb{R}^n) \subset C^\infty(\mathbb{R}^n),$$

and correspondingly, for the dual spaces, it holds

$$D^*(\mathbb{R}^n) \supset S^*(\mathbb{R}^n) \supset E^*(\mathbb{R}^n).$$

Theorem 5.3.6 (construction theorem of $E^*(\mathbb{R}^n)$-distribution) For a distribution space $E^*(\mathbb{R}^n)$, we have

(1) $E^*(\mathbb{R}^n) = \{S \in D^*(\mathbb{R}^n) : \text{supp } S \text{ is compact set}\}$;

(2) $S \in E^*(\mathbb{R}^n) \Leftrightarrow S = \sum_{|p| \leqslant r} \partial^p f, f \in C_C^r(\mathbb{R}^n), p \in \mathbb{N}^n, r \in \mathbb{N}$

with $|p| = p_1 + \cdots + p_n$, and supp $f \subset (\text{supp } S)_\varepsilon$, here $(\text{supp } S)_\varepsilon$ is an ε-neighborhood of supp S;

(3) If supp $S = \{0\}$, then $S = \sum_{|p| \leqslant k} c_p \delta^{(p)}$; i.e. the distribution S with support $\{0\}$ can be expressed by a finite linear combinations of Dirac distribution $\delta \in E^*(\mathbb{R}^n)$ and derivatives $\delta^{(p)}$ $(p \in \mathbb{N}^n)$ with $k \in \mathbb{N}$, $c_p \in \mathbb{R}$.

Theorem 5.3.7 (construction theorem of $S^*(\mathbb{R}^n)$-distribution) For a distribution $T \in S^*(\mathbb{R}^n)$, we have

$$T \in S^*(\mathbb{R}^n) \Leftrightarrow T = \partial^p [(1 + |x|^2)^{\frac{k}{2}} f(x)],$$
$$f \in C(\mathbb{R}^n) \cap B(\mathbb{R}^n), \quad p \in \mathbb{N}^n, \quad k \in \mathbb{N},$$

with $B(\mathbb{R}^n)$, the bounded function class on $\mathbb{R}^n$.

We refer to [5] for the proofs of these two theorems.

5.3.4 Fourier transformations of convolutions of Schwartz distributions

1. Classical results

Definition 5.3.3 (softened kernel, Fredrish softened operator) Let $\alpha \in C_C^\infty(\mathbb{R}^n)$ be a function on $\mathbb{R}^n$ with $\alpha \geqslant 0$, $\int_{\mathbf{R}^n} \alpha(x)\mathrm{d}x = 1$, and $\operatorname{supp}\alpha = \overline{B_1(0)}$, the closed unit ball. Then $\alpha(x)$ is said to be a **Fredrish softened kernel.** Let $\alpha_\varepsilon(x) = \varepsilon^{-n}\alpha\left(\frac{x}{\varepsilon}\right)$, $\varepsilon > 0$. It is clear that:

$$\alpha_\varepsilon \in C_C^\infty(\mathbb{R}^n), \alpha_\varepsilon \geqslant 0;\quad \int_{\mathbf{R}^n} \alpha_\varepsilon(x)\mathrm{d}x = 1;\ \operatorname{supp}\alpha_\varepsilon = \overline{B_\varepsilon(0)};\ \lim_{\varepsilon\to 0}\operatorname{supp}\alpha_\varepsilon(x) = \{0\}.$$

Take $\varepsilon = j^{-1}$ $(j = 1, 2, \cdots)$, and define a number sequence by

$$\alpha_j(x) \equiv \alpha_{\varepsilon = j^{-1}}(x) = j^n \alpha(jx),\quad j = 1, 2, \cdots,$$

it is said to be **a regularized sequence**; and the convolution

$$J_\varepsilon f(x) = f * \alpha_\varepsilon(x) = \int_{\mathbf{R}^n} f(t)\,\alpha_\varepsilon(x - t)\,\mathrm{d}t$$

is said to be **a Fredrish softened operator of function** $f(x)$.

Theorem 5.3.8 For $J_\varepsilon f(x) = f * \alpha_\varepsilon(x)$, we have

(1) If $f \in L_{\mathrm{loc}}(\mathbb{R}^n)$, then $f * \alpha_\varepsilon \in C^\infty(\mathbb{R}^n)$;

(2) If $f \in L_{\mathrm{loc}}(\mathbb{R}^n)$, and $\operatorname{supp} f = K \subset \mathbb{R}^n$ is a compact set, then

$$\operatorname{supp} f * \alpha_\varepsilon = K_\varepsilon = \bigcup_{x\in K} B_\varepsilon(x);$$

(3) If $f \in C(\mathbb{R}^n)$, then $\lim\limits_{\varepsilon\to 0} f * \alpha_\varepsilon(x) = f(x)$, uniformly on any compact set $L \subset \mathbb{R}^n$;

(4) If $f \in L^p(\mathbb{R}^n)$, $1 \leqslant p < +\infty$, then $\lim\limits_{\varepsilon\to 0} f * \alpha_\varepsilon \overset{L^p}{=} f$, that is,

$$\lim_{\varepsilon\to 0} \| f * \alpha_\varepsilon - f \|_{L^p(\mathbf{R}^n)} = 0;$$

(5) If $f \in C_C^\infty(\mathbb{R}^n)$, then $f * \alpha_\varepsilon \xrightarrow{D^*(\mathbf{R}^n)} f\ (\varepsilon \to 0)$, that is,

$$\lim_{\varepsilon\to 0} f * \alpha_\varepsilon = f,\ \text{uniformly on any bounded set } B \subset C_C^\infty(\mathbb{R}^n).$$

Example 5.3.7 The sense of "$\lim\limits_{\varepsilon\to 0} f * \alpha_\varepsilon = f$" in Theorem 5.3.8(5)— the term $f * \alpha_\varepsilon$ is regarded as a result of the action of convolution operator "$J_\varepsilon \equiv \alpha_\varepsilon *$" with $J_\varepsilon f \equiv \alpha_\varepsilon * f \equiv f * \alpha_\varepsilon$, its limit is regarded as a result of the action of unit operator I with $If \equiv f$. In this point of view, **the convolution operator** $J_\varepsilon \equiv \alpha_\varepsilon *$ **has limit unit operator** I **in any bounded set uniformly when** $\varepsilon \to 0$; Similarly, "$\lim\limits_{\varepsilon\to 0} \| f * \alpha_\varepsilon - f \|_{L^p(\mathbf{R}^n)} = 0$" in Theorem

5.3.8(4) tells that: **the convolution operator $J_\varepsilon \equiv \alpha_\varepsilon *$ has limit unit operator I in $L^p(\mathbb{R}^n)$-mean sense when** $\varepsilon \to 0$.

Example 5.3.8 Let $h(t)$ be the **pulse response function** of a linear system, $f(t)$ be an input signal (Fig. 5.3.1). What is the output signal?

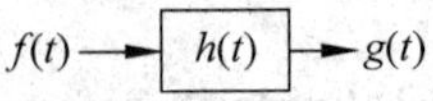

Fig. 5.3.1

If an input $f(t)$ is coming into the linear system, since the time retards at $t-\tau$, and the frequency is mixed, then, the output signal is

$$g(t)=f*h(t)=\int f(\tau-t)\,h(\tau)\,\mathrm{d}\tau,$$

thus, the concept of convolutions of functions possesses very important and realistic senses in the signal analysis and physics fields.

2. Convolutions of distributions

If $f, g \in D(\mathbb{R}) \subset D^*(\mathbb{R})$, are two any-order derivable functions with compact supports, then the convolution of f and g

$$f*g(x)=\int_{\mathbf{R}} f(x-t)\,g(t)\,\mathrm{d}t=\int_{\mathbf{R}} f(y)\,g(x-y)\,\mathrm{d}y=\langle f(y), g(x-y)\rangle$$

can be regarded as a convolution operator $f*g$, and can act on $\varphi \in D(\mathbb{R})$, denoted by

$$\langle f*g, \varphi\rangle=\left\langle \int_{\mathbf{R}} f(t)\,g(x-t)\,\mathrm{d}t, \varphi\right\rangle=\int_{\mathbf{R}} \varphi(x)\left(\int_{\mathbf{R}} f(t) g(x-t)\,\mathrm{d}t\right)\mathrm{d}x$$

$$=\int_{\mathbf{R}} f(t)\left(\int_{\mathbf{R}} \varphi(t+y) g(y)\,\mathrm{d}y\right)\mathrm{d}t=\int_{\mathbf{R}} f(x)\left(\int_{\mathbf{R}} \varphi(x+y) g(y)\,\mathrm{d}y\right)\mathrm{d}x$$

$$=\langle f(x), \langle g(y), \varphi(x+y)\rangle\rangle.$$

This motivates for defining convolutions of distributions "f" and "g" as

$$\langle f*g, \varphi\rangle=\langle f_{(x)}, \langle g_{(y)}, \varphi(x+y)\rangle\rangle.$$

Definition 5.3.4 (convolution of distributions) Consider two cases.

1) **Convolution of distribution** $T \in D^*(\mathbb{R}^n)$ **and function** $f \in D(\mathbb{R}^n)$

For $T \in D^*(\mathbb{R}^n)$, $f \in D(\mathbb{R}^n)$, **the convolution** $T*f$ **is a distribution satisfying**

$$\langle T*f, \varphi\rangle=\langle T_x, \langle f_y, \varphi(x+y)\rangle\rangle, \quad \forall \varphi \in D(\mathbb{R}^n),$$

denoted by $\langle T*f, \varphi\rangle=\langle T_x, \langle f_y, \varphi_{x+y}\rangle\rangle$, $\forall \varphi \in D(\mathbb{R}^n)$.

Note The definition makes sense, the reason is: since $f \in D(\mathbb{R}^n)$, and $\forall \varphi \in D(\mathbb{R}^n)$, it holds

$$\psi(x) \equiv \langle f(y), \varphi(x+y)\rangle=\int_{\mathbf{R}^n} f(y)\varphi(x+y)\,\mathrm{d}y \in C^\infty(\mathbb{R}^n);$$

because it can be deduced as

$$\psi(x)=\int_{\mathbb{R}^n} f(-y)\varphi(x-y)\,\mathrm{d}y=\int_{\mathbb{R}^n}\tilde{f}(y)\,\varphi(x-y)\,\mathrm{d}y=\tilde{f}*\varphi(x) \quad (5.3.1)$$

with $\tilde{f}(x)=f(-x)$, and $\psi\in D(\mathbb{R}^n)$ (since $\mathrm{supp}\psi\subset\mathrm{supp}\tilde{f}+\mathrm{supp}\varphi$ is a compact set for $\tilde{f},\varphi\in D(\mathbb{R}^n)$).

On the other hand, $T*f$ is a continuous linear operator on $D(\mathbb{R}^n)$, since $\langle T*f,\varphi\rangle=\langle T_x,\langle f_{(y)},\varphi(x+y)\rangle\rangle$ is linear, firstly; about continuity, for $\{\varphi_j\}_{j=1}^{+\infty}\subset D(\mathbb{R}^n)$, we have "$\varphi_j\xrightarrow{D(\mathbb{R}^n)}0$ implies $\langle f_{(y)},\varphi_j(x+y)\rangle\xrightarrow{\mathbb{R}}0$", then this implies $\langle T*f,\varphi_j\rangle=\langle T_x,\langle f_{(y)},\varphi_j(x+y)\rangle\rangle\xrightarrow{\mathbb{R}}0$. So that $T*f$ is continuous operator. This tells that by Definition 5.3.4, $\langle T*f,\varphi\rangle=\langle T_x,\langle f_y,\varphi(x+y)\rangle\rangle$, $\forall\varphi\in D(\mathbb{R}^n)$ makes sense.

2) **Convolution of distribution** $T\in D^*(\mathbb{R}^n)$ **and distribution** $S\in E^*(\mathbb{R}^n)$

For $T\in D^*(\mathbb{R}^n)$, $S\in E^*(\mathbb{R}^n)$, i. e., supp S is compact, then **the convolution** $T*S$ is defined as **a distribution satisfying**

$$\langle T*S,\varphi\rangle=\langle T_x\otimes S_y,\varphi(x+y)\rangle,\ \forall\varphi\in D(\mathbb{R}^n),$$

where

$$\langle T_x\otimes S_y,u(x,y)\rangle=\langle T_x,\langle S_y,u(x,y)\rangle\rangle=\langle S_y,\langle T_x,u(x,y)\rangle\rangle,$$
$$\forall u(x,y)\in D(\mathbb{R}^n\times\mathbb{R}^n),$$

and $T\otimes S$ is with $\langle T\otimes S,\varphi\otimes\psi\rangle=\langle T,\varphi\rangle\langle S,\psi\rangle$, it is **a tensor product of** T **and** S.

Note The definition of $T*S$ is only for compact support supp$S$, and it makes sense by tensor theory; if supp T is compact, $T*S$ is also well-defined. However, if the condition "one support is compact" is removed, that is

$$T,S\in D^*(\mathbb{R}^n)\backslash E^*(\mathbb{R}^n),$$

then more theories are needed, we will omit this case, and refer to [5], [8].

Theorem 5.3.9 We have the following results:

(1) If $T\in D^*(\mathbb{R}^n)$, and $f\in D(\mathbb{R}^n)$, then $T*f(x)$ is a function

$$T*f(x)=\langle T_y,f(x-y)\rangle,\quad x\in\mathbb{R}^n, \quad (5.3.2)$$

and satisfies

① $T*f\in C^\infty(\mathbb{R}^n)\subset S^*(\mathbb{R}^n)$, and holds $(T*f)^\wedge=T^\wedge f^\wedge$;

② $\mathrm{supp}(T*f)\subset\mathrm{supp}T+\mathrm{supp}f$;

(2) If $T\in D^*(\mathbb{R}^n)$, and $S\in E^*(\mathbb{R}^n)$, then $T*S$ is a distribution

$$\langle T*S,\varphi\rangle=\langle T_x,\langle S_y,\varphi(x+y)\rangle\rangle,\quad\forall\varphi\in D(\mathbb{R}^n), \quad (5.3.3)$$

and satisfies

① $T*S\in D^*(\mathbb{R}^n)$;

② $\mathrm{supp}(T*S)\subset\mathrm{supp}T+\mathrm{supp}S$;

(3) If $T \in \mathbb{S}^*(\mathbb{R}^n)$, and $\varphi \in \mathbb{S}(\mathbb{R}^n)$, then $T * \varphi$ is a function

$$T * \varphi(x) = \langle T_y, \varphi(x-y) \rangle, \quad x \in \mathbb{R}^n, \tag{5.3.4}$$

and satisfies

① $T * \varphi \in \mathbb{O}_M^n(\mathbb{R}^n)$, where $\mathbb{O}_M^n(\mathbb{R}^n) \subset \mathbb{S}^*(\mathbb{R}^n)$ is the function space

$\mathbb{O}_M^n(\mathbb{R}^n) = \{f \in C^\infty(\mathbb{R}^n): \forall \alpha, \exists c_\alpha > 0, \exists N_\alpha, \text{s.t. } |D^\alpha f(x)| \leqslant c_\alpha (1+|x|^2)^{N_\alpha}\}$;

and holds $(T * \varphi)^\wedge = T^\wedge \varphi^\wedge$;

② $\operatorname{supp}(T * \varphi) \subset \operatorname{supp} T + \operatorname{supp} \varphi$;

(4) If $T \in \mathbb{S}^*(\mathbb{R}^n)$, and $S \in E^*(\mathbb{R}^n)$, then

① $S^\wedge(\xi) = \langle S_x, e^{-i(x \cdot \xi)} \rangle, \xi \in \mathbb{R}^n$; and $T * S \in \mathbb{S}^*(\mathbb{R}^n)$;

② $\operatorname{supp}(T * S) \subset \operatorname{supp} T + \operatorname{supp} S$.

Proof We only prove (1), (4), and refer to [5] for the others.

For (1) ① $\forall T \in D^*(\mathbb{R}^n), f \in D(\mathbb{R}^n)$, by Definition 5.3.4(1), it holds

$$\langle T * f, \varphi \rangle = \langle T_x, \langle f(y), \varphi(x+y) \rangle \rangle, \quad \forall \varphi \in D(\mathbb{R}^n)$$

and it makes sense. On the other hand, $f \in D(\mathbb{R}^n)$ and $\varphi \in D(\mathbb{R}^n)$ imply that

$$\langle f(y), \varphi(x+y) \rangle = \int_{\mathbb{R}^n} f(y) \varphi(x+y) \mathrm{d}y = \int_{\mathbb{R}^n} f(\tau - x) \varphi(\tau) \mathrm{d}\tau;$$

Thus

$$\langle T * f, \varphi \rangle = \langle T_x, \langle f(y), \varphi(x+y) \rangle \rangle = \langle T_x, \int_{\mathbb{R}^n} f(\tau - x) \varphi(\tau) \mathrm{d}\tau \rangle. \tag{5.3.5}$$

Since φ and f are in $D(\mathbb{R}^n)$ with compact supports, and infinitely derivable, the term $\int_{\mathbb{R}^n} f(\tau - x) \varphi(\tau) \mathrm{d}\tau$ is

$$\int_{\mathbb{R}^n} f(\tau - x) \varphi(\tau) \mathrm{d}\tau = \Sigma_\tau f(\tau - x) \varphi(\tau), \tag{5.3.6}$$

here the right-hand side is a convergent series. Take "action of T_x" on both sides of (5.3.6), we get

$$\left\langle T_x, \int_{\mathbb{R}^n} f(\tau - x) \varphi(\tau) \mathrm{d}\tau \right\rangle = \langle T_x, \Sigma_\tau f(\tau - x) \varphi(\tau) \rangle. \tag{5.3.7}$$

Again, by the compact supports and infinitely derivable of φ and f, the order of "action of T_x" and "summation" in (5.3.7) can be exchanged, such that it holds

$$\langle T_x, \Sigma_\tau f(\tau - x) \varphi(\tau) \rangle = \Sigma_\tau \langle T_x, f(\tau - x) \rangle \varphi(\tau).$$

The right-hand side of the above equality is an integral summation of

$$\int_{\mathbb{R}^n} \langle T_x, f(\tau - x) \rangle \varphi(\tau) \mathrm{d}\tau,$$

so we have

$$\Sigma_\tau \langle T_x, f(\tau - x) \rangle \varphi(\tau) = \int_{\mathbb{R}^n} \langle T_x, f(\tau - x) \rangle \varphi(\tau) \mathrm{d}\tau. \tag{5.3.8}$$

Since the right-hand side of (5.3.8) is $\langle\langle T_x, f(\tau-x)\rangle, \varphi(\tau)\rangle$, then combine with (5.3.5), it holds $\langle\langle T_x, f(\tau-x)\rangle, \varphi(\tau)\rangle = \langle T * f, \varphi\rangle$; change x to y, and τ to x, finally, we get $T * f(x) = \langle T_y, f(x-y)\rangle, x \in \mathbb{R}^n$. This is (5.3.2).

To prove $T * f \in C^\infty(\mathbb{R}^n)$, take any $\alpha, \beta \in \mathbb{N}^n$, then $f \in D(\mathbb{R}^n) \Rightarrow \partial_x^\alpha \partial_\xi^\beta f(x-y) \in D(\mathbb{R}^n)$. Thus, say, $n=1$, for $T \in D^*(\mathbb{R})$, and $h \in \mathbb{R}$ with $h \to 0$, it follows

$$\frac{\mathrm{d}}{\mathrm{d}x}[T * f(x)] = \lim_{h\to 0}\frac{1}{h}[T * f(x+h) - T * f(x)]$$

$$= \lim_{h\to 0}\frac{1}{h}\{\langle T_y, f(x+h-y)\rangle - \langle T_y, f(x-y)\rangle\} \quad (\text{in } D^*(\mathbb{R}))$$

$$= \lim_{h\to 0}\left\langle T_y, \frac{f(x-y+h)-f(x-y)}{h}\right\rangle = \left\langle T_y, \partial_x f(x-y)\right\rangle.$$

For $p \in \mathbb{N}^n$, we can evaluate $D^p[T * f(x)]$, similarly. Hence, $T * f \in C^\infty(\mathbb{R}^n)$. This is the first assertion in(1) ①.

Then, we prove the formula of convolution $(T * f)^\wedge = T^\wedge f^\wedge$ in(1) ①. That is, we evaluate $(T * f)^\wedge$ for $T \in D^*(\mathbb{R}^n)$, $f \in D(\mathbb{R}^n)$— since by $T * f \in C^\infty(\mathbb{R}^n) \subset \mathbb{S}^*(\mathbb{R}^n)$ and (5.3.3), we have

$$\langle (T * f)^\wedge, \varphi\rangle = \langle T * f, \varphi^\wedge\rangle = \langle T_x, \langle f_y, \varphi^\wedge(x+y)\rangle\rangle, \quad \varphi \in \mathbb{S}(\mathbb{R}^n). \tag{5.3.9}$$

For $\langle f_y, \varphi^\wedge(x+y)\rangle$, we evaluate

$$(f^\wedge \varphi)^\wedge(x) = \int_{\mathbb{R}^n} \mathrm{e}^{-\mathrm{i}(x\cdot\xi)}[f^\wedge(\xi)\varphi(\xi)]\,\mathrm{d}\xi$$

$$= \int_{\mathbb{R}^n}\langle f_y, \mathrm{e}^{-\mathrm{i}(y\cdot\xi)}\rangle\varphi(\xi)\mathrm{e}^{-\mathrm{i}(x\cdot\xi)}\,\mathrm{d}\xi$$

$$= \int_{\mathbb{R}^n}\langle f_y, \mathrm{e}^{-\mathrm{i}(x+y)\cdot\xi}\rangle\varphi(\xi)\,\mathrm{d}\xi$$

$$= \left\langle f_y, \int_{\mathbb{R}^n}\mathrm{e}^{-\mathrm{i}(x+y)\cdot\xi}\varphi(\xi)\,\mathrm{d}\xi\right\rangle = \langle f_y, \varphi^\wedge(x+y)\rangle,$$

this shows that

$$(f^\wedge\varphi)^\wedge = \langle f_y, \varphi^\wedge(x+y)\rangle;$$

substitute into (5.3.9), it follows

$$\langle (T * f)^\wedge, \varphi\rangle = \langle T_x, \langle f_y, \varphi^\wedge(x+y)\rangle\rangle = \langle T, (f^\wedge\varphi)^\wedge\rangle = \langle T^\wedge, f^\wedge\varphi\rangle$$

$$= \langle T^\wedge \cdot f^\wedge, \varphi\rangle, \quad \varphi \in \mathbb{S}(\mathbb{R}^n).$$

This implies the formula of convolution: $(T * f)^\wedge = T^\wedge f^\wedge$.

For (1)②: Prove relation $\operatorname{supp}(T * S) \subset \operatorname{supp}T + \operatorname{supp}S$—

Take $T \in D^*(\mathbb{R}^n)$, $f \in D(\mathbb{R}^n)$, suppose that $x \notin \operatorname{supp}T + \operatorname{supp}f$, then $y \in \operatorname{supp}f$ implies $x - y \notin \operatorname{supp}f$ (otherwise, it holds $x \in \{y\} + \operatorname{supp}f \subset \operatorname{supp}T + \operatorname{supp}f$); this means that "$x \notin \text{supp}T + \text{supp}f \Leftrightarrow \text{supp}T \cap \text{supp}f(x - \cdot) = \varnothing$".

Hence, $T * f = 0$ in a neighborhood of x. So that $x \notin \text{supp}(T * f)$. Thus we assert that "$x \notin \text{supp}T + \text{supp}f$" implies "$x \notin \text{supp}(T * f)$". This implies $\text{supp}(T * f) \subset \text{supp}T + \text{supp}f$.

Next, we prove (4).

For (4)①: Prove the first formula $S^\wedge(\xi) = \langle S_x, e^{-i(x \cdot \xi)} \rangle, \xi \in \mathbb{R}^n$ in (4)①—since $S \in E^*(\mathbb{R}^n)$, then

$$\begin{aligned}\langle S^\wedge, \varphi \rangle &= \langle S, \varphi^\wedge \rangle = \langle S_x, \langle \varphi_\xi, e^{-i(x\cdot\xi)} \rangle\rangle = \langle S_x \otimes \varphi_\xi, e^{-i(x\cdot\xi)} \rangle = \langle \varphi_\xi \otimes S_x, e^{-i(x\cdot\xi)} \rangle \\ &= \langle \varphi_\xi, \langle S_x, e^{-i(x\cdot\xi)} \rangle\rangle = \langle \langle S_x, e^{-i(x\cdot\xi)} \rangle, \varphi_\xi \rangle, \quad \forall \varphi \in \mathbb{S}(\mathbb{R}^n)\end{aligned}$$

this implies $S^\wedge(\xi) = \langle S_x, e^{-i(x \cdot \xi)} \rangle$.

Prove the second formula $T * S \in \mathbb{S}^*(\mathbb{R}^n)$ in (4)① — take $T \in \mathbb{S}^*(\mathbb{R}^n), S \in E^*(\mathbb{R}^n)$, by $\mathbb{S}^*(\mathbb{R}^n) \subset D^*(\mathbb{R}^n)$, it holds $\langle T * S, \varphi \rangle = \langle T_x, \langle S_y, \varphi_{x+y} \rangle\rangle$ (Definition 5. 3. 4 (2)). The Theorem 5. 3. 6 (2) gives

$$S \in E^*(\mathbb{R}^n) \Leftrightarrow S = \sum_{|p| \leqslant r} \partial^p f, \quad f \in C_C^r(\mathbb{R}^n),$$

where f is an r-order continuous derivable function with compact support, and $\text{supp}f \subset (\text{supp } S)_\varepsilon$.

Let $S_y = g(y) = \partial^p f(y)$, then

$$\psi(x) \equiv \langle S_y, \varphi(x+y) \rangle = \langle g(y), \varphi(x+y) \rangle = (-1)^{|p|} \int f(y) \partial_x^p \varphi(x+y) dy.$$

By using Peetre inequality[5], it follows that

$$\begin{aligned}|(1+|x|^2)^{\frac{k}{2}} \partial_x^q \psi(x)| &\leqslant C \int |f(y)(1+|y|^2)^{\frac{k}{2}}| \, |(1+|x+y|^2)^{\frac{k}{2}} \partial_x^{p+q} \varphi(x+y)| dy \\ &\leqslant C \sup_{\mathbb{R}^n} |(1+|x|^2)^{\frac{k}{2}} \partial_x^{p+q} \varphi(x)|,\end{aligned}$$

then, $\varphi \in \mathbb{S}(\mathbb{R}^n)$ implies $\psi(x) \equiv \langle g(y), \varphi(x+y) \rangle \in \mathbb{S}(\mathbb{R}^n)$. We rewrite $\psi(x) \equiv \langle S_y, \varphi(x+y) \rangle$, then

$$\langle T_x, \langle S(y), \varphi(x+y) \rangle\rangle = \langle T_x * S_y, \varphi(x+y) \rangle, \quad \forall T \in \mathbb{S}^*(\mathbb{R}^n)$$

makes sense, and $T * S \in \mathbb{S}^*(\mathbb{R}^n)$.

For (4) ② Prove the relation $\text{supp}(T * S) \subset \text{supp}T + \text{supp}S$ —

Take $T \in D^*(\mathbb{R}^n), S \in E^*(\mathbb{R}^n)$, suppose that $x \notin \text{supp}T + \text{supp}S$, then $y \in \text{supp}S$ implies $x - y \notin \text{supp}S$ (otherwise, it holds $x \in \{y\} + \text{supp}S \subset \text{supp}T + \text{supp}S$); this means that "$x \notin \text{supp}T + \text{supp}S \Leftrightarrow \text{supp}T \cap \text{supp}f(x - \cdot) = \varnothing$".

Hence, $T * f = 0$ in a neighborhood of x. So that $x \notin \text{supp}(T * S)$. Thus we assert that "$x \notin \text{supp}T + \text{supp}S$" implies "$x \notin \text{supp}(T * S)$". This implies $\text{supp}(T * S) \subset$

supp$T$+supp$S$. The proofs are complete.

Note In Theorem 5.3.9 above, the multiplication formula of Fourier transformations of convolutions of Schwartz distributions holds only in the cases (1),(3), but not in that of (2),(4). However, when does the multiplication formula in cases (2),(4) hold ?

In fact, we have proved in Theorem 5.3.9,(2),(4):

$$T \in D^*(\mathbb{R}^n),\ S \in E^*(\mathbb{R}^n) \Rightarrow T * S \in D^*(\mathbb{R}^n);$$

$$T \in \mathbb{S}^*(\mathbb{R}^n),\ S \in E^*(\mathbb{R}^n) \Rightarrow T * S \in \mathbb{S}^*(\mathbb{R}^n).$$

On the other hand, we have by Theorem 5.3.4,

$$T * S \in \mathbb{S}^*(\mathbb{R}^n) \Leftrightarrow (T * S)^\wedge \in \mathbb{S}^*(\mathbb{R}^n).$$

In the case (2), the condition $T \in D^*(\mathbb{R}^n)$ and result $T * S \in D^*(\mathbb{R}^n)$ both can not guarantee the existences of Fourier transformations $T^\wedge$, $(T * S)^\wedge$. So that for this case we do not have, in generally, the formula of convolution $(T * S)^\wedge = T^\wedge S^\wedge$.

In the case (4), the conditions $T \in \mathbb{S}^*(\mathbb{R}^n)$, $S \in E^*(\mathbb{R}^n)$ and result $T * S \in D^*(\mathbb{R}^n)$, all guarantee the existences of Fourier transformations $T^\wedge$, $S^\wedge$, $(T * S)^\wedge$. So that it is reasonable to ask whether we have the formula $(T * S)^\wedge = T^\wedge S^\wedge$.

We evaluate $(T * S)^\wedge$ in this case, firstly. Take $\varphi \in D(\mathbb{R}^n) \subset \mathbb{S}(\mathbb{R}^n)$, then by $T * S \in \mathbb{S}^*(\mathbb{R}^n) \subset D^*(\mathbb{R}^n)$ and (5.3.3), it holds

$$\langle (T * S)^\wedge, \varphi \rangle = \langle T * S, \varphi^\wedge \rangle = \langle T_x, \langle S_y, \varphi^\wedge(x+y) \rangle \rangle, \quad \varphi \in \mathbb{S}(\mathbb{R}^n). \tag{5.3.10}$$

Then, evaluate $\langle S_y, \varphi^\wedge(x+y) \rangle$: by (4)①, we have

$$\begin{aligned}\int_{\mathbb{R}^n} \mathrm{e}^{-\mathrm{i}(x\cdot\xi)} [(S_\xi^\wedge)(\xi) \cdot \varphi(\xi)]\,\mathrm{d}\xi &= \int_{\mathbb{R}^n} \langle S_y, \mathrm{e}^{-\mathrm{i}(y\cdot\xi)} \rangle \varphi(\xi) \mathrm{e}^{-\mathrm{i}(x\cdot\xi)}\,\mathrm{d}\xi \\ &= \int_{\mathbb{R}^n} \langle S_y, \mathrm{e}^{-\mathrm{i}(x+y)\cdot\xi} \rangle \varphi(\xi)\,\mathrm{d}\xi \\ &= \left\langle S_y, \int_{\mathbb{R}^n} \mathrm{e}^{-\mathrm{i}(x+y)\cdot\xi} \varphi(\xi)\,\mathrm{d}\xi \right\rangle = \langle S_y, \varphi^\wedge(x+y) \rangle,\end{aligned}$$

this shows that

$$\langle S_y, \varphi^\wedge(x+y) \rangle = (S^\wedge \cdot \varphi)^\wedge \equiv (S^\wedge \varphi)^\wedge;$$

substitute into (5.3.9), it follows

$$\langle (T * S)^\wedge, \varphi \rangle = \langle T * S, \varphi^\wedge \rangle = \langle T_x, \langle S_y, \varphi^\wedge(x+y) \rangle \rangle = \langle T, (S^\wedge \cdot \varphi)^\wedge \rangle. \tag{5.3.11}$$

We have a formula $\langle (T * S)^\wedge, \varphi \rangle = \langle T, (S^\wedge \cdot \varphi)^\wedge \rangle \equiv \langle T, (S^\wedge \varphi)^\wedge \rangle$ which can be used to evaluate $(T * S)^\wedge$.

Here in (5.3.11), $S^\wedge \varphi$ is the product of Schwartz distribution $S^\wedge \in \mathbb{S}^*(\mathbb{R}^n)$ with Schwartz function $\varphi \in \mathbb{S}(\mathbb{R}^n)$, it is defined as a distribution satisfying

$$\langle S^\wedge \varphi, \psi \rangle = \langle S^\wedge, \varphi\psi \rangle, \quad \forall \psi \in \mathbb{S}^*(\mathbb{R}^n),$$

and makes sense $S^\wedge \varphi \in \mathbb{S}^*(\mathbb{R}^n)$.

Unfortunately, $S \in E^*(\mathbb{R}^n) \subset \mathbb{S}^*(\mathbb{R}^n)$ only implies $S^\wedge \varphi \in \mathbb{S}^*(\mathbb{R}^n)$, but not guarantees $S^\wedge \varphi \in \mathbb{S}(\mathbb{R}^n)$. So that, we only have formula (5.3.11) $\langle (T * S)^\wedge, \varphi \rangle = \langle T, (S^\wedge \varphi)^\wedge \rangle$, but " Does $\langle T, (S^\wedge \varphi)^\wedge \rangle = \langle T^\wedge, S^\wedge \varphi \rangle$ hold ? " and " Does $\langle T^\wedge, S^\wedge \varphi \rangle = \langle T^\wedge S^\wedge, \varphi \rangle$ hold ?" These problems make the formula $(T * S)^\wedge = T^\wedge S^\wedge$ may fail.

3. Properties of convolutions of distributions

Theorem 5.3.10 For $T \in D^*(\mathbb{R}^n)$, $f, g \in D(\mathbb{R}^n)$, $\alpha \in \mathbb{P}^n$, $\delta \in E^*(\mathbb{R}^n)$, then

(1) $(T * f) * g = T * (f * g)$; $(T * \delta) * g = T * (\delta * g)$; (combination law)

(2) $T * f = f * T$; (exchange law)

(3) $T * \delta = \delta * T = T$; (unit element)

(4) $\tau_h T = (\tau_h \delta) * T$; (translation of distribution)

(5) $\tau_h (T * f) = (\tau_h T) * f = T * (\tau_h f)$; (translation of convolution)

(6) $D^\alpha (T * f) = D^\alpha T * f = T * D^\alpha f$. (derivative of convolution)

Proof For (1), by definition,

$$\langle T * f, \varphi \rangle = \langle T_x, \langle f_y, \varphi_{x+y} \rangle \rangle, \forall \varphi \in D(\mathbb{R}^n), x, y \in \mathbb{R}^n.$$

Take $\varphi \in D(\mathbb{R}^n)$, *then the action of* $(T * f) * g$ *on* $\forall \varphi \in D(\mathbb{R}^n)$ is

$$\langle (T * f) * g, \varphi \rangle = \langle (T * f)_x, \langle g_y, \varphi_{x+y} \rangle \rangle = \langle T_x, \langle f_y, \langle g_y, \varphi_{x+y} \rangle \rangle \rangle$$
$$= \langle T_x, \langle (f * g)_y, \varphi_{x+y} \rangle \rangle = \langle T_x * (f * g)_y, \varphi_{x+y} \rangle = \langle T * (f * g), \varphi \rangle;$$

this implies the first formula in (1); the second one $(T * \delta) * g = T * (\delta * g)$ can be shown, similarly.

For (2), by two steps:

First step—If $T_1, T_2 \in D^*(\mathbb{R}^n)$ and $T_1 * \varphi = T_2 * \varphi$, $\forall \varphi \in D(\mathbb{R}^n)$, then $T_1 = T_2$.

In fact, Theorem 5.3.9 shows that for $\varphi \in D(\mathbb{R}^n)$,

$$T * f(x) = \langle T_y, f(x - y) \rangle = \langle T_y, \tilde{f}(y - x) \rangle, \quad x \in \mathbb{R}^n;$$

let $x = 0$ in above equality, then $T * f(0) = \langle T_y, \tilde{f}(y) \rangle$, also, $T * \tilde{f}(0) = \langle T_y, f(y) \rangle = \langle T, f \rangle$. Hence,

$$\langle T_1, \varphi \rangle = (T_1 * \tilde{\varphi})(0) = (T_2 * \tilde{\varphi})(0) = \langle T_2, \varphi \rangle, \quad \forall \varphi \in D(\mathbb{R}^n).$$

So, $T_1 = T_2$.

Second step—If $T \in D^*(\mathbb{R}^n)$, $f \in D(\mathbb{R}^n)$, then $T * f = f * T$.

In fact, take $\varphi, \psi \in D(\mathbb{R}^n)$, then $\varphi * \psi = \psi * \varphi \in D(\mathbb{R}^n)$. Thus,

$$(T*f)*(\varphi*\psi)=(T*f)*(\psi*\varphi)=T*(f*\psi*\varphi)$$

combination in $D(\mathbb{R}^n)$ (first equality) combination low in (1) (second equality)

$$=T*(f*\psi)*\varphi=T*\varphi*(f*\psi)$$
$$=(T*\varphi)*(f*\psi)=(f*\psi)*(T*\varphi)$$
$$=f*\psi*T*\varphi=f*T*\psi*\varphi=(f*T)*(\varphi*\psi)$$

$\psi*T=T*\psi$ (second equality) $\psi*\varphi=\varphi*\psi$ (third equality)

the (2) is proved by the first step result.

For (3), by two steps:

First step—If $T=f\in D(\mathbb{R}^n)$, then $\delta*f=f$.

We have by definition that $\delta*f(x)=\langle\delta,f(y+x)\rangle=f(x+y)|_{y=0}=f(x)$; then regard δ as $\delta\in D^*(\mathbb{R}^n)$ and $f\in D(\mathbb{R}^n)$, then $\delta*f=f*\delta$.

Second step— If $T\in D^*(\mathbb{R}^n)$, then $(T*\delta)*\varphi=T*(\delta*\varphi)=T*\varphi$, thus $T*\delta=\delta*T=T$.

The property (3) means that $\delta\in D^*(\mathbb{R}^n)$ is the unit element of convolution operation "$*$" in $D^*(\mathbb{R}^n)$.

For (4), to prove $\tau_h T=(\tau_h\delta)*T$, take $\varphi\in D(\mathbb{R}^n)$, then, $\delta*f=f*\delta$. And

$$\langle(\tau_h\delta)*T,\varphi\rangle=\langle T*(\tau_h\delta),\varphi\rangle=\langle T_x,\langle(\tau_h\delta)_y,\varphi_{x+y}\rangle\rangle=\langle T_x,(\tau_h\delta)_y*\varphi_{x+y}\rangle$$
$$=\langle T_x,\varphi_{x+h}\rangle=\langle T,\tau_{-h}\varphi\rangle=\langle\tau_h T,\varphi\rangle.$$

For (5), to prove $\tau_h(T*f)=(\tau_h T)*f=T*(\tau_h f)$, take $\varphi\in D(\mathbb{R}^n)$, then

$$\langle\tau_h(T*f),\varphi\rangle=\langle T*f,\tau_{-h}\varphi\rangle=\langle(T*f)_x,\varphi(x+h)\rangle=\langle T_x,\langle f_y,\varphi_{x+y+h}\rangle\rangle$$
$$=\langle T_x,\langle f_y,\tau_{-h}\varphi_{x+y}\rangle\rangle=\langle T_x,\langle\tau_h f_y,\varphi_{x+y}\rangle\rangle$$
$$=\langle T_x*\tau_h f_y,\varphi_{x+y}\rangle=\langle T*\tau_h f,\varphi\rangle;$$

moreover, it holds $\tau_h(T*f)=\tau_h(f*T)$ by similar deduction and commutative law, and then it follows

$$\tau_h(f*T)=(\tau_h f)*T=T*(\tau_h f),$$

this implies (5).

The property (5) means: The convolution operation has **invariance under the translations.**

For (6), to prove $D^\alpha(T*f)=D^\alpha T*f=T*D^\alpha f$, take $\varphi\in D(\mathbb{R}^n)$, then for any $p=\alpha\in\mathbb{N}^n$,

$$\langle\partial^\alpha(T*f),\varphi\rangle=(-1)^{|\alpha|}\langle T*f,\partial^\alpha\varphi\rangle=\langle T_x,(-1)^{|\alpha|}\langle f_y,\partial_y^\alpha\varphi_{x+y}\rangle\rangle$$
$$=\langle T_x,\langle\partial_x^\alpha f_y,\varphi_{x+y}\rangle\rangle=\langle T*\partial^\alpha f,\varphi\rangle;$$

this is $D^\alpha(T*f)=T*D^\alpha f$; similarly, by $T*f=f*T$ and $D^\alpha(T*f)=D^\alpha(f*T)$, it holds

$$D^\alpha(T*f)=f*D^\alpha T=D^\alpha T*f.$$

The proofs are complete.

Theorem 5.3.11 If $T\in D^*(\mathbb{R}^n)$, $\varphi\in D(\mathbb{R}^n)$, then $\langle T,\varphi\rangle=(T_\xi*\tilde{\varphi})(0)$.

4. Semi-normed algebra on $E^*(\mathbb{R}^n)$

Theorem 5.3.12 (1) If $T*\alpha_\varepsilon(x)$ with $T\in D^*(\mathbb{R}^n)$, then it holds

$$\lim_{\varepsilon\to 0} T*\alpha_\varepsilon=T,\quad \text{in the strong topology of } D^*(\mathbb{R}^n).$$

(2) Distribution space $(E^*(\mathbb{R}^n),+,\alpha\cdot,\{p_{m,j}\},*)$ is a combinative, commutative, semi-normed algebra with unit element $\delta\in E^*(\mathbb{R}^n)$, and holds

$$\delta*T=T*\delta,\quad \forall T\in E^*(\mathbb{R}^n).$$

5.4 Wavelet Analysis

5.4.1 Introduction

Fourier transform theory plays leading role in the scientific areas almost 2 hundred years, such that lots of natural sciences have great developments, particularly, for the signal analysis, such as, radios, televisions, radars, and communication techniques, … . However, it has some disadvantages, which could not be satisfactory in applications. At about 1984, American mathematicians Grossmann, A. and Morlet, J. introduced so-called **wavelet transform.** Then, foundation works, research results, and application jobs concerned appear in most of all scientific areas and technological fields. Hence, "wavelet analysis" as a new field of harmonic analysis in mathematics has been listed, also as a cross branch of pure and applied mathematics, standing on the forward position of mathematic science.

The theory of wavelet analysis is including on one-dimensional case and on higher-dimensional case, we only concentrate our mind on the first case in this section.

1. Gabor transformations

Recall sections 5.2 and 5.3, Fourier transformations of $f\in L^p(\mathbb{R})$ $(1\leqslant p\leqslant 2)$-functions, Schwartz functions as well as Schwartz distributions are defined. Regard $f\in L^1(\mathbb{R})$ as a signal, denoted by $f(t)$ with **time** t **in "time field"** $\mathbb{R}$, then the Fourier transformation $f^\wedge(\xi)=\int_{\mathbb{R}} f(t)\,\mathrm{e}^{-\mathrm{i}t\xi}\,\mathrm{d}t$ is **the continuous frequency spectrum** of $f(t)$ with

frequency ξ in "frequency field" $\mathbb{R}$. We see that: to determine the spectrum $f^\wedge(\xi)$ of $f(t)$, we need all information (i.e. all values) of $f(t)$ on $t\in\mathbb{R}=(-\infty,+\infty)$. So that, to obtain the frequency spectrum $f^\wedge(\xi)$ on the frequency field of a signal $f(t)$, we have to supply all values of $f(t)$ in time field $\mathbb{R}$; otherwise, the frequency spectrum could not be determined.

On the other hand, a little bit variation of signal $f(t)$ on a small neighborhood $(t_0-\Delta t, t_0+\Delta t)$ in time field, may affect the values of frequency spectrum. For example, Dirac distribution $\delta_{t_0}=\tau_{t_0}\delta$ supported at t_0 has $(\delta_{t_0})^\wedge=\mathrm{e}^{-\mathrm{i}t_0\xi}$, $\xi\in\mathbb{R}$ by Theorem 5.3.3(1), this shows that a signal $\delta_{t_0}=\tau_{t_0}\delta$ supported only on one point t_0 may affect whole frequency spectrum.

To collect a local information of Fourier transformation $f^\wedge(\xi)$ of signal $f(t)$, Gabor, B. introduced **Gabor transformation of "localization of time field"** in 1946,

$$G_b^a f(\xi)=\int_{\mathbf{R}} f(t)\mathrm{e}^{-\mathrm{i}\xi t} g_a(t-b)\mathrm{d}t,$$

with so-called "weight function" $g_a(t)=\dfrac{1}{2\sqrt{\pi a}}\mathrm{e}^{-\frac{t^2}{4a}}$, $a>0$. Thus, the frequency spectrum $f^\wedge$ has been localized by Gaussian function $\tilde{g}(x)=\mathrm{e}^{-x^2}$ and its Fourier transformation $\tilde{g}^\wedge(\xi)=\int_{\mathbf{R}}\mathrm{e}^{-x^2}\mathrm{e}^{-\mathrm{i}x\xi}\mathrm{d}x=\sqrt{\dfrac{\pi}{a}}\mathrm{e}^{-\frac{\xi^2}{4a}}$, $\xi\in\mathbb{R}$.

For $g_a(t)$ and $G_b^a f(\xi)$, we have

$$\int_{\mathbf{R}} g_a(t-b)\mathrm{d}b=\frac{1}{2\sqrt{\pi a}}\int_{\mathbf{R}}\exp\left\{-\frac{(t-b)^2}{4a}\right\}\mathrm{d}b=\frac{1}{\sqrt{\pi}}\int_{\mathbf{R}}\mathrm{e}^{-\xi^2}\mathrm{d}\xi=1,$$

and

$$\begin{aligned}\int_{\mathbf{R}}(G_b^a f)(\xi)\mathrm{d}b&=\int_{\mathbf{R}}\left(\int_{\mathbf{R}} f(t)\,\mathrm{e}^{-\mathrm{i}\xi t} g_a(t-b)\mathrm{d}t\right)\mathrm{d}b\\&=\int_{\mathbf{R}}\left(\int_{\mathbf{R}}[f(t)g_a(t-b)]\,\mathrm{e}^{-\mathrm{i}\xi t}\mathrm{d}t\right)\mathrm{d}b\\&=\int_{\mathbf{R}} f(t)\,\mathrm{e}^{-\mathrm{i}\xi t}\left(\int_{\mathbf{R}} g_a(t-b)\mathrm{d}b\right)\mathrm{d}t=\int_{\mathbf{R}} f(t)\,\mathrm{e}^{-\mathrm{i}\xi t}\cdot 1\mathrm{d}t\\&=\int_{\mathbf{R}} f(t)\,\mathrm{e}^{-\mathrm{i}\xi t}\mathrm{d}t=f^\wedge(\xi),\quad \xi\in\mathbb{R}.\end{aligned}$$

This shows that $\int_{\mathbf{R}} G_a^b f(\xi)\mathrm{d}b = f^\wedge(\xi)$ is independent of parameter a in Gabor transformation $G_b^a f(\xi)$, and $f^\wedge(\xi)$ has been localized at $t=b$. That is, $f^\wedge(\xi)$ can be regarded as the frequency spectrum of signal function $f(t)g_a(t-b)$ which are the values of $f(t)$ in a neighborhood of $t=b$, and the local information of the frequency spectrum is obtained. Gabor transformation set $\{G_b^a f: b \in \mathbb{R}\}$ gives a decomposition of $f^\wedge(\xi)$, and some information of a local spectrum of $f(t)$.

Let $G_{b,\xi}^a(t) = e^{-i\xi t} g_a(t-b)$, rewrite $G_b^a f$ as

$$(G_b^a f)(\xi) = \int_{\mathbf{R}} f(t) e^{-i\xi t} g_a(t-b) \mathrm{d}t \equiv \int_{\mathbf{R}} f(t) G_{b,\xi}^a(t) \mathrm{d}t = \langle G_{b,\xi}^a, f \rangle,$$

it shows a new sense of $G_b^a f$: open a "window" for $f(t)$, use a "window function" $G_{b,\xi}^a(t)$ to restrict $f^\wedge$ in a local scope.

New problems for this Gabor transformation are: (1) how to determine the sizes of "window" by g_a? (2) how to overcome the restriction of properties of Gaussian function, such that "window function" could not be restricted? exactly to solve these theoretical and technical problems, further promotes the flourishing developments of the study of wavelet analysis.

2. Window functions

We use the following notations: t is time; ω is the frequency of a signal $g(t)$; Fourier transformation $g^\wedge(\omega)$ is the frequency spectrum of $g(t)$; and (t, ω) is the **phase space**, i. e. **time-frequency space.**

Definition 5.4.1 (window function) If $g(t) \in L^2(\mathbb{R})$ and $tg(t) \in L^2(\mathbb{R})$, then $g(t)$ is said to be **a window function.** If both $g(t)$ and $g^\wedge(\omega)$ are window functions, then

$$(t_0, \omega_0) = \left(\frac{1}{\| g \|_2^2} \int_{\mathbf{R}} t \mid g(t) \mid^2 \mathrm{d}t, \frac{1}{\| g^\wedge \|_2^2} \int_{\mathbf{R}} \omega \mid g^\wedge(\omega) \mid^2 \mathrm{d}\omega \right)$$

is said to be **a center of phase space**, with $\| g \|_2 = \| g \|_{L^2(\mathbf{R})}$; and

$$\Delta_g = \left\{ \frac{\int_{\mathbf{R}} (t - t_0)^2 \mid g(t) \mid^2 \mathrm{d}t}{\| g \|_2^2} \right\}^{\frac{1}{2}}, \quad \Delta_{g^\wedge} = \left\{ \frac{\int_{\mathbf{R}} (\omega - \omega_0)^2 \mid g(\omega) \mid^2 \mathrm{d}\omega}{\| g^\wedge \|_2^2} \right\}^{\frac{1}{2}}$$

are said to be **time width** and **frequency width** of $g(t)$, respectively; the rectangle

$$[T_g^0; \Omega_{g^\wedge}^0] \equiv T_g^0 \times \Omega_{g^\wedge}^0 = [t_0 - \Delta_g, t_0 + \Delta_g; \omega_0 - \Delta_{g^\wedge}, \omega_0 + \Delta_{g^\wedge}]$$

is said to be **a time-frequency window in phase space** (t, ω) **with center** (t_0, ω_0), **side lengths** $2\Delta_g$ **and** $2\Delta_{g^\wedge}$ **of** $g(t)$.

3. Properties of window functions

Proposition 5.4.1 (uncertainty principle) If $g \in L^2(\mathbb{R})$, and $g(t)$, $g^\wedge(\omega)$ are window functions, then

$$\Delta_g \cdot \Delta_{g^\wedge} \geqslant \frac{1}{2};$$

moreover, the equal sign holds if and only if $g(t)=c\dfrac{e^{iat}}{2\sqrt{\pi a}}\exp\left\{-\dfrac{(t-b)^2}{4a}\right\}$, $c\neq 0, a>0$, $a, b\in\mathbb{R}$.

Note The area of time-frequency window is $4\Delta_g\Delta_{g^\wedge}$ by definition; so that the uncertainty principle tells that: the area of time-frequency window of every window function is greater than or equal to 2; the area is equal to 2 if and only if for the Gauss-type function. The proof of this principle is not diffical, however, its sense in physics and its applications is very important.

Proposition 5.4.2 (properties of translation-dilation) If $g, g^\wedge$ are window functions, then the translation-dilation function $g^{ba}(x)=\dfrac{1}{\sqrt{a}}g\left(\dfrac{x-b}{a}\right)$ with $a>0, b\in\mathbb{R}$ is a window function. Moreover, the Fourier transformation $(g^{ba})^\wedge$ of g^{ba}, i.e. $(g^{ba}(\circ))^\wedge(\omega)=\sqrt{|a|}\cdot e^{-ib\omega}g^\wedge(a\omega)$ with $a>0, b\in\mathbb{R}, \omega\in\mathbb{R}$ is a window function; where the translation-dilation function $g^{ba}(x)$ is with center $b+at_0$, radius $a\Delta_g$; and $(g^{ba})^\wedge(\omega)$ is with center $\dfrac{\omega_0}{a}$, radius $\dfrac{\Delta_{g^\wedge}}{a}$; the time-frequency window is

$$[T^0_{g^{ba}}; \Omega^0_{(g^{ba})^\wedge}]=\left[b+at_0-a\Delta_g, b+at_0+a\Delta_g; \frac{\omega_0}{a}-\frac{\Delta_{g^\wedge}}{a}, \frac{\omega_0}{a}+\frac{\Delta_{g^\wedge}}{a}\right];$$

the area of time-frequency window is $4\Delta_g\Delta_{g^\wedge}$.

The restriction of a signal in time window is called **time localization of the signal**; the restriction of a frequency spectrum of a signal in frequency window is called **frequency localization of the signal.**

5.4.2 Continuous wavelet transformations

1. Definition of continuous wavelet transformations

Definition 5.4.2 (admissible condition, base wavelet) If $\psi\in L^2(\mathbb{R})$ satisfies $0<c_\psi=\int_{\mathbb{R}}\dfrac{|\psi^\wedge(\omega)|^2}{|\omega|}d\omega<+\infty$, then ψ is said to satisfy **the admissible condition**, and $\psi\in L^2(\mathbb{R})$ is said to be **a base wavelet.** We suppose $\|\psi\|_2=1$ without loss of generality.

For any base wavelet $\psi\in L^2(\mathbb{R})$, the translation-dilation function with $a, b\in\mathbb{R}, a\neq 0$, is

$$\psi^{ba}(x)=\frac{1}{\sqrt{|a|}}\psi\left(\frac{x-b}{a}\right). \tag{5.4.1}$$

$\{\psi^{ba}(x)\}_{a,b\in\mathbf{R},a\neq 0}$ is **a bi-index function family**, called **a wavelet family generalized by base wavelet** ψ. Clearly, it holds

$$(\psi^{ba})^{\wedge}(\omega)=\sqrt{|a|}\,\mathrm{e}^{-\mathrm{i}b\omega}\psi^{\wedge}(a\omega). \tag{5.4.2}$$

Definition 5.4.3 (continuous wavelet transformation) The operator W_ψ determined by wavelet family (5.4.1)

$$(W_\psi f)(b,a)=\frac{1}{\sqrt{|a|}}\int_{\mathbf{R}} f(t)\,\overline{\psi\left(\frac{t-b}{a}\right)}\,\mathrm{d}t\equiv\langle f,\psi^{ba}\rangle,\quad f\in L^2(\mathbb{R}) \tag{5.4.3}$$

with $a,b\in\mathbb{R}$, $a\neq 0$, is said to be **the continuous wavelet transformation**, or simply, **the wavelet transformation, of** f.

Theorem 5.4.1 (equivalent representation) The continuous wavelet transformation of f in (5.4.3) is a convolution operator

$$(W_\psi f)(b,a)=\langle f,\psi^{ba}\rangle=\frac{1}{\sqrt{|a|}}f*\overline{\psi(a^{-1}\circ)}(b),\quad f\in L^2(\mathbb{R})$$

with $a,b\in\mathbb{R}$, $a\neq 0$.

Theorem 5.4.2 (time-frequency window for a wavelet family) The time-frequency window of wavelet family (5.4.1) with $a>0$, $b\in\mathbb{R}$, is

$$T^0_{\psi^{ba}}\times\Omega^0_{(\psi^{ba})^\wedge}=\left[b+at_0-a\Delta_\psi,\,b+at_0+a\Delta_\psi;\ \frac{\omega_0}{a}-\frac{1}{a}\Delta_{\psi^\wedge},\,\frac{\omega_0}{a}+\frac{1}{a}\Delta_{\psi^\wedge}\right].$$

Proof Since the time-frequency window of ψ is

$$T^0_\psi\times\Omega^0_{\psi^\wedge}=[t_0-\Delta_\psi,t_0+\Delta_\psi;\ \omega_0-\Delta_{\psi^\wedge},\omega_0+\Delta_{\psi^\wedge}],$$

then by Proposition 5.4.2 and formula (5.4.2), the result of this theorem can be obtained.

2. Sense of admissible condition

Theorem 5.4.3 If ψ is a base wavelet, then the continuous wavelet transformation (5.4.3) determined by ψ satisfies

$$J=\int_{\mathbf{R}}\int_{\mathbf{R}}(W_\psi f)(b,a)\cdot\overline{(W_\psi g)(b,a)}\,\frac{\mathrm{d}a}{a^2}\mathrm{d}b=c_\psi\langle f,g\rangle,\quad\forall f,g\in L^2(\mathbb{R}),$$

where c_ψ is, by Definition 5.4.2, satisfies $0<c_\psi=\int_{\mathbf{R}}\frac{|\psi^\wedge(\omega)|^2}{|\omega|}\mathrm{d}\omega<+\infty$.

Proof Denote $J=\int_{\mathbf{R}}I\,\frac{\mathrm{d}a}{a^2}$. We deduce I, firstly.

$$I\equiv\int_{\mathbf{R}}(W_\psi f)(b,a)\cdot\overline{(W_\psi g)(b,a)}\,\mathrm{d}b \qquad\text{(definition)}$$

$$=\frac{1}{|a|}\int_{\mathbf{R}}\left\{\int_{\mathbf{R}}f(t)\,\overline{\psi\left(\frac{t-b}{a}\right)}\mathrm{d}t\cdot\overline{\int_{\mathbf{R}}g(s)\,\overline{\psi\left(\frac{s-b}{a}\right)}\mathrm{d}s}\right\}\mathrm{d}b$$

$$=\frac{1}{|a|}\int_{\mathbf{R}}\left\{\int_{\mathbf{R}}f(t)\,\overline{\psi\left(\frac{t-b}{a}\right)}\,\mathrm{d}t\cdot\int_{\mathbf{R}}\overline{g(s)}\,\psi\left(\frac{s-b}{a}\right)\mathrm{d}s\right\}\mathrm{d}b$$

(Plancherel-type formula and (5.4.2))

$$=\int_{\mathbf{R}}\left\{\frac{|a|^{\frac{1}{2}}}{\sqrt{2\pi}}\int_{\mathbf{R}}f^{\wedge}(\omega)\,\mathrm{e}^{ib\omega}\overline{\psi^{\wedge}(a\omega)}\mathrm{d}\omega\cdot\overline{\frac{|a|^{\frac{1}{2}}}{\sqrt{2\pi}}\int_{\mathbf{R}}g^{\wedge}(\tau)\,\mathrm{e}^{ib\tau}\overline{\psi^{\wedge}(a\tau)}\,\mathrm{d}\tau}\right\}\mathrm{d}b$$

$$=\int_{\mathbf{R}}\left\{\frac{|a|^{\frac{1}{2}}}{\sqrt{2\pi}}\int_{\mathbf{R}}f^{\wedge}(\omega)\,\mathrm{e}^{ib\omega}\overline{\psi^{\wedge}(a\omega)}\mathrm{d}\omega\cdot\frac{|a|^{\frac{1}{2}}}{\sqrt{2\pi}}\int_{\mathbf{R}}\overline{g^{\wedge}(\tau)}\,\mathrm{e}^{-ib\tau}\psi^{\wedge}(a\tau)\,\mathrm{d}\tau\right\}\mathrm{d}b$$

$$=|a|\int_{\mathbf{R}}\left\{\frac{1}{\sqrt{2\pi}}\int_{\mathbf{R}}F(\omega)\,\mathrm{e}^{ib\omega}\,\mathrm{d}\omega\cdot\frac{1}{\sqrt{2\pi}}\int_{\mathbf{R}}\overline{G(\tau)}\,\mathrm{e}^{-ib\tau}\,\mathrm{d}\tau\right\}\mathrm{d}b$$

$$\left(\text{let } F(\omega)=f^{\wedge}(\omega)\overline{\psi^{\wedge}(a\omega)},\quad G(\tau)=g^{\wedge}(\tau)\overline{\psi^{\wedge}(a\tau)}\right)$$

$$=|a|\int_{\mathbf{R}}\left\{\overline{\frac{1}{\sqrt{2\pi}}\int_{\mathbf{R}}\overline{F(\omega)}\,\mathrm{e}^{-ib\omega}\,\mathrm{d}\omega}\cdot\frac{1}{\sqrt{2\pi}}\int_{\mathbf{R}}\overline{G(\tau)}\,\mathrm{e}^{-ib\tau}\,\mathrm{d}\tau\right\}\mathrm{d}b$$

$$=\frac{|a|}{2\pi}\int_{\mathbf{R}}\overline{\overline{F(\circ)}^{\wedge}(b)}\,\overline{G(\circ)}^{\wedge}(b)\mathrm{d}b=|a|\int_{\mathbf{R}}F(x)\overline{G(x)}\mathrm{d}x\,.$$

(Plancherel type formula)

Secondly,

$$J=\int_{\mathbf{R}}I\,\frac{\mathrm{d}a}{a^2}=\int_{\mathbf{R}}\int_{\mathbf{R}}(W_{\psi}f)(b,a)\cdot\overline{(W_{\psi}g)(b,a)}\,\mathrm{d}b\cdot\frac{\mathrm{d}a}{a^2}$$

$$=\int_{\mathbf{R}}\frac{a^2}{|a|}\int_{\mathbf{R}}\overline{G(x)}\,F(x)\mathrm{d}x\cdot\frac{\mathrm{d}a}{a^2}$$

$$=\int_{\mathbf{R}}\frac{1}{|a|}\left\{\int_{\mathbf{R}}\overline{g^{\wedge}(x)}\,\psi^{\wedge}(ax)\cdot f^{\wedge}(x)\overline{\psi^{\wedge}(ax)}\mathrm{d}x\right\}\mathrm{d}a.$$

Finally, by Fubini theorem, we have

$$J=\int_{\mathbf{R}}f^{\wedge}(x)\,\overline{g^{\wedge}(x)}\left\{\int_{\mathbf{R}}\frac{|\psi^{\wedge}(ax)|^2}{|a|}\mathrm{d}a\right\}\mathrm{d}x=c_{\psi}\langle f^{\wedge},g^{\wedge}\rangle=c_{\psi}\langle f,g\rangle,$$

the proof is complete.

Note The result of this theorem can be written as

$$\langle W_{\psi}f,W_{\psi}g\rangle=c_{\psi}\langle f,g\rangle,$$

the left-hand side is the "inner product" of functions $W_{\psi}f$ and $W_{\psi}g$ with "weight element" $\frac{1}{a^2}$; **the constant c_{ψ} in admissible condition is a finite positive ratio factor of the inner product** depending on base wavelet ψ in wavelet transformation. The constant $c_{\psi}\neq 0, \infty$; otherwise, the wavelet transformation does not make sense.

3. Inversion formulas

Theorem 5.4.4 If ψ is a base wavelet, then the continuous wavelet transformation

(5.4.3) satisfies inversion formula

$$f(x)=\frac{1}{c_\psi}\int_{\mathbf{R}}\int_{\mathbf{R}}(W_\psi f)(b,a)\,\psi^{ba}(x)\,\frac{\mathrm{d}a}{a^2}\mathrm{d}b,\quad x\in\mathbb{R},\forall f\in L^2(\mathbb{R}).$$

Proof Take $g_a(t-b)=\frac{1}{2\sqrt{\pi a}}\mathrm{e}^{-\frac{(t-b)^2}{4a}}$, $a>0, b\in\mathbb{R}$, in Theorem 5.4.3 and in (5.4.3), then

$$\begin{aligned}I&=\frac{1}{c_\psi}\lim_{a\to0+}\int_{\mathbf{R}}\int_{\mathbf{R}}(W_\psi f)(b,a)\cdot\overline{(W_\psi g_a(\circ-b))}\,\frac{\mathrm{d}a}{a^2}\mathrm{d}b\\&=\frac{1}{c_\psi}\lim_{a\to0+}\int_{\mathbf{R}}\int_{\mathbf{R}}(W_\psi f)(b,a)\cdot\overline{\langle g_a(\circ-b),\psi^{ba}\rangle}\,\frac{\mathrm{d}a}{a^2}\mathrm{d}b\\&=\frac{1}{c_\psi}\int_{\mathbf{R}}\int_{\mathbf{R}}(W_\psi f)(b,a)\cdot\overline{\lim_{a\to0+}g_a*\psi^{ba}(x)}\,\frac{\mathrm{d}a}{a^2}\mathrm{d}b.\end{aligned}$$

Substitute $\lim\limits_{a\to0+}g_a*\psi^{ba}(x)=\psi^{ba}(x)$ in the last form, it follows

$$I=\frac{1}{c_\psi}\int_{\mathbf{R}}\int_{\mathbf{R}}(W_\psi f)(b,a)\cdot\psi^{ba}(x)\,\frac{\mathrm{d}a}{a^2}\mathrm{d}b.$$

Finally, the Theorem 5.4.3 shows that

$$I=\frac{1}{c_\psi}\lim_{a\to0+}\{c_\psi\langle f,g\rangle\}=\lim_{a\to0+}\langle f,g_a(\circ-x)\rangle=\lim_{a\to0+}g_a*f(x)=f(x),$$

then the proof is complete.

Theorem 5.4.5 (criteria for a base wavelet, inversion formula) If $\psi\in L^2(\mathbb{R})$, $\|\psi\|_2=1$, satisfies

$$\int_{(0,+\infty)}\frac{|\psi^\wedge(\omega)|^2}{\omega}\mathrm{d}\omega=\int_{(0,+\infty)}\frac{|\psi^\wedge(-\omega)|^2}{|-\omega|}\mathrm{d}\omega=\frac{1}{2}c_\psi<+\infty,$$

then ψ is a base wavelet, and the continuous wavelet transformation (5.4.3) determined by ψ satisfies the inversion formula

$$f(x)=\frac{2}{c_\psi}\int_{(0,+\infty)}\left[\int_{\mathbf{R}}(W_\psi f)(b,a)\left\{\frac{1}{\sqrt{a}}\psi\left(\frac{x-b}{a}\right)\right\}\mathrm{d}b\right]\frac{\mathrm{d}a}{a^2},\quad a>0, b\in\mathbb{R};$$

it holds for $\forall f\in L^2(\mathbb{R})$.

Theorem 5.4.6 If ψ is a base wavelet, i.e., $\psi(t)\in L^2(\mathbb{R})$ with $0<c_\psi<+\infty$; moreover, if $\psi^\wedge$ is continuous at $\omega=0$, then $\int_{\mathbf{R}}\psi(t)\mathrm{d}t=0$.

Proof We deduce

$$0<c_\psi=\int_{\mathbf{R}}\frac{|\psi^\wedge(\omega)|^2}{|\omega|}\mathrm{d}\omega=\int_{(-\infty,0)}\frac{|\psi^\wedge(\omega)|^2}{|\omega|}\mathrm{d}\omega+\int_{(0,+\infty)}\frac{|\psi^\wedge(\omega)|^2}{|\omega|}\mathrm{d}\omega<+\infty\Rightarrow$$

$\dfrac{|\psi^\wedge(\omega)|^2}{|\omega|} \in L^1(\mathbb{R}^+) \Rightarrow \dfrac{|\psi^\wedge(\omega)|}{|\omega|^{\frac{1}{2}}} \in L^2(\mathbb{R}^+) \Rightarrow \dfrac{|\psi^\wedge(\omega)|}{|\omega|^{\frac{1}{2}}}$ is a. e. finite in a neighborhood of $\omega=0 \Rightarrow \lim\limits_{|\omega|\to 0} \dfrac{|\psi^\wedge(\omega)|}{|\omega|^{\frac{1}{2}}} < +\infty$, $\lim\limits_{|\omega|\to 0} |\psi^\wedge(\omega)| = 0 \Rightarrow \psi^\wedge(0)=0$ (since $\psi^\wedge(\omega)$ is continuous at $\omega=0$) $\Rightarrow \int_{\mathbb{R}} \psi(t)\mathrm{d}t = 0 \Rightarrow$ The proof is complete.

4. Examples of base wavelets

Example 5. 4. 1 Mexican hat $\varphi(x)$ with $\varphi(x)=\begin{cases}1-|x|, & |x|\leqslant 1\\ 0, & |x|>1\end{cases}$ is a base wavelet.

Example 5. 4. 2 Haar function $\psi(t)=\begin{cases}1, & 0\leqslant t<\dfrac{1}{2},\\ -1, & \dfrac{1}{2}\leqslant t\leqslant 1,\\ 0, & t<\dfrac{1}{2}, t>1\end{cases}$ is a base wavelet.

Example 5. 4. 3 B-spline function $\varphi(x)=\begin{cases}\dfrac{1}{2}(x+1)^2, & -1\leqslant x\leqslant 0\\ \dfrac{3}{4}-\left(x-\dfrac{1}{2}\right)^2, & 0\leqslant x\leqslant 1\\ \dfrac{1}{2}(x-1)^2, & 1\leqslant x\leqslant 2\\ 0, & x<-1, x>2\end{cases}$ is a base wavelet.

5. 4. 3 Discrete wavelet transformations

1. Orthogonal wavelets

Definition 5. 4. 4 (orthogonal wavelet base) Let $\psi \in L^2(\mathbb{R})$, $\|\psi\|_2 = 1$, and denoted by

$$\psi_{j,k}(x) = 2^{\frac{j}{2}}\psi(2^j - k), \quad j,k \in \mathbb{Z}.$$

If

(1) $\langle \psi_{j,k}, \psi_{l,m}\rangle = \delta_{j,l}\delta_{k,m}$, $\delta_{j,k}=\begin{cases}1, & j=k\\ 0, & j\neq k\end{cases}$;

(2) $\forall f\in L^2(\mathbb{R})$, the series

$$f(x)=\sum_{j,k=-\infty}^{+\infty} c_{j,k}\psi_{j,k}(x) \tag{5.4.4}$$

converges in $L^2(\mathbb{R})$-norm sense, i. e.

$$\lim_{N,M\to+\infty}\left\| f(x)-\sum_{j=-M}^{+M}\sum_{k=-N}^{+N} c_{j,k}\psi_{j,k}(\circ)\right\|_{L^2(\mathbf{R})}=0, \tag{5.4.5}$$

then $\{\psi_{j,k}\}_{j,k\in\mathbf{Z}}$ is **an orthonormal base** in $L^2(\mathbb{R})$. This ψ is said to be **an orthogonal base wavelet**, or **a discrete base wavelet.** The series $\sum_{j,k=-\infty}^{+\infty} c_{j,k}\psi_{j,k}(x)$ is said to be **a wavelet series**, or **a discrete wavelet series**, and

$$c_{j,k}=\langle f,\psi_{j,k}\rangle=\int_{\mathbf{R}} f(x)\overline{\psi_{j,k}(x)}\,\mathrm{d}x, \quad j,k\in\mathbb{Z} \tag{5.4.6}$$

are said to be **the coefficients of discrete wavelet series**, or **the finite wavelet transformation.**

2. Stability condition, the most stability condition

Definition 5.4.5 (stability condition) If $\psi\in L^2(\mathbb{R})$ satisfies **condition**: there exist constants A,B $(0<A\leqslant B<+\infty)$, such that

$$A\leqslant\sum_{j=-\infty}^{+\infty}|\psi^\wedge(2^j\omega)|^2\leqslant B, \quad \text{a. e. } \omega\in\mathbb{R},$$

then the condition is said to be **a stability condition**; if $0<A=B<+\infty$, then it is said to be **the most stability condition.**

We have the equivalent theorem for stability condition.

Theorem 5.4.7 The stability condition of ψ is equivalent to: there exist constants $A,B(0<A\leqslant B<+\infty)$, such that

$$A\|f\|^2_{L^2(\mathbf{R})}\leqslant\sum_{j=-\infty}^{+\infty}\|W_\psi f(j,\circ)\|^2_{L^2(\mathbf{R})}\leqslant B\|f\|^2_{L^2(\mathbf{R})}, \quad \forall f\in L^2(\mathbb{R}),$$

where $W_\psi f$ is given in (5.4.3).

Theorem 5.4.8 (property of the discrete base wavelet) If ψ is a discrete base wavelet with stability condition, then it is a discrete base wavelet with the admissible condition, and there exist constants $A,B(0<A\leqslant B<+\infty)$, such that

$$A\ln 2\leqslant\int_{(0,+\infty)}\frac{|\psi^\wedge(\omega)|^2}{\omega}\mathrm{d}\omega\leqslant B\ln 2, \quad A\ln 2\leqslant\int_{(0,+\infty)}\frac{|\psi^\wedge(-\omega)|^2}{\omega}\mathrm{d}\omega\leqslant B\ln 2.$$

Proof For integral $\int_{(1,2)}\frac{|\psi^\wedge(2^j\omega)|^2}{\omega}\mathrm{d}\omega$, we deduce by the substitution method of variables, then

$$\int_{(1,2)}\frac{|\psi^\wedge(2^j\omega)|^2}{\omega}\mathrm{d}\omega=\int_{(2^{-j},2^{-j+1})}\frac{|\psi^\wedge(t)|^2}{t}\mathrm{d}t.$$

On the other hand, from the stability condition, $\frac{A}{\omega} \leqslant \sum_{j=-\infty}^{+\infty} \frac{|\psi^\wedge(2^{-j}\omega)|^2}{\omega} \leqslant \frac{B}{\omega}$, then, integrate both sides on $[1,2]$, it follows

$$\int_{[1,2]} \frac{A}{\omega} d\omega \leqslant \sum_{j=-\infty}^{+\infty} \int_{[1,2]} \frac{|\psi^\wedge(2^{-j}\omega)|^2}{\omega} d\omega \leqslant \int_{[1,2]} \frac{B}{\omega} d\omega,$$

i. e., $A\ln 2 \leqslant \sum_{j=-\infty}^{+\infty} \int_{[2^{-j},2^{-j+1})} \frac{|\psi^\wedge(\omega)|^2}{\omega} d\omega \leqslant B\ln 2$, so that we have $A\ln 2 \leqslant \int_{(0,+\infty)} \frac{|\psi^\wedge(\omega)|^2}{\omega} d\omega \leqslant B\ln 2$. Another inequality can be obtained similarly.

3. Discrete wavelet transformations

Definition 5. 4. 6 (discrete wavelet transformation) If a base wavelet ψ satisfies the stability condition, then **a discrete wavelet transformation** of $f \in L^2(\mathbb{R})$ is defined by

$$(W_\psi f)\left(\frac{1}{2^j}, \frac{k}{2^j}\right) = \int_{\mathbf{R}} f(x)\, \overline{\psi_{j,k}(x)}\, dx = \langle f, \psi_{j,k} \rangle, \quad j, k \in \mathbb{Z}, \tag{5.4.7}$$

or, it is said to be **a dyadic wavelet transformation.**

Compare with (5. 4. 6), the discrete wavelet transformation is, as a matter of fact, the coefficients of discrete wavelet series of function $f \in L^2(\mathbb{R})$ expanded by orthonormal system $\{\psi_{j,k}\}_{j,k \in \mathbf{Z}}$.

Moreover, if we set $b = \frac{1}{2^j}$, $a = \frac{k}{2^j}$, $j, k \in \mathbb{Z}$, in (5. 4. 3), then

$$(W_\psi f)(b, a) = \frac{1}{\sqrt{|a|}} \int_{\mathbf{R}} f(t)\, \overline{\psi\left(\frac{t-b}{a}\right)}\, dt \equiv \langle f, \psi^{ba} \rangle$$

is as in Definition 5. 4. 6.

4. Inversion formula of discrete wavelet transformation

The inversion formulas in discrete wavelet transformation are important, similar to the case of Fourier series.

Theorem 5. 4. 9 If a discrete base wavelet ψ satisfies the stability condition, then the discrete wavelet transformation

$$(W_\psi f)\left(\frac{1}{2^j}, \frac{k}{2^j}\right) = \int_{\mathbf{R}} f(x)\, \overline{\psi_{j,k}(x)}\, dx = \langle f, \psi_{j,k} \rangle, \quad j, k \in \mathbb{Z}$$

of $f \in L^2(\mathbb{R})$ has the inversion formula

$$f(x) = \sum_{j,k=-\infty}^{+\infty} d_{j,k} \tilde{\psi}_{j,k}(x), \tag{5.4.8}$$

with $c_{j,k}=\langle f,\tilde{\psi}_{j,k}\rangle=\int_{\mathbf{R}} f(x)\overline{\tilde{\psi}_{j,k}(x)}\,dx$, $j,k\in\mathbb{Z}$, and $\tilde{\psi}_{j,k}(x)=2^{\frac{j}{2}}\tilde{\psi}(2^j x-k)$, $j,k\in\mathbb{Z}$, the notation $\tilde{\psi}$ is the "dual" of ψ defined by its Fourier transformation:

$$(\tilde{\psi})^\wedge(\omega)=\frac{\psi^\wedge(\omega)}{\sum_{k=-\infty}^{+\infty}|\psi^\wedge(\omega+2k\pi)|^2}.$$

5.4.4 Applications of wavelet transformations

Wavelet transformations have widely applications, such as, in the areas of multi-resolution analysis, coded data compression, signal-noise separation and filter. We only give a simply example.

1. Multi-resolution analysis

Definition 5.4.7 (multi-resolution analysis) Let $\{V_j\}_{j\in\mathbf{Z}}$ be a sequence of closed subspaces with $V_j\subset L^2(\mathbb{R})$, $j\in\mathbb{Z}$, satisfy

(1) $V_j\subset V_{j-1}$, $\forall j\in\mathbb{Z}$;

(2) $\overline{\bigcup_{j\in\mathbf{Z}} V_j}=L^2(\mathbb{R})$, $\overline{\bigcap_{j\in\mathbf{Z}} V_j}=\{0\}$;

(3) $u(x)\in V_j\Leftrightarrow u(2x)\in V_{j-1}$, $j\in\mathbb{Z}$;

(4) $u(x)\in V_0\Rightarrow u(x-k)\in V_0$, $k\in\mathbb{Z}$;

(5) $\exists g(x)\in V_0$, s. t. $\{g(x-k): k\in\mathbb{Z}\}$, such that it is **a Riesz base of** V_0, i. e.

$$\forall u(x)\in V_0,\exists!\ \{\alpha_k: k\in\mathbb{Z}\}\subset l^2,\text{s. t. } u(x)=\sum_{k\in\mathbf{Z}}\alpha_k g(x-k);$$

and conversely, $\forall\{\alpha_k: k\in\mathbb{Z}\}\subset l^2$, $\exists A,B$, $0<A\leqslant B<+\infty$, such that

$$A\|u\|^2_{L^2(\mathbf{R})}\leqslant\sum_{k\in\mathbf{Z}}|\alpha_k|^2\leqslant\|u\|^2_{L^2(\mathbf{R})},\quad\forall u(x)\in V_0.$$

Then, $\{V_j\}_{j\in\mathbf{Z}}$ is said to be **a multi-resolution analysis sequence**, and $g(x)$ is said to be **a generator.**

The process for constructing multi-resolution analysis: For a given subspace sequence $\{V_j\}_{j\in\mathbf{Z}}$, $V_j\subset L^2(\mathbb{R})$, as **a multi-resolution analysis sequence**, and select a function $g(x)\in V_0$ as **a generator**, we construct **a scale function** $\varphi(x)\in V_0$, such that $\{\varphi(x-k)\}_{k\in\mathbf{Z}}$ is an orthogonal base of V_0, for example, by setting

$$\varphi(x)=\frac{1}{\sqrt{2\pi}}\left(\left(\sum_{k\in\mathbf{Z}}|g^\wedge(\omega+2k\pi)|^2\right)^{-1/2}g^\wedge(\omega)\right)^\vee.$$

Then $\{V_j\}_{j\in\mathbf{Z}} \longrightarrow g(x)\in V_0$

generator

$\varphi(x)\in V_0 \longrightarrow \{\varphi(x-k):k\in\mathbb{Z}\} \longrightarrow \{\varphi_{jk}(x)=2^{-j/2}\varphi(2^{-j}x-k):j,k\in\mathbb{Z}\}$

scale function orthogonal base of V_0 orthogonal base of V_j

$\frac{1}{\sqrt{2}}\varphi\left(\frac{x}{2}\right)\in V_1\subset V_0$

$$\frac{1}{\sqrt{2}}\varphi\left(\frac{x}{2}\right)=\sum_{k\in\mathbf{Z}}h_k\,\varphi(x-k), \tag{5.4.9}$$

with

$$h_k=\left\langle\frac{1}{\sqrt{2}}\varphi\left(\frac{x}{2}\right),\varphi(x-k)\right\rangle=\frac{1}{\sqrt{2}}\int_{\mathbf{R}}\varphi\left(\frac{x}{2}\right)\bar{\varphi}(x-k)\,\mathrm{d}x. \tag{5.4.10}$$

$$\varphi^\wedge(2\omega)=H(\omega)\varphi^\wedge(\omega), \tag{5.4.11}$$

and

$$H(\omega)=\frac{1}{\sqrt{2}}\sum_{k\in\mathbf{Z}}h_k\mathrm{e}^{-\mathrm{i}k\omega},\quad H\in L^2([0,2\pi]). \tag{5.4.12}$$

$\{h_k\}_{k\in\mathbf{Z}}$ is called **the response of frequency**, and $H(\omega)$ is called **the transfer function of** $\{h_k\}_{k\in\mathbf{Z}}$.

Theorem 5.4.10 For the generator $\varphi(x)$ of multi-resolution analysis sequence $\{V_j\}_{j\in\mathbf{Z}}$, its response frequency $\{h_k\}_{k\in\mathbf{Z}}$ and transfer function $H(\omega)$ of $\{h_k\}_{k\in\mathbf{Z}}$ satisfy

(1) $|H(\omega)|^2+|H(\omega+\pi)|^2=1$; (5.4.13)

(2) if $\{h_k\}_{k\in\mathbf{Z}}\in l^1$, and $\varphi^\wedge(\omega)$ is continuous with $\varphi^\wedge(0)=1$, then $H(0)=1$.

2. Orthogonal complement of multi-resolution analysis sequences

For given multi-resolution analysis sequence $\{V_j\}_{j\in\mathbf{Z}}$, we define new closed subspaces W_j, $j\in\mathbb{Z}$, such that

$$V_j\oplus W_j=V_{j-1},\quad j\in\mathbb{Z},$$

where W_j is called the orthogonal complement (or ortho-complement) of V_j, and holds

$$V_j\oplus W_j\oplus W_{j-1}\oplus\cdots\oplus W_{j-m+1}=V_{j-m};$$

$$\bigoplus_{j=-\infty}^{J}W_j=V_{J-1},\quad J\in\mathbb{Z};$$

$$\bigoplus_{j=-\infty}^{+\infty}W_j=L^2(\mathbb{R}^n);$$

$$V_J\oplus\bigoplus_{j=J}^{+\infty}W_j=L^2(\mathbb{R}^n),\quad J\in\mathbb{Z}.$$

$\{W_j\}_{j\in\mathbf{Z}}$ has the following properties.

Theorem 5. 4. 11 For the orthogonal complement $\{W_j\}_{j\in\mathbf{Z}}$ of $\{V_j\}_{j\in\mathbf{Z}}$, see Fig. 5. 4. 1, we have

(1) $u(x)\in W_j \Rightarrow u(x-2^j k)\in W_j, j, k\in\mathbb{Z}$;

(2) $u(x)\in W_j \Leftrightarrow u(2x)\in W_{j-1}, j\in\mathbb{Z}$;

(3) $P_{W_j}u(x)\to 0, |j|\to+\infty, \forall u\in L^2(\mathbb{R}^n)$,

where $P_{W_j}u(x)$ are the orthogonal projections of $u\in L^2(\mathbb{R}^n)$ on space W_j.

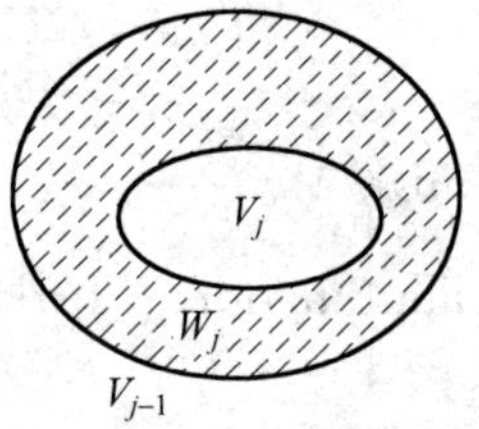

Fig. 5. 4. 1 Ortho-complement

We will analyze the signals in both multi-resolution analysis sequence $\{V_j\}_{j\in\mathbf{Z}}$ and orthogonal complement $\{W_j\}_{j\in\mathbf{Z}}$.

3. Wavelet function ψ

From scale function $\varphi\in V_0$, we construct a wavelet function $\psi\in W_0$, such that the set $\{\psi_{j,k}: k\in\mathbb{Z}\}, j\in\mathbb{Z}$, is an orthogonal base of W_j, with $\psi_{j,k}(x)=2^{-j/2}\psi(2^{-j}x-k), j, k\in\mathbb{Z}$. By

$\varphi(x)\in V_0 \longrightarrow \{\varphi(x-k): k\in\mathbb{Z}\} \longrightarrow \{\varphi_{jk}(x)=2^{-j/2}\varphi(2^{-j}x-k): j,k\in\mathbb{Z}\}$

scale function — orthogonal base of V_0 — orthogonal base of V_j

$$\varphi_{-1,k}(x)=2^{1/2}\varphi(2x-k)$$

$$\psi(x)=\sum_{k\in\mathbf{Z}}g_k\varphi_{-1,k}(x)=2^{1/2}\sum_{k\in\mathbf{Z}}g_k\varphi(2x-k), \tag{5.4.14}$$

with $g_k=\langle\psi,\varphi_{-1,k}\rangle$. And ψ defined in (5. 4. 14) is called **a wavelet function.** Let

$$G(\omega)=\frac{1}{\sqrt{2}}\sum_{k\in\mathbf{Z}}g_k e^{-ik\omega},\quad G(\omega)\in L^2([0,2\pi]).$$

Then we can prove that the wavelet function ψ has the following "symmetric properties":

(1) $\frac{1}{\sqrt{2}}\psi\left(\frac{x}{2}\right)=\sum_{k\in\mathbf{Z}}g_k\varphi(x-k)$, with $g_k=\left\langle\frac{1}{\sqrt{2}}\psi\left(\frac{x}{2}\right),\varphi(x-k)\right\rangle$;

$\left(\text{compare with (5. 4. 9): } \frac{1}{\sqrt{2}}\varphi\left(\frac{x}{2}\right)=\sum_{k\in\mathbf{Z}}h_k\varphi(x-k), \text{ with } h_k=\left\langle\frac{1}{\sqrt{2}}\varphi\left(\frac{x}{2}\right),\varphi(x-k)\right\rangle\right)$

(2) $\psi^\wedge(2\omega)=G(\omega)\varphi^\wedge(\omega)$, with $G(\omega)=\frac{1}{\sqrt{2}}\sum_{k\in\mathbf{Z}}g_k e^{-ik\omega}\in L^2([0,2\pi])$,

$\left(\text{compare with (5. 4. 11): } \varphi^\wedge(2\omega)=H(\omega)\varphi^\wedge(\omega), \text{ with } H(\omega)=\frac{1}{\sqrt{2}}\sum_{k\in\mathbf{Z}}h_k e^{-ik\omega}\right).$

Theorem 5. 4. 12 For the wavelet function $\psi(x)$ generalized from scale function

$\varphi(x)$, we have:

(1) $\psi \in W_0 \Leftrightarrow H(\omega)\overline{G(\omega)} + H(\omega+\pi)\overline{G(\omega+\pi)} = 0$;

(2) $\{\psi(x-k)\}_{k\in\mathbf{Z}}$ is orthogonal system $\Leftrightarrow$ $|G(\omega)|^2 + |G(\omega+\pi)|^2 = 1$;

(3) $\{\psi(x-k)\}_{k\in\mathbf{Z}}$ is orthogonal system of $W_0 \Leftrightarrow$ $|G(\omega)|^2 + |G(\omega+\pi)|^2 = 1$ and

$$H(\omega)\overline{G(\omega)} + H(\omega+\pi)\overline{G(\omega+\pi)} = 0$$

holds. Thus,

$$(\{V_j\}_{j\in\mathbf{Z}}, g) \Rightarrow (\{V_j\}_{j\in\mathbf{Z}}, \varphi) \Rightarrow (\{W_j\}_{j\in\mathbf{Z}}, \psi).$$

The following Fig. 5.4.2 is a draft for establishing of the multi-resolution analysis.

4. Signal analysis, Mallat algorithm

For a signal $f(t)$ with time variable t, Mallat algorithm is shown, by virtue of the notations in Fig. 5.4.2.

If $\{V_j\}_{j\in\mathbf{Z}}$ is a multi-resolution sequence and $\{W_j\}_{j\in\mathbf{Z}}$ is the ortho-complement. Suppose $f \in V_{J_1}$, $J_1 \in \mathbb{Z}$, is as

$$f(t) \equiv A_{J_1} f(t) = \sum_{k\in\mathbf{Z}} c_{J_1,k}\varphi_{J_1,k}(t), \tag{5.4.15}$$

with $\langle \varphi_{J_1,k}, \varphi_{J_1+1,m}\rangle = \overline{h_{k-2m}}$, $\langle \varphi_{J_1,k}, \psi_{J_1+1,m}\rangle = \overline{g_{k-2m}}$, $k, m \in \mathbb{Z}$. Decompose $f(t)$ as

$$f(t) = A_{J_1} f(t) = A_{J_1+1} f(t) + B_{J_1+1} f(t), \tag{5.4.16}$$

with $A_{J_1+1} f(t) = \sum_{m\in\mathbf{Z}} C_{J_1+1,m}\varphi_{J_1+1,m}(t)$ possessing coefficients

$$C_{J_1+1,m} = \sum_{k\in\mathbf{Z}} \overline{h_{k-2m}}\, C_{J_1,k}; \tag{5.4.17}$$

and $B_{J_1+1} f(t) = \sum_{m\in\mathbf{Z}} D_{J_1+1,m}\psi_{J_1+1,m}(t)$ possessing coefficients

$$D_{J_1+1,m} = \sum_{k\in\mathbf{Z}} \overline{g_{k-2m}}\, C_{J_1,k}. \tag{5.4.18}$$

Thus, the signal $f(t)$ has been decomposed into multi-resolution sequence $\{V_j\}_{j\in\mathbf{Z}}$ and the ortho-complement $\{W_j\}_{j\in\mathbf{Z}}$.

Introduce new notations in (5.4.17), (5.4.18)

$$C_{J_1+1} \equiv C_{J_1+1,m}, \quad \widetilde{H}C_{J_1} \equiv \sum_{k\in\mathbf{Z}} \overline{h_{k-2m}}\, C_{J_1,k},$$

$$B_{J_1+1} \equiv D_{J_1+1,m}, \quad \widetilde{G}C_{J_1,k} \equiv \sum_{k\in\mathbf{Z}} \overline{g_{k-2m}}\, C_{J_1,k},$$

then $C_{J_1+1} = \widetilde{H}C_{J_1}$, $B_{J_1+1} = \widetilde{G}C_{J_1}$; where $\widetilde{H} = [H_{m,k}] = [\bar{h}_{k-2m}]$, $\widetilde{G} = [G_{m,k}] = [\bar{g}_{k-2m}]$ are infinite matrix.

By the new notations, we have

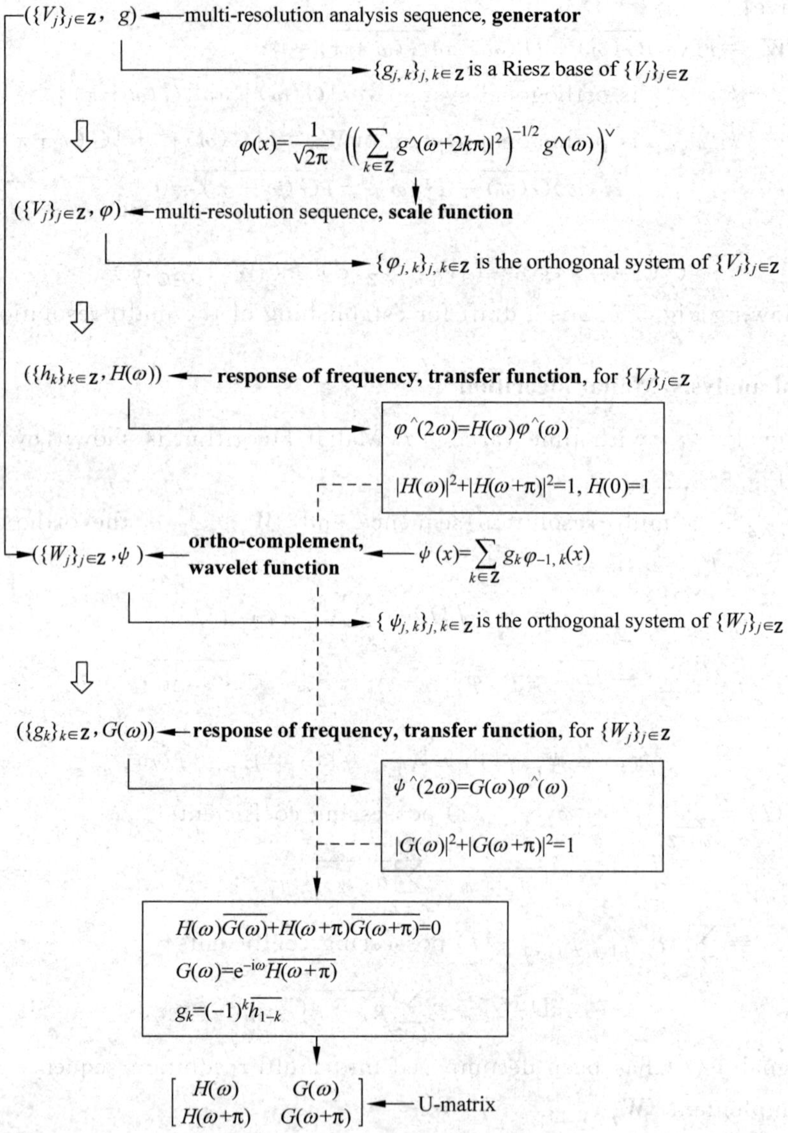

Fig. 5. 4. 2 Multi-resolution analysis

$$f(t)=A_{J_2}f(t)+\sum_{j=J_1+1}^{J_2}B_jf(t), \tag{5.4.19}$$

$$A_jf(t)=\sum_{k\in\mathbf{Z}}C_{j,k}\varphi_{j,k}(t),\quad B_jf(t)=\sum_{k\in\mathbf{Z}}D_{j,k}\psi_{j,k}(t), \tag{5.4.20}$$

$$C_{j+1}=\widetilde{H}C_j,\quad B_{j+1}=\widetilde{G}C_j,\quad j=J_1,J_1+1,\cdots,J_2-1, \tag{5.4.21}$$

The formulas (5.4.20) and (5.4.21) are called **Mallat decomposition formulas of signal** $f(t)$. And we have:

(1) $A_j f$ is **a continuous approximation of** $f(t)$ in the resolution ratio lower than 2^j; it is the part of $f(t)$ with the frequency lower than 2^{-j}; C_j is **a discrete approximation of** $f(t)$ in the resolution ratio lower than 2^j;

(2) $B_j f$ is **the continuous details of** $f(t)$ in the resolution ratio lower than 2^j; it is a part of $f(t)$ with the frequency between 2^{-j} and 2^{-j+1}; D_j is the discrete details of $f(t)$ in the resolution ratio lower than 2^j.

Let $\widetilde{H}^*$, $\widetilde{G}^*$ be the conjugate operators of $\widetilde{H}, \widetilde{G}$ in space l^2, respectively, or say, $\widetilde{H}^*$ and $\widetilde{G}^*$ be the conjugate transformations of $\widetilde{H}$ and $\widetilde{G}$, respectively. Thus, a signal $f(t)$ can be decomposed into different frequencies $A_j f$ and $D_j f$, then, the higher frequency part and detail part can be determined by the requirement of the problem.

We have a draft in Fig. 5.4.3.

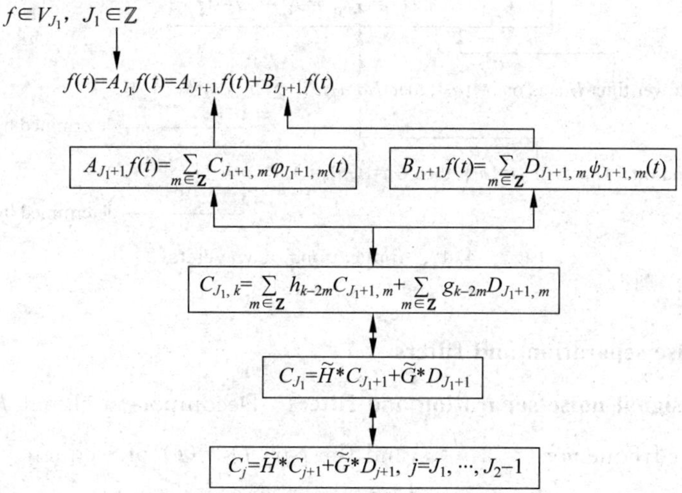

Fig. 5.4.3 Applications of wavelets(1)

5. Coded data compressions

Take $f(t)$ as an input signal, and let

$$f(t)\in V_{J_1}, \tag{5.4.22}$$

$$f(t)=A_{J_1}f(t)+\sum_{k\in} C_{J_1,k}\varphi_{J_1,k}(t), \tag{5.4.23}$$

where $\{V_j\}_{j\in\mathbf{Z}}$ is a multi-resolution analysis sequence in $L^2(\mathbb{R})$, and $V_j\subset L^2(\mathbb{R}), j\in\mathbf{Z}$;

coefficient sequence $\{C_{J_1,k}\}_{k\in\mathbf{Z}}$ has finite length (finite terms). To recover the signal, we use the inductive algorithm. Let

$$C_j = H^* C_{j+1} + G^* D_{j+1},\quad j = J_2 - 1, \cdots, J_1.$$

By Mallat algorithm, decompose signal $f(t)$ into different frequencies. Suppose that for $\{h_k\}$, $\{g_k\}$, the length $\ll N_{J_1}$. When C_{J_1}, D_{J_1} take a half, then C_j, D_j take $\frac{N_{J_1}}{2^{j-J_1}}$, $j = J_1+1, \cdots, J_2$, such that the data are compressed.

We have the Fig. 5. 4. 4.

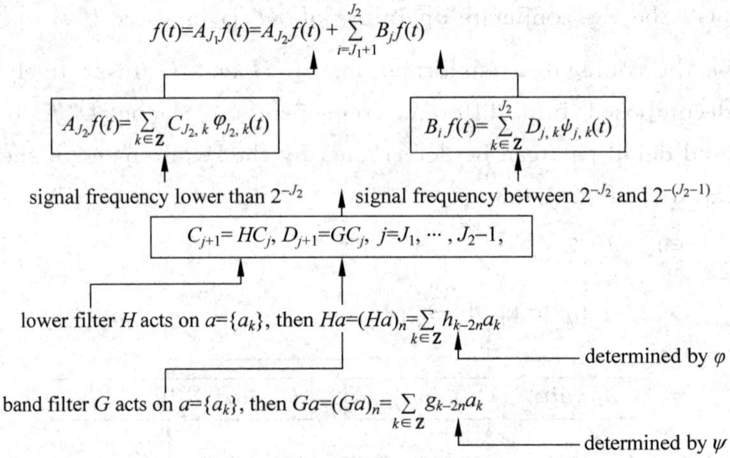

Fig. 5. 4. 4 Applications of wavelets(2)

6. Signal-noise separation and filters

Principle of signal-noise separation and filters Decompose a signal $f(t)$ into the part $A_{J_2}f(t)$ of lower frequency ($<2^{-J_2}$), and the part $D_jf(t)$ of frequency between 2^{-j} and $2^{-(j-1)}$, with $J_1-1<j<J_2$; then, each part is decomposed into the "sub-parts" with the same "range" possessing the same frequency but with different phases

$$A_{J_2}f(t) = \sum_{k\in\mathbf{Z}} C_{J_2,k}\varphi_{J_2,k}(t) \quad \text{and} \quad D_jf(t) = \sum_{k\in\mathbf{Z}}^{J_2} D_{j,k}\psi_{j,k}(t).$$

Analysis: If the energies of the scale function and wavelet function concentrate at the neighborhoods of $t=a$ and $t=b$, then

$$A_{J_2}f(t) = \sum_{k\in\mathbf{Z}} C_{J_2,k}\varphi_{J_2,k}(t) \quad \text{and} \quad B_jf(t) = \sum_{k\in\mathbf{Z}}^{J_2} D_{j,k}\psi_{j,k}(t),$$

where the energies of $C_{J_2,k}\varphi_{J_2,k}(t)$ and $D_{j,k}\psi_{j,k}(t)$ are concentrated at the neighborhoods of $t=2^j(k+a)$ and $t=2^j(t+b)$, respectively. We see that: the resolution ratios of time (spaces) fields are changed with the frequencies; as results, are accurate at higher frequencies, whereas, are coarse at lower frequencies.

By Mallat decomposition algorithm, it follows

$$C_{j+1}=HC_j,\quad D_{j+1}=GC_j,\quad j=J_1,\cdots,J_2-1.$$

Decompose a signal $f(t)$ by Fig. 5.4.4; then, distinguish the effect part of signal and noise by certain priori knowledge; and then, take zero values for $C_{J_2,k}$ and $D_{j,k}$, $j=J_1+1,\cdots,J_2$, so that a new sequence $\tilde{C}_{J_2,k}$, $\tilde{D}_{j,k}$, $j=J_1+1,\cdots,J_2$, are determined with $\tilde{C}_{j-1}=H^*\tilde{C}_j+G^*\tilde{D}_j$, $j=J_1+1,\cdots,J_2$; finally, the signal without noise is obtained

$$\tilde{f}(t)=A_{J_1}\tilde{f}(t)=\sum_{k\in\mathbf{Z}}\tilde{C}_{J_1,k}\,\varphi_{J_1,k}(t).$$

Exercise 5

1. Prove: $\mathbb{S}(\mathbb{R}^n)\subset L^p(\mathbb{R}^n)$, $1\leqslant p<+\infty$.

2. Prove: $\mathrm{e}^{-x^2}\in\mathbb{S}(\mathbb{R}^n)$.

3. Prove: $|x^\beta\partial^p\varphi(x)|\leqslant M_{\beta,p}$, $\forall\beta,p\in\mathbb{N}^n\Leftrightarrow|(1+|x|^2)^{\beta/2}\partial^p\varphi(x)|\leqslant M_{\beta,p}$, $\forall\beta,p\in\mathbb{N}^n\Leftrightarrow\lim\limits_{|x|\to+\infty}|x|^\beta\partial^p\varphi(x)=0$, $\forall\beta,p\in\mathbb{N}^n$.

4. Generalize various operations of Schwartz distributions to spaces $E^*(\mathbb{R}^n)$ and $D^*(\mathbb{R}^n)$.

5. Prove: $\dfrac{\partial^2 u}{\partial x_i\partial x_j}=\dfrac{\partial^2 u}{\partial x_j\partial x_i}$ holds in distribution sense in $\mathbb{S}^*(\mathbb{R}^n)$, $E^*(\mathbb{R}^n)$, $D^*(\mathbb{R}^n)$, respectively.

6. Prove: $\dfrac{\partial}{\partial x_i}(a\,u)=a\dfrac{\partial u}{\partial x_i}+\dfrac{\partial a}{\partial x_i}u$, for $a\in C^\infty(\mathbb{R}^n)$, $u\in D^*(\mathbb{R}^n)$, holds in distribution sense.

7. Prove: $D^*(\mathbb{R}^n)\supset\mathbb{S}^*(\mathbb{R}^n)\supset E^*(\mathbb{R}^n)$ for $D(\mathbb{R}^n)\subset\mathbb{S}(\mathbb{R}^n)\subset E(\mathbb{R}^n)$.

8. Prove: $\operatorname{supp}\delta=\{0\}$ and $\delta\in E^*(\mathbb{R}^n)$.

9. Prove: for $f\in L^1(\mathbb{R})$, Fourier transformation $f^\wedge(\xi)$ is an even function on $\mathbb{R}$, if and only if $f(x)$ is even function on $\mathbb{R}$; and it holds $[f(-\circ)]^\wedge(\xi)=f^\wedge(-\xi)$.

10. Prove: Theorems 5.2.1, 5.2.7 and 5.2.8.

11. Prove: $(L^1([a,b]),+,\alpha\cdot,\|f\|_{L^1([a,b])},f*g)$ is a Banach algebra without unit, by Theorem 5.2.4.

12. Prove: the properties of $L^2(\mathbb{R})$-Fourier transformations (similar to those of $L^1(\mathbb{R})$-FT).

13. Prove: the properties of $L^p(\mathbb{R}^n)$ $(1<p<2)$-Fourier transformations (similar to those of $L^1(\mathbb{R}^n)$-FT).

14. Could we define $L^p(\mathbb{R}^n)$-Fourier transformation for $p>2$, similar to that of for $2<p<\infty$?

15. Prove: the unique theorem of Fourier transformation.

16. Let $\Omega_1\subset\mathbb{R}^n$, $\Omega_2\subset\mathbb{R}^m$ be open sets, $K_1\subset\mathbb{R}^n$, $K_2\subset\mathbb{R}^m$ be compact sets. Prove that $\int\langle T,\varphi(\circ,y)\rangle\mathrm{d}y=\langle T,\int\varphi(\circ,y)\,\mathrm{d}y\rangle$ holds for $T\in D^*(\Omega_1\times\Omega_2)$, $\varphi\in D(\Omega_1\times\Omega_2)$, with $\operatorname{supp}\varphi\in K_1\times K_2$.

17. Let $\Omega_1\subset\mathbb{R}^n$, $\Omega_2\subset\mathbb{R}^m$ be open sets. Prove: integral summation $\sum_j f(\xi_j,y)\Delta_j x$ converges into $\int f(x,y)\mathrm{d}x$ for $f\in D(\Omega_1\times\Omega_2)$ in space $D(\Omega_2)$.

18. Laplace transformation of $S\in E^*(\mathbb{R})$ is defined by $L_S(\lambda)=\langle S(\cdot),\mathrm{e}^{-\lambda\cdot}\rangle$. Prove: for $T\in E(\mathbb{R})$, it holds Laplace transformation formula $L_{S*T}(\lambda)=L_S(\lambda)L_T(\lambda)$ with convolution $S*T$.

19. Prove: mapping $\langle S,\varphi\rangle\to S*\varphi$ of $S\in D^*(\mathbb{R}^n)$ and $\varphi\in D(\mathbb{R}^n)$ is bilinear; and is continuous mapping about S and φ, respectively.

20. Evaluate convolutions: $\mathrm{e}^{-|x|}*\mathrm{e}^{-|x|}$; $\mathrm{e}^{-ax^2}*\mathrm{e}^{-ax^2}$, $a>0$; $x\mathrm{e}^{-ax^2}*x\mathrm{e}^{-ax^2}$, $a>0$.

21. Prove: translation operator τ_h and convolution operator $*$ are exchanged, i. e.

$$\tau_h(T*S)=T*\tau_h S,\quad \forall S\in D^*(\mathbb{R}^n), T\in E^*(\mathbb{R}^n);$$

and prove: $\tau_h T=\delta(x-h)*T$, for $T\in E^*(\mathbb{R}^n)$.

22. Prove: convolution operation $(T,S)\to T*S$ is bilinear, and continuous about T and S, respectively.

23. Prove: if I: $\mathbb{S}^*(\mathbb{R}^n)\to D^*(\mathbb{R}^n)$ is the embedding mapping from $\mathbb{S}^*(\mathbb{R}^n)$ into $D^*(\mathbb{R}^n)$, and if it embeds zero of $\mathbb{S}^*(\mathbb{R}^n)$ to zero of $D^*(\mathbb{R}^n)$, then $\mathbb{S}^*(\mathbb{R}^n)\to D^*(\mathbb{R}^n)$ is an injection mapping.

24. Prove: $L^p(\mathbb{R}^n)\subset\mathbb{S}^*(\mathbb{R}^n)$, $1\leqslant p<+\infty$.

25. Define a multiplication of two distributions $T\in\mathbb{S}^*(\mathbb{R}^n)$, $S\in E^*(\mathbb{R}^n)$, by "$\langle T^\wedge S^\wedge,\varphi\rangle=\langle T^\wedge,S^\wedge\varphi\rangle$, $\forall\varphi\in\mathbb{S}(\mathbb{R}^n)$", and give an interpretation: multiplication $T^\wedge S^\wedge$ defined above makes sense under the assumption of Theorem 5. 3. 9 (4).

26. Evaluate Fourier transformation of $x\,\delta$ for δ-distribution.

27. Prove: Theorem 5.3.11.

28. Prove: the functions in Examples 5.4.1, 5.4.2, 5.4.3 are base wavelet.

29. Show a comparison between Fourier analysis and wavelet analysis (including Fourier series and discrete wavelet series; Fourier transformation and continuous wavelet transformation).

30. Show an example by using wavelet analysis to deal with a signal $f(t)$, and give its Mallat algorithm.

Chapter 6

Calculus on Manifolds

S. S. Chern, the great master of mathematics, once predicted: manifold will be the main research object in modern mathematics. That is true, "analysis on manifolds" is the most important research branch nowadays, calculus on manifolds, Riemann geometry and certain non-Euclidean geometries were bringing into physics, astronomy, biochemistry, biology, geology, even for medicine science, social science, and humanity, because these branches contain abundant modern ideology and advanced thought, as well as many useful theories and technologies.

The main contents of this chapter are selected from [3], written by S. S. Chern, et al., other references are [7], [9], [15], [16].

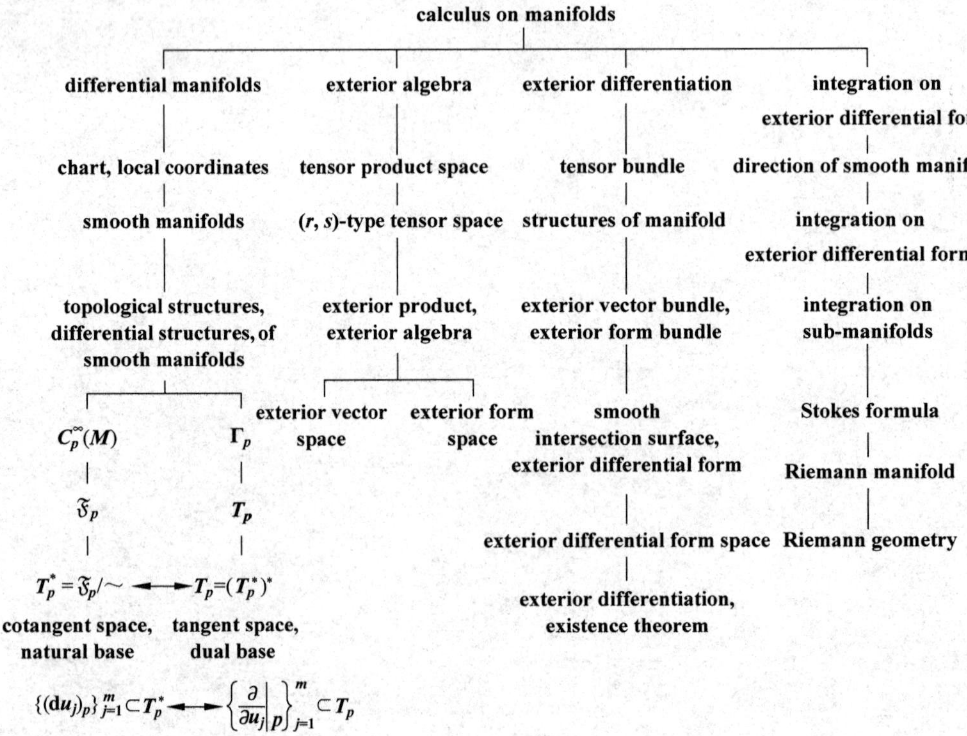

6.1 Basic Concepts

The originator of differential geometry is German mathematician C. F. Gauss (1777—1855). He introduced the first fundamental form of surfaces, established the theory of them, and founded "bending" geometry. Then, B. Riemann (1826—1866) generalized the theory built by Gauss to n-dimensional spaces, and established the Riemann geometry. The importance of Riemann geometry is that it formed the theoretical foundation of generalized relativity theory of Einstein, there, Einstein interpreted the gravitation phenomena as a curvature property of Riemann space, such that "gravitation wave" has a mathematical footing; Moreover, the development of the gauge field theory of Yang-Mills is a canonical example of combining mathematics with physics.

Nowadays, from Euclidean geometry to non-Euclidean geometry, from projective, affine, and differential geometries to Riemann geometry, it has been built a rigor, exquisite frame by virtue of exterior algebra, exterior differentiation, manifold, curvature, connection, etc. ; and has been applied to many natural science fields, successfully.

We introduce in this chapter the calculus on manifolds as basic knowledge and a starting point for going to higher and deeper theories, as well as going to science research, for our undergraduate students to prepare their accomplishments of mathematics and abilities of creativity.

6.1.1 Structures of differential manifolds

We use the notations in this section: Let $\mathbb{R}^m=\{x=(x_1,\cdots,x_m): x_j\in\mathbb{R}, j=1,2,\cdots,m\}$ be m-dimensional Euclidean space; $\Omega\subseteq\mathbb{R}^m$ be an open subset; $f: \Omega\to\mathbb{R}$ be a real-valued function on Ω. And let

$$C(\Omega)\equiv C^0(\Omega)=\{f: \Omega\to\mathbb{R}, f \text{ is continuous on } \Omega\};$$

$$C^r(\Omega)=\{f: \Omega\to\mathbb{R}, \partial^k f\in C(\Omega), k=0,1,\cdots,r\},\quad r\in\mathbb{N};$$

$$C^\infty(\Omega)=\{f: \Omega\to\mathbb{R}, \partial^k f\in C(\Omega), \forall k=0,1,\cdots\}.$$

1. Differential manifolds

We construct topology structure on a T_2-type (Hausdorff) topological space (X,τ), such that it becomes a topological manifold, firstly.

Definition 6.1.1 (topological manifold, chart, and atlas) Let (X,τ) be a T_2-type topological space. If $\forall p\in X$, there exists an open neighborhood $U_p\in\tau$ with $p\in U_p$, such that U_p is homeomorphic with an open set in $\mathbb{R}^m$, then X is said to be an m-**dimensional topological manifold**, denoted by M, and the homeomorphism mapping denoted by $\varphi_{U_p} \equiv \varphi: U_p \to \varphi(U_p)$.

It is easy to see that $U_p \subset X$ and $\varphi(U_p) \subset \mathbb{R}^m$, the mapping φ is bisingle-valued and bicontinuous, so that $\varphi_{U_p}(U_p) \subset \mathbb{R}^m$ is an open set in $\mathbb{R}^m$. Omit the sub-index, denoted by $(U, \varphi_U) \equiv (U_p, \varphi_{U_p})$, then, (U, φ_U) is said to be **a chart of** M **at** $p \in M$; the collection $\mathbb{A} = \{(U, \varphi_U): U \in \tau\}$ is said to be **an atlas of** M.

A chart (U, φ_U) of $p \in M$ on M is shown in Fig. 6.1.1.

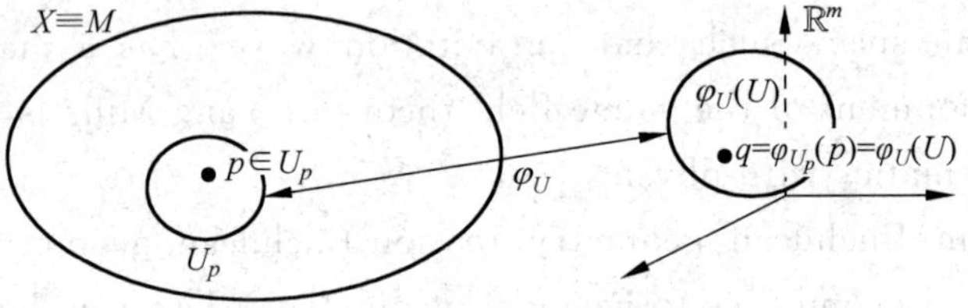

Fig. 6.1.1 Chart $(U, \varphi_U) \equiv (U_p, \varphi_{U_p})$

Note, for one point $p \in M$, there might be not unique $U \in \tau$ with $p \in U$ such that (U, φ_U) is a chart of M at $p \in M$. So that, we need to define "compatibility" for charts.

Compatibility conditions: Let (U, φ_U) and (V, φ_V) be two charts of M at $p \in M$. We distinguish two cases:

(1) for $U \cap V = \varnothing$, and $p \in U \setminus V$, take (U, φ_U) as a chart of M at $p \in M$;

(2) for $U \cap V \neq \varnothing$, and $p \in U \cap V$, then $\varphi_U(U \cap V)$ and $\varphi_V(U \cap V)$ are nonempty open sets in $\mathbb{R}^m \Rightarrow \varphi_U(p) \in \varphi_U(U \cap V)$, $\varphi_V(p) \in \varphi_V(U \cap V)$ should satisfy the compounded conditions:

$\varphi_V \circ \varphi_U^{-1}|_{\varphi_U(U \cap V)}: \varphi_U(U \cap V) \to \varphi_V(U \cap V)$ is homeomorphic from $\mathbb{R}^m$ to $\mathbb{R}^m$;

$(\varphi_V \circ \varphi_U^{-1}|_{\varphi_U(U \cap V)})^{-1}: \varphi_V(U \cap V) \to \varphi_U(U \cap V)$ is homeomorphic from $\mathbb{R}^m$ to $\mathbb{R}^m$.

Suppose that

$$F \equiv \varphi_V \circ \varphi_U^{-1}|_{\varphi_U(U \cap V)},$$

it is a homeomorphism from open set $\varphi_U(U \cap V)$ onto open set $\varphi_V(U \cap V)$. Then, the two charts (U, φ_U) and (V, φ_V) satisfying (1) and (2) are called **compatible.**

Let $F(z) = (f_1(z), \cdots, f_m(z)) \in \mathbb{R}^m$, $\forall z \in \varphi_U(U \cap V)$, then

$$\varphi_U(p) = ((\varphi_U(p))_1, \cdots, (\varphi_U(p))_m) \equiv (u_1, \cdots, u_m).$$

Let $u_j = (\varphi_U(p))_j$, $p \in U$, $j = 1, 2, \cdots, m$.

Definition 6.1.2 (local coordinates of $p \in U$, local coordinate system on M) For $p \in U$,

$\varphi_U(p)\in\mathbb{R}^m$, then, $u_j=(\varphi_U(p))_j$, $1\leqslant j\leqslant m$, are said to be **local coordinates of** $p\in U$, and

$$\{(u_1,u_2,\cdots,u_m): u_j=(\varphi_U(p))_j, j=1,2,\cdots,m,(U,\varphi_U)\in\mathbb{A}\}$$

is said to be **a local coordinate system of** M, or for simply, denoted by

$$\{(u_j=(\varphi_U(p))_j)_{j=1}^m: (U,\varphi_U)\in\mathbb{A}\},$$

or $\{(U,u_j)\}$ (Fig. 6.1.2).

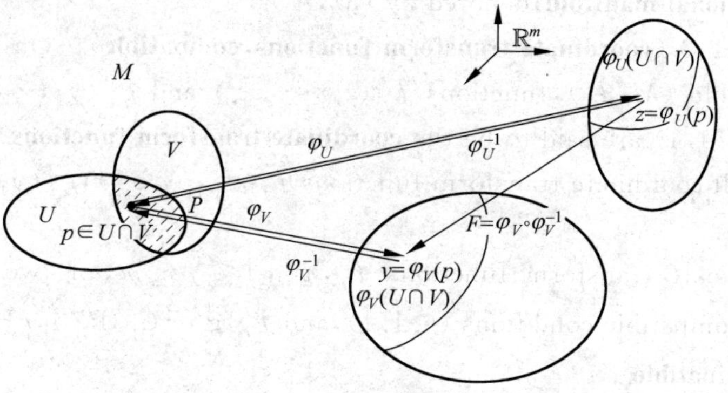

Fig. 6.1.2 Compatibility of local coordinates of $p\in U$

For the local coordinates, we deduce

$\forall z\in\varphi_U(U\cap V)\subset\mathbb{R}^m\Rightarrow y\in F(z)\in\varphi_V(U\cap V)\subset\mathbb{R}^m$

$\Rightarrow F(z)=(f_1(z),\cdots,f_m(z))\equiv(y_1,\cdots,y_m)=y\in\mathbb{R}^m\Rightarrow y=F(z)$

$\Rightarrow y_j=f_j(z)=f_j(z_1,\cdots,z_m),\quad j=1,2,\cdots,m$

$\Rightarrow z=F^{-1}(y)=(g_1(y),\cdots,g_m(y))$

$\Rightarrow z_k=g_k(y)=g_k(y_1,\cdots,y_m),\quad k=1,2,\cdots,m$

$\Rightarrow f_j, g_k$ are continuous functions, and satisfy

$y_j=f_j(z)=f_j(z_1,\cdots,z_m)=f_j(g_1(y_1,\cdots,y_m),\cdots,g_m(y_1,\cdots,y_m)),\quad j=1,2,\cdots,m,$

$z_j=g_j(y)=g_j(y_1,\cdots,y_m)=g_j(f_1(z_1,\cdots,z_m),\cdots,f_m(z_1,\cdots,z_m)),\quad j=1,2,\cdots,m;$

or for simply, rewrite as

$$\begin{cases} y_j=f_j(z_1,\cdots,z_m)=(\varphi_V\circ\varphi_U^{-1}(z_1,\cdots,z_m))_j, & (z_1,\cdots,z_m)\in\varphi_U(U\cap V) \\ z_j=g_j(y_1,\cdots,y_m)=(\varphi_U\circ\varphi_V^{-1}(y_1,\cdots,y_m))_j, & (y_1,\cdots,y_m)\in\varphi_V(U\cap V) \end{cases},\quad j=1,2,\cdots,m.$$

The above formulas can be deduced as

$$\begin{cases} f_j(g_1(y_1,\cdots,y_m),\cdots,g_m(y_1,\cdots,y_m))=y_j, \\ g_j(f_1(z_1,\cdots,z_m),\cdots,f_m(z_1,\cdots,z_m))=z_j, \end{cases}\quad j=1,2,\cdots,m,\qquad (6.1.1)$$

the formula (6.1.1) is said to be the **compatible conditions.**

Local coordinate system $\{(u_j=(\varphi_U(p))_j)_{j=1}^{m}:\ (U,\varphi_U)\in\mathbb{A}\}$ has been constructed on a T_2-type topological space (X,τ).

A T_2-type topological space (M,τ) with atlas $\mathbb{A}=\{(U,\varphi_U):U\in\tau\}$ satisfying the compatibility conditions (1) and (2) is called m-**dimensional topological manifold**, or simply, m-**dimensional manifold**, denoted by $(M,\mathbb{A})$.

Definition 6.1.3 (coordinate transform functions, compatible) On an m-dimensional topological manifold $(M,\mathbb{A})$, functions $f_j(z_1,\cdots,z_m)$ and $g_j(y_1,\cdots,y_m)$, $1\leqslant j\leqslant m$, determined by (6.1.1) are said to be the **coordinate transform functions.**

As we see, all coordinate transform functions $f_j(z_1,\cdots,z_m)$, $g_j(y_1,\cdots,y_m)$, $1\leqslant j\leqslant m$, are continuous.

If the coordinate transform functions f_j, g_j, $1\leqslant j\leqslant m$, of two charts (U,φ_U), (V,φ_V) satisfy compatible conditions (6.1.1), and $f_j, g_j\in C^r(\mathbb{R}^m)$, $r\in\mathbb{N}$, then they are said to be C^r-**compatible.**

Definition 6.1.4 (C^r-differential manifold, smooth manifold) If $(M,\mathbb{A})$ is an m-dimensional topological manifold, $\mathbb{A}=\{(U,\varphi_U)\}$ is an atlas. If $\mathbb{A}$ satisfies

(1) $\mathbb{B}=\{U\}$ is **an open covering** of M;

(2) Any two charts in $\mathbb{A}=\{(U,\varphi_U)\}$ are C^r-**compatible**;

(3) $\mathbb{A}=\{(U,\varphi_U)\}$ is **maximum**, i.e. if a chart $(\widetilde{U},\varphi_{\widetilde{U}})$ of M is C^r-compatible with any chart in $\mathbb{A}=\{(U,\varphi_U)\}$, then $(\widetilde{U},\varphi_{\widetilde{U}})\in\mathbb{A}$;

Then, $\mathbb{A}=\{(U,\varphi_U)\}$ is said to be a C^r-**differential construction** on M.

If there is a C^r-differential construction $\mathbb{A}$ on M, then $(M,\mathbb{A})$ is said to be a C^r-**differential manifold**, and $\mathbb{A}=\{(U,\varphi_U)\}$ is said to be **a permissible chart system** on M.

A chart system of a topological manifold is C^0-compatible, by definition. If a topological manifold M with C^∞-**differential structure** $\mathbb{A}$, then $(M,\mathbb{A})$ is said to be **a smooth manifold.**

Example 6.1.1 $M=\mathbb{R}^m$

Take $U=M$, and φ_U an identity mapping, then $\mathbb{A}=\{(U,\varphi_U)\}$ is an open covering of $\mathbb{R}^m$, and $(\mathbb{R}^m,\mathbb{A})$ is a smooth manifold, called **a standard differential structure** of $\mathbb{R}^m$.

Example 6.1.2 Construct a smooth manifold M on $S^1=\{(x_1,x_2)\in\mathbb{R}^2:\ x_1^2+x_2^2=1\}$, here S^1 is the unit circle in $\mathbb{R}^2$ (Fig. 6.1.3).

Solve On $S^1=\{(x_1,x_2)\in\mathbb{R}^2:\ x_1^2+x_2^2=1\}$, denote $x=(x_1,x_2)$, and let

$$U_1=\{(x_1,x_2)\in S^1:\ x_2>0\}\ \text{red (real line)},\quad \varphi_{U_1}(x)=x_1;$$

$U_2 = \{(x_1, x_2) \in S^1 : x_2 < 0\}$ blue (real line), $\varphi_{U_2}(x) = x_1$;

$U_3 = \{(x_1, x_2) \in S^1 : x_1 > 0\}$ green (dotted line), $\varphi_{U_3}(x) = x_2$;

$U_4 = \{(x_1, x_2) \in S^1 : x_1 < 0\}$ purple (dotted line), $\varphi_{U_4}(x) = x_2$.

The sets U_1, U_2, U_3, U_4 are open in S^1, since in sub-topology on S^1, the set

$$U_1 = \{(x_1, x_2) \in S^1 : x_2 > 0\}$$

is regarded as an open set in $\mathbb{R}^2$

$$O = \{(x_1, x_2) \in \mathbb{R}^2 : x_2 > 0\}$$

intersecting with S^1, then the intersection is

$$U_1 = \{(x_1, x_2) \in S^1 : x_2 > 0\} = O \cap S^1.$$

Similar for U_2, U_3, U_4. And it holds $S^1 \subset U_1 \cup U_2 \cup U_3 \cup U_4$.

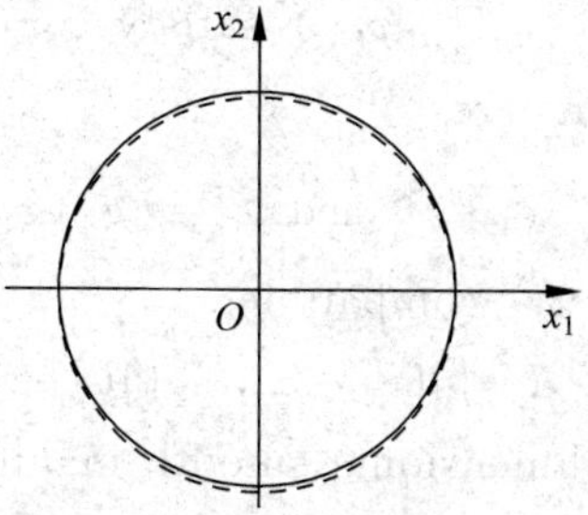

Fig. 6. 1. 3 One-dimensional smooth manifold on S^1

Proof $\mathbb{A} = \{(U_j, \varphi_{U_j}) : j = 1, 2, 3, 4\}$ is a one-dimensional smooth manifold structure on S^1.

In fact, $U_1 \cap U_2 = \varnothing$, $U_3 \cap U_4 = \varnothing$. We only need to deal with cases $U_1 \cap U_3 \neq \varnothing$, $U_1 \cap U_4 \neq \varnothing$, $U_2 \cap U_3 \neq \varnothing$, $U_2 \cap U_4 = \varnothing$. Take $U_1 \cap U_4 = \{(x_1, x_2) \in S^1 : x_1 < 0, x_2 > 0\}$ as example. By $\varphi_{U_1}(x) = x_1$, $\varphi_{U_4}(x) = x_2$, then

$$\varphi_{U_1}(U_1) = \{z \in \mathbb{R}^1 : z = (x_1, 0), -1 < x_1 < 1\},$$

$$\varphi_{U_4}(U_4) = \{z \in \mathbb{R}^1 : z = (0, x_2), -1 < x_2 < 1\};$$

thus, $\varphi_{U_1}(U_1 \cap U_4) = \{x_1 : -1 < x_1 < 0\} \subset \mathbb{R}^1$ and $\varphi_{U_4}(U_1 \cap U_4) = \{x_2 : 0 < x_2 < 1\} \subset \mathbb{R}^1$; moreover, $\varphi_{U_1}(U_1 \cap U_4)$ and $\varphi_{U_4}(U_1 \cap U_4)$ are non empty open sets in $\mathbb{R}^1$. The mapping

$$F = \varphi_{U_4} \circ \varphi_{U_1}^{-1}|_{\varphi_{U_1}(U_1 \cap U_4)} : \varphi_{U_1}(U_1 \cap U_4) \to \varphi_{U_4}(U_1 \cap U_4)$$

is a homeomorphism, such that

$$F = \varphi_{U_4} \circ \varphi_{U_1}^{-1}|_{\varphi_{U_1}(U_1 \cap U_4)} : \{x_1 : -1 < x_1 < 0\} \subset \mathbb{R}^1 \to \{x_2 : 0 < x_2 < 1\} \subset \mathbb{R}^1.$$

This implies

$$z=(x_1,0)\in\varphi_{U_1}(U_1\cap U_4)\xrightarrow{F}y=(0,x_2)\in\varphi_{U_4}(U_1\cap U_4).\qquad(*)$$

We determine functions in (6.1.1). Denote $z=(z_1,z_2)$, $y=(y_1,y_2)$, then

$$\begin{cases}y_1=f_1(z_1,z_2)\\ y_2=f_2(z_1,z_2)\end{cases},\quad\begin{cases}z_1=g_1(y_1,y_2)\\ z_2=g_2(y_1,y_2)\end{cases},$$

and deduces by(*), it follows

$$\begin{cases}y_1=f_1(z_1,z_2)=0,\\ y_2=f_2(z_1,z_2)=x_2,\end{cases}\quad\begin{cases}z_1=g_1(y_1,y_2)=x_1\\ z_2=g_2(y_1,y_2)=0\end{cases};$$

thus,

$$x_2=f_2(z_1,z_2)=f_2(x_1,0)\equiv f_2(x_1)=\sqrt{1-x_1^2},\quad\text{with}\quad -1<x_1<0;$$

$$x_1=g_1(y_1,y_2)=g_1(0,x_2)\equiv g_1(x_2)=-\sqrt{1-x_2^2},\quad\text{with}\quad 0<x_2<1.$$

This shows that charts (U_1,φ_{U_1}), (U_4,φ_{U_4}) satisfy compatible condition (6.1.1), and the coordinate transform functions are

$x_2=f_2(x_1)=\sqrt{1-x_1^2}$, $-1<x_1<0$ and $x_1=g_1(x_2)=-\sqrt{1-x_2^2}$, $0<x_2<1$, they are C^∞-functions, thus, are C^∞-compatible.

For other cases $U_1\cap U_3\neq\varnothing$, $U_2\cap U_3\neq\varnothing$, $U_2\cap U_4=\varnothing$, can be proved, similarly.

We have determined a one-dimensional smooth manifold structure $\mathbb{A}=\{U_1,U_2,U_3,U_4\}$ on S^1, such that $S^1=\{(x_1,x_2)\in\mathbb{R}^2: x_1^2+x_2^2=1\}$ is a one-dimensional smooth manifold $M=S^1$ of $\mathbb{R}^2$.

Note We can endow two different differential manifold structures on one topological manifold. Mathematician J. Milnor gave a famous "Milnor strange sphere" in 1956, as an example, it shows that there exist different differential manifold structures on two homeomorphic topological manifolds, and the differential manifold structures on the sphere are independent of the topological structures.

2. Smoothness of mappings on differential manifolds

1) Smooth functions on smooth manifolds

Definition 6.1.5 (smooth function on a smooth manifold) Let $(M,\mathbb{A})$ be an m-dimensional smooth manifold, and $f: M\to\mathbb{R}$ be a real-valued function on (M,τ). For a point $p\in M$, if $(U,\varphi_U)\in\mathbb{A}$ is a permissible chart contained p, then compound function $f\circ\varphi_U^{-1}$ is defined on open set $\varphi_U(U)\subset\mathbb{R}^m$ and with values in $\mathbb{R}$, i.e. $f\circ\varphi_U^{-1}: \varphi_U(U)\to\mathbb{R}$. If the real-valued function $f\circ\varphi_U^{-1}: \varphi_U(U)\to\mathbb{R}$ on open set $\varphi_U(U)\subset\mathbb{R}^m$ is C^∞-continuous and differentiable at point $\varphi_U(p)\in\varphi_U(U)\subset\mathbb{R}^m$, then $f: M\to\mathbb{R}$ on smooth manifold $(M,\mathbb{A})$

is said to be **a C^{∞}-continuous differentiable function at point** $\varphi_U(p)$. If $f: M\to\mathbb{R}$ is a C^{∞}-continuous differentiable function at every point on $(M, \mathbb{A})$, then it is said to be **a C^{∞}-continuous differentiable function on** $(M, \mathbb{A})$. or simply, **smooth function on** M(Fig. 6. 1. 4).

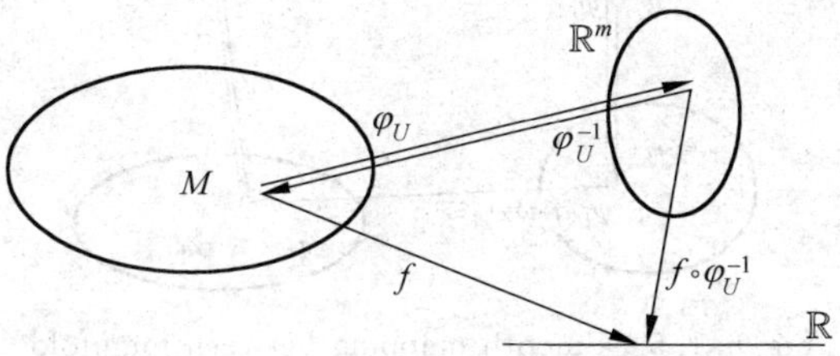

Fig. 6. 1. 4 Smooth function on manifold M

Sign $C^{\infty}(M)=\{f: f$ is C^{∞}-continuous diffetentiable on $M\}$.

Then, a smooth function $f: M\to\mathbb{R}$ on manifold $(M, \mathbb{A})$ can be expressed by

$f: M\to\mathbb{R}$ **is a smooth function on manifold** $(M, \mathbb{A}) \Leftrightarrow$

$f\circ\varphi_U^{-1}: \varphi_U(U)\to\mathbb{R}$ **is an** m-**variable smooth function on** $\mathbb{R}^m$

2) Smooth mappings between smooth manifolds

Definition 6. 1. 6 (smooth mapping between two smooth manifolds) Let $(M, \mathbb{A})$ and $(N, \mathbb{B})$ be m-dimensional and n-dimensional smooth manifolds, $f: M\to N$ be a continuous mapping from (M, τ) to (N, ν), respectively. If there exist charts $(U, \varphi_U)\in\mathbb{A}$ of $p\in M$ and $(V, \psi_V)\in\mathbb{B}$ of $f(p)\in N$, respectively, such that the function

$$\psi_V\circ f\circ\varphi_U^{-1}: \varphi_U(U)\subset\mathbb{R}^m \to \psi_V(V)\subset\mathbb{R}^n$$

from $\mathbb{R}^m$ to $\mathbb{R}^n$ is a C^{∞}-continuous differentiable vector-valued function at point $\varphi_U(p)\in\varphi_U(U)\subset\mathbb{R}^m$, then $f: M\to N$ is said to be C^{∞}-**mapping at** $p\in M$; If $f: M\to N$ is C^{∞}-mapping at every point $p\in M$, then $f: M\to N$ is said to be C^{∞}-**mapping on** M; or, $f: M\to N$ is **a smooth mapping from smooth manifolds** $(M, \mathbb{A})$ to $(N, \mathbb{B})$.

Note (1) the smooth function is a special example of smooth mappings;

(2) the smoothness of mappings is independent of the selection of permissible charts.

If $\dim M=\dim N$, and $f: M\to N$ is a homeomorphism mapping (bisingle-valued, bicontinuous); if f and f^{-1} are smooth mappings, then, $f: M\to N$ is said to be **a differential homeomorphism**, and two manifolds $(M, \mathbb{A})$, $(N, \mathbb{B})$ are called **differentiable homeomorphic**, or, the smooth manifold structures of $(M, \mathbb{A})$, $(N, \mathbb{B})$ are **isomorphic**(Fig. 6. 1. 5).

Then, a smooth mapping from manifold to manifold can be expressed by:

$f: M\to N$ **is a smooth mapping from smooth manifolds** $(M, \mathbb{A})$ to $(N, \mathbb{B}) \Leftrightarrow \psi_V\circ f\circ\varphi_U^{-1}: \varphi_U(U)\to\psi_V(V)$ **is an** m-**variable smooth vector-valued mapping from** $\mathbb{R}^m$ **onto** $\mathbb{R}^n$

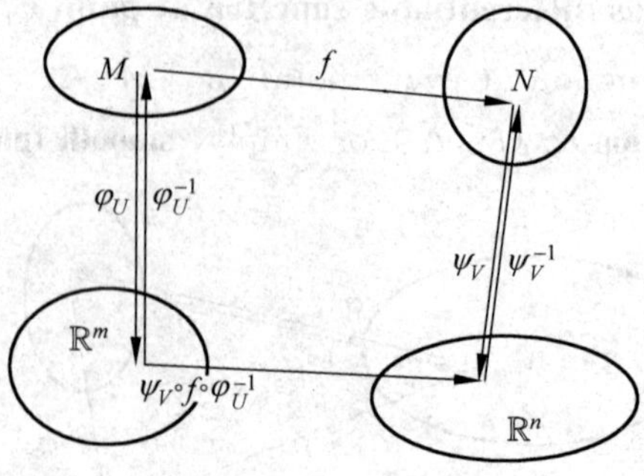

Fig. 6. 1. 5 Smooth mapping between manifolds

3) Smooth parameter curves on smooth manifolds

Definition 6. 1. 7 (smooth parameter curve on smooth manifold) Let $M=(a,b)$, and τ be as a usual topology on $\mathbb{R}^1$; and $(M,\mathbb{A})=((a,b),\tau)$, $(N,\mathbb{B})$ be one-dimensional smooth manifold, n-dimensional smooth manifold, respectively. If $f:(a,b)\to N$ is a smooth mapping from (a,b) to N, then $f:(a,b)\to N$ is said to be **a smooth parameter curve on manifold**, or simply, **a smooth curve, on** $(N,\mathbb{B})$ (Fig. 6. 1. 6).

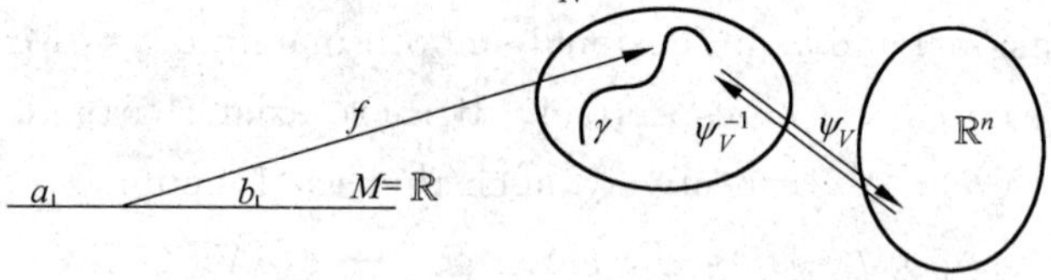

Fig. 6. 1. 6 Smooth curve on manifold

In fact, $(M,\mathbb{A})=((a,b),\tau)$ is regarded as one-dimensional manifold with τ as a usual topology on $\mathbb{R}^1$; and $(N,\mathbb{B})$ is an n-dimensional smooth manifold with permissible chart system $\mathbb{B}=\{(V,\psi_V)\}$. The domain of mapping $f: M\to N$ is $\mathfrak{D}_f=(a,b)\subset M$, and the range is $\mathfrak{R}_f\subset N$. If $f:(a,b)\to N$ is a smooth mapping from $\mathfrak{D}_f\subset M$ to N, then $f:(a,b)\to N$ is called **a smooth curve on** N.

Then, a smooth curve on smooth manifold $(N,\mathbb{B})$ can be expressed by:

$f:(a,b)\to N$ **is a smooth curve on smooth manifold** $(N,\mathbb{B})\Leftrightarrow$

$\psi_V\circ f:(a,b)\to\mathbb{R}^n$ **is a smooth n-dimensional vector-valued function on** (a,b)

where $f:\mathfrak{D}_f=(a,b)\subset M\to\mathfrak{R}_f=N$ is from one-dimensional manifold M to n-dimensional manifold, and $g=\psi_V\circ f:(a,b)\to\mathbb{R}^n$ is an n-dimensional vector valued function with one-dimensional variable in (a,b).

Note The smooth parameter curve is another special example of the smooth mappings from smooth manifold to smooth manifold.

We denote for smooth curve $f:(a,b)\to N$ by $\gamma:(a,b)\to N$, in the sequel.

6.1.2 Cotangent spaces, tangent spaces

The geometric visualization in 3-dimensional Euclidean space tells that: there exists a tangential line at each point on a smooth curve, and exists a tangent plane at each point on a smooth surface; moreover, the tangent plane could be understood as being consisted of all tangential lines of smooth curves on the smooth surface passing through the same point. We now generalize these concepts on a smooth manifold $(M,\mathbb{A})$, $\mathbb{A}=\{(U,\varphi_U)\}$.

Suppose $(M,\mathbb{A})$, $\mathbb{A}=\{(U,\varphi_U)\}$ is an m-dimensional smooth manifold in this section.

1. Basic space $C_p^\infty(M)$

The set

$$C_p^\infty(M)=\{f:U\to\mathbb{R},f\in C^\infty(U),p\in U,U\in\tau\}$$

is all C^∞-real-valued functions $f:U\to\mathbb{R}$ from a neighborhood $U\equiv U_p\in\tau$ of point $p\in M$ to $\mathbb{R}$. The domain of f is U, denoted by $\mathfrak{D}_f=U$.

Let $f,g\in C_p^\infty(M)$, and $\mathfrak{D}_f=U$, $\mathfrak{D}_g=V$; the algebraic operations on $C_p^\infty(M)$ are defined as:

addition $\quad (f+g)(x)=f(x)+g(x),\quad \forall x\in\mathfrak{D}_f\cap\mathfrak{D}_g$;

number product $\quad (\alpha f)(x)=\alpha f(x),\quad \forall x\in\mathfrak{D}_f,\alpha\in\mathbb{R}$;

thus, $(C_p^\infty(M),+,\alpha\cdot)$ is a linear space on $\mathbb{R}$; moreover,

multiplication $\quad (f\cdot g)(x)=f(x)g(x),\quad \forall x\in\mathfrak{D}_f\cap\mathfrak{D}_g$.

We emphasize: the definition of C^∞-smoothness of function $f:U\to\mathbb{R}$ in linear space $C_p^\infty(M)$ is in the sense of Definition 6.1.5.

2. Function germ space $\mathfrak{F}_p$

① **Equivalent relation $\sim$ in $C_p^\infty(M)$** for $f,g\in C_p^\infty(M)$, if $\exists$ a neighborhood H of p with $p\in H\subset M$, s. t. $f|_H=g|_H$, then f and g are said to be **equivalent**, denoted by $f\sim g$.

The set of all $C_p^\infty(M)$-functions equivalent to $f\in C_p^\infty(M)$, denoted by $[f]$, i. e. $[f]=\{g\in C_p^\infty(M):g\sim f\}$, is called **a germ of manifold M at point $p\in M$**.

The sense of equivalent relation $\sim$ in $C_p^\infty(M)$: All functions are regarded as in "one germ", if they are equal to each other in a neighborhood H of point $p\in M$, i. e., $f,g\in C_p^\infty(M)$, if $\exists H$, s. t. $f|_H=g|_H$, then $f\sim g$ are in one germ.

② **Equivalent class space** The C^∞-function set (quotient set $\mathfrak{F}_p$)

$$\mathfrak{F}_p = C_p^\infty(M)/\sim \equiv \{[f]: f \in C_p^\infty(M)\}$$

is a germ set of functions, called **a function germ set**, it is a quotient set of $C_p^\infty(M)$ with respect to equivalent relation $\sim$.

Define algebraic operations in function germ set $\mathfrak{F}_p = C_p^\infty(M)/\sim$:

addition $[f],[g] \in \mathfrak{F}_p \Rightarrow [f]+[g]=[f+g]$;

number product $[f] \in \mathfrak{F}_p, \alpha \in \mathbb{R} \Rightarrow \alpha[f] = [\alpha f]$;

then, function germ set $\mathfrak{F}_p = C_p^\infty(M)/\sim$ becomes a real linear space, called **a function germ space**, or simply, **a germ space.** Certainly, it is not a finite-dimensional space.

The sense of germ space $\mathfrak{F}_p = C_p^\infty(M)/\sim$: each element $[f]$ in space $\mathfrak{F}_p$ is an equivalent class, or a germ, however, each germ $[f] \in \mathfrak{F}_p$ contends infinite functions $g \in [f]$, they have same values $g|_H = f|_H$ in a neighborhood H of $p \in M$ for $g \in [f]$. The f is regarded as a representative member for all $g \in [f]$, it is as **"a germ"**. Here provides a most important idea to study the properties of a function space.

3. Smooth parameter curve space Γ_p

Smooth parameter curve passing through point $p \in M$ on manifold $(M, \mathbb{A})$ For a point $p \in M$, if $\exists \delta > 0$, such that one-one C^∞-mapping $\gamma: (-\delta, \delta) \to M$, such that $0 \leftrightarrow p$, i. e., $\gamma(0) = p$, then γ is called **a smooth parameter curve passing through $p \in M$ on manifold** $(M, \mathbb{A})$; i. e., $g = \varphi_U \circ \gamma: (-\delta, \delta) \to \mathbb{R}^m$ is a smooth mapping from $(-\delta, \delta)$ to $\mathbb{R}^m$. Denote the set of all smooth parameter curves passing through $p \in M$ on manifold $(M, \mathbb{A})$ by

$$\Gamma_p = \{\gamma: (-\delta, \delta) \to M, \quad \gamma \text{ is } C^\infty\text{-smooth curve}\},$$

and Γ_p is called **a smooth parameter curve set passing through p**, or simply, **a parameter curve set.**

Define the algebraic operations in Γ_p:

addition $(\gamma_1 + \gamma_2)(t) = \gamma_1(t) + \gamma_2(t), \quad \forall t \in (-\delta, \delta)$;

number product $(\alpha\gamma)(t) = \alpha\gamma(t), \quad \forall t \in (-\delta, \delta), \alpha \in \mathbb{R}$;

then Γ_p becomes a real linear space, called **a parameter curve space.**

The sense of parameter curve space Γ_p: there are infinite smooth parameter curves passing through $p \in M$, so that we have to restrict parameter t in $(-\delta, \delta)$, here $\exists \delta > 0$ is a common value, but not interval (a, b), such that $t = 0$ is correspondent to point p, denoted by $\gamma(0) = p$.

4. zero-match subspace $\mathfrak{I}_p$

The match of function germ $[f]\in\mathfrak{F}_p$ and parameter curve $\gamma\in\Gamma_p$ For $[f]\in\mathfrak{F}_p$ and $\gamma\in\Gamma_p$, **the match** $\langle\langle\gamma,[f]\rangle\rangle$ **of** $[f]\in\mathfrak{F}_p$ **and** $\gamma\in\Gamma_p$ is defined by:

$$\langle\langle\gamma,[f]\rangle\rangle=\left.\frac{\mathrm{d}(f\circ\gamma)}{\mathrm{d}t}\right|_{t=0},\quad t\in(-\delta,\delta),\tag{6.1.2}$$

here the smoothness of function $f\circ\gamma:(-\delta,\delta)\to\mathbb{R}$ makes sense.

The sense of "match": we have

$[f]\in\mathfrak{F}_p,\gamma\in\Gamma_p\Rightarrow f:U_p\to\mathbb{R}^m,\gamma:(-\delta,\delta)\to M\Rightarrow f\circ\gamma:(-\delta,\delta)\to\mathbb{R}^m$ has derivative $\frac{\mathrm{d}(f\circ\gamma)}{\mathrm{d}t}$ at $t=0$ for $t\in(-\delta,\delta)\Rightarrow$ value of the derivative at $t=0$ is $\left.\frac{\mathrm{d}(f\circ\gamma)}{\mathrm{d}t}\right|_{t=0}=$ $\langle\langle\gamma,[f]\rangle\rangle$, it is the tangent vector of mapping $f\circ\gamma$ at point p ($\leftrightarrow t=0$).

Since the match $\langle\langle\gamma,[f]\rangle\rangle$ is independent of the choice of f (why?), then we may define the subspace of the function germ space $\mathfrak{F}_p$ by

$$\mathfrak{I}_p=\{[f]\in\mathfrak{F}_p:\langle\langle\gamma,[f]\rangle\rangle=0,\forall\gamma\in\Gamma_p\}\subset\mathfrak{F}_p,$$

$\mathfrak{I}_p$ is called **the zero-match set** (Fig. 6.1.7).

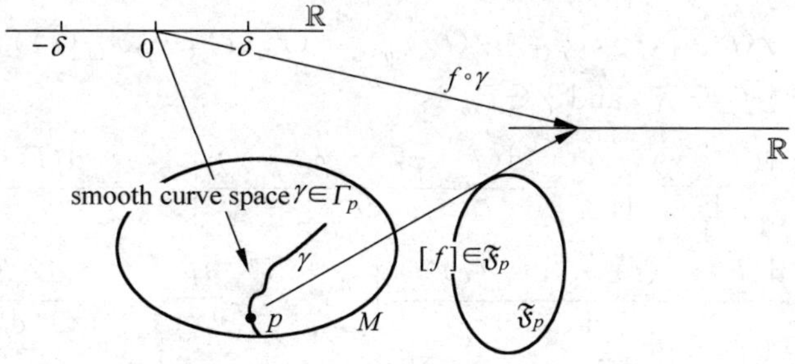

Fig. 6.1.7 $f\circ\gamma$

The algebraic operations on zero-match set $\mathfrak{I}_p\subset\mathfrak{F}_p$:

addition $[f],[g]\in\mathfrak{I}_p\Rightarrow\forall\gamma\in\Gamma_p,[f]+[g]=[f+g]$;

number product $[f]\in\mathfrak{I}_p,\alpha\in\mathbb{R},\Rightarrow\forall\gamma\in\Gamma_p,\alpha[f]=[\alpha f]$.

We check that the above operations are closed in zero-match set $\mathfrak{I}_p$:

$$[f],[g]\in\mathfrak{I}_p\Rightarrow\forall\gamma\in\Gamma_p,\langle\langle\gamma,[f+g]\rangle\rangle=\left.\frac{\mathrm{d}(f+g)\circ\gamma}{\mathrm{d}x}\right|_{t=0}$$

$$=\left.\frac{\mathrm{d}(f\circ\gamma)}{\mathrm{d}x}\right|_{t=0}+\left.\frac{\mathrm{d}(g\circ\gamma)}{\mathrm{d}x}\right|_{t=0}=\langle\langle\gamma,[f]\rangle\rangle+\langle\langle\gamma,[g]\rangle\rangle=0$$

$$\Rightarrow\langle\langle\gamma,[f+g]\rangle\rangle=0\Rightarrow[f+g]\in\mathfrak{I}_p;$$

$$[f]\in\mathfrak{I}_p,\alpha\in\mathbb{R}\Rightarrow\forall\gamma\in\Gamma_p,\langle\langle\gamma,\alpha[f]\rangle\rangle=\left.\frac{\mathrm{d}(\alpha f\circ\gamma)}{\mathrm{d}t}\right|_{t=0}=\alpha\left.\frac{\mathrm{d}(f\circ\gamma)}{\mathrm{d}t}\right|_{t=0}$$

$$=\alpha\langle\langle\gamma,[f]\rangle\rangle=0$$
$$\Rightarrow \alpha[f]\in\mathfrak{J}_p \Rightarrow \langle\langle\gamma,\alpha[f]\rangle\rangle=0 \Rightarrow \alpha[f]\in\mathfrak{J}_p.$$

Thus,

$$\mathfrak{J}_p=\{[f]\in\mathfrak{F}_p:\langle\langle\gamma,[f]\rangle\rangle=0,\quad \forall\gamma\in\Gamma_p\}$$

is a real linear subspace of $\mathfrak{F}_p$, and $\mathfrak{J}_p$ is called **the zero-match subspace of** $\mathfrak{F}_p$.

The following is the representation theorem of zero-match subspace.

Theorem 6.1.1 Let $(M,\mathcal{A})$ be an m-dimensional smooth manifold, $\mathfrak{F}_p$ be the C^∞-function germ space, $\mathfrak{J}_p$ be zero-match subspace. For $[f]\in\mathfrak{F}_p$, then $[f]\in\mathfrak{J}_p$, if and only if any permissible chart (U,φ_U) containing $p\in M$ satisfies $\left.\frac{\partial(f\circ\varphi_U^{-1})}{\partial x_j}\right|_{\varphi_U(p)}=0, 1\leqslant j\leqslant m$.

Proof Let (U,φ_U) be a permissible chart containing $p\in M$. Then, for any $[f]\in\mathfrak{F}_p$, let

$$f\circ\varphi_U^{-1}(x_1,\cdots,x_m)\equiv F(x_1,\cdots,x_m),\quad (x_1,\cdots,x_m)\in\varphi_U(U).$$

On the other hand, for $\gamma\in\Gamma_p$ with $\gamma:(-\delta,\delta)\to U\subset M$, its coordinates in the permissible chart (U,φ_U) are

$$((\varphi_U\circ\gamma(t))_1,\cdots,(\varphi_U\circ\gamma(t))_m)=(x_1(t),\cdots,x_m(t))\in\mathbb{R}^m.$$

Thus, the match for $[f]\in\mathfrak{F}_p$ and $\gamma\in\Gamma_p$ is

$$\langle\langle\gamma,[f]\rangle\rangle=\left.\frac{\mathrm{d}(f\circ\gamma)}{\mathrm{d}t}\right|_{t=0}=\left.\frac{\mathrm{d}(f\circ\varphi_U^{-1}\circ\varphi_U\circ\gamma)}{\mathrm{d}t}\right|_{t=0}=\left.\frac{\mathrm{d}(F\circ\varphi_U\circ\gamma)}{\mathrm{d}t}\right|_{t=0}$$
$$=\left.\frac{\mathrm{d}(F(x_1(t),\cdots,x_m(t)))}{\mathrm{d}t}\right|_{t=0}=\sum_{j=1}^m\left.\frac{\partial F}{\partial x_j}\right|_{\varphi_U(p)}\left.\frac{\mathrm{d}x_j(t)}{\mathrm{d}t}\right|_{t=0}.$$

Then, by $\langle\langle\gamma,[f]\rangle\rangle=\sum_{j=1}^m\left.\frac{\partial F}{\partial x_j}\right|_{\varphi_U(p)}\left.\frac{\mathrm{d}x_j(t)}{\mathrm{d}t}\right|_{t=0}$, we get

$$[f]\in\mathfrak{J}_p\Leftrightarrow\forall\gamma\in\Gamma_p,\langle\langle\gamma,[f]\rangle\rangle=0\Leftrightarrow\forall\gamma\in\Gamma_p,\sum_{j=1}^m\left.\frac{\partial F}{\partial x_j}\right|_{\varphi_U(p)}\left.\frac{\mathrm{d}x_j(t)}{\mathrm{d}t}\right|_{t=0}=0\Leftrightarrow$$
$$\left.\frac{\partial F}{\partial x_j}\right|_{\varphi_U(p)}=0, 1\leqslant j\leqslant m\Leftrightarrow\left.\frac{\partial(f\circ\varphi_U^{-1})}{\partial x_j}\right|_{\varphi_U(p)}=0, 1\leqslant j\leqslant m.$$

The proof is complete.

The sense of zero-match subspace $\mathfrak{J}_p=\{[f]\in\mathfrak{F}_p:\langle\langle\gamma,[f]\rangle\rangle=0,\forall\gamma\in\Gamma_p\}$: any element $[f]$ in zero-match subspace $\mathfrak{J}_p$ is just the smooth function germ which all partial derivatives with respect to local coordinates (Definition 6.1.2) are equal to zero at point $p\in M$.

The match defined in (6.1.2) plays very important role: by virtue of this concept, we may introduce two equivalent relationships in **function germ space** $\mathfrak{F}_p$ and in **smooth parameter curve space** Γ_p, respectively.

For $[f]\in\mathfrak{F}_p$ and $\gamma\in\Gamma_p$, from their match $\langle\langle\gamma,[f]\rangle\rangle=\left.\frac{\mathrm{d}(f\circ\gamma)}{\mathrm{d}t}\right|_{t=0}$, $t\in(-\delta,\delta)$,

we define two equivalent relationships:

① In $\mathfrak{F}_p$—for $[f_1],[f_2]\in\mathfrak{F}_p$

$$[f_1]\sim[f_2]\Leftrightarrow\langle\langle\gamma,[f_1]\rangle\rangle=\langle\langle\gamma,[f_2]\rangle\rangle,\quad\forall\gamma\in\Gamma_p;\tag{6.1.3}$$

(later, we will see that the cotangent space can be defined by $T_p^*=\mathfrak{F}_p/\sim$);

② In Γ_p— for $\gamma,\gamma'\in\Gamma_p$

$$\gamma\sim\gamma'\Leftrightarrow\langle\langle\gamma,[f]\rangle\rangle=\langle\langle\gamma',[f]\rangle\rangle,\quad\forall[f]\in\mathfrak{F}_p;\tag{6.1.4}$$

(later, we will see that the tangent space can be defined by $T_p=\Gamma_p/\sim$).

5. Cotangent space $T_p^*=\mathfrak{F}_p/\sim$—equivalent class in $\mathfrak{F}_p$

Definition 6.1.8 (**cotangent space**) Let $\mathfrak{F}_p$ be the function germ space. By (6.1.3), the equivalent relation $\sim$ in $\mathfrak{F}_p$ for $[f],[g]\in\mathfrak{F}_p$ is defined. We see that,

$$[f_1]\sim[f_2]\Leftrightarrow\langle\langle\gamma,[f_1]\rangle\rangle=\langle\langle\gamma,[f_2]\rangle\rangle,\quad\forall\gamma\in\Gamma_p\Leftrightarrow[f]-[g]\in\mathfrak{J}_p.$$

The set of all $[g]\in\mathfrak{F}_p$ which are equivalent to $[f]\in\mathfrak{F}_p$, denoted by $[f]^*$, i. e.

$$[f]^*=\{[g]\in\mathfrak{F}_p:[g]\sim[f]\},$$

it is an equivalent class; The set of all these equivalent classes, denoted by $\mathfrak{F}_p/\sim$, i. e.

$$\mathfrak{F}_p/\sim\ \equiv\{[f]^*:[f]\in\mathfrak{F}_p\},$$

is **the quotient set of function germ space $\mathfrak{F}_p$ with respect to equivalent relation** $\sim$, endowed the algebraic operations by

addition $\quad [f]^*,[g]^*\in T_p^*\Rightarrow[f]^*+[g]^*=[f+g]^*$;

number product $\quad [f]^*\in T_p^*,\alpha\in\mathbb{R}\Rightarrow\alpha[f]^*=[\alpha f]^*$;

Then, quotient set $\mathfrak{F}_p/\sim$ becomes a real linear space, called **the cotangent space**, and denoted by $T_p^*=\mathfrak{F}_p/\sim$.

In cotangent space $T_p^*=\mathfrak{F}_p/\sim$, an element $[f]^*\in T_p^*$ is denoted by $(\mathrm{d}f)_p\equiv[f]^*$, and called **a cotangent vector of manifold M at point $p\in M$**, or simply, **a cotangent vector.**

6. A formula of cotangent vector $(\mathrm{d}f)_p$

Let $f^1,\cdots,f^s\in C_p^\infty(M)$ be s smooth functions on smooth manifold M, and $F(y^1,\cdots,y^s):\mathbb{R}^s\to\mathbb{R}$ be a smooth function on $\mathbb{R}^s$. Then, $F(f^1,\cdots,f^s):\mathbb{R}^s\to\mathbb{R}$ is a smooth function in a neighborhood of point $(f^1(p),\cdots,f^s(p))\in\mathbb{R}^s$. Compound function $f(q)\equiv F(f^1(q),\cdots,f^s(q))$, $q\in U_p$, is coming from $f^j:M\to\mathbb{R}$, $j=1,2,\cdots,s$, and $F:\mathbb{R}^s\to\mathbb{R}$, i. e.

$$f^1,\cdots,f^s \in C_p^\infty(M) \quad F(f^1,\cdots,f^s):\mathbb{R}^s \to \mathbb{R}$$

$$\downarrow$$

$$f(q) \equiv F(f^1(q),\cdots,f^s(q)), \quad q \in U_p$$

then the cotangent vector $(\mathrm{d}f)_p = [f]^*$ of $f(q) \equiv F(f^1(q),\cdots,f^s(q))$, $q \in U_p$, has the following formula.

Theorem 6.1.2 Let M be an m-dimensional smooth manifold, and functions $f^1,\cdots,f^s \in C_p^\infty(M)$. For a smooth function $F(y^1,\cdots,y^s):\mathbb{R}^s \to \mathbb{R}$ on $\mathbb{R}^s$, let

$$f(q) \equiv F(f^1(q),\cdots,f^s(q)), \quad q \in U_p.$$

Then $f \in C_p^\infty(M)$, and holds

$$(\mathrm{d}f)_p = \sum_{k=1}^{s} \left(\frac{\partial F}{\partial f^k}\right)_{(f^1(p),\cdots,f^s(p))} (\mathrm{d}f^k)_p. \tag{6.1.5}$$

Proof Let the domains of $f^1,\cdots,f^s \in C_p^\infty(M)$ be U_k $(p \in U_k, 1 \leqslant k \leqslant s)$, respectively. Then, the domain of f is $U \equiv \bigcap_{k=1}^{s} U_k$. So that for $q \in U \equiv \bigcap_{k=1}^{s} U_k$, we have

$$f(q) \equiv F(f^1(q),\cdots,f^s(q)), \quad q \in U_p,$$

and $F \in C^\infty(\mathbb{R}^s)$ implies $f \in C_p^\infty(M)$ (Fig. 6.1.8).

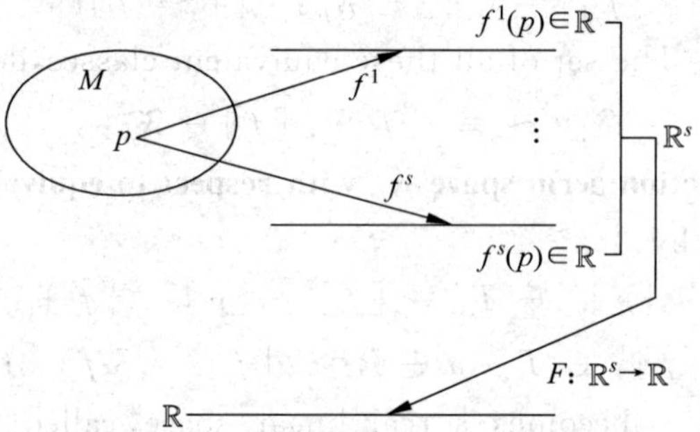

Fig. 6.1.8 Diagram of theorem 6.1.2

We evaluate for $\forall \gamma \in \Gamma_p$,

$$\begin{aligned}\langle\langle \gamma,[f]\rangle\rangle &= \left.\frac{\mathrm{d}(f \circ \gamma)}{\mathrm{d}t}\right|_{t=0} \left.\frac{\mathrm{d}F(f^1 \circ \gamma(t),\cdots,f^s \circ \gamma(t))}{\mathrm{d}t}\right|_{t=0} \\ &= \sum_{k=1}^{s} \left(\frac{\partial F}{\partial f^k}\right)_{(f^1(p),\cdots,f^s(p))} \left.\frac{\mathrm{d}(f^k \circ \gamma(t))}{\mathrm{d}t}\right|_{t=0} \\ &= \sum_{k=1}^{s} \left(\frac{\partial F}{\partial f^k}\right)_{(f^1(p),\cdots,f^s(p))} \langle\langle \gamma,[f^k]\rangle\rangle\end{aligned}$$

$$=\left\langle\left\langle\gamma,\sum_{k=1}^{s}\left(\frac{\partial F}{\partial f^{k}}\right)_{(f^{1}(p),\cdots,f^{s}(p))}[f^{k}]\right\rangle\right\rangle.$$

Then,

$$\langle\langle\gamma,[f]\rangle\rangle=\left\langle\left\langle\gamma,\sum_{k=1}^{s}\left(\frac{\partial F}{\partial f^{k}}\right)_{(f^{1}(p),\cdots,f^{s}(p))}[f^{k}]\right\rangle\right\rangle,$$

this implies

$$[f]-\sum_{k=1}^{s}\left(\frac{\partial F}{\partial f^{k}}\right)_{(f^{1}(p),\cdots,f^{s}(p))}[f^{k}]\in\mathfrak{I}_p,$$

i. e.

$$(\mathrm{d}f)_p=\sum_{k=1}^{s}\left(\frac{\partial F}{\partial f^{k}}\right)_{(f^{1}(p),\cdots,f^{s}(p))}(\mathrm{d}f^{k})_p.$$

The proof is complete.

The important properties of cotangent vector $(\mathrm{d}f)_p=[f]^*$:

Theorem 6. 1. 3 Suppose $f,g\in C_p^{\infty}(M),\alpha\in\mathbb{R}$, then $(\mathrm{d}f)_p=[f]^*\in T_p^*$ satisfies

$$(\mathrm{d}(\alpha f+\beta g))_p=\alpha(\mathrm{d}f)_p+\beta(\mathrm{d}g)_p;$$

$$(\mathrm{d}(f\cdot g))_p=(\mathrm{d}f)_p g(p)+(\mathrm{d}g)_p f(p).$$

The proof can be obtained by the basic formulas of the calculus in *Advance Mathematics*.

Theorem 6. 1. 4 (dimension theorem of T_p^*) If $(M,\mathbb{A})$ is an m-dimensional smooth manifold, then $\dim T_p^*=m$.

Proof Take a permissible chart (U,φ_U) of point $p\in M$. Each $q\in U$ has local coordinate $(u_1,\cdots,u_m)$ with

$$(u_1(q),\cdots,u_m(q))=((\varphi_U(q))_1,\cdots,(\varphi_U(q))_m)=((e_1\circ\varphi_U)(q),\cdots,(e_m\circ\varphi_U)(q)),\tag{6. 1. 6}$$

and $\{e_1,\cdots,e_m\}\subset\mathbb{R}^m$ is a given coordinate system in $\mathbb{R}^m$, such that $\forall q\in U\subset M\Rightarrow\varphi_U(q)\in\mathbb{R}^m$.

For instance, $m=3$ (Fig. 6. 1. 9): $\varphi_U(q)=(\varphi_U(q))_1e_1+(\varphi_U(q))_2e_2+(\varphi_U(q))_3e_3$, i. e.

$$q\in U\Rightarrow\varphi_U(q)\in\mathbb{R}^3\Rightarrow((\varphi_U(q))_1,(\varphi_U(q))_2,(\varphi_U(q))_3).$$

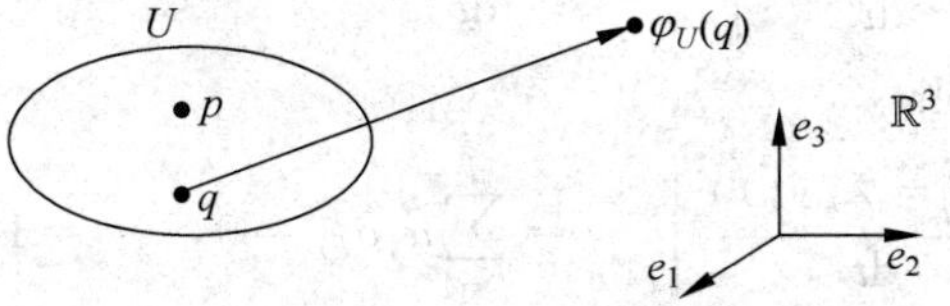

Fig. 6. 1. 9 Diagram of Theorem 6. 1. 4

In general, denote $u_j(q)\equiv(\varphi_U(q))_j e_j\equiv(e_j\circ\varphi_U)(q), j=1,2,\cdots,m$. Thus,

$$((\varphi_U(q))_1,\cdots,(\varphi_U(q))_m)=\sum_{j=1}^{m}(\varphi_U(q))_j e_j=\sum_{j=1}^{m}(e_j\circ\varphi_U)(q). \qquad (6.1.7)$$

Note Here $\{e_1,\cdots,e_m\}\subset\mathbb{R}^m$ is a selected coordinate system in $\mathbb{R}^m$, but is not necessarily an orthonormal system. By the smoothness of a manifold, we have $u_j\in C_p^\infty(M)$ in (6.1.6). Then,

$$u_j\in C_p^\infty(M)\Rightarrow[u_j]\in\mathfrak{F}_p\Rightarrow[u_j]^*\in\mathfrak{F}_p/\sim\Rightarrow(\mathrm{d}u_j)\equiv[u_j]^*\in T_p^*,$$

this implies $(\mathrm{d}u_j)_p=[u_j]^*\in T_p^*$. We next prove

$$\{(\mathrm{d}u_j)_p, 1\leqslant j\leqslant m\} \qquad (6.1.8)$$

is a base of T_p^*, i. e., prove the following:

① $\forall(\mathrm{d}f)_p\in T_p^*$ can be expressed as a unique linear combination by $(\mathrm{d}u_j)_p$, $1\leqslant j\leqslant m$;

② $\{(\mathrm{d}u_j)_p: 1\leqslant j\leqslant m\}$ is an independent set.

Firstly, $\forall(\mathrm{d}f)_p\in T_p^*\Rightarrow f\circ\varphi_U^{-1}$ is a smooth function on an open set V with $\varphi_U(q)\in V\Rightarrow$ for $(x_1,\cdots,x_m)\in V\subset\mathbb{R}^m$, let $F(x_1,\cdots,x_m)\equiv f\circ\varphi_U^{-1}(x_1,\cdots,x_m)$, then $f(q)=F(u_1,\cdots,u_m)\Rightarrow$ by Theorem 6.1.2, we have $(\mathrm{d}f)_p=\sum_{k=1}^{m}\left(\frac{\partial F}{\partial u_k}\right)_{(u_1(p),\cdots,u_m(p))}(\mathrm{d}u_k)_p$, this means that $(\mathrm{d}f)_p$ is a linear combination of $(\mathrm{d}u_j)_p$, $1\leqslant j\leqslant m$, thus, ① is proved.

Secondly, if $(\mathrm{d}u_j)_p$, $1\leqslant j\leqslant m$ would not be linear independent $\Rightarrow\exists\alpha_j$, $1\leqslant j\leqslant m$, with $\alpha_j\in\mathbb{R}$, and at least one $\alpha_j\neq0$, such that $\sum_{j=1}^{m}\alpha_j(\mathrm{d}u_j)_p=0\Rightarrow\sum_{j=1}^{m}\alpha_j[u_j]\in\mathfrak{I}_p\Rightarrow$ $\forall\gamma\in\Gamma_p$ implies that

$$\left\langle\left\langle\gamma,\sum_{j=1}^{m}\alpha_j[u_j]\right\rangle\right\rangle=0\Rightarrow\sum_{j=1}^{m}\alpha_j\left.\frac{\mathrm{d}(u_j\circ\gamma(t))}{\mathrm{d}t}\right|_{t=0}=0 \qquad (6.1.9)$$

$\Rightarrow$ take $\lambda_k\in\Gamma_p$, $1\leqslant k\leqslant m$ with $u_j\circ\lambda_k(t)=u_j(p)+t\delta_{kj}$, and $\delta_{kj}=\begin{cases}1, & j=k,\\0, & j\neq k;\end{cases}$ in fact, $\lambda_k\in\Gamma_p$, $1\leqslant k\leqslant m$, are correspondent to the straight line passing through point p.

$\Rightarrow\left.\frac{\mathrm{d}(u_j\circ\lambda_k(t))}{\mathrm{d}t}\right|_{t=0}=\left.\frac{\mathrm{d}(u_j(p))}{\mathrm{d}t}\right|_{t=0}+\frac{\mathrm{d}(t\delta_{kj})}{\mathrm{d}t}=0+\delta_{kj}\Rightarrow$ put $\gamma(t)=\lambda_k(t)$, in (6.1.9) we have: $\gamma=\lambda_k\in\Gamma_p$ implies

$$\sum_{j=1}^{m}\alpha_j\left.\frac{\mathrm{d}(u_j\circ\lambda_k(t))}{\mathrm{d}t}\right|_{t=0}=\sum_{j=1}^{m}\alpha_j\delta_{kj}=0,\quad k=1,2,\cdots,m.$$

$\Rightarrow\alpha_j=0, 1\leqslant j\leqslant m\Rightarrow$ this contradicts "at least one $\alpha_j\neq0$" $\Rightarrow(\mathrm{d}u_j)_p$ $(1\leqslant j\leqslant m)$ are linear independent, thus, ② holds. The proof is complete.

$\{(\mathrm{d}u_j)_p: 1\leqslant j\leqslant m\}$ in (6.1.8) is called **the natural base of T_p^* with respect to local coordinate system $\{(U,u_j)\}$ of smooth manifold M**; or simply, **natural base of T_p^***.

7. Tangent space $T_p=\Gamma_p/\sim$—equivalent class in Γ_p

Definition 6.1.9 (tangent space) Let $T_p^*=\mathfrak{F}_p/\sim$ be the cotangent space.

(1) **Equivalent relation in parameter curve space** $\Gamma_p=\{\gamma: (-\delta,\delta)\to M\}$ By (6.1.4), the equivalent relation $\sim$ on Γ_p is defined as

$$\gamma_1,\gamma_2\in\Gamma_p, \gamma_1\sim\gamma_2\Leftrightarrow\langle\langle\gamma_1-\gamma_2,(\mathrm{d}f)_p\rangle\rangle=0,\quad \forall(\mathrm{d}f)_p\in T_p^*.$$

(2) **Tangent space** $T_p=\Gamma_p/\sim$ Denote the equivalent class by $[\gamma]=\{\gamma'\in\Gamma_p: \gamma'\sim\gamma\}$, and the set of all equivalent classes by

$$T_p=\Gamma_p/\sim=\{[\gamma]: \gamma\in\Gamma_p\}.$$

Endow algebraic operations on T_p,

addition $\qquad \gamma_1,\gamma_2\in T_p\Rightarrow\gamma_1+\gamma_2\in T_p$;

number product $\qquad \gamma\in T_p, \alpha\in\mathbb{R}\Rightarrow\alpha\gamma\in T_p$;

then, quotient set $T_p=\Gamma_p/\sim$ is a real linear space, and $T_p=\{[\gamma]: \gamma\in\Gamma_p\}$ is called **the tangent space of manifold M at point $p\in M$**, or simply, **the tangent space** at p.

In tangent space $T_p=\Gamma_p/\sim$, an element $[\gamma]\in T_p$ is called **a tangent vector of manifold M at point $p\in M$**, or simply, **a tangent vector**, see Definition 6.1.10.

Theorem 6.1.5 (properties of the tangent space) Suppose that M is an m-dimensional smooth manifold, then

(1) $T_p=\{[\gamma]: \gamma\in\Gamma_p\}$ is a linear space;

(2) T_p is the dual space of cotangent space T_p^*, i.e., T_p is the linear functional space of T_p^*, or $(T_p^*)^*=T_p$, and $\dim T_p^*=m=\dim T_p$;

(3) "the match" of $[\gamma]\in T_p$ with $(\mathrm{d}f)_p\in T^*$ can be defined by

$$\langle[\gamma],(\mathrm{d}f)_p\rangle\equiv\langle\langle\gamma,(\mathrm{d}f)_p\rangle\rangle,\quad \forall\gamma\in[\gamma]\in T_p, \forall(\mathrm{d}f)_p\in T_p^*,$$

and $\langle[\gamma],(\mathrm{d}f)_p\rangle$ is a "bilinear form" on $T_p\times T_p^*$.

Proof (1) $T_p=\{[\gamma]: \gamma\in\Gamma_p\}$ is a linear space, since the algebraic operations are

$$[\gamma_1], [\gamma_2]\in T_p\Rightarrow[\gamma_1]+[\gamma_2]=[\gamma_1+\gamma_2];$$

$$\alpha\in\mathbb{R}, [\gamma]\in T_p\Rightarrow\alpha[\gamma]=[\alpha\gamma];$$

such that T_p is a real linear space.

(2) To prove $(T_p^*)^*=T_p$, we take the local coordinate system of smooth manifold M in (6.1.6).

Suppose that a smooth parameter curve $\gamma\in\Gamma_p$ on M is given by $u_j=u_j(t), 1\leqslant j\leqslant m$; and $p\in M$ has local coordinates $p=(u_j(0))_{1\leqslant j\leqslant m}$. By similar idea for evaluate "match" $\langle\langle \gamma, [f] \rangle\rangle$ in Theorem 6.1.1, and by Theorem 6.1.2, we deduce as follows: $\forall [\gamma] \in T_p, \forall (\mathrm{d}f)_p \in T_p^*$, $\langle [\gamma], (\mathrm{d}f)_p \rangle \equiv \langle\langle \gamma, (\mathrm{d}f)_p \rangle\rangle$, By (6.1.2) $\Rightarrow \langle [\gamma], (\mathrm{d}f)_p \rangle = \sum_{j=1}^{m} a_j \xi_j$, with a_j, ξ_j in the following (6.1.10)

$$a_j = \left(\frac{\partial (f \circ \varphi_U^{-1})}{\partial u_j}\right)_{\varphi_U(p)}, \quad \xi_j = \left(\frac{\mathrm{d}u_j}{\mathrm{d}t}\right)_{t=0} = \left(\frac{\mathrm{d}(u_j \circ \gamma(t))}{\mathrm{d}t}\right)_{t=0}, \quad 1 \leqslant j \leqslant m \tag{6.1.10}$$

$\Rightarrow a_j = \left(\frac{\partial (f \circ \varphi_U^{-1})}{\partial u_j}\right)_{\varphi_U(p)}$ is the component of cotangent vector $(\mathrm{d}f)_p \in T_p^*$ with respect to the natural base $\{\xi_j : 1 \leqslant j \leqslant m\} \equiv \{(\mathrm{d}u_j)_p, 1 \leqslant j \leqslant m\}$ in (6.1.8)

$\Rightarrow \langle [\gamma], (\mathrm{d}f)_p \rangle = \sum_{j=1}^{m} a_j \xi_j$ is a linear mapping from T_p^* to $\mathbb{R}$, i. e., a linear functional. Then, we conclude:

For $\gamma \in [\gamma] \in T_p$, system $\{\xi_j : 1 \leqslant j \leqslant m\}$ is determined as a linear functional, denoted by

$$F_{[\gamma]} \leftrightarrow \xi_j = \left(\frac{\mathrm{d}(u_j \circ \gamma(t))}{\mathrm{d}t}\right)_{t=0}, \quad 1 \leqslant j \leqslant m.$$

For $(\mathrm{d}f)_p \in T_p^*$, by the natural base $\{\xi_j : 1 \leqslant j \leqslant m\}$ with $\xi_j = \left(\frac{\mathrm{d}u_j}{\mathrm{d}t}\right)_{t=0}$, $1 \leqslant j \leqslant m$, then

$a_j = \left(\frac{\partial (f \circ \varphi_U^{-1})}{\partial u_j}\right)_{\varphi_U(p)}$, $1 \leqslant j \leqslant m$, are the components of $(\mathrm{d}f)_p \in T_p^*$ in the natural base.

$\Rightarrow \left\{F_{[\gamma]}(\cdot) = \sum_{j=1}^{m} a_j \xi_j = \langle\langle [\gamma], \cdot \rangle\rangle : \forall [\gamma] \in T_p\right\} \leftrightarrow T_p$ is the set of all linear functionals on T_p^*, i. e., $(T_p^*)^* = T_p$.

(3) is evident.

The proof is complete.

Definition 6.1.10 (tangent vector) Space $T_p = \{[\gamma] : \gamma \in \Gamma_p\}$ is **the tangent space of smooth manifold** $(M, \mathbb{A})$ **at point** $p \in M$; And an element $[\gamma]$ in T_p is said to be **a tangent vector passing through** $p \in M$, or simply, **tangent vector.**

Geometry sense of the tangent vector: for two tangent vectors $[\gamma], [\gamma'] \in T_p$, if $\gamma, \gamma' \in \Gamma_p$ have local coordinates, respectively:

$$\gamma \in \Gamma_p : (-\delta, \delta) \to M : (u_j(t)) = (u_1(t), \cdots, u_m(t)), \quad t \in (-\delta, \delta),$$

$$\gamma' \in \Gamma_p : (-\delta, \delta) \to M : (u'_j(t)) = (u'_1(t), \cdots, u'_m(t)), \quad t \in (-\delta, \delta),$$

then, $[\gamma]\sim[\gamma'] \Leftrightarrow \left(\frac{\mathrm{d}u_j}{\mathrm{d}t}\right)_{t=0} = \left(\frac{\mathrm{d}u'_j}{\mathrm{d}t}\right)_{t=0}$, $1\leqslant j\leqslant m$; so that $[\gamma]\sim[\gamma']$ means:

two smooth parameter curves $\gamma, \gamma' \in \Gamma_p$ have the same tangent vector at point $p \in M$ (Fig. 6.1.10).

Thus, **the tangent vector of manifold M at point $p \in M$** is an equivalent class (**the set of all smooth parameter curves**) in which each element has the same tangent vector at point p.

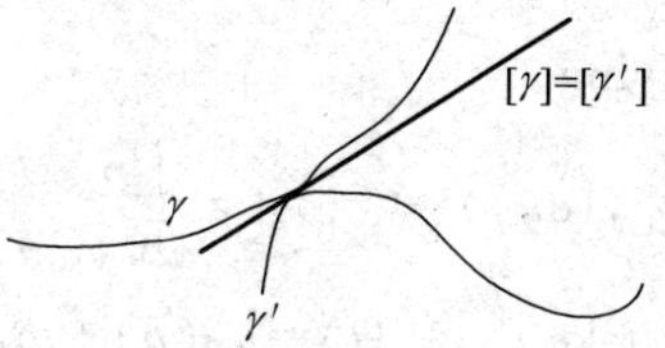

Fig. 6.1.10 Tangent vector $[\gamma]\sim[\gamma']$

We denote by $X=X_p=[\gamma]$, then, Theorem 6.1.5(3) shows that

$$\langle[\gamma],(\mathrm{d}f)_p\rangle \equiv \langle X,(\mathrm{d}f)_p\rangle, \quad X=[\gamma]\in T_p, \quad (\mathrm{d}f)_p \in T_p^*,$$

is a bilinear form.

The sense of bilinear form $\langle[\gamma],(\mathrm{d}f)_p\rangle\equiv\langle X,(\mathrm{d}f)_p\rangle$ is given as follows.

Definition 6.1.11 (differentiation of a smooth function) Let $(M,\mathbb{A})$ be a smooth manifold, T_p be the tangent space at point $p\in M$, and $[\gamma]\in T_p$ be tangent vector passing through p, denoted by $X\equiv X_p=[\gamma]$, $X\in T_p$; T_p^* be the cotangent space at point $p\in M$, and $[f]^*\in T_p^*$ be the cotangent vector passing though p, denoted by $(\mathrm{d}f)_p=[f]^*\in T_p^*$. Then, $(\mathrm{d}f)_p\in T_p^*$ is said to be **the differentiation of smooth function** $f\in C_p^\infty(M)$ at point p, and the bilinear form

$$Xf \equiv \langle X,(\mathrm{d}f)_p\rangle, \quad X\in T_p, \quad (\mathrm{d}f)_p\in T_p^*$$

is said to be **the directional derivative of f along tangent vector X**, or simply, **directional derivative of f**.

If $(\mathrm{d}f)_p=0$, then, point $p\in M$ is said to be **a critical point of f**.

Theorem 6.1.6 The directional derivatives have the following properties: If $X\in T_p$, and $f,g\in C_p^\infty(M)$, $\alpha,\beta\in\mathbb{R}$, then

(1) $X(\alpha f+\beta g)=\alpha Xf+\beta Xg$;

(2) $X(f\cdot g)=f(p)Xg+g(p)Xf$.

Thus, $X=X_p: C_p^\infty(M)\to\mathbb{R}$ is a real function on $C_p^\infty(M)$.

Proof By Theorems 6.1.2 and 6.1.3.

Note A tangent vector $X\in T_p$ can be regarded as an operator — directional derivative operator, then, $X\in T_p$ is a linear operator from $C_p^\infty(M)$ to $C_p^\infty(M)$, by Theorem 6.1.6.

8. Natural bases of T_p^* and T_p

Since $Xf \equiv \langle X, (\mathrm{d}f)_p \rangle$ with $X=[\gamma]\in T_p, (\mathrm{d}f)_p \in T_p^*$, is a bilinear form, take a local coordinate system $\{u_j(q): 1\leqslant j\leqslant m\}, q\in U_p\subset M$ of smooth manifold $(M, \mathbb{A})$: $(u_1(q), \cdots, u_m(q))=((\varphi_U(q))_1, \cdots, (\varphi_U(q))_m)=((e_1\circ\varphi_U)(q), \cdots, (e_m\circ\varphi_U)(q))$, the natural base $\{(\mathrm{d}u_j)_p: 1\leqslant j\leqslant m\}$ of T_p^* is obtained; then, take smooth parameter curves $\lambda_k\in\Gamma_p, 1\leqslant k\leqslant m$, satisfying

$$\langle [\lambda_k], (\mathrm{d}u_j)_p \rangle = \delta_{kj} = \begin{cases} 1, & j=k \\ 0, & j\neq k \end{cases}, \tag{6.1.11}$$

(see the proof of Theorem 6.1.4, take $u_j\circ\lambda_k(t)=u_j(p)+t\delta_{kj}$, where δ_{kj} is in (6.1.11)). We obtain:

$\{[\lambda_k], 1\leqslant k\leqslant m\}\subset T_p$ **is the dual base of natural base** $\{(\mathrm{d}u_j)_p: 1\leqslant j\leqslant m\}\subset T_p^*$.

For tangent vector system $[\lambda_k]\in T_p, 1\leqslant k\leqslant m$ satisfying (6.1.11), we have the following interpretation.

In the equality

$$\langle[\gamma], (\mathrm{d}f)_p\rangle \equiv \left\langle\left\langle \gamma, (\mathrm{d}f)_p \right\rangle\right\rangle,$$

we take $\langle[\gamma], (\mathrm{d}f)_p\rangle = \sum_{j=1}^m a_j\xi_j$ by Theorem 6.1.5(3), with

$$a_j = \left(\frac{\partial(f\circ\varphi_U^{-1})}{\partial u_j}\right)_{\varphi_U(p)}, \quad \xi_j = \left(\frac{\mathrm{d}u_j}{\mathrm{d}t}\right)_{t=0} = \left(\frac{\mathrm{d}(u_j\circ\gamma(t))}{\mathrm{d}t}\right)_{t=0}, \quad 1\leqslant j\leqslant m.$$

On the other hand, $(\mathrm{d}f)_p = \sum_{j=1}^m \left(\frac{\partial f}{\partial u_j}\right)_p \cdot (\mathrm{d}u_j)_p$, with $\left(\frac{\partial f}{\partial u_j}\right)_p = \left(\frac{\partial(f\circ\varphi_U^{-1})}{\partial u_j}\right)_p = a_j$, $1\leqslant j\leqslant m$.

Moreover, by (6.1.11), it follows that

$$\begin{aligned}\langle [\lambda_k], (\mathrm{d}f)_p\rangle &= \left\langle [\lambda_k], \sum_{j=1}^m \left(\frac{\partial f}{\partial u_j}\right)_p \cdot (\mathrm{d}u_j)_p \right\rangle \\ &= \left(\frac{\partial f}{\partial u_k}\right)_p = \left(\frac{\partial(f\circ\varphi_U^{-1})}{\partial u_k}\right)_p, \quad 1\leqslant k\leqslant m.\end{aligned}$$

This means: tangent vector $[\lambda_k]\in T_p$ acts on cotangent vector $[f]^* = (\mathrm{d}f)_p$ generalized by function germ $[f]$, in fact, it is the action on $(\mathrm{d}f)_p$ of partial differential operator $[\lambda_k] = \left(\frac{\partial}{\partial u_k}\right)_p$, so that formula(6.1.11) can be rewritten as

$$\left\langle \frac{\partial}{\partial u_k}\bigg|_p, (\mathrm{d}u_j)_p \right\rangle = \delta_{kj} = \begin{cases} 1, & j=k. \\ 0, & j\neq k. \end{cases} \tag{6.1.12}$$

Now, we conclude that: for smooth manifold $(M, \mathbb{A})$,

in cotangent space T_p^*, **the natural base** is $\{(\mathrm{d}u_j)_p, 1\leqslant j\leqslant m\}\subset T_p^*$;

in tangent space T_p, **the dual base** is $\left\{\frac{\partial}{\partial u_j}\bigg|_p, 1\leqslant j\leqslant m\right\}\subset T_p$.

9. Expressions of elements in T_p and T_p^*

Usually, by natural base $\{(\mathrm{d}u_j)_p, 1\leqslant j\leqslant m\}\subset T_p^*$, we have

$$\forall X\in T_p \Rightarrow X=\sum_{j=1}^{m}\xi_j\frac{\partial}{\partial u_j}, \quad \text{with } \xi_j=\frac{\mathrm{d}(u_j\circ\gamma)}{\mathrm{d}t}, \quad X=[\gamma];$$

$$\forall \alpha\in T_p^* \Rightarrow \alpha=(\mathrm{d}f)_p=\sum_{k=1}^{m}\alpha_k\,\mathrm{d}u_k, \quad \text{with } \alpha_k=\frac{\partial f}{\partial u_k}.$$

If $\{(\mathrm{d}\tilde{u}_j)_p, 1\leqslant j\leqslant m\}\subset T_p^*$ is another local coordinate system of T_p^*, the expressions of X and α are

$$\forall X\in T_p \Rightarrow X=\sum_{j=1}^{m}\tilde{\xi}_j\frac{\partial}{\partial \tilde{u}_j}, \quad \text{with } \tilde{\xi}_j=\frac{\mathrm{d}(\tilde{u}_j\circ\gamma)}{\mathrm{d}t}, X=[\gamma],$$

$$\forall \alpha\in T_p^* \Rightarrow \alpha=(\mathrm{d}f)_p=\sum_{k=1}^{m}\tilde{\alpha}_k\,\mathrm{d}\tilde{u}_k, \quad \text{with } \tilde{\alpha}_k=\frac{\partial f}{\partial \tilde{u}_k};$$

respectively, Then, there are transformation formulas

$$\tilde{\xi}_j=\sum_{k=1}^{m}\xi_k\frac{\partial \tilde{u}_j}{\partial u_k} \quad \text{and} \quad \alpha_k=\sum_{j=1}^{m}\tilde{\alpha}_j\frac{\partial \tilde{u}_j}{\partial u_k}, \tag{6.1.13}$$

where

$$\frac{\partial \tilde{u}_j}{\partial u_k}=\frac{\partial(\varphi_{\tilde{U}}\circ\varphi_U^{-1})_j}{\partial u_k}, \tag{6.1.14}$$

they are elements of Jacobi matrix of coordinate transformation $\varphi_{\tilde{U}}\circ\varphi_U^{-1}$.

We emphasize: $\{(U, u_j)\}$ is a local coordinate system of smooth manifold $(M, \mathbb{A})$.

10. Cotangent space and tangent space of mappings between smooth manifolds

In the above subsections, the cotangent space T_p^* and tangent space T_p of smooth manifold $(M, \mathbb{A})$ at one point $p\in M$ are studied. We now turn to study these concepts at all points on $(M, \mathbb{A})$.

1) **Cotangent mapping F^* of smooth mapping $F: M\to N$**

Definition 6.1.12 (**cotangent mapping** (**differentiation**) **of mapping** $F: M\to N$)

Let $(M,\mathbb{A})$ and $(N,\mathbb{B})$ be m-dimensional and n-dimensional smooth manifolds, and let $F: M\to N$ be a smooth mapping; let $p\in M$ and $q=F(p)\in N$. Denote by $T_p^*(M)$ the cotangent spaces for $\forall p\in M$, and $T_q^*(N)$ for $q=F(p)\in N$. We define a mapping

$$F^*: T_q^*(N)\to T_p^*(M),\quad p\in M、\ q=F(p)\in N, \tag{6.1.15}$$

satisfying

$$F^*((\mathrm{d}f)_q)=\mathrm{d}(f\circ F)_p,\quad (\mathrm{d}f)_q\in T_q^*(N), \tag{6.1.16}$$

then, $F^*: T_q^*(N)\to T_p^*(M)$ is said to be **a cotangent mapping of smooth mapping** $F: M\to N$ **on** M, or simply, F^* **is a cotangent mapping of** F.

Note $\forall(\mathrm{d}f)_q\in T_q^*(N)\Rightarrow F^*((\mathrm{d}f)_q)=\mathrm{d}(f\circ F)_p\in T_p^*(M)$, i. e., F^* maps **a differentiation** $(\mathrm{d}f)_q$ in $T_q^*(N)$ onto **differentiation** $\mathrm{d}(f\circ F)\equiv(\mathrm{d}(f\circ F))_p$ in $T_p^*(M)$, here $\mathrm{d}(f\circ F)$ makes sense, because $F: M\to N$ and $(\mathrm{d}f)_q\in T_q^*(N)$, thus $f\in C_q^\infty(N)$, so that $f\circ F: M\to\mathbb{R}$ is C^∞-function on M. Then, the cotangent mapping $F^*: T_q^*(N)\to T_p^*(M)$ of mapping $F: M\to N$ defined in (6.1.15), (6.1.16) makes sense, and it is reasonable that F^* is also called **differentiation** (or "differentiation operator") **of mapping** $F: M\to N$.

2) **Tangent mapping F_* of mapping $F: M\to N$**

Definition 6.1.13 (tangent mapping of mapping $F: M\to N$) Let $(M,\mathbb{A})$ and $(N,\mathbb{B})$ be m-dimensional and n-dimensional smooth manifolds, respectively; $F: M\to N$ be a smooth mapping with $p\in M$, $q=F(p)\in N$; the cotangent spaces be $T_p^*(M)$ and $T_q^*(N)$, respectively. Let $F^*: T_q^*(N)\to T_p^*(M)$ be the cotangent mapping of $F: M\to N$ in Definition 6.1.12. Denote **the conjugate mapping of** $F^*: T_q^*(N)\to T_p^*(M)$ by

$$F_*: T_p(M)=(T_p^*(M))^*\to T_q(N)=(T_q^*(N))^*, \tag{6.1.17}$$

i. e.

$$\langle F_*X,\alpha\rangle=\langle X,F^*\alpha\rangle,\quad X\in T_p(M),\quad \alpha\in T_q^*(N).$$

Then, conjugate mapping $F_*: T_p(M)\to T_q(N)$ of cotangent mapping $F^*: T_q^*(N)\to T_p^*(M)$ (differentiation) is said to be **tangent mapping induced by mapping**, or simply, **the tangent mapping of** $F: M\to N$.

Note Interpretations of "cotangent mapping" and "tangent mapping".

(1) mapping $F: M\to N\Rightarrow$ differentiation (cotangent mapping) $F^*: T_q^*(N)\to T_p^*(M)$ satisfying

$$F^*((\mathrm{d}f)_q)=(\mathrm{d}(f\circ F))_p,\quad (\mathrm{d}f)_q\in T_q^*(N).$$

Since $(T_p^*(M))^*=T_p(M)$ and $(T_q^*(N))^*=T_q(N)$, then, for $F^*: T_q^*(N)\to T_p^*(M)$, we have

$$F^*((df)_q) = (d(f \circ F))_p, \quad \forall (df)_q \in T_q^*(N),$$

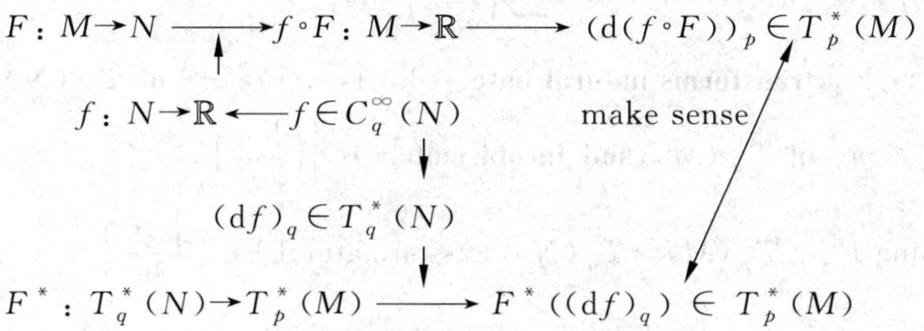

F : M→N
f∘F : M→ℝ
(d(f∘F))_p ∈ T_p^*(M)
f : N→ℝ
f ∈ C_q^∞(N)
make sense
(df)_q ∈ T_q^*(N)
F^* : T_q^*(N)→T_p^*(M)
F^*((df)_q) ∈ T_p^*(M)

The above diagram shows that the definition of cotangent mapping (differentiation) is reasonable and makes sense.

(2) cotangent mapping $F^*: T_q^*(N) \to T_p^*(M) \Rightarrow$ tangent mapping $F_*: T_p(M) \to T_q(N)$ with

$$\langle F_* X, \alpha \rangle = \langle X, F^* \alpha \rangle, \quad X \in T_p(M), \quad \alpha \in T_q^*(N).$$

By (1), $F^*: T_q^*(N) \to T_p^*(M)$, then, F^* maps $\forall \alpha \in T_q^*(N)$ onto $F^* \alpha \in T_p^*(M)$, thus,

$$\forall X \in T_p(M), \quad \forall F^* \alpha \in T_p^*(M) \Rightarrow \langle X, F^* \alpha \rangle \in \mathbb{R}.$$

This guarantees the operator F_* in (2) makes sense: for $F_*: T_p(M) \to T_q(N)$, each $X \in T_p(M)$ guarantees $F_* X$ is in $T_q(N)$. Moreover, for element $\alpha \in T_q^*(N)$ in space $T_q^*(N) = (T_q(N))^*$, the action $\langle F_* X, \alpha \rangle$ makes sense, and it is a real number in $\mathbb{R}$; So that, equality $\langle F_* X, \alpha \rangle = \langle X, F^* \alpha \rangle$ gives the definition of tangent operator F_* with the tangent mapping $F_*: T_p(M) \to T_q(N)$.

Let $(M, \mathbb{A})$, $(N, \mathbb{B})$ be m-dimensional, n-dimensional smooth manifolds, respectively; $F: M \to N$ be a smooth mapping with $p \in M$, $q = F(p) \in N$. We now determine the transformation matrices of cotangent mapping $F^*: T_q^*(N) \to T_p^*(M)$ and tangent mapping $F_*: T_p(M) \to T_q(N)$ in the natural base.

Let $(u_j)_{1 \leqslant j \leqslant m}$ be a local coordinate system in an open neighborhood of $p \in M$, and $(v_k)_{1 \leqslant k \leqslant n}$ be a local coordinate system in an open neighborhood of $f(p) = q \in N$. Correspondingly, the natural base $\{(du_j): 1 \leqslant j \leqslant m\}$ of $T_p^*(M)$ and natural base $\{(dv_k): 1 \leqslant k \leqslant n\}$ of $T_q^*(N)$ are determined, respectively.

Then, smooth mapping $F: M \to N$ can be expressed in the open neighborhood of $p \in M$ as

$$v_k = F_k(u_1, \cdots, u_m), \quad 1 \leqslant k \leqslant n;$$

the action of cotangent mapping (differentiation) $F^*: T_q^*(N) \to T_p^*(M)$ on natural base

$\{(dv_k): 1\leqslant k\leqslant n\}$ of $T_q^*(N)$ is denoted by

$$F^*((dv_k)_q)=d((v_k\circ F)_p)=\sum_{j=1}^{m}\left(\frac{\partial F_k}{\partial u_j}\right)_p du_j.$$

This formula shows that: F^* **transforms natural base** $\{(dv_k): 1\leqslant k\leqslant n\}$ **of** $T_q^*(N)$ **to natural base** $\{(du_j): 1\leqslant j\leqslant m\}$ **of** $T_p^*(M)$**, and Jacobi matrix is** $\left[\left(\frac{\partial F_k}{\partial u_j}\right)_p\right]_{n\times m}$.

Similarly, tangent mapping $F_*: T_p(M)\to T_q(N)$ acts on natural base $\left\{\frac{\partial}{\partial u_j}\right\}_{1\leqslant j\leqslant m}$ of $T_p(M)$, expressed by

$$\left\langle F_*\left(\frac{\partial}{\partial u_j}\right), dv_k\right\rangle=\left\langle\frac{\partial}{\partial u_j}, F^*(dv_k)\right\rangle=\sum_{s=1}^{m}\left\langle\frac{\partial}{\partial u_j}, du_s\right\rangle\left(\frac{\partial F_k}{\partial u_s}\right)_p$$
$$=\left\langle\sum_{t=1}^{n}\left(\frac{\partial F_t}{\partial u_j}\right)_p\frac{\partial}{\partial v_t}, dv_k\right\rangle,$$

i. e.

$$F_*\left(\frac{\partial}{\partial u_j}\right)=\sum_{t=1}^{n}\left(\frac{\partial F_t}{\partial u_j}\right)_p\frac{\partial}{\partial v_t}.$$

Thus, F_* **transforms natural base** $\left\{\frac{\partial}{\partial u_j}\right\}_{1\leqslant j\leqslant m}$ **of** $T_p(M)$ **to natural base** $\left\{\frac{\partial}{\partial v_k}\right\}_{1\leqslant k\leqslant n}$ **of** $T_q(N)$**, and Jacobi matrix is** $\left[\left(\frac{\partial F_k}{\partial u_j}\right)_p\right]_{m\times n}$.

Fig. 6. 1. 11 is a simple diagram of tangent space and tangent mapping, as well as cotangent space and cotangent mapping(Fig. 6. 1. 11).

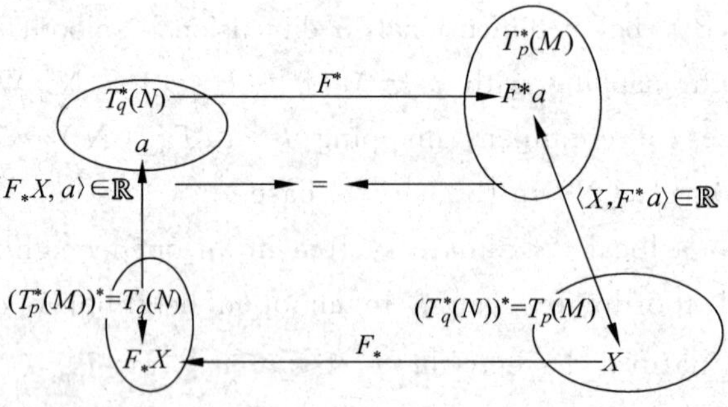

Fig. 6. 1. 11 Tangent space, tangent mapping; cotangent space, cotangent mapping

11. Tangent vector filed

We suppose that $(M,\mathbb{A})\equiv(M,\tau,\mathbb{A})$ is a C^∞-smooth manifold defined on T_2-type topological space (M,τ), and $T_p^*(M)$ is a cotangent space on M, $T_p(M)$ is a tangent space on M. Then, Theorem 6.1.6 shows that tangent vector $X_p \in T_p$ is a real linear operator defined on $C_p^\infty(M)$, and $X_p: C_p^\infty(M) \to C_p^\infty(M)$, thus, $f \in C_p^\infty(M) \Rightarrow X_p f \in C_p^\infty(M)$. Denoted by

$$X_p f \equiv (Xf)(p),$$

this means: Xf is a real function on M, i. e. , $Xf: M \to \mathbb{R}$.

For every point $p \in M$, if a tangent vector X_p is assigned, then the set

$$X = \{X_p: p \in M\}$$

is called **a tangent vector field on manifold** M.

Definition 6.1.14 (smooth tangent vector field) Let $X = \{X_p: p \in M\}$ be a tangent vector field on smooth manifold M. If for any $f \in C^\infty(M)$, it holds $Xf \in C^\infty(M)$, then, X is said to be **a smooth tangent vector on** M.

If in tangent vector field $X = \{X_p: p \in M\}$, every tangent vector is smooth, then $X = \{X_p: p \in M\}$ is said to be **a smooth tangent vector field.**

We conclude that: a smooth tangent vector field $X = \{X_p: p \in M\}$ is an operator from $C^\infty(M)$ to $C^\infty(M)$.

The properties of smooth tangent vector field are: for $f, g \in C^\infty(M)$, $\alpha, \beta \in \mathbb{R}$,

(1) $X(\alpha f + \beta g) = \alpha(Xf) + \beta(Xg)$; (linearity)

(2) $X(f \cdot g) = f \cdot (Xg) + g \cdot (Xf)$. (multiplication formula)

Lemma 6.1.1 If $X = \{X_p: p \in M\}$ is a smooth tangent vector field on smooth manifold M, then for any nonempty open subset $U \subset M$, the restriction of X on U, i. e. , $X|_U$, is a smooth tangent vector field on open submanifold U.

Proof Only need to prove that $X|_U f$ is a smooth function on U for each $f \in C^\infty(U)$.

(1) Take a point $p \in U$, then there exists a coordinate field V of p with $p \in V$, such that $\bar{V}$ is compact, and $\bar{V} \subset U$. Then, there exists smooth function $h: M \to \mathbb{R}$ on M, such that

① $0 \leqslant h \leqslant 1$; ② $h(q) = \begin{cases} 1, & p \in V \\ 0, & p \notin V \end{cases}$.

(the proof is similar to that of in Uryshohn lemma).

(2) Let $F(x) = \begin{cases} f(x) \cdot h(x), & x \in U, \\ 0, & x \notin U, \end{cases}$ then the smoothness of f and h guarantees $F \in C^\infty(M)$ with $F|_V = f|_V$. Hence, $XF \in C^\infty(M)$ by the assumption of smooth tangent vector field of X on M. Furthermore,

$$(X|_U f)(x) = X|_X f(x) = (XF)(x), \quad \forall x \in V,$$

then, function $X|_U f$ is smooth at $q \in U$, that is, $X|_U f$ is a smooth function on U.

For a smooth tangent vector field, we have the coordinate representation:

Theorem 6.1.7 A tangent vector field $X = \{X_p : p \in M\}$ on smooth manifold M is smooth tangent vector field, if and only if for every point $p \in M$, there exists a local coordinate system (U, u_j) of p, such that the restriction of X on U can be expressed as $X|_U = \sum_{j=1}^{m} \xi_j \frac{\partial}{\partial u_j}$ with smooth functions ξ_j $(1 \leqslant j \leqslant m)$ on U.

We omit the proof, and refer to [3].

Definition 6.1.15 (Poisson brackets) Let X, Y be smooth tangent vector fields on smooth manifold M. **Poisson brackets of** X **and** Y is defined by

$$[X, Y] = XY - YX,$$

i. e., Poisson brackets $[X, Y]$ consist of an operator on $C^\infty(M)$ satisfying

$$[X, Y]f = X(Yf) - Y(Xf), \quad f \in C^\infty(M).$$

Theorem 6.1.8 (properties of Poisson brackets) Let X, Y, Z be smooth tangent vector fields on smooth manifold M. Then, for $f, g \in C^\infty(M)$,

(1) $[X, Y](\alpha f + \beta g) = \alpha [X, Y] f + \beta [X, Y] g, \alpha, \beta \in \mathbb{R}$;

(2) $[X, Y](f \cdot g) = f \cdot [X, Y] g + g \cdot [X, Y] f$;

(3) $[X, Y] = -[Y, X]$;

(4) $[X + Y, Z] = [X, Z] + [Y, Z]$;

(5) $[fX, gY] = f \cdot (Xg) Y - g \cdot (Yf) X + f \cdot g [X, Y]$;

(6) $[X, [Y, Z]] + [Y, [Z, X]] + [Z, [X, Y]] = 0$.

Proof We only prove for (5), the others are left as exercises. Take $h \in C^\infty(M)$, by definition,

$$\begin{aligned}
[fX, gY]h &= (fX)((gY)h) - (gY)((fX)h) \\
&= f \cdot X(g \cdot Yh) - g \cdot Y(f \cdot Xh) \\
&= f \cdot (Xg)(Yh) + f \cdot g \cdot X(Yh) - g \cdot (Yf)(Xh) - g \cdot f \cdot Y(Xh) \\
&= (f (Xg) \cdot Y - g(Yf) \cdot X + f \cdot g[X, Y]) h.
\end{aligned}$$

Theorem 6.1.9 (local coordinate expression of Poisson brackets) Let X, Y be two smooth tangent vector fields on smooth manifold $(M, \mathbb{A})$; and $\{(U, u_j)\}$ be a local coordinate system on smooth manifold. If two smooth tangent vector fields have expressions

$$X|_U = \sum_{j=1}^{m} \xi_j \frac{\partial}{\partial u_j}, \quad Y|_U = \sum_{j=1}^{m} \eta_j \frac{\partial}{\partial u_j},$$

then

$$[X,Y]|_U = [X|_U, Y|_U] = \sum_{j=1}^{m}\sum_{k=1}^{m}\left(\frac{\partial \eta_j}{\partial u_k} - \frac{\partial \xi_j}{\partial u_k}\right)\frac{\partial}{\partial u_j}.$$

Proof By evaluating $X = \frac{\partial}{\partial u_j}$, we have $\left[\frac{\partial}{\partial u_j}, \frac{\partial}{\partial u_k}\right] = \delta_{kj} = \begin{cases} 1, & j = k \\ 0, & j \neq k \end{cases}$.

Definition 6.1.16 (singular point) Let $(M, \mathbb{A})$ be a smooth manifold, X be a smooth tangent vector field on $(M, \mathbb{A})$. If $X_p = 0$ at point $p \in M$, then the point p is said to be **a singular point of smooth tangent vector field X.**

The properties of the singular points of smooth tangent vector field X on M are very complex. However, non-singular points are rather simple, but more important and useful. Our mind will be concentrated at the non-singular points in our course.

Theorem 6.1.10 Let M be a smooth manifold, X be a smooth tangent vector field on $(M, \mathbb{A})$. If $X_p \neq 0$ at $p \in M$, then there exists a local coordinate system $\{(W, w_j)\}$, such that $X|_W = \frac{\partial}{\partial w_j}$.

12. Diagram for establishing cotangent mappings and tangent mappings

smooth mapping $F: M \to N$ between smooth manifolds M and N

↓

cotangent mapping (differentiation) of $F: M \to N$—$F^*: T_q^*(N) \to T_p^*(M)$ by

$$F^*((\mathrm{d}f)) = \mathrm{d}(f \circ F), \quad (\mathrm{d}f) \in T_q^*(N)$$

↓

tangent mapping of $F: M \to N$— $F_*: T_p(M) \to T_q(N)$ by

$$\langle F_* X, \alpha \rangle = \langle X, F^* \alpha \rangle, \quad X \in T_p(M), \alpha \in T_q^*(N).$$

6.1.3 Submanifolds

1. Existence theorem of the inverse functions in $\mathbb{R}^n$

In *Advanced Calculus*, a familiar result is: if a function $y = f(x)$ has positive derivative at a point $x = a$, that is, $f'(a) > 0$, then $y = f(x)$ is increasing at a neighborhood of a; this means "the properties of differentiation at a point can affect the ones of a function in some neighborhood". In case of manifold, do we have similar result? The answer is "positive".

In the classical case, we have the following theorem.

Theorem 6.1.11 (existence theorem of the inverse function in $\mathbb{R}^n$) Let $W \subset \mathbb{R}^n$ be an open set, $f: W \to \mathbb{R}^n$ be a smooth mapping from W to $\mathbb{R}^n$. If $\det\left[\frac{\partial f_j}{\partial x_k}\bigg|_{x_0}\right] \neq 0$ at $x_0 \in W$, then, there exists a neighborhood $U \subset W$ of x_0, such that $V = f(U)$ is a neighborhood of $f(x_0)$ in $\mathbb{R}^n$, and f has smooth inverse function $g = f^{-1}: V \to U$ on V.

We may interpret the above theorem in "manifold language":

Jacobi matrix $\left[\frac{\partial f_j}{\partial x_k}\right]$ of mapping $f: W \to \mathbb{R}^n$ is just the matrix of tangent mapping $f_*: T_p(\mathbb{R}^n) \to T_q(\mathbb{R}^n)$ of f in natural base $\left\{\frac{\partial}{\partial u_k}\right\}_{1 \leqslant k \leqslant n} = \left\{\frac{\partial}{\partial x_k}\right\}_{1 \leqslant k \leqslant n}$. Thus, the determinate is not equal to zero, $\det \left[\frac{\partial f_j}{\partial x_k}\bigg|_{x_0}\right] \neq 0$, and means that: the tangent mapping (linear) satisfies the following relations:

$$\begin{array}{ccc} f_*: T_p(W \subset \mathbb{R}^n) \to T_q(\mathbb{R}^n) & \Leftrightarrow & f_*: \mathbb{R}^n \to \mathbb{R}^n \\ \updownarrow & & \updownarrow \\ T_p(W \subset \mathbb{R}^n) \approx \mathbb{R}^n & & T_q(\mathbb{R}^n) \approx \mathbb{R}^n \end{array}$$

i. e., f_* is an isomorphism from $\mathbb{R}^n$ to $\mathbb{R}^n$.

Moreover, "$g = f^{-1}: V \to U$ is the inverse function of $f: W \to f(W) \subseteq \mathbb{R}^n$ that" implies "two compound mappings $g \circ f = I: U \to U$, $f \circ g = I: V \to V$, and two mappings f, g all are smooth". Thus, the restriction function $f|_U: U \to V$ is a differentiable homeomorphism from U to V.

The sense of "existence theorem of the inverse function": If tangent mapping $f_*: T_p(\mathbb{R}^n) \to T_q(\mathbb{R}^n)$ of $f: W \to \mathbb{R}^n$ at $p \in W$ is an isomorphism mapping, then, mapping f is a differentiable homeomorphism in U of the neighborhood of p with $U \subset W$.

We emphasize that: the properties of tangent mapping $F_*: T_p(M) \to T_q(N)$ at $p \in M$ could determine the ones of mapping $F: M \to N$ in a neighborhood of $p \in M$.

2. Properties of tangent mappings

The following is a generalization of Theorem 6. 1. 11.

Theorem 6. 1. 12 Suppose that $(M, \mathbb{A})$ and $(N, \mathbb{B})$ are two n-dimensional smooth manifolds, $f: M \to N$ is a smooth mapping. If tangent mapping $f_*: T_p(M) \to T_{f(p)}(N)$ at point $p \in M$ is a topological isomorphism, then there exists a neighborhood U of $p \in M$ with $f(U) = W \subset N$ being neighborhood of $f(p) \in N$, and the restriction mapping $f|_U: U \to W$ is a differentiable homeomorphism.

Proof Since $f: M \to N$ is a smooth mapping, we take local coordinate chart (U_0, φ) of point $p \in M$ in M, and take local coordinate chart (V_0, ψ) of point $q = f(p) \in N$ in N with $f(U_0) \subset V_0$, such that

$$\tilde{f}=\psi\circ f\circ\varphi^{-1}:\varphi(U_0)\to\psi(V_0)\subset\mathbb{R}^n$$

is a smooth mapping. It is clear that Jacobi determinate of $\tilde{f}$ at point $\varphi(p)$ is not zero, then, there exist a neighborhoods $\tilde{U}\subset\varphi(U_0)$ of $\varphi(p)$, and $\tilde{V}\subset\varphi(V_0)$ of $\psi(q)$ in $\mathbb{R}^n$, respectively, such that $f|_{\tilde{U}}:\tilde{U}\to\tilde{V}$ is a differentiable homeomorphism, by Theorem 6.1.11.

Let $U=\varphi^{-1}(\tilde{U})$, $V=\psi^{-1}(\tilde{V})$, then $U\subset M$, $V\subset N$ are neighborhoods of p, q, respectively, such that $f=\psi^{-1}\circ\tilde{f}\circ\varphi$: $U\to V$ is a differentiable homeomorphism mapping. The proof is complete.

Note If the dimensions of smooth manifolds $(M,\mathbb{A})$ and $(N,\mathbb{B})$ are equal each other, then "**tangent mapping f_* is a homeomorphism at a point**" is equivalent to "**f_* is single at the point**". However, if $(M,\mathbb{A})$ is an m-dimensional smooth manifold, $(N,\mathbb{B})$ is an n-dimensional smooth manifold, Theorem 6.1.12 holds too.

If $f: M\to N$ is a smooth mapping, and tangent mapping f_* is single at point $p\in M$, then f_* is said to be non-degenerative at p.

If f_* is non-degenerate, then it holds $m\leqslant n$, and the rank of Jacobi matrix of f is m at $p\in M$ (as exercise).

Theorem 6.1.13 Let $(M,\mathbb{A})$ be m-dimensional smooth manifold, $(N,\mathbb{B})$ be n-dimensional smooth manifold, and $m<n$; let $f: M\to N$ be a smooth mapping. If tangent mapping f_* is non-degenerative at $p\in M$, then, there exist a local coordinate system $\{(U,u_j)\}$ of $p\in M$, and a local coordinate system $\{(V,v_k)\}$ of $q=f(p)\in N$, such that $f(U)\subset V$, and $f|_U$ has expression: for any $x\in U$,

$$\begin{cases} v_j(f(x))=u_j(x), & 1\leqslant j\leqslant m \\ v_s(f(x))=0, & m+1\leqslant s\leqslant n \end{cases}. \tag{6.1.18}$$

We omit the proof, and refer to [3].

The sense of Theorem 6.1.13: If the tangent mapping f_* is non-degenerative at $p\in M$, we certainly can obtain a local coordinate system $\{(U,u_j)\}$ of p with $u_j(p)=0$, such that image $q=f(p)$ of $p\in M$ for the mapping f: $M\to N$ has $(n-m)$-zero local coordinates, i.e., $f(u_1,\cdots,u_m)=(u_1,\cdots,u_m,0,\cdots,0)$. Or, if tangent mapping f_* of $f: M\to N$ is non-degenerate, then f maps the manifold M onto an m-dimensional subspace $f(M)$ of manifold N, and the points in $f(M)$ have $(n-m)$-zero local coordinates (Fig. 6.1.12).

3. Embedding submanifolds, immersion submanifolds

Definition 6.1.17 (embedding and immersion submanifolds) Let $(M,\mathbb{A})$, $(N,\mathbb{B})$ be smooth manifolds. If a smooth mapping φ: $M\to N$ satisfies:

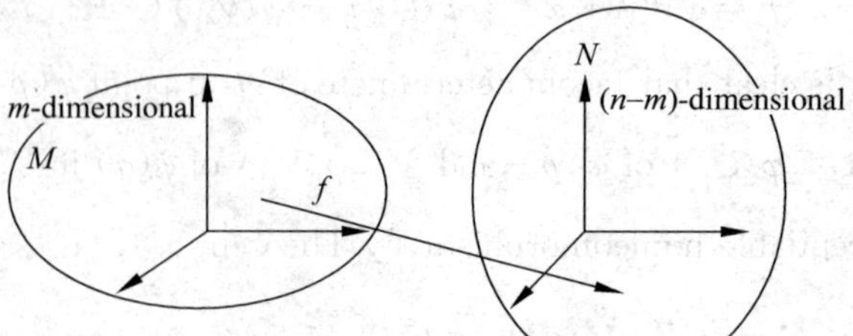

Fig. 6. 1. 12 Diagram of Theorem 6. 1. 13

(1) φ is one-one;

(2) tangent mapping φ_* : $T_p(M) \to T_{q=\varphi(p)}(N)$ is non-degenerative at every $p \in M$;

then $(M,\varphi) \equiv \varphi(M)$ is said to be **an embedding smooth submanifold of** N; and φ is said to be **an embedding mapping from** M **to** N. If mapping φ: $M \to N$ satisfies condition(2), then $(M,\varphi) \equiv \varphi(M)$ is said to be **an immersion smooth submanifold of** N; and φ is said to be **an immersion mapping from** M **to** N.

The immersion mapping is single, locally but not in global area; The difference of graphs between "immersion" and "embedding" mappings is that whether the image set $\varphi(M)$ has self-intersect points.

Example 6. 1. 3 (open submanifold)

Let $(N,\mathbb{B})$ be an n-dimensional smooth manifold, $U \subset N$ be an open subset of N. Take the subspace structure on U of N, then $(U,\mathbb{B})$ is a smooth submanifold with the same dimension of N.

Let I: $U \to N$ be the identity mapping. Then, $I(U) = (U,I)$ is an embedding submanifold of N, also, $I(U)$ is **an open submanifold of** N.

As we see, $\mathbb{R}^n \subset \mathbb{R}^{n+1}$ with identity mapping I: $\mathbb{R}^n \to \mathbb{R}^{n+1}$ gives an n-dimensional smooth open submanifold $\mathbb{R}^n$ of $\mathbb{R}^{n+1}$; $\mathbb{R}^n$ is an embedding submanifold.

Example 6. 1. 4 (closed submanifold)

Let $(N,\mathbb{B})$ be an n-dimensional smooth manifold, $(M,\mathbb{A})$ be an m-dimensional smooth manifold, and φ: $M \to N$ be a smooth mapping. Then, $\varphi(M)$ is a smooth submanifold of N.

If (1) $\varphi(M) \subset N$ is a closed subset of N; (2) $\forall q \in \varphi(M)$, there exists a local coordinate system $\{(V,v_k)\}$, such that $\varphi(M) \cap V$ is determined by equations $v_{m+1} = \cdots = v_n = 0$;

Then $\varphi(M) \equiv (M,\varphi)$ is called **a closed submanifold of** N.

As we see, unit ball $S^n \subset \mathbb{R}^{n+1}$ with identity mapping I: $S^n \to \mathbb{R}^{n+1}$ gives an n-dimensional closed sub-manifold S^n of $\mathbb{R}^{n+1}$. S^n is an embedding submanifold.

Example 6.1.5 (immersion submanifold)

Let $f: \mathbb{R}^1 \to \mathbb{R}^2$ be mapping $f(t)=\left(2\cos\left(t-\frac{\pi}{2}\right), \sin 2\left(t-\frac{\pi}{2}\right)\right)$. Then $f(\mathbb{R})\subset\mathbb{R}^2$ is an immersion submanifold of $\mathbb{R}^2$, but is not embedding submanifold.

Example 6.1.6 (embedding submanifold)

Let $g: \mathbb{R}^1 \to \mathbb{R}^2$ be a mapping $g(t)=\left(2\cos\left(2\arctan t+\frac{\pi}{2}\right), \sin 2\left(2\arctan t+\frac{\pi}{2}\right)\right)$. Then $g(\mathbb{R})\subset\mathbb{R}^2$ is an embedding submanifold. It is clear that, $g(0)=(0,0)$ for $t=0$; and $g(t)\to(0,0)$ as $t\to\pm\infty$ (Fig. 6.1.13).

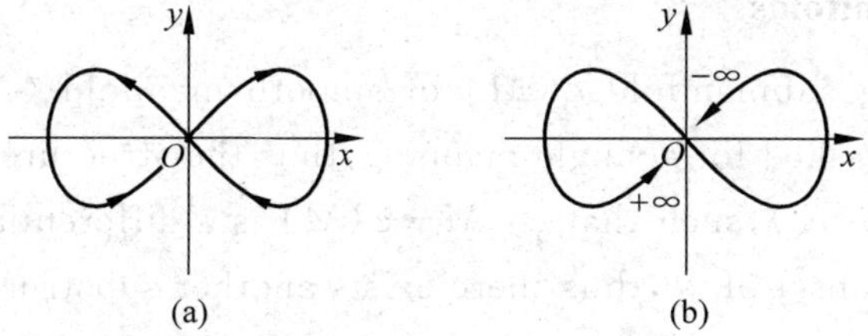

Fig. 6.1.13 Example 6.1.5, Example 6.1.6

Example 6.1.7 (embedding submanifold)

Let $F(t)=\begin{cases}\left(\frac{3}{t^2}, \sin\pi t\right), & 1\leqslant t<+\infty \\ f(t), & 0\leqslant t\leqslant 1 \\ f(t), & -1\leqslant t\leqslant 0 \\ (0, t+2), & -\infty<t\leqslant -1\end{cases}$; where $f(t)$ be a smooth curve in $t\in[-1,1]$ connected points $(3,0)$ and $(0,1)$; moreover, $f(t)$ tends to the part of itself on $t\in[-3,-1]$ as $t\to+\infty$ (Fig. 6.1.14(a)). Then, $F(\mathbb{R})=(\mathbb{R}, F)$ is an embedding mapping from $\mathbb{R}$ onto $\mathbb{R}^2$.

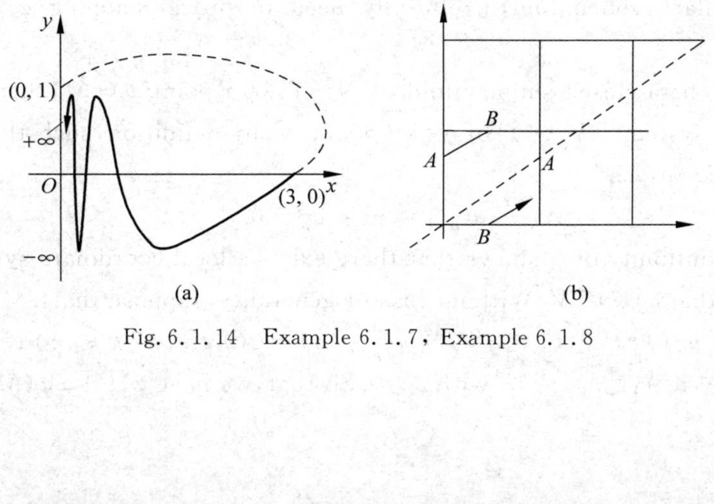

Fig. 6.1.14 Example 6.1.7, Example 6.1.8

Example 6.1.8 (toroid)

A toroid $\mathbb{T}^2 = S^1 \times S^1$ can be regarded as a 2-dimensional manifold generated by a unit square in the following way(Fig. 6.1.14(b)):

Let (x, y) be a point on toroid $\mathbb{T}^2$ with $x, y \in \mathbb{R}$, and x, y be mod 1 real numbers. Or, take real numbers a, b, such that $a : b$ is irrational number. Then, define a mapping φ: $\mathbb{R}^1 \to \mathbb{T}^2$ by $\varphi(t) = (x, y) = (at \operatorname{mod} 1, bt \operatorname{mod} 1)$. It is clear that $(\mathbb{R}, \varphi) \subset \mathbb{T}^2$ is an embedding submanifold of $\mathbb{T}^2$, image set $\varphi(\mathbb{R}) \subset \mathbb{T}^2$ is a dense subset in $\mathbb{T}^2$. If $a : b$ is a rational number, then $\varphi(\mathbb{R}) \subset \mathbb{T}^2$ is an embedding submanifold of $\mathbb{T}^2$.

4. Regular submanifolds

For an embedding submanifold $\varphi(M)$ of smooth manifold $(N, \mathbb{B})$, the embedding mapping φ: $M \to N$ is asked for a single mapping, thus the structure of submanifold $\varphi(M)$ agrees with that of $(M, \mathbb{A})$, such that φ: $M \to \varphi(M)$ is a differentiable homeomorphism. However, $\varphi(M)$ is a subset of N, thus there exists another substructure on $\varphi(M)$ induced from that of N, so that, there are two structures on $\varphi(M)$: one is from M, the other is from N. The two structures could be the same, or not.

Definition 6.1.18 (regular submanifold) Let $(M, \mathbb{A})$, $(N, \mathbb{B})$ be smooth manifolds. If a smooth mapping φ: $M \to N$ is a homeomorphism from M to N, then image set $\varphi(M) = (M, \varphi) \subset N$ is said to be **a regular submanifold in** N, and φ: $M \to N$ is said to be **a regular mapping.**

Theorem 6.1.14 (characteristic property of regular submanifold) Let $(M, \mathbb{A})$ be m-dimensional smooth manifold, $(N, \mathbb{B})$ be n-dimensional smooth manifold with $m \leqslant n$. Then (M, φ) is a regular submanifold of N, if and only if $(M, \varphi) \subset N$ is a closed submanifold of some open submanifold of N.

Proof **Sufficiency** We only need to prove: if (M, φ) is a closed submanifold, then it must be a regular submanifold, i.e., only need to prove mapping φ is one-one and bicontinuous.

Let (M, φ) be a closed submanifold of N. Take a point $p \in M$, then, there exists a local coordinate system $\{(V, v_k)\}$ of $q = \varphi(p)$ in N, by definition, such that $\varphi(M) \cap V$ is determined by equations

$$v_{m+1} = \cdots = v_n = 0. \tag{6.1.19}$$

Moreover, the continuity of φ shows that there exists a local coordinate system $\{(U, u_j)\}$ of $p \in M$, such that $\varphi(U) \subset V$. Without loss of generality, suppose that:

$$u_j(p) = 0, (1 \leqslant j \leqslant m);\ v_k(q) = 0 (m+1 \leqslant k \leqslant n);$$

and $V = \{(v_1, \cdots, v_n): |v_k| < \delta\}$ with $\delta > 0$. So that, we have $\varphi(U) \subset \varphi(M) \cap V$.

Next, prove $\varphi^{-1}: \varphi(M)\subset N\to M$ is a continuous mapping, i. e., there exists $\delta_1>0$, such that for $\varphi(M)\cap V_1$, it holds $\varphi^{-1}(\varphi(M)\cap V_1)\subset U$, with V_1 being determined later.

By (6.1.18), the local coordinate system of mapping $\varphi|_U$ is $\begin{cases} v_j=\varphi_j(u_1,\cdots,u_m), & 1\leqslant j\leqslant m, \\ v_k=0, & m+1\leqslant k\leqslant n, \end{cases}$ thus Jacobi determinate $\left.\dfrac{\partial(\varphi_1,\cdots,\varphi_m)}{\partial(u_1,\cdots,u_m)}\right|_{u_j}\neq 0$. Then by Theorem 6.1.11, there exists a δ_1 with $0<\delta_1<\delta$, such that the function system $(\varphi_1,\cdots,\varphi_m)$ has inverse functions $u_j=\psi_j(v_1,\cdots,v_m)$, $|v_j|<\delta_1$.

Take $V_1=\{(v_1,\cdots,v_m,v_{m+1},\cdots,v_n): |v_k|<\delta_1\}$, then the inverse image of $\varphi(M)\cap V_1$ in mapping φ (i. e., the set $\varphi^{-1}(\varphi(M)\cap V_1)$) is in U. This implies φ^{-1} is continuous.

Necessity Let (M,φ) be a regular submanifold of N. We prove: there exists an open submanifold $W\subset N$ of N, such that (M,φ) is a closed submanifold of W.

In fact, take $p\in M$. Then for any neighborhood $U\subset M$ of point $p\in M$, there exists a neighborhood V of point $q=\varphi(p)$ in N, such that it holds $\varphi(U)=\varphi(M)\cap V$. Thus, Theorem 6.1.13 shows that, there exist local coordinate system $\{(U_1,u_j)\}$ of p, and local coordinate system $\{(V_1,v_k)\}$ of q, respectively, such that $\varphi(U_1)\subset V_1$, and $\varphi|_{U_1}$ has a local coordinate expression

$$\varphi(u_1,\cdots,u_m)=(u_1,\cdots,u_m,\underbrace{0,\cdots,0}_{n-m}). \tag{6.1.20}$$

Without loss of generality, let $U_1\subset U$, and take $V_1\subset V$. Thus, $\varphi(U_1)=\varphi(M)\cap V_1$. Expression (6.1.20) shows that $\varphi(M)\cap V_1$ is determined by equations

$$v_{m+1}=\cdots=v_n=0. \tag{6.1.21}$$

Now we construct an open submanifold $W\subset N$: firstly, construct $V_q\equiv V_1$ as above, it is a coordinate field of point q in N with (6.1.21). Let $W=\bigcup\limits_{q\in\varphi(M)}V_q$, then W is an open submanifold of N containing $\varphi(M)$.

We only prove: regarded as a topological subspace, $\varphi(M)\subset N$ satisfies $W\cap\overline{\varphi(M)}=\varphi(M)$ with $\overline{\varphi(M)}$: the closure of $\varphi(M)$ in N (a set $\varphi(M)$ is called **a relatively closed set**, if holds $W\cap\overline{\varphi(M)}=\varphi(M)$).

In fact, take $s\in W\cap\overline{\varphi(M)}$, there exists $q\in\varphi(M)$, such that $s\in V_q$, by $W=\bigcup\limits_{q\in\varphi(M)}V_q$. Then, (6.1.20) shows that $\varphi(M)\cap V_q$ is an m-dimensional coordinate surface in V_q, so it is a relatively closed set in V_q.

On the other hand, $s\in V_q\cap\overline{\varphi(M)}\subset W\cap\overline{\varphi(M)}$ means that "s belongs to the relative closure of $V_q\cap\varphi(M)$ in V_q", and this implies $s\in V_q\cap\varphi(M)$, then it implies

$$W \cap \overline{\varphi(M)} \subset \varphi(M). \qquad (*)$$

Combine the above relationship (*) with $W \cap \overline{\varphi(M)} \supset \varphi(M)$, we have $W \cap \overline{\varphi(M)} = \varphi(M)$.

The proof is complete.

Corollary (M,φ) is a regular submanifold of smooth manifold $(N, \mathbb{B})$, if and only if for every point $p \in M$, there exists a local coordinate system $\{(V, v_k)\}$ of $q=\varphi(p)$ in N with $v_k(q)=0$, such that set $\varphi(M) \cap V$ is determined by equations $v_{m+1}=\cdots=v_n=0$.

Theorem 6.1.15 Let (M,φ) be a submanifold of smooth manifold $(N,\mathbb{B})$. If $(M,\mathbb{A})$ is a compact manifold, then $\varphi: M \to N$ is a regular embedding mapping.

Proof $\varphi(M) \subset N$, regarded as a subspace of N, is T_2-type, and mapping $\varphi: M \to \varphi(M) \subset N$ is one-one, and continuous from compact space M to T_2-type subspace $\varphi(M)$, thus, $\varphi: M \to \varphi(M) \subset N$ is a homeomorphism. Then, (M, φ) is a regular submanifold of N.

Theorem 6.1.16 Let $(M, \mathbb{A})$ be an m-dimensional compact smooth manifold. Then, there exist integer $n \in \mathbb{N}$ and smooth mapping $\varphi: M \to \mathbb{R}^n$, such that (M,φ) is a regular sub-manifold of $\mathbb{R}^n$.

5. Frobenius theorem

Recall Theorem 6.1.10, it is a simplified theorem of the local coordinate system for smooth tangent vector field X on a smooth manifold M: If $(M,\mathbb{A})$ is a smooth manifold, X is a smooth tangent vector field on M. Then, for a point $p \in M$ with $X_p \neq 0$, there exists a local coordinate system $\{(W, w_j)\}$ of p, such that $X|_W = \dfrac{\partial}{\partial w_j}$.

An important problem is: on a smooth manifold $(M, \mathbb{A})$, smooth tangent vector fields $X_1, \cdots, X_s$ are linear independent in a neighborhood $U \subset M$. Does there exist a local coordinate system $\{(W, w_j)\}$, satisfying

$$X_j|_W = \frac{\partial}{\partial w_j}, \quad 1 \leqslant j \leqslant s$$

at every point $p \in U$? i. e. , does there is a local coordinate system in U, on which linear independent smooth tangent vector fields $X_1, \cdots, X_s$ could be simplified, simultaneously (not only at some points but at whole U)?

We have the following theorem.

Theorem 6.1.17 Let $(M, \mathbb{A})$ be a smooth manifold, $X_1, \cdots, X_s$ be linearly independent smooth tangent vector fields in a neighborhood $U \subset M$. Then, there exists a local coordinate system $\{(W, w_j)\}$, satisfying $X_j|_W = \dfrac{\partial}{\partial w_j}$, $1 \leqslant j \leqslant s$ at each point $p \in U$, if and only if $[X_j, X_k]=0, 1\leqslant j,k\leqslant s$.

However, the following is a weaker condition, called **Frobenius condition.**

Definition 6.1.19 (s-dimensional tangent subspace field) Let $(M, \mathbb{A})$ be a smooth manifold. For each point $p\in M$, assign an s-dimensional subspace $L^s(p)$ of tangent space T_p, i.e., L^s is an s-dimensional tangent subspace field on M.

If for each $p\in M$, there exist linearly independent smooth tangent vector fields $X_1, \cdots, X_s$ on a neighborhood U of p, such that for every $w\in U$, subspace $L^s(w)$ is spanned by tangent vectors $X_1(w), \cdots, X_s(w)$, then L^s is said to be **an s-dimensional smooth tangent subspace filed**, or said to be **s-dimensional smooth distribution on smooth manifold** M, denoted by $L^s|_U=\{X_1, \cdots, X_s\}$.

Definition 6.1.20 (Frobenius condition) Let $(M, \mathbb{A})$ be a smooth manifold, L^s be an s-dimensional smooth tangent subspace field on M. Smooth tangent subspace field L^s is said to **satisfy the Frobenius condition**, if on any coordinate field U, when L^s is spanned by linearly independent smooth tangent vector fields $X_1, \cdots, X_s$, then every $[X_j, X_k]$, $1\leqslant j$, $k\leqslant s$, can be expressed by the linear combination of X_j, $1\leqslant j\leqslant s$.

Theorem 6.1.18 (Frobenius Theorem) Let $(M, \mathbb{A})$ be a smooth manifold, L^s be an s-dimensional smooth tangent subspace field on an open set U of M. Then, at any point $p\in U$, there exists a local coordinate system $\{(W, w_j)\}$ with $W\subset U$ of p, such that $L^s|_U=\left\{\frac{\partial}{\partial w_1}, \cdots, \frac{\partial}{\partial w_s}\right\}$, if and only if L^s satisfies Frobenius condition.

The submanifolds play essential roles in the integration on manifolds, in section 6.4.2.

We close this subsection by the following summation.

Submanifold: Let $(M, \mathbb{A})$ be an m-dimensional smooth manifold, $N\subset M$ be a subset of M. If for any point $p\in N$, there exists a chart (U, φ_U) of p in M, such that $\varphi_U(N\cap U)=(\mathbb{R}^k)\cap\varphi_U(U)$, then N is called a **k-dimensional smooth submanifold of** M, or simply, submanifold.

Regular submanifold: If N is a k-dimensional smooth submanifold of M, a mapping Φ: $N\to\varphi(N)$ with $\varphi(N)\subset M$ is a homeomorphism, then (N, Φ) is called **a regular submanifold of** M, and Φ is called **a regular mapping** from smooth manifold N to smooth manifold M.

Embedding mapping: A mapping $h: N\to M$ is called **an embedding mapping** from N to M, if it satisfies ① $h: N\to M$ is a smooth, one-one mapping, and $q=h(p)\in M$ for $p\in N$; its cotangent mapping (differentiation) $h^*: T_q^*(M)\to T_p^*(N)$ is defined by $h^*((df))=d(f\circ F)$ for $(df)_q\in T_q^*(M)$; and its tangent mapping $h_*: T_p(N)\to T_q(M)$ is defined by $\langle h_* X, \alpha\rangle = \langle X, h^* \alpha\rangle$ for $X \in T_p(N), \alpha \in T_q^*(M)$; ② $h_*: T_p(N) \to T_q(M)$ is non-degeneration at $\forall p \in N$.

Embedding submanifold: Let N be a k-dimensional smooth submanifold of M. If there exists **embedding mapping** $h: N \to M$, then N is called **an embedding submanifold of** M with $h(N) \subset M$.

Immersion mapping: A mapping $h: N \to M$ is called **immersion mapping** from N to M, if it satisfies the above ②; i. e., $h: N \to M$ is a smooth mapping (not necessarily one-one), and its differentiation is $h^*: T_q^*(M) \to T_p^*(N)$, its tangent mapping is $h_*: T_p(N) \to T_q(M)$. So that, $h: N \to M$ is called **an immersion mapping**, if its tangent mapping $h_*: T_p(N) \to T_q(M)$ is non-degenerative at every $p \in N$.

Immersion submanifold: Let N be a k-dimensional submanifold of M. If there exists an immersion mapping $h: N \to M$ from N to M, then N is called **an immersion submanifold of** M with $h(N) \subset M$.

Regular embedding submanifold: If N is a k-dimensional smooth embedding submanifold of M, and the embedding mapping $h: N \to M$ is a homeomorphism from N to $h(N)$, then N is called **a regular embedding submanifold of** M.

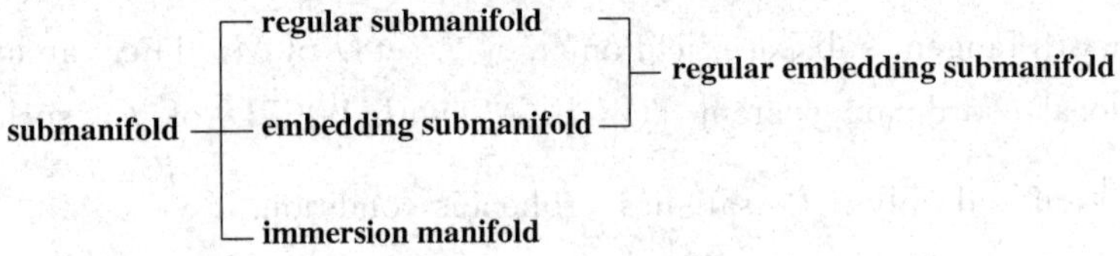

6.2 External Algebra

6.2.1 (r,s)-type tensors, (r,s)-type tensor spaces

1. Tensor product, tensor product space

Definition 6.2.1 (tensor product, tensor product space) Suppose that V is an n-dimensional linear space on $\mathbb{F}$, and W is m-dimensional linear space on $\mathbb{F}$ ($\mathbb{R}$ or $\mathbb{C}$); V^*, W^* are the dual spaces of V, W, respectively.

For $v^* \in V^* = L(V, \mathbb{F})$, $w^* \in W^* = L(W, \mathbb{F})$, **a tensor product** $v^* \otimes w^*$ of v^*, w^* is defined by mapping $v^* \otimes w^*: V \times W \to \mathbb{F}$ satisfying the following form (i. e., form (2.2.8))

$$v^* \otimes w^*(v, w) = v^*(v) \cdot w^*(w) = \langle v, v^*\rangle \cdot \langle w, w^*\rangle, \quad v \in V, w \in W, \tag{6.2.1}$$

where $\langle v, v^* \rangle \in \mathbb{F}$, $\langle w, w^* \rangle \in \mathbb{F}$ are the action of $v^* \in V^*$ on $v \in V$ and the action of $w^* \in W^*$ on $w \in W$, respectively (see Remark 1 of Definition 2.2.7). Thus, the tensor product $v^* \otimes w^*$ is a bilinear functional on $V \times W$, i.e., $v^* \otimes w^* \in L(V \times W; \mathbb{F})$. (compare with the definition of tensor product $v^* \otimes w^*$ of $v^* \in V^*$ and $w^* \in W^*$ in Definition 2.2.9).

The tensor product space $V^* \otimes W^*$ of V^*, W^* — Let

$$V^* \otimes W^* \equiv \operatorname{span}\{v^* \otimes w^* : v \in V^*, w^* \in W^*\}, \tag{6.2.2}$$

with operations: $\forall f^*, g^* \in V^* \otimes W^*$, $\alpha \in \mathbb{F}$,

addition $(f^* + g^*)((v,w)) = f^*((v,w)) + g^*((v,w))$, $(v,w) \in V \times W$;

number product $(\alpha f^*)((v,w)) = \alpha f^*((v,w))$, $(v,w) \in V \times W$;

where $f^* \in V^* \otimes W^*$ is a linear combination of $v^* \otimes w^* \in V^* \otimes W^*$, and $v^* \otimes w^*(v, w)$ has been defined in (6.2.1) or (2.2.8). Then $V^* \otimes W^*$ is said to be **a tensor product space.**

If $\mathfrak{A} = \{a_1, \cdots, a_n\}$ is a base of n-dimensional linear space V, and $\mathfrak{A}^* = \{a_j^*, 1 \leqslant j \leqslant n\}$ is the dual base of V^* with respect to base $\mathfrak{A}$, satisfying

$$a_j^*(a_k) = \delta_{kj}, \quad 1 \leqslant j, k \leqslant n, \tag{6.2.3}$$

where δ_{kj}, $1 \leqslant j, k \leqslant n$ is a Kronecker symbol. Correspondingly, $\mathfrak{B} = \{b_1, \cdots, b_m\}$ is a base of m-dimensional linear space W, and $\mathfrak{B}^* = \{b_k^*, 1 \leqslant k \leqslant m\}$ is the dual base of W^* with respect to base $\mathfrak{B}$, satisfying

$$b_j^*(b_k) = \delta_{kj}, \quad 1 \leqslant j, k \leqslant m. \tag{6.2.4}$$

Then, the tensor product $\otimes$ can be regarded as an operator from $V^* \times W^*$ to $L(V \times W; \mathbb{F})$,

$$\otimes: (v^*, w^*) \in V^* \times W^* \to v^* \otimes w^* \in L(V \times W; \mathbb{F}),$$

and $v^* \otimes w^* = \sum_{j,k} (v^*(a_j) w^*(b_k))(a_j^* \otimes b_k^*)$ by bilinear. Then, we conclude that

$$\{a_j^* \otimes b_k^* : 1 \leqslant j \leqslant n, 1 \leqslant k \leqslant m\}$$

is a base of $V^* \otimes W^*$ such that $V^* \otimes W^*$ is an $n \times m$-dimensional linear space on $\mathbb{F}$, and

$$V^* \otimes W^* \leftrightarrow L(V \times W; \mathbb{F}).$$

Moreover, we have

$$\langle a_j \otimes b_k, a_l^* \otimes b_s^* \rangle = \delta_{jl}\delta_{ks} = \begin{cases} 1, & (j,k) = (l,s), \\ 0, & (j,k) \neq (l,s), \end{cases}$$

thus $\{a_j \otimes b_k : 1 \leqslant j \leqslant n, 1 \leqslant k \leqslant m\}$ is a base of $V \otimes W$, and $\{a_j^* \otimes b_k^* : 1 \leqslant j \leqslant n, 1 \leqslant k \leqslant m\}$ is a base of $V^* \otimes W^*$, respectively; they are dual bases each other. So that

$$V^* \otimes W^* = (V \otimes W)^*.$$

2. tensor spaces of (r,s)-type and their elements

Definition 6.2.2 (tensor space of (r,s)-type) For n-dimensional linear space V on $\mathbb{F}$, **the tensor space of (r,s)-type** is defined as

$$V_s^r = \underbrace{V\otimes\cdots\otimes V}_{r}\otimes\underbrace{V^*\otimes\cdots\otimes V^*}_{s}, \tag{6.2.5}$$

with operations $+$ and $\alpha\cdot$, where $(r,s)\in\mathbb{P}\times\mathbb{P}$. It is a linear space on $\mathbb{F}$, and **we agree on** $V_0^0=\mathbb{F}$; an element $x\in V_s^r$ is said to be a **tensor of (r,s)-type**; r is called the **degree of contra-variant of** x, and s is called the **degree of covariant of** x.

Particularly, $V_0^r=\underbrace{V\otimes\cdots\otimes V}_{r}$ is said to be a **contra-variant tensor space of r-degree**; $x\in V_0^r$ is called a **contra-variant tensor of r-degree**; $x\in V_0^1$ is called **a contra-variant tensor.** The set $V_s^0=\underbrace{V^*\otimes\cdots\otimes V^*}_{s}$ is said to be **a covariant tensor space of s-degree**; $x\in V_s^0$ is called **a covariant tensor of s-degree**; $x\in V_1^0$ is called a **covariant tensor.**

The following conclusions are very important and useful:

(1) $\dim V_s^r=n^{r+s}$;

(2) $V_s^r \leftrightarrow L(\underbrace{V^*\times\cdots\times V^*}_{r}\times\underbrace{V\times\cdots\times V}_{s};\mathbb{F})$, thus, (r,s)-type tensor $x\in V_s^r=\underbrace{V\otimes\cdots\otimes V}_{r}\otimes\underbrace{V^*\otimes\cdots\otimes V^*}_{s}$ is an $\mathbb{F}$**-valued $(r+s)$-multiple linear function on tensor space** $\underbrace{V^*\times\cdots\times V^*}_{r}\times\underbrace{V\times\cdots\times V}_{s}$.

(3) For n-dimensional linear space V with base $\mathfrak{A}=\{a_j,1\leqslant j\leqslant n\}$, *and dual space* V^* *with base* $\mathfrak{A}^*=\{a_j^*,1\leqslant j\leqslant n\}$, then, **the base of tensor space of (r,s)-type** $V_s^r=\underbrace{V\otimes\cdots\otimes V}_{r}\otimes\underbrace{V^*\otimes\cdots\otimes V^*}_{s}$ is

$$\{a_{j_1}\otimes\cdots\otimes a_{j_r}\otimes a_{k_1}^*\cdots\otimes a_{k_s}^*,1\leqslant j_1,\cdots,j_r,k_1,\cdots,k_s\leqslant n\};$$

A tensor of (r,s)-type $x\in V_s^r$ can be expressed uniquely by

$$x=\sum_{1\leqslant j_1,\cdots,j_r,k_1,\cdots,k_s\leqslant n}x_{k_1,\cdots,k_s}^{j_1,\cdots,j_r}(a_{j_1}\otimes\cdots\otimes a_{j_r})\otimes(a_{k_1}^*\otimes\cdots\otimes a_{k_s}^*)$$

with

$$x_{k_1,\cdots,k_s}^{j_1,\cdots,j_r}=x(a_{j_1},\cdots,a_{j_r},a_{k_1}^*,\cdots,a_{k_s}^*)=\langle a_{j_1}\otimes\cdots\otimes a_{j_r}\otimes a_{k_1}^*\otimes\cdots\otimes a_{k_s}^*,x\rangle.$$

Note we use **Einstein convention of sum** (or simply, **Einstein convention**): a sum

$$x=\sum_{1\leqslant j_1,\cdots,j_r,k_1,\cdots,k_s\leqslant n}x_{k_1,\cdots,k_s}^{j_1,\cdots,j_r}(a_{j_1}\otimes\cdots\otimes a_{j_r})\otimes(a_{k_1}^*\otimes\cdots\otimes a_{k_s}^*),$$

denoted by

$$x = x^{j_1,\cdots,j_r}_{k_1,\cdots,k_s}(a_{j_1}\otimes\cdots\otimes a_{j_r})\otimes(a^*_{k_1}\otimes\cdots\otimes a^*_{k_s}), \tag{6.2.6}$$

or simply, $x = x^{j_1,\cdots,j_r}_{k_1,\cdots,k_s}$.

3. Operations on tensor spaces of (r,s)-type

Definition 6.2.3 (addition, number product, and tensor product on V^r_s) Let V be an n-dimensional linear space on number field $\mathbb{F}$, and $V^r_s = \underbrace{V\otimes\cdots\otimes V}_{r}\otimes\underbrace{V^*\otimes\cdots\otimes V^*}_{s}$ be **a tensor space of** (r,s)**-type**. The **addition** and **number product** of tensors of (r,s)-type $x\in V^r_s$ and $y\in V^r_s$ with $\alpha\in\mathbb{F}$, are expressed by the notations of Einstein convention: if

$$x = x^{j_1,\cdots,j_r}_{k_1,\cdots,k_s}(a_{j_1}\otimes\cdots\otimes a_{j_r})\otimes(a^*_{k_1}\otimes\cdots\otimes a^*_{k_s}),$$

and

$$y = y^{j_1,\cdots,j_r}_{k_1,\cdots,k_s}(a_{j_1}\otimes\cdots\otimes a_{j_r})\otimes(a^*_{k_1}\otimes\cdots\otimes a^*_{k_s}),$$

then

$$x+y = (x^{j_1,\cdots,j_r}_{k_1,\cdots,k_s}+y^{j_1,\cdots,j_r}_{k_1,\cdots,k_s})(a_{j_1}\otimes\cdots\otimes a_{j_r})\otimes(a^*_{k_1}\otimes\cdots\otimes a^*_{k_s}), \tag{6.2.7}$$

$$\alpha x = (\alpha x^{j_1,\cdots,j_r}_{k_1,\cdots,k_s})(a_{j_1}\otimes\cdots\otimes a_{j_r})\otimes(a^*_{k_1}\otimes\cdots\otimes a^*_{k_s}). \tag{6.2.8}$$

The tensor product Let $x\in V^{r_1}_{s_1}$ be a tensor of (r_1,s_1)-type, $y\in V^{r_2}_{s_2}$ be a tensor of (r_2,s_2)-type. Then **the tensor product** $x\otimes y$ of x and y is defined by

$$x\otimes y(v^*_1,\cdots,v^*_{r_1+r_2},v_1,\cdots,v_{s_1+s_2})$$
$$= x(v^*_1,\cdots,v^*_{r_1},v_1,\cdots,v_{s_1})\cdot y(v^*_{r_1+1},\cdots,v^*_{r_1+r_2},v_{s_1+1},\cdots,v_{s_1+s_2}). \tag{6.2.9}$$

And in certain bases, we have:

$$(x\otimes y)^{j_1,\cdots,j_{r_1+r_2}}_{k_1,\cdots,k_{s_1+s_2}} = x^{j_1,\cdots,j_{r_1}}_{k_1,\cdots,k_{s_1}}\cdot y^{j_{r_1+1},\cdots,j_{r_1+r_2}}_{k_{s_1+1},\cdots,k_{s_1+s_2}}. \tag{6.2.10}$$

It is easy to prove that, the tensor product satisfies **combination law**

$$(x\otimes y)\otimes z = x\otimes(y\otimes z);$$

and **distribution law**: if x and y are the same type tensor, then

$$z\otimes(x+y) = (z\otimes x)+(z\otimes y),$$
$$(x+y)\otimes z = (x\otimes z)+(y\otimes z).$$

Definition 6.2.4 (tensor spaces of $(r,0)$-type and $(0,s)$-type) Denote by $T^r(V)\equiv V^r_0 = \underbrace{V\otimes\cdots\otimes V}_{r}$ the tensor space of $(r,0)$-type. The operations addition and number product are defined: for $x, y\in T^r(V)$ with $x = x^{j_1,\cdots,j_r}(a_{j_1}\otimes\cdots\otimes a_{j_r})$, $y = y^{j_1,\cdots,j_r}(a_{j_1}\otimes\cdots\otimes a_{j_r})$,

$$x+y = (x^{j_1,\cdots,j_r}+y^{j_1,\cdots,j_r})(a_{j_1}\otimes\cdots\otimes a_{j_r}),$$

$$\alpha x=(\alpha x^{j_1,\cdots,j_r})(a_{j_1}\otimes\cdots\otimes a_{j_r}).$$

Similar for the tensor space of $(0,s)$-type $T^s(V^*)=(V^*)_s^0=\underbrace{V^*\otimes\cdots\otimes V^*}_{s}$, and addition, number product: $x,y\in T^s(V^*)$ with $x=x_{j_1,\cdots,j_s}(a_{j_1}^*\otimes\cdots\otimes a_{j_s}^*)$, $y=y_{j_1,\cdots,j_s}(a_{j_1}^*\otimes\cdots\otimes a_{j_s}^*)$,

$$x+y=(x_{j_1,\cdots,j_s}+y_{j_1,\cdots,j_s})(a_{j_1}^*\otimes\cdots\otimes a_{j_s}^*),$$

$$\alpha x=(\alpha x_{j_1,\cdots,j_s})(a_{j_1}^*\otimes\cdots\otimes a_{j_s}^*).$$

The duality theorem of the tensor space of $(r,0)$-type and tensor space of $(0,r)$-type is:

Theorem 6.2.1 The tensor space of $(r,0)$-type $T^r(V)$ and tensor space of $(0,r)$-type $T^r(V^*)$ are dual each other, $(T^r(V))^*=T^r(V^*)$; and $\dim T^r(V)=\dim T^r(V^*)=n^r$.

Definition 6.2.5 (matching of elements in tensor spaces $T^r(V)$ and $T^r(V^*)$) For the elements in tensor spaces $T^r(V^*)$ and $T^r(V)$, **the number in** $\mathbb{F}$ defined by

$$\langle v_1\otimes\cdots\otimes v_r, v_1^*\otimes\cdots\otimes v_r^*\rangle=\langle v_1,v_1^*\rangle\cdots\cdots\langle v_r,v_r^*\rangle,$$

$$v_j\in V,\quad v_k^*\in V^*,\quad 1\leqslant j,k\leqslant n$$

is said to be **"a matching" of elements** $v_1\otimes\cdots\otimes v_r\in T^r(V)$ **and** $v_1^*\otimes\cdots\otimes v_r^*\in T^r(V^*)$, or simply, **matching of** $v_1\otimes\cdots\otimes v_r$ **and** $v_1^*\otimes\cdots\otimes v_r^*$; it is the action of $v_1\otimes\cdots\otimes v_r$ on $v_1^*\otimes\cdots\otimes v_r^*$, since tensor space of $(r,0)$-type $T^r(V)$ and tensor space of $(0,r)$-type $T^r(V^*)$ are dual each other.

4. self-homomorphism σ on $T^r(V)$, permutation group of r-degree $\mathfrak{T}(r)$

Recall the permutation group in section 1.2: for a set $S=\{1,2,\cdots,r\}$ with r elements, $r\in\mathbb{Z}^+$, a one-one mapping $\sigma: S\to S$ from S onto itself is called **a permutation.** Denote the set of all permutations on S by $\mathfrak{T}(r)=\{\sigma:\sigma\{1,\cdots,r\}=\{\sigma(1),\cdots,\sigma(r)\}\}$, and $\sigma\in\mathfrak{T}(r)$ by $\sigma=\begin{pmatrix}1 & 2 & \cdots & r\\ \sigma(1) & \sigma(2) & \cdots & \sigma(r)\end{pmatrix}$. There are $r!$ elements in $\mathfrak{T}(r)$.

The operation on $\mathfrak{T}(r)$ is defined by "compound", $\sigma,\tau\in\mathfrak{T}(r)\Rightarrow\tau\circ\sigma$, such that

$$(\tau\circ\sigma)(s)=\tau(\sigma(s)),\quad\forall s\in S.$$

Then, $(\mathfrak{T}(r),\circ)$ is a group with operation "$\circ$", the unit is identity $\sigma\equiv I: \sigma(s)=s, s\in S$, and the inverse element of $\sigma\in\mathfrak{T}(r)$ is $\tau^{-1}\in\mathfrak{T}(r)$.

The multiplication value $\upsilon(\sigma)\equiv\prod\limits_{1\leqslant k<j\leqslant r}(\sigma(j)-\sigma(k))$ of $\sigma\in\mathfrak{T}(r)$ plays important role. For $\forall\sigma\in\mathfrak{T}(r)$ with $\sigma=\begin{pmatrix}1 & 2 & \cdots & r\\ \sigma(1) & \sigma(2) & \cdots & \sigma(r)\end{pmatrix}$, the value $\upsilon(\sigma)$ has a formula

$$\prod_{1\leqslant k<j\leqslant r}(\sigma(j)-\sigma(k))=\pm(2!)(3!)\cdots((n-1)!). \tag{6.2.11}$$

If $\upsilon(\sigma)>0$, then σ is said to be an **even permutation**; if $\upsilon(\sigma)<0$, then σ is sad to be an **odd permutation.**

Since $\sigma\{1,\cdots,r\}=\{\sigma(1),\cdots,\sigma(r)\}=\{j_1,\cdots,j_r\}$, $1\leqslant j_1,\cdots,j_r\leqslant r$, thus $\sigma\in\mathfrak{T}(r)$ is **a self-homomorphism** $\sigma: T^r(V)\to T^r(V)$ from a tensor space $T^r(V)$ to itself. So that, for $x\in T^r(V)=V_0^r=\underbrace{V\otimes\cdots\otimes V}_{r}$, define

$$\sigma(x(v_1^*,\cdots,v_r^*))=x(v_{\sigma(1)}^*,\cdots,v_{\sigma(r)}^*),\quad v_j^*\in V^*, j=1,\cdots,r. \tag{6.2.12}$$

Note 1 The sense of (6.2.12) is: $T^r(V)$ and $T^r(V^*)$ are dual each other, $(T^r(V))^*=T^r(V^*)$, so that

$$\begin{aligned}\forall x\in T^r(V)&\Rightarrow x=x^{j_1,\cdots,j_r}(a_{j_1}\otimes\cdots\otimes a_{j_r})\\&\Rightarrow x(v_1^*,\cdots,v_r^*)\in\mathbb{F},\quad\forall(v_1^*,\cdots,v_r^*)\in T^r(V^*).\end{aligned}$$

Then, $\forall\sigma\in\mathfrak{T}(r)$, $\sigma: T^r(V)\to T^r(V)$ maps $x\in T^r(V)$ onto $\sigma x\in T^r(V)$ satisfying

$$(\sigma x)(v_1^*,\cdots,v_r^*)=\sigma(x(v_1^*,\cdots,v_r^*))=x(v_{\sigma(1)}^*,\cdots,v_{\sigma(r)}^*),\quad v_j^*\in V^*.$$

Note 2 Self-homomorphism $\sigma: T^r(V)\to T^r(V)$ satisfies

$$\sigma(x+y)=\sigma(x)+\sigma(y),\quad x,y\in T^r(V);$$

$$\sigma(\alpha x)=\alpha\sigma(x),\quad x\in T^r(V),\alpha\in\mathbb{F}.$$

Theorem 6.2.2 **If $x\in T^r(V)$, $x=v_1\otimes\cdots\otimes v_r$, then**

$$\sigma x=v_{\sigma^{-1}(1)}\otimes\cdots\otimes v_{\sigma^{-1}(r)},\quad\forall\sigma\in\mathfrak{T}(r),$$

with inverse element σ^{-1} of σ.

5. Contra-variant tensors in $T^r(V)$

Definition 6.2.6 (**symmetric and skew-symmetric contra-variant tensors**) Let $x\in T^r(V)$. If for any $\sigma\in\mathfrak{T}(r)$, it holds

$$\sigma x=x, \tag{6.2.13}$$

then x is said to be **a symmetric contra-variant tensor of r-degree**; If for any $\sigma\in\mathfrak{T}(r)$, it holds

$$\sigma x=(\operatorname{sgn}\sigma)\cdot x, \tag{6.2.14}$$

then x is said to be **a skew-symmetric contra-variant tensor of r-degree**; where $\operatorname{sgn}\sigma$ is

$$\operatorname{sgn}\sigma=\begin{cases}1, & \sigma \text{ is even permutation}\\-1, & \sigma\text{ is odd permutation}\end{cases}.$$

Theorem 6.2.3 If $x\in T^r(V)$, **then**

(1) x is a symmetric contra-variant tensor, if and only if the components of x are

symmetric with respect to each index;

(2) x is a skew-symmetric contra-variant tensor, if and only if the components of x are skew-symmetric with respect to each index.

Proof Let the base of V be $\{a_1,\cdots,a_n\}$, and the dual base of V^* be $\{a_1^*,\cdots,a_n^*\}$.

(1) if x is a symmetric contra-variant tensor, then, for any $\sigma\in\mathfrak{T}(r)$, it holds

$$x^{j_1,\cdots,j_r}(a_{j_1}^*,\cdots,a_{j_r}^*)=x(a_{j_1}^*,\cdots,a_{j_r}^*)=\sigma x(a_{j_1}^*,\cdots,a_{j_r}^*)$$
$$=x(a_{j_{\sigma(1)}}^*,\cdots,a_{j_{\sigma(r)}}^*)=x^{j_{\sigma(1)},\cdots,j_{\sigma(r)}}(a_{j_1}^*,\cdots,a_{j_r}^*),$$

and vice versa.

(2) If x is a skew-symmetric contra-variant tensor, then, for any $\sigma\in\mathfrak{T}(r)$, it holds

$$x^{j_1,\cdots,j_r}(a_{j_1}^*,\cdots,a_{j_r}^*)=x(a_{j_1}^*,\cdots,a_{j_r}^*)=(\mathrm{sgn}\sigma)\cdot\sigma x(a_{j_1}^*,\cdots,a_{j_r}^*)$$
$$=(\mathrm{sgn}\sigma)\cdot x(a_{j_{\sigma(1)}}^*,\cdots,a_{j_{\sigma(r)}}^*)=(\mathrm{sgn}\sigma)\cdot x^{j_{\sigma(1)},\cdots,j_{\sigma(r)}}(a_{j_1}^*,\cdots,a_{j_r}^*),$$

and vice versa. The proof is complete.

The set of symmetric contra-variant tensors of r-degree in $T^r(V)$ is denoted by $P^r(V)$; the set of skew-symmetric contra-variant tensors of r-degree in $T^r(V)$ is denoted by $\Lambda^r(V)$.

It is clear that, $P^r(V)$ and $\Lambda^r(V)$ are two sublinear spaces of $T^r(V)$, since permutation $\sigma\in\mathfrak{T}(r)$ is a self-homomorphism on $T^r(V)$.

Definition 6.2.7 (symmetrizing and skew-symmetrizing operators) For $x\in T^r(V)$, define

$$S_r(x)=\frac{1}{r!}\sum_{\sigma\in\mathfrak{T}(r)}\sigma x \quad\text{and}\quad A_r(x)=\frac{1}{r!}\sum_{\sigma\in\mathfrak{T}(r)}(\mathrm{sgn}\sigma)\cdot\sigma x. \tag{6.2.15}$$

Then, $S_r: T^r(V)\to T^r(V)$, $A_r: T^r(V)\to T^r(V)$, and $S_r(x), A_r(x)\in T^r(V)$. Thus, operators S_r and A_r are self-homomorphic on $T^r(V)$.

S_r is said to be **the symmetrizing operator** on $T^r(V)$; A_r is said to be **the skew-symmetrizing operator on** $T^r(V)$.

Theorem 6.2.4 $P^r(V)=S_r(T^r(V))$ and $\Lambda^r(V)=A_r(T^r(V))$; moreover, $P^r(V)$ and $\Lambda^r(V)$ are sublinear spaces of $T^r(V)$.

Proof Firstly, we prove: the symmetrizing of $x\in T^r(V)$ is a symmetric tensor, and the skew-symmetrizing of $x\in T^r(V)$ is a skew-symmetric tensor. Since $\forall x\in T^r(V)$, $\forall \tau\in\mathfrak{T}(r)$, it holds

$$\tau(S_r(x))=\frac{1}{r!}\sum_{\sigma\in\mathfrak{T}(r)}\tau(\sigma(x))=S_r(x);$$

and

$$\tau(A_r(x)) = \frac{1}{r!}\sum_{\sigma\in\mathfrak{T}(r)}(\mathrm{sgn}\sigma)\cdot\tau(\sigma(x))$$

$$= (\mathrm{sgn}\tau)\cdot\frac{1}{r!}\sum_{\sigma\in\mathfrak{T}(r)}(\mathrm{sgn}(\tau\circ\sigma))\cdot(\tau\circ\sigma)(x) = (\mathrm{sgn}\tau)\cdot A_r(x).$$

Thus, $S_r(T^r(V))\subset P^r(V)$ and $A_r(T^r(V))\subset\Lambda^r(V)$.

Secondly, we prove: a symmetric tensor is invariant under the action of symmetrizing operator; so that we have $P^r(V)\subset S_r(P^r(V))$; moreover, a skew-symmetric tensor is invariant under the action of skew-symmetrizing operator (left for exercises); so that we have $\Lambda^r(V)\subset A_r(\Lambda^r(V))$. Combine the above relationships proved in the first step, we have

$$P^r(V) = S_r(T^r(V)) \text{ and } \Lambda^r(V) = A_r(T^r(V)). \tag{6.2.16}$$

It is clear that, $P^r(V)$ and $\Lambda^r(V)$ are sublinear spaces of $T^r(V)$, since permutation $\sigma\in\mathfrak{T}(r)$ is a self-homomorphism on $T^r(V)$.

Similarly, the above results hold for **covariant tensors.** We list the results below.

The set of the symmetric covariant tensors of r-degree in $T^r(V^*)$ is denoted by $P^r(V^*)$; the set of the skew-symmetric covariant tensors of r-degree in $T^r(V^*)$ is denoted by $\Lambda^r(V^*)$.

By symmetrizing operator S_r, it holds $P^r(V^*) = S_r(T^r(V^*))$; By **skew-symmetrizing operator** A_r, it holds $\Lambda^r(V^*) = A_r(T^r(V^*))$; It is clear that, $P^r(V^*)$ and $\Lambda^r(V^*)$ are sublinear spaces of $T^r(V^*)$.

Askew-symmetric contra-variant tensor space of r-degree $\Lambda^r(V)$ is also said to be **an exterior vector space of r-degree**; and **a skew-symmetric contra-variant tensor of r-degree in** $\Lambda^r(V)$ is called **an exterior vector of r-degree.** We agree on: $\Lambda^0(V)=\mathbb{F}$, $\Lambda^1(V)=V$.

Askew-symmetric covariant tensor space of r-degree $\Lambda^r(V^*)$ is said to be **an exterior form space of r-degree**; and **a skew-symmetric covariant tensor of r-degree in** $\Lambda^r(V^*)$ is called **an exterior form of r-degree.** We agree on: $\Lambda^0(V^*)=\mathbb{F}$, $\Lambda^1(V^*)=V^*$.

We emphasize that: the skew-symmetric tensors, including the skew-symmetric contra-variant tensors and skew-symmetric covariant tensors, play very important and essential roles in the study of calculus on differential manifolds, specially, of **the exterior differential forms.**

6.2.2 Tensor algebra

1. Graded sum of tensor space of $(r,0)$-type $T^r(V)=V_0^r$, $r\geqslant 0$

Let $T^r(V)=V_0^r=\underbrace{V\otimes\cdots\otimes V}_{r}$, $r\geqslant 0$, be a tensor space of $(r,0)$-type. Consider

$$T(V)=\sum_{r\geqslant 0}T^r(V), \tag{6.2.17}$$

the sum in (6.2.17) is called **a graded sum**, if its element $x\in T(V)$ can be expressed by "a direct sum" as

$$x=\sum_{r\geqslant 0}x^r,\quad x^r\in T^r(V), \tag{6.2.18}$$

with finite terms in the summation. Then, tensor space $T(V)$ defined in (6.2.17) is regarded as a "graded sum set"; operations "addition $+$" and "number product $\alpha\cdot$" are introduced for the elements in $T(V)$, such that $(T(V),+,\alpha\cdot)$ becomes a linear space with infinite dimension.

2. Tensor algebra

Tensor product $\otimes$, i.e., multiplication operation of two tensors, can be generalized to the graded sum space $(T(V),+,\alpha\cdot)=\left(\sum_{r\geqslant 0}T^r(V),+,\alpha\cdot\right)$: for $x,y\in T(V)=\sum_{r\geqslant 0}T^r(V)$,

① if $x\in T^{r_1}(V), y\in T^{r_2}(V), r_1\geqslant 0, r_2\geqslant 0$, then

$$x\otimes y(v_1^*,\cdots,v_{r_1+r_2}^*)=x(v_1^*,\cdots,v_{r_1}^*)\cdot y(v_{r_1+1}^*,\cdots,v_{r_1+r_2}^*).$$

In certain selected base of V, the components of $x\otimes y$ satisfy

$$(x\otimes y)^{j_1,\cdots,j_{r_1+r_2}}=x^{j_1,\cdots,j_{r_1}}\cdot y^{j_{r_1+1},\cdots,j_{r_1+r_2}}; \tag{6.2.19}$$

It is clear, $x\in T^{r_1}(V), y\in T^{r_2}(V)\Rightarrow x\otimes y\in T^{r_1+r_2}(V)$; thus, operation $\otimes$ is closed in linear space $(T(V),+,\alpha\cdot)=\left(\sum_{r\geqslant 0}T^r(V),+,\alpha\cdot\right)$.

② If $x=\sum_{r\geqslant 0}x_r$, $y=\sum_{r\geqslant 0}y_r$, then

$$x\otimes y=\left(\sum_{r\geqslant 0}x_r\right)\otimes\left(\sum_{r\geqslant 0}y_r\right) \tag{6.2.20}$$

evaluated by the distribution law (since both summations have finite terms).

Definition 6.2.8 (tensor algebra) Tensor space $T(V)$ with operations $+,\alpha\cdot,\otimes$ becomes an algebra (recall the definition of "algebra" in Chapter 2, section 2.2.2), denoted by $(T(V),+,\alpha\cdot,\otimes)=\left(\sum_{r\geqslant 0}T^r(V),+,\alpha\cdot,\otimes\right)$, called **a tensor algebra on linear space** V, or simply, $(T(V),+,\alpha\cdot,\otimes)$ is said to be **a tensor algebra.**

Similarly, **a tensor algebra** $(T(V^*),+,\alpha\cdot,\otimes)=\left(\sum_{r\geqslant 0}T^r(V^*),+,\alpha\cdot,\otimes\right)$ on dual space V^* of linear space V can be defined.

Note, that the graded sums $T(V)=\sum_{r\geqslant 0} T^r(V)=\sum_{r\geqslant 0} V_0^r$ and $T(V^*)=\sum_{r\geqslant 0} T^r(V^*)=\sum_{r\geqslant 0} V_r^0$ are algebras with respect to multiplication operation $\otimes$; however, tensor spaces of $(r,0)$-type $T^r(V)$, $T^r(V^*)$ are not, since $\otimes$ is not closed in them. We now introduce so-called "exterior product" $\wedge$ in both $T^r(V)$ and $T^r(V^*)$, such that they become "exterior algebra".

3. Exterior product

1) Exterior product in skew-symmetric contra-variant tensor spaces

Definition 6.2.9 (exterior product in $\Lambda^k(V)$) Let ξ be an exterior vector of k-degree, $\xi\in\Lambda^k(V)$, η be an exterior vector of l-degree, $\eta\in\Lambda^l(V)$. Denote

$$\xi\wedge\eta=A_{k+l}(\xi\otimes\eta),$$

with A_r, **the skew-symmetrizing operator** $A_r(\xi)=\frac{1}{r!}\sum_{\sigma\in\mathfrak{T}(r)}(\mathrm{sgn}\sigma)\cdot\sigma\xi$. Then, $\xi\wedge\eta$ (a $k+l$-degree exterior vector) is said to be **an exterior product of** ξ and η.

Theorem 6.2.5 For exterior vectors of k-degree $\xi,\xi_1,\xi_2\in\Lambda^k(V)$, exterior vectors of l-degree $\eta,\eta_1,\eta_2\in\Lambda^l(V)$ and exterior vector of h-degree $\zeta\in\Lambda^h(V)$, we have

(1) Combination law $(\xi\wedge\eta)\wedge\zeta=\xi\wedge(\eta\wedge\zeta)$;

(2) Distribution law $(\xi_1+\xi_2)\wedge\eta=\xi_1\wedge\eta+\xi_2\wedge\eta$;

$$\xi\wedge(\eta_1+\eta_2)=\xi\wedge\eta_1+\xi\wedge\eta_2;$$

(3) Contra-exchange law $\xi\wedge\eta=(-1)^{kl}\eta\wedge\xi$.

Proof The distribution law (2) is clear by the linearity of a tensor product, and linearity of a skew-symmetrizing operator.

Then, prove the contra-exchange law (3): for exterior vector of k-degree $\xi\in\Lambda^k(V)$ and exterior vector of l-degree $\eta\in\Lambda^l(V)$, exterior product $\xi\wedge\eta=A_{k+l}(\xi\otimes\eta)\in\Lambda^{k+l}(V)$ is a skew-symmetric tensor, so that for any permutation $\tau\in\mathfrak{T}(k+l)$, it holds $\tau(\xi\wedge\eta)=\mathrm{sgn}\tau\cdot(\xi\wedge\eta)$. Take permutation

$$\tau=\begin{pmatrix}1 & \cdots & k & k+1 & \cdots & k+l\\ 1+l & \cdots & k+l & 1 & \cdots & l\end{pmatrix},$$

with $\mathrm{sgn}\tau=(-1)^{kl}$, then for $\forall v_1^*,\cdots,v_k^*,v_{k+1}^*,\cdots,v_{k+l}^*\in V^*$, we deduce

$$\begin{aligned}\xi\wedge\eta(v_1^*,\cdots,v_{k+l}^*)&=(-1)^{kl}\xi\wedge\eta(v_{\tau(1)}^*,\cdots,v_{\tau(k+l)}^*)\\&=\frac{(-1)^{kl}}{(k+l)!}\sum_{\sigma\in\mathfrak{T}(k+l)}\mathrm{sgn}\sigma\cdot\sigma(\xi\otimes\eta)(v_{\tau(1)}^*,\cdots,v_{\tau(k+l)}^*)\\&=\frac{(-1)^{kl}}{(k+l)!}\sum_{\sigma\in\mathfrak{T}(k+l)}\mathrm{sgn}\sigma\cdot\xi(v_{\sigma\circ\tau(1)}^*,\cdots,v_{\sigma\circ\tau(k)}^*)\cdot\end{aligned}$$

$$\eta(v^*_{\sigma\circ\tau(k+1)},\cdots,v^*_{\sigma\circ\tau(k+l)})$$

$$=\frac{(-1)^{kl}}{(k+l)!}\sum_{\sigma\in\mathfrak{T}(k+l)}\operatorname{sgn}\sigma\cdot\eta(v^*_{\sigma(1)},\cdots,v^*_{\sigma(l)})\cdot$$

$$\xi(v^*_{\sigma(l+1)},\cdots,v^*_{\sigma(l+k)})$$

$$=\frac{(-1)^{kl}}{(k+l)!}\sum_{\sigma\in\mathfrak{T}(k+l)}\operatorname{sgn}\sigma\cdot\sigma(\eta\otimes\xi)(v^*_1,\cdots,v^*_l,v^*_{l+1},\cdots,v^*_{l+k})$$

$$=(-1)^{kl}A_{l+k}(\eta\otimes\xi)(v^*_1,\cdots,v^*_{l+k})$$

$$=(-1)^{kl}\eta\wedge\xi(v^*_1,\cdots,v^*_{l+k}).$$

This implies (3).

Prove the combination law (1): for $\xi\in\Lambda^k(V),\eta\in\Lambda^l(V),\zeta\in\Lambda^h(V)$, take

$$v^*_1,\cdots,v^*_k,v^*_{k+1},\cdots,v^*_{k+l},v^*_{k+l+1},\cdots,v^*_{k+l+h}\in V^*,$$

then

$$((\xi\wedge\eta)\wedge\zeta)(v^*_1,\cdots,v^*_k,v^*_{k+1},\cdots,v^*_{k+l},v^*_{k+l+1},\cdots,v^*_{k+l+h})$$

$$=A_{(k+l)+h}((\xi\wedge\eta)\otimes\zeta)(v^*_1,\cdots,v^*_k,v^*_{k+1},\cdots,v^*_{k+l},v^*_{k+l+1},\cdots,v^*_{k+l+h})$$

$$=\frac{1}{((k+l)+h)!}\sum_{\sigma\in\mathfrak{T}(k+l+h)}(\operatorname{sgn}\sigma)\cdot$$

$$\sigma((\xi\wedge\eta)\otimes\zeta)(v^*_1,\cdots,v^*_k,v^*_{k+1},\cdots,v^*_{k+l},v^*_{k+l+1},\cdots,v^*_{k+l+h})$$

$$=\frac{1}{((k+l)+h)!}\sum_{\sigma\in\mathfrak{T}(k+l+h)}\operatorname{sgn}\sigma\cdot(\xi\wedge\eta)(v^*_{\sigma(1)},\cdots,v^*_{\sigma(k)},v^*_{\sigma(k+1)},\cdots,v^*_{\sigma(k+l)})\cdot$$

$$\zeta(v^*_{\sigma(k+l+1)},\cdots,v^*_{\sigma(k+l+h)})$$

$$=\frac{1}{((k+l)+h)!}\sum_{\sigma\in\mathfrak{T}(k+l+h)}\operatorname{sgn}\sigma\cdot\frac{1}{(k+l)!}\sum_{\tau\in\mathfrak{T}(k+l)}\operatorname{sgn}\tau\cdot\xi(v^*_{\sigma\circ\tau(1)},\cdots,v^*_{\sigma\circ\tau(k)})\cdot$$

$$\eta(v^*_{\sigma\circ\tau(k+1)},\cdots,v^*_{\sigma\circ\tau(k+l)})\cdot\zeta(v^*_{\sigma\circ\tau(k+l+1)},\cdots,v^*_{\sigma\circ\tau(k+l+h)})$$

$$=\frac{1}{(k+l+h)!}\cdot\frac{1}{(k+l)!}\sum_{\sigma\in\mathfrak{T}(k+l+h)}\sum_{\tau\in\mathfrak{T}(k+l)}\operatorname{sgn}\sigma\cdot\operatorname{sgn}\tau\cdot\xi(v^*_{\sigma\circ\tau(1)},\cdots,v^*_{\sigma\circ\tau(k)})\cdot$$

$$\eta(v^*_{\sigma\circ\tau(k+1)},\cdots,v^*_{\sigma\circ\tau(k+l)})\cdot\zeta(v^*_{\sigma\circ\tau(k+l+1)},\cdots,v^*_{\sigma\circ\tau(k+l+h)})$$

$$=A_{k+l+h}(\xi\otimes\eta\otimes\zeta)(v^*_1,\cdots,v^*_k,v^*_{k+1},\cdots,v^*_{k+l},v^*_{k+l+1},\cdots,v^*_{k+l+h});$$

on the other hand,

$$(\xi\wedge(\eta\wedge\zeta))(v^*_1,\cdots,v^*_k,v^*_{k+1},\cdots,v^*_{k+l},v^*_{k+l+1},\cdots,v^*_{k+l+h})$$

$$=A_{k+(l+h)}(\xi\otimes(\eta\wedge\zeta))(v^*_1,\cdots,v^*_k,v^*_{k+1},\cdots,v^*_{k+l},v^*_{k+l+1},\cdots,v^*_{k+l+h})$$

$$=\frac{1}{(k+(l+h))!}\sum_{\sigma\in\mathfrak{T}(k+l+h)}(\operatorname{sgn}\sigma)\cdot$$

$$\sigma(\xi\otimes(\eta\wedge\zeta))(v_1^*,\cdots,v_k^*,v_{k+1}^*,\cdots,v_{k+l}^*,v_{k+l+1}^*,\cdots,v_{k+l+h}^*)$$
$$=\frac{1}{(k+l+h)!}\sum_{\sigma\in\mathfrak{T}(k+l+h)}\operatorname{sgn}\sigma\cdot\xi(v_{\sigma(1)}^*,\cdots,v_{\sigma(k)}^*)\cdot$$
$$(\eta\wedge\zeta)(v_{\sigma(k+1)}^*,\cdots,v_{\sigma(k+l)}^*v_{\sigma(k+l+1)}^*,\cdots,v_{\sigma(k+l+h)}^*)$$
$$=\frac{1}{((k+l)+h)!}\sum_{\sigma\in\mathfrak{T}(k+l+h)}\operatorname{sgn}\sigma\cdot\frac{1}{(k+l)!}\sum_{\tau\in\mathfrak{T}(k+l)}\operatorname{sgn}\tau\cdot\xi(v_{\sigma\circ\tau(1)}^*,\cdots,v_{\sigma\circ\tau(k)}^*)\cdot$$
$$\eta(v_{\sigma\circ\tau(k+1)}^*,\cdots,v_{\sigma\circ\tau(k+l)}^*)\cdot\zeta(v_{\sigma\circ\tau(k+l+1)}^*,\cdots,v_{\sigma\circ\tau(k+l+h)}^*)$$
$$=\frac{1}{((k+l)+h)!}\cdot\frac{1}{(k+l)!}\sum_{\sigma\in\mathfrak{T}(k+l+h)}\sum_{\tau\in\mathfrak{T}(k+l)}\operatorname{sgn}\sigma\cdot\operatorname{sgn}\tau\cdot\xi(v_{\sigma\circ\tau(1)}^*,\cdots,v_{\sigma\circ\tau(k)}^*)\cdot$$
$$\eta(v_{\sigma\circ\tau(k+1)}^*,\cdots,v_{\sigma\circ\tau(k+l)}^*)\cdot\zeta(v_{\sigma\circ\tau(k+l+1)}^*,\cdots,v_{\sigma\circ\tau(k+l+h)}^*)$$
$$=A_{k+l+h}(\xi\otimes\eta\otimes\zeta)(v_1^*,\cdots,v_k^*,v_{k+1}^*,\cdots,v_{k+l}^*,v_{k+l+1}^*,\cdots,v_{k+l+h}^*);$$

this implies combination law

$$(\xi\wedge\eta)\wedge\zeta=A_{k+l+h}(\xi\otimes\eta\otimes\zeta)=\xi\wedge(\eta\wedge\zeta).$$

Theorem 6.2.6 The exterior product has the following properties

(1) If $\xi,\eta\in V=\Lambda^1(V)$, then $\xi\wedge\xi=0$, $\xi\wedge\eta=-\eta\wedge\xi$;

Generally, if a polynomial of exterior product has more than 2 same 1-degree factors, then it must be 0.

(2) If $\{e_1,\cdots,e_n\}$ is a base of V, then

$$e_{i_1}\wedge\cdots\wedge e_{i_r}=A_r(e_{i_1}\otimes\cdots\otimes e_{i_r}),\quad 1\leqslant i_1,\cdots,i_r\leqslant n$$

is a base of $\Lambda^r(V)$ with $r\leqslant n$. (note that, exterior vector $e_{i_1}\wedge\cdots\wedge e_{i_r}$ is not zero, only for all $i_1,\cdots,i_r$ are different each other; so that, it is 0 for $r>n$).

(3) $\Lambda^r(V)=\{0\}$, $\forall r>n$.

(4) $\forall r\leqslant n$, in base $\{e_{j_1}\wedge\cdots\wedge e_{j_r},1\leqslant j_1<\cdots<j_r\leqslant n\}$ of $\Lambda^r(V)$, it holds

$$\xi=r!\sum_{1\leqslant i_1<\cdots<i_r\leqslant n}\xi^{j_1,\cdots,j_r}e_{j_1}\wedge\cdots\wedge e_{j_r},\quad\forall\xi\in\Lambda^r(V).$$

Proof Only need to prove: in $\{e_{j_1}\wedge\cdots\wedge e_{j_r}:1\leqslant j_1<\cdots<j_r\leqslant n\}$, there are $\binom{n}{r}=\frac{n!}{r!(n-r)!}$ linearly independent exterior vectors. By three steps.

First the formula of $e_{j_1}\wedge\cdots\wedge e_{j_r}$: take $v_1^*,\cdots,v_r^*\in V^*$,

$$e_{j_1}\wedge\cdots\wedge e_{j_r}(v_1^*,\cdots,v_r^*)=\frac{1}{r!}\sum_{\sigma\in\mathfrak{T}(r)}\operatorname{sgn}\sigma\cdot\langle e_{j_1},v_{\sigma(1)}^*\rangle\cdots\langle e_{j_r},v_{\sigma(r)}^*\rangle$$

$$=\frac{1}{r!}\begin{vmatrix}\langle e_{j_1},v_1^*\rangle & \cdots & \langle e_{j_1},v_r^*\rangle\\ \langle e_{j_2},v_1^*\rangle & \cdots & \langle e_{j_2},v_r^*\rangle\\ \vdots & \ddots & \vdots\\ \langle e_{j_r},v_1^*\rangle & \cdots & \langle e_{j_r},v_r^*\rangle\end{vmatrix}. \tag{6.2.21}$$

This is the evaluation formula of $e_{j_1}\wedge\cdots\wedge e_{j_r}$; specially,

$$\begin{aligned}e_{j_1}\wedge\cdots\wedge e_{j_r}(e_{k_1}^*,\cdots,e_{k_r}^*)&=\frac{1}{r!}\langle e_{j_1}\wedge\cdots\wedge e_{j_r},e_{k_1}^*\wedge\cdots\wedge e_{k_r}^*\rangle\\ &=\frac{1}{r!}\det(\langle e_{j_\alpha},e_{k_\beta}^*\rangle)=\frac{1}{r!}\delta_{k_1\cdots k_r}^{j_1\cdots j_r},\end{aligned} \tag{6.2.22}$$

with $r\leqslant n$, and

$$\delta_{k_1\cdots k_r}^{j_1\cdots j_r}=\begin{cases}1, & 1\leqslant j_1<\cdots<j_r\leqslant n,\{k_1,\cdots,k_r\}\text{ is }\{j_1,\cdots,j_r\}\text{ even},\\ -1, & 1\leqslant j_1<\cdots<j_r\leqslant n,\{k_1,\cdots,k_r\}\text{ is }\{j_1,\cdots,j_r\}\text{ odd},\\ 0, & \text{others},\end{cases}$$

called **a general Kronecker symbol**.

Second by the first step, it holds $e_1\wedge\cdots\wedge e_n\neq 0$.

Third for $r\leqslant n$, if $\{e_{j_1}\wedge\cdots\wedge e_{j_r}: 1\leqslant j_1<\cdots<j_r\leqslant n\}$ would be a linearly dependent set, then, $\exists\,\alpha^{j_1,\cdots,j_r}\in\mathbb{F}$ with not all zero, such that

$$\sum_{1\leqslant j_1<\cdots<j_r\leqslant n}\alpha^{j_1,\cdots,j_r}e_{j_1}\wedge\cdots\wedge e_{j_r}=0.$$

Suppose, say, $\alpha^{\overline{j_1,\cdots,j_r}}\neq 0$, $1\leqslant j_1<\cdots<j_r\leqslant n$, add coordinates $k_1,\cdots,k_{n-r}$, such that $\{j_1,\cdots j_k,k_1,\cdots,k_{n-r}\}\in\mathfrak{T}(n)$. Then we evaluate the exterior product with $e_{k_1}\wedge\cdots\wedge e_{k_{n-r}}$, and obtain

$$\alpha^{\overline{j_1,\cdots,j_r}}e_{j_1}\wedge\cdots\wedge e_{j_r}\wedge e_{k_1}\wedge\cdots\wedge e_{k_{n-k}}=\pm\alpha^{\overline{j_1,\cdots,j_r}}e_1\wedge\cdots\wedge e_n=0;$$

then $e_1\wedge\cdots\wedge e_n\neq 0$ implies $\alpha^{\overline{j_1,\cdots,j_r}}=0$, this contradicts $\alpha^{\overline{j_1,\cdots,j_r}}\neq 0$. So that,

$$\{e_{j_1}\wedge\cdots\wedge e_{j_r}: 1\leqslant j_1<\cdots<j_r\leqslant n\}$$

is independent. We obtain that $\{e_{j_1}\wedge\cdots\wedge e_{j_r}: 1\leqslant j_1<\cdots<j_r\leqslant n\}$ is a base of $\Lambda^r(V)$, and $\Lambda^r(V)$ is a $\binom{n}{r}=\frac{n!}{r!(n-r)!}$-dimensional linear space.

Now, (1), (2), (4) can be proved by the three steps respectively.

Finally, for (3): suppose the components of exterior vector of r-degree ξ is $\xi=\xi^{j_1,\cdots,j_r}e_{j_1}\otimes\cdots\otimes e_{j_r}$. Since the linearity of the skew-symmetrizing operator, we obtain

$$\xi=A_r\xi=\xi^{j_1,\cdots,j_r}A_r(e_{j_1}\otimes\cdots\otimes e_{j_r})=\xi^{j_1,\cdots,j_r}e_{j_1}\wedge\cdots\wedge e_{j_r},$$

By (2), $\forall r>n \Rightarrow \Lambda^r(V)=0$. This is (3).

The proof is complete.

2) Exterior product in skew-symmetric covariant tensor spaces

The definition of exterior product in $\Lambda^r(V^*)$ is similar to that in $\Lambda^r(V)$, and a similar theorem is as follows.

The exterior product in $\Lambda^k(V^*)$—Let ξ be an exterior form of k-degree, $\xi\in\Lambda^k(V^*)$, η be an exterior form of l-degree, $\eta\in\Lambda^l(V^*)$. Denote $\xi\wedge\eta=A_{k+l}(\xi\otimes\eta)$, with A_{k+l} skew-symmetrizing operator $A_r(\xi)=\dfrac{1}{r!}\sum\limits_{\sigma\in\mathfrak{T}(r)}(\text{sgn}\sigma)\cdot\sigma\xi$. Then, $\xi\wedge\eta$ (an exterior form of $k+l$-degree) is called **an exterior product of ξ and η**.

Theorem 6.2.7 For the exterior forms of k-degree $\xi,\xi_1,\xi_2\in\Lambda^k(V^*)$, the exterior forms of l-degree $\eta,\eta_1,\eta_2\in\Lambda^l(V^*)$ and exterior form of h-degree $\zeta\in\Lambda^h(V^*)$, we have

(1) Combination law $(\xi\wedge\eta)\wedge\zeta=\xi\wedge(\eta\wedge\zeta)$;

(2) Distribution law $(\xi_1+\xi_2)\wedge\eta=\xi_1\wedge\eta+\xi_2\wedge\eta$;

$$\xi\wedge(\eta_1+\eta_2)=\xi\wedge\eta_1+\xi\wedge\eta_2;$$

(3) Contra-commutative law $\xi\wedge\eta=(-1)^{kl}\eta\wedge\xi$.

Theorem 6.2.8 The exterior product has the following properties:

(1) If $\xi,\eta\in V^*=\Lambda^1(V^*)$, then $\xi\wedge\xi=0,\xi\wedge\eta=-\eta\wedge\xi$;

Generally, if a polynomial of exterior product has more than 2 same 1-degree factors, then it must be 0.

(2) If $\{e_1^*,\cdots,e_n^*\}$ is a base of V^*, then

$$e_{i_1}^*\wedge\cdots\wedge e_{i_r}^*=A_r(e_{i_1}^*\otimes\cdots\otimes e_{i_r}^*),\quad 1\leqslant i_1,\cdots,i_r\leqslant n$$

is a base of $\Lambda^r(V^*)$ with $r\leqslant n$. (Note that, the exterior form $e_{i_1}^*\wedge\cdots\wedge e_{i_r}^*$ is not zero, only for all $i_1,\cdots,i_r$ are different each other; so that, it is 0 for $r>n$.)

(3) $\Lambda^r(V^*)=\{0\}$, $\forall r>n$;

(4) $\forall r\leqslant n$, in base $\{e_{j_1}^*\wedge\cdots\wedge e_{j_r}^*: 1\leqslant j_1<\cdots<j_r\leqslant n\}$ of $\Lambda^r(V^*)$, it holds

$$\eta=r!\sum_{1\leqslant k_1<\cdots<k_r\leqslant n}\eta_{k_1,\cdots,k_r}e_{k_1}^*\wedge\cdots\wedge e_{k_r}^*,\quad \forall\eta\in\Lambda^r(V^*).$$

6.2.3 Grassmann algebra (exterior algebra)

Let V be an n-dimensional linear space on number field $\mathbb{F}$, linear space V^* be dual of V, and $\{e_1,\cdots,e_n\}$, $\{e_1^*,\cdots,e_n^*\}$ be the bases of V,V^*, respectively.

1. Grassmann algebras $\Lambda(V)$ and $\Lambda(V^*)$

Definition 6.2.10 (exterior vector space $\Lambda(V)$) Suppose that **the graded sum** of **exterior vector spaces of r-degree** $\Lambda^r(V)$, $r=0,1,2,\cdots,n$, is denoted by

$$\Lambda(V)=\sum_{r=0}^{n}\Lambda^r(V), \tag{6.2.23}$$

here $\Lambda(V)$ is a 2^n-dimensional linear space on $\mathbb{F}$ (compare with(6.2.17)). For $\xi,\eta\in\Lambda(V)$ with

$$\xi=\sum_{r=0}^{n}\xi_r,\quad \xi_r\in\Lambda^r(V)\quad\text{and}\quad \eta=\sum_{s=0}^{n}\eta_s,\quad \eta_s\in\Lambda^s(V),$$

the exterior product is $\xi\wedge\eta=\sum_{r,s=0}^{n}\xi_r\wedge\eta_s$. Then, $(\Lambda(V),+,\alpha\cdot,\wedge)$ is an algebra, called **a Grassmann algebra on** V, or **an exterior algebra on** V; space $\Lambda(V)$ is said to be **an exterior vector space on** V; an element in $\Lambda(V)$ is said to be an **exterior vector.** The base of exterior vector space $\Lambda(V)$ is

$$\left\{\begin{matrix}\{1\};\ \{e_j:1\leqslant j\leqslant n\};\ \{e_{j_1}\wedge e_{j_2}:1\leqslant j_1<j_2\leqslant n\};\ \cdots;\\ \{e_{j_1}\wedge\cdots\wedge e_{j_r}:1\leqslant j_1<\cdots<j_r\leqslant n\};\ \cdots;\ \{e_1\wedge\cdots\wedge e_n\}\end{matrix}\right\}. \tag{6.2.24}$$

Definition 6.2.11 (exterior form space $\Lambda(V^*)$) Suppose that **the graded sum** of **exterior form spaces of r-degree** $\Lambda^r(V^*)$, $r=0,1,2,\cdots,n$, is denoted by

$$\Lambda(V^*)=\sum_{r=0}^{n}\Lambda^r(V^*) \tag{6.2.25}$$

with $\Lambda(V^*)$, a 2^n-dimensional linear space on $\mathbb{F}$. For $\xi,\eta\in\Lambda(V^*)$ with

$$\xi=\sum_{r=0}^{n}\xi_r,\quad \xi_r\in\Lambda^r(V^*)\quad\text{and}\quad \eta=\sum_{s=0}^{n}\eta_s,\quad \eta_s\in\Lambda^s(V^*),$$

the exterior product is $\xi\wedge\eta=\sum_{r,s=0}^{n}\xi_r\wedge\eta_s$. Then, $(\Lambda(V^*),+,\alpha\cdot,\wedge)$ is an algebra, called **a Grassmann algebra on** V^*, or **an exterior algebra on** V^*; space $\Lambda(V^*)$ is said to be **an exterior form space on** V^*; an element in $\Lambda(V^*)$ is said to be **an exterior form.** The base of exterior form space $\Lambda(V^*)$ is

$$\left\{\begin{matrix}\{1\};\ \{e_j^*:1\leqslant j\leqslant n\};\ \{e_{j_1}^*\wedge e_{j_2}^*:1\leqslant j_1<j_2\leqslant n\};\ \cdots;\\ \{e_{j_1}^*\wedge\cdots\wedge e_{j_r}^*:1\leqslant j_1<\cdots<j_r\leqslant n\};\ \cdots;\ \{e_1^*\wedge\cdots\wedge e_n^*\}\end{matrix}\right\}. \tag{6.2.26}$$

2. Matching of elements in $\Lambda^r(V)$ and $\Lambda^r(V^*)$ with $0\leqslant r\leqslant n$

Definition 6.2.12 (matching of elements in $\Lambda^r(V)$ and $\Lambda^r(V^*)$) Exterior vector space of r-degree $\Lambda^r(V)$ and exterior form space of r-degree $\Lambda^r(V^*)$ are dual each other; **the matching** for elements $v_1\wedge\cdots\wedge v_r\in\Lambda^r(V)$ and $v_1^*\wedge\cdots\wedge v_r^*\in\Lambda^r(V^*)$ is defined by

$$\langle v_1\wedge\cdots\wedge v_r, v_1^*\wedge\cdots\wedge v_r^*\rangle=\det\left[\langle v_\alpha, v_\beta^*\rangle\right]. \tag{6.2.27}$$

Since the bases of $\Lambda^r(V)$ and $\Lambda^r(V^*)$ are $\{e_{j_1}\wedge\cdots\wedge e_{j_r}: 1\leqslant j_1<\cdots<j_r\leqslant n\}$ and $\{e_{k_1}^*\wedge\cdots\wedge e_{k_r}^*: 1\leqslant k_1<\cdots<k_r\leqslant n\}$, respectively, then the matching for them is expressed as

$$\langle e_{j_1}\wedge\cdots\wedge v_{j_r}, e_{k_1}^*\wedge\cdots\wedge e_{k_r}^*\rangle=\det\left[\langle e_{j_\alpha}, e_{j_\beta}^*\rangle\right]$$
$$=\delta_{j_1,\cdots,j_r}^{k_1,\cdots,k_r}=\begin{cases}1, & \{k_1,\cdots,k_r\}=\{j_1,\cdots,j_r\},\\ 0, & \{k_1,\cdots,k_r\}\neq\{j_1,\cdots,j_r\},\end{cases}$$

with $1\leqslant k_1<\cdots<k_r\leqslant n$ and $1\leqslant j_1<\cdots<j_r\leqslant n, 0\leqslant r\leqslant n$.

Note Compare with Definition 6.2.5, two definitions of **matching** are different with a factor $(r!)$ for $\Lambda^r(V), \Lambda^r(V^*)$. That is, the matching of the elements in $\Lambda^r(V), \Lambda^r(V^*)$ is defined by Definition 6.2.12, not by Definition 6.2.5; whereas the matching of the elements in $T^r(V), T^r(V^*)$ is given by Definition 6.2.5, even though we have $\Lambda^r(V)\subset T^r(V)$ and $\Lambda^r(V^*)\subset T^r(V^*)$. However, this is only for convenience of computation.

3. Mappings between exterior form spaces on smooth manifolds V and W

Recall that for function $f: M\to\mathbb{R}$ with $f\in C_p^\infty(M)$ on m-dimensional smooth manifold M, and $p\in U\subset M$, cotangent space T_p^* and tangent space T_p at $p\in M$ are defined; then cotangent vector $(\mathrm{d}f)_p\in T_p^*$ (i.e. the differentiation of p), tangent vector $X_p\in T_p$, directional derivative $Xf=\langle X, (\mathrm{d}f)_p\rangle$ at p, as well as smooth tangent vector field $X\equiv\{X_p: p\in M\}$ (i.e., an smooth operator $X: C^\infty(M)\to C^\infty(M)$) are defined in section 6.1.2. Then, the generalizations of above definitions for two manifolds, m-dimensional smooth manifold M, n-dimensional smooth manifold N, i.e. spaces $T_p(M)$, $T_p^*(M), T_q(N)\equiv T_{f(q)}(N), T_q^*(N)\equiv T_{f(q)}^*(N)$ at points $p\in U\subset M, q=f(p)\in N$, respectively, are given. Moreover, for a smooth mapping $F: M\to N$, the cotangent mapping $F^*: T_q^*(N)\to T_p^*(M)$ (i.e., differentiation) in Definition 6.1.12, tangent mapping $F_*: T_p(M)\to T_q(N)$ in Definition 6.1.3 (i.e., $F_*: T_p(M)\equiv(T_p^*(M))^*\to T_q(N)\equiv T_q^*(N)$) are defined; In fact,

smooth mapping $F: M\to N$ of M and N

↓

cotangent mapping F^* of $F: M\to N$ (differentiation)

$F^*: T_q^*(N) \to T_p^*(M)$ with $F^*((df)_q) = d(f \circ F)$, $\quad (df)_q \in T_q^*(N)$

$\downarrow$

tangent mapping F_* of $F: M \to N$

$F_*: T_p(M) \to T_q(N)$ with $\langle F_* X, \alpha \rangle = \langle X, F^* \alpha \rangle$, $\quad X \in T_p(M), \alpha \in T_q^*(N)$.

Now, for two exterior form spaces of r-degree $\Lambda^r(V^*)$ and $\Lambda^r(W^*)$ related to two smooth manifolds V, W, respectively, for a linear mapping $G: V \to W$, mapping $G^*: \Lambda^r(W^*) \to \Lambda^r(V^*)$ can be induced, satisfying for any $v_1, \cdots, v_r \in V$ and $\varphi \in \Lambda^r(W^*)$, it holds

$$G^* \circ \varphi(v_1, \cdots, v_r) = \varphi(G(v_1), \cdots, G(v_r)). \tag{6.2.28}$$

Then, G^* is linear, clearly; moreover, it has the following important property.

Theorem 6.2.9 If $G: V \to W$ is a smooth mapping, then the linear mapping

$$G^*: \Lambda^r(W^*) \to \Lambda^r(V^*)$$

defined in (6.2.28) can be commutated with exterior product $\wedge$, i.e., for any $\varphi \in \Lambda^r(W^*)$ and $\psi \in \Lambda^r(W^*)$, it holds $G^*(\varphi \wedge \psi) = (G^* \varphi) \wedge (G^* \psi)$.

Proof Take $v_1, \cdots, v_r, v_{r+1}, \cdots, v_{r+s} \in V$, then, we have

$$\begin{aligned}
& G^*(\varphi \wedge \psi)(v_1, \cdots, v_r, v_{r+1}, \cdots, v_{r+s}) \\
&= \varphi \wedge \psi(G(v_1), \cdots, G(v_{r+s})) \\
&= \frac{1}{(r+s)!} \sum_{\sigma \in \mathfrak{T}(r+s)} \operatorname{sgn}\sigma \cdot \varphi(G(v_{\sigma(1)}), \cdots, G(v_{\sigma(r)})) \cdot \psi(G(v_{\sigma(r+1)}), \cdots, G(v_{\sigma(r+s)})) \\
&= \frac{1}{(r+s)!} \sum_{\sigma \in \mathfrak{T}(r+s)} \operatorname{sgn}\sigma \cdot G^* \varphi(G(v_{\sigma(1)}), \cdots, G(v_{\sigma(r)})) \cdot G^* \psi(G(v_{\sigma(r+1)}), \cdots, G(v_{\sigma(r+s)})) \\
&= G^* \varphi \wedge G^* \psi(v_1, \cdots, v_r, v_{r+1}, \cdots, v_{r+s}),
\end{aligned}$$

thus, $G^*(\varphi \wedge \psi) = (G^* \varphi) \wedge (G^* \psi)$. The proof is complete.

Hence, G^* is a homomorphism from Grassmann algebra $(\Lambda(W^*), +, \alpha \cdot, \wedge)$ to $(\Lambda(V^*), +, \alpha \cdot, \wedge)$.

4. Four important theorems for Grassmann algebra

Theorem 6.2.10 Vectors $v_1, \cdots, v_r \in V$ are linearly dependent on each other, if and only if $v_1 \wedge \cdots \wedge v_r = 0$.

Proof **Necessity** If $v_1, \cdots, v_r \in V$ are linearly dependent on each other, without loss of generality, suppose that $v_r = \alpha_1 v_1 + \cdots + \alpha_{r-1} v_{r-1}$. Then $v_1 \wedge \cdots \wedge v_{r-1} \wedge (\alpha_1 v_1 + \cdots + \alpha_{r-1} v_{r-1})$ must be 0.

Sufficiency Conversely, if $v_1, \cdots, v_r$ would be independent each other, then it could be extended as a base of V, say, $\{v_1, \cdots, v_r, u_{r+1}, \cdots, u_n\}$ with $v_1 \wedge \cdots \wedge v_r \wedge u_{r+1} \wedge \cdots \wedge$

$u_n \neq 0$. This implies $v_1 \wedge \cdots \wedge v_r \neq 0$. Then it contradicts $v_1 \wedge \cdots \wedge v_r = 0$. The proof is complete.

Theorem 6.2.11 (Cartan lemma) Let vectors $v_1, \cdots, v_r, w_1, \cdots, w_r \in V$ be with $\sum_{k=1}^{r} v_k \wedge w_k = 0$. If $v_1, \cdots, v_r$ are linearly independent, then w_k can be expressed as a linear combination, $w_k = \sum_{j=1}^{r} a_{kj} v_j$, $1 \leqslant k \leqslant r$, with the symmetric formula $a_{kj} = a_{jk}$.

Proof Since $v_1, \cdots, v_r$ are independent, it can be extended as a base $\{v_1, \cdots, v_r, v_{r+1}, \cdots, v_n\}$ of V, and holds $w_k = \sum_{j=1}^{r} a_{kj} v_j + \sum_{j=r+1}^{n} b_{kj} v_j, k = 1, \cdots, r$. Substitute into $\sum_{k=1}^{r} v_k \wedge w_k = 0$, we obtain that

$$\begin{aligned} 0 &= \sum_{k,j=1}^{r} a_{kj} v_k \wedge v_j + \sum_{k=1}^{r} \sum_{i=r+1}^{n} b_{ki} v_k \wedge v_i \\ &= \sum_{1 \leqslant k < j \leqslant 1} (a_{kj} - a_{jk}) v_k \wedge v_j + \sum_{k=1}^{r} \sum_{i=r+1}^{n} b_{ki} v_k \wedge v_i. \end{aligned}$$

Since $\{v_k \wedge v_j, v_k \wedge u_j, u_k \wedge u_j : 1 \leqslant j < k \leqslant n\}$ is a base of $\Lambda^2(V)$, the above form implies $a_{kj} - a_{jk} = 0$ and $b_{ki} = 0$, this is $w_k = \sum_{j=1}^{r} a_{kj} v_j$, $k = 1, \cdots, r$, and $a_{kj} = a_{jk}$. The proof is complete.

Theorem 6.2.12 If vectors $v_1, \cdots, v_r \in V$ are linearly independent, w is an exterior vector of s-degree in V. Then there exist $\psi_1, \cdots, \psi_r \in \Lambda^{s-1}(V)$, such that $w = \sum_{k=1}^{r} v_k \wedge \psi_k$, if and only if $v_1 \wedge \cdots \wedge v_r \wedge w = 0$. This result is expressed by $w \equiv 0 \mod (v_1, \cdots, v_r)$.

Theorem 6.2.13 If vectors $v_j, w_j, v'_j, w'_j, 1 \leqslant j \leqslant k$, are in space V; the vector system $\{v_j, w_j, 1 \leqslant j \leqslant k\}$ is a linearly independent system, and

$$\sum_{j=1}^{k} v_j \wedge w_j = \sum_{j=1}^{k} v'_j \wedge w'_j, \tag{6.2.29}$$

then $v'_j, w'_j \in V$ are the combinations of $v_1, \cdots, v_k, w_1, \cdots, w_k$, and they are linearly independent.

Proof By (6.2.29), evaluating $J \equiv \underbrace{(\sum_{j=1}^{k} v_j \wedge w_j) \wedge \cdots \wedge (\sum_{j=1}^{k} v_j \wedge w_j)}_{k}$, we get

$$J = k!(v_1 \wedge w_1 \wedge \cdots \wedge v_k \wedge w_k) = k!(v'_1 \wedge w'_1 \wedge \cdots \wedge v'_k \wedge w'_k); \tag{6.2.30}$$

and $k!(v_1 \wedge w_1 \wedge \cdots \wedge v_k \wedge w_k) \neq 0$ since $\{v_j, w_j, 1 \leqslant j \leqslant k\}$ is independent. This implies $\{v_j', w_j', 1 \leqslant j \leqslant k\}$ is independent by Theorem 6.2.10, so that the conclusion of Theorem holds.

Note The exterior product and determinate have very closed relation. Since by (6.2.21), we have:

$$e_{j_1} \wedge \cdots \wedge e_{j_r}(v_1^*, \cdots, v_r^*) = \langle e_{j_1} \wedge \cdots \wedge e_{j_r}, v_{\sigma(1)}^* \wedge \cdots \wedge v_{\sigma(r)}^* \rangle$$

$$= \frac{1}{r!} \sum_{\sigma \in \mathfrak{T}(r)} \mathrm{sgn}\sigma \cdot \langle e_{j_1}, v_{\sigma(1)}^* \rangle \cdots \langle e_{j_r}, v_{\sigma(r)}^* \rangle$$

$$= \frac{1}{r!} \det(\langle e_{j_\alpha}, v_{\sigma(j_\beta)}^* \rangle).$$

Moreover, if $v_1, \cdots, v_k \in V$, and $w_1, \cdots, w_k \in V$ are linear combinations of $\{v_1, \cdots, v_k\}$, i.e., $w_\alpha = \sum_{\beta=1}^{k} t_\alpha^\beta v_\beta$, $1 \leqslant \alpha \leqslant k$. Then,

$$w_1 \wedge \cdots \wedge w_k = \det[t_\alpha^\beta] v_1 \wedge \cdots \wedge v_k. \tag{6.2.31}$$

Thus, exterior vectors $w_1 \wedge \cdots \wedge w_k$ and $v_1 \wedge \cdots \wedge v_k$ are different only with a scale factor by the determinate.

Grassmann manifold $G(k,n)$ In an n-dimensional vector space V, if L^k is a k-dimensional linear subspace, and $G(k,n)$ is the set of all these L^k with $k \leqslant n$. Then, on $G(k,n)$, there exists a natural differential structure, such that $G(k,n)$ becomes a $k(n-k)$-dimensional differential manifold, called a Grassmann manifold.

Grassmann manifold is a very important concept, and has very practical applications.

6.3 Exterior Differentiation of Exterior Differential Forms

6.3.1 Tensor bundles and vector bundles

1. Tensor space of (r,s)-type $T_s^r(p)$

Tensor space of (r,s)-type $V_s^r = \underbrace{V \otimes \cdots \otimes V}_{r} \otimes \underbrace{V^* \otimes \cdots \otimes V^*}_{s}$ has been defined in section 6.2. We now use those concepts there in tensor theory to smooth manifolds in this section.

Given an m-dimensional smooth manifold $M \equiv (M, \mathbb{A})$, take $p \in M$, and take V, V^* as tangent space T_p, cotangent space T_p^* of p, respectively: $V = T_p$, $V^* = T_p^*$. Then, we have a vector space of (r,s)-type $T_s^r(p) = \underbrace{T_p \otimes \cdots \otimes T_p}_{r} \otimes \underbrace{T_p^* \otimes \cdots \otimes T_p^*}_{s}$ at each point

$p\in M$, it is m^{r+s}-dimensional linear space. The following spaces are used here.

V: m-dimensional real linear space on $\mathbb{R}$ with a base $\{e_1,\cdots,e_m\}$, such that

$$y=(y_1,\cdots,y_m)\in V\leftrightarrow\mathbb{R}^m.$$

$GL(V)$: linear self-isomorphic group on V, such that

$$GL(V)\leftrightarrow GL(m;\mathbb{R})\leftrightarrow\mathfrak{M}_{m\times m}=\{A=(a_{jk})_{m\times m}:\det[a_{jk}]\neq 0\},$$

and $A\in GL(m;\mathbb{R})$ actsing on $y\in V$ is denoted by

$$y\cdot A=(y_1,\cdots,y_m)\cdot[a_{jk}],\quad \det[a_{jk}]\neq 0.$$

V_s^r: tensor space of (r,s)-type $V_s^r\equiv\underbrace{V\otimes\cdots\otimes V}_{r}\otimes\underbrace{V^*\otimes\cdots\otimes V^*}_{s}$ on V, with base

$$e_{j_1}\otimes\cdots\otimes e_{j_r}\otimes e_{k_1}^*\otimes\cdots\otimes e_{k_s}^*,\quad 1\leqslant j_\alpha,k_\beta\leqslant m.$$

T_p^*: cotangent space of m-dimensional smooth manifold M at point $p\in M$, with local coordinate system $\{(U,u_j)\}$, natural base $\{(\mathrm{d}u_k)_p,1\leqslant k\leqslant m\}\subset T_p^*$ of cotangent space T_p^*, and dual base $\left\{\left.\frac{\partial}{\partial u_j}\right|_p,1\leqslant j\leqslant m\right\}\subset T_p$ of tangent space T_p.

$T_s^r(p)$: tensor space of (r,s)-type $T_s^r(p)=\underbrace{T_p\otimes\cdots\otimes T_p}_{r}\otimes\underbrace{T_p^*\otimes\cdots\otimes T_p^*}_{s}$ at $p\in M$ with base

$$\left\{\left(\frac{\partial}{\partial u_{j_1}}\right)_p\otimes\cdots\otimes\left(\frac{\partial}{\partial u_{j_r}}\right)_p\otimes(\mathrm{d}u_{k_1})_p\otimes\cdots\otimes(\mathrm{d}u_{k_s})_p,1\leqslant j_\alpha,k_\beta\leqslant m,1\leqslant\alpha\leqslant r,1\leqslant\beta\leqslant s\right\}.$$

2. Tensor bundle of (r,s)-type $T_s^r\equiv T_s^r(M)$

Let $T_s^r\equiv T_s^r(M)=\bigcup_{p\in M}T_s^r(p)$ with tensor space of (r,s)-type $T_s^r(p)$, here notation "$\bigcup_{p\in M}$" is called **an isolate union.** We will endow topological structure, such that T_s^r is a T_2-type topological space; moreover, endow C^∞-differential manifold structure, such that T_s^r is a smooth manifold. Then, we show that T_s^r is differentiation homeomorphic with product mani-fold $M\times\mathbb{R}^\mu$; this T_s^r is called **a tensor bundle of (r,s)-type on** M.

1) Topological structure on T_s^r Let M be m-dimensional smooth manifold on real field $\mathbb{R}$, and $\{(U,u_j)\}$ be a local coordinate system. Take m-dimensional real linear space V, say, $\mathbb{R}^m$, and take $\{e_1,\cdots,e_m\}$ a base of V.

Define a mapping φ_U with

$$\varphi_U:U\times V_s^r\to\bigcup_{p\in U}T_s^r(p),\tag{6.3.1}$$

satisfies"$\forall\ (p,y)\in U\times V_s^r\Rightarrow\varphi_U(p,y)\in T_s^r(p)$", such that the expression

$$\varphi_U(p,y)=z^{j_1,\cdots,j_r}_{k_1,\cdots,k_s}\left(\left(\frac{\partial}{\partial u_{j_1}}\right)_p\otimes\cdots\otimes\left(\frac{\partial}{\partial u_{j_r}}\right)_p\otimes(\mathrm{d}u_{k_1})_p\otimes\cdots\otimes(\mathrm{d}u_{k_s})_p\right)$$

of point $\varphi_U(p,y)\in T^r_s(p)$ with respect to the base

$$\left\{\left(\frac{\partial}{\partial u_{j_1}}\right)_p\otimes\cdots\otimes\left(\frac{\partial}{\partial u_{j_r}}\right)_p\otimes(\mathrm{d}u_{k_1})_p\otimes\cdots\otimes(\mathrm{d}u_{k_s})_p, 1\leqslant j_\alpha,k_\beta\leqslant m\right\}$$

of $T^r_s(p)$ is equal to the expression $y=y^{j_1,\cdots,j_r}_{k_1,\cdots,k_s}(e_{j_1}\otimes\cdots\otimes e_{j_r}\otimes e^*_{k_1}\otimes\cdots\otimes e^*_{k_s})$ of $y\in V^r_s$ with respect to the base $e_{j_1}\otimes\cdots\otimes e_{j_r}\otimes e^*_{k_1}\otimes\cdots\otimes e^*_{k_s}$, $1\leqslant j_\alpha,k_\beta\leqslant m$, i. e.

$$z^{j_1,\cdots,j_r}_{k_1,\cdots,k_s}=y^{j_1,\cdots,j_r}_{k_1,\cdots,k_s}.$$

So that, $(p,y)\xleftrightarrow{\text{one to one}}\varphi_U(p,y)$.

Take an open covering $\{U,W,\cdots\}$ of M, such that each local coordinate system defines a mapping as in (6.3.1), then we get

$$\{\varphi_U,\varphi_W,\cdots\}. \tag{6.3.2}$$

Take the set of all images of open subsets as $U\times V^r_s$ under mapping φ_U, then this set can be taken as a topological base of T^r_s, since it satisfies the conditions of topological base (see Chapter. 3, 3.2.2). So far, we endow a topology, such that T^r_s is a T_2-type (Hausdorff) topological space.

2) C^∞-differential manifold structure on T^r_s For a fixed point $p\in U$, we define a mapping $\varphi_{U,p}: V^r_s\to T^r_s(p)$, by φ_U in (6.3.1), satisfying

$$\varphi_{U,p}(y)=\varphi_U(p,y),\quad y\in V^r_s. \tag{6.3.3}$$

Clearly, $\varphi_{U,p}$ is a linear isomorphism from V^r_s to $T^r_s(p)$, i. e., it is linear, bisingle, and bi-continuous.

For the open covering $\{U,W,\cdots\}$ taken in 1), if $U\cap W\neq\varnothing$ with $p\in U\cap W$, then

$$V^r_s\underset{\varphi_{U,p}}{\rightarrow}T^r_s(p)\underset{\varphi^{-1}_{W,p}}{\rightarrow}V^r_s.$$

We define a mapping

$$g_{UW}(p)=\varphi^{-1}_{W,p}\circ\varphi_{U,p}: V^r_s\to V^r_s, \tag{6.3.4}$$

then, g_{UW} is a self-isomorphism from V^r_s to V^r_s, and thus $g_{UW}(p)\in GL(V^r_s)$ satisfies

$$y,y'\in V^r_s\Rightarrow\varphi_U(p,y)=\varphi_W(p,y');$$

also, it holds

$$y,y'\in V^r_s\Rightarrow\varphi_U(p,y)=\varphi_W(p,y')\Leftrightarrow\varphi_{U,p}(y)=\varphi_{W,p}(y')$$
$$\Leftrightarrow y'=\varphi^{-1}_{W,p}\circ\varphi_{U,p}(y)\Leftrightarrow y'=y\cdot g_{UW}(p),$$

the last step is because $g_{UW}(p)\in GL(V^r_s)$, where $y\equiv y_U\in U$, $y'\equiv y_W\in W$.

On the other hand, for the mapping g_{UW}, it holds

$$g_{UW}: U \cap W \to GL(V_s^r). \tag{6.3.5}$$

In fact, we define a new mapping $\varphi_U(p, y) \equiv \varphi_{U,p}(y)$ by $\varphi_U: U \times V_s^r \to \bigcup_{p \in U} T_s^r(p)$, such that $\varphi_{U,p}: V_s^r \to T_s^r(p)$, thus, $g_{UW}(p) = \varphi_{W,p}^{-1} \circ \varphi_{U,p}: V_s^r \to V_s^r$ (see Fig. 6.3.1 and Fig. 6.3.2).

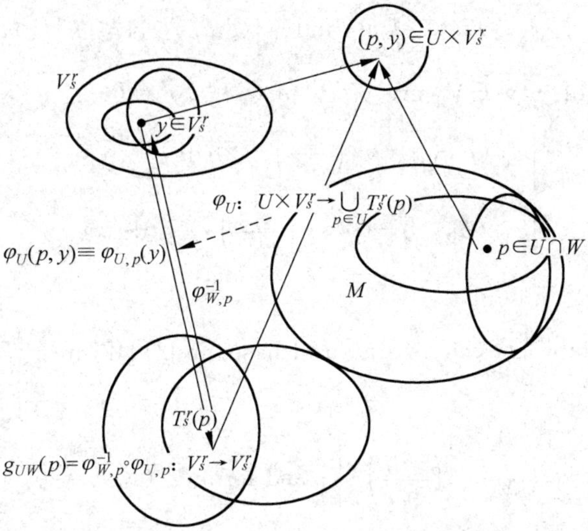

Fig. 6.3.1 Mapping $g_{UW}(p)$

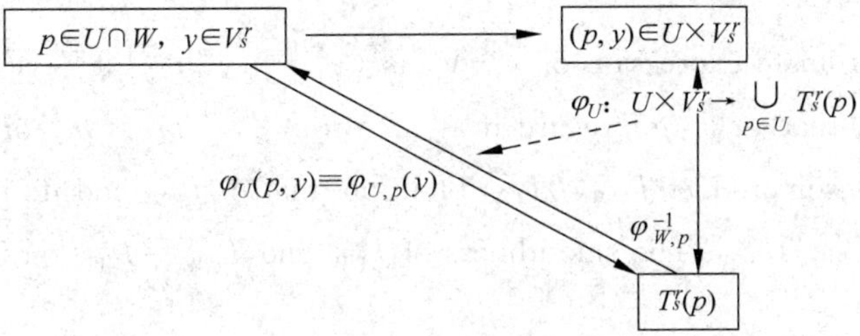

Fig. 6.3.2 C^∞-differentiation manifold structure on T_s^r

Hence, we have

$$p \in U \cap W \neq \varnothing \Rightarrow y, y' \in V_s^r, (p, y), (p, y') \in U \times V_s^r$$
$$\Rightarrow \varphi_U(p, y) = \varphi_W(p, y') \Leftrightarrow y' = y \cdot g_{UW}(p). \tag{6.3.6}$$

Next, we need to prove:

(1) g_{UW} **is a smooth mapping** the proof of the smoothness of g_{UW} is more complex, we only prove for $r=1, s=1$.

Let $\{(U, u_j)\}$, $\{(W, w_j)\}$ be a local coordinate systems of M, and $\{u_1, \cdots, u_m\}$, $\{w_1, \cdots, w_m\}$ be the corresponding natural bases, respectively; the tangent space T_p and cotangent space T_p^* have bases with respect to $\{u_1,\cdots,u_m\}$

$$\left\{\left.\frac{\partial}{\partial u_j}\right|_p, 1\leqslant j\leqslant m\right\}\subset T_p \quad \text{and} \quad \{(\mathrm{d}u_k)_p, 1\leqslant k\leqslant m\}\subset T_p^*;$$

T_p and T_p^* have bases with respect to $\{w_1,\cdots,w_m\}$

$$\left\{\left.\frac{\partial}{\partial w_j}\right|_p, 1\leqslant j\leqslant m\right\}\subset T_p \quad \text{and} \quad \{(\mathrm{d}w_k)_p, 1\leqslant k\leqslant m\}\subset T_p^*.$$

If the expresses of $y, y'\in V_1^1$ are $y=y_k^j e_j\otimes e_k^*$, $y'=(y')_k^j e_j\otimes e_k^*$; then

$$\begin{cases}\varphi_U(p,y)=y_k^j\left(\dfrac{\partial}{\partial u_j}\right)_p\otimes(\mathrm{d}u_k)_p\\ \varphi_W(p,y')=(y')_k^j\left(\dfrac{\partial}{\partial w_j}\right)_p\otimes(\mathrm{d}w_k)_p\end{cases}.$$

So that, the relationship between two natural bases in $U\cap W$ are: $\begin{cases}\mathrm{d}u_k=\dfrac{\partial u_k}{\partial w_k}\mathrm{d}w_k,\\ \dfrac{\partial}{\partial u_j}=\dfrac{\partial w_j}{\partial u_j}\dfrac{\partial}{\partial w_j},\end{cases}$ $1\leqslant j$, $k\leqslant m$ with Jacobi matrix $J_{UW}=\left[\left(\dfrac{\partial w_j}{\partial u_k}\right)_p\right]$; and formula $(y')_{k'}^{j'}=y_k^j\left[\left(\dfrac{\partial w_{j'}}{\partial u_j}\right)_p\right]\left[\left(\dfrac{\partial u_k}{\partial w_{k'}}\right)_p\right]$ holds, i. e.

$$(y\cdot g_{UW}(p))_{k'}^{j'}=y_k^j\left(\frac{\partial w_{j'}}{\partial u_j}\right)_p\left(\frac{\partial u_k}{\partial w_{k'}}\right)_p. \tag{6.3.7}$$

In fact, the coordinate expression of $y\in V_1^1$ is $(y_1^1,\cdots,y_1^m;\ y_2^1,\cdots,y_2^m;\ \cdots;\ y_m^1,\cdots,y_m^m)$, and (6.3.6) shows that $g_{UW}(p)$ regarded as an $(m^{r+s})^2=m^2\times m^2$-order non-degenerated matrix, is just the tensor product $J_{UW}\otimes J_{UW}^{-1}$ of Jacobi matrix J_{UW} and its inverse matrix J_{UW}^{-1}, denoted by $g_{UW}=J_{UW}\otimes J_{UW}^{-1}$. The smoothness of J_{UW} and $J_{UW}^{-1}=J_{WU}$ on $U\cap W$ guarantee the smoothness of g_{UW}.

(2) **T_s^r is a smooth manifold**

Coordinate open covering system $\{\varphi_U(U\times V_s^r), \varphi_W(W\times V_s^r),\cdots\}$ of T_s^r is C^∞-compatible, thus, each point (p,y) in $\varphi_U(U\times V_s^r)$ has coordinate $(u_j(p), y_{k_1,\cdots,k_s}^{j_1,\cdots,j_r})$, where u_j is a local coordinate in coordinate field U of M, and $y_{k_1,\cdots,k_s}^{j_1,\cdots,j_r}$ is the component with respect to base $e_{j_1}\otimes\cdots\otimes e_{j_r}\otimes e_{k_1}^*\otimes\cdots\otimes e_{k_s}^*$, $1\leqslant j_\alpha, k_\beta\leqslant m$.

If $U\cap W\neq\varnothing$, for $g_{UW}=J_{UW}\otimes J_{UW}^{-1}$, we see that mapping

$$g_{UW}=J_{UW}\otimes J_{UW}^{-1}: U\cap W\to GL(V_s^r)$$

is smooth. The relation (6.3.6) shows that coordinate open covering $\{\varphi_U(U\times V_s^r), \varphi_W(W\times V_s^r),\cdots\}$ of T_s^r is C^∞-compatible, then, T_s^r is a smooth manifold, and $\mathbb{A}=$

$\{(U,\varphi_U)\}\equiv\{\varphi_U(U\times V_s^r),h\circ g_{UW}\}$ is its C^∞-differential manifold structure, where g_{UW} determined by (6.3.4) or (6.3.5), and $h: GL(V_s^r)\to\mathbb{R}^\mu$ with $\mu=(m^{r+s})^2$ is a topological isomorphism from $GL(V_s^r)$ to $\mathbb{R}^\mu$, with $\dim GL(V_s^r)=(m^{r+s})^2$.

Definition 6.3.1 (tensor bundle of (r,s)-type) Set $T_s^r\equiv T_s^r(M)=\bigcup_{p\in M}T_s^r(p)$ with C^∞-differential manifold structure $\mathbb{A}=\{(U,\varphi_U)\}\equiv\{\varphi_U(U\times V_s^r),h\circ g_{UW}\}$ is said to be **a tensor bundle of (r,s)-type on manifold M**, or simply, **a tensor bundle of (r,s)-type.**

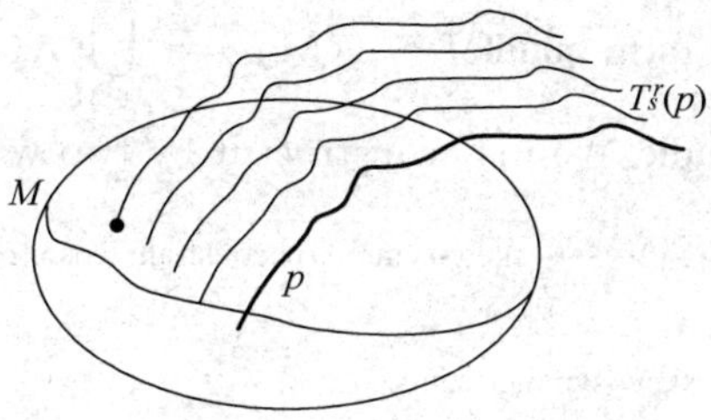

Fig. 6.3.3 Tensor bundle

A natural project mapping $\pi: T_s^r\to M$, from tensor bundle of (r,s)-type T_s^r to smooth manifold M, maps all elements on tensor space of (r,s)-type $T_s^r(p)$ at $p\in M$ to point p itself, and it is a smooth surjective mapping, then $\pi: T_s^r\to M$ is said to be **a bundle project mapping**, or simply, **a bundle project**; here tensor space of (r,s)-type $T_s^r(p)=\underbrace{T_p\otimes\cdots\otimes T_p}_{r}\otimes\underbrace{T_p^*\otimes\cdots\otimes T_p^*}_{s}$ is said to be **a fiber of (r,s)-type** at $p\in M$ on manifold M, or simply, **a fiber at** $p\in M$.

Note 1 Regard a smooth manifold M and a tensor bundle of (r,s)-type T_s^r on M as a whole body, then they look like a head with hairs of human being, that is, M is a head, bundle is its hair.

Note 2 We consider two smooth manifolds M, N, and define "vector bundle of (r,s)-type" as generalized concept of "tensor bundle of (r,s)-type", moreover, combine the concept of "connection", it will forms a fundamental theory for the gauge field and plays essential role in Riemann manifold theory.

3. Tangent bundle and cotangent bundle

Definition 6.3.2 In tensor bundle of (r,s)-type $T_s^r\equiv T_s^r(M)=\bigcup_{p\in M}T_s^r(p)$, take $r=1,s=0$, then a tensor bundle of $(1,0)$-type on M is said to be **a tangent bundle**, denoted by $T(M)\equiv T_0^1(M)=\bigcup_{p\in M}T_0^1(p)=\bigcup_{p\in M}T_p$; take $r=0,s=1$, then a tensor bundle of $(0,1)$-type on M is said to be **a cotangent bundle**, denoted by $T^*(M) \equiv T_1^0(M) = \bigcup_{p \in M} T_1^0(p) = \bigcup_{p \in M} T_p^*$.

4. Exterior vector bundle and exterior form bundle

Take $V = T_p$, $V^* = T_p^*$, we define and construct **the exterior vector bundle** and **exterior form bundle** on smooth manifold M. The definition and process of constructing for exterior form bundle are as follows, for instance.

Definition 6.3.3 (exterior form bundle) $\Lambda(M^*) = \bigcup_{p \in M} \Lambda(T_p^*) = \bigcup_{p \in M} \left(\sum_{r=0}^{m} \Lambda^r(T_p^*)\right)$ is said to be **the exterior form bundle.** We may construct it by two ways:

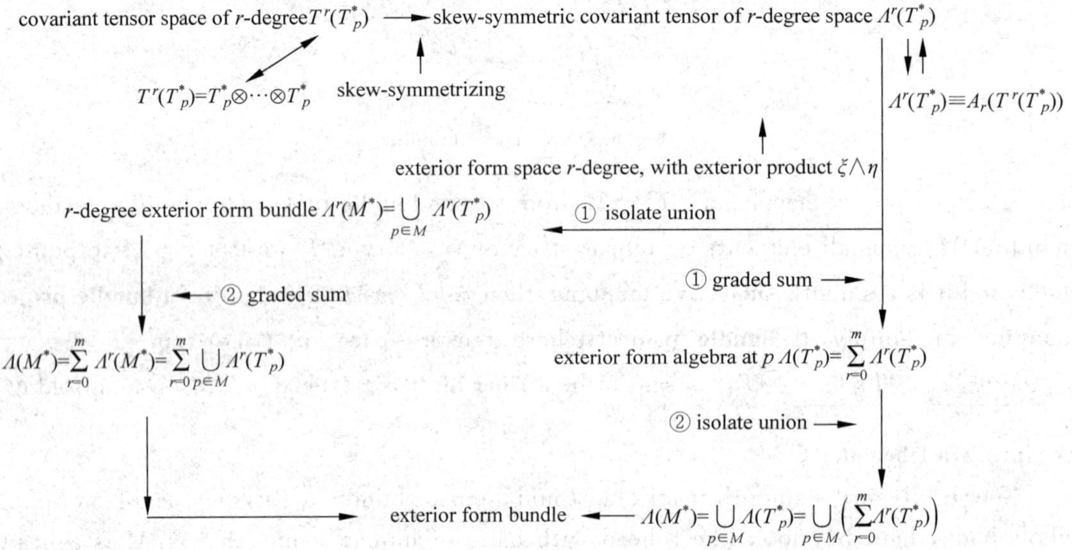

The above processes show that we have

$$\Lambda(M^*) = \sum_{r=0}^{m} \Lambda^r(M^*) = \sum_{r=0}^{m} \bigcup_{p \in M} \Lambda^r(T_p^*) \text{ and } \Lambda(M^*) = \bigcup_{p \in M} \Lambda(T_p^*) = \bigcup_{p \in M} \left(\sum_{r=0}^{m} \Lambda^r(T_p^*)\right),$$

remember: **"bundle"** is to construct **"an isolate union"** on the manifold, and **"algebra"** is to construct **"a graded sum"** on the manifold, respectively.

The construction of exterior vector bundle $\Lambda(M)$ on M is similar to that of exterior form bundle $\Lambda(M^*)$ on M, left to readers.

5. Smooth intersecting surfaces

1) Smooth intersecting surface of a tensor bundle T_s^r

Definition 6.3.4 (smooth intersecting surface of a tensor bundle of (r,s)-type) Let

T_s^r be a tensor bundle of (r,s)-type, $T_s^r \equiv T_s^r(M) = \bigcup_{p\in M} T_s^r(p)$, and $g: M \to T_s^r$ be a smooth mapping on to M. If a compound mapping $\pi \circ g$ of $g: M \to T_s^r$ and a bundle project $\pi: T_s^r \to M$ is the identity mapping, i. e. $\pi \circ g = I: M \to M$, then $g(M)$ is said to be **a smooth intersecting surface of tensor bundle of (r,s)-type T_s^r**, or simply, **a smooth intersecting surface of T_s^r**; or, mapping $g(M)$ is said to be **a smooth tensor field of (r,s)-type on M**, or simply, **smooth tensor field of (r,s)-type.**

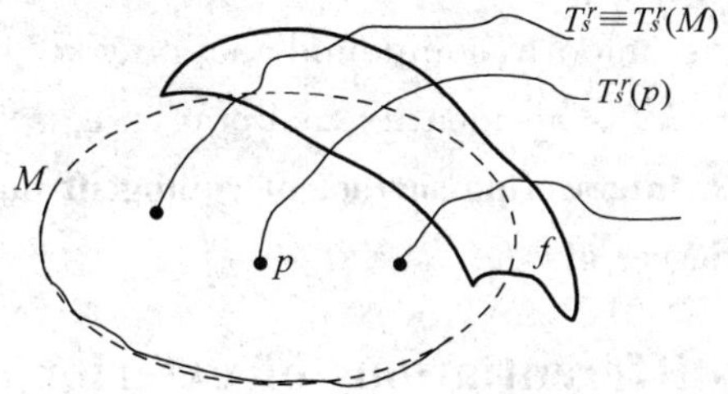

Fig. 6. 3. 4 Smooth section on tensor bundle

In Fig. 6. 3. 4, we show that: smooth manifold M; fiber $T_s^r(p)$ at point $p \in M$; smooth intersecting surface $g(M)$; tensor bundle of (r,s)-type $T_s^r = \bigcup_{p\in M} T_s^r(p)$. As we see, the image point of g on $T_s^r(p)$ is $g(p)$; the image set $g(M)$ is the smooth intersecting surface, $g(M) \subset T_s^r$.

2) Smooth intersecting surface of a tangent bundle $T(M)$

Definition 6. 3. 5 (smooth intersecting surface of a tangent bundle) For tangent bundle $T(M) \equiv T_0^1(M) = \bigcup_{p\in M} T_p$ constructed by tangent space T_p at point $p \in M$, if mapping $g: M \to T(M)$ is smooth, compound mapping $\pi \circ g$ of $g: M \to T(M)$ and bundle project $\pi: T(M) \to M$ is an identity mapping, i. e. $\pi \circ g = I: M \to M$, then image $g(M)$ is said to be **a smooth intersecting surface of tangent bundle $T(M)$**, or, is said to be **a tangent vector field on M.**

Recall the definition of "tangent vector field" in the subsection 11 of section 6. 1. 2, and compare two definitions. Now, we see that **a smooth intersecting surface of tangent bundle $T(M)$ is, in fact, a tangent vector field on M.**

Definition 6. 3. 6 (smooth tangent vector field) Let $X = \{X_p: p \in M\}$ be a tangent vector field on smooth manifold M. If for any $f \in C^\infty(M)$, it holds $Xf \in C^\infty(M)$, then X is said to be **a smooth tangent vector field** on M.

In this opinion, smooth tangent vector field $X = \{X_p: p \in M\}$ is an operator from space $C^\infty(M)$ to it-self. And we have: $\forall f, g \in C^\infty(M), \alpha, \beta \in \mathbb{R}$,

(1) $X(\alpha f+\beta g)=\alpha(Xf)+\beta(Xg)$;

(2) $X(f\cdot g)=f\cdot(Xg)+g\cdot(Xf)$.

Recall Definition 6.1.14 of "smooth tangent vector field" in the subsection 11 of section 6.1.2.

3) Smooth intersecting surface of a cotangent bundle $T^*(M)$

Definition 6.3.7 (smooth intersecting surface of a cotangent bundle) For cotangent bundle $T^*(M)\equiv T_1^0(M)=\bigcup_{p\in M} T_p^*$ constructed by cotangent space T_p^* at point $p\in M$, if mapping $g: M\to T^*(M)$ is smooth, compound mapping $\pi\circ g$ of $g: M\to T^*(M)$ and bundle project $\pi: T^*(M)\to M$ is an identity mapping, i. e. $\pi\circ g=I: M\to M$, then image $g(M)$ is said to be **a smooth intersecting surface of cotangent bundle $T^*(M)$**, or, is said to be **a differential form of 1-degree of M**.

6.3.2 Exterior differentiations of exterior differential form

We start from an m-dimensional smooth manifold $M\equiv(M,\mathbb{A})$.

1. Exterior differential form space $A(M)$ on M

1) Exterior differential form space of r-degree $A^r(M)$

Definition 6.3.3 gives **exterior form bundle of** r-degree $\Lambda^r(M^*)=\bigcup_{p\in M}\Lambda^r(T_p^*)$ on M, it is a kind of tensor bundle on M. Denote by $A^r(M)=\Gamma(\Lambda^r(M^*))$ the set of all smooth intersecting surfaces of $\Lambda^r(M^*)$ with certain operations, then $A^r(M)$ is called **the exterior differential form space of r-degree on M**; an element in $A^r(M)$ is called **an exterior differential form of r-degree on M** (an exterior differential form of r-degree on M, in fact, is a smooth skew-symmetric covariant tensor field of r-degree).

2) Exterior differential form space $A(M)$

Definition 6.3.3 gives exterior form bundle $\Lambda(M^*)=\bigcup_{p\in M}\Lambda(T_p^*)=\bigcup_{p\in M}(\sum_{r=0}^{m}\Lambda^r(T_p^*))$ on M, it is a kind of tensor bundle on M. Denote by $A(M)=\Gamma(\Lambda(M^*))$ the set of all smooth intersecting surfaces of $\Lambda(M^*)$ with certain operations, then $A(M)$ is called **the exterior differential form space on M**. An element in $A(M)$ is called **an exterior differential form on M** (an exterior differential form on M, in fact, is a smooth skew-symmetric covariant tensor field).

Note 1 $\Lambda^r(M^*)$ is the exterior form bundle of r-degree, without graded sum, so that the notation of smooth intersecting surface set $A^r(M)=\Gamma(\Lambda^r(M^*))$ contains index r.

Note 2 $\Lambda(M^*)$ is the exterior form bundle, with graded sum, so that the notation of smooth intersecting surface set $A(M)=\Gamma(\Lambda(M^*))$ does not contain index r.

3) Construction of exterior differential form space $A(M)$ on M

We have shown the construction of exterior differential form space $A(M)$ on M in 2). We see that

$$A(M)=\sum_{r=0}^{m}A^r(M)=\sum_{r=0}^{m}\Gamma(\Lambda^r(M^*))=\sum_{r=0}^{m}\Gamma(\bigcup_{p\in M}\Lambda^r(T_p^*)).$$

However, by the following way, we have

$$A(M)=\Gamma(\Lambda(M^*))=\Gamma(\bigcup_{p\in M}\Lambda(T_p^*))=\Gamma(\bigcup_{p\in M}(\sum_{r=0}^{m}\Lambda^r(T_p^*)))$$

$$=\Gamma(\sum_{r=0}^{m}(\bigcup_{p\in M}\Lambda^r(T_p^*)))=\sum_{r=0}^{m}\Gamma(\bigcup_{p\in M}\Lambda^r(T_p^*)).$$

That is,

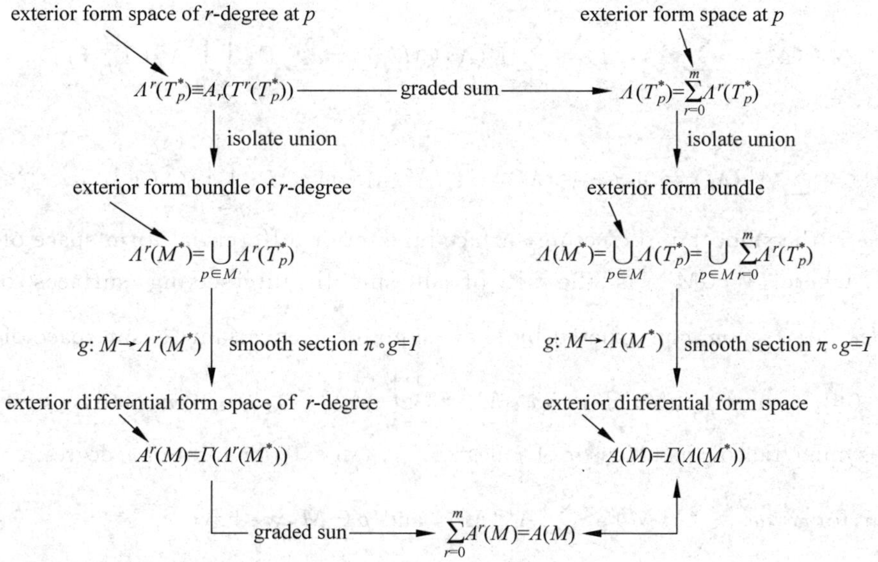

We need the bases of two spaces:

(1) **base of exterior form space of r-degree** $\Lambda^r(T_p^*)\equiv A_r(T^r(T_p^*))$ **at** p:

$$(du_{k_1})_p\wedge\cdots\wedge(du_{k_r})_p,\quad 1\leqslant k_\beta\leqslant m, 1\leqslant\beta\leqslant r;$$

exterior algebra $(\sum_{r=0}^{m}\Lambda^r(T_p^*), +, \alpha\cdot, \wedge)$ is obtained by taking graded sum of $\Lambda^r(T_p^*)$;

(2) **base of exterior vector space of r-degree** $\Lambda^r(T_p)\equiv A_r(T^r(T_p))$ **at** p:

$$\left(\frac{\partial}{\partial u_{j_1}}\right)_p \wedge \cdots \wedge \left(\frac{\partial}{\partial u_{j_r}}\right)_p, \quad 1 \leqslant j_\alpha \leqslant m, 1 \leqslant \alpha \leqslant r;$$

exterior algebra $(\sum_{r=0}^{m} \Lambda^r(T_p), +, \alpha \cdot, \wedge)$ is obtained by taking graded sum of $\Lambda^r(T_p)$.

4) Expression of exterior differential form in an exterior differential form space

Each exterior differential form $\omega \in A(M) = \sum_{r=0}^{m} A^r(M)$ in an exterior differential form space $A(M)$ can be expressed uniquely as

$$\omega = \omega^0 + \omega^1 + \cdots + \omega^m = \sum_{r=0}^{m} \omega^r,$$

where $\omega^r \in A^r(M)$ is the exterior differential form of r-degree, $r \in \{0, 1, \cdots, m\}$.

5) Exterior product operation in an exterior differential form space

We now generalize the operation of exterior product $\wedge$ to exterior differential form space

$$A(M) = \sum_{r=0}^{m} A^r(M) = \sum_{r=0}^{m} \Gamma(\Lambda^r(M^*)) = \sum_{r=0}^{m} \Gamma(\bigcup_{p \in M} \Lambda^r(T_p^*)).$$

In fact,

$$\omega \in A(M) = \sum_{r=0}^{m} A^r(M) \Rightarrow \omega \in A^j(M) = \Gamma(\Lambda^r(M^*)) = \Gamma(\bigcup_{p \in M} \Lambda^j(T_p^*)), \quad 0 \leqslant j \leqslant m.$$

i. e. each ω is the smooth intersecting surface in exterior differential form space of j-degree $A^j(M)$, where $A^j(M)$ is the set of all smooth intersecting surfaces of bundle $\bigcup_{p \in M} \Lambda^j(T_p^*)$, $0 \leqslant j \leqslant m$, constructed by skew-symmetric covariant tensor space of j-degree $\Lambda^j(T_p^*)$, $0 \leqslant j \leqslant m$ at $p \in M$. Thus, $\omega: M \to \bigcup_{p \in M} \Lambda^j(T_p^*)$, $0 \leqslant j \leqslant m$, and $\omega(p) \in \Lambda^j(T_p^*)$ is a skew-symmetric covariant tensor of j-degree, i. e., an exterior form of j-degree.

Then, for $\omega_1, \omega_2 \in A(M) = \sum_{r=0}^{m} A^r(M)$, and $p \in M$, we have

$$\omega_1(p) \in \Lambda^j(T_p^*), \quad \omega_2(p) \in \Lambda^k(T_p^*), \quad 0 \leqslant j, k \leqslant m,$$

exterior product $\omega_1(p) \wedge \omega_2(p)$ of two exterior forms $\omega_1(p), \omega_2(p)$ makes sense.

We define **an exterior product** $\omega_1 \wedge \omega_2$ **of two exterior forms** $\omega_1(p), \omega_2(p)$ by

$$(\omega_1 \wedge \omega_2)(p) \equiv \omega_1(p) \wedge \omega_2(p),$$

and $\omega_1 \wedge \omega_2 \in A(M)$. Thus, $(A(M) = \sum_{r=0}^{m} A^r(M), +, \alpha \cdot, \tau, \wedge)$ becomes **a graded algebra**; $A(M) = \sum_{r=0}^{m} A^r(M)$ is a graded sum of exterior differential form space of r-degree

$A^r(M), 0\leqslant r\leqslant n$, with exterior product $\wedge$, and

$$\wedge : A^r(M)\times A^s(M)\to A^{r+s}(M)\equiv\begin{cases}A^{r+s}(M), & r+s\leqslant m\\ \{0\}, & r+s>m\end{cases}.$$

6) Matching in an exterior vector bundle and an exterior form bundle of r-degree

Exterior vector bundle of r-degree $\Lambda^r(M)=\bigcup_{p\in M}\Lambda^r(T_p)$ and exterior form bundle of r-degree $\Lambda^r(M^*)=\bigcup_{p\in M}\Lambda^r(T_p^*)$ are dual bundles, thus **a matching of two fibers** at point $p\in M$ on manifold M can be induced from the matching of $\Lambda^r(T_p)$ and $\Lambda^r(T_p^*)$.

We only need to define the **matching** of bases

$$\left(\frac{\partial}{\partial u_{j_1}}\right)_p\wedge\cdots\wedge\left(\frac{\partial}{\partial u_{j_r}}\right)_p,\quad 1\leqslant j_\alpha\leqslant m, 1\leqslant\alpha\leqslant r$$

of space $\Lambda^r(T_p)$ and bases

$$(\mathrm{d}u_{k_1})_p\wedge\cdots\wedge(\mathrm{d}u_{k_s})_p,\quad 1\leqslant k_\beta\leqslant m, 1\leqslant\beta\leqslant r$$

of space $\Lambda^r(T_p^*)$ by

$$\left\langle\left(\frac{\partial}{\partial u_{j_1}}\right)_p\wedge\cdots\wedge\left(\frac{\partial}{\partial u_{j_r}}\right)_p, (\mathrm{d}u_{k_1})_p\wedge\cdots\wedge(\mathrm{d}u_{k_r})_p\right\rangle=\delta_{j_1\cdots j_r}^{k_1\cdots k_r};$$

then component expressions $\omega_{k_1\cdots k_r}(\mathrm{d}u_{k_1}\wedge\cdots\wedge\mathrm{d}u_{k_r})$ of $\omega(p)\in\Lambda^r(T_p^*)$ in local coordinate system $\{u_j\}_{1\leqslant j\leqslant m}$ are

$$\omega_{k_1\cdots k_r}=\frac{1}{r!}\left\langle\left(\frac{\partial}{\partial u_{k_1}}\right)_p\wedge\cdots\wedge\left(\frac{\partial}{\partial u_{k_r}}\right)_p,\omega\right\rangle.$$

2. Exterior differentiation operation on exterior differential form space $A(M)$

From graded algebra $\left(A(M)=\sum_{r=0}^m A^r(M), +, \alpha\cdot, \tau, \wedge\right)$ with

$$A(M)=\sum_{r=0}^m A^r(M)=\sum_{r=0}^m\Gamma(\Lambda^r(M^*))=\sum_{r=0}^m\Gamma\left(\bigcup_{p\in M}\Lambda^r(T_p^*)\right),$$

and with C^∞-smooth differential manifold structure, then we have the exterior differential form space $(A(M), +, \alpha\cdot, \mathbb{A}, \wedge)$. Introduce "exterior differentiation", a new operation, on this space, we have $(A(M), +, \alpha\cdot, \mathbb{A}, \wedge, d)$.

1) Definition of exterior differentiation

Definition 6.3.8 (exterior differentiation) Let M be m-dimensional smooth manifold, $A(M)$ be exterior differential form space on M. If there exists a mapping d: $A(M)\to A(M)$, satisfying $\omega\in A(M)\to\mathrm{d}\omega\in A(M)$, and

(1) $\mathrm{d}(A^r(M))\subset A^{r+1}(M)$;

(2) $\omega_1, \omega_2 \in A(M)$ implies $d(\omega_1 + \omega_2) = d\omega_1 + d\omega_2$;

(3) $\omega_1 \in A^r(M)$ (an exterior differential form of r-degree) and $\omega_2 \in A(M)$, implies $d(\omega_1 \wedge \omega_2) = (d\omega_1) \wedge \omega_2 + (-1)^r \omega_1 \wedge d\omega_2$;

(4) $f \in A^0(M) = C^\infty(M)$ (f is a smooth function on M) implies df is the differentiation of f in classical sense (in Newton calculus sense);

(5) $\omega \in A^0(M)$ implies $d(d\omega) = 0$.

Then, the operator $d: A(M) \to A(M)$ is said to be **an exterior differential operator on exterior differential form space** $A(M)$, or simply, **an exterior differential operator**; and $d\omega \in A(M)$ is said to be the **exterior differentiation of** ω.

2) Existence and uniqueness of exterior differentiation

The following is the existence and uniqueness theorem of exterior differentiation on exterior differential form space $(A(M), +, \alpha \cdot, \tau, \wedge)$.

Theorem 6.3.1 Let M be an m-dimensional smooth manifold. Then, there exists unique mapping $d: A(M) \to A(M)$, satisfying the conditions (1)-(5) in definition **6.3.8.**

Proof We define $d: A(M) \to A(M)$, and prove it satisfies (1)-(5) by following steps.

Firstly: operator d satisfying (1)-(5) is a **local operator**, i. e.

$\omega_1, \omega_2 \in A(M), \omega_1|_U = \omega_2|_U, U \in \tau$ implies $d\omega_1|_U = d\omega_2|_U$;

$\omega \in A(M), \omega|_U = \omega|_W, U, W \in \tau$ implies $d(\omega|_U)|_{U\cap W} = d(\omega|_W)|_{U\cap W}$.

These two results mean: d is a local operator.

Secondly: define "an exterior differentiation operator d", and show it is linear. By the following deduction:

$\omega \in A(M)$, then $\exists r\ (0 \leqslant r \leqslant m)$, s. t. $\omega \in A^r(M) \Rightarrow \omega|_U = a \cdot du_{k_1} \wedge \cdots \wedge du_{k_r}, a \in A^0(M)$, is smooth function on $U \in \tau \Rightarrow$ **define**

$$d\omega = d(a \cdot du_{k_1} \wedge \cdots \wedge du_{k_r}) \equiv da \wedge du_{k_1} \wedge \cdots \wedge du_{k_r}, \tag{$*$}$$

with

$$da = \frac{\partial a}{\partial u_j} du_j = \sum_{j=1}^{m} \frac{\partial a}{\partial u_j} du_j = \frac{\partial a}{\partial u_1} du_1 + \cdots + \frac{\partial a}{\partial u_m} du_m, \tag{$**$}$$

and $a = a(u_1, \cdots, u_m) \in C^\infty(M)$.

We show that ($*$) and ($**$) $\Leftrightarrow$ (4): Clearly, $a \in A^0(M) = C^\infty(M)$ and ($**$) implies $da \in A^1(M)$. The base of $A^1(M)$ is $\{du_1, \cdots, du_m\}$, then

$$da = \frac{\partial a}{\partial u_k} du_k = \sum_{k=1}^{m} \frac{\partial a}{\partial u_j} du_j$$

↑ Einstein notation ↑ one-degree differentiation

Thus (4) holds; and vice versa.

By (*) and (**), we see (2) is clear. Moreover, $\forall \alpha \in \mathbb{R}$ implies $\mathrm{d}(\alpha\omega)=\alpha \mathrm{d}\omega$, then the d defined in (*) and (**) is a linear operator.

Till now, the operator d is well-defined and satisfies (2) and (4).

Thirdly: prove (1), (3), (5) by the local coordinate of $\omega \in A(M)$.

For (1) By $\omega \in A^r(M)$, $\omega = a \cdot \mathrm{d}u_{k_1} \wedge \cdots \wedge \mathrm{d}u_{k_r}$ with $a \in A^0(M)=C^\infty(M)$, and by (**), the differentiation of smooth function a with m-variable is $\mathrm{d}a = \frac{\partial a}{\partial u_k}\mathrm{d}u_k = \sum_{k=1}^{m}\frac{\partial a}{\partial u_j}\mathrm{d}u_j$; then, by definition in (*), for $\omega \in A(M)$ with $\omega = a \cdot \mathrm{d}u_{k_1} \wedge \cdots \wedge \mathrm{d}u_{k_r}$, it holds

$$\begin{aligned}
\mathrm{d}\omega &= \mathrm{d}(a \cdot \mathrm{d}u_{k_1} \wedge \cdots \wedge \mathrm{d}u_{k_r}) = \mathrm{d}a \wedge \mathrm{d}u_{k_1} \wedge \cdots \wedge \mathrm{d}u_{k_r} \\
&= \left(\frac{\partial a}{\partial u_1}\mathrm{d}u_1 + \cdots + \frac{\partial a}{\partial u_m}\mathrm{d}u_m\right) \wedge \mathrm{d}u_{k_1} \wedge \cdots \wedge \mathrm{d}u_{k_r} \\
&= \frac{\partial a}{\partial u_1}\mathrm{d}u_1 \wedge (\mathrm{d}u_{k_1} \wedge \cdots \wedge \mathrm{d}u_{k_r}) + \cdots + \frac{\partial a}{\partial u_m}\mathrm{d}u_m \wedge (\mathrm{d}u_{k_1} \wedge \cdots \wedge \mathrm{d}u_{k_r}).
\end{aligned}$$

$\uparrow$

$\mathrm{d}u_1 \wedge (\mathrm{d}u_{k_1} \wedge \cdots \wedge \mathrm{d}u_{k_r})$, (k_r+1)-terms, corresponding to $\mathrm{d}u_{k_1} \wedge \cdots \wedge \mathrm{d}u_{k_r+1}$ so that,

$$\frac{\partial a}{\partial u_j}\mathrm{d}u_j \wedge (\mathrm{d}u_{k_1} \wedge \cdots \wedge \mathrm{d}u_{k_r}) \in A^{r+1}(M), \quad j \neq k_1, \cdots, k_r;$$

other terms are similar. Thus, (1) holds, and $\mathrm{d}(A^r(M)) \subset A^{r+1}(M)$.

For (3) If $\omega_1 = \mathrm{a} \cdot d\mathrm{u}_{\mathrm{k}_1} \wedge \cdots \wedge d\mathrm{u}_{\mathrm{k}_r} \in \mathrm{A}^{\mathrm{r}}(\mathrm{M})$, $\omega_2 = \mathrm{b} \cdot d\mathrm{u}_{\mathrm{j}_1} \wedge \cdots \wedge d\mathrm{u}_{\mathrm{j}_s} \in \mathrm{A}^{\mathrm{s}}(\mathrm{M})$, then

$$\begin{aligned}
\omega_1 \wedge \omega_2 &= (a \cdot \mathrm{d}u_{k_1} \wedge \cdots \wedge \mathrm{d}u_{k_r}) \wedge (b \cdot \mathrm{d}u_{j_1} \wedge \cdots \wedge \mathrm{d}u_{j_s}) \\
&= (a \cdot b)(\mathrm{d}u_{k_1} \wedge \cdots \wedge \mathrm{d}u_{k_r} \wedge \mathrm{d}u_{j_1} \wedge \cdots \wedge \mathrm{d}u_{j_s}),
\end{aligned}$$

here $a \cdot b \in C^\infty(M)$, by (*), and

$$\begin{aligned}
\mathrm{d}(\omega_1 \wedge \omega_2) &= \mathrm{d}\{(a \cdot \mathrm{d}u_{k_1} \wedge \cdots \wedge \mathrm{d}u_{k_r}) \wedge (b \cdot \mathrm{d}u_{j_1} \wedge \cdots \wedge \mathrm{d}u_{j_s})\} \\
&= \mathrm{d}\{(a \cdot b)(\mathrm{d}u_{k_1} \wedge \cdots \wedge \mathrm{d}u_{k_r} \wedge \mathrm{d}u_{j_1} \wedge \cdots \wedge \mathrm{d}u_{j_s})\} \\
&= \{b(\mathrm{d}a) + a(\mathrm{d}b)\}(\mathrm{d}u_{k_1} \wedge \cdots \wedge \mathrm{d}u_{k_r} \wedge \mathrm{d}u_{j_1} \wedge \cdots \wedge \mathrm{d}u_{j_s}) \\
&= (b(\mathrm{d}a) \wedge \mathrm{d}u_{k_1} \wedge \cdots \wedge \mathrm{d}u_{k_r} \wedge \mathrm{d}u_{j_1} \wedge \cdots \wedge \mathrm{d}u_{j_s}) + \\
&\quad (a(\mathrm{d}b) \wedge \mathrm{d}u_{k_1} \wedge \cdots \wedge \mathrm{d}u_{k_r} \wedge \mathrm{d}u_{j_1} \wedge \cdots \wedge \mathrm{d}u_{j_s}) \\
&= ((\mathrm{d}a) \wedge \mathrm{d}u_{k_1} \wedge \cdots \wedge \mathrm{d}u_{k_r}) \wedge (b \wedge \mathrm{d}u_{j_1} \wedge \cdots \wedge \mathrm{d}u_{j_s}) +
\end{aligned}$$

$$(-1)^r (a \wedge du_{k_1} \wedge \cdots \wedge du_{k_r})((db) \wedge du_{j_1} \wedge \cdots \wedge du_{j_s})$$

$$= d\omega_1 \wedge \omega_2 + (-1)^r \omega_1 \wedge d\omega_2,$$

this is (3),

$$\omega_1 \in A^r(M), \omega_2 \in A(M) \Rightarrow d(\omega_1 \wedge \omega_2) = (d\omega_1) \wedge \omega_2 + (-1)^r \omega_1 \wedge d\omega_2.$$

For (5) If $\omega = f \in A^0(M) = C^\infty(M)$, then f is a smooth function on M, and (**) implies that

$$df = \sum_{j=1}^{m} \frac{\partial f}{\partial u_j} du_j \Rightarrow$$

$$d(df) = d\left(\sum_{j=1}^{m} \frac{\partial f}{\partial u_j} du_j\right) \wedge (du_k) = \sum_{k=1}^{m}\sum_{j=1}^{m} \frac{\partial^2 f}{\partial u_j \partial u_k}(du_j \wedge du_k)$$

$$= \sum_{k,j=1}^{m} \frac{1}{2}\left(\frac{\partial^2 f}{\partial u_j \partial u_k} + \frac{\partial^2 f}{\partial u_k \partial u_j}\right) du_j \wedge du_k$$

$$= \frac{1}{2}\left(\sum_{j,k=1}^{m} \frac{\partial^2 f}{\partial u_j \partial u_k} du_j \wedge du_k - \sum_{k,j=1}^{m} \frac{\partial^2 f}{\partial u_k \partial u_j} du_k \wedge du_j\right) \underset{\uparrow}{=} 0,$$

since $f \in C^\infty(M)$ implies $\dfrac{\partial^2 f}{\partial u_j \partial u_k} = \dfrac{\partial^2 f}{\partial u_k \partial u_j}$,

and $\displaystyle\sum_{j,k=1}^{m} \frac{\partial^2 f}{\partial u_j \partial u_k} du_j \wedge du_k = \sum_{k,j=1}^{m} \frac{\partial^2 f}{\partial u_k \partial u_j} du_k \wedge du_j$

thus $d(df)=0$. This implies $d^2\omega = 0$ for $\omega \in A^0(M)$, and (5) holds.

Fourthly: Prove the uniqueness of d

The uniqueness can be obtained, since by $d(\omega|_U)|_{U\cap W} = d(\omega|_{U\cap W})| = d(\omega|_W)|_{U\cap W}$, the definition of differential operator d: $A(M) \to A(M)$ by (*), (**) are coincided on $U \cap W$, then d is well-defined and unique on smooth manifold M.

Finally, for $\omega \in A^r(M)$, $\omega = f \cdot du_1 \wedge \cdots \wedge du_r$, $f \in C^\infty(M)$

$$\Rightarrow d\omega = d(f \cdot du_1 \wedge \cdots \wedge du_r) = df \wedge du_1 \wedge \cdots \wedge du_r \equiv df \wedge (du_1 \wedge \cdots \wedge du_r)$$

$$\Rightarrow d(d\omega) = d\{df \wedge d(du_1 \wedge \cdots \wedge du_r)\}$$

$$= \underset{\substack{\uparrow \\ d(df)=0}}{d(df)} \wedge (du_1 \wedge \cdots \wedge du_r) - (-1)^1 df \underset{\substack{\uparrow \\ d(du_1)\wedge(du_2\wedge\cdots\wedge du_r)=0}}{\wedge} d\{(du_1 \wedge \cdots \wedge du_r)\}$$

the last step is by $d(du_1) = d(1 \cdot du_1) = d(1) \wedge du_1 = 0$. Then, $\forall \omega \in A^r(M)$, it holds $d^2\omega = 0$.

The proof is complete.

The following example is a particular case for the differential operator; It helps us to understand the sense of exterior differentiation.

Example 6.3.1 $M=\mathbb{R}^3$

Let $M=\mathbb{R}^3$ be 3-dimensional Euclidean space $\{u_1,u_2,u_3\}$ with $u_1=x, u_2=y, u_3=z$, be a local coordinate system.

(1) **cotangent space at** p: $T_p^*=\mathfrak{F}_p/\sim=\{[f]^*:[f]\in\mathfrak{F}_p\}$

The natural base $\{(\mathrm{d}u_k)_p, 1\leqslant k\leqslant 3\}=\{(\mathrm{d}u_1)_p,(\mathrm{d}u_2)_p,(\mathrm{d}u_3)_p\}$ with respect to the local coordinate system $\{u_1,u_2,u_3\}=\{x,y,z\}$, i.e. the natural base is

$$\{(\mathrm{d}x)_p,(\mathrm{d}y)_p,(\mathrm{d}z)_p\}\equiv\{\mathrm{d}x,\mathrm{d}y,\mathrm{d}z\};$$

differentiation of f **at** p: $f\in C_p^\infty(\mathbb{R}^3)\Rightarrow(\mathrm{d}f)_p\in T_p^*$ with

$$(\mathrm{d}f)_p=\sum_{k=1}^3\alpha_k(\mathrm{d}u_k)_p=\alpha_1\mathrm{d}x+\alpha_2\mathrm{d}y+\alpha_3\mathrm{d}z,$$

where $\alpha_1=\frac{\partial f}{\partial x}$, $\alpha_2=\frac{\partial f}{\partial y}$, $\alpha_3=\frac{\partial f}{\partial z}\in C_p^\infty(\mathbb{R}^3)$;

(2) **tangent space at** p: $T_p=(T_p^*)^*$

dual base $\left\{\left.\frac{\partial}{\partial u_j}\right|_p, 1\leqslant j\leqslant 3\right\}=\left\{\left.\frac{\partial}{\partial u_1}\right|_p,\left.\frac{\partial}{\partial u_2}\right|_p,\left.\frac{\partial}{\partial u_3}\right|_p\right\}$, i.e.

$$\left\{\left.\frac{\partial}{\partial x}\right|_p,\left.\frac{\partial}{\partial y}\right|_p,\left.\frac{\partial}{\partial z}\right|_p\right\}\equiv\left\{\frac{\partial}{\partial x},\frac{\partial}{\partial y},\frac{\partial}{\partial z}\right\};$$

tangent vector at p: $[\gamma]=X_p\in T_p$ with

$$X_p=\sum_{j=1}^3\xi_j\left.\frac{\partial}{\partial u_j}\right|_p=\xi_1\frac{\partial}{\partial x}+\xi_2\frac{\partial}{\partial y}+\xi_3\frac{\partial}{\partial z},$$

where $\xi_1=\frac{\mathrm{d}(x\circ\gamma)}{\mathrm{d}t}$, $\xi_2=\frac{\mathrm{d}(y\circ\gamma)}{\mathrm{d}t}$, $\xi_3=\frac{\mathrm{d}(z\circ\gamma)}{\mathrm{d}t}$, here $\gamma=\gamma(t)$ is a smooth curve passing p.

(3) **covariant tensor space of** r**-degree at** p: $T^r(T_p^*)\equiv\underbrace{T_p^*\otimes\cdots\otimes T_p^*}_{r}, 0\leqslant r\leqslant 3$

(4) **exterior form space of** r**-degree at** p: $\Lambda^r(T_p^*)\equiv A_r(T^r(T_p^*)), 0\leqslant r\leqslant 3$ with base $\{(\mathrm{d}u_{k_1})_p\wedge\cdots\wedge(\mathrm{d}u_{k_s})_p: 1\leqslant k_\beta\leqslant 3, 1\leqslant\beta\leqslant r\}$:

① $\Lambda^0(T_p^*)$: base is $\{1\}$;

② $\Lambda^1(T_p^*)$: base is $\{\mathrm{d}u_1,\mathrm{d}u_2,\mathrm{d}u_3\}$;

③ $\Lambda^2(T_p^*)$: base is $\{\mathrm{d}u_1\wedge\mathrm{d}u_2,\mathrm{d}u_2\wedge\mathrm{d}u_3,\mathrm{d}u_3\wedge\mathrm{d}u_1\}$;

④ $\Lambda^3(T_p^*)$: base is $\{\mathrm{d}u_1\wedge\mathrm{d}u_2\wedge\mathrm{d}u_3\}$.

(5) **exterior form bundle of** r**-degree**: $\Lambda^r((\mathbb{R}^3)^*)=\bigcup_{p\in\mathbf{R}^3}\Lambda^r(T_p^*), 0\leqslant r\leqslant 3$

(6) **exterior form bundle (the graded sum of exterior form bundle of r-degree):**

$$\Lambda(\mathbb{R}^3)=\sum_{r=0}^{3}\Lambda^r((\mathbb{R}^3)^*)=\sum_{r=0}^{3}\Big(\bigcup_{p\in\mathbf{R}^3}\Lambda^r(T_p^*)\Big)$$

(7) **exterior differential form space (smooth section space of exterior form bundle)**

$$A(\mathbb{R}^3)=\Gamma(\Lambda(\mathbb{R}^3))=\Gamma\Big(\sum_{r=0}^{3}\bigcup_{p\in\mathbf{R}^3}\Lambda^r(T_p^*)\Big)$$

base $\{(\mathrm{d}u_{k_1})_p\wedge\cdots\wedge(\mathrm{d}u_{k_s})_p:\ 1\leqslant k_\beta\leqslant 3,1\leqslant\beta\leqslant r\}$ is

$$\{\{1\},\{\mathrm{d}x,\mathrm{d}y,\mathrm{d}z\},\{\mathrm{d}x\wedge\mathrm{d}y,\mathrm{d}y\wedge\mathrm{d}z,\mathrm{d}z\wedge\mathrm{d}x\},\{\mathrm{d}x\wedge\mathrm{d}y\wedge\mathrm{d}z\}\}.$$

Exterior differential operator d: $A(\mathbb{R}^3)\to A(\mathbb{R}^3)$ **for** $\forall\omega\in A(\mathbb{R}^3)$:

define exterior differential form of one degree:

$$P\mathrm{d}x+Q\mathrm{d}y+R\ \mathrm{d}z,$$

with $P,Q,R\in A^0(\mathbb{R}^3)\equiv C_p^\infty(\mathbb{R}^3)$; then $\omega\in A^0(\mathbb{R}^3)\Rightarrow\mathrm{d}\omega\in A^1(\mathbb{R}^3)$.

define exterior differential form of two degree:

$$P\mathrm{d}x\wedge\mathrm{d}y+Q\mathrm{d}y\wedge\mathrm{d}z+R\mathrm{d}z\wedge\mathrm{d}x,$$

with $P,Q,R\in A^0(\mathbb{R}^3)$; then $\omega\in A^1(\mathbb{R}^3)\Rightarrow\mathrm{d}\omega\in A^2(\mathbb{R}^3)$.

define exterior differential form of three degree:

$$P\mathrm{d}x\wedge\mathrm{d}y\wedge\mathrm{d}z,$$

with $P\in A^0(\mathbb{R}^3)$; then $\omega\in A^1(\mathbb{R}^3)\Rightarrow\mathrm{d}\omega\in A^2(\mathbb{R}^3)$; $\omega\in A^2(\mathbb{R}^3)\Rightarrow\mathrm{d}\omega\in A^3(\mathbb{R}^3)$.

The above definitions of exterior differential operator d: $A(\mathbb{R}^3)\to A(\mathbb{R}^3)$ and exterior differential forms of one-, two-, and three-degrees satisfy

(1) $\mathrm{d}(A^r(M))\subset A^{r+1}(M)$; $r=0,1,2,3$;

(2) $\omega_1,\omega_2\in A(M)\Rightarrow\mathrm{d}(\omega_1+\omega_2)=\mathrm{d}\omega_1+\mathrm{d}\omega_2$;

(3) ω_1 is an exterior differential form of r-degree, then

$$\mathrm{d}(\omega_1\wedge\omega_2)=(\mathrm{d}\omega_1)\wedge\omega_2+(-1)^r\omega_1\wedge\mathrm{d}\omega_2;$$

(4) $f\in A^0(M)$, then $\mathrm{d}f$ is the differentiation of f in classical sense;

(5) $\omega\in A^0(M)\Rightarrow\mathrm{d}(\mathrm{d}\omega)=0$;

(6) $\mathrm{d}x\wedge\mathrm{d}x=0,\mathrm{d}x\wedge\mathrm{d}y=-\mathrm{d}y\wedge\mathrm{d}x,\mathrm{d}y\wedge\mathrm{d}z=-\mathrm{d}z\wedge\mathrm{d}y,\mathrm{d}z\wedge\mathrm{d}x=-\mathrm{d}x\wedge\mathrm{d}z$.

Example 6. 3. 2 Evaluation of exterior differentiation on exterior differential form space $A(\mathbb{R}^3)$

$$\forall\omega\in A(\mathbb{R}^3)=\Gamma(\Lambda((\mathbb{R}^3)^*))\Rightarrow\omega(p)\in\Lambda((\mathbb{R}^3)^*)=\sum_{r=0}^{3}\Lambda^r((\mathbb{R}^3)^*),$$

$$\Rightarrow\omega=\omega^0+\omega^1+\omega^2+\omega^3,\quad\omega^j\in\Lambda^r((\mathbb{R}^3)^*),j=0,1,2,3.$$

(1) $\omega^0\in\Lambda^0((\mathbb{R}^3)^*)\Rightarrow\omega=f\in C_p^\infty(\mathbb{R}^3)\Rightarrow$

$$\mathrm{d}\omega^0=(\mathrm{d}f)_p=P\mathrm{d}x+Q\mathrm{d}y+R\mathrm{d}z=\left(\frac{\partial f}{\partial x}\right)_p\mathrm{d}x+\left(\frac{\partial f}{\partial y}\right)_p\mathrm{d}y+\left(\frac{\partial f}{\partial z}\right)_p\mathrm{d}z,$$

where $\left(\frac{\partial f}{\partial x}\right)_p, \left(\frac{\partial f}{\partial x}\right)_p, \left(\frac{\partial f}{\partial x}\right)_p \in A^0(\mathbb{R}^3) \equiv C_p^\infty(\mathbb{R}^3)$; then

$$\omega^0 \in A^0(\mathbb{R}^3) \Rightarrow \mathrm{d}\omega^0 \in A^1(\mathbb{R}^3) \Rightarrow$$

$$\mathrm{d}\omega^0 = (\mathrm{d}f)_p = \left(\frac{\partial f}{\partial x}\right)_p \mathrm{d}x + \left(\frac{\partial f}{\partial y}\right)_p \mathrm{d}y + \left(\frac{\partial f}{\partial z}\right)_p \mathrm{d}z = \mathrm{grad} f \cdot (\mathrm{d}x, \mathrm{d}y, \mathrm{d}z);$$

thus, exterior differentiation $\mathrm{d}f \in A^1(\mathbb{R}^3)$ of smooth function $f \in A^0(\mathbb{R}^3)$ is the coefficients of $\mathrm{grad} f$ with respect to base vector $(\mathrm{d}x, \mathrm{d}y, \mathrm{d}z)$.

(2) $\omega^1 \in \Lambda^1((\mathbb{R}^3)^*) \Rightarrow \omega^1 = a\,\mathrm{d}x + b\,\mathrm{d}y + c\,\mathrm{d}z \Rightarrow$

$$\begin{aligned}\mathrm{d}\omega^1 &= \mathrm{d}(a\,\mathrm{d}x + b\,\mathrm{d}y + c\,\mathrm{d}z) = \mathrm{d}a \wedge \mathrm{d}x + \mathrm{d}b \wedge \mathrm{d}y + \mathrm{d}c \wedge \mathrm{d}z \\ &= \left(\frac{\partial c}{\partial y} - \frac{\partial b}{\partial z}\right)\mathrm{d}y \wedge \mathrm{d}z + \left(\frac{\partial a}{\partial z} - \frac{\partial c}{\partial x}\right)\mathrm{d}z \wedge \mathrm{d}x + \left(\frac{\partial b}{\partial x} - \frac{\partial a}{\partial y}\right)\mathrm{d}x \wedge \mathrm{d}y,\end{aligned}$$

where $a, b, c \in A^0(\mathbb{R}^3)$; then

$$\omega^1 \in A^1(\mathbb{R}^3) \Rightarrow \mathrm{d}\omega^1 \in A^2(\mathbb{R}^3) \Rightarrow \mathrm{d}\omega^1 = \mathrm{rot}(a, b, c) \cdot (\mathrm{d}y \wedge \mathrm{d}z, \mathrm{d}z \wedge \mathrm{d}x, \mathrm{d}x \wedge \mathrm{d}y)$$

thus, exterior differentiation $\mathrm{d}\omega^1 \in A^2(\mathbb{R}^3)$ of exterior differential form of one-degree $\omega^1 = (a, b, c) \in A^1(\mathbb{R}^3)$ is the coefficients of $\mathrm{rot}(a, b, c)$ with respect to base vector $(\mathrm{d}y \wedge \mathrm{d}z, \mathrm{d}z \wedge \mathrm{d}x, \mathrm{d}x \wedge \mathrm{d}y)$.

(3) $\omega^2 \in \Lambda^2((\mathbb{R}^3)^*) \Rightarrow \omega^2 = a\,\mathrm{d}y \wedge \mathrm{d}z + b\,\mathrm{d}z \wedge \mathrm{d}x + c\,\mathrm{d}x \wedge \mathrm{d}y \Rightarrow$

$$\begin{aligned}\mathrm{d}\omega^2 &= d(a\,\mathrm{d}y \wedge \mathrm{d}z + b\,\mathrm{d}z \wedge \mathrm{d}x + c\,\mathrm{d}x \wedge \mathrm{d}y) \\ &= \mathrm{d}a \wedge \mathrm{d}y \wedge \mathrm{d}z + \mathrm{d}b \wedge \mathrm{d}z \wedge \mathrm{d}x + \mathrm{d}c \wedge \mathrm{d}x \wedge \mathrm{d}y \\ &= \left(\frac{\partial a}{\partial x}\mathrm{d}x + \frac{\partial a}{\partial y}\mathrm{d}y + \frac{\partial a}{\partial z}\mathrm{d}z\right) \wedge \mathrm{d}y \wedge \mathrm{d}z + \left(\frac{\partial b}{\partial x}\mathrm{d}x + \frac{\partial b}{\partial y}\mathrm{d}y + \frac{\partial b}{\partial z}\mathrm{d}z\right) \wedge \mathrm{d}z \wedge \mathrm{d}x + \\ &\quad \left(\frac{\partial c}{\partial x}\mathrm{d}x + \frac{\partial c}{\partial y}\mathrm{d}y + \frac{\partial c}{\partial z}\mathrm{d}z\right) \wedge \mathrm{d}x \wedge \mathrm{d}y \\ &= \frac{\partial a}{\partial x}\mathrm{d}x \wedge \mathrm{d}y \wedge \mathrm{d}z + \frac{\partial b}{\partial y}\mathrm{d}y \wedge \mathrm{d}z \wedge \mathrm{d}x + \frac{\partial c}{\partial z}\mathrm{d}z \wedge \mathrm{d}x \wedge \mathrm{d}y \\ &= \frac{\partial a}{\partial x}\mathrm{d}x \wedge \mathrm{d}y \wedge \mathrm{d}z - \frac{\partial b}{\partial y}\mathrm{d}y \wedge \mathrm{d}x \wedge \mathrm{d}z - \frac{\partial c}{\partial z}\mathrm{d}x \wedge \mathrm{d}z \wedge \mathrm{d}y \\ &= \frac{\partial a}{\partial x}\mathrm{d}x \wedge \mathrm{d}y \wedge \mathrm{d}z - (-1)\frac{\partial b}{\partial y}\mathrm{d}x \wedge \mathrm{d}y \wedge \mathrm{d}z - (-1)\frac{\partial c}{\partial z}\mathrm{d}x \wedge \mathrm{d}y \wedge \mathrm{d}z \\ &= \left(\frac{\partial a}{\partial x} + \frac{\partial b}{\partial y} + \frac{\partial c}{\partial z}\right)\mathrm{d}x \wedge \mathrm{d}y \wedge \mathrm{d}z,\end{aligned}$$

where $a, b, c \in A^0(\mathbb{R}^3)$; then

$$\omega^2 \in A^2(\mathbb{R}^3) \Rightarrow \mathrm{d}\omega^2 \in A^3(\mathbb{R}^3) \Rightarrow$$

$$\mathrm{d}\omega^2 = \left(\frac{\partial a}{\partial x} + \frac{\partial b}{\partial y} + \frac{\partial c}{\partial z}\right)\mathrm{d}x \wedge \mathrm{d}y \wedge \mathrm{d}z = \mathrm{div}(a, b, c)(\mathrm{d}x \wedge \mathrm{d}y \wedge \mathrm{d}z)$$

thus, exterior differentiation $\mathrm{d}\omega^2 \in A^3(\mathbb{R}^3)$ of exterior differential form of two-degree $\omega^2 =$

$(a, b, c) \in A^2(\mathbb{R}^3)$ is the coefficients of div(a, b, c) with respect to base vector $(\mathrm{d}x \wedge \mathrm{d}y \wedge \mathrm{d}z)$.

(4) $\omega^3 \in \Lambda^3((\mathbb{R}^3)^*) \Rightarrow \omega^3 = a\,\mathrm{d}x \wedge \mathrm{d}y \wedge \mathrm{d}z$, with $a \in A^0(\mathbb{R}^3)$, then

$$\omega^3 \in A^3(\mathbb{R}^3) \Rightarrow \mathrm{d}\omega^3 = 0.$$

3. Properties of exterior differentiation on exterior differential form space $A(M)$

Theorem 6.3.2 (Poincare) Let M be m-dimensional smooth manifold. Then for any exterior differential form $\omega \in A(M)$, its exterior differentiation of two-degree $\mathrm{d}^2\omega = \mathrm{d}(\mathrm{d}\omega)$ holds $\mathrm{d}^2\omega = 0$.

Proof Since exterior differential operator d is linear, we only need to prove for single item of $\omega \in A(M)$. On the other hand, by the locality of the exterior differential operator, we only need to prove for $\omega = a(\mathrm{d}u_1 \wedge \mathrm{d}u_2 \wedge \cdots \wedge \mathrm{d}u_r)$. So that, we deduce:

$$\omega = a(\mathrm{d}u_1 \wedge \mathrm{d}u_2 \wedge \cdots \wedge \mathrm{d}u_r) \Rightarrow \mathrm{d}\omega = \mathrm{d}a \wedge \mathrm{d}u_1 \wedge \mathrm{d}u_2 \wedge \cdots \wedge \mathrm{d}u_r \Rightarrow$$

$\mathrm{d}^2\omega = \mathrm{d}(\mathrm{d}a \wedge \mathrm{d}u_1 \wedge \mathrm{d}u_2 \wedge \cdots \wedge \mathrm{d}u_r) \Rightarrow$ for exterior differential form of r-degree ω_1, it holds $\mathrm{d}(\omega_1 \wedge \omega_2) = (\mathrm{d}\omega_1) \wedge \omega_2 + (-1)^r \omega_1 \wedge \mathrm{d}\omega_2 \Rightarrow \omega \in A^0(M)$ implies $\mathrm{d}(\mathrm{d}\omega) = 0 \Rightarrow \mathrm{d}^2\omega = \mathrm{d}(\mathrm{d}a) \wedge (\mathrm{d}u_1 \wedge \mathrm{d}u_2 \wedge \cdots \wedge \mathrm{d}u_r) - \mathrm{d}a \wedge \mathrm{d}(\mathrm{d}u_1) \wedge \cdots \wedge \mathrm{d}u_r + \cdots = 0$, the proof is complete.

By Theorem 6.3.3, the basic formulas in the field theory of *Advanced Calculus* are obtained:

$$\mathrm{rot}(\mathrm{grad} f) = 0, \quad \mathrm{div}(\mathrm{rot} X) = 0. \tag{6.3.8}$$

Theorem 6.3.3 Let ω be an exterior differential form of one-degree on smooth manifold M, and X, Y be smooth tangent vector fields on M. Then

$$\langle X \wedge Y, \mathrm{d}\omega \rangle = X\langle Y, \omega \rangle - Y\langle X, \omega \rangle - \langle [X, Y], \omega \rangle \tag{6.3.9}$$

with $[X, Y] = XY - YX: C^\infty(M) \to C^\infty(M)$ (*Poisson brackets see* Definition 6.1.15.),

$$[X, Y](f) = X(Yf) - Y(Xf), \quad f \in C^\infty(M).$$

Proof We only prove for $\omega = g\,\mathrm{d}f$ with $f, g \in C^\infty(M)$ since both sides in (6.3.9) are linear with respect to ω. Deduce as follows.

$$\omega = g\,\mathrm{d}f \Rightarrow \mathrm{d}\omega = \mathrm{d}g \wedge \mathrm{d}f \Rightarrow \langle X \wedge Y, \mathrm{d}\omega \rangle = \langle X \wedge Y, \mathrm{d}g \wedge \mathrm{d}f \rangle$$

$\Rightarrow$by (6.2.26), the left-hand side of (6.3.9) is $\langle X \wedge Y, \mathrm{d}g \wedge \mathrm{d}f \rangle = \begin{vmatrix} \langle X, \mathrm{d}g \rangle & \langle X, \mathrm{d}f \rangle \\ \langle Y, \mathrm{d}g \rangle & \langle Y, \mathrm{d}f \rangle \end{vmatrix} = Xg \cdot Yf - Xf \cdot Yg$.

On the other hand, since

$$\langle X, \omega \rangle = \langle X, g\,\mathrm{d}f \rangle = g \cdot Xf \Rightarrow Y\langle X, \omega \rangle = Yg \cdot Xf + g \cdot Y(Xf);$$

$$\langle Y, \omega \rangle = \langle Y, g\,\mathrm{d}f \rangle = g \cdot Yf \Rightarrow X\langle Y, \omega \rangle = Xg \cdot Yf + g \cdot X(Yf);$$

$\Rightarrow$ the right-hand side of (6.3.9) is $X\langle Y, \omega \rangle - Y\langle X, \omega \rangle - \langle [X, Y], \omega \rangle$, then

$$
\begin{aligned}
& X\langle Y,\omega\rangle - Y\langle X,\omega\rangle - \langle [X,Y],\omega\rangle \\
&= Xg\cdot Yf + g\cdot X(Yf) - Yg\cdot Xf - g\cdot Y(Xf) + \langle [X,Y],\omega\rangle \\
&= Xg\cdot Yf - Yg\cdot Xf + g\cdot (X(Yf) - Y(Xf)) - g\langle [X,Y],\mathrm{d}f\rangle \\
&= Xg\cdot Yf - Yg\cdot Xf,
\end{aligned}
$$

this is the right-hand side of (6.3.9).

The formula of exterior form of r-degree $\omega\in A^r(M)$ Let $X_1,\cdots,X_{r+1}$ be $(r+1)$ smooth tangent vector fields. Then we have

$$
\begin{aligned}
\langle X_1\wedge\cdots\wedge X_{r+1},\mathrm{d}\omega\rangle = & \sum_{j=1}^{r+1}(-1)^{j+1}X_j(\langle X_1\wedge\cdots\wedge\hat{X}_j\wedge\cdots\wedge X_{r+1},\mathrm{d}\omega\rangle) + \\
& \sum_{1\leqslant j<k\leqslant r+1}(-1)^{j+k}\langle [X_j,X_k]\wedge\cdots\wedge\hat{X}_j\wedge\cdots\wedge\hat{X}_k\wedge\cdots\wedge X_{r+1},\omega\rangle.
\end{aligned}
\tag{6.3.10}
$$

Theorem 6.3.4 Let $F: M\to N$ be a smooth mapping from manifolds M on to N. Then the linear mapping $F^*: A(N)\to A(M)$ induced by F in (6.2.27) is a smooth mapping from exterior differential form spaces $A(N)$ to $A(M)$, and F^* can be commutated with exterior differential operator d,

$$F^*\circ \mathrm{d} = \mathrm{d}\circ F^*: A(N)\to A(M),$$

so that it holds the interchanged graph Fig. 6.3.5:

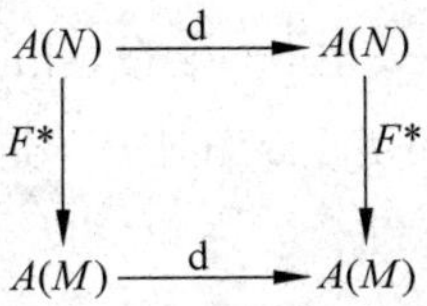

Fig. 6.3.5 Interchanged graph

Proof We interpret the mapping F^*, firstly: $F: M\to N$ is a smooth mapping $\Rightarrow$ induced mapping $F^*: A(N)\to A(M)$ by F, with $\left(A(N)=\sum_{r=0}^{m}A^r(N),+,\alpha\cdot,\tau,\wedge\right)$, $\left(A(M)=\sum_{r=0}^{m}A^r(M),+,\alpha\cdot,\tau,\wedge\right)$, and $\forall\beta\in A^r(N)$, implies $F^*\beta\in A^r(M)$.

① for $r\geqslant 1$, any r smooth tangent vector fields $X_1,\cdots,X_r$ on M, such that the matching is determined by

$$\langle X_1\wedge\cdots\wedge X_r,F^*\beta\rangle_p = \langle F_*X_1\wedge\cdots\wedge F_*X_r,\beta\rangle_{F(p)},\quad p\in M; \tag{6.3.11}$$

② for $r=0$ by $F^*\beta=\beta\circ F\in A^0(M)$ in two exterior differential form spaces of r-degree $A^r(N)$ and $A^r(M)$. By Theorem 6.2.9, mapping F^* can be commutated with the exterior product, i.e. for any $\omega,\eta\in A(N)$, it holds $F^*(\omega\wedge\eta)=F^*\omega\wedge F^*\eta$ (note:

$A(N)=\sum_{r=0}^{n}A^r(N)=\sum_{r=0}^{n}A^r(N)=\sum_{r=0}^{n}\Gamma(\Lambda^r(N^*))$, so that Theorem 6.2.9. works).

Then, turn to prove the theorem: the commutative formula of F^* with exterior differential operator d.

First step if $\beta\in A^0(N)$, take a smooth tangent vector field X on M, then (6.3.11) shows

$$\langle X, F^*(\mathrm{d}\beta)\rangle=\langle F_*X, \mathrm{d}\beta\rangle=F_*X(\mathrm{d}\beta)=X(\beta\circ F)=\langle X, \mathrm{d}(F^*\beta)\rangle,$$

this implies $F^*(\mathrm{d}\beta)=\mathrm{d}(F^*\beta)$.

Second step if $\beta=u\mathrm{d}v$, with u, v smooth functions on N, then

$$F^*(\mathrm{d}\beta)=F^*(\mathrm{d}u\wedge\mathrm{d}v)=F^*(\mathrm{d}u)\wedge F^*(\mathrm{d}v)=\mathrm{d}(F^*u)\wedge\mathrm{d}(F^*v)=\mathrm{d}(F^*\beta).$$

Third step if $F^*\circ\mathrm{d}=\mathrm{d}\circ F^*$ holds for all exterior differential forms of k-degree with $k<r$, then it holds for r.

In fact, let β be a simple term form of r-degree, $\beta\in A^r(N)$ with $\beta=\beta_1\wedge\beta_2$, where β_1 is exterior differential form of one-degree on N, and β_2 is exterior differential form of $(r-1)$-degree on N, thus by induction, it follows that

$$\begin{aligned}\mathrm{d}\circ F^*(\beta_1\wedge\beta_2)&=\mathrm{d}(F^*\beta_1\wedge F^*\beta_2)=\mathrm{d}(F^*\beta_1)\wedge F^*\beta_2-F^*\beta_1\wedge\mathrm{d}(F^*\beta_2)\\&=F^*(\mathrm{d}\beta_1\wedge\beta_2)-F^*(\beta_1\wedge\mathrm{d}\beta_2)=F^*\circ\mathrm{d}(\beta_1\wedge\beta_2),\end{aligned}$$

this implies $\mathrm{d}\circ F^*=F^*\circ\mathrm{d}$. The proof is complete.

Example 6.3.3 Let $\omega=xy\mathrm{d}x+z\mathrm{d}y-yz\mathrm{d}z$, smooth mapping $F:\mathbb{R}^2\to\mathbb{R}^3$ defined by $F(u,v)=(uv,u^2,3u+v)$. Evaluate $F^*\omega$ and $F^*\mathrm{d}\omega$.

Solve Evaluate $\mathrm{d}\omega$—

$$\begin{aligned}\mathrm{d}\omega&=(x\mathrm{d}y+y\mathrm{d}x)\wedge\mathrm{d}x+\mathrm{d}z\wedge\mathrm{d}z-(y\mathrm{d}x+z\mathrm{d}y)\wedge\mathrm{d}z\\&=-x\mathrm{d}x\wedge\mathrm{d}y-(1+z)\mathrm{d}y\wedge\mathrm{d}z.\end{aligned}$$

To evaluate $F^*\mathrm{d}x$, by $x=uv, y=u^2, z=3u+v$ from $F(u,v)=(uv,u^2,3u+v)$, then

$$F^*\mathrm{d}x=v\mathrm{d}u+u\mathrm{d}v;\quad F^*\mathrm{d}y=2u\mathrm{d}u;\quad F^*\mathrm{d}z=3\mathrm{d}u+\mathrm{d}v;$$

moreover,

$$F^*(\mathrm{d}x\wedge\mathrm{d}y)=F^*\mathrm{d}x\wedge F^*\mathrm{d}y=(v\mathrm{d}u+u\mathrm{d}v)\wedge(2u\mathrm{d}u)=-2u^2\mathrm{d}u\wedge\mathrm{d}v;$$

$$F^*(\mathrm{d}y\wedge\mathrm{d}z)=F^*\mathrm{d}y\wedge F^*\mathrm{d}z=(2u\mathrm{d}u)\wedge(3\mathrm{d}u+\mathrm{d}v)=2u\mathrm{d}u\wedge\mathrm{d}v;$$

$$F^*(\mathrm{d}z\wedge\mathrm{d}x)=F^*\mathrm{d}z\wedge F^*\mathrm{d}x=(3\mathrm{d}u+\mathrm{d}v)\wedge(v\mathrm{d}u+u\mathrm{d}v)=(3u-v)\mathrm{d}u\wedge\mathrm{d}v.$$

By $\omega=xy\mathrm{d}x+z\mathrm{d}y-yz\mathrm{d}z$, then

$$\begin{aligned}F^*\omega&=(xy\mathrm{d}x+z\mathrm{d}y-yz\mathrm{d}z)\big|_{x=uv,y=u^2,z=3u+v}\\&=u^3v(v\mathrm{d}u+u\mathrm{d}v)+2u(3u+v)\mathrm{d}u-u^2(3u+v)(3\mathrm{d}u+\mathrm{d}v).\end{aligned}$$

And deduce it to $F^*\omega=(u^3v^2+6u^2+2uv-9u^3-3u^2v)\mathrm{d}u+(u^4v-3u^3-u^2v)\mathrm{d}v$.

Similarly, $F^*\mathrm{d}\omega=-uv(-2u^2\mathrm{d}u\wedge\mathrm{d}v)-2(1+3u+v)u\mathrm{d}u\wedge\mathrm{d}v=(2u^3v-6u^2-2uv-2u)\mathrm{d}u\wedge\mathrm{d}v$.

Example 6.3.4 If $\Omega\equiv P\mathrm{d}x+Q\mathrm{d}y+R\mathrm{d}z=0$ is a total differential equation with P, $Q, R\in C^\infty(\mathbb{R}^3)$. It is said to be **totally integrable**, if there exists a smooth function F in a (small enough) neighborhood U, s. t. $F=$ const. is an initial integration of the equation $\Omega=0$. Prove that: equation $\Omega\equiv P\mathrm{d}x+Q\mathrm{d}y+R\mathrm{d}z=0$ is totally integrable, if and only if $\mathrm{d}\Omega\wedge\Omega=0$; or

$$P\left(\frac{\partial R}{\partial y}-\frac{\partial Q}{\partial z}\right)+Q\left(\frac{\partial P}{\partial z}-\frac{\partial R}{\partial x}\right)+R\left(\frac{\partial Q}{\partial x}-\frac{\partial P}{\partial y}\right)=0. \tag{6.3.12}$$

In other words, equation $\Omega\equiv P\mathrm{d}x+Q\mathrm{d}y+R\mathrm{d}z=0$ has integrating factor, if and only if (6.3.12) holds.

6.4 Integration of Exterior Differential Forms

We turn to the integration of exterior differential forms (global property), after studying the exterior differentiation of exterior differential forms (local property). Important topics are constructing integration of exterior differential forms on so-called "directed smooth manifold", and evaluating that of exterior differential forms. The following is the frame of this section.

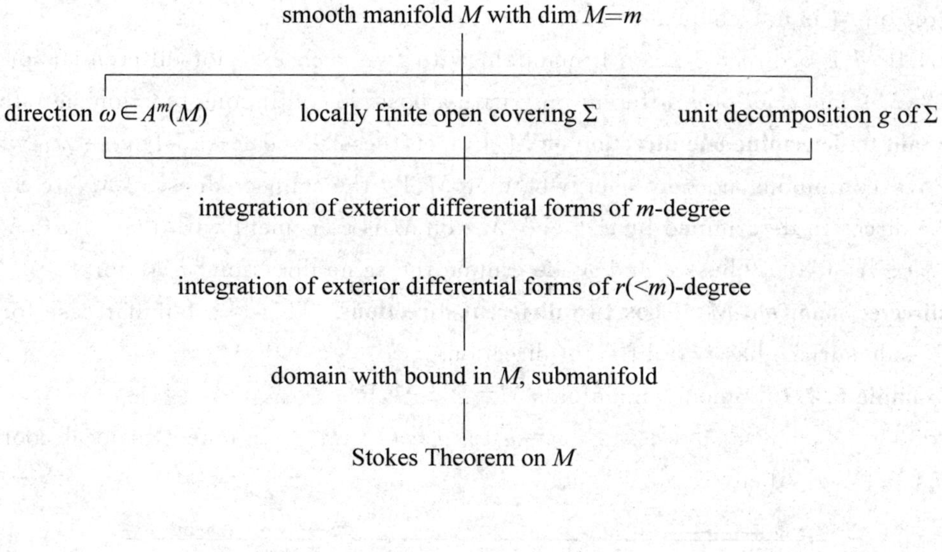

6.4.1 Directions of smooth manifolds

1. Definition of direction

Definition 6.4.1 (directed manifold) m-dimensional smooth manifold $M \equiv (M, \mathbb{A})$ is said to be **directed**, if there exists a continuous, nonzero everywhere, exterior differential form of m-degree $\omega \in A^m(M)$, $\omega \neq 0$, on M. If such a nonzero, exterior differential form of m-degree $\omega \neq 0$ with $\omega \in A^m(M)$ given on M, then M is said to be **a directed smooth manifold**, or simply, **a directed manifold.**

Note The following remarks about directed manifold are important.

(1) **"direction" described by local coordinate system**—For a smooth manifold M, suppose that

$$\Sigma_0 = \{(U_\alpha, x_\alpha^j) : \alpha \in I, 1 \leqslant j \leqslant m\}$$

is **a coordinate open covering** satisfying: if $U_\alpha \cap U_\beta \neq \varnothing$, $\forall \alpha, \beta \in I$, and Jacobi determinate $\det \left[\frac{\partial x_\alpha^j}{\partial x_\beta^k}\right]\Bigg|_{U_\alpha \cap U_\beta}$ keeps sign on $U_\alpha \cap U_\beta$ everywhere, say positive, then M is a directed manifold. This covering Σ_0 is called **a directed smooth coordinate open-covering system of** M.

(2) If M is a connected smooth manifold satisfying the second axiom of countability, then M is directed manifold, if and only if there exists an nonzero everywhere, exterior differential form of m-degree $\omega \in A^m(M)$ on M. However, this exterior differential form of m-degree on M is not unique.

(3) If M is a directed smooth manifold with two such exterior differential forms ω, $\omega' \in A^m(M)$, and ω, ω' are different only with a positive continuous function sign, then ω, ω' **are said to determine one direction on** M. In fact, these ω and ω' satisfy $\omega' = f\omega$, where f is positive, continuous, nonzero everywhere on M. By the connectedness of M, we conclude that the direction determined by $\omega' \in A^m(M)$ on M is coincident with that of $\omega \in A^m(M)$ or $(-\omega) \in A^m(M)$; Thus, ω and ω' determine the same direction on M for $f > 0$. Then, **for a directed manifold** M, **it has two different directions.** This is a familiar case for us in $\mathbb{R}^3$, a bi-side surface has two different directions.

Example 6.4.1 Smooth manifolds $M = \mathbb{R}^1, \mathbb{R}^2, \mathbb{R}^3$

For $M = \mathbb{R}^1$— $\omega \in A^1(\mathbb{R}^1) \Rightarrow \omega = a\,\mathrm{d}x$, $a \in C^\infty(\mathbb{R}^3) \Rightarrow$ take two local coordinate systems $\{u\}$, $\{v\}$, then,

Direction of $\mathbb{R}^1$: take base $\{\{1\}; \{\mathrm{d}x\}\}$ of $A(\mathbb{R}^1)$, then $\{\mathrm{d}x\}$ is a base of $A^1(\mathbb{R}^1)$, it

can be used to represent direction of $\mathbb{R}^1$. Clearly, $\exists\, \omega = a\,\mathrm{d}x \in A^1(\mathbb{R}^1), a \in C^\infty(\mathbb{R}^1)$, say, $a=1$, then $\omega = a\,\mathrm{d}x = \mathrm{d}x$ determines a direction of $\mathbb{R}^1$.

For $M=\mathbb{R}^2$—$\omega \in A^2(\mathbb{R}^2) \Rightarrow \omega = a\,\mathrm{d}x \wedge \mathrm{d}y, a \in C^\infty(\mathbb{R}^3) \Rightarrow$ take two local coordinate systems $\{u_1, u_2\}$, $\{v_1, v_2\}$, then:

Direction of $\mathbb{R}^2$: take base $\{1; \{\mathrm{d}x, \mathrm{d}y\}; \{\mathrm{d}x \wedge \mathrm{d}y\}\}$ of $A(\mathbb{R}^2)$, then $\{\mathrm{d}x \wedge \mathrm{d}y\}$ is a base of $A^2(\mathbb{R}^2)$, it can be used to represent direction of $\mathbb{R}^2$. Clearly, $\exists\, \omega = a\,\mathrm{d}x \wedge \mathrm{d}y, \omega \neq 0, a \in C^\infty(\mathbb{R}^2)$ to determine direction of $\mathbb{R}^2$, say $a=1$. Since $\mathrm{d}x \wedge \mathrm{d}y = -\mathrm{d}y \wedge \mathrm{d}x$, then $\mathrm{d}x \wedge \mathrm{d}y$ and $\mathrm{d}y \wedge \mathrm{d}x$ are just two different directions of $\mathbb{R}^2$.

For $M=\mathbb{R}^3$— $\omega \in A^3(\mathbb{R}^3) \Rightarrow \omega = a\,\mathrm{d}x \wedge \mathrm{d}y \wedge \mathrm{d}z, a \in C^\infty(\mathbb{R}^3)$.

Direction of $\mathbb{R}^3$: take base $\{1; \{\mathrm{d}x, \mathrm{d}y, \mathrm{d}z\}; \{\mathrm{d}y \wedge \mathrm{d}z, \mathrm{d}z \wedge \mathrm{d}x, \mathrm{d}z \wedge \mathrm{d}y\}; \{\mathrm{d}x \wedge \mathrm{d}y \wedge \mathrm{d}z\}\}$ of $A(\mathbb{R}^3)$, then $\{\mathrm{d}x \wedge \mathrm{d}y \wedge \mathrm{d}z\}$ is a base of $A^3(\mathbb{R}^3)$, it can be used to represent direction of $\mathbb{R}^3$. Clearly, $\exists\, \omega = a\,\mathrm{d}x \wedge \mathrm{d}y \wedge \mathrm{d}z, \omega \neq 0, a \in C^\infty(\mathbb{R}^3)$, say $a=1$. If we take $\{u_1, u_2, u_3\}$, $\{v_1, v_2, v_3\}$, then $\mathrm{d}u_1 \wedge \mathrm{d}u_2 \wedge \mathrm{d}u_3$ and $\mathrm{d}v_1 \wedge \mathrm{d}v_2 \wedge \mathrm{d}v_3$ determine two directions

$$\begin{cases} \mathrm{d}u_1 \wedge \mathrm{d}u_2 \wedge \mathrm{d}u_3 \\ \mathrm{d}v_1 \wedge \mathrm{d}v_2 \wedge \mathrm{d}v_3 \end{cases} \qquad \begin{cases} \mathrm{d}u_1 \wedge \mathrm{d}u_2 \wedge \mathrm{d}u_3 \\ \mathrm{d}v_2 \wedge \mathrm{d}v_1 \wedge \mathrm{d}v_3 \end{cases}$$

2. Unit decomposition theorem on smooth manifold M

Definition 6.4.2 (support of exterior differential form) Let M be an m-dimensional smooth manifold, $f: M \to \mathbb{R}$ be a real function on M with support $\operatorname{supp} f$ defined by $\operatorname{supp} f = \overline{\{p \in M: f(p) \neq 0\}}$.

The support $\operatorname{supp}\omega$ **of an exterior differential form** $\omega \in A(M)$ is defined by

$$\operatorname{supp}\omega = \overline{\{p \in M: \omega(p) \neq 0\}}.$$

Thus, the complimentary set of $\operatorname{supp}\omega$ is the largest set on which $\omega = 0$ i. e., $\omega|_{\complement(\operatorname{supp}\omega)} = 0$.

Definition 6.4.3 (local finite open covering of smooth manifold M**)** Let Σ_0 be an open

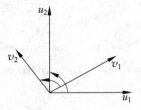

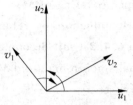

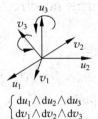

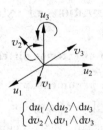

covering of smooth manifold M. If any compact subset $N \subset M$ of M intersects with only a finite open set in Σ_0, then Σ_0 is said to be **a local finite open covering of** M.

Theorem 6.4.1 (existence of local finite open covering) Let Σ be a topological base of m-dimensional smooth manifold M. Then there exists a subset Σ_0 of Σ, such that Σ_0 is a local finite open covering of M.

Proof By induction, we refer to [3].

Theorem 6.4.2 (unit decomposition theorem) Let $\Sigma = \{W_j\}$ be an open covering of m-dimensional smooth manifold M. Then there exists a system $\{g_\alpha\}$ of smooth functions on M, satisfying:

(1) $\forall \alpha$, there are ① $0 \leqslant g_\alpha \leqslant 1$; ② supp g_α is a compact set in M; ③ there exists an open set $W_j \in \Sigma$ with supp $g_\alpha \subset W_j$; ④ $\sum_\alpha g_\alpha = 1$;

(2) $\forall p \in M$, there exists an open neighborhood U, such that U intersects only with finite supports supp g_α.

Proof We refer to [3].

Note From (2), the $\sum_\alpha g_\alpha$ in ④ is a finite summation. The system $\{g_\alpha\}$ of smooth functions is called **a unit decomposition subordinate to open covering** Σ, denoted by $(\Sigma = \{W_j\}, g = \{g_\alpha\})$. "unit decomposition" means 1 can be decomposed as a summation $\sum_\alpha g_\alpha$.

6.4.2 Integrations of exterior differential forms on directed manifold M

1. Integration of exterior differential form of m-degree

Let M be an m-dimensional directed smooth manifold, $\omega \in A^m(M)$ be an exterior differential form of m-degree on M with compact support supp ω; let $\Sigma = \{W_j\}$ be coordinate covering of M coincident with the direction of M, and $\{g_\alpha\}$ be the unit decomposition subordinate to Σ. Then, it holds

$$\omega = \left(\sum_\alpha g_\alpha\right) \cdot \omega = \sum_\alpha (g_\alpha \cdot \omega),$$

where the sum is with finite terms.

Since supp ω is compact, then for each term $g_\alpha \cdot \omega$ in summation $\sum_\alpha (g_\alpha \cdot \omega)$, there exists $W_j \in \Sigma$, such that $\operatorname{supp}(g_\alpha \cdot \omega) \subset W_j$.

We may define **the integration of $g_\alpha \cdot \omega$ on M** by

$$\int_M g_\alpha \cdot \omega = \int_{W_j} g_\alpha \cdot \omega, \tag{6.4.1}$$

the right-hand side in (6.4.1) makes sense, it is a Riemann integration, because $g_\alpha \cdot \omega$, as an exterior differential form, can be expressed by the local coordinate $u_1, \cdots, u_m$ of W_j, i.e.

$$g_\alpha \cdot \omega du_1 \wedge \cdots \wedge du_m = f(u_1, \cdots, u_m)\, du_1 \cdots du_m,$$

so that it is an m-multiple Riemann integration on $W_j \subset \mathbb{R}^m$

$$\int_{W_j} g_\alpha \cdot \omega \equiv \int_{W_j} f(u_1, \cdots, u_m)\, du_1 \cdots du_m. \tag{6.4.2}$$

Note For the integration in (6.4.2), we have to give the following two explanations:

Firstly, $(\Sigma = \{W_j\}, g = \{g_\alpha\})$, **a unit decomposition subordinate to open covering** Σ, is given, then if we take $W_j \cap W_k \neq \varnothing, j \neq k$, in (6.4.2), values $\int_{W_j} g_\alpha \cdot \omega$ and $\int_{W_k} g_\alpha \cdot \omega$ are equal to each other on $W_j \cap W_k$, i.e., the value of integration $\int_{W_j} g_\alpha \cdot \omega$ is unique, independent of choice of local coordinate system.

In fact, suppose that compact support $\text{supp}(g_\alpha \cdot \omega)$ is contained in W_j and W_k for $W_j \cap W_k \neq \varnothing, j \neq k$, simultaneously; and $W_j \leftrightarrow u_1, \cdots, u_m, W_k \leftrightarrow v_1, \cdots, v_m$, with Jacobi determinate of coordinate transformation $J = J = \dfrac{\partial(v_1 \cdots, v_m)}{\partial(u_1, \cdots, u_m)} > 0$. Let $J > 0$, and

$$g_\alpha \cdot \omega du_1 \wedge \cdots \wedge du_m \equiv f(u_1, \cdots, u_m)\, du_1 \cdots du_m, \quad \text{in} \quad W_j,$$

$$g_\alpha \cdot \omega dv_1 \wedge \cdots \wedge dv_m \equiv h(v_1, \cdots, v_m)\, dv_1 \cdots dv_m, \quad \text{in} \quad W_k.$$

Then $f(u_1, \cdots, u_m) = h(v_1, \cdots, v_m) \cdot |J|$. By the formula of variable substitution in Riemann Integration, we have

$$\begin{aligned}\int_{W_j \cap W_k} g_\alpha \cdot \omega &= \int_{W_j \cap W_k} h(v_1, \cdots, v_m)\, dv_1 \cdots dv_m \\ &= \int_{W_j \cap W_k} h(v_1, \cdots, v_m)\, |J|\, du_1 \cdots du_m \\ &= \int_{W_j \cap W_k} f(u_1, \cdots, u_m)\, du_1 \cdots du_m,\end{aligned}$$

thus, (6.4.2) is independent of choice of local coordinate system. (for $J < 0$, the proof is similar.)

Moreover, by the compactness of $\text{supp}\,\omega$, and by the definition of unit decomposition $\{g_\alpha\}$ subordinate to open covering Σ, then $\text{supp}\,\omega$ intersects with W_j in $\Sigma = \{W_j\}$ only finite. Thus we may define the integration of ω on M by

$$\int_M \omega = \int_M \sum_\alpha g_\alpha \cdot \omega = \sum_\alpha \int_M g_\alpha \cdot \omega, \tag{6.4.3}$$

and we conclude that for a given pair $(\Sigma = \{W_j\}, g = \{g_\alpha\})$, the definition of (6.4.3) is well-defined.

If the compact support is contained in W_j, i.e. $\text{supp}(g_\alpha \cdot \omega) \subset W_j$, so that

$$g_\alpha \cdot \omega du_1 \wedge \cdots \wedge du_m = f(u_1, \cdots, u_m) du_1 \cdots du_m,$$

then, integration $\int_M \omega = \int_{W_j} g_\alpha \cdot \omega du_1 \wedge \cdots \wedge du_m = \int_{W_j} f(u_1, \cdots, u_m) du_1 \cdots du_m$ is a Riemann integration.

Secondly, let an open covering $\Sigma = \{W_j\}$ is given. If $(\Sigma = \{W_j\}, g = \{g_\alpha\})$, $(\Sigma = \{W_j\}, g' = \{g'_\beta\})$ are with different unit decompositions $g = \{g_\alpha\}$, $g' = \{g'_\alpha\}$, respectively. Then the values of (6. 4. 3) for g and g' are equal to each other. In fact, decompose g'_β as $g'_\beta = \sum_\alpha g_\alpha \cdot g'_\beta$, we have

$$\sum_\beta \int_M g'_\beta \cdot \omega = \sum_\beta \int_M g'_\beta \left(\sum_\alpha g_\alpha \cdot \omega\right) = \sum_\beta \Sigma_\alpha \int_M g'_\beta \cdot g_\alpha \cdot \omega$$

$$= \sum_\alpha \sum_\beta \int_M g_\alpha \cdot g'_\beta \cdot \omega = \sum_\alpha \int_M g_\alpha \left(\sum_\beta g'_\beta \cdot \omega\right) = \sum_\alpha \int_M g_\alpha \cdot \omega.$$

Now, we may define the integration of exterior differential form of m-degree ω on m-dimensional smooth manifold M.

Definition 6. 4. 4 (integration of an exterior differential form of m-degree) Let M be an m-dimensional smooth manifold, $\omega \in A^m(M)$ be exterior differential form of m-degree on M with compact support supp ω. The integration $\int_M \omega$ of ω on M is defined as

$$\int_M \omega = \sum_\alpha \int_M g_\alpha \cdot \omega, \tag{6. 4. 4}$$

the value in (6. 4. 4) is independent of choice of $(\Sigma = \{W_j\}, g = \{g_\alpha\})$.

Theorem 6. 4. 3 (linearity of integration) For exterior differential forms of m-degree $\omega_1, \omega_2 \in A^m(M)$ on smooth manifold M with compact supports, and any real numbers c_1, $c_2 \in \mathbb{R}$, the integrations on M satisfy $\int_M c_1\omega_1 + c_2\omega_2 = c_1\int_M \omega_1 + c_2\int_M \omega_2$; i. e., the integration of exterior differential forms is linear.

2. Integration of exterior differential forms of r-degree ($r<m$)

If $\omega \in A^r(M)$, $r<m$, is an exterior differential form of r-degree with compact support supp $\omega \subset N$, where N is an r-dimensional smooth submanifold of m-dimensional smooth manifold M. If $h: N \to M$ is **an embedding mapping** from N to M, such that N is r-dimensional embedding submanifold of M, then mapping $h^*: A(M) \to A(N)$ induced by $h: N \to M$ satisfying "$\omega \in A^r(M)$ $(r<m)$ implies $h^*\omega \in A^r(N)$", (see the definition in (6. 2. 27)), where $h^*\omega$ is an exterior differential form of r-degree on r-dimensional smooth manifold N with compact support supp $h^*\omega \subset N$. Hence, the following integrals

$$\int_M \omega = \int_{h(N)} \omega = \int_N h^* \omega$$

make sense. Thus, the integration of $\omega \in A^r(M), r < m$ can be defined.

Definition 6.4.5 (integration of an exterior differential form of r-degree, $r < m$) For exterior differential form of r-degree $\omega \in A^r(M)$, $r < m$, with compact support, **integration** $\int_M \omega$ **on an m-dimensional smooth manifold** M is defined as

$$\int_M \omega = \int_{h(N)} \omega = \int_N h^* \omega. \tag{6.4.5}$$

The integrations $\int_{h(N)} \omega$ and $\int_N h^* \omega$ in (6.4.5) make sense, that is: integration of exterior differential form of r-degree $\omega \in A^r(M)$ on r-dimensional submanifold N, dim $N = r < m$ (by Definition 6.4.4). Usually, we take identity mapping $I: N \to M$ with $I(x) = x$, then $I^*: A(M) \to A(N)$ satisfies (6.2.27).

Example 6.4.2 Integrations of exterior differential form $\omega \in A^r(M)$ on $M = \mathbb{R}^3$

Integration of exterior differential form of three-degree $\omega \in A^3(\mathbb{R}^3)$ on $M = \mathbb{R}^3$. Take local coordinate system $\{u_1, u_2, u_3\}$ as $u_1 = x, u_2 = y, u_3 = z$, the base of $A^3(\mathbb{R}^3)$ is $\{du_1 \wedge du_2 \wedge du_3\} = \{dx, dy, dz\}$,

$\omega \in A^3(\mathbb{R}^3) \Rightarrow \omega = H dx \wedge dy \wedge dz$, with $H \in A^0(\mathbb{R}^3) \Rightarrow$

$$\int_{\mathbb{R}^3} \omega = \int_{\mathbb{R}^3} H(u_1, u_2, u_3)\, du_1 \wedge du_2 \wedge du_3 = \int_{\mathbb{R}^3} H(x, y, z)\, dx dy dz,$$

the integration in right-hand side is a 3-multiple integration on $\mathbb{R}^3$, denoted by

$$\int_{\mathbb{R}^3} \omega dx \wedge dy \wedge dz = \int_{\mathbb{R}^3} H(x, y, z)\, dx dy dz.$$

Integration of exterior differential form of two-degree $\omega \in A^2(\mathbb{R}^3)$ on $M = \mathbb{R}^3$. Take local coordinate system $\{u_1, u_2, u_3\}$ as $u_1 = x, u_2 = y, u_3 = z$, the base of $A^2(\mathbb{R}^3)$ is $\{dy \wedge dz, dz \wedge dx, dx \wedge dy\}$, $\omega \in A^2(\mathbb{R}^3) \Rightarrow \omega = P dy \wedge dz + Q dz \wedge dx + R dx \wedge dy$, with $P, Q, R \in A^0(\mathbb{R}^3)$. Take $h: N \to M$ as identity mapping $I: \mathbb{R}^2 \to \mathbb{R}^3$, then $\int_{\mathbb{R}^3} \omega = \int_{\mathbb{R}^2} h^* \circ \omega$ with $h^* \circ \omega \in A^2(\mathbb{R}^2)$, and $h^* \circ \omega = h^*(\omega) = I^*(\omega)$.

By $\omega = P dy \wedge dz + Q dz \wedge dx + R dx \wedge dy$, we evaluate $I^*(\omega) = \omega|_U$,

$$\begin{aligned} I^*(\omega) &= I^*(P dy \wedge dz + Q dz \wedge dx + R dx \wedge dy) \\ &= P dy \wedge dz + Q dz \wedge dx + R dx \wedge dy, \end{aligned}$$

then $h^* \circ \omega|_U = P dy \wedge dz + Q dz \wedge dx + R dx \wedge dy$, and

$$\int_{\mathbb{R}^2} h^* \circ \omega = \int_{\mathbb{R}^2} P du_2 \wedge du_3 + Q du_3 \wedge du_1 + R du_1 \wedge du_2$$

$$=\int_{\mathbb{R}^2} P\,dy \wedge dz + Q\,dz \wedge dx + R\,dx \wedge dy$$

$$=\int_{\mathbb{R}^2} P\,dy\,dz + Q\,dz\,dx + R\,dx\,dy.$$

Integration of exterior differential form of one-degree $\omega \in A^1(\mathbb{R}^3)$ on $M=\mathbb{R}^3$. Take local coordinate system $\{u_1, u_2, u_3\}$ as $u_1 = x, u_2 = y, u_3 = z$, then $N=\mathbb{R}^1$, the base of $A^1(\mathbb{R}^3)$ is $\{dx, dy, dz\}$,

$$\omega \in A^1(\mathbb{R}^3) \Rightarrow \omega = P\,dx + Q\,dy + R\,dz, \quad \text{with } P, Q, R \in A^0(\mathbb{R}^3).$$

Take $h: N \to M$ as identity mapping $I: \mathbb{R}^1 \to \mathbb{R}^3$, then $\int_{\mathbb{R}^3} \omega = \int_{\mathbb{R}^1} h^* \circ \omega$ with $h^* \circ \omega \in A^1(\mathbb{R}^1)$, and $h^* \circ \omega = h^*(\omega) = I^*(\omega)$, by $I^*(\omega) = \omega|_U$ with $\omega = P\,dx + Q\,dy + R\,dz$, then

$$I^*(\omega) = I^*(P\,dx + Q\,dy + R\,dz) = P\,dx + Q\,dy + R\,dz,$$

then, $h^* \circ \omega|_U = P\,dx + Q\,dy + R\,dz$, and

$$\int_{\mathbb{R}^1} h^* \circ \omega = \int_{\mathbb{R}^1} P\,du_1 + Q\,du_2 + R\,du_3 = \int_{\mathbb{R}^1} P\,dx + Q\,dy + R\,dz.$$

Example 6.4.3 Integrations on submanifolds:

(1) Evaluate integration of exterior differential form of two-degree $\omega \in A^2(\mathbb{R}^3)$ on two-dimensional smooth surface $N=S$ of $M=\mathbb{R}^3$.

Suppose that a parameter equation of smooth surface is

$$S: \begin{cases} x = x(u,v), \\ y = y(u,v), \quad (u,v) \in D \subset \mathbb{R}^2, \\ z = z(u,v), \end{cases}$$

the normal vector of S is

$$\boldsymbol{n} = (\cos\alpha, \cos\beta, \cos\gamma) = \frac{\left(\dfrac{\partial(y,z)}{\partial(u,v)}, \dfrac{\partial(z,x)}{\partial(u,v)}, \dfrac{\partial(x,y)}{\partial(u,v)}\right)}{\sqrt{\left(\dfrac{\partial(y,z)}{\partial(u,v)}\right)^2 + \left(\dfrac{\partial(z,x)}{\partial(u,v)}\right)^2 + \left(\dfrac{\partial(x,y)}{\partial(u,v)}\right)^2}}.$$

Let $\omega \in A^2(\mathbb{R}^3)$, $\omega = P\,dy \wedge dz + Q\,dz \wedge dx + R\,dx \wedge dy$, take $h: S \to \mathbb{R}^3$ as identity mapping $I: S \to \mathbb{R}^3$, then $h^* \circ \omega = h^*(\omega) = I^*(\omega)$ is

$$\begin{aligned} I^*(\omega) &= I^*(P\,dy \wedge dz + Q\,dz \wedge dx + R\,dx \wedge dy) \\ &= P\left(\frac{\partial y}{\partial u}du + \frac{\partial y}{\partial v}dv\right) \wedge \left(\frac{\partial z}{\partial u}du + \frac{\partial z}{\partial v}dv\right) + Q\left(\frac{\partial z}{\partial u}du + \frac{\partial z}{\partial v}dv\right) \wedge \left(\frac{\partial x}{\partial u}du + \frac{\partial x}{\partial v}dv\right) + \\ &\quad R\left(\frac{\partial x}{\partial u}du + \frac{\partial x}{\partial v}dv\right) \wedge \left(\frac{\partial y}{\partial u}du + \frac{\partial y}{\partial v}dv\right) \\ &= P\frac{\partial(y,z)}{\partial(u,v)}du \wedge dv + Q\frac{\partial(z,x)}{\partial(u,v)}du \wedge dv + R\frac{\partial(x,y)}{\partial(u,v)}du \wedge dv. \end{aligned}$$

Thus,

$$\int_{\mathbb{R}^3}\omega=\int_S I^*\circ\omega=\int_S\left\{P\frac{\partial(y,z)}{\partial(u,v)}+Q\frac{\partial(z,x)}{\partial(u,v)}+R\frac{\partial(x,y)}{\partial(u,v)}\right\}\mathrm{d}u\wedge\mathrm{d}v$$

$$=\int_D\left\{P\frac{\partial(y,z)}{\partial(u,v)}+Q\frac{\partial(z,x)}{\partial(u,v)}+R\frac{\partial(x,y)}{\partial(u,v)}\right\}\mathrm{d}u\mathrm{d}v;$$

Denote $\mathrm{d}S=\sqrt{\left(\frac{\partial(y,z)}{\partial(u,v)}\right)^2+\left(\frac{\partial(z,x)}{\partial(u,v)}\right)^2+\left(\frac{\partial(x,y)}{\partial(u,v)}\right)^2}\,\mathrm{d}u\mathrm{d}v$, then

$$\int_{\mathbb{R}^3}\omega=\int_S\{P\cos\alpha+Q\cos\beta+R\cos\gamma\}\,\mathrm{d}S.$$

(2) Evaluate integration of exterior differential form of one-degree $\omega\in A^1(\mathbb{R}^3)$ on one-dimensional smooth curve $N=\gamma$ of $M=\mathbb{R}^3$.

Suppose that a parameter equation of a smooth curve is $\gamma:\begin{cases}x=x(t)\\ y=y(t),t\in[a,b]\subset\mathbb{R}^1.\\ z=z(t)\end{cases}$

By $\omega\in A^1(\mathbb{R}^3)$ with $\omega=P\mathrm{d}x+Q\mathrm{d}y+R\mathrm{d}z$, and $h:\gamma\to\mathbb{R}^3$ is identity mapping $I:\gamma\to\mathbb{R}^3$, then $h^*\circ\omega=h^*(\omega)=I^*(\omega)$,

$$I^*(\omega)=I^*(P\mathrm{d}x+Q\mathrm{d}y+R\mathrm{d}z)=P\left(\frac{\mathrm{d}x}{\mathrm{d}t}\mathrm{d}t\right)+Q\left(\frac{\mathrm{d}y}{\mathrm{d}t}\mathrm{d}t\right)+R\left(\frac{\mathrm{d}z}{\mathrm{d}t}\mathrm{d}t\right)$$

$$=\left(P\frac{\mathrm{d}x}{\mathrm{d}t}+Q\frac{\mathrm{d}y}{\mathrm{d}t}+R\frac{\mathrm{d}z}{\mathrm{d}t}\right)\mathrm{d}t;$$

it follows that

$$\int_{\mathbb{R}^3}\omega=\int_\gamma I^*\circ\omega=\int_\gamma I^*(P\mathrm{d}x+Q\mathrm{d}y+R\mathrm{d}z)=\int_a^b\left\{P\frac{\mathrm{d}x}{\mathrm{d}t}+Q\frac{\mathrm{d}y}{\mathrm{d}t}+R\frac{\mathrm{d}z}{\mathrm{d}t}\right\}\mathrm{d}t.$$

6.4.3 Stokes formula

We prove Stokes formula on smooth manifolds in this section.

Connection between the integrations on domain and that on boundary of domain is a basic content in *Advanced Calculus*. Recall Newton-Leibniz, Green, Gauss and Stokes formulas:

Newton-Leibniz formula $\int_{[a,b]}f\,\mathrm{d}x=f(b)-f(a);$

Green formula $\iint_D\left(\frac{\partial Q}{\partial x}-\frac{\partial P}{\partial y}\right)\mathrm{d}x\mathrm{d}y=\int_{\partial D}P\mathrm{d}x+Q\mathrm{d}y;$

Gauss formula $\iiint_\Omega\left(\frac{\partial P}{\partial x}+\frac{\partial Q}{\partial y}+\frac{\partial R}{\partial z}\right)\mathrm{d}x\mathrm{d}y\mathrm{d}z=\iint_{\partial\Omega}P\mathrm{d}y\mathrm{d}z+Q\mathrm{d}z\mathrm{d}x+R\mathrm{d}x\mathrm{d}y;$

Stokes formula $\iint_\Sigma\left(\frac{\partial R}{\partial y}-\frac{\partial Q}{\partial z}\right)\mathrm{d}y\mathrm{d}z+\left(\frac{\partial P}{\partial z}-\frac{\partial R}{\partial x}\right)\mathrm{d}z\mathrm{d}x+\left(\frac{\partial Q}{\partial x}-\frac{\partial P}{\partial y}\right)\mathrm{d}x\mathrm{d}y$

$$=\int_{\partial\Sigma} P\mathrm{d}x+Q\mathrm{d}y+R\mathrm{d}z.$$

Unite the above formulas by virtue of exterior differentiation, we have:

Newton-Leibniz formula: denote $[a,b]=D$, boundary $\partial D=\{\{a\},\{b\}\}$; the formula becomes

$$\int_D \mathrm{d}f=\int_{\partial D} f;$$

Green formula: let $\omega=P\mathrm{d}x+Q\mathrm{d}y$, then $\left(\frac{\partial Q}{\partial x}-\frac{\partial P}{\partial y}\right)\mathrm{d}x\wedge\mathrm{d}y=\mathrm{d}\omega$; the formula becomes

$$\int_D \mathrm{d}\omega=\int_{\partial D}\omega$$

with domain $D\subset\mathbb{R}^2$ and piece-wise smooth boundary of curve ∂D;

Gauss formula: let $\zeta=P\mathrm{d}y\mathrm{d}z+Q\mathrm{d}z\mathrm{d}x+R\mathrm{d}x\mathrm{d}y$, then $\left(\frac{\partial P}{\partial x}+\frac{\partial Q}{\partial y}+\frac{\partial R}{\partial z}\right)\mathrm{d}x\wedge\mathrm{d}y\wedge\mathrm{d}z=\mathrm{d}\zeta$; the formula becomes

$$\int_D \mathrm{d}\zeta=\int_{\partial D}\zeta$$

with domain $D\subset\mathbb{R}^3$ and piece-wise smooth boundary of surfaces ∂D;

Stokes formula: let $\eta=P\mathrm{d}x+Q\mathrm{d}y+R\mathrm{d}z$, then

$$\left(\frac{\partial R}{\partial y}-\frac{\partial Q}{\partial z}\right)\mathrm{d}y\wedge\mathrm{d}z+\left(\frac{\partial P}{\partial z}-\frac{\partial R}{\partial x}\right)\mathrm{d}z\wedge\mathrm{d}x+\left(\frac{\partial Q}{\partial x}-\frac{\partial P}{\partial y}\right)\mathrm{d}x\wedge\mathrm{d}y=\mathrm{d}\eta;$$

the formula becomes

$$\int_\Sigma \mathrm{d}\eta=\int_{\partial\Sigma}\eta$$

with domain of smooth surface $\Sigma\subset\mathbb{R}^3$ and piece-wise smooth boundary of curves ∂D.

We now describe the integration formula — Stokes formula — on smooth manifold M.

1. Domain with boundary on smooth manifold M

Definition 6.4.6 (domain with boundary) Let $D\subset M$ be a subset of m-dimensional smooth manifold M. If the points in D can be classified into two classes:

(1) inner point $p\in D$—if there exists an open neighborhood U of p, satisfying $p\in U\subset D$;

(2) boundary point $s\in M$— if there exists local coordinate system $\{(U,u_j)\}$ of s, satisfying $u_j(s)=0$, and $U\cap D=\{q\in U: u_m(q)\geqslant 0\}$;

then subset D is said to be **a domain with boundary**; local coordinate system $\{(U,u_j)\}$ in

(2) is said to be **an adapted coordinate system of boundary point** $s \in M$, or, simply, **an adapted coordinate system.**

The set of all boundary points of domain D, denoted by ∂D, is said to be **boundary of** D.

Note Let $\Sigma_0 = \{(U_\alpha, x_\alpha^j): \alpha \in I, 1 \leqslant j \leqslant m\}$ be a coordinate covering system. A point $p \in D$ is called **an inner point**, if there exists $U_\alpha \in \Sigma_0$, such that $p \in U_\alpha \subset D$; A point $s \in M$ is called **a boundary point**, if there exists $U_\alpha \in \Sigma_0$, such that $\forall j, 1 \leqslant j \leqslant m$, it holds $x_\alpha^j(\varphi(s)) = 0$ with $(U_\alpha, \varphi_\alpha)$, a chart of s; then, D is called **a domain with boundary ∂D and boundary point $s \in \partial D$ of smooth manifold M.** Moreover, we have

$$\varphi(U \cap D) = \varphi(U) \cap \{(x^1, \cdots, x^m) \in \mathbb{R}^m : x^m \geqslant 0\} = \varphi(U) \cap \mathbb{R}^m;$$

$$\varphi(U \cap \partial D) = \varphi(U) \cap \{(x^1, \cdots, x^m) \in \mathbb{R}^m : x^m = 0\}.$$

Theorem 6.4.4 (direction of a domain with boundary) Let D be a domain with boundary on smooth manifold M. Then boundary ∂D is a regular embedding closed submanifold of M; If M is directed manifold, then ∂D is a directed submanifold of M.

Proof Clearly, boundary ∂D of D is a closed subset of M. Let $\{(U, u_j)\}$ be **an adapted coordinate system of** $s \in M$, then, $U \cap \partial D = \{q \in U: u_m(q) = 0\}$. Thus ∂D is a regular embedding closed submanifold of M, by definition.

The direction of ∂D determined by that of M—for any $s \in \partial D$, take adapted coordinate system $\{(U, u_j)\}$ of $s \in \partial D$, such that it agrees with the direction of M, i. e., $du_1 \wedge \cdots \wedge du_{m-1} \wedge du_m$ is the direction of M (Fig. 6.4.1). Then $(u_1, \cdots, u_{m-1})$ is local coordinate of ∂D at point $s \in \partial D$. Thus, take

$$(-1)^m du_1 \wedge \cdots \wedge du_{m-1} \tag{6.4.6}$$

as a direction of coordinate domain $U \cap \partial D = \{q \in U: u_m(q) = 0\}$ of boundary ∂D with $(m-1)$ dimensions at point $s \in \partial D$.

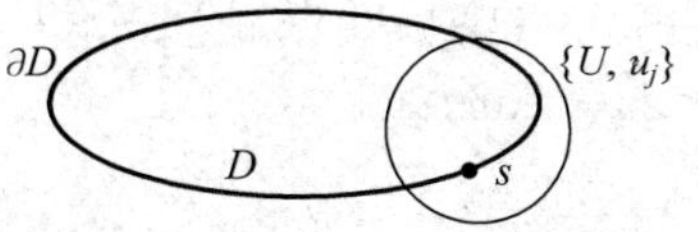

Fig. 6.4.1 Domain and boundary

We prove that the above direction of $U \cap \partial D = \{q \in U: u_m(q) = 0\}$ is compatible.

Let $\{(W, w_k)\}$ be an other adapted coordinate system of $s \in \partial D$. If it agrees with the direction of M, then

$$\frac{\partial(w_1, \cdots, w_{m-1}, w_m)}{\partial(u_1, \cdots, u_{m-1}, u_m)} > 0. \tag{6.4.7}$$

Suppose that $w_m = f_m(u_1, \cdots, u_{m-1}, u_m)$, then for fixed $u_1, \cdots, u_{m-1}$, the signs of variables w_m and u_m are the same, and $u_m=0 \Rightarrow w_m=0$. Then, $\dfrac{\partial w_m}{\partial u_m}>0$ at $s\in\partial D$. Without loss of generality, suppose that $w_m=u_m$. So that, (6.4.7) implies

$$\frac{\partial(w_1,\cdots,w_{m-1})}{\partial(u_1,\cdots,u_{m-1})}>0. \tag{6.4.8}$$

Then, two directions on $U\cap W\cap\partial D$ given by $(-1)^m\mathrm{d}u_1\wedge\cdots\wedge\mathrm{d}u_{m-1}$ and $(-1)^m\mathrm{d}w_1\wedge\cdots\wedge\mathrm{d}w_{m-1}$, respectively, are agreed with each other. This shows that ∂D is directed.

Definition 6.4.7 (induced direction on boundary) Let M be an m-dimensional directed smooth manifold; $D\subset M$ be a domain with boundary ∂D, and direction $(-1)^m\mathrm{d}u_1\wedge\cdots\wedge\mathrm{d}u_{m-1}$ of ∂D, here $\{(U,u_j)\}$ be adapted coordinate system of $s\in\partial D$, and $(u_1,\cdots,u_{m-1})$ be the local coordinate system of boundary ∂D at $s\in\partial D$. Then, direction $(-1)^m\mathrm{d}u_1\wedge\cdots\wedge\mathrm{d}u_{m-1}$ in (6.4.6) is said to be **an induced direction on boundary ∂D of directed smooth manifold M**, or, simply, **an induced direction on ∂D**.

Usually, the direction of manifold M is regarded as the direction of domain D with boundary ∂D, thus, the induced direction of domain D and induced direction of boundary ∂D are totally determined.

In $\mathbb{R}^3$ case, domain $D\subset\mathbb{R}^2$ and its boundary ∂D (curve in $\mathbb{R}^2$), domain $\Omega\subset\mathbb{R}^3$ and its boundary $\partial\Omega$ (surface in space $\mathbb{R}^3$), surface $\Sigma\subset\mathbb{R}^3$ and its boundary $\partial\Sigma$ (curve in space $\mathbb{R}^3$), are special cases of Definition 6.4.7.

Example 6.4.4 ① Unit closed disc $B_0^{(2)}(1)=\{(x,y)\in\mathbb{R}^2: x^2+y^2\leqslant 1\}$ is a domain with boundary in $\mathbb{R}^2$, and its boundary is unit circle $S^1(1)=\{(x,y)\in\mathbb{R}^2: x^2+y^2=1\}=\partial B_0^{(2)}(1)$.

$M=\mathbb{R}^2$, $\omega\in A^2(\mathbb{R}^2)\Rightarrow\omega=a\,\mathrm{d}x\wedge\mathrm{d}y$, $a\in C^\infty(\mathbb{R}^3)\Rightarrow$ take $a=1$, then $\omega=\mathrm{d}x\wedge\mathrm{d}y$ (**right-hand rule**) is the direction of $\mathbb{R}^2$. Take adapted local coordinate system $\{(U,(u_1,u_2))\}$ of $B_0^{(2)}(1)$, the direction is $\mathrm{d}u_1\wedge\mathrm{d}u_2$. By Definition 6.4.7, the induced direction on $\partial B_0^{(2)}(1)$ is $(-1)^2\mathrm{d}u_1=\mathrm{d}u_1$.

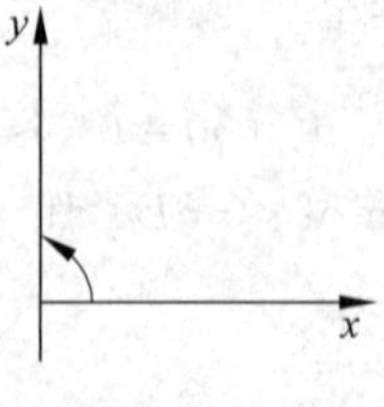

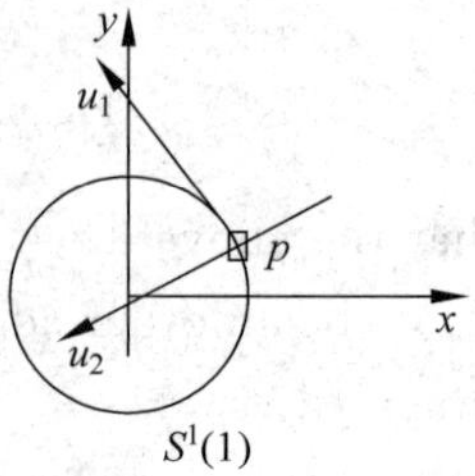

The above definition on $\partial B_0^{(2)}(1)$ is agreed with usual direction of $\partial B_0^{(2)}(1)$. Since the usual direction of $\partial B_0^{(2)}(1)$ is defined as: when a walker walks along the positive direction (anticlockwise) of boundary $\partial B_0^{(2)}(1)$, then domain $B_0^{(2)}(1)$ itself is on left-hand side of the walker. That is, in fact, the relationship between usual direction of $\partial B_0^{(2)}(1)$ and adapted coordinate system $\{(U,(u_1,u_2))\}$ is: $\frac{\partial}{\partial u_1}$ and $\partial B_0^{(2)}(1)$ are tangent each other at point p, and thus $\frac{\partial}{\partial u_1}$ is the tangent frame of p, such that $\frac{\partial}{\partial u_2}$ points the inner part of $B_0^{(2)}(1)$, that is, $\frac{\partial}{\partial u_2}$ points left-hand side of the walker who walks along the positive direction of $\partial B_0^{(2)}(1)$. Then, take $(-1)^2 \mathrm{d}u_1 = \mathrm{d}u_1$ as the direction of $\partial B_0^{(2)}(1)$, it is agreed with usual direction of it.

② Unit closed ball $B_0^{(3)}(1) = \{(x,y,z) \in \mathbb{R}^3 : x^2+y^2+z^2 \leqslant 1\}$ is a domain with boundary in $\mathbb{R}^3$, its boundary is a unit sphere $\partial B_0^{(3)}(1) = S^2(1) = \{(x,y,z) \in \mathbb{R}^3 : x^2+y^2+z^2=1\}$.

$M=\mathbb{R}^3$, $\omega \in A^3(\mathbb{R}^3) \Rightarrow \omega = a\,\mathrm{d}x \wedge \mathrm{d}y \wedge \mathrm{d}z$, $a \in C^\infty(\mathbb{R}^3)$, take $a=1$, then $\omega = 1 \cdot \mathrm{d}x \wedge \mathrm{d}y \wedge \mathrm{d}z = \mathrm{d}x \wedge \mathrm{d}y \wedge \mathrm{d}z$ is the direction of $\mathbb{R}^3$ (**right-hand rule**). Take an orthonormal adapted local coordinate system $(U_p,(e_1,e_2,e_3))$ at point $p \in \partial B_0^{(3)}(1)$, such that the direction determined by (e_1,e_2,e_3) is agreed with that of by $\omega = \mathrm{d}x \wedge \mathrm{d}y \wedge \mathrm{d}z$; i. e., (e_1,e_2) is the tangent vector of boundary $\partial B_0^{(3)}(1)$, and e_3 is the direction of outer normal line of $B_0^{(3)}(1)$. Hence, tangent frame $(p;\ e_1,e_2)$ gives the direction of $\partial B_0^{(3)}(1)$ on tangent space at point $p \in \partial B_0^{(3)}(1)$, so that, the direction of $\partial B_0^{(3)}(1)$ is given by the outer normal line at $p \in \partial B_0^{(3)}(1)$.

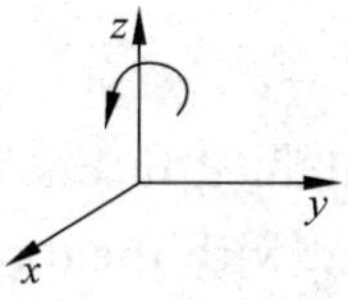

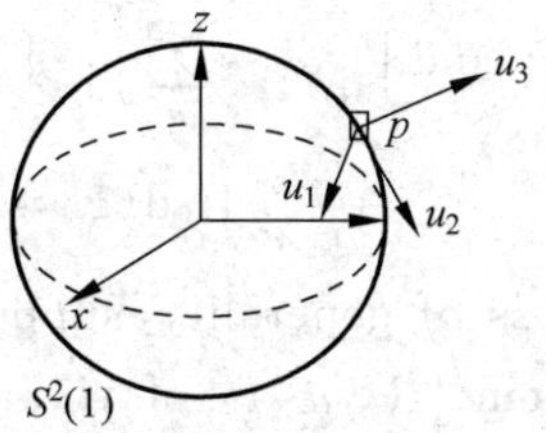

Next, we determine the direction of $\partial B_0^{(3)}(1)$ by the method of manifolds.

Take an adapted local coordinate system $\{(U_p,(u_1,u_2,u_3))\}$ of $B_0^{(3)}(1)$, its direction $\mathrm{d}u_1 \wedge \mathrm{d}u_2 \wedge \mathrm{d}u_3$ is agreed with $\omega = \mathrm{d}x \wedge \mathrm{d}y \wedge \mathrm{d}z$. The direction of $\partial B_0^{(3)}(1)$ is given by $(-1)^3 \mathrm{d}u_1 \wedge \mathrm{d}u_2 = -\mathrm{d}u_1 \wedge \mathrm{d}u_2$. This direction is agreed with usual outer normal line direction, because: The adapted local coordinate system $\{(U_p,(u_1,u_2,u_3))\}$ of $B_0^{(3)}(1)$ generates local coordinate system $\{(p,(\mathrm{d}u_1)_p,(\mathrm{d}u_2)_p,(\mathrm{d}u_3)_p)\}$ of cotangent space, and local coordinate system $\left\{\left(p, \left.\frac{\partial}{\partial u_1}\right|_p, \left.\frac{\partial}{\partial u_2}\right|_p, \left.\frac{\partial}{\partial u_3}\right|_p\right)\right\}$, respectively. Note that, $\left\{\left(p, \left.\frac{\partial}{\partial u_1}\right|_p, \left.\frac{\partial}{\partial u_2}\right|_p\right)\right\}$ is the tangent frame of $\partial B_0^{(3)}(1)$, and $\left.\frac{\partial}{\partial u_3}\right|_p$ intersects with $\partial B_0^{(3)}(1)$; moreover, it points to the inner part of

$B_0^{(3)}(1)$. If we take anti-directions of $\left.\frac{\partial}{\partial u_1}\right|_p$ and $\left.\frac{\partial}{\partial u_3}\right|_p$, then frames $\left\{p, \left.\frac{\partial}{\partial u_1}\right|_p, \left.\frac{\partial}{\partial u_2}\right|_p, \left.\frac{\partial}{\partial u_3}\right|_p\right\}$ and $\left\{p, -\left.\frac{\partial}{\partial u_1}\right|_p, \left.\frac{\partial}{\partial u_2}\right|_p, -\left.\frac{\partial}{\partial u_3}\right|_p\right\}$ are agreed with the direction of $\mathbb{R}^3$, i. e., $\omega = \mathrm{d}x \wedge \mathrm{d}y \wedge \mathrm{d}z$, and $\left.\frac{\partial}{\partial u_3}\right|_p$ intersects with $\partial B_0^{(3)}(1)$, but it points to the outer part of $B_0^{(3)}(1)$. Thus, we see that, on tangent space, the direction of boundary given by $-\mathrm{d}u_1 \wedge \mathrm{d}u_2$ is that given by tangent frame $\left\{p, -\left.\frac{\partial}{\partial u_1}\right|_p, \left.\frac{\partial}{\partial u_2}\right|_p\right\}$, then " the outer normal line direction of $\partial B_0^{(3)}(1)$ is its positive direction".

2. Stokes formula on manifold

Theorem 6.4.5 Let M be an m-dimensional directed smooth manifold, $D \subset M$ be a domain, its boundary ∂D with the induced direction; ω be exterior differential form of $(m-1)$-degree on M with compact support. Then

$$\int_D \mathrm{d}\omega = \int_{\partial D} \omega. \tag{6.4.9}$$

Proof Let $\{U_j\}$ be the coordinate covering agreed with the direction of M; and $\{g_\alpha\}$ be the unit decomposition subordinate to the covering $\{U_j\}$; denoted by $\{\{U_j\}, \{g_\alpha\}\}$. Then, $\omega = \sum_\alpha g_\alpha \cdot \omega$, with compact support, and the sum is finite terms. Since $\int_D \mathrm{d}\omega = \sum_\alpha \int_D \mathrm{d}(g_\alpha \cdot \omega)$ and $\int_{\partial D} \omega = \sum_\alpha \int_{\partial D} g_\alpha \cdot \omega$, we only need to prove

$$\int_D \mathrm{d}(g_\alpha \cdot \omega) = \int_{\partial D} g_\alpha \cdot \omega, \quad \forall \alpha.$$

Without loss of generality, suppose that supp ω is contained in one coordinate field (U, u_j) of system $\{(U, u_j)\}$ of M, and this (U, u_j) is agreed with the direction of M. Let

$$\omega = \sum_{j=1}^m (-1)^j a_j \mathrm{d}u_1 \wedge \cdots \wedge \mathrm{d}\hat{u}_j \wedge \cdots \wedge \mathrm{d}u_m,$$

where $a_j \in A^0(M)$, and $\mathrm{d}u_1 \wedge \cdots \wedge \mathrm{d}\hat{u}_j \wedge \cdots \wedge \mathrm{d}u_m = \mathrm{d}u_1 \wedge \cdots \wedge \mathrm{d}u_{j-1} \wedge \mathrm{d}u_{j+1} \wedge \cdots \wedge \mathrm{d}u_m$. Then

$$\mathrm{d}\omega = \left(\sum_{j=1}^m \frac{\partial a_j}{\partial u_j}\right) \mathrm{d}u_1 \wedge \cdots \wedge \mathrm{d}u_m.$$

Two cases are discussed:

(1) If $U \cap \partial D = \varnothing$, then the right-hand side of (6.4.9) is zero, $\int_{\partial D} \omega = 0$.

For left-hand side $\int_D \mathrm{d}\omega$: when $U \subset M \setminus D$, it is clear that $\int_D \mathrm{d}\omega = 0$; when $U \subset D$, it holds

$$\int_{\partial D}\omega=\int_D d\omega=\int_D\left(\sum_{j=1}^m\frac{\partial a_j}{\partial u_j}\right)du_1\wedge\cdots\wedge du_m.$$

Consider a cube $K=\{|u_j|\leqslant c,1\leqslant j\leqslant m\}$ in $\mathbb{R}^m$ with $U\subset K$. Then, extend functions a_j, $1\leqslant j\leqslant m$, to K, such that each a_j is continuous differential on K, and zero on $K\backslash U$. Thus,

$$\int_U\frac{\partial a_j}{\partial u_j}du_1\cdots du_m=\int_K\frac{\partial a_j}{\partial u_j}du_1\cdots du_m=\int_U\frac{\partial a_j}{\partial u_j}du_1\cdots du_m$$

$$=\int_{|u_j|\leqslant c}\left(\int_{-c}^c\frac{\partial a_j}{\partial u_j}du_j\right)du_1\cdots du_{j-1}du_{j+1}\cdots du_m=0;$$

the last term is zero, since

$$\int_{-c}^c\frac{\partial a_j}{\partial u_j}du_j=a_j(u_1,\cdots,u_{j-1},c,u_{j+1},\cdots,u_m)-$$

$$a_j(u_1,\cdots,u_{j-1},-c,u_{j+1},\cdots,u_m)=0.$$

(2) If $U\cap\partial D\neq\varnothing$ with U the adapted coordinate field agreed with direction of M, i. e.

$$U\cap D=\{q\in U:\ u_m(q)\geqslant 0\}\quad\text{and}\quad U\cap\partial D=\{q\in U:\ u_m(q)=0\}.$$

Similarly, take a cube in $\mathbb{R}^m$

$$K=\{|u_j|\leqslant c,1\leqslant j\leqslant m-1;\ 0\leqslant u_m\leqslant c\},$$

then, for an enough large constant $c>0$, set $U\cap D$ is contained in $\mathring{K}\cup\{u_m=0\}$. Similar as in (1), extend a_j on K keeping continuous differentiability, and zero on the outside of $U\cap D$.

The right-hand side of (6.4.9)—

$$\int_{\partial D}\omega=\int_{U\cap\partial D}\omega=\sum_{j=1}^m(-1)^{j-1}\int_{U\cap\partial D}a_j\,du_1\wedge\cdots\wedge du_{j-1}\wedge du_{j+1}\wedge\cdots\wedge du_m$$

$$=(-1)^{m-1}\int_{U\cap\partial D}a_m\,du_1\wedge\cdots\wedge du_{m-1}=-\int_{|u_j|\leqslant c,1\leqslant j\leqslant m-1}a_m(u_1,\cdots,u_{m-1},0)\,du_1\cdots du_{m-1}.$$

↑ $du_m=0$ on $U\cap\partial D$ ↑ induced direction on ∂D of M

The left-hand side of (6.4.9)—

$$\int_D d\omega=\int_{D\cap U}d\omega=\sum_{j=1}^m\int_{D\cap U}\frac{\partial a_j}{\partial u_j}du_1\wedge\cdots\wedge du_m,\tag{6.4.10}$$

for $1\leqslant j\leqslant m-1$, then

$$\int_{D\cap U}\frac{\partial a_j}{\partial u_j}du_1\wedge\cdots\wedge du_m=\int_{\substack{|u_j|\leqslant c,\\ j\neq m,\\ 0\leqslant u_m\leqslant c}}\left(\int_{-c}^c\frac{\partial a_j}{\partial u_j}du_j\right)du_1\cdots du_{j-1}du_{j+1}\cdots du_m=0.$$

The right-hand side of (6.4.10) only has one nonzero term, i. e.

$$\int_{D\cap U}\frac{\partial a_m}{\partial u_m}du_1\wedge\cdots\wedge du_m=\int_{\substack{|u_j|\leqslant c,\\ j\neq m}}(a_m(u_1,\cdots,u_{m-1},c)-a_m(u_1,\cdots,u_{m-1},0))\,du_1\cdots du_{m-1}-\int_{\substack{|u_j|\leqslant c,\\ j\neq m}}a_m(u_1,\cdots,u_{m-1},0)\,du_1\cdots du_{m-1}.$$

So that (6.2.9) holds, the proof is complete.

Note Usually, a domain D is closed or compact, so that Stokes formula still holds without assumption "exterior differential form of $(m-1)$-degree is with compact support".

Stokes formula is applied to mathematics, physics, astronomy, mechanics, and lots of other scientific areas.

Example 6.4.5 Evaluate $\int_{\mathbb{R}^3}\omega=\int_{S^2(1)}I^*\circ\omega$, where $S^2(1)=\{(x,y,z)\in\mathbb{R}^3: x^2+y^2+z^2=1\}$, and $\omega=x\mathrm{d}y\wedge\mathrm{d}z+y\mathrm{d}z\wedge\mathrm{d}x+z\mathrm{d}x\wedge\mathrm{d}y$.

Solve (1) Evaluate by taking local coordinate system—Suppose that: on $S^2(1)$, we take local coordinate systems

$$U=\{(x,y,z)\in S^2(1): z\geqslant 0\},\quad \{U,(u_1,u_2)\},$$

$$V=\{(x,y,z)\in S^2(1): z\leqslant 0\},\quad \{V,(v_1,v_2)\},$$

and let

$$\varphi_U(u_1,u_2)=\left(u_1,u_2,\sqrt{1-u_1^2-u_2^2}\right): \varphi_U: U\rightarrow\varphi_U(U), \varphi_U(U)\subset\mathbb{R}^3,$$

$$\varphi_V(v_1,v_2)=\left(v_1,v_2,-\sqrt{1-v_1^2-v_2^2}\right): \varphi_V: V\rightarrow\varphi_V(V), \varphi_V(V)\subset\mathbb{R}^3.$$

Then, $S^2(1)\subset U\cup V$; and

$$I^*(x\mathrm{d}y\wedge\mathrm{d}z+y\mathrm{d}z\wedge\mathrm{d}x+z\mathrm{d}x\wedge\mathrm{d}y)|_U$$

$$=u_1\mathrm{d}u_2\wedge\left(\frac{\partial z}{\partial u_1}\mathrm{d}u_1+\frac{\partial z}{\partial u_2}\mathrm{d}u_2\right)+u_2\left(\frac{\partial z}{\partial u_1}\mathrm{d}u_1+\frac{\partial z}{\partial u_2}\mathrm{d}u_2\right)\wedge\mathrm{d}u_1+\sqrt{1-u_1^2-u_2^2}\,\mathrm{d}u_1\wedge\mathrm{d}u_2$$

$$=\mathrm{d}u_2\wedge\left(\frac{-u_1^2}{\sqrt{1-u_1^2-u_2^2}}\mathrm{d}u_1+\frac{-u_1u_2}{\sqrt{1-u_1^2-u_2^2}}\mathrm{d}u_2\right)+\left(\frac{-u_2u_1}{\sqrt{1-u_1^2-u_2^2}}\mathrm{d}u_1+\frac{-u_2^2}{\sqrt{1-u_1^2-u_2^2}}\mathrm{d}u_2\right)\wedge\mathrm{d}u_1+\sqrt{1-u_1^2-u_2^2}\,\mathrm{d}u_1\wedge\mathrm{d}u_2$$

$$=\frac{u_1^2}{\sqrt{1-u_1^2-u_2^2}}\mathrm{d}u_1\wedge\mathrm{d}u_2+\frac{u_2^2}{\sqrt{1-u_1^2-u_2^2}}\mathrm{d}u_2\wedge\mathrm{d}u_1+\frac{1-u_1^2-u_2^2}{\sqrt{1-u_1^2-u_2^2}}\mathrm{d}u_1\wedge\mathrm{d}u_2+$$

$$\frac{u_2^2}{\sqrt{1-u_1^2-u_2^2}}\mathrm{d}u_2 \wedge \mathrm{d}u_1 + \frac{1-u_1^2-u_2^2}{\sqrt{1-u_1^2-u_2^2}}\mathrm{d}u_1 \wedge \mathrm{d}u_2.$$

So that

$$I^*(x\mathrm{d}y \wedge \mathrm{d}z + y\mathrm{d}z \wedge \mathrm{d}x + z\mathrm{d}x \wedge \mathrm{d}y)|_U = \frac{1}{\sqrt{1-u_1^2-u_2^2}}\mathrm{d}u_1 \wedge \mathrm{d}u_2;$$

$$I^*(x\mathrm{d}y \wedge \mathrm{d}z + y\mathrm{d}z \wedge \mathrm{d}x + z\mathrm{d}x \wedge \mathrm{d}y)|_V = \frac{1}{\sqrt{1-v_1^2-v_2^2}}\mathrm{d}v_1 \wedge \mathrm{d}v_2.$$

Set $U \cap V = \{(x,y,z) \in \mathbb{R}^3: z=0\}$ is the celestial equator, its Lebesgue measure is 0; we get

$$\int_{S^2(1)} I^* \circ \omega = \int_U I^* \circ \omega + \int_V I^* \circ \omega.$$

Take the polar coordinate system on U by $\begin{cases} u_1 = r\cos\theta, \\ u_2 = r\sin\theta, \end{cases}$ $0 \leqslant r \leqslant 1, -\pi \leqslant \theta \leqslant \pi$, then

$$\mathrm{d}u_1 \wedge \mathrm{d}u_2 = \left(\frac{\partial u_1}{\partial r}\mathrm{d}r + \frac{\partial u_1}{\partial \theta}\mathrm{d}\theta\right) \wedge \left(\frac{\partial u_2}{\partial r}\mathrm{d}r + \frac{\partial u_2}{\partial \theta}\mathrm{d}\theta\right) = r\mathrm{d}r \wedge \mathrm{d}\theta,$$

since $U \leftrightarrow D_0 = D \setminus \{(u,0) \in D: -1 \leqslant u \leqslant 0\}$, it follows that

$$\int_U I^* \circ \omega = \int_U \frac{1}{\sqrt{1-u_1^2-u_2^2}}\mathrm{d}u_1 \wedge \mathrm{d}u_2 = \int_{D_0} \frac{r}{\sqrt{1-r^2}}\mathrm{d}r \wedge \mathrm{d}\theta$$

$$= \int_{-\pi}^{\pi}\int_0^1 \frac{r}{\sqrt{1-r^2}}\mathrm{d}r\,\mathrm{d}\theta = 2\pi.$$

Similar for V. Finally, we get

$$\int_{S^2(1)} I^* \circ \omega = \int_U I^* \circ \omega + \int_V I^* \circ \omega = 4\pi.$$

(2) Evaluate by Stokes formula— Let $S^2(1) = \{(x,y,z) \in \mathbb{R}^3: x^2+y^2+z^2=1\}$, and the domain be $B_0^{(3)}(1) = \{(x,y,z) \in \mathbb{R}^3: x^2+y^2+z^2 \leqslant 1\}$. We evaluate $\int_{S^2(1)} I^* \circ \omega = \int_{B_0^{(3)}(1)} \mathrm{d}\omega$.

By $\omega = x\mathrm{d}y \wedge \mathrm{d}z + y\mathrm{d}z \wedge \mathrm{d}x + z\mathrm{d}x \wedge \mathrm{d}y$, then

$$\begin{aligned}\mathrm{d}\omega &= \mathrm{d}(x\mathrm{d}y \wedge \mathrm{d}z + y\mathrm{d}z \wedge \mathrm{d}x + z\mathrm{d}x \wedge \mathrm{d}y) \\ &= \mathrm{d}x \wedge \mathrm{d}y \wedge \mathrm{d}z + \mathrm{d}y \wedge \mathrm{d}z \wedge \mathrm{d}x + \mathrm{d}z \wedge \mathrm{d}x \wedge \mathrm{d}y = 3\mathrm{d}x \wedge \mathrm{d}y \wedge \mathrm{d}z,\end{aligned}$$

and

$$\int_{S^2(1)} I^* \circ \omega = \int_{B_0^{(3)}(1)} \mathrm{d}\omega = \int_{B_0^{(3)}(1)} 3\mathrm{d}x \wedge \mathrm{d}y \wedge \mathrm{d}z = 3\int_{B_0^{(3)}(1)} \mathrm{d}x \wedge \mathrm{d}y \wedge \mathrm{d}z.$$

Take ball coordinate system

$$\begin{cases} x = r\cos\varphi\cos\theta, \\ y = r\cos\varphi\sin\theta, \quad 0 \leqslant r \leqslant 1, 0 \leqslant \theta \leqslant 2\pi, -\dfrac{\pi}{2} \leqslant \varphi \leqslant \dfrac{\pi}{2}, \\ z = r\sin\varphi, \end{cases}$$

then

$$\mathrm{d}x \wedge \mathrm{d}y \wedge \mathrm{d}z = \begin{vmatrix} \cos\varphi\cos\theta & -r\cos\varphi\sin\theta & -r\sin\varphi\cos\theta \\ \cos\varphi\sin\theta & r\cos\varphi\cos\theta & -r\sin\varphi\sin\theta \\ \sin\varphi & 0 & r\cos\varphi \end{vmatrix} \mathrm{d}r \wedge \mathrm{d}\theta \wedge \mathrm{d}\varphi$$

$$= r^2 \cos\varphi \mathrm{d}r \wedge \mathrm{d}\theta \wedge \mathrm{d}\varphi;$$

thus

$$\int_{S^2(1)} I^* \circ \omega = \int_{B_0^{(3)}(1)} \mathrm{d}\omega = 3\int_{B_0^{(3)}(1)} \mathrm{d}x \wedge \mathrm{d}y \wedge \mathrm{d}z = 3\int_{B_0^{(3)}(1)} r^2 \cos\varphi \, \mathrm{d}r \wedge \mathrm{d}\theta \wedge \mathrm{d}\varphi$$

$$= 3\int_{B_0^{(3)}(1)} r^2 \cos\varphi \mathrm{d}r \mathrm{d}\theta \mathrm{d}\varphi = 3\int_0^1 r^2 \mathrm{d}r \int_0^{2\pi} \mathrm{d}\theta \int_{-\frac{\pi}{2}}^{\frac{\pi}{2}} \cos\varphi \mathrm{d}\varphi = 3 \times \frac{4\pi}{3} = 4\pi.$$

Note By

$$\mathrm{d}x \wedge \mathrm{d}y \wedge \mathrm{d}z = \left(\frac{\partial x}{\partial r}\mathrm{d}r + \frac{\partial x}{\partial \theta}\mathrm{d}\theta + \frac{\partial x}{\partial \varphi}\mathrm{d}\varphi\right) \wedge \left(\frac{\partial y}{\partial r}\mathrm{d}r + \frac{\partial y}{\partial \theta}\mathrm{d}\theta + \frac{\partial y}{\partial \varphi}\mathrm{d}\varphi\right) \wedge$$

$$\left(\frac{\partial z}{\partial r}\mathrm{d}r + \frac{\partial z}{\partial \theta}\mathrm{d}\theta + \frac{\partial z}{\partial \varphi}\mathrm{d}\varphi\right),$$

we have the same result. $\left\{\left(\frac{\partial}{\partial r}, \frac{\partial}{\partial \theta}, \frac{\partial}{\partial \varphi}\right)\right\}$ is by right-hand rule, the direction is the same with $\mathrm{d}x \wedge \mathrm{d}y \wedge \mathrm{d}z$ of $\mathbb{R}^3$.

3. Integral operator on a smooth manifold M

If $(M, \mathcal{A})$ is a locally compact, directed, and m-dimensional smooth manifold satisfying second countable axiom, then integral operator $\int_M : A^m(M) \to \mathbb{R}$ is a functional on $A^m(M)$ with the following properties:

Theorem 6.4.6 Integral operator $\int_M : A^m(M) \to \mathbb{R}$ is a linear functional on $A^m(M)$

(1) $\int_M \omega^1 + \omega^2 = \int_M \omega^1 + \int_M \omega^2, \quad \omega^1, \omega^2 \in A^m(M)$;

(2) $\int_M \alpha\omega = \alpha \int_M \omega, \quad \omega \in A^m(M), \alpha \in \mathbb{R}$.

6.5 Riemann Manifolds, Mathematics, and Modern Physics

6.5.1 Riemann manifolds

We establish the theory of Riemann manifolds, now.

1. Non-degenerate tensor field, positive definite tensor field

Let $M \equiv (M, \mathbb{A})$ be an m-dimensional smooth manifold, G be a symmetric covariant tensor field of two-degree on M. If $\{(U, u_j)\}$ is a local coordinate system of M, then G can be expressed as

$$G = g_{ij} \mathrm{d}u_i \otimes \mathrm{d}u_j, \tag{6.5.1}$$

where $g_{ij} = g_{ji}$ are smooth functions on U. Then, G gives a two-multilinear function on $T_p(M)$ at point $p \in M$, i. e., for $X = X_i \dfrac{\partial}{\partial u_i}$, $Y = Y_i \dfrac{\partial}{\partial u_i}$, it holds

$$G(X, Y) = g_{ij} X_i X_j, \tag{6.5.2}$$

with the **convention of Einstein summation.**

Definition 6.5.1 (non-degenerate tensor field) A tensor field G is said to be **non-degenerate at** $p \in M$, if a tangent vector $X \in T_p(M)$ satisfies $G(X, Y) = 0$, $\forall Y \in T_p(M)$, implying $X = 0$.

Proposition 6.5.1 A tensor field G is non-degenerate at p, if and only if the equation system

$$g_{ij}(p) X_i = 0, \quad 1 \leqslant j \leqslant m$$

only has zero solution; also, if and only if $\det[g_{ij}(p)] \neq 0$.

Definition 6.5.2 (positive definite tensor field) for $G(X, Y) = g_{ij} X_i X_j$ in (6.5.2), if $G(X, X) \geqslant 0$, $\forall X \in T_p(M)$; and "=" holds, if and only if $X = 0$; then tensor field G is said to be **positive and definite at** $p \in M$ or, positive definite at $p \in M$.

A tensor field G is positive definite, if and only if matrix $[g_{ij}]$ is positive and definite, by linear algebra knowledge. So that, a positive definite tensor field must be non-degenerate.

2. Generalized Riemann manifolds, Riemann manifolds

Definition 6.5.3 (generalized Riemann manifold, Riemann manifold) If on an m-dimensional smooth manifold M, a smooth, non-degenerate everywhere, symmetric covariant tensor field of two-degree G is given, then M is said to be **a generalized Riemann manifold**; and G is said to be **a basic tensor field on** M, or **a metric tensor field.**

If G is a smooth, non-degenerate everywhere, symmetric, positive definite covariant tensor field of two-degree, then M is said to be **a Riemann manifold**; and G is called **a basic tensor field on Riemann manifold M.**

A generalized Riemann manifold, or Riemann manifold, is denoted by (M,G).

In section 6.1, a differential manifold M is defined, it has topological structure and differential structure, but it does not have metric. However, we may endow a metric on a Riemann manifold (M,G), such that there is a "distance" on it, so it has metric structure.

We introduce the inner product, length on **a Riemann manifold** (M,G), firstly.

Inner product For two tangent vectors $X, Y \in T_p(M)$ on a Riemann manifold (M,G),

$$X \cdot Y \equiv G(X,Y) = g_{ij}(p) X_i Y_j$$

defined in (6.5.2) is said to be an **inner product of X and Y.**

Length For a tangent vector $X \in T_p(M)$ on Riemann manifold (M,G),

$$\| X \| = \sqrt{X \cdot X} \tag{6.5.3}$$

is said to be **a length** of X. Since G is positive and definite, the length makes sense. Moreover, **the cotangent of angle** $\theta \equiv \angle(X,Y)$ of two tangent vectors $X, Y \in T_p(M)$ at point $p \in M$ is defined by

$$\cos\theta = \cos\angle(X,Y) = \frac{X \cdot Y}{\| X \| \; \| Y \|}, \tag{6.5.4}$$

where $0 \leqslant \theta = \angle(X,Y) \leqslant \pi$.

Thus, **a Riemann manifold (M,G) is a differential manifold on which the positive definite inner product $X \cdot Y$ on tangent space $T_p(M)$ at every point $p \in M$ is determined**; Moreover, if X, Y are smooth tangent vector fields, then inner product $X \cdot Y$ is a smooth function on M.

3. Riemann matric, differential of arc

Definition 6.5.4 (Riemann matric, differential of arc) The differential form of two-degree

$$ds^2 = g_{ij}\, du_i\, du_j \equiv g_{ij}\, du_i \otimes du_j \tag{6.5.5}$$

is said to be **a Riemann metric**, or **a metric form.**

A Riemann metric $ds^2 = g_{ij}\, du_i\, du_j$ is independent of the choice of local coordinate system $\{(U, u_i)\}$. And $ds = \sqrt{g_{ij}\, du_i\, du_j}$ is just a length of infinitesimal tangent vector, called **differential of** arc, or **infinitesimal of arc.**

Let Γ: $u_i = u_i(t), t_0 \leqslant t \leqslant T$, be a continuous, piecewise smooth, parameter curve on smooth manifold M.

Then the length of arc of Γ is

$$s=\int_{t_0}^{T}\sqrt{g_{ij}\frac{du_i}{dt}\frac{du_j}{dt}}\,dt. \tag{6.5.6}$$

Theorem 6.5.1 (existence theorem of Riemann metric) There exists Riemann metric on an m-dimensional smooth manifold $(M,\mathbb{A})$.

Proof Take a local finite coordinate covering $\{(U_\alpha,u_i^\alpha)\}$ of M, and suppose that $\{h_\alpha\}$ is the corresponding unit decomposition, satisfying supp $h_\alpha \subset U_\alpha$. Denote $ds_\alpha^2=\sum_{i=1}^{m}(du_i^\alpha)^2$, then

$$ds^2=\sum_\alpha h_\alpha\cdot ds_\alpha^2, \tag{6.5.7}$$

where $h_\alpha\cdot ds_\alpha^2$ is $(h_\alpha\cdot ds_\alpha^2)(p)=\begin{cases}h_\alpha(p)\cdot ds_\alpha^2, & p\in U\\ 0, & p\notin U\end{cases}$, they are smooth differential forms of 2-degree on M.

The summation in (6.5.7) makes sense at each point $p\in M$, it is, in fact, a finite sum, because there exists a coordinate system $\{(U,u_j)\}$ of p, such that $\overline{U}$ is compact; and the local finiteness of $\{U_\alpha\}$ guarantees that U intersects with finite $U_{\alpha_1},\cdots,U_{\alpha_r}$; then the restriction on U of $ds^2=\sum_\alpha h_\alpha\cdot ds_\alpha^2$, thus (6.5.7) becomes

$$ds^2=\sum_{j=1}^{r}h_{\alpha_j}\cdot ds_{\alpha_j}^2=g_{ij}\,du_i\,du_j,$$

with $g_{ij}=\sum_{\lambda=1}^{r}\sum_{\mu=1}^{m}h_{\alpha_\lambda}\frac{\partial u_{\alpha_\lambda}^\mu}{\partial u_i}\frac{\partial u_{\alpha_\lambda}^\mu}{\partial u_j}$. Since $0\leqslant h_\alpha\leqslant 1$, $\sum_\alpha h_\alpha=1$, then there exists an index β satisfying $h_\beta(p)>0$, and thus $ds^2(p)\geqslant h_\beta\cdot ds_\beta^2$. This implies the positive definiteness of ds^2 everywhere on M.

The proof is complete.

The transform formula of component g_{ij} of $G=g_{ij}\,du_i\otimes du_j$ for local coordinate transformation $(U,u_j)\rightarrow(W,w_j)$ is

$$g'_{ij}=g_{i'j'}\frac{\partial u_{i'}}{\partial w_i}\frac{\partial u_{j'}}{\partial w_j}. \tag{6.5.8}$$

Example 6.5.1 The differential of arc in $\mathbb{R}^n$

Take local coordinate $(\mathbb{R}^n,(x_1,\cdots,x_n))$ in $\mathbb{R}^n$, then the differential of arc is

$$ds^2=(dx_1)^2+\cdots+(dx_n)^2.$$

6.5.2 Connections

In order to study "concept of differentiation" on vector field, we must introduce a new structure on "vector bundle", so-called "connection structure", including "affine connection", "connection on frame bundle", "Riemann connection", and so on. By the concept of connection, we may introduce differential structure on a vector field.

1. Vector bundle (E, M, π) on a smooth manifold

The vector bundle is a generalization of tensor bundle.

1) Definition of vector bundle

Definition 6.5.5 (vector bundle) Let M be a smooth manifold with $\dim M = m$; and $\{U, W, \cdots\}$ be an open covering of M, as well as $\{\varphi_U, \varphi_W, \cdots\}$ be smooth mappings with respect to $\{U, W, \cdots\}$. Let V be l-dimensional linear space, $\dim V = l$. If E is a smooth manifold, it satisfies that there exists smooth mapping $\pi: E \to M$, such that

(1) $\forall \varphi_U: U \times \mathbb{R}^l \to \pi^{-1}(U)$ is a differential homeomorphism; and $\forall (p, y) \in U \times \mathbb{R}^l$, it holds $\pi \circ \varphi_U(p, y) = p$;

(2) $\forall p \in U$, let $\varphi_{U,p}(y) = \varphi_U(p, y)$, $y \in \mathbb{R}^l$. Then $\varphi_{U,p}: \mathbb{R}^l \to \pi^{-1}(p)$ is a homeomorphism. If $U, W \in \{U, W, \cdots\}$, $U \cap W \neq \varnothing$, then mapping $\varphi_{U,W}(p) \equiv \varphi_{W,p}^{-1} \circ \varphi_{U,p}: \mathbb{R}^l \to \mathbb{R}^l$ is a linear self-isomorphism on linear space $V (\leftrightarrow \mathbb{R}^l)$, i. e.

$$\varphi_{U,W}(p) \equiv \varphi_{W,p}^{-1} \circ \varphi_{U,p} \in GL(V) = GL(\mathbb{R}^l); \qquad (*)$$

(3) For $U \cap W \neq \varnothing$, the mapping $\varphi_{U,W}: U \cap W \to GL(V)$ in $(*)$ is smooth;

Then, (E, M, π) (simply denoted by E) is said to be **an l-dimensional (real) vector bundle on manifold** M; and E is said to be **a bundle space**; M is called **the underlying space**; π is called **the bundle projection**; and V is called **a fiber form of** E. Moreover, $\forall p \in M$, let $E_p \equiv \pi^{-1}(p)$. Then, E_p is said to be **a fiber of vector bundle E at point $p \in M$.**

The tensor bundle $T_s^r = \bigcup_{p \in M} T_s^r(p)$ at $p \in M$ is a special case of vector bundle (E, M, π), and $E \leftrightarrow T_s^r$. The product manifold $M \times \mathbb{R}^l = E$ is a simple example of vector bundle, it is called **the trivial vector bundle.**

Note We give explanations about the conditions (2) and (3):

For (2) Here, the elements y_U, y_W on a fiber form V satisfy the following "$\Leftrightarrow$" condition

$$\varphi_U(p, y_U) = \varphi_W(p, Y_W) \Leftrightarrow y_U \circ g_{U,W}(p) = y_W,$$

where $g_{U,W}(p) \in L(\mathbb{R}^l)$ is regarded as a non-degenerate $(l \times l)$ matrix. Thus, if U is a coordinate field containing $p \in M$ in M, then the structure on fiber form V may transfer to that on fiber E_p by mapping $g_{U,W}(p) \in L(\mathbb{R}^l)$, such that fiber E_p becomes an l-dimensional linear space, and the structure of E_p is independent of choice of chart

(U, φ_U). Thus, a vector bundle (E, M, π) can be regarded as a glutinous body of product manifolds, as $U \times \mathbb{R}^l$, along fibers, as E_p at the same point $p \in M$, keeping the linear relation on fibers.

For (3) Now, mapping $\varphi_{U,W}: U \cap W \to GL(V)$ satisfies **the following compatible conditions ① and ②:**

① $\forall p \in U, \varphi_{U,U}(p) = I: V \to V$;

② if $p \in U \cap W \cap Q \neq \varnothing$, then $\varphi_{U,W}(p) \circ \varphi_{W,Q}(p) \circ \varphi_{Q,U}(p) = I: V \to V$.

Thus, $\{\varphi_{U,W}\}$ is called **a transition function family of vector bundle** (E, M, π), and the compatible conditions ①, ② are sufficient condition such that $\{\varphi_{U,W}\}$ is a transition function family. That is, we have the following theorem:

Theorem 6.5.2 (existence theorem of vector bundle) Let M be an m-dimensional smooth manifold, $\{U_\alpha\}_{\alpha \in \Lambda}$ be an open covering of M, and V be an l-dimensional linear space. If for any index pair $\alpha, \beta \in \Lambda$ with $U_\alpha \cap U_\beta \neq \varnothing$, the smooth mapping $g_{\alpha,\beta}: U_\alpha \cap U_\beta \to GL(V)$ satisfying compatible conditions ①, ② is determined, then there exists an l-dimensional vector bundle (E, M, π) on M with transition function family $\{g_{\alpha,\beta}\}$.

2) Examples of vector bundles

Example 6.5.2 (dual bundle of a vector bundle) Let M be an m-dimensional smooth manifold; $\{U, W, \cdots\}$ be a given local coordinate system of M with a function system $\{\varphi_U, \varphi_W, \cdots\}$. Let (E, M, π) be a vector bundle on M, where the vector bundle is defined in Definition 6.5.4. Then the local product structure of E is $\{(U, \varphi_U)\}$.

Let V^* be the dual space of l-dimensional linear space V; and E^* be the vector bundle on M with V^* the fiber form, π^* be the bundle projection (the senses of E^*, π^* are similar as in Definition 6.5.4).

Suppose that the local product structure of E^* is $\{(U, \psi_U)\}$. If for any point $p \in U \cap W \neq \varnothing$, it always holds

$$\langle y_U, \lambda_U \rangle = \langle y_W, \lambda_W \rangle \tag{**}$$

at $y_U, y_W \in V, \lambda_U, \lambda_W \in V^*$ with

$$\varphi_U(p, y_U) = \varphi_W(p, y_W) \quad \text{and} \quad \psi_U(p, \lambda_U) = \varphi_W(p, \lambda_W).$$

Then, **the matching of fibres $\pi^{-1}(p)$ and $(\pi^*)^{-1}(p)$** is defined by

$$\langle \varphi_U(p, y_U), \psi_U(p, \lambda_U) \rangle = \langle y_U, \lambda_U \rangle, \tag{6.5.9}$$

it is independent of the choice of U, such that two fibres $\pi^{-1}(p)$ and $(\pi^*)^{-1}(p)$ are dual spaces.

So far, the vector bundle (E^*, M, π^*) is well-defined with π^* satisfying (**), such that E^* and E are dual bundles each other, and (E^*, M, π^*) is called **a dual bundle**

of (E, M, π).

For l-dimensional linear space V with base $\{b_1, \cdots, b_l\}$ and its dual space V^* with dual base $\{b_1^*, \cdots, b_l^*\}$, an $y \in V$ is an l-dimensional row vector, and $\lambda \in V^*$ is an l-dimensional line vector, then **the matching** $\langle y, \lambda \rangle$ of $y \in V$ and $\lambda \in V^*$ in (6.5.9) can be expressed by multiplication $\langle y, \lambda \rangle = y \cdot \lambda$ of matrices with "$\cdot$" in $GL(V)$.

Since $y_W = y_U \cdot g_{U,W}(p)$ and $\langle y_U, \lambda_U \rangle = \langle y_W, \lambda_W \rangle \Leftrightarrow y_U \cdot \lambda_U = y_W \cdot \lambda_W$, by (6.3.6), then $y_U \cdot \lambda_U = y_U \cdot g_{U,W}(p) \cdot \lambda_W$, this implies that $\lambda_U = g_{U,W}(p) \cdot \lambda_W$, where $\{g_{U,W}\}$ is the **transition function family of vector bundle** (E, M, π).

Similarly, the transition function family $\{h_{U,W}\}$ of dual bundle (E^*, M, π^*) can be determined, and we have relationship

$$[h_{U,W}] = ([g_{U,W}]^{-1})^{\mathrm{T}} = [g_{W,U}]^{\mathrm{T}}.$$

Tangent bundle and cotangent bundle are special cases of vector bundle (see Definition 6.3.2); The transition function family $\{g_{U,W}\}$ of tangent bundle consists of Jacobi matrix $[J_{U,W}]$ of coordinate transformation, and transition function family $\{h_{U,W}\}$ of cotangent bundle consists of the transposition of inverse Jacobi matrix $([J_{U,W}]^{-1})^{\mathrm{T}}$.

Example 6.5.3 Direct sum $E \oplus E'$ of vector bundles E and E'

Let E and E' be vector bundles on smooth manifold M with fiber forms V and V', with transition function families $\{g_{U,W}\}$ and $\{g'_{U,W}\}$, respectively. Let

$$G_{U,W} = \begin{bmatrix} g_{U,W} & 0 \\ 0 & g'_{U,W} \end{bmatrix}.$$

Then, $\{G_{U,W}\}$ is a linear self-isomorphism on $V \oplus V'$, and satisfies **the compatible conditions ① and ②**.

The vector bundle on smooth manifold M, denoted by $E \oplus E'$, with fiber form $V \oplus V'$, with $\{G_{U,W}\}$ transition function family, is called **the direct sum of vector bundles** E and E'.

Example 6.5.4 Tensor product $E \otimes E'$ of vector bundles E and E'

Let E, E' be vector bundles on smooth manifold M with fiber types V, V', with transition function families $\{g_{U,W}\}, \{g'_{U,W}\}$, respectively. Let $k_{U,W} \equiv g_{U,W} \otimes g'_{U,W}$ be the tensor product of $g_{U,W}$ and $g'_{U,W}$, i. e.

$$(v \otimes v') \cdot k_{U,W} = (v \cdot g_{U,W}) \otimes (v' \cdot g'_{U,W}), v \in V, v' \in V'.$$

Then, the vector bundle on smooth manifold M, denoted by $E \otimes E'$, with fiber form $V \otimes V'$, with transition function family $\{k_{U,W}\} = \{g_{U,W} \otimes g'_{U,W}\}$, is called **a tensor product of vector bundles** E **and** E'.

3) Smooth intersecting surface of vector bundle

Similar to Definitions 6.3.5~6.3.7, we discuss smooth intersecting surface of a vector bundle.

Definition 6.5.6 (smooth intersecting surface of a vector bundle) Let M be a smooth manifold with $\dim M = m$, and (E, M, π) be a vector bundle on M. If a smooth mapping t: $M \to E$ satisfies $\pi \circ t \equiv I$: $M \to M$ with bundle projection π: $E \to M$, then t: $M \to E$ is said to be **a smooth intersecting surface of vector bundle** (E, M, π). The set of all smooth sections of vector bundle (E, M, π) is denoted by $\Gamma(E)$.

Since each **fiber** $E_p = \pi^{-1}(p)$ of vector bundle (E, M, π) is a linear spaces, isomorphic with fiber form V, thus, the "addition" and "number product" of intersecting surfaces on set $\Gamma(E)$ can be defined by:

$$(t_1 + t_2)(p) = t_1(p) + t_2(p), \quad t_1, t_2 \in \Gamma(E);$$

$$(\alpha t)(p) = \alpha(p)\, t(p), \quad t \in \Gamma(E), \alpha \in C^\infty(M).$$

Thus, $\Gamma(E)$ is a real linear space, called **the space of smooth intersecting surface of** E.

If E is tangent bundle $E = T(M)$, then $\Gamma(T(M))$ is the linear space consisted by all smooth tangent vector field X (Definitions 6.3.5, 6.3.6), and "$X \in \Gamma(T(M)) \Leftrightarrow X \in T_p(M)$, $\forall p \in M$".

Note that, smooth intersecting surface of vector bundle, everywhere non-zero (E, M, π) does not always exist. In fact, the existence of such smooth intersecting surfaces represents certain topological properties of smooth manifolds.

2. Connection on vector bundle

Let E be an l-dimensional real vector bundle on smooth manifold M, and $\Gamma(E)$ be the space of smooth intersecting surfaces of E on M, it is a real linear space, and a $C^\infty(M)$-model (see Definition 1.2.4).

Definition 6.5.7 (connection) Let E be a vector bundle. A mapping D on $\Gamma(E)$ is said to be **a connection**, if

$$D: \Gamma(E) \to \Gamma(T^*(M) \otimes E) \tag{6.5.10}$$

is a mapping from $\Gamma(E)$ on to $\Gamma(T^*(M) \otimes E)$, satisfying

(1) $\forall s_1, s_2 \in \Gamma(E)$, it holds $D(s_1 + s_2) = Ds_1 + Ds_2$;

(2) $\forall s \in \Gamma(E)$, $\forall \alpha \in C^\infty(M)$, it holds $D(\alpha s) = d\alpha \otimes s + \alpha Ds$.

Note 1 If $\alpha = -1$, then $D(-s) = -Ds$, that is, D maps the zero-section on to itself.

Note 2 D is a linear operator acting on intersecting surfaces of E.

Note 3 D is an operator, however, it is with **locality**: if $s_1, s_2 \in \Gamma(E)$ are intersecting surfaces, and $s_1|_U = s_2|_U$ on an open set $U \subseteq M$ of M, then $Ds_1|_U = Ds_2|_U$. Thus, by the locality of D, we may regard D as an operator acting on local intersecting surfaces, such that Ds is a totally determined intersecting surface on U.

Theorem 6.5.3 (existence theorem of connection) There exists a connection on any vector bundle (E, M, π) of smooth manifold M. Moreover, on the tangent bundle $T(M)$ of smooth manifold M, as a special vector bundle, there exists a connection.

For the proof, we refer to [3].

3. Absolute derivative of a smooth intersecting surface

From the concept of connection, we may define "absolute derivative of smooth intersecting surfaces along a tangent vector field".

Definition 6.5.8 (absolute derivative) Let X be a smooth tangent vector field on smooth manifold M, i.e., $X \in \Gamma(T(M))$; let (E, M, π) be a vector bundle on M; $\Gamma(E)$ be the space of smooth intersecting surfaces of vector bundle E; and D be a connection on E. For an intersecting surface $s \in \Gamma(E)$, let

$$D_X s \equiv \langle X, Ds \rangle, \tag{6.5.11}$$

where $\langle \cdot, \cdot \rangle$ is the matching of $T(M)$ and $T^*(M)$. Then, $D_X s$ is a smooth intersecting surface of E, i.e. $D_X s \in \Gamma(E)$. Thus, $D_X s$ is said to be **an absolute derivative of intersecting surface s along tangent vector field X with respect to connection D**, or simply, **an absolute derivative of intersecting surface s**.

Theorem 6.5.4 (properties of absolute derivative) Let M be an m-dimensional smooth manifold. A connection D with $D_X s = \langle X, Ds \rangle$ can be regarded as a mapping of two-variables:

$$\forall X \in \Gamma(T(M)), \ \forall s \in \Gamma(E), \text{ implies } D: \Gamma(T(M)) \times \Gamma(E) \to \Gamma(E).$$

Then, $\mathfrak{D}_X$ has the following properties: for $X, Y \in T(M)$, $s, s_1, s_2 \in \Gamma(E)$, $\alpha \in C^\infty(M)$, hold

(1) $D_{X+Y} s = D_X s + D_Y s$;

(2) $D_{\alpha X} s = \alpha D_X s$;

(3) $D_X(s_1 + s_2) = D_X s_1 + D_X s_2$;

(4) $D_X(\alpha s) = X\alpha + \alpha D_X s$.

Theorem 6.5.5 (locality of abstract derivative) Let X_1, X_2 be smooth tangent vector fields on smooth manifold M, such that they take the same value at point $p \in M$. Then, absolute derivatives $D_{X_1} s, D_{X_2} s$ on any intersecting surface $s \in \Gamma(E)$ of the vector bundle E take the same value at p.

By this theorem, we conclude that:

① **absolute derivative operator** D_X about tangent vector field X on the space of smooth intersecting surfaces $\Gamma(E)$ of vector bundle E of smooth manifold M at point $p \in M$ can be defined as

$$D_X : \Gamma(E) \to E_p, \quad \forall X \in T_p(M). \tag{6.5.12}$$

② for a point $p \in M$, a mapping D_X in (6.5.12) satisfies: if intersecting surfaces s_1, s_2 take the same value at those parameter curves on M which are tangent with X and take the same value at p, then it holds $D_X s_1 = D_X s_2$, i. e. absolute derivative operator D_X possesses locality.

4. Local frame field, connection matrix and curvature matrix

By the concept of connection and its locality, we may define "local frame field".

Definition 6.5.9 (local frame field) Let $\{(U, u_j)\}$ be a local coordinate system on smooth manifold M. Take smooth intersecting surfaces s_α, $1 \leqslant \alpha \leqslant q$, denoted by $\{s_\alpha, 1 \leqslant \alpha \leqslant q\}$, the linearly independent everywhere on U of vector bundle E, and called **a local frame field on U of E**, or simply, **a local frame field.**

At every point $p \in U$, by local frame field $\{s_\alpha, 1 \leqslant \alpha \leqslant q\}$, the base of tensor space $T_p^* \otimes E_p$ is

$$\{du_j \otimes s_\alpha : 1 \leqslant j \leqslant m, 1 \leqslant \alpha \leqslant q\}.$$

Since Ds_α is a (local) intersecting surface on U of the vector bundle $T^*(M) \otimes E$, we express it as

$$Ds_\alpha = \sum_{i=1}^{m} \sum_{\beta=1}^{q} \Gamma_{\alpha,i}^{\beta} \, du_i \otimes s_\beta, \quad 1 \leqslant \alpha \leqslant q \tag{6.5.13}$$

with $\Gamma_{\alpha,i}^{\beta}$ smooth function on U. Denoted by $\omega_\alpha^\beta = \sum_{i=1}^{m} \Gamma_{\alpha,i}^{\beta} \, du_i$, (6.5.13) turns out

$$Ds_\alpha = \sum_{\beta=1}^{q} \omega_\alpha^\beta \otimes s_\beta, \quad 1 \leqslant \alpha \leqslant q, \tag{6.5.14}$$

if $S = \begin{bmatrix} s_1 \\ \vdots \\ s_q \end{bmatrix}$ is the line matrix form of local frame field $\{s_\alpha, 1 \leqslant \alpha \leqslant q\}$, and $\omega = [\omega_\alpha^\beta]_{1 \leqslant \alpha, \beta \leqslant q} =$

$\begin{bmatrix} \omega_1^1 & \cdots & \omega_1^q \\ \vdots & \ddots & \vdots \\ \omega_q^1 & \cdots & \omega_q^q \end{bmatrix}$ is with elements ω_α^β. Then, $\omega = [\omega_\alpha^\beta]_{1 \leqslant \alpha, \beta \leqslant q}$ is said to be **the connection matrix.** Formula (6.5.14) can be expressed by tensor product $DS = \omega \otimes S$, where **"tensor product" of matrices** $A = [a_{ij}]_{m \times n}$ and $B = [b_{kl}]_{r \times s}$ is defined by

$$A \otimes B = \begin{bmatrix} a_{11}B & \cdots & a_{1n}B \\ \vdots & \ddots & \vdots \\ a_{m1}B & \cdots & a_{mn}B \end{bmatrix}.$$

If two local frame fields $S'=\begin{bmatrix} s'_1 \\ \vdots \\ s'_q \end{bmatrix}$ and $S=\begin{bmatrix} s_1 \\ \vdots \\ s_q \end{bmatrix}$ satisfy: $S'=A\ S$ with $\det A\neq 0$, and the transform matrix $A=\begin{bmatrix} a_1^1 & \cdots & a_1^q \\ \vdots & \ddots & \vdots \\ a_q^1 & \cdots & a_q^q \end{bmatrix}$ with a_j^k smooth functions on U, then the connection matrix ω has **a transform formula**

$$\omega' = \mathrm{d}A \cdot A^{-1} + A \cdot \omega \cdot A^{-1} \tag{6.5.15}$$

of $\omega \rightarrow \omega'$. It is an important formula in differential geometry (for the proof of it, we refer to [3]).

Rewrite (6.5.15) as $\omega' \cdot A = \mathrm{d}A + A \cdot \omega$, take the exterior differentiation of 1-degree both sides, then we have

$$\mathrm{d}\omega' \cdot A - \omega' \wedge \mathrm{d}A = \mathrm{d}A \wedge \omega + A \cdot \mathrm{d}\omega, \tag{6.5.16}$$

where the elements of exterior product $\wedge$ of matrices are exterior products of the elements of matrices. Substitute $\mathrm{d}A = \omega' \cdot A - A \cdot \omega$ into (6.5.16), then, it holds

$$(\mathrm{d}\omega' - \omega' \wedge \omega') \cdot A = A \cdot (\mathrm{d}\omega - \omega \wedge \omega). \tag{6.5.17}$$

Theorem 6.5.6 (existence theorem of a local frame field) Let D be a connection on vector bundle E, and $p \in M$. Then, there exists a local frame field S on coordinate field U of point p, such that the corresponding connection matrix ω equals zero at p.

We emphasize that: the Theorems 6.5.2, 6.5.3, 6.5.6 are three important existence theorems in differential geometry.

Definition 6.5.10 (curvature matrix) The matrix $\Omega = \mathrm{d}\omega - \omega \wedge \omega$ is said to be **a curvature matrix of connection** D at U.

The transform formula (6.5.15) can be rewritten by curvature matrices as

$$\Omega' = A \cdot \Omega \cdot A^{-1}. \tag{6.5.18}$$

Note The transform formula (6.5.18) of curvature matrix $\Omega = [\Omega_\alpha^\beta]_{1\leqslant \alpha,\beta \leqslant q}$ is homogeneous, whereas the transform formula (6.5.17) of connection matrix $\omega = [\omega_\alpha^\beta]_{1\leqslant \alpha,\beta \leqslant q}$ is not homogeneous.

Definition 6.5.11 (curvature operator) For two smooth tangent vector fields $X, Y \in \Gamma(T(M))$ on smooth manifold M, if $\forall s \in \Gamma(E)$, $\forall p \in M$, the operator $R(X,Y)$: $\Gamma(E) \rightarrow \Gamma(E)$ satisfies

$$(R(X,Y)s)(p) = R(X_p, Y_p)s_p, \tag{6.5.19}$$

then, $R(X,Y)$ is said to be **a curvature operator of connection** D.

The expression of curvature operator $R(X,Y)$ **of connection** D: Let $X, Y \in \Gamma(T(M))$

be smooth tangent vector fields on smooth manifold M. For a connection D on M and its curvature matrix $\Omega=[\Omega_\alpha^\beta]_{1\leqslant\alpha,\beta\leqslant q}$, take any $p\in U\subset M$, then $X,Y\in T_p(M)$, and $\langle X\wedge Y, \Omega_\alpha^\beta\rangle$ is an (1,1)-type tensor on fiber $\pi^{-1}(p)$ (linear space). So that $s\in\pi^{-1}(p)$ has representation $s=\sum_{\alpha=1}^{q}\lambda_\alpha(s_\alpha)|_p$ with $\lambda_\alpha\in\mathbb{R}$ by local frame field $S_U=\begin{bmatrix} s_1\\ \vdots \\ s_q\end{bmatrix}$ on U of vector bundle E. Furthermore, operator $R(X,Y)$ acts on s is

$$R(X,Y)s=\sum_{\alpha=1}^{q}\sum_{\beta=1}^{q}\lambda_\alpha\langle X\wedge Y,\Omega_\alpha^\beta\rangle(s_\beta)|_p, \tag{6.5.20}$$

it is independent of the choice of local frame field, whereas is determined by connection D and curvature matrix Ω , and it is a linear mapping from $\pi^{-1}(p)$ on to itself. This tells that the definition of curvature operator $R(X,Y)$ makes sense.

Theorem 6.5.7 The curvature operator $R(X,Y)$ has the following properties:

(1) $R(X,Y)=-R(Y,X)$;

(2) $R(fX,Y)=f\cdot R(X,Y)$;

(3) $R(X,Y)(fs)=f\cdot(R(X,Y)s)$;

where $X,Y\in\Gamma(T(M))$, $f\in C^\infty(M)$, $s\in\Gamma(E)$.

Theorem 6.5.8 The curvature operator $R(X,Y)$ is a local operator. By connection D and absolute derivative, it can be expressed as

$$R(X,Y)=D_XD_Y-D_YD_X-D_{[X,Y]}.$$

Concept "parallel" in elementary geometry can be generalized to manifolds.

Definition 6.5.12 (parallel intersecting surface) If an intersecting surface $s\in\Gamma(E)$ of vector bundle E satisfies $Ds=0$, then s is said to be **a parallel intersecting surface.**

Let $\gamma\subset M$ be a smooth parameter curve on smooth manifold M, and X be a tangent vector field of γ. If the smooth intersecting surface s of vector bundle E on γ satisfies $D_Xs=0$, then s is said to be **parallel along curve γ.**

5. Affine connection

The tangent bundle $T(M)$ on m-dimensional smooth manifold M is an m-dimensional vector bundle determined by differential structure of M itself; thus, the connection on tangent bundle $T(M)$ exists, certainly, by Theorem 6.5.3. A connection on tangent bundle is called an affine connection defined in the following definition.

Definition 6.5.13 (affine connection, admissible affine connection) Let M be an m-dimensional smooth manifold, and $T(M)$ be an tangent bundle of M. A connection of tangent bundle $T(M)$ is said to be **an affine connection on smooth manifold M**, or simply,

an affine connection on manifold M. A smooth manifold M is said to be **an affine connection space**, if an affine connection on M is given.

Let (M,G) be an m-dimensional generalized Riemann manifold, D be an affine connection on M. If $DG=0$, then D is said to be **an admissible affine connection of generalized Riemann manifold** (M,G), or simply, **admissible affine connection.**

6. Absolute differential of smooth tangent vector field

To define and evaluate "absolute differential" of tensor field, the "affine connection" on manifold has defined in Definition 6.5.13. Now, we introduce the concept of absolute differential of vector field.

Let M be an m-dimensional affine connection space, D be a given affine connection on M. Take a local coordinate system $\{(U,u_j)\}$ of M, then the natural base of tangent space T_p at $p\in U$ is $\left\{\frac{\partial}{\partial u_i},1\leqslant i\leqslant m\right\}$, and it generates a local frame field $\left\{s_i=\frac{\partial}{\partial u_i},1\leqslant i\leqslant m\right\}$ on U of tangent bundle $T(M)$. Suppose that

$$Ds_i=\sum_{j=1}^{m}\omega_i^j\otimes s_j\equiv\sum_{k=1}^{m}\sum_{j=1}^{m}\Gamma_{i,k}^j\,\mathrm{d}u_k\otimes s_j,\quad 1\leqslant i\leqslant m, \tag{6.5.21}$$

where $\Gamma_{i,k}^j$ is smooth function on U. Then, $\Gamma_{i,k}^j$, $1\leqslant i,j,k\leqslant m$, are called **the connection coefficients of connection** D **with respect to local coordinate of** u_i, or simply, **the connection coefficients of** D.

Theorem 6.5.9 If a transformation of local coordinate systems is $\{(U,u_j)\}\to\{(W,w_j)\}$, and the corresponding local frame fields are transferred by $S=\left\{s_i=\frac{\partial}{\partial u_i},1\leqslant i\leqslant m\right\}\to S'=\left\{s_i'=\frac{\partial}{\partial w_i},1\leqslant i\leqslant m\right\}$, then, we have $S'=J_{WU}\cdot S$ with Jacobi determinant $J_{WU}=\left[\frac{\partial u_i}{\partial w_j}\right]_{1\leqslant i,j\leqslant m}$. Moreover, the formula of coordinate transformation of connection coefficients $\Gamma_{i,k}^j$, $1\leqslant i,j,k\leqslant m$, are

$$\Gamma'^{j}_{i,k}=\sum_{i',j',k'=1}^{m}\Gamma_{i',k'}^{j'}\frac{\partial w_j}{\partial u_{j'}}\frac{\partial u_{i'}}{\partial w_i}\frac{\partial u_{k'}}{\partial w_k}+\sum_{i'=1}^{m}\frac{\partial^2 u_{i'}}{\partial w_i\partial w_k}\cdot\frac{\partial w_j}{\partial u_{i'}}. \tag{6.5.22}$$

Note **The formula** (6.5.22) tells that: connection coefficients $\Gamma_{i,k}^j$ do not satisfy the rule of tensor components on M.

Definition 6.5.14 (**absolute differential of a smooth tangent vector field**) Let M be m-dimensional affine connection space, X be a smooth tangent vector field on M with $X=\sum_{j=1}^{m}x_j\frac{\partial}{\partial u_j}$; and D be a given affine connection on M. Then

$$DX = \sum_{i,j=1}^{m} (\mathrm{d}x_i + x_j \omega_j^i) \otimes \frac{\partial}{\partial u_j}$$

is said to be **the absolute differential of** X with connection matrix $\omega = [\omega_\alpha^\beta]_{1\leqslant \alpha,\beta \leqslant q}$, or simply, **absolute differential of** X.

Absolute differential DX of X has the form

$$DX = \sum_{i,j=1}^{m} (\mathrm{d}x_i + x_j \omega_j^i) \otimes \frac{\partial}{\partial u_j} = \left(\frac{\partial x_i}{\partial u_j} + x_k \Gamma_{k,j}^i\right) \mathrm{d}u_j \otimes \frac{\partial}{\partial u_j},$$

it is a intersecting surface of vector bundle $T^*(M) \otimes T(M)$, and a tensor field of (1,1)-type on M. Particularly, we agree on: The absolute differential of a number field is the differential of it in the classical sense of Newton calculus.

7. Curvature tensor, torsion tensor

For an affine connection D on affine connection space M, and curvature matrix $\Omega = [\Omega_i^j]_{1\leqslant i,j\leqslant q}$, we have

$$\Omega_i^j = \sum_{k=1}^{m} \Gamma_{i,k}^j \mathrm{d}u_k, \quad 1 \leqslant i,j \leqslant q, \tag{6.5.23}$$

where the coefficients $\Gamma_{i,k}^j$ of D with respect to local coordinate u_j are smooth functions on U. Thus, curvature matrix $\Omega = \mathrm{d}\omega - \omega \wedge \omega$ with $\omega = [\omega_i^j]_{1\leqslant i,j\leqslant q}$ can be expressed as

$$\begin{aligned}
\mathrm{d}\omega_i^j - \omega_i^h \wedge \omega_h^j &= \frac{\partial \Gamma_{i,k}^j}{\partial u_l} \mathrm{d}u_l \wedge \mathrm{d}u_k - \Gamma_{i,l}^h \Gamma_{h,k}^j \mathrm{d}u_l \wedge \mathrm{d}u_k \\
&= \frac{1}{2}\left(\frac{\partial \Gamma_{i,l}^j}{\partial u_k} - \frac{\partial \Gamma_{i,k}^j}{\partial u_l} + \Gamma_{i,l}^h \Gamma_{h,k}^j - \Gamma_{i,k}^h \Gamma_{h,l}^j\right) \mathrm{d}u_k \wedge \mathrm{d}u_l \\
&= \frac{1}{2} R_{i,k,l}^j \mathrm{d}u_k \wedge \mathrm{d}u_l,
\end{aligned}$$

with $R_{i,k,l}^j = \dfrac{\partial \Gamma_{i,l}^j}{\partial u_k} - \dfrac{\partial \Gamma_{i,k}^j}{\partial u_l} + \Gamma_{i,l}^h \Gamma_{h,k}^j - \Gamma_{i,k}^h \Gamma_{h,l}^j$.

Definition 6.5.15 (curvature tensor) $R = R_{i,k,l}^j \dfrac{\partial}{\partial u_j} \otimes \mathrm{d}u_i \otimes \mathrm{d}u_k \otimes \mathrm{d}u_l$ is said to be **a curvature tensor of affine connection** D **of** m**-dimensional affine connection space** M, or simply, **a curvature tensor of affine connection** D; and $R_{i,k,l}^j$ is called **the coefficients of curvature tensor** R.

A curvature tensor is independent of the choice of local coordinate system, then if $\{(U, u_j)\} \to \{(W, w_j)\}$, we have

$$R_{i,k,l}^j = R_{i',k',l'}^{j'} \frac{\partial w_j}{\partial u_{j'}} \frac{\partial u_{i'}}{\partial w_i} \frac{\partial u_{k'}}{\partial w_k} \frac{\partial u_{l'}}{\partial w_l}, \tag{6.5.24}$$

it satisfies the rule of components of tensor of (1,3)-type, thus, the curvature tensor is a tensor field of (1,3)-type on manifold M.

Theorem 6.5.10 Let X, Y, Z be smooth tangent vector fields with local expressions $X=X_i \frac{\partial}{\partial u_i}, Y=Y_i \frac{\partial}{\partial u_i}, Z=Z_i \frac{\partial}{\partial u_i}$, respectively. Then, curvature operator $R(X, Y)$ can be represented by curvature tensor as

$$R(X,Y)Z=Z_i\langle X \wedge Y, \Omega_i^j\rangle \frac{\partial}{\partial u_i}=R^j_{i,k,l}Z_iX_kY_l \frac{\partial}{\partial u_i}.$$

Thus, coefficients $R^j_{i,k,l}$ of curvature tensor R can be expressed by

$$R^j_{i,k,l}=\left\langle R\left(\frac{\partial}{\partial u_k},\frac{\partial}{\partial u_l}\right)\frac{\partial}{\partial u_i}, du_j\right\rangle. \tag{6.5.25}$$

Definition 6.5.16 (torsion tensor) Let $\Gamma^j_{i,k}$ be the coefficients of curvature tensor R, and $T^j_{i,k} \equiv \Gamma^j_{k,i}-\Gamma^j_{i,k}$. Then, $T=T^j_{i,k} \frac{\partial}{\partial u_j} \otimes du_i \otimes du_k$ is said to be **a torsion tensor of affine connection D on m-dimensional affine connection space** M, or simply, **a torsion tensor of affine connection D**.

Torsion tensor is independent of the choice of local coordinate system, then if $\{(U, u_j)\} \to \{(W, w_j)\}$, we have

$$T^j_{i,k}=T^{j'}_{i',k'} \frac{\partial w_j}{\partial u_{j'}} \frac{\partial u_{i'}}{\partial w_i} \frac{\partial u_{k'}}{\partial w_k},$$

it satisfies the rule of components of tensor of (1,2)-type, thus, the torsion tensor is a tensor field of (1,2)-type on manifold M.

Definition 6.5.17 (torsion-free affine connection) If torsion tensor T of affine connection D is zero, then D is said to be **a torsion-free affine connection.**

8. Important existence theorems

Recall that: ① Theorem 6.5.1 is existence theorem of Riemann metric on m-dimensional smooth manifold M; ② Theorem 6.5.2 is existence theorem of vector bundle E on smooth manifold M; ③ Theorem 6.5.3 is existence theorem of connection on a vector bundle E; ④ Theorem 6.5.6 is existence theorem of local frame field on which connection matrix is zero.

We have more two important existence theorems.

Theorem 6.5.11 (existence theorem of torsion-free affine connection) There exists torsion-free affine connection on m-dimensional smooth manifold M, certainly. Moreover, if D is a torsion-free affine connection of M, then there exists a local coordinate system

$\{(U,u_j)\}$ at any point $p\in M$, such that corresponding connection coefficients $\Gamma^j_{i,k}=0$ at p.

Theorem 6.5.12 (fundamental theorem of Riemann geometry) If M is an m-dimensional generalized Riemann manifold, then there exists unique torsion-free admissible affine connection on M.

A torsion-free admissible affine connection is called **a Riemann connection**, or **Levi-Civita connection.**

9. Geodesic, geodesic normal coordinate

Definition 6.5.18 (parallel along curve, geodesic) Let M be m-dimensional smooth manifold; γ: $u_i=u_i(t)$ with $t\in[t_0,T]$ be smooth parameter curve on M; and $X(t)$ be a tangent vector field defined on γ with expression $X(t)=\sum_{j=1}^{m}x_j(t)\left(\frac{\partial}{\partial u_j}\right)_{\Gamma(t)}$. A tangent vector field $X(t)$ defined on γ is said to be **parallel along curve** γ, if the absolute differential of $X(t)$ along γ is zero; i.e., $DX|_\gamma=0$. Or equivalently, $\frac{\mathrm{d}x_i(t)}{\mathrm{d}t}+x_j(t)\Gamma^i_{j,k}\frac{\mathrm{d}u_k(t)}{\mathrm{d}t}=0$, this is an one-order differential equation system, so that if we give a tangent vector X_p at every point $p\in\gamma$, then tangent vector field X on γ is generated and determined, it is parallel along γ.

If a tangent vector field $X(t)$ defined on smooth curve γ is parallel along γ, then γ is said to be **a geodesic on** M.

For a geodesic γ, its tangent vector field $X(t)$ can be expressed by $X(t)=\sum_{j=1}^{m}\frac{\mathrm{d}u_j(t)}{\mathrm{d}t}\left(\frac{\partial}{\partial u_j}\right)_{\gamma(t)}$, since $X(t)$ is parallel along γ, so that γ satisfies the two-order differential equation system

$$\frac{\mathrm{d}^2u_i}{\mathrm{d}t^2}+\sum_{j,k=1}^{m}\Gamma^i_{j,k}\frac{\mathrm{d}u_j}{\mathrm{d}t}\frac{\mathrm{d}u_k}{\mathrm{d}t}=0,\quad i=1,2,\cdots,m,\tag{6.5.26}$$

then, we conclude that: there is just one geodesic γ passing through a point $p\in M$ on manifold M, which is tangent with the given tangent vector at $p\in M$.

Definition 6.5.19 (geodesic on a Riemann manifold) If M is an m-dimensional Riemann manifold, and smooth parameter curve γ: $u_j=u_j(t)$, $1\leqslant j\leqslant m$, $t\in[t_0,T]$, is a geodesic with respect to Riemann connection (Levi-Civita connection, it is torsion-free admissible affine connection) on M, then γ is said to be **a geodesic of Riemann manifold** M.

In Euclidean space $\mathbb{R}^3$, the geodesic of the surface of earth is the equator.

Theorem 6.5.13 If M is an m-dimensional Riemann manifold, then parameter t of a geodesic γ on M must be a linear function $t=\lambda s+\mu$ of the arc length parameter s with $\lambda\neq 0$, and λ, μ constants.

Proof In fact, if $\{(U,u_j)\}$ is a local coordinate system of M, and the connection coefficients of Riemann connection D are $\Gamma^j_{i,k}$. If a smooth curve γ: $u_i=u_i(t), 1\leqslant i\leqslant m$, is a geodesic, then its coordinates satisfy equation system (6.5.26), or by Einstein convention, it is

$$\frac{\mathrm{d}^2 u_i}{\mathrm{d}t^2}+\Gamma^i_{j,k}\frac{\mathrm{d}u_j}{\mathrm{d}t}\frac{\mathrm{d}u_k}{\mathrm{d}t}=0, \tag{6.5.27}$$

where $X_i=\dfrac{\mathrm{d}u_i}{\mathrm{d}t}$ is a tangent vector of γ. Since the tangent vector field of geodesic is parallel along curve γ itself with respect to Riemann connection; and Riemann connection keeps invariance of metric properties under translation, thus the length of tangent vector field is a constant, i. e., $\dfrac{\mathrm{d}s}{\mathrm{d}t}=$const., so that $t=\lambda s+\mu$. The proof is complete.

Hence, we may establish a particular **coordinate system** $\{(U,u_j)\}$ at point $p\in M$ on the manifold M, such that in this coordinate system, the parameter of any geodesic started from p is linear function of arc length parameter, then this special coordinate system $\{(U,u_j)\}$ is called **the geodesic normal coordinate system.**

Construction of geodesic normal coordinate system: (1) Take a local coordinate system $\{(U,u_j)\}$ at $p\in M$ in affine connection space M. Give initial values $\left(u_i(0),\dfrac{\mathrm{d}u_i(0)}{\mathrm{d}t}\right)=(x_i,\alpha_i)$ at any point $x\in U$ with $t=0$, and solve the equation system of geodesic

$$\begin{cases}\dfrac{\mathrm{d}^2 u_i}{\mathrm{d}t^2}+\Gamma^i_{j,k}\dfrac{\mathrm{d}u_j}{\mathrm{d}t}\dfrac{\mathrm{d}u_k}{\mathrm{d}t}=0,\\ u_i(0)=x_i, \qquad\qquad 1\leqslant i\leqslant m.\\ \dfrac{\mathrm{d}u_i}{\mathrm{d}t}(0)=\alpha_i,\end{cases} \tag{6.5.28}$$

By the theory in ordinary differential equations, for any point $y\in U$, there exist a neighborhood $W\subset U$ and two positive numbers r and δ, such that for any initial point $x\in W$, and any $\alpha\in\mathbb{R}^m$ with $\|\alpha\|_{\mathbb{R}^m}\leqslant r$, unique solution

$$u_i=f_i(t,x_k,\alpha_k),\quad |t|<\delta, \tag{6.5.29}$$

can be obtained with smooth functions f_i depending on t, x_k, α_k, satisfying initial conditions $u_i(0)=f_i(0,x_k,\alpha_k)=x_i$, and $\dfrac{\mathrm{d}u_i}{\mathrm{d}t}(0)=\dfrac{\partial f_i(0,x_k,\alpha_k)}{\partial t}=\alpha_i$.

(2) Take constant $c\neq 0$, then function $\tilde{u}_i=f_i(ct,x_k,\alpha_k)$ with $x\in W$, $\|\alpha\|_{\mathbf{R}^m}\leqslant r$, $|t|<\dfrac{\delta}{|c|}$, still satisfies equation $\dfrac{\mathrm{d}^2u_i}{\mathrm{d}t^2}+\Gamma^i_{j,k}\dfrac{\mathrm{d}u_j}{\mathrm{d}t}\dfrac{\mathrm{d}u_k}{\mathrm{d}t}=0$, and conditions $f_i(ct,x_k,\alpha_k)|_{t=0}=x_i$, $\left.\dfrac{\partial f_i(ct,x_k,\alpha_k)}{\partial t}\right|_{t=0}=c\alpha_i$. However, (6.5.28) has unique solution, thus, it always holds $f_i(ct,x_k,\alpha_k)=f_i(t,x_k,c\alpha_k)$ when $\|\alpha\|_{\mathbf{R}^m}\leqslant r$, $\|c\alpha\|_{\mathbf{R}^m}\leqslant r$, $|t|<\delta$, $|ct|<\delta$. The function $f_i(ct,x_k,\alpha_k)$ is well-defined since $f_i(ct,x_k,\alpha_k)$ makes sense for $x\in W$, $\|\alpha\|_{\mathbf{R}^m}\leqslant r$, $|t|<\dfrac{\delta}{|c|}$. Hence, the function $f_i(t,x_k,\alpha_k)$ is well-defined for $x\in W$, $\|\alpha\|_{\mathbf{R}^m}\leqslant|c|r$, $|t|<\dfrac{\delta}{|c|}$. Particularly, take $|c|<\delta$, then function $f_i(t,x_k,\alpha_k)$ is well-defined on $x\in W$, $\|\alpha\|_{\mathbf{R}^m}\leqslant|c|r$, $|t|\leqslant 1$.

Let $u_i=f_i(1,x_k,\alpha_k)$, then $f_i(1,x_k,\alpha_k)=f_i(0,x_k,\alpha_k)=x_i$. Hence, "the function" $u_i=f_i(1,x_k,\alpha_k)$ gives a smooth mapping from a neighborhood of the origin of tangent space $T_x(M)(=\mathbb{R}^m)$ for fixed $x\in W$, on to a neighborhood of $p\in M$ on manifold M. Since

$$\left.\frac{\partial f_i(1,x_k,t\alpha_k)}{\partial t}\right|_{t=0}=\left.\frac{\partial f_i(1,x_k,\alpha_k)}{\partial\alpha_i}\right|_{\alpha=0}\cdot\alpha_i$$

and

$$\left.\frac{\partial f_i(1,x_k,t\alpha_k)}{\partial t}\right|_{t=0}=\left.\frac{\partial f_i(t,x_k,\alpha_k)}{\partial t}\right|_{t=0}=\alpha_i;$$

then, we have $\left.\dfrac{\partial u_i}{\partial\alpha_j}\right|_{\alpha=0}=\delta^i_j$. This means that the smooth mapping $f_i(1,x_k,\alpha_k)$ is normal at origin; i.e., α_i can be taken as a local coordinate system.

Then, we continue: take $\alpha_k=\alpha_k^0$ as a local coordinate of point $x\in W\subset U$, such that $\{(W,\alpha_k)\}$ is a local coordinate system of x, then $\{(W,\alpha_k)\}$ is called **the geodesic normal coordinate system of** x; and α_k is called **the geodesic normal coordinate of** x, or simply, **a geodesic normal coordinate.**

Since the tangent space of a smooth manifold is a linear space, thus two coordinate systems on tangent space are different only up to a non-degenerate linear transformation, so that the normal coordinate systems are different only up to a non-degenerate linear transformation. Hence, for a fixed system $\alpha_k=\alpha_k^0$, then $t\alpha_k^0$ is the straight line started from origin in $T_x(M)$ for variable t; and correspondingly, a geodesic started from x, tangent with tangent vector (α_k^0) is appeared on manifold M. Thus, in geodesic normal coordinate system $\{(W,\alpha_k)\}$, the equation of geodesics is $\alpha_k = t\,\alpha_k^0$ with constants α_k^0.

Theorem 6.5.14 Let M be m-dimensional torsion-free affine connection space. Then, for geodesic normal coordinate system $\{(W,\alpha_j)\}$ at every point $x \in M$, its connection coefficients $\Gamma^i_{j,k}$ is zero at x, i.e.

$$\Gamma^i_{j,k}(0) = 0, \quad 1 \leqslant i,j,k \leqslant m.$$

Theorem 6.5.15 (existence theorem of geodesic normal coordinate system) There exists a neighborhood W at every point $x \in M$ on m-dimensional affine connection space M, such that there exists a geodesic normal coordinate system $\{(W,\alpha_j)\}$ of each point in W, and this W is a normal coordinate field in the $\{(W,\alpha_j)\}$.

The establishment of geodesic normal coordinate system is very beneficial to the research of Riemann manifolds.

6.5.3 Lie group and moving-frame method

1. Lie groups

The operation, topological, differential and other structures, on Euclidean space $\mathbb{R}^n$ supply very abundant models for structures of spaces, and if there are various structures on the same space, then there are various different, useful and deep properties present for us. Lie group is an example; There are operation and differential structures on it.

Definition 6.5.20 (Lie group) Let G be a nonempty set. If it satisfies

(1) G is a group $(G,\cdot)$;

(2) G is an r-dimensional smooth manifold;

(3) $\tau: G \to G, \tau(g) = g^{-1}$ and $\varphi: G \times G \to G, \varphi(g_1,g_2) = g_1 g_2$ are smooth mappings;

Then, G is said to be an r-**dimensional Lie group.**

Example 6.5.5 $\mathbb{R}^n$ is an n-dimensional Lie group with the addition of vectors, usual topology and differential structure on $\mathbb{R}^n$.

Example 6.5.6 Ring group $\mathbb{T}^n$ is an n-dimensional Lie group.

In fact, take base $\{e_i: 1 \leqslant i \leqslant n\}$ in $\mathbb{R}^n$. Let $L = \left\{x = \sum_{i=1}^{n} a_i e_i : a_i \in \mathbb{Z}\right\} = \mathbb{Z}^n$, called **"a lattice"**, be a subgroup of $\mathbb{R}^n$. Then, quotient group $\mathbb{T}^n = \mathbb{R}^n / L$ is called **a ring group.** Its topological structure group $\mathbb{T}^n$ is an n-dimensional toroid, thus a compact Lie group.

Example 6.5.7 Let G_1, G_2 be Lie groups. Define an operation $\cdot$ on product manifold $G_1 \times G_2$ by

$$(a_1, a_2), (b_1, b_2) \in G_1 \times G_2 \Rightarrow (a_1, a_2) \cdot (b_1, b_2) = (a_1 b_1, a_2 b_2) \in G_1 \times G_2,$$

then $(G_1 \times G_2, \cdot)$ is a Lie group, called **direct product of two Lie groups.**

Example 6.5.8 Linear groups $GL(n, \mathbb{R})$, $GL(n, \mathbb{C})$ (see Chapter 1)

Linear group $GL(n, \mathbb{R})$ is the set of all $n \times n$-degree non-degenerate real matrices, with the operation structure of the multiplication of matrix, topological structure and differential structure of $\mathbb{R}^{n^2}$, then it is **a non-interchanged Lie group.** Similar for the $GL(n, \mathbb{C})$.

Example 6.5.9 $SL(n, \mathbb{R}) = \{A \in GL(n, \mathbb{R}) : \det A = 1\}$, $O(n, \mathbb{R}) = \{A \in GL(n, \mathbb{R}) : AA^{\mathrm{T}} = I\}$

$SL(n, \mathbb{R})$, $O(n, \mathbb{R})$ are regular sub-manifolds of $GL(n, \mathbb{R})$, thus are Lie subgroups; $SL(n, \mathbb{R})$ is called **a unit modular Lie group**, and $O(n, \mathbb{R})$ is called **a real orthogonal Lie group.**

Definition 6.5.21 (right and left translations of a group) Let $(G, \cdot)$ be a group. If a mapping $R_g : G \to G$ is given by $R_g(x) = x \cdot g$, then it is said to be **a right translation on** G; if a mapping $L_g : G \to G$ is given by $L_g(x) = g \cdot x$, then it is said to be **a left translation on** G. If a tangent vector field X is invariant after right (left) translation, then it is called **a right (left) invariant tangent vector field.**

Theorem 6.5.16 Let G be an r-dimensional Lie group. If X, Y are right invariant tangent vector fields on G, then Poisson bracket $[X, Y] = XY - YX$ consists of a right invariant vector field on G; and so is for the left invariant vector field.

Definition 6.5.22 (Lie algebra) If an r-dimensional real linear space has a multiplication operation satisfying distribution law, anti-interchange law, and the law of Jacobi identity, then it is said to be an r-**dimensional Lie algebra.**

Definition 6.5.23 (Lie algebra of a Lie group) Let G be a Lie group. The set of all right invariant smooth tangent vector fields on G is denoted by $\mathfrak{A}$. If Poisson brackets $[X, Y]$ of invariant smooth tangent vector fields $X, Y \in \mathfrak{A}$ satisfys

(1) $[c_1 X_1 + c_2 X_2, Y] = c_1 [X_1, Y] + c_2 [X_2, Y]$, $c_1, c_2 \in \mathbb{R}$; (distribution law)

(2) $[X, Y] = -[Y, X]$; (anti-interchange law)

(3) $[X, [Y, Z]] + [Y, [Z, X]] + [Z, [X, Y]] = 0$; (Jacobi identity)

Then $\mathfrak{A}$ is called **a Lie algebra of Lie group** G.

In fact, the linear space of all right invariant smooth tangent vector fields of G, denoted by $\mathfrak{A}$, then $\mathfrak{A}$ is a Lie algebra of Lie group G.

2. Constructive constants of a Lie group

Let G be an r-dimensional Lie group, e be the unit of G. The right translation $R_{a^{-1}}$:

$G \to G$ is defined by $R_{a^{-1}}(x) = x \cdot a^{-1}$, then it satisfies

(1) $R_{a^{-1}}(e) = e \cdot a^{-1} = a^{-1}$;

(2) $R_{a^{-1}}(a) = a \cdot a^{-1} = e$;

(3) $R_{a^{-1}}: G \to G$ is a differentiable homeomorphism form G to G itself.

Thus, the tangent mapping $(R_{a^{-1}})_*: T_a(G) \to T_e(G)$ of $R_{a^{-1}}$ is a linear isomorphism from tangent space $T_a(G)$ of manifold G at point $a \in G$ to tangent space $T_e(G)$ of G at unit $e \in G$. Hence, for $X \in T_a(G)$, let

$$\omega(X) = (R_{a^{-1}})_*(X). \tag{6.5.30}$$

Then, ω is a differential form of one-degree defined on $T_a(G)$ and taken value in $T_e(G)$, we call this ω **a fundamental differential form of Lie group** G, or, **Maurer-Cartan form.**

Take a base of $T_e(G)$, denoted by $\{\delta_i, 1 \leqslant i \leqslant r\}$, then

$$\omega = \sum_{i=1}^{r} \omega^i \delta_i, \tag{6.5.31}$$

where ω^i, $i = 1, 2, \cdots, r$, are independent everywhere differential forms of one-degree of Lie group G.

Evaluate exterior differentiation of ω^i in (6.5.31). We have

$$\begin{cases} \mathrm{d}\omega^i = -\dfrac{1}{2} \displaystyle\sum_{j,k=1}^{r} c_{jk}^i \omega^j \wedge \omega^k \\ c_{jk}^i + c_{kj}^i = 0 \end{cases}. \tag{6.5.32}$$

It is easy to prove that ω^i, $\mathrm{d}\omega^i$ are right invariant on Lie group G, and coefficients c_{jk}^i are constant.

Definition 6.5.24 (constructive constants of a Lie group) The coefficients c_{jk}^i in (6.5.32) is said to be the **constructive constants of Lie group** G.

Theorem 6.5.17 Let G be an r-dimensional Lie group. Then the constrictive constants of G satisfies

$$\sum_{j=1}^{r} (c_{jk}^i c_{hl}^j + c_{jh}^i c_{lk}^j + c_{jl}^i c_{kh}^j) = 0. \tag{6.5.33}$$

Note that the constructive constants play important role, since they determine the structure of **local Lie group**, denoted by $\widetilde{G}$, for so-called local Lie group, we refer to [3].

3. Moving-frame method

Denote by $\mathbb{E}(n)$, the moving group of rigid body of n-dimensional Euclidean space $\mathbb{R}^n$, and take an orthogonal frame $(\mathbb{R}^n, O; \delta_1, \cdots, \delta_n)$ on $\mathbb{R}^n$. Denote the (right) action of each motion of rigid body $F \in \mathbb{E}(n)$ at $x \in \mathbb{R}^n$ by

$$x \cdot F \equiv F(x) = x \cdot A + a,$$

where $x = (x_1, \cdots, x_n) = \sum_{\alpha=1}^{n} x_\alpha \delta_\alpha \in \mathbb{R}^n$, and $a = (a_1, \cdots, a_n) = \sum_{\alpha=1}^{n} a_\alpha \delta_\alpha \in \mathbb{R}^n$, orthogonal matrix $A = \begin{bmatrix} \beta_{11} & \cdots & \beta_{1n} \\ \vdots & \ddots & \vdots \\ \beta_{n1} & \cdots & \beta_{nn} \end{bmatrix}$ satisfies $AA^{\mathrm{T}} = I$ with $\det A = 1$. Then, any motion of rigid body $F \in \mathbb{E}(n)$ can be corresponded by a "pair" (A, a):

$$F \in \mathbb{E}(n) \leftrightarrow (A, a) \in (O(n, \mathbb{R}), \mathbb{R}^n).$$

Moreover, an operation $\cdot$ can be introduced in $\mathbb{E}(n)$: if $F_1, F_2 \in \mathbb{E}(n) \leftrightarrow (A_1, a_1)$, (A_2, a_2), then

$$F_1 \cdot F_2 = (A_1 \cdot A_2, a_1 A_2 + a_2),$$

the inverse of $F \in \mathbb{E}(n) \leftrightarrow (A, a)$ is defined by

$$(F)^{-1} = ((A)^{-1}, -a(A)^{-1}).$$

Till now, $\mathbb{E}(n)$ is a $\frac{1}{2}n(n+1)$-dimensional Lie group.

Denote the set of all orthogonal frames on $\mathbb{R}^n$ by $\mathfrak{F}(\mathbb{R}^n) = \{(\mathbb{R}^n, O; \delta_1, \cdots, \delta_n)\}$, and denote by $\mathfrak{F}_+(\mathbb{R}^n)$ the set of all orthogonal frames having the same direction with a fixed frame on $\mathbb{R}^n$. Then, both $\mathfrak{F}(\mathbb{R}^n)$ and $\mathfrak{F}_+(\mathbb{R}^n)$ are main bundles on $\mathbb{R}^n$, and $O(n, \mathbb{R})$, $SO(n, \mathbb{R})$ are their structure groups, respectively, so that, $\mathfrak{F}_+(\mathbb{R}^n) \leftrightarrow \mathbb{E}(n)$.

Hence, we establish the moving frame $(\mathbb{E}(n), p; e_1, \cdots, e_n) \leftrightarrow (\mathbb{R}^n, O; \delta_1, \cdots, \delta_n)$ on $\mathbb{E}(n)$, such that $F \in \mathbb{E}(n) \leftrightarrow (A, a)$ can be expressed in $(\mathbb{E}(n), p; e_1, \cdots, e_n)$ by

$$\begin{cases} \mathrm{d}p = \sum_{\alpha=1}^{n} \omega^\alpha e_\alpha, \\ \mathrm{d}e_\alpha = \sum_{\gamma=1}^{n} \omega_\gamma^\alpha e_\gamma, \end{cases} \tag{6.5.34}$$

and ω^α, ω_γ^α, $1 \leqslant \alpha, \gamma \leqslant n$, are called **relative components of the moving frame.**

The concept of moving frame comes from mechanics, when physicists study the motion of rigid body, they fix an orthogonal frame on rigid body; then the orthogonal frame is moving as rigid body moves. After, a family of orthogonal frames depending on time are generated, and this family totally describes the rigid motion of rigid body. French mathematicians Cotton and Darboux generalized the moving frames with single parameter to those of multi-parameters. After these, moving-frame method combines with the exterior differentiation, play tremendous role in the study of geometry and physics.

6.5.4 Mathematics and modern physics

Mathematician Master S. S. Chern has analyzed incisively and deeply the relationship between morden physics and mathematics in point of view of theoretical physics and differential geometry. He points out that: Both theoretical physics and differential geometry use calculus as tool, the former studies physical phenomena, and the later studies geometric objects. However, the scope of the former is wider than that of the later since any physical phenomena are happened in spaces, but the later is a foundation of the former. Although the thinking and method of research in both subjects are the ratiocination, but theoretical physics needs more support of experiments, whereas differential geometry is not restricted by this, so that its research topics can be selected widely and freely.

We will state some important examples to show the closed relationship of these two scientific fields.

The first example is **dynamics and moving frame.**

In dynamics, to describe the motion of a solid, physicists set up a moving frame on a solid, and observe the motion of moving frame to obtain information. For instance, in $\mathbb{R}^3$, moving frame is a system with point $x \in \mathbb{R}^3$, three unit vectors e_1, e_2, e_3 being perpendicular to each other and passing through x. If x represents coordinate vector of a point, then

$$\begin{cases} \dfrac{\mathrm{d}x}{\mathrm{d}t} = \displaystyle\sum_{\alpha=1}^{3} p_\alpha(t)\, e_\alpha, \\ \dfrac{\mathrm{d}e_\alpha}{\mathrm{d}t} = \displaystyle\sum_{\gamma=1}^{3} q_{\gamma\alpha}(t)\, e_\gamma, \end{cases}$$

with $\alpha = 1, 2, 3$, this is just (6.5.34) with time t, the components $q_{\gamma\alpha}$ of moving frame satisfy $q_{\gamma\alpha}(t) + q_{\alpha\gamma}(t) = 0$, then, functions $p_\alpha(t)$, $q_{\gamma\alpha}(t)$ describe motion of moving frame and solid, fully.

On the other hand, dynamics are closely connected with curve and surface theories of space. Moreover, biparameter moving-frame method is used in research of dynamics, then, (6.5.34) is interpreted as a partial differential equation system, expressed by exterior differentiation

$$\begin{cases} \mathrm{d}x = \displaystyle\sum_{i} \omega_i e_i \\ \mathrm{d}e_i = \displaystyle\sum_{j} \omega_{ji} e_i \end{cases} \quad \text{and} \quad \omega_{ij} + \omega_{ji} = 0, \quad 1 \leqslant i, j \leqslant 3,$$

where ω_i and ω_{ij} are exterior differential forms of one-degree in parameter spaces, take exterior differentiation, then we get

$$\begin{cases} d\omega_i = \sum_j \omega_j \wedge \omega_{ji}, \\ d\omega_{ij} = \sum_k \omega_{ik} \wedge \omega_{jk}, \end{cases} \quad 1 \leqslant i, j \leqslant 3.$$

This is a famous Maurer-Cartan equation system of Lie group G. This equation system is the dual with respect to multiplication equation system of Lie algebra of Lie group G. Thus, from dynamics to moving frame, and then to basic equation of Lie group, it is a very natural process.

Further, the above process is continued: After publishing of **the generalized relativity theory of Einstein**, Elie Cartan (one of the three greatest differential geometricians, the other two are Gauss and Riemann) published papers continuously, stated and developed **the generalized affine space theory**, pointed out applications of this theory in **relativity theory**, as well as a generalization of Maurer-Cartan equation system:

$$\begin{cases} d\omega_i = \sum_{j=1}^{3} \omega_j \wedge \omega_{ji}, & 1 \leqslant i \leqslant 3, \\ d\omega_{ij} = \sum_{k=1}^{3} \omega_{ik} \wedge \omega_{jk} + \Omega_{ij}, & 1 \leqslant i, j \leqslant 3, \end{cases}$$

where Ω_{ij} is differential form of two-degree, called **the curvature form**, and the above equation system is **the basic equations of three-dimensional Riemann geometry.**

The second example **is the surface theory and soliton, as well as σ-pattern.**

In 3-dimensional Euclidean space $\mathbb{R}^3$, take a surface S. Let x be a coordinate vector of S, and ξ be the unit normal vector. Then, invariant variables on S are two differential forms of two-degree

$$\mathrm{I} = (dx, dx) \quad \text{and} \quad \mathrm{II} = -(dx, d\xi),$$

called **the first** and, **the second basic form, of surface** S, respectively.

The first basic form $\mathrm{I} = (dx, dx)$ of S is positive and definite.

The second basic form $\mathrm{II} = -(dx, d\xi)$ of S has **two characteristic values,** κ_i, $i = 1, 2$, called **main curvatures** of surface S; and symmetric function $H = \frac{1}{2}(\kappa_1 + \kappa_2)$ is called **the middle curvature of** S, symmetric function $K = \kappa_1 \kappa_2$ is called **the total curvature** (or **Gauss curvature**) of S.

These curvatures have their senses of geometry: for instance, a surface with $H =$

$\frac{1}{2}(\kappa_1+\kappa_2)=0$ is the minimum surface; and a surface with $K=-1$ is that one, having real asymptotic curves and not coincident. Suppose that the angle of asymptotic curves is φ, then selecting parameters u, t on S, such that $\varphi_{ut}=\sin\varphi$, this is the famous **Sine-Gordon** (SG) **equation.** If $\varphi_{ut}=\varphi(u,t)$ is interpreted as a wave on straight line u, and t is time, then SG equation has soliton solution.

Moreover, the case of surface with constant middle curvature $H=\frac{1}{2}(\kappa_1+\kappa_2)=c$, including minimum surface, has lots of applications in theoretical physics, such as, one can define **"energy"** $E(f)$ of a mapping $f: X \to Y$ from Riemann manifold X to Riemann manifold Y; and the critical mapping of this functional is called **a harmonic mapping.** As is well-known, harmonic mapping plays active role in physics science. In 1980, some physicists determined the set of all harmonic mappings from two-dimensional sphere S^2 to n-dimensional complex projective space $P_n(\mathbb{C})$, called the σ-pattern. More and more open problems are raised, nowadays.

The third example is **the gauge field.**

The foundation of mathematics of gauge field is the concept of vector bundle.

The vector bundle plays essential role in modern mathematics, the key idea is to connect product space $X \times Y$ with vector bundle E. Consider mapping $\pi: E \to X$, satisfying: $\forall x \in X$, there exists a $U \subset X$, such that $\pi^{-1}(U)$ and $U \times Y$ are equal to each other, topologically. Clearly, $X \times Y$ and E are equal, locally. So that, an important problem is: Whether this locality is global, certainly? In fact, this is a very delicate and pretty question which introduces a concept of **characteristic class**, and pushes the development of differential geometry. Further, for vector bundle E and bundle projective mapping $\pi: E \to X$, if a mapping $F: X \to E$ satisfies $\pi \circ F(x)=x, x \in X$, then F is called **a intersecting surface.** To define the differentiation of a intersecting surface, we need a concept of **connection**, and so-called **the gauge field**, it is a connection of vector bundle.

In physics, all theory will be induced to "**quantization**"; correspondingly, in mathematics, the **infinite-dimensional spaces and discrete phenomena** have to be studied. Physics needs mathematics, whereas mathematics can not depart from physics, they are connected closely.

Exercise 6

1. Determine $U_2 \cap U_3 \neq \varnothing$ and corresponding $\varphi_{U_2}, \varphi_{U_3}$ in Example 6. 1. 2, and show

smoothness of them.

2. Let $U \subset \mathbb{R}^3$ be an open set in $\mathbb{R}^3$, a mapping $F: U \to \mathbb{R}$ be continuous differentiable function $F \in C^1(\mathbb{R}^3)$. If $F(x_1, x_2, x_3) = ax_1 + bx_2 + cx_3$ with $(x_1, x_2, x_3) \in \mathbb{R}^3$ and real a, b, c with $a^2 + b^2 + c^2 > 0$, such that $F(x_1, x_2, x_3) = 0$ is a plane in $\mathbb{R}^3$ passing through the origin, and $F^{-1}(0) \subset \mathbb{R}^3$. Prove that $F^{-1}(0)$ is a two-dimensional linear subspace of $\mathbb{R}^3$.

3. Prove: the differentiability of real function $f: M \to \mathbb{R}$ at point $p \in M$ on m-dimensional smooth manifold M is independent of the choice of permissible chart system.

4. Let $e_1 = (1,0,0)$, $e_2 = (1,1,0)$, $e_3 = (1,1,1)$ form be a base of $\mathbb{R}^3$. Find the dual base e_1^*, e_2^*, e_3^*.

5. Prove: the differentiability of a mapping $f: M \to N$ at point $p \in M$ from m-dimensional smooth manifold M onto n-dimensional smooth manifold N is independent of the choice of permissible chart system.

6. Prove Theorem 6.1.6.

7. Prove relationships (6.1.13), (6.1.14).

8. Let M, N be m-dimensional, n-dimensional manifolds, respectively, $F: M \to N$ be a smooth mapping. Find the transform matrix between differential mapping $F^*: T_q^*(N) \to T_p^*(M)$ and tangent mapping $F_*: T_p(M) \to T_q(N)$ with respect to natural bases.

9. Prove Theorem 6.2.2.

10. Interpret the sense of $\sigma(x(v_1^*, \cdots, v_r^*)) = x(v_{\sigma(1)}^*, \cdots, v_{\sigma(r)}^*)$, $v_j^* \in V^*$, and show an example.

11. Prove: the mapping $\sigma: T^r(V) \to T^r(V)$, $\forall \sigma \in \mathfrak{T}(r)$, is a self-homomorphism.

12. Prove: symmetrical tensor is invariant with respect to the action of symmetric operator; and skew-symmetrical tensor is invariant with respect to the action of skew-symmetric operator.

13. Why we take the permutation $\tau = \begin{pmatrix} 1 & \cdots & k & k+1 & \cdots & k+l \\ 1+l & \cdots & k+l & 1 & \cdots & l \end{pmatrix}$ in the proof of Theorem 6.2.5?

14. Prove: Theorems 6.2.7, 6.2.8, 6.2.9.

15. Prove: Theorems 6.2.12, 6.2.13.

16. Prove: Formulas (6.2.24) and (6.2.25) are the bases of spaces $\Lambda(V)$ and $\Lambda(V^*)$, respectively.

17. Give and prove the counterpart in four-dimensional case of Theorem 6.3.2.

18. Give the representation of exterior differential form of four-degree.

19. Prove: the anti-Leibniz formula of exterior differential form ω of p-degree and

exterior differential form of q-degree

$$d(\omega \wedge \theta) = d\omega \wedge \theta + (-1)^p \omega \wedge d\theta.$$

20. Evaluate exterior differential form of one-degree and two-degree of

$$\omega = (y^3 - z^3)\,dx + (z^3 - x^3)\,dy + (x^3 - y^3)\,dz.$$

21. Evaluate $d\omega$ of $\omega = (x^2 + y^2 + z^2)dx + (2xy - z^2)dy + (xy + yz + zx)dz$.

22. Evaluate $d\omega$ of exterior differential form of two-degree $\omega = 2x\,dy \wedge dz + 2y\,dz \wedge dx + 3z\,dx \wedge dy$ in the exterior differential form space $A(\mathbb{R}^3)$.

23. Let V be two-dimensional linear space with base $\{e_1, e_2\}$. Show the expression of skew-symmetric covariate tensor of one-degree; If $\omega \in A(M)$, give the expression of $d\omega$.

24. Let $\omega = -y\,dx + x\,dy$ be exterior differential form of one-degree on $\mathbb{R}^3$, and $\alpha = \omega|_{S^2}$ be the restriction of ω on the unit sphere $S^2 = \{(x, y, z) \in \mathbb{R}^3 : x^2 + y^2 + z^2 = 1\}$ in $\mathbb{R}^3$. Evaluate $d(\alpha|_U)$ and $d(\alpha|_V)$ with respect to local coordinate system $\{(U; u, v)\}$ on S^2 with $U = \left\{(x, y, z) \in S^2 : x = u, y = v, z = \sqrt{1 - u^2 - v^2}\right\}$.

25. Evaluate $f^*\omega, f^*d\omega$, if $\omega = xy\,dx + z\,dy - yz\,dz$, and smooth mapping $f: \mathbb{R}^2 \to \mathbb{R}^3$ is given by $f(u, v) = (uv, u^2, 3u + v)$.

26. Prove: Theorems 6.4.1 and 6.4.2.

27. Let $T^2 = \left\{(x, y, z) \in \mathbb{R}^3 : \left(\sqrt{x^2 + y^2} - R\right)^2 + z^2 = r^2\right\}, 0 < r < R$. If $I: T^2 \to \mathbb{R}^3$ is an embedding mapping, and positive direction of T^2 is the outer normal direction. Evaluate $\int_{T^2} I^*\omega$ for $\omega = x\,dy \wedge dz + y\,dz \wedge dx + z\,dx \wedge dy$.

28. Let $S^n(1) = \{(x_1, \cdots, x_n, x_{n+1}) \in \mathbb{R}^{n+1} : x_1^2 + \cdots + x_n^2 + x_{n+1}^2 = 1\}$ be the unit sphere in $\mathbb{R}^{n+1}$. Prove: $S^n(1)$ is an n-dimensional Riemann manifold in $\mathbb{R}^{n+1}$.

29. Evaluate the first basic form of the following surfaces in $\mathbb{R}^3$:

(1) $r = (u\cos v, u\sin v, av), a > 0$; (2) $r = (\cosh u \cos v, \cosh u \sin v, u)$.

30. Interpret the theory of Riemann geometry in case $n = 5, 6$.

Chapter 7

Complimentary Knowledge

The complimentary knowledge is the self-reading contents for readers.

7.1 Variational Calculus

Variational approach is a mathematical method to deal with the maximum and minimum problems. The extremal problems with finite variables are dealt with in *advanced calculus*, whereas for extreme value problems of functionals are studied in variational calculus.

7.1.1 Variation and variation problems

1. Examples

We show some examples, firstly.

Example 7.1.1 Extreme value problem of 2-degree polynomial

Consider the two-degree polynomial of n real variables $u_1, \cdots, u_n \in \mathbb{R}$

$$J(u) = \sum_{j,k=1}^{n} a_{jk} u_j u_k + 2\sum_{j=1}^{n} a_j u_j + a_0 \tag{7.1.1}$$

with real coefficients a_{jk}, a_j, a_0, and real symmetric matrix $[a_{jk}]_{n\times n}, a_{jk} = a_{kj}$.

To find u on which $J(u)$ takes extreme value, we search out the "**static point**" by the method in *Advanced Calculus*, firstly, i. e. solve equation system

$$\frac{1}{2}\frac{\partial J}{\partial u_j} = \sum_{k=1}^{n} a_{jk} u_k + a_j = 0, \quad j = 1, 2, \cdots, n. \tag{7.1.2}$$

Let $A = [a_{jk}]$, $A^{-1} = [a_{jk}]^{-1} = [a'_{jk}]$, $b = \begin{bmatrix} a_1 \\ \vdots \\ a_n \end{bmatrix}$, $u = \begin{bmatrix} u_1 \\ \vdots \\ u_n \end{bmatrix}$, then (7.1.2) deduces to $Au + b = 0$. Solve it, we have solution

$$\tilde{u} = -A^{-1}b = -[a'_{jk}]\begin{bmatrix} a_1 \\ \vdots \\ a_n \end{bmatrix}, \tag{7.1.3}$$

where $\tilde{u}$ is called **a static point of** $J(u)$, and $J(\tilde{u})$ is called **a static value.** Substitute $\tilde{u}$ into $J(\tilde{u})$,

$$\begin{aligned} J(\tilde{u}) &= (\tilde{u})^{\mathrm{T}}A\tilde{u} + 2b^{\mathrm{T}}\tilde{u} + a_0 = \{-b^{\mathrm{T}}(A^{-1})^{\mathrm{T}}\}A\{-A^{-1}b\} + 2b^{\mathrm{T}}\{-A^{-1}b\} + a_0 \\ &= b^{\mathrm{T}}\{(A^{-1})^{\mathrm{T}}\}AA^{-1}b - 2b^{\mathrm{T}}A^{-1}b + a_0 = b^{\mathrm{T}}A^{-1}b - 2b^{\mathrm{T}}A^{-1}b + a_0 \\ &= -b^{\mathrm{T}}A^{-1}b + a_0 = -\sum_{j,k=1}^{n} a'_{jk}a_j a_k + a_0, \end{aligned}$$

by the symmetry of A^{-1}. Let $u_j - \tilde{u}_j = \delta\tilde{u}_j$ $(j=1,2,\cdots,n)$, and by

$$\delta\tilde{u}_j\delta\tilde{u}_k = (u_j - \tilde{u}_j)(u_k - \tilde{u}_k) = u_j u_k - \tilde{u}_j u_k - u_j\tilde{u}_k + \tilde{u}_j\tilde{u}_k,$$

it follows

$$\begin{aligned} J(u) - J(\tilde{u}) &= u^{\mathrm{T}}Au + 2b^{\mathrm{T}}u + a_0 - (\tilde{u})^{\mathrm{T}}A\tilde{u} - 2b^{\mathrm{T}}\tilde{u} - a_0 \\ &= u^{\mathrm{T}}Au + 2b^{\mathrm{T}}u - (\tilde{u})^{\mathrm{T}}A\tilde{u} - 2b^{\mathrm{T}}\tilde{u} \\ &= \sum_{j,k=1}^{n} a_{jk}(u_j u_k - \tilde{u}_j\tilde{u}_k) + 2\sum_{j=1}^{n} a_j(u_j - \tilde{u}_j) = \sum_{j,k=1}^{n} a_{jk}\delta\tilde{u}_j\delta\tilde{u}_k. \end{aligned}$$

Thus, the problem turns to discussing quadratic form $x^{\mathrm{T}}Ax = \sum_{j,k=1}^{n} a_{jk}x_j x_k$ with matrix $A = [a_{jk}]$ and $x = \begin{bmatrix} x_1 \\ \vdots \\ x_n \end{bmatrix}$. Hence,

$J(\tilde{u})$ is a minimum $\Leftrightarrow$ quadratic form $(\delta\tilde{u})^{\mathrm{T}}A(\delta\tilde{u})$ is positive and definite;

$J(\tilde{u})$ is a maximum $\Leftrightarrow$ quadratic form $(\delta\tilde{u})^{\mathrm{T}}A(\delta\tilde{u})$ is negative and definite.

We call $\tilde{u}$ **the extreme point of** $J(u)$, and $J(\tilde{u})$ **the extreme value of** $J(u)$.

Example 7.1.2 Characteristic problems

Consider a quotient form of n real variables $u_1, \cdots, u_n \in \mathbb{R}$

$$J(u) = \frac{\sum_{j,k=1}^{n} a_{jk}u_j u_k}{\sum_{j,k=1}^{n} b_{jk}u_j u_k} = \frac{u^{\mathrm{T}}Au}{u^{\mathrm{T}}Bu} = \frac{(u, Au)}{(u, Bu)}, \tag{7.1.4}$$

where coefficients a_{jk}, b_{jk} are real numbers; $A = [a_{jk}]_{n\times n}$, $B = [b_{jk}]_{n\times n}$ are real symmetric matrices with $a_{jk} = a_{kj}$, $b_{jk} = b_{kj}$, respectively; $|B| \neq 0$ and u^{T} is the transposed matrix of u. Find the extreme value and extreme point of $J(u)$ in (7.1.4).

Since $J(u)=\dfrac{u^{\mathrm{T}}Au}{u^{\mathrm{T}}Bu}=\dfrac{(u,Au)}{(u,Bu)}$ is a **zero-degree homogeneous form**, i. e. , $J(\lambda u)=J(u)$, then, we may suppose that "the inner product" of u and Bu satisfies $(u,Bu)=1$ without loss of generality. This problem has an application to find the minimum (maximum) radius of an ellipse.

To find minimum value of $J(u)$, we write out the "**static condition**"

$$\frac{\partial J}{\partial u_j}=0, \quad j=1,2,\cdots,n. \tag{7.1.5}$$

Rewrite (7. 1. 4) as $(u^{\mathrm{T}}Bu)J(u)=u^{\mathrm{T}}Au$, and let $\lambda=J(u)$. Then, (7. 1. 5) deduces to

$$\begin{cases}\displaystyle\sum_{j,k=1}^{n}(a_{jk}-\lambda b_{jk})u_k=0,\\ \lambda=J(u),\end{cases}$$

or simply, $Au=\lambda Bu$, thus the problem deduces to an algebraic problem, i. e. find **the characteristic values** (eigenvalues) $\lambda_1,\cdots,\lambda_n$ of matrix A, or, find the solutions of **the characteristic equation** (eigenequation) $|A-\lambda I|=0$; correspondingly, find **the characteristic function** (eigenfunction, or eigenvector) u satisfying $J(u)=\lambda$.

The more general case is to find the extreme values and extreme points of

$$J(u)=J(u_1,\cdots,u_n), \quad (u_1,\cdots,u_n)\in D\subset\mathbb{R}^n.$$

And we have the definitions:

Minimal point—if $\tilde{u}\in D$ satisfies $J(u)\geqslant J(\tilde{u})$, then it is called **a generalized minimal point of** $J(u)$; if $\tilde{u}\in D$ satisfies $J(u)>J(\tilde{u})$, then it is called **a minimal point of** $J(u)$. Similarly, we may define **maximal point.**

About static point, we have:

Static point—two necessary conditions: ① if $\tilde{u}\in D, \tilde{u}\notin\partial D$, such that $\dfrac{\partial J}{\partial u_j}=0, j=1,2,\cdots,n$; ② $\tilde{u}\in\partial D$. After finding static points, we must use the sufficient conditions to check whether the static points are extreme points.

Example 7. 1. 3 Shortest-path problem (Fig. 7. 1. 1)

A particle P moves along a smooth curve from point P_0 to P_1 by gravity. Find a smooth curve γ along which P spends the shortest time moving from P_0 to P_1 by gravity ?

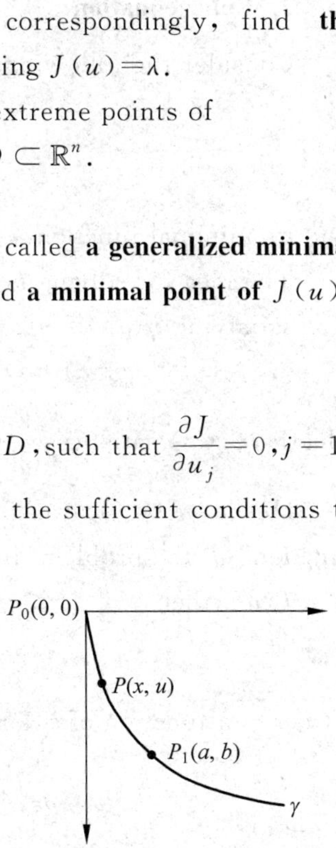

Fig. 7. 1. 1 Shortest path

Solve Suppose the equation of curve γ is $u=u(x)$, and gravity is g; the arc differential on γ is

$$ds = \sqrt{1+[u'(x)]^2}\ dx.$$

For any point $P(x,u)$ on γ, the velocity along γ is $\sqrt{2gu}$, and the time along ds is

$$dt = v^{-1} ds = \frac{1}{\sqrt{2gu}}\sqrt{1+[u'(x)]^2}\ dx.$$

Without loss of generality, let $2g=1$. Then the moving time from P_0 to P_1 is $J(u) = \int_0^a \frac{\sqrt{1+[u'(x)]^2}}{\sqrt{u(x)}} dx$, and the problem induces to find a minimal function $u(x)$ satisfying $u(0)=0, u(a)=b$, such that $J(u)$ takes the minimal value, i. e. find the minimal function $u(x)$ of problem

$$\begin{cases} J(u) = \int_0^a \frac{\sqrt{1+[u'(x)]^2}}{\sqrt{u(x)}} dx, \\ u(0)=0, \quad u(a)=b. \end{cases} \tag{7.1.6}$$

We will find solutions of (7.1.6) in subsection 3, Example 7.1.5, later.

2. Euler equation

Consider the following problem

$$\begin{cases} J(u) = \int_{a_0}^{a_1} F(x,u,u')\, dx \\ u(a_0)=b_0, u(a_1)=b_1 \end{cases}, \tag{7.1.7}$$

find its minimal function $u=\tilde{u}(x)$, see Fig. 7.1.2.

The idea for solving is take $\eta(x)$ satisfying $\eta(a_0)=\eta(a_1)=0$; then, $\forall \varepsilon \in \mathbb{R}$, let

$$u_\varepsilon(x) = \tilde{u}(x) + \varepsilon\eta(x). \tag{7.1.8}$$

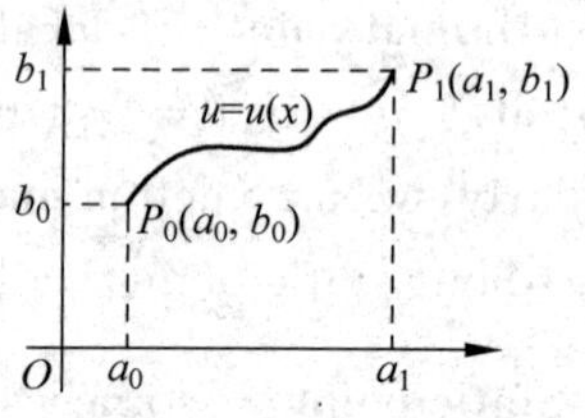

Fig. 7.1.2 $u=u(x)$

Clearly, $u_\varepsilon(x)$ satisfies $\begin{cases} u_\varepsilon(a_0)=\tilde{u}(a_0)=b_0 \\ u_\varepsilon(a_1)=\tilde{u}(a_1)=b_1 \end{cases}$.

Hence, $u_\varepsilon(x)$ can be regarded as a "**permissible function**" of the problem. In fact, it is the function from variation of $\tilde{u}(x)$. Substitute it into $J(u)$, then

$$J(u_\varepsilon) = \int_{a_0}^{a_1} F(x,u_\varepsilon,u'_\varepsilon)\ dx.$$

$J(u_\varepsilon)$ is a function of ε, and it takes minimal value when $\varepsilon=0, u_0(x)=\tilde{u}(x)$. So that

$$\frac{\partial}{\partial\varepsilon} J(u_\varepsilon)\Big|_{\varepsilon=0} = \frac{\partial}{\partial\varepsilon}\int_{a_0}^{a_1} F(x,u_\varepsilon,u'_\varepsilon)\ dx\Big|_{\varepsilon=0}$$

$$=\frac{\partial}{\partial \varepsilon}\int_{a_0}^{a_1} F(x,\tilde{u}(x)+\varepsilon\eta(x),(\tilde{u})'(x)+\varepsilon\eta'(x))\,\mathrm{d}x\Big|_{\varepsilon=0}$$

$$=\int_{a_0}^{a_1}\left\{\frac{\partial}{\partial u}F(x,\tilde{u}(x),(\tilde{u})'(x))\,\eta(x)+\frac{\partial}{\partial u'}F(x,\tilde{u}(x),(\tilde{u})'(x))\,\eta'(x)\right\}\mathrm{d}x$$

$$=\int_{a_0}^{a_1}\widetilde{F}_u(x)\eta(x)\mathrm{d}x+\widetilde{F}_{u'}(x)\eta(x)\big|_{a_0}^{a_1}-\int_{a_0}^{a_1}\left(\frac{\mathrm{d}}{\mathrm{d}x}\widetilde{F}_{u'}(x)\right)\eta(x)\,\mathrm{d}x$$

$$=\int_{a_0}^{a_1}\left(\widetilde{F}_u-\frac{\mathrm{d}\widetilde{F}_{u'}}{\mathrm{d}x}\right)\eta(x)\mathrm{d}x=0.$$

This means that $u=\tilde{u}(x)$ must satisfy differential equation

$$\widetilde{F}_u-\frac{\mathrm{d}\widetilde{F}_{u'}}{\mathrm{d}x}=0, \tag{7.1.9}$$

or rewrite as

$$\widetilde{F}_u-\frac{\mathrm{d}\widetilde{F}_{u'}}{\mathrm{d}x}=F_u(x,u,u'(x))-\frac{\mathrm{d}}{\mathrm{d}x}F_{u'}(x,u,u'(x))=0. \tag{7.1.10}$$

The equation (7.1.9) or (7.1.10) is called **Euler equation**, it is **a necessary condition** for which $J(u)$ takes extreme value.

In the above deduction, we have used the "**basic lemma of variation**": if a continuous function $f(x)$ satisfies integral equation $\int_{a_0}^{a_1} f(x)\eta(x)\mathrm{d}x=0$ for any function $\eta(x)$ with some smoothness and boundary conditions $\eta(a_0)=\eta(a_1)=0$, then it holds $f(x)\equiv 0$. In section 7.1.2, we will be back to this lemma.

3. Integration method of Euler equation

Euler equation (7.1.10) is a differential one, we expound its integrals in some special cases of F.

1) If F does not contain u, i.e., $F(x,u')=0$: (7.1.10) presents as $\frac{\mathrm{d}F_{u'}}{\mathrm{d}x}=0$, then we get a first integral

$$F_{u'}(x,u'(x))=C.$$

2) If F does not contain x, i.e., $F(u,u')=0$: (7.1.10) presents as

$$\frac{\mathrm{d}}{\mathrm{d}x}(F-u'F_{u'})=0, \tag{7.1.11}$$

because

$$\frac{\mathrm{d}}{\mathrm{d}x}(F-u'F_{u'})=F_x+F_u\cdot u'+F_{u'}\cdot u''-u''F_{u'}-u'\frac{\mathrm{d}}{\mathrm{d}x}F_{u'}$$

$$=F_u \cdot u' - u'\frac{d}{dx}F_{u'} = u'\left(F_u - \frac{d}{dx}F_{u'}\right) = 0.$$

Solve (7.1.11), we get a first integral $F - u'F_{u'} = C$.

Example 7.1.4 Generalized shortest-path problem: $J(u) = \int_{a_0}^{a_1} f(u)\sqrt{1+[u'(x)]^2}\,dx$

For this $J(u)$, we have $F(x,u,u') = f(u)\sqrt{1+[u'(x)]^2}$, and F does not contain x. Thus, by

$$\frac{d}{dx}(F - u'F_{u'}) = \frac{d}{dx}\left[f(u)\sqrt{1+(u')^2} - u'f(u)\frac{u'}{\sqrt{1+(u')^2}}\right] = \frac{d}{dx}\left[\frac{f(u)}{\sqrt{1+(u')^2}}\right] = 0,$$

the first integral $\dfrac{f(u)}{\sqrt{1+(u')^2}} = c$ is obtained. So that, $\dfrac{c^2(u')^2}{f^2(u) - c^2} = 1$, and for the extreme problem

$$J(u) = \int_{a_0}^{a_1} f(u)\sqrt{1+[u'(x)]^2}\,dx,$$

the minimal function $u(x)$ satisfies

$$c\int [f^2(u) - c^2]^{-\frac{1}{2}}\,du = \pm(x - \xi). \tag{7.1.12}$$

Note, that $u = \text{const}$ maybe a solution of the problem.

Example 7.1.5 Solutions of the shortest-path problem in Example 7.1.3.

For $J(u) = \int_0^a \dfrac{\sqrt{1+[u'(x)]^2}}{\sqrt{u(x)}}dx$ with $u(0)=0, u(a)=b$ in (7.1.6), we may use the result in Example 7.1.4 by setting $f(u) = \dfrac{1}{\sqrt{u}}$, then get $c\int\left[\dfrac{u}{1-c^2u}\right]^{\frac{1}{2}}du = \pm(x-\xi)$.

Suppose that $u = c^{-2}\sin^2\theta$, then the solutions are

$$\begin{cases} x = \dfrac{1}{2c^2}(2\theta - \sin 2\theta) + \xi \\ u = \dfrac{1}{2c^2}(1 - \cos 2\theta) \end{cases}, \tag{7.1.13}$$

this is **a pendulum line**, it is **the locus** of curves which describe by a point on the circle with radius $\dfrac{1}{2c^2}$, turning along Ox axis.

Note that: $u = b$ is not extreme point, since $f'(b) \neq 0$. Although $u = c^{-2}$ is a singular point of function equation $F - u'F_{u'} = C$, but it is not solution of Euler equation. Hence, only (7.1.13) is an extreme curve of our problem.

To find the solution with boundary conditions, we take, for instance, $\theta = 0$ as a starting point, then $\xi = 0$. Constant c^2 can be determined by a point (a_1, b_1) which the curve passes.

In Fig. 7.1.3, the center of circle is A, a point on pendulum line is B, tangent point cutting with Ox axis is C, and $AC = \frac{1}{2c^2}$, $\angle BAC = 2\theta$.

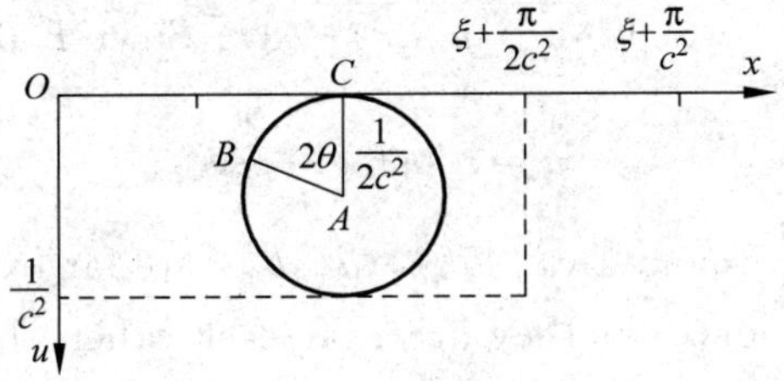

Fig. 7.1.3 Pendulum line

The existence of c^2 is by audiovisual geometry, and it is unique. If $\frac{b_1}{a_1} \geqslant \frac{2}{\pi}$, then the particle P falls down to the earth surface along the pendulum line; If $\frac{b_1}{a_1} < \frac{2}{\pi}$, then P falls down on the earth surface, after a while jumps up, and then falls down again.

More generally, for given points (a_0, b_0), (a_1, b_1), if $b_0 \geqslant 0, b_1 \geqslant 0$, then the minimal curve exists and is unique, passing through these two points.

Example 7.1.6 Another extreme value problem (Fig. 7.1.4)

In Example 7.1.4, take $f(u) = \sqrt{u}$, then it is an application of "the minimal action principle" to the throwing motion in mechanics. We solve this problem for $J(u) = \int_{a_0}^{a_1} \sqrt{u(x)}\sqrt{1+[u'(x)]^2}\,dx$, and get a first integral by (7.1.12)

$$u = \frac{1}{4c^2}(x-\xi)^2 + c^2. \qquad (7.1.14)$$

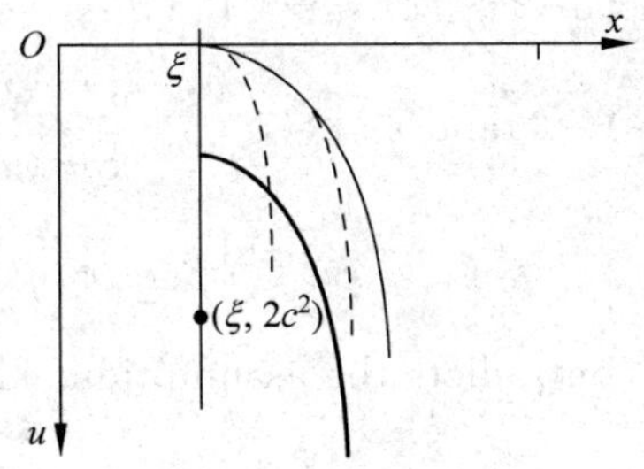

Fig. 7.1.4 Parabola

By similar discussion in Example 7.1.4, the above u in (7.1.14) is the unique extreme curve, a parabola with the director line $u=0$, and axis $x=\xi$, which is parallel to u-axis, its focus is point $(\xi, 2c^2)$.

Parameters (integral constants) ξ, c can be used to determine boundary conditions. In fact, there is one and only one extreme curve passing through given point (a_0, b_0), it is a parabola $u=\frac{(x-a_0)^2}{4b_0}$.

Example 7.1.7 In Example 7.1.4, take $f(u)=u$, it become a minimum problem: find a surface Σ generated by curve $u=u(x)$ passing through given points $P_0(a_0,b_0)$, $P_1(a_1,b_1)$, and turning around Ox axis such that the area of Σ is minimum; i.e. find the extreme value of $J(u)=\int_{a_0}^{a_1} u(x)\sqrt{1+[u'(x)]^2}\,dx$. Since F does not contain x, then,

$$u=c\cosh\frac{x-\xi}{c},$$

it is called **a catenary**. If two points (a_0,b_0), (a_1,b_1) are far from each other on Ox axis, then there is no extreme curve; if they are near each other, then maybe there exist two extreme curves.

7.1.2 Variation principle

1. Variation basic lemma

Now we prove "basic lemma of the variation" which is the variation principle.

"Basic lemma of the variation": If a continuous function $f(x)$ satisfies integral equation $\int_{a_0}^{a_1} f(x)\eta(x)\,dx=0$ for any function $\eta(x)$ with certain smoothness and boundary conditions $\eta(a_0)=\eta(a_1)=0$, then $f(x)\equiv 0$.

Proof Suppose that this would be not true, then, $\exists\, x_0\in(a_0,a_1)$, such that $f(x_0)\neq 0$, say $f(x_0)>0$. By the continuity of $f(x)$, there exists $\zeta>0$, such that $f(x)>0$ on interval $(x_0-\zeta,x_0+\zeta)$. Thus, we may take a smooth function $\eta(x)$ satisfying $\eta(a_0)=\eta(a_1)=0$, and $\eta(x)\begin{cases}>0, & x\in(x_0-\zeta,x_0+\zeta)\\ =0, & x\notin(x_0-\zeta,x_0+\zeta)\end{cases}$. Then,

$$\int_{a_0}^{a_1} f(x)\eta(x)\,dx=\int_{x_0-\zeta}^{x_0+\zeta} f(x)\eta(x)\,dx>0,$$

this contradicts the assumption. The proof is complete.

2. The lowest value, minimal value

For the extreme problem $\begin{cases}J(u)=\int_{a_0}^{a_1} F(x,u,u')\,dx\\ u(a_0)=b_0, u(a_1)=b_1\end{cases}$ in (7.1.7), function $F(x,u,u')$ is with certain smoothness on domain $D\subset\mathbb{R}^3$, and $J(u)=\int_{a_0}^{a_1} F(x,u,u')\,dx$ is a functional with piecewise smooth function $u = u(x)$. If the $u = u(x)$ satisfies (7.1.7), then is called **a suitable function.** Denote the set of all suitable functions by $\Psi = \{u: u(x)$ is suitable function of $J(u)\}$.

1) The smallest value of functional $J(u)$

If there exists a suitable function $u = \tilde{u}(x) \in \Psi$ such that it holds

$$J(\tilde{u}) \leqslant J(u), \quad \forall u \in \Psi, \tag{7.1.15}$$

then $J(\tilde{u})$ is **the smallest value of** $J(u)$, and $\tilde{u}$ is called **the smallest function of** $J(u)$. Moreover, if

$$J(\tilde{u}) < J(u), \quad u \neq \tilde{u}, \forall u \in \Psi,$$

then $J(\tilde{u})$ is called **the strictly smallest value of** $J(u)$; otherwise, called **the generalized smallest one.**

2) Minimal value of functional $J(u)$

① **Strong minimal value**

For a positive number $\delta > 0$, the set Ψ_δ contained in Ψ satisfying

$$|u(x) - \tilde{u}(x)| < \delta, \quad a_0 \leqslant x \leqslant a_1 \tag{7.1.16}$$

is called **a δ-near neighbor of** $\tilde{u}$.

If there exists a suitable $\delta > 0$, such that it holds

$$J(\tilde{u}) \leqslant J(u), \quad \forall u \in \Psi_\delta,$$

then $J(\tilde{u})$ is called **the strong minimal value of functional** $J(u)$; Also may define the **strictly strong** minimal value and **generalized strong** minimal value of $J(u)$.

② **Weak minimal value**

For a positive number $\delta > 0$, the set Φ_δ contained in Ψ satisfying

$$|u(x) - \tilde{u}(x)| < \delta, \quad |u'(x) - (\tilde{u})'(x)| < \delta, \quad a_0 \leqslant x \leqslant a_1 \tag{7.1.17}$$

is called **a δ-weak near neighbor of** $\tilde{u}$.

If there exists a suitable $\delta > 0$, such that it holds

$$J(\tilde{u}) \leqslant J(u), \quad \forall u \in \Phi_\delta,$$

then $J(\tilde{u})$ is called **the weak minimal value of functional** $J(u)$; Also may define the **strictly weak** minimal value and **generalized weak** minimal value of $J(u)$.

The largest value of functional $J(u)$ **and the maximal value of functional** $J(u)$ can be defined, similarly.

7.1.3 More general variation problems

1. Case of several unknown functions

Consider k unknown functions $u_1(x), \cdots, u_k(x)$, the extreme problem is presented by

$$\begin{cases} J(u)=\int_{a_0}^{a_1} F(x, u_1, \cdots, u_k, u_1', \cdots, u_k')\,dx, \\ u_j(a_0)=b_{0j}, u_j(a_1)=b_{1j}, 1 \leqslant j \leqslant k; \end{cases}$$

then, the corresponding Euler equation system is

$$F_{u_j}-\frac{dF_{u_j'}}{dx}=0, \quad j=1,2,\cdots,k, \tag{7.1.18}$$

with **regular condition** $\det\left[\frac{\partial^2 F}{\partial u_j' \partial u_s'}\right]_{1 \leqslant j, s \leqslant k} \neq 0$.

In the analytic mechanic, physicists take x as time variable, $u_1(x), \cdots, u_k(x)$ as the general coordinates, and F as a Lagrange function which represents the difference between the kinetic and potential energy, then, the system of this motion is determined with J, Euler equation system (7.1.18) is the moving equation.

2. Case of higher-order derivatives

Consider the extreme problem contained unknown function $u(x), x \in [a_0, a_1]$, and higher-order derivatives $u'(x), \cdots, u^{(k)}(x)$, it is

$$\begin{cases} J(u)=\int_{a_0}^{a_1} F(x, u, u', \cdots, u^{(k)})\,dx, \\ u(a_0)=b_0, \cdots, u^{(k-1)}(a_0)=b_0^{(n-1)}, \\ u(a_1)=b_1, \cdots, u^{(k-1)}(a_1)=b_1^{(n-1)}, \end{cases}$$

the corresponding Euler equation is

$$F_u-\frac{d}{dx}F_{u'}+\frac{d^2}{dx^2}F_{u''}-\cdots+(-1)^k \frac{d^k}{dx^k}F_{u^{(k)}}=0, \tag{7.1.19}$$

it is a $2k$-order ordinary differential equation.

3. Case of several variables

For instance, two variables x, y and unknown function $u(x, y)$, as well as u_x, u_y, we have

$$\begin{cases} J(u)=\iint_G F(x, y, u, u_x, u_y)\,dx\,dy, \\ u|_{\partial G}=f(s), \quad s \in \partial G, \end{cases}$$

the corresponding Euler equation is

$$F_u-\frac{\partial}{\partial x}F_{u_x}-\frac{\partial}{\partial y}F_{u_y}=0, \tag{7.1.20}$$

it is a two-order partial differential equation.

4. Isoperimetric problem

The isoperimetric problem is an extreme problem by $J(u)=\int_{a_0}^{a_1} F(x,u,u')\,du$ with $K(u)=\int_{a_0}^{a_1} G(x,u,u')\,dx=c$, i. e. find a closed curve $u=\tilde{u}(x)$ satisfying $K(u)=c$, such that $J(u)$ arrives minimal value, and the area of domain with boundary $u=\tilde{u}(x)$ takes maximal value.

Suppose that $u=\tilde{u}$ is the minimal function. Consider function

$$u=\tilde{u}(x)+\varepsilon_1\eta_1(x)+\varepsilon_2\eta_2(x),$$

substitute $J(u)=\int_{a_0}^{a_1} F(x,u,u')\,du$, $K(u)=\int_{a_0}^{a_1} G(x,u,u')\,dx$, then we get two functions $J(\varepsilon_1,\varepsilon_2)$, $K(\varepsilon_1,\varepsilon_2)$, respectively. By assumption $K(\varepsilon_1,\varepsilon_2)=c$, then $J(u)$ takes minimum at $\varepsilon_1=\varepsilon_2=0$. Thus,

$$\left(\frac{\partial J}{\partial \varepsilon_1}\right)\bigg|_{\varepsilon_1=\varepsilon_2=0}-\lambda\left(\frac{\partial K}{\partial \varepsilon_1}\right)\bigg|_{\varepsilon_1=\varepsilon_2=0}=0,\quad \left(\frac{\partial J}{\partial \varepsilon_2}\right)\bigg|_{\varepsilon_1=\varepsilon_2=0}-\lambda\left(\frac{\partial K}{\partial \varepsilon_2}\right)\bigg|_{\varepsilon_1=\varepsilon_2=0}=0,$$

then Euler equation is

$$(F-\lambda G)_u-\frac{d}{dx}(F-\lambda G)_{u'}=0, \tag{7.1.21}$$

Example 7.1.8 An isoperimetric problem

For the minimal problem $J(u)=\int_{a_0}^{a_1} u(x)\sqrt{1+[u'(x)]^2}\,dx$ with $K(u)=\int_{a_0}^{a_1}\sqrt{1+(u')^2}\,dx=l$, take $f(u)=u-\lambda$ in (7.1.12), then we get the integral of Euler equation

$$u=\lambda+c\cosh\frac{x-\xi}{c}, \tag{7.1.22}$$

this is the minimal curve. Substitute it into $K(u)=l$, then we have

$$\sinh\frac{a_1-\xi}{c}-\sinh\frac{a_0-\xi}{c}=\frac{1}{c}. \tag{7.1.23}$$

The condition of curve passing through points $P_0(a_0,b_0)$, $P_1(a_1,b_1)$ is

$$\lambda+c\cosh\frac{a_0-\xi}{c}=b_0,\quad \lambda+c\cosh\frac{a_1-\xi}{c}=b_1. \tag{7.1.24}$$

By (7.1.22)-(7.1.24), constants ξ, c, λ can be determined, and when $l^2\geqslant(a_1-a_0)^2+(b_1-b_0)^2$, there exists unique solution, it is the curve with the length greater then the distance between P_0 and P_1.

7.2 Some Important Theorems in Banach Spaces

In Chapter 6 of [11], the calculus of Banach spaces is introduced. For a mapping $f: X \to Y$ of Banach spaces X on to Y, we define limits, derivatives, series, etc. About integral, we only define the integral of a mapping $f: [a,b] \to Y$ from real interval $[a,b]$ to Banach space Y, however, it is essentially a Riemann integral, we refer to [10]. Now we introduce four important theorems: Stone-Weierstrass, implicit-mapping, inverse-mapping and fixed-point theorems.

7.2.1 Stone-Weierstrass theorems

Recall the classical Weierstrass theorem:

Theorem 7.2.1 Suppose that $f(x)$ is a real continuous function on $[a,b]$, then there exists a real polynomial sequence $\{p_n(x)\}_{n=1}^{+\infty}$, such that $p_n(x)$ converges to $f(x)$ on $[a,b]$, uniformly.

We refer to [6] for the proof.

To show this Stone-Weierstrass theorem, we introduce the concept of "distinguish points".

Definition 7.2.1 (set A distinguishes points of set X) Let X be a set, $\mathfrak{B}_{\mathbf{R}}(X)$ be the set of all real bounded functions on X, i. e., $\mathfrak{B}_{\mathbf{R}}(X) = \{f: X \to \mathbb{R}, f$ is real and bounded on $X\}$, and $A \subset \mathfrak{B}_{\mathbf{R}}(X)$. The set A **is said to distinguish points of** X, if for any points $x, y \in X$ with $x \neq y$, there exists a function $f \in A$, such that $f(x) \neq f(y)$.

In the following theorem, suppose that X is a T_2-type compact topological space, $\mathfrak{B}_{\mathbf{R}}(X)$ is the space of all bounded functions on X, thus space $C_{\mathbf{R}}(X)$, i. e. the space of all real continuous functions on X, is a subspace of $\mathfrak{B}_{\mathbf{R}}(X)$. Moreover, $(C_{\mathbf{R}}(X), +, \alpha \cdot, \|f\|_{C_{\mathbf{R}}(X)}, f \cdot g)$ with addition $+$, number product $\alpha \cdot$, norm $\|\cdot\|_{C_{\mathbf{R}}(X)}$, and multiplication $\cdot$, becomes a Banach algebra with unit $e_{C_{\mathbf{R}}(X)} = 1$.

Theorem 7.2.2 (Stone-Weierstrass Theorem) Let X be a T_2-type compact topological space, $C_{\mathbf{R}}(X)$ be a Banach algebra on X. If a subset $A \subset C_{\mathbf{R}}(X)$ is a sub-algebra of $C_{\mathbf{R}}(X)$, and contains unit e of the multiplication of $C_{\mathbf{R}}(X)$, then $\overline{A} = C_{\mathbf{R}}(X)$, if and only if A distinguishes points of X.

Proof **Necessity** If $\overline{A} = C_{\mathbf{R}}(X)$, we prove that A distinguishes points of X.

Since X is a T_2-type space, then it is T_4-type. Take $x, y \in X$ with $x \neq y$, by Urysohn Lemma, there exists a continuous function $f \in C_{\mathbf{R}}(X)$, such that

$$f(x)=0,\quad f(y)=1. \tag{7.2.1}$$

By necessary condition $\overline{A}=C_{\mathbf{R}}(X)$, for the above $f\in C_{\mathbf{R}}(X)=\overline{A}$, and any given $\varepsilon>0$, there exists $g\in A$, such that

$$|f(z)-g(z)|<\varepsilon,\quad \forall z\in X. \tag{7.2.2}$$

Hence, take $\varepsilon=\frac{1}{2}$, combining (7.2.1) and (7.2.2), it holds

$$-\frac{1}{2}<g(x)<\frac{1}{2},\quad \frac{1}{2}<g(y)<\frac{3}{2}. \tag{7.2.3}$$

This implies $g(x)\neq g(y)$. So that, A distinguishes points of X.

Sufficiency If subalgebra $A\subset C_{\mathbf{R}}(X)$ distinguishes points of X, we prove $\overline{A}=C_{\mathbf{R}}(X)$ by 3 steps.

First step Prove: the operations $\vee$, $\wedge$ defined as

$$(f\vee g)(x)=\max\{f(x),g(x)\},\ (f\wedge g)(x)=\min\{f(x),g(x)\}$$

are closed in subalgebra $\overline{A}$.

In fact, by

$$\max\{f(x),g(x)\}=\frac{1}{2}\{f(x)+g(x)\}+\frac{1}{2}|f(x)-g(x)|,$$
$$\min\{f(x),g(x)\}=\frac{1}{2}\{f(x)+g(x)\}-\frac{1}{2}|f(x)-g(x)|,$$

we only need to prove that the mapping $\varphi: f\to|f|$ is closed in $\overline{A}$.

For $f\in\overline{A}\subset C_{\mathbf{R}}(X)\subset\mathfrak{B}_{\mathbf{R}}(X)$, there exists c with $0<c<+\infty$, such that $|f|<c$. So that,

$$|f|=\sqrt{c^2-(c^2-f^2)}=c\sqrt{1-\left(1-\frac{f^2}{c^2}\right)}=c\left\{1-\sum_{n=1}^{+\infty}\frac{(2n-3)!!}{(2n)!!}\left(1-\frac{f^2}{c^2}\right)^n\right\},$$

the above series is convergent uniformly with respect to $x\in X$, and f is continuous since $f\in\overline{A}\subset C_{\mathbf{R}}(X)$, thus each term $\left(1-\frac{f^2}{c^2}\right)^n$ is continuous, this implies the sum of series in $\overline{A}$, i.e., $|f|\in\overline{A}$. The first step is proved.

Second step Prove: $\forall f\in C_{\mathbf{R}}(X)$ and $\forall x,y\in X$ with $x\neq y$, there exists $g_{xy}^0\in A$, such that

$$g_{xy}^0(x)=f(x),\quad g_{xy}^0(y)=f(y).$$

Suppose that $x,y\in X, x\neq y$. Then assumption "A distinguishes points of X" implies that there exists $g^0\in A$ satisfying $g^0(x)\neq g^0(y)$, i.e., $\begin{vmatrix} g^0(x) & 1\\ g^0(y) & 1\end{vmatrix}\neq 0$. So that there exists unique solution pair α, β in real field $\mathbb{R}$ of $\begin{cases} g^0(x)\cdot\alpha+1\cdot\beta=f(x), \\ g^0(y)\cdot\alpha+1\cdot\beta=f(y). \end{cases}$ Then, function $g_{xy}^0=\alpha g^0+\beta e$ with $g^0\in A$ satisfying $g_{xy}^0\in A$ is needed.

Third step Prove: $\overline{A}=C_{\mathbf{R}}(X)$, i. e., only need to prove $C_{\mathbf{R}}(X)\subset\overline{A}$.

In fact, for given $\varepsilon>0$, and $x\in X$, take $y\in X$ with $y\neq x$. By the second step, $\exists g_{xy}\in A$, such that

$$g_{xy}(x)=f(x), \quad g_{xy}(y)=f(y).$$

Then, there exists an open neighborhood $U_x\subset X$ of $x\in X$ by the continuity of g_{xy}, such that

$$-\varepsilon<g_{xy}(z)-f(z)<\varepsilon, \quad z\in U_x.$$

Clearly, it holds $\bigcup\limits_{x\in X}U_x=X$.

By the compactness of X, there exist finite neighborhoods $U_{x_1},\cdots,U_{x_n}$, such that $\bigcup\limits_{j=1}^{n}U_{x_j}=X$. Then,

$$g_{x_jy}(z)>f(z)-\varepsilon, \quad z\in U_{x_j}, \quad j=1,2,\cdots,n.$$

Let $g_y=g_{x_1y}\vee\cdots\vee g_{x_ny}$. Then, $g_y\in A$, by the second step. It follows that

$$g_y(z)>f(z)-\varepsilon, \quad \forall z\in X=\bigcup_{j=1}^{n}U_{x_j}. \tag{7.2.4}$$

Furthermore, by the continuity of g_y, there exists an open neighborhood $V_y\subset X$ of $y\in X$, such that

$$-\varepsilon<g_y(z)-f(z)<\varepsilon, \quad z\in V_y.$$

Clearly, it holds $\bigcup\limits_{y\in X}V_y=X$. Thus, the compactness of X implies that there exist finite neighborhoods $V_{y_1},\cdots,V_{y_m}$, such that $\bigcup\limits_{k=1}^{m}V_{y_k}=X$. So that, it follows

$$g_{y_k}(z)<f(z)+\varepsilon, \quad z\in V_{y_k}, \quad k=1,2,\cdots,m.$$

Let $g=g_{y_1}\wedge\cdots\wedge g_{y_m}$, then $g\in A$, by the second step. It follows that

$$g(z)<f(z)+\varepsilon, \quad \forall z\in X=\bigcup_{j=1}^{m}V_{y_j}. \tag{7.2.5}$$

Combining (7.2.4) and (7.2.5), it follows that for any $f\in C_{\mathbf{R}}(X)$, and any given $\varepsilon>0$, there exists $g\in A$, such that

$$\|f-g\|_X=\max_{z\in X}|f(z)-g(z)|<\varepsilon.$$

The proof is complete.

Theorem 7.2.3 **(Stone-Weierstrass Theorem on complex field)** Let X be a T_2-type compact topological space, and $\mathfrak{C}_C(X)$ be a complex Banach algebra on X. If $A \subset \mathfrak{C}_C(X)$ is a subalgebra of $\mathfrak{C}_C(X)$ containing the unit e of multiplication of $\mathfrak{C}_C(X)$, then $\overline{A} = \mathfrak{C}_C(X)$, if and only if A distinguishes points of X.

7.2.2 Implicit- and inverse-mapping theorems

Theorem 7.2.4 (classical existence theorem of implicit function) Suppose that:

(1) Real function $F(x,y)$ is defined and continuous on $\mathfrak{D} = [x_0 - \delta, x_0 + \delta; y_0 - \delta_1, y_0 + \delta_1]$;

(2) F'_x, F'_y are continuous on $\mathfrak{D}$;

(3) $F(x_0, y_0) = 0, F'_y(x_0, y_0) \neq 0$.

Then, we have:

(1) There exists some neighborhood $U \subset \mathfrak{D}$ of (x_0, y_0), such that y is determined as a function $y = f(x)$ of x (single value) by $F(x,y) = 0$ in U;

(2) $y_0 = f(x_0)$;

(3) $y = f(x)$ is continuously derivable in U, and $\dfrac{dy}{dx} = -\dfrac{\frac{\partial F}{\partial x}}{\frac{\partial F}{\partial y}}$.

Theorem 7.2.5 (Existence theorem of implicit mapping in Banach space) Let X, Y, Z be Banach spaces, and $A \subset X \times Y$ be an open subset of $X \times Y$. If a mapping $f: A \to Z$ satisfies:

(1) f is continuously derivable on A;

(2) $f(x_0, y_0) = 0$ at $(x_0, y_0) \in A$;

(3) $\partial_y f(x_0, y_0)$ is a linear homeomorphism from Y to Z.

Then, there exists an open neighborhood $U_0 \subset X$ of x_0, such that

(1) There exists unique continuous mapping $u: U \to Y$ in each connected open subset $U \subset U_0$ with $x_0 \in U$ satisfying $u(x_0) = y_0$, and holding $(x, u(x)) \in A, f(x, u(x)) = 0$ at each $x \in U$;

(2) u is continuously derivable in U, and holds

$$u'(x) = -(\partial_y f(x, u(x)))^{-1} \circ \partial_x f(x, u(x)).$$

Theorem 7.2.6 (inverse mapping theorem in Banach space) Let X, Y be Banach spaces, $x_0 \in X$, and $V \subset X$ be an open neighborhood of x_0. If $f: V \to Y$ is a continuously derivable mapping, and $f'(x_0)$ is a linear homeomorphism from X to Y. Then, there exists an open neighborhood $U \subset V$ of x_0, such that $f|_U: U \to W \subset Y$ is a homeomorphism, where W is an open neighborhood of $y_0 = f(x_0)$; Moreover, if $f: U \to$

W is p-order continuously derivable in U, then the inverse mapping $g: f(U) \to U$ of $f: U \to f(U) \subset W$ is p-order continuously derivable in $f(U)$.

Note For the definition of continuously derivable in Banach space, we refer to Chapter 6 of [11].

7.2.3 Fixed point theorems

Theorem 7.2.7 (compression mapping principle) Let X, Y be Banach spaces, $U = B_a(0) \subset X$ be a ball with center zero of X and radius $a > 0$; correspondingly, $V = B_b(0) \subset Y$ be a ball with center zero of Y and radius $b > 0$. If $u: U \times V \to Y$ is a continuous mapping satisfying:

(1) $\| u(x, y_1) - u(x, y_2) \|_{X \times Y} \leqslant k \| y_1 - y_2 \|_Y$, where $0 < k < 1$, $x \in U$, $y_1, y_2 \in V$;

(2) $\| u(x, 0) \|_{X \times Y} < \beta(1-k)$, where $x \in U$;

Then, there exists unique mapping $f: U \to V$, satisfying $f(x) = u(x, f(x))$.

The existence and uniqueness theorem of the solution for initial-valued problem of ordinary differential equations in *Advanced Calculus* (refer to [11], Ch. 5) is a special case of this theorem.

Theorem 7.2.8 (Brouwer fixed-point theorem) Let $B_1(0) = \{x \in \mathbb{R}^n : \| x \|_{\mathbb{R}^n} \leqslant 1\}$ be a unit ball in $\mathbb{R}^n$. If a mapping $f: B_1(0) \to B_1(0)$ is continuous, then there exists $x_0 \in B(0,1)$ satisfying $f(x_0) = x_0$.

Schauder fixed-point theorem has two forms, usually as follows:

Theorem 7.2.9 (fixed-point theorem of compact mapping for bounded closed convex set) Let X be a Banach space, $M \subset X$ be a bounded closed convex subset in X. If $f: M \to M$ is a compact mapping, then there exists $x_0 \in M$ satisfying $f(x_0) = x_0$.

Note 1 Let M be a subset in Banach space X. If the **"line"** from any points $x_1 \in M$ to $x_2 \in M$, denoted by $L = \{x \in X: x = \alpha x_1 + (1-\alpha) x_2, 0 \leqslant \alpha \leqslant 1\}$, contains in M, i. e. $L \subset M$, then M is called **a convex set.**

Note 2 Let $f: M \to M$ be a mapping on subset M of Banach space X to M. If for any sequence $\{x_n\} \subset M$, the corresponding mapping sequence $\{f(x_n)\} \subset M$ has the subsequence that converges to an element $x \in M$, then $f: M \to M$ is said to be a compact mapping.

Theorem 7.2.10 (fixed-point theorem of continuous mapping for compact convex set) Let X be a Banach space, $M \subset X$ be a compact convex subset of X. If $f: M \to M$ is a continuous mapping, then there exists $x_0 \in M$, satisfying $f(x_0) = x_0$.

7.3 Haar Integrals on Locally Compact Groups

Haar measure, a generalization of Lebesgue measure, was established in 1933 by Germen mathematician A. Haar; then, correspondingly, Haar integral appeared. This kind of measure and integer has invariance of translation, and summarizes a class of metrics with invariant "volume" of rigidity motion bodes, so that Haar integral plays important role in mathematics, physics and many other natural science fields.

1. Locally compact groups, compact groups

Definition 7.3.1 (topological group) Let G be a group with multiplication operation $\cdot$, and be a topological space with topology τ. If mapping $(x, y) \to x \cdot y$ from $G \times G$ onto G is continuous, and mapping $x \to x^{-1}$ from G onto G is continuous, then G is called **a topological group**, denoted by $(G, \cdot, \tau)$, or simply, by G.

There are lots of examples of topological groups, such as, real number set $\mathbb{R}$ with usual addition $+$ and usual topology τ, then, $(\mathbb{R}, +, \tau)$ is a topological group; the positive real number set $\mathbb{R}^+ = (0, +\infty)$ with usual multiplication $\times$ and usual topology τ is a topological group $(\mathbb{R}^+, \times, \tau)$.

Definition 7.3.2 (locally compact topological group, compact topological group) Let $G \equiv (G, \cdot, \tau)$ be a topological group. If (G, τ) is a locally compact topological space, then G is called **a locally compact topological group**, or simply, **a locally compact group**; if (G, τ) is a compact topological group, then G is called **a compact topological group**, or simply, **a compact group**.

2. Haar integral on locally compact groups

Definition 7.3.3 (positive functional on G, Haar integral) Let G be a T_2-type locally compact group, $\mathfrak{C}_C(G)$ be the set of all complex-valued continuous functions with compact support on G. Then a linear functional I on $\mathfrak{C}_C(G)$ is called **a left Haar integral on** G, if it satisfies:

(1) "$f \in \mathfrak{C}_C(G), f > 0$" implies $I(f) > 0$;

(2) "$f \in \mathfrak{C}_C(G), f \neq 0, \forall a \in G$" implies $I({}_a f) = I(f)$;

Where ${}_a f(x) = f(ax)$; Similar definition for **a right Haar integral on** G.

If a linear functional I is both left and right Haar integral, then it is called **a Haar integral on** G. Clearly, if G is a commutative group, then the left and right Haar integrals are the same, being a Haar integral.

Theorem 7.3.1 (existence theorem of Haar integral on a locally compact group) Let G be a T_2-type locally compact group. Then, there exists left (right) Haar integral on G, and this left (right) Haar integral is unique except a constant factor.

For a T_2-type locally compact group G, Haar integral of a complex-valued function $f: G \to \mathbb{C}$ is denoted by

$$\int_G f(x)\mathrm{d}x. \tag{7.3.1}$$

3. Examples

Example 7.3.1 Let $(\mathbb{R}^n, +, \tau)$ be a T_2-type locally compact Abel group on Euclidean space $\mathbb{R}^n$ with usual addition $+$ and usual topology τ. Thus Haar integral on $(\mathbb{R}^n, +, \tau)$ is an n-multiple Lebesgue integral $\int_{\mathbb{R}^n} f(x)\mathrm{d}m$ with $f: \mathbb{R}^n \to \mathbb{C}$.

Example 7.3.2 Let $(G, \cdot) = (\{x_1, \cdots, x_r\}, \cdot)$ be an r-order finite Abel group. Then, the linear functional $I(f) = \frac{1}{r}\sum_{j=1}^{r} f(x_j)$ with $f: G \to \mathbb{C}$ is a Haar integral on G.

Example 7.3.3 Let $(\mathbb{R}^+, \times, \tau) = ((0, +\infty), \times, \tau)$ be the T_2-type locally compact Abel group on positive real axis $\mathbb{R}^+$ with usual multiplication $\times$ and usual topology. Then, $\int_{\mathbb{R}^+} f(x)\frac{\mathrm{d}x}{x}$ is a Haar integral on $\mathbb{R}^+$.

4. Haar measure on locally compact groups

Definition 7.3.4 (Borel set class) Let G be a T_2-type locally compact group. The set of all open subsets, closed subsets, unions of closed subsets, and intersections of open subsets, denoted by $\mathfrak{R}(G)$, is called **a Borel set class on** G. A set in $\mathfrak{R}(G)$ is called **a Borel set.**

Definition 7.3.5 (Haar measure) Let G be a T_2-type locally compact group, $\mathfrak{R}(G)$ be a Borel set class on G. If a non-negative real-valued set function $\mu: \mathfrak{R}(G) \to [0, +\infty)$ satisfies:

(1) $\mu(\varnothing) = 0$;

(2) for $E_j \in \mathfrak{R}(G)$, $j = 1, 2, \cdots$, with $E_j \cap E_k = \varnothing$, $j \neq k$, it holds

$$\mu\left(\bigcup_{j=1}^{+\infty} E_j\right) = \sum_{j=1}^{+\infty} \mu(E_j), \tag{7.3.2}$$

then μ is called **a Borel measure on** $\mathfrak{R}(G)$.

If a Borel measure μ satisfies: for any $x \in G$ and $E \in \mathfrak{R}(G)$, it holds

$$\mu(xE)=\mu(E) \quad (\text{or}, \mu(Ex)=\mu(E)),$$

then μ is called **an left** (or **right**) **translation invariant measure on** $\mathfrak{R}(G)$, or called **a left** (**or right**) **Haar measure.** If μ is both the left and right Haar measures on $\mathfrak{R}(G)$, then called **a Haar measure on** $\mathfrak{R}(G)$.

5. Relationship between Haar measure and Haar integral on locally compact groups

Let G be a T_2-type locally compact group, $\int_G f(x)\mathrm{d}x$ with $f: G \rightarrow \mathbb{C}$ be a Haar integral on G. If a set $E \in \mathfrak{R}(G)$ is a Borel measurable set, and $\mu(E)$ is a Haar measure on E, then the "characteristic function" of set E is defined by $\chi_E(x)=\begin{cases}1, & x \in E, \\ 0, & x \notin E.\end{cases}$

Thus, **a Haar measure** ν **on** $\mathfrak{R}(G)$ **induced from the Haar integral** $\int_G f(x)\mathrm{d}x$ is defined by

$$\nu(E)=\int_G \chi_E(x)\mathrm{d}x;$$

moreover, this measure ν satisfies $\mu(E)=c\nu(E)$, $\forall E \in \mathfrak{R}(G)$ with constant factor c.

Conversely, for any Haar measure μ on T_2-type locally compact group G, a Haar integral can be defined by

$$I(f)=\int_G f(x)\mathrm{d}\mu(x),$$

where $\mathrm{d}\mu$ is the infinitesimal element with respect to measure μ.

6. Some properties of Haar integrals

Theorem 7.3.2 Let G be a T_2-type locally compact group. If μ is a Haar measure on G, then

(1) for any nonempty open set $O \subset G$, it holds $\mu(O)>0$;

(2) for a positive continuous $f>0$ with compact support on G, it holds $\int_G f \,\mathrm{d}\mu > 0$;

(3) $\mu(G)<+\infty$, if and only if G is a compact group.

Definition 7.3.6 (**modulus function**) Let G be a T_2-type locally compact group, I be a Haar integral on G. For $f \in \mathfrak{C}_{\mathrm{C}}(G)$, $f>0$ and $x \in G$, then, $\Delta(x) \equiv \dfrac{I(f_{x^{-1}})}{I(f)}$ is called **a modulus function of** x on G.

Definition 7.3.7 (**properties of modulus function**) Let G be a T_2-type locally compact group. If $\Delta(x)=\dfrac{I(f_{x^{-1}})}{I(f)}$ is the modulus function on of x on G, then

(1) $\Delta(x)$ is only dependent on x, whereas, it is independent of f and I;

(2) $\Delta(x)$ is a positive continuous function on G, satisfying $\Delta(x \cdot y)=\Delta(x)\Delta(y)$;

(3) every compact group G is a unit modulus, i. e. $\Delta(x)=1$.

Exercise 7

1. Fermat principle is: if the refracted ratio of medium at point (x, u) is $n(x, u)$, then when the light passes through two points, the path on which light needs minimal time is the "ray of light". Rewrite it as a variation problem.

2. Deduce this Euler equation of $J(u)=\int_{a_0}^{a_1} F(x, u_1, \cdots, u_k, u_1', \cdots, u_k')\,dx$.

3. Deduce this Euler equation of $J(u)=\int_{a_0}^{a_1} F(x, u, u', \cdots, u^{(k)})\,dx$.

4. Deduce this Euler equation of $J(u)=\iint_G F(x, y, u, u_x, u_y)\,dx\,dy$.

5. In the shortest-path problem, evaluate the minimal value $J(\tilde{u})$.

6. Evaluate the minimal value and minimal function $J(u)=\int u^k (u^2+(u')^2)^{\frac{1}{2}}\,dx, u>0$.

7. Prove: $u=c\varphi\left(\frac{x-\xi}{c}\right)$ is the minimal function of $J(u)=\int u^k g(u')\,dx$ with constants c, ξ.

8. Consider extreme problem of $J(u)$, and give an idea to solve it, where

$$J(u)=\int_{a_0}^{a_1} [p(x)u^2+2q(x)uu'+r(x)u'^2+2f(x)u+2g(x)u']\,dx.$$

9. Consider extreme problem of $J(u)=\iint_\Omega [u_x^2+u_y^2+2f(x, y)u]\,dx\,dy$.

10. Consider extreme problem $J(u)=\iint_\Omega [u_x^2+u_y^2+2f(u)]\,dx\,dy+\int_{\partial\Omega} [\rho(s)u^2+2\sigma(s)u]\,ds$.

11. Solve eigenvalue problem $J(u)=\int_{a_0}^{a_1} [p(x)u^2+2q(x)uu'+r(x)(u')^2]\,dx$ with condition $K(u)=\int_{a_0}^{a_1} k(x)u^2\,dx=1$, and find its minimal value.

12. Deduce Euler equation of $J(u)=\iint_\Omega (u_x^2-u_y^2)\,dx\,dy$.

References

[1] 张禾瑞. 近世代数基础[M]. 北京：高等教育出版社，1978.

[2] DEODONATE J. Foundation of Modern Analysis[M]. Beijing：Science Press，1983.

[3] 陈省身，陈维恒. 微分几何讲义[M]. 北京：北京大学出版社，1984.

[4] BUTZER P L， Nessel R J. Fourier Analysis and Approximation Theory（Vol. I）[M]. Beijing：Advance Education Press，1985.

[5] 齐民友. 线性偏微分算子引论(上)[M]. 北京：科学出版社，1986.

[6] 郑维行，王声望. 实变函数与泛函分析概要(一)、(二)[M]. 北京：高等教育出版社，1992.

[7] 吴时敏. 广义相对论教程[M]. 北京：北京师范大学出版社，1998.

[8] 齐民友，吴同方. 广义函数与数学物理方程[M]. 北京：高等教育出版社，1999.

[9] 陈维桓. 流形上的微积分[M]. 北京：高等教育出版社，2000.

[10] 苏维宜. 近代分析引论[M]. 北京：北京大学出版社，2000.

[11] 仇庆久. 高等数学[M]. 北京：高等教育出版社，2003.

[12] 熊金城. 点集拓扑讲义[M]. 北京：高等教育出版社，2003.

[13] 龚昇，线性代数五讲[M]. 北京：科学出版社，2005.

[14] 徐森林，胡自胜，金亚东，等. 点集拓扑学[M]. 北京：高等教育出版社，2007.

[15] 曾谨言. 量子力学教程[M]. 北京：科学出版社，2008.

[16] 徐森林，薛春华. 近代微分几何[M]. 北京：中国科学技术大学出版社，2009.

[17] 徐森林，薛春华. 实变函数论[M]. 北京：清华大学出版社，2009.

[18] 苏维宜. 局部域上的调和分析与分形分析[M]. 北京：科学出版社，2011.

Index

A

B

C

D

E

I

J

K

L

M

N

O

P

Q

R

S

U